34th International DAAAM Symposium on Intelligent Manufacturing and Automation 2023

Vienna, Austria
26-27 October 2023

Editor:

B. Katalinic

ISBN: 978-1-7138-8616-7

Printed from e-media with permission by:

Curran Associates, Inc.
57 Morehouse Lane
Red Hook, NY 12571

Some format issues inherent in the e-media version may also appear in this print version.

Printed with permission by Curran Associates, Inc. (2024)

For permission requests, please contact DAAAM International Vienna
at the address below.

DAAAM International Vienna
TU Wien, Getreidemarkt 9
1060 Wien
Austria, EU

Phone: +436646424398

http://www.daaam.info

Additional copies of this publication are available from:

Curran Associates, Inc.
57 Morehouse Lane
Red Hook, NY 12571 USA
Phone: 845-758-0400
Fax: 845-758-2633
Email: curran@proceedings.com
Web: www.proceedings.com

ISSN 1726-9679

Danube Adria Association for Automation & Manufacturing

DAAAM International Vienna

www.daaam.info

PROCEEDINGS

OF THE 34TH INTERNATIONAL DAAAM SYMPOSIUM
"INTELLIGENT MANUFACTURING & AUTOMATION"
26-27TH OCTOBER 2023
HOSTED FROM VIENNA UNIVERSITY OF TECHNOLOGY, VIENNA, AUSTRIA, EU

ORGANIZED BY:
DAAAM INTERNATIONAL VIENNA & UNIVERSITY OF MOSTAR

IN COOPERATION WITH:
ÖIAV 1848, VIENNA UNIVERSITY OF TECHNOLOGY, INTERNATIONAL ACADEMY OF ENGINEERING
AND UNIVERSITY OF APPLIED SCIENCES – TECHNIKUM WIEN

UNDER THE AUSPICES OF THE DANUBE RECTORS' CONFERENCE &
RECTORS' AND PRESIDENTS' HONOR COMMITTEE OF DAAAM INTERNATIONAL

EDITOR: B.[RANKO] KATALINIC

IMPRESSUM
2023

Proceedings of the 34th International DAAAM Symposium "Intelligent Manufacturing & Automation" 26-27th October 2023, hosted from Vienna University of Technology, Vienna, Austria, EU, Volume 34, No.1,
Abbreviated key title: ***Ann. DAAAM 2023***

ISSN 1726-9679

Ed. B. Katalinic,

Published by DAAAM International, Vienna, Austria, EU, 2023
Layout & Design: B[ranko] Katalinic
Technical Editors: I[lija] Zec / L[uka] Zec

Publisher:

DAAAM International Vienna
TU Wien, Getreidemarkt 9, 1060 Wien, Austria
E-mails: president@daaam.info / branko.katalinic@tuwien.ac.at
DAAAM homepage: http://www.daaam.info

CONTENTS

AUTHOR INDEX

ORGANIZING COMMITTEE FOR 2023

KATALINIC, BRANKO – PRESIDENT OF DAAAM INTERNATIONAL
COVIC, DRAGAN – HONOR PRESIDENT OF ORGANIZING COMMITTEE
GUSEV, BORIS – PRESIDENT OF INTERNATIONAL ACADEMY OF ENGINEERING
ZEC, ILIJA - SECRETARY

ABURIA, MOHAMED
BEVANDA, IVAN
BIELIC, TONI
BOSNJAK, IGOR
BUCHMEISTER, BORUT
COLAK, IVO
COLAK, STANKO
COSIC, ILIJA
CVETKOVSKA, ROBERTA
DURAKBASA, NUMAN
FUERST, LAURENZ
GODEC, DAMIR
HOELLER, LIANE
KALENIC, MIODRAG
KNEZOVIC, NIKOLA
KOSTOLANI, DAVID
KUKUSHKIN, ILYA
MARIC, PETAR
NAUMBURGER, JOERG
PANFILOV, PETER
PARK, HONG-SEOK
PRYANICHNIKOV, VALENTIN
RAKIC, KRESIMIR
SARAVANJA, LUKA
SCHMOELLEBECK, FRITZ
SEREMET, ZELJKO
STAZHKOV, SERGEY
STAROVESKI, TOMISLAV
STOJKIC, INJA
STOJKIC, ZELJKO
STUJA, KEMAJL
TEKIC, ZELJKO
TOMIC, ZORAN
UDILJAK, TOMA
UGLESIC, ANTE
VLADOV, MAREK
ZEC, ANTE
ZEC, LUKA

RECTORS' AND PRESIDENTS' HONOR COMMITTEE

ABDRASILOV, BOLATBEK	RECTOR, KARAGANDA STATE INDUSTRIAL UNIVERSITY
ABRUDAN, IOAN VASILE	RECTOR, TRANSILVANIA UNIVERSITY OF BRASOV
ALEXANDROV, ANATOLIY	RECTOR, BAUMAN MOSCOW STATE TECHNICAL UNIVERSITY
BAAIJENS, FRANK	RECTOR, EINDHOVEN UNIVERSITY OF TECHNOLOGY
BIAŁKIEWICZ, ANDRZEJ ANTONI	RECTOR, CRACOW UNIVERSITY OF TECHNOLOGY
BONDREA, IOAN	PRESIDENT, LUCIAN BLAGA UNIVERSITY OF SIBIU
BORAS, DAMIR	RECTOR, UNIVERSITY OF ZAGREB
BOROVIKOV, YURY	RECTOR, THE YAROSLAV-THE-WISE NOVGOROD STATE UNIVERSITY
CHEON, YEON	PRESIDENT, UNIVERSITY OF ULSAN
CISNEROS PÉREZ, GUILLERMO	RECTOR, UNIVERSIDAD POLITECNICA DE MADRID
COSTOIU, MIHNEA	RECTOR, UNIVERSITY POLITEHNICA OF BUCHAREST
CZIGANY, TIBOR	RECTOR, BUDAPEST UNIV. OF TECHNOLOGY AND ECONOMICS
DAM, HENRIK	DIRECTOR, UNIVERSITY OF SOUTHERN DENMARK
DANIELS, MITCHELL	PRESIDENT, PURDUE UNIVERSITY
DI SCIASCIO, EUGENIO	RECTOR, POLYTECHNIC OF BARI
DRĂGAN, FLORIN	RECTOR, POLYTECHNIC UNIVERSITY OF TIMISOARA
G. BOUDOUVIS, ANDREAS	RECTOR, NATIONAL TECHNICAL UNIVERSITY OF ATHENS
GALIJEV, ILJHAM	PRESIDENT, OMSK STATE TRANSPORT UNIVERSITY
GARCÍA MARÍN, JUAN FRANCISCO	RECTOR, UNIVERSITY OF LEON
GARCIA, EVA FERREIRA	RECTOR, UNIVERSITY OF THE BASQUE COUNTRY
GELENCSER, ANDRAS	RECTOR, UNIVERSITY OF PANNONIA
GEYER, SYLVIA	RECTOR, UNIV. OF APPLIED SCIENCES TECHNIKUM, VIENNA
GONOKAMI, MAKOTO	PRESIDENT EMERITUS, UNIVERSITY OF TOKYO
GUBERAC, VLADO	RECTOR, STROSSMAYER UNIVERSITY OF OSIJEK

INTERNATIONAL SCIENTIFIC COMMITTEE

INTERNATIONAL PROGRAM COMMITTEE

BLEY, HELMUT
BOTEZATU, CORNELIA PAULINA
BOTEZATU, MIHAI
BRAJLIH, TOMAZ
BREZAK, DANKO
BREZOCNIK, MIRAN
BUCHMEISTER, BORUT
BUCURESTEANU, ANCA
CAMPBELL, R. IAN
CEKIC, AHMET
CELAR, STIPE
CHAN, CHEE SENG
CHAUS, ALEXANDER
CHRAMCOV, BRONISLAV
COHODAR, MAIDA
COLEMAN, J. J.
COSIC, ILIJA PETAR
COSMA, SMARANDA
COTET, COSTEL EMIL
CUBONOVA, NADEZDA
CURTU, IOAN
DAMIC, VJEKOSLAV
DENKENA, BEREND
DEVAHASTIN, SAKAMON
DOLGA, LIA
DOLGA, VALER
DOLINAY, JAN
DOLINAY, VILIAM
DOSTAL, PETR
DOSTALEK, PETR
DRAGCEVIC, ZVONKO
DRAGICEVIC, MARIJA
DRSTVENSEK, IGOR
DUHOVNIK, JOZE
EJUB DZAFEROVIC,
FIRST ROGALE, SNJEZANA
FUJII, NORIO
GHEORGHIU, HORIA
GODEC, DAMIR
GOMEZ, ANTONIO
GORDIN, DOINA MARGARETA
HLAVA, JAROSLAV
HOFREITER, MILAN
HRUBINA, KAMIL
HRUSKA, FRANTISEK
HRUSKOVA, ERIKA
IBISHI, ISMET M.
JANACOVA, DAGMAR
JANSSENS, THIERRY
JARVIS, CHRISTINE W.
JAVOROVA, ANGELA
JERBIC, BOJAN
JURAGA, IVAN
JURUM-KIPKE, JASNA
KARAULOVA, TATYANA
KATALINIC, BRANKO
KAWALEC, ANDRZEJ
KHANNA, ASHUTOSH
KHOMCHENKO, VASILY
KIRK, CHRIS. S.
KISS, IMRE
KITTL, BURKHARD
KLIMENKO, S. A.
KNEGO, NIKOLA
KNEZEVIC, PETAR
KOHLERT, HELMUT
LUKIC, IVANA
LUPEA, IULIAN
LUPU, CIPRIAN
LUPULESCU, NOURAS BARBU
MACHEDON PISU, TEODOR
MACKU, LUBOMIR
MACUROVA, ANNA
MAGHETI, IOAN
MAJETIC, DUBRAVKO
MALISA, VIKTORIO
MALOPOLSKI, WALDEMAR
MAN, MARIANA
MANDIC, LIDIJA
MANDRU, DAN
MANESCU, TIBERIU STEFAN
MARASCU KLEIN, VLADIMIR
MARCOS, MARIANO
MARCU, MARIUS
MAREK, KUBALCIK
MARIAN TOLNAY
MARIN, DUMITRU
MARIN, MIHNEA
MARINESCU, NICULAE ION
MARN, FERDINAND
MARONEK, MILAN
MARSLAND, DAVID
MARTA, CONSTANTIN
MARTINEZ, SUSANA
MASTACAN, LUCIAN
MATUSOVA, MIRIAM
MATUSU, RADEK
MEHTA, NARINDER KUMAR
MEIER, HORST
MENZ, PETER
MICHAL, BLIZNAK
MICIETA, BRANISLAV
MICLOSI, VIOREL
MOHORA, CRISTINA
MOK, H.
MORARU, SORIN AUREL
MORIWAKI, T.
MORTOIU, DOINA
MOTOMANCEA, ADRIAN
MUNTEAN, MIHAELA
MUNTEAN, SEBASTIAN
MUNTEANU, ANCA
MUSCA, GAVRIL
MUSTAFIZUL, KARIM
MYSHLYAVTSEV, ALEXANDR
NAGI, MIHAI
NAGLA, K. S.
NAKAMOTO, NOBUYUKI
NEUMANN, PETR
NICA COTET, GABRIELA BEATRICE
NIKOLIC, GOJKO
NKUMBWA, RODGER
NOH, SANG DO
NOVAC, OVIDIU
NOVOSELOVIC, DANIEL
NURUL AMIN A.K.M.
OANCEA, GHEORGHE I.
OKAMOTO, J.
OLARU, ADRIAN, DUMITRU
OLARU, ONISIFOR
OLEKSIK, VALENTIN
OLTEANU, CIPRIAN
PLESA, ANGELA
PODURAEV, JURY
POKORNY, PAVEL
POMAZAN, VALENTINA
POPA, CICERONE LAURENTIU
POPPEOVA, VIERA
PRODAN, DAN
PROKOP, ROMAN
PRYANICHNIKOV, VALENTIN
PYTKO, STANISLAW
QUINN, ROGER D.
RAMAJ, VEHBI XH.
RAVAS, RUDOLF
RAY, T. S.
RIKALOVIC, ALKSANDAR
RIZESCU, SABIN VASILE
ROGALE, DUBRAVKO
ROHDE, ULRICH L.
ROMERO, FERNANDO
ROS, OLIMPIA
ROSADO, V.O.G.
RUDAS, IMRE J.
RUGESCU, DRAGOS RADU DAN
RUJA, IOAN
RUSU ANGHEL, NICOLAE
RUSU, LIGIA
RUSU-CASANDRA, AURELIA
RUTEMI, GANI BRAHIM
RUZIC, DRAGO
SAKULSKI, DUSAN
SALIHU, AVDI HAJDIN
SAMEK, DAVID
SCHREIBER, PETER
SELVAKUMAR, ARUL LAWRENCE
SEONG, POONG HYUN
SERCER, MLADEN
SHIBASAKA, TOSHIROH
SHIH, MING CHANG
SIEMIATKOWSKI, M.
SIHN, WILFRIED
SINAY, JURAJ
SISEK, BORIS
SITUM, ZELJKO
SKOCZYNSKI, WACLAW
SKOLUD, BOZENA
SLUGA, ALOJZ
SMIRNOV, ALEXANDER
STANKOVSKI, STEVAN
STAROVEŠKI, TOMISLAV
STEFANIC, NEDELJKO
STOPPER, MARKUS
STRGAR KURECIC, MAJA
SUBRAMANIAN, ARUNKUMAR
SUGAR, PETER
SYROVA, LIVIA
SYSALA, TOMAS
TADAMURA, KATSUMI
TAKALA, JOSU
TAMASILA, MATEI
TAN, KOK KIONG
THIERHEIMER, WALTER WILHELM
TORIMS, TOMS
ULRICH, GUENTHER
VAARIO, JARI
VASEK, VLADIMIR
VELISEK, KAROL

Reviewing Committee

CULJAK, EVA
CUS, FRANCI
DAR-EL, E.
DE FILIPPI, A.
DEVAHASTIN, SAKAMON
DINCMEN, M.
DINIZ, ANSELMO EDUARDO
DOHDA, KUNIAKI
DOINA MARGARETA
DONG, JIANHUA
DOVE, RICK
DURAKBASA, NUMAN
EGOROV, SERGEY
ENDTER, R.
EVERSHEIM, W.
GALANTUCCI, LUIGI MARIA
GAWLIK, JOZEF
GIACAGLIA, G. E.
GIBSON, I.
GIROT, FRANK
GOLDFARB, VENYAMIN Y.
GOMEZ, ANTONIO
GRABENWEGER, JOHANN
GUSEV, BORIS V.
HABCHI, GEORGES
HAN, QING YAO
HOBLE, DOREL ANTON
HOCK, SEO SWEE
HOFMANN, PETER
HRYU, JEE-HWAN
ILTSCHEV, VELKO
MACUTA, SILVIU-DANUTI
MARCOS, MARIANO
MARCOS, MARIANO
MARIC, PETAR
MATUSU, RADEK
MEGHDARI, ALI
MERTINS, KAI
MICHELINI, R. C.
MILOS, LAURA RAISA
MIRONEASA, SILVIA
MOMETE, DANIELA CRISTINA
MUNTEANU RADU
NASTAC, SILVIU
NEUMANN, H. J.
NICOLETTO, G.
NOVAC, CORNELIA MIHAELA
NOVAC, OVIDIU CONSTANTIN
OCZOS, KAZIMIERZ E.
OTTO, TAUNO
PALCIC, IZTOK
PANDA, ANTON
PANFILOV, PETER
PARK, HONG-SEOK
PEGDEN, DENIS
PERAKOVIC, DRAGAN
PERUTKA, KAREL
PETERKA, JOZEF
PIVONKA, PETR
PODURAEV, JURY
POP, PETRU ADRIAN
POPESCU, DANIEL
PORTER, BRIAN
TOPALOVA, IRINA
TOPLICEANU, LILIANA
TORIMS, TOMS
TORRES-LEZA, F.
TORVINEN, S.
TSAI, B. S. Y.
UDILJAK, TOMA
UHLMANN, ECKART
UYAR, EROL
VALMOHAMMADI, CHANGIZ
VAN BRUSSEL, HENDRIK
VAN OUDHEUSDEN, DIRK L.
VARGA, GYULA
VAVRIN, PETR
VENKATESH, V. C.
VETTURI, DAVID
VITECEK, ANTONIN
VOSNIAKOS, G. CHRISTOPHER
VOUK, RUDOLF
WALDIR, P. O.
WERTHEIM, RAFI
WESSELY, EMIL
WESTKAEMPER, ENGELBERT
WIENDAHL, HANS-PETER
WILLIAMS, D. J.
WINSTON, PATRICK H.
WOZNIAK, ADAM
ZALZALA, ALI M. S.
ZEC, ILIJA
ZEC, LUKA
ZHAO, DAOZHI
ZLOTO, TADEUSZ

34TH DAAAM INTERNATIONAL SYMPOSIUM ON INTELLIGENT MANUFACTURING AND AUTOMATION

Preface

The 34th DAAAM International Symposium on Intelligent Manufacturing and Automation was organised as virtual online conference hosted by the Vienna University of Technology, Vienna, Austria, between the 26th and 27th October 2023, during the DAAAM International Week. The Symposium was organized by DAAAM International Vienna and University of Mostar in cooperation with ÖIAV 1848, Vienna University of Technology, International Academy of Engineering and University of Applied Sciences – Technikum Wien and Under the Auspices of the Danube Rectors' Conference & Rectors' and Presidents' Honor Committee of DAAAM International for 2023. The Symposium was hosted from Vienna, Austria. This year's symposium aimed at continuing the success of the previous years, focusing on the five-fold traditional objectives of the symposium: the presentation of the most recent high-quality results, support of development of young scientists and researchers, organization of international (summer) doctoral school, inauguration of new members of Central European Branch of International Academy of Engineering and the provision of the necessary setting for stimulating discussions, brainstorming and networking among European and international researchers coming both from the academia government agencies and industry

DAAAM International Symposium has been established as the leading Central European conference devoted to the cutting-edge advancements in the field of modern production and automation. Symposium provides an international forum for engineers, scientists and industrial managers to present and discuss the current status and impact of advanced manufacturing and automation. The international series of Symposium was initiated by the first meeting in Vienna (1990) and continued in Štrbske Pleso (1991), Budapest (1992), Brno (1993), Maribor (1994), Cracow (1995), Vienna (1996), Dubrovnik (1997), Cluj - Napoca (1998), Vienna (1999), Opatija (2000), Jena (2001), Vienna (2002), Sarajevo (2003), Vienna (2004), Opatija (2005), Vienna (2006), Zadar (2007), Trnava (2008), Vienna (2009), Zadar (2010), Vienna (2011), Zadar (2012), Zadar (2013), Vienna (2014), Zadar (2015), Mostar (2016), Zadar (2017), Zadar (2018), Zadar (2019), Mostar (2020), Vienna (2021), Vienna (2022), Vienna (2023).

Regular organization of the annual Symposium is one of the central activities of the DAAAM International association. The association was founded on 5. November 1990 in Vienna during a meeting organised by Professor B. Katalinic in order to celebrate the 175 Years of Vienna University of Technology. The main idea of the association was to establish a network for a new kind of international scientific and academic cooperation in the fields of intelligent manufacturing and automation. Our primary focus is on: all aspects of production of technical products made out of parts and corresponding automation and knowledge, trans-disciplinary application of methods and all kinds of education.

DAAAM International activities, during the last thirty-four years, have been supported by scientists and experts from over fifty countries all over the world. In the course of the last decade, DAAAM International has focused its activities on the regional and global long-term co-operation between (technical) universities. The result of this initiative is the founding of the University Network of DAAAM International Vienna and DAAAM International Network for Advanced Technologies which was founded on 2002-10-25 in Vienna.

The main activities of DAAAM are:

- Organisation of annual DAAAM International Symposium.
- To maintain and improve the University Network of DAAAM International Vienna as a strategic alliance and permanent international open platform for long-term co-operation and partnership in all fields of technology, research and education.
- Organisation of long-term projects, initially known as Inter-University Doctoral Studies.
- Organisation of long-term projects, initially known as the Virtual University.
- Publishing of international journals, books, and proceedings.
- Promotion of the exchange of people, scientific, and technical information.
- Promotion of international co-operation among scientists and engineers.
- Promotion of R & D, education, and industrial application.
- To co-operate with other national and international organisations in matters of common interest.
- Organisation of International Doctoral School as a platform for the support of the next generation of engineers at the highest level.

This proceeding contains 73 refereed papers from 210 authors presented at the 34th DAAAM 2023. In DAAAM proceedings, since the first symposium in 1990, we published papers from 9755 authors.

All the papers included in these proceedings are reviewed by experts in double blind peer review way, and finally revised by the authors for the improvement according to referee's reports and advices.

The papers discuss many aspects of modern manufacturing and automation such as: Algorithms, Artificial Intelligence, CAX, Computer Integration, Control, Cutting Tools, Design, FEM, Invited Lectures, Knowledge, Management, Manufacturing System, Mechatronics, Methodology, Methods, Modelling, Optimization, Robotics, Simulation, Technical Solutions, Technology and Trends.

Appreciation is forwarded to reviewers who contributed to improve the quality of papers reviewing the papers, providing constructive critical comments, improvements, and corrections to the authors. We also recognize the support of reviewers, who were in charge of acceptance/rejection process. We wish to thank the Scientific, Program and Organizing Committees, Session Chairs, student helpers and administrative assistants, for contributing to the success of this Conference. They did an excellent job.

Finally, we wish to thank all the authors who submitted papers, making this conference possible and the authors of accepted papers for making appropriate modifications and corrections with care and patience and for preparing the final manuscript. It is the quality of their presentations and their passion to communicate with other participants that really makes this conference series a success.

We would like to thank all members of the International Academy of Engineering – Central European Branch, Rector of the Vienna University of Technology Professor Dr Sabine Seidler, and Rector of the University of Mostar Professor Dr. Sc. Zoran Tomić and his team, management and all those nice colleagues for their excellent support in the organization of 34th DAAAM International Symposium and the 12th DAAAM International Doctoral School, which was held parallel to the symposium and was positively integrated.

Vienna, 2023-12-07
Editor
Chair of the 34th DAAAM 2023

Branko Katalinic

Professor rtd Dr.sc. Dr.mult.h.c.Prof.h.c. Branko Katalinic
President of DAAAM International
TU Wien, IFLT-IMS, Getreidemarkt 9
1060 Wien, Austria, Europe
URL: http://www.daaam.info
E-mail: president@daaam.info
branko.katalinic@tuwien.ac.at

DOI: 10.2507/34th.daaam.proceedings.001

PRODUCTIVITY FACTORS IN AGILE SOFTWARE DEVELOPMENT PROJECTS

Mili Turić, Stipe Čelar & Srđana Dragičević

This Publication has to be referred as: Turic, M[ili]; Celar, S[tipe] & Dragicevic, S[rdana] (2023). Productivity Factors in Agile Software Development Projects, Proceedings of the 34th DAAAM International Symposium, pp.0004-0008, B. Katalinic (Ed.), Published by DAAAM International, ISBN 978-3-902734-41-9, ISSN 1726-9679, Vienna, Austria
DOI: 10.2507/34th.daaam.proceedings.001

Abstract

Agile methodology is the state-of-the-art methodology for software development projects. Important point for successful managing of agile software development project is precise project effort estimation. Productivity of agile team and team member significantly affect the accuracy of the estimation. In the literature, there are works dealing with even over 200 factors that can affect the productivity. This paper gives a literature overview of productivity factors. These works are mostly team productivity oriented and classified team productivity into different areas. Further, the paper suggests 18 factors as personal productivity factors for developers in agile software development teams. Each of these 18 factors has one value on a three-level weight scale.

Keywords: agile software development; software effort estimation; productivity factors; team and personal productivity.

1. Introduction

To adapt to constantly changed environment conditions, companies utilized lean management as a systematic approach for identifying and eliminating waste in their core processes. [1], [2] As a result of successful implementation of lean management in software processes, Agile Software Development (ASD) methodologies were created. Agile methodologies are the state-of-the-art methodologies for software development projects. However, the estimation of project effort, duration, cost, and product quality is a big challenge for ASD. [3] Using traditional estimation techniques (e.g. functional analysis, planning poker, etc.) can produce imprecise estimates. [4] Some recent studies show that estimation methods can successfully introduce Artificial Intelligence (AI) techniques for improvement of these methods [5], [6], [7].

An important element to assess in today's software estimations is productivity – both team productivity and individual team member productivity. It is not easy to identify factors that affect a productivity. And when they are even recognized, it is difficult to measure them.

Matos et al. [8] include 'team productivity' on the comprehensive list of 90 factors influencing effort estimation. Recognising important productivity factors and their interrelationships will enable project team management to identify areas where more effort is needed to improve effort estimates and overall project goals. Even when productivity factors are identified, it is difficult to set measurable criteria for their impact on productivity because such 'measurement' is largely subjective.

The rest of this paper is structured as follows: a literature overview regarding productivity factors (Section 2); a suggestion of personal productivity factors in agile software development projects (Section 3); and the presentation of conclusions and the outlines of future work (Section 4).

2. Related work

2.1. Productivity in general

Productivity generally indicates success in carrying out some work in relation to the resources used. [9] Jørgensen [10] emphasizes importance of 'productivity of the individuals or teams completing the work'. Jones [11] identified over 200 factors affecting the productivity of developers and software quality, and lists the primary factors that have the greatest impact:

- the skill and experience of staff,
- the cooperation of users during requirements and design,
- schedule or resource constraints,
- methods employed on the project,
- tools available,
- appropriate choice of programming language(s),
- problem complexity,
- code complexity,
- data complexity,
- project organization structures, and
- physical environment.

2.2. Team productivity factors

There is more research in the literature on factors affecting team productivity than on individual productivity. Factors related to developer productivity are rarely highlighted or the developer is seen as part of a team.

The study [12] describes research made in middle (> 30 developers) and large companies put also 'agile teamwork productivity' into the focus. It also claims that, from the perspective of agile team members, the four most perceived factors impacting on their productivity are:

- team effectiveness,
- team management,
- motivation, and
- customer satisfaction.

A study [13] identified 101 human factors and 79 influences influencing the individuals, the development team, and the software project activities. The authors concluded that the most investigated human factors in software development teams are:

- communication,
- collaboration,
- knowledge, and
- motivation.

A systematic mapping study [14] included 616 studies and identified 63 productivity factors that contribute to the measurement of team productivity in ASD. A systematic literature review [15] included 53 primary studies with the aim to identify and classify the factors influencing teamwork productivity in ASD. As a result, review identified 77 influential factors and classified them into six groups that affect teamwork productivity in ASD:

- technical,
- non-technical,
- organizational,
- environmental,
- project management, and
- user requirements.

2.3. Personal productivity factors

Some developers in a team can be multiple times more productive than the others with similar background and experience. The productivity of developer can enhance the productivity of the whole software development team.

Interesting correlation between needed soft skills (s. Table 1), big-five human traits, and developer's tasks through the whole software life cycle (SLC). These matrices can be good template for managing developers in the teams because they commonly work in a team. [16]

Big-Five Personality Traits	Phases of SLC	Skill Description
1. Openness	1. Requirement Analysis	1. Communication Skill
2. Conscientiousness	2. System Design	2. Interpersonal Skills
3. Extraversion	3. Implementation	3. Analytical & Problem-Solving Skills
4. Agreeableness	4. Testing	4. Open and Adaptable to Changes Skills
5. Neuroticism	5. Deployment	5. Organizational Skills
	6. Maintenance	6. Team Player Skills
		7. Ability to Work Independently Skills

Table 1. Personality traits, phases of SLC, and developer soft skills [16]

In the study made across three large USA companies researchers surveyed 622 developers about personal productivity factors and about self-rated productivity. They identified 48 factors, and the three most statistically significant among them in all three companies are [17]:

- job enthusiasm,
- peer support for new ideas, and
- useful feedback about job performance.

3. Productivity in small agile software development projects

Individual developer productivity is especially important in smaller teams and micro/small software companies because it has a greater impact on the project than in larger teams and large companies. An important node in the models for task effort estimation in ASD is 'Developer skills.' [18], [19] This node represents productivity of each team member involved in the project and whose productivity the project manager must evaluate when planning and managing the project. Even when such models are reliable, this node can be a source of instability, i.e. unreliability because it is a 'human' factor in the model.

Parameters used for calculation of an individual team member personal capability (s. Table 2) can be candidate for factors influencing above mentioned node. [9] They are, in fact, *personal productivity* factors. These factors are focused either on the experience or the personality of a developer. They can be used both as personal productivity factors and as team productivity factors.

Factor	*Factor description*
F1 – Work experience	Number of years of experience in similar jobs.
F2 – Knowledge of technology	Level of knowledge and handling of technology in which the software will be developed.
F3 – The level of learning and work habits	Adoption of new working procedures as part of the work being performed. It is reflected in self-acceptance and performance of the procedures.
F4 – Ability to work autonomously	Extent to which tasks are performed in accordance with the general or specific guidelines and instructions of his superiors and the scope of control superiors required in performing a particular job.
F5 – Work complexity	Level of complexity of the tasks performed in the workplace and the complexity of the procedures that apply in their resolution, the required level of personal contribution of employees and the scope of the workplace
F6 – Attitude towards work	Extent to which the employee identifies with the project in which it participates, how relaxed or seriously certain situations, and how it relates to the daily work in terms of their engagement.
F7 – Concentration	Degree of person's presence in the work that is done. Manifested by jumping from one job (or other content) to another or the commitment to carrying out the task.
F8 – Skill of performing of work types	For each task defines the kind of work it belongs, and each person has evaluated the performance of skill for any type of work.
F9 – Knowledge of the project	Knowledge of the project plan and involves participation in the planning of the project from the very beginning.
F10 – Knowledge of the product	Knowledge of the entire system, not just the knowledge of the individual segments.
F11 – Responsibility and influence on decision making	Extent to which the tasks performed in the workplace have an impact on the implementation of the project objectives.
F12 – Communication	It reflects the type and frequency of contacts that are achieved when performing tasks, and their importance to the project.

Factor	*Factor description*
F13 – Agility	Agility reflects the degree of dexterity and diligence in the performance of individual tasks.
F14 – Producing bugs	It manifests itself as a number of bugs produced during development but also in the time spent on the removal of bugs produced.
F15 – Knowledge of the area	Extent to which person needs consulting about business area.
F16 – Testing during development	Degree of effort required to test components developed by person, in order to reduce the number of bugs produced.
F17 – Experience in similar tasks	A measure that shows how person is experienced in performing similar tasks.
F18 – Need for supervision	Extent to which person needs daily mentoring by the project manager and / or head of software product development.

Table 2. Personal productivity factors, adapted from [9]

A general difficulty in working with productivity factors is the difficulty in measuring them. These factors do not contribute equally to productivity. A three-levels scale proved to be sufficient for the quantifying the capabilities of developers [9]:

- f = 10 – very important factor with the greatest impact on productivity,
- f = 8 – medium important factor that has a pretty big impact, but not decisive,
- f = 5 – factor that is not crucial but it would be good to have it.

Three-levels scale of weight factors and their numerical values (in this case: 5, 8, 10) are based on the author's practical experience in 13 years capability and productivity monitoring in small agile teams. [9], [20], [21] The number of levels and values of weighting factors should be adjusted by the company to its specific needs and environment.

4. Conclusion

An important step in the management of agile software development projects is the estimation of project efforts, costs, and quality. An important component of the estimation is the productivity of the team and project team members. When the factors affecting productivity are recognized, management can try to quantify them and then effectively manage them and the entire project.

This paper gives an overview of literature of productivity factors. These works describe more team productivity factors then personal productivity factors. In this paper 18 factors influencing personal productivity in small agile software development projects are described and suggested for use in such projects. A three-level weight scale based on experience for these 18 factors is also described and suggested.

In their future work, the authors will engage in deeper research into the connection between factors of personal productivity and estimation models in agile software development projects.

5. Acknowledgments

This research is a result of research in the frame of cooperation between University of Split and company Venio Indicium Ltd, Split, Croatia.

6. References

[1] Softic, A., Basic, H.,, Lulic, H. & Nakic, A. (2020). Application of the Lean Concept for Analysis and Optimization of the Automobile Filter Production Company, Proceedings of the 31st DAAAM International Symposium, pp.0708-0716, B. Katalinic (Ed.), Published by DAAAM International, ISBN 978-3-902734-29-7, ISSN 1726-9679, Vienna, Austria DOI: 10.2507/31st.daaam.proceedings.098

[2] Trebuna, P.; Pekarcikova, M.; Kliment, M.; Kopec, J. & Svantner, T. Online e-Kanban System Implementation in a Manufacturing Company, International Journal of Simulation Modelling Volume 22: Number 1: March 2023, pp. 5-16

[3] B. W. Boehm, "Software Engineering Economics," in IEEE Transactions on Software Engineering, vol. SE-10, no. 1, pp. 4-21, Jan. 1984, doi: 10.1109/TSE.1984.5010193.

[4] Canedo, E.D., Da Costa, R.P. Methods and metrics for estimating and planning agile software projects, (2018) 24th Americas Conference on Information Systems 2018: Digital Disruption, AMCIS, New Orleans, 2018, ISBN: 978-099668316-6

[5] Rodríguez Sánchez, E.; Vázquez Santacruz, E.F.; Cervantes Maceda, H. Effort and Cost Estimation Using Decision Tree Techniques and Story Points in Agile Software Development. Mathematics 2023, 11, 1477.

[6] BaniMustafa, A. Predicting Software Effort Estimation Using Machine Learning Techniques. In Proceedings of the 2018 8th International Conference on Computer Science and Information Technology (CSIT), Amman, Jordan, 11–12 July 2018; pp. 249–256.

[7] Cabral, J.T.H.; Oliveira, A.L.I. Ensemble Effort Estimation using dynamic selection. J. Syst. Softw. 2021, 175, 110904.
[8] O. Matos, T. Conte, E. Mendes, „Is there a Place for Qualitative Studies when Identifying Effort Predictors? A case in Web Effort Estimation,“ Proc. of the 18th Int. Conf. EASE '14 40, May 13 – 14 2014, London, England, ACM, Article No. 40
[9] Celar, Stipo : Turic, Mili : Vickovic, Linda, Method for personal capability assessment in agile teams using personal points // 22nd Telecommunications Forum/ / Paunović, George (ur.). Beograd: Institute of Electrical and Electronics Engineers (IEEE), 2014. str. 1134-1137.
[10] M. Jørgensen, "What We Do and Don't Know about Software Development Effort Estimation,“ IEEE Software, vol 31(2), 2014, pp. 37-40
[11] C. Jones, "Applied Software Measurement: Global Analysis of Productivity and Quality," 3rd ed, McGraw-Hill, Inc. New York, USA, 2008
[12] I. Fatema and K. Sakib, Factors Influencing Productivity of Agile Software Development Teamwork: A Qualitative System Dynamics Approach, 2017 24th Asia-Pacific Software Engineering Conference (APSEC), Nanjing, China, 2017, pp. 737-742, doi: 10.1109/APSEC.2017.95.
[13] E. Dutra, B. Diirr, and G. Santos, "Human Factors and their Influence on Software Development Teams - A Tertiary Study," Sep. 2021, pp. 442–451. doi: 10.1145/3474624.3474625
[14] Guerrero-Calvache, M., Hernández, G. (2022). Team Productivity in Agile Software Development: A Systematic Mapping Study. In: Florez, H., Gomez, H. (eds) Applied Informatics. ICAI 2022. Communications in Computer and Information Science, vol 1643. Springer, Cham. https://doi.org/10.1007/978-3-031-19647-8_32
[15] Javed Iqbal, Azman Yasin and Mazni Omar, "Defining Teamwork Productivity Factors in Agile Software Development," International Journal on Advanced Science, Engineering and Information Technology, vol. 12, no. 3, pp. 1160-1172, 2022. [Online]. Available: http://dx.doi.org/10.18517/ijaseit.12.3.13648.
[16] M. A. Iqbal, A. R. Aldaihani, A. Shah, "Big-Five Personality Traits Mapped with Software Development Tasks to Find Most Productive Software Development Teams," IJITEE, vol. 8, no. 12, pp. 965–971, Oct. 2019, doi: 10.35940/ijitee.J9755.1081219.
[17] E. Murphy-Hill et al., "What Predicts Software Developers' Productivity?," in IEEE Transactions on Software Engineering, vol. 47, no. 3, pp. 582-594, 1 March 2021, doi: 10.1109/TSE.2019.2900308.
[18] Dragicevic, S.; Celar, S.; Turic, M. "Bayesian Network Model for Task Effort Estimation in Agile Software Development." J. Syst. Softw.-Elsevier 2017, 127, 109–119.
[19] Turic, Mili, Stipe Celar, Srdjana Dragicevic, and Linda Vickovic. 2023. "Advanced Bayesian Network for Task Effort Estimation in Agile Software Development" Applied Sciences 13, no. 16: 9465. https://doi.org/10.3390/app13169465
[20] S. Čelar, Ž. Šeremet, Ž. Marušić, M. Turić, "Using of Web Objects Method in Agile Web Software Projects," Proceedings of 21st TELFOR, Beograd, IEEE, 2013, pp. 873-876
[21] S. Čelar, V. Vicković, E. Mudnić, "Evolutionary Measurement-Estimation Method for Micro, Small and Medium-Sized Enterprises Based on Estimation Objects," APEM, vol 7(2), 2012, pp. 81-92

DOI: 10.2507/34th.daaam.proceedings.002

ARTIFICIAL INTELLIGENCE AS AN ESSENTIAL FACTOR OF SUPPORT TO CORPORATE COMMUNICATION IN BUSINESS DECISION PROCESS

Berislav Andrlić, Krešimir Lacković & Ines Resler

This Publication has to be referred as: Andrlic, B[erislav]; Lackovic, K[resimir] & Resler, I[nes] (2023). Artificial Intelligence as an Essential Factor of Support to Corporate Communication in Business Decision Process, Proceedings of the 34th DAAAM International Symposium, pp.0009-0015, B. Katalinic (Ed.), Published by DAAAM International, ISBN 978-3-902734-41-9, ISSN 1726-9679, Vienna, Austria
DOI: 10.2507/34th.daaam.proceedings.002

Abstract

In more and more situations, information technologies are successfully replacing physical work, but also some mental human activities where artificial intelligence is increasingly relevant. In this paper, the goal of the research is determined, which is the use of artificial intelligence in defining corporate strategic decisions. The reason for this is because the decisions determine the long-term direction and intensity of the corporation as a complex economic system. In this sense, the hypothesis was put forward that artificial intelligence is a great support for the process of creating strategic decisions. That is why the features, experiences and possibilities of IT technologies, especially artificial intelligence, were investigated. Based on these findings, a communication model was set up where all participants in the process of creating artificial intelligence can use its capabilities. The discussion and concluding consideration confirm the goal and hypothesis, but also the need for research into innovations in IT technologies, especially artificial intelligence, with the goal of its even more effective use.

Keywords: artificial intelligence; process; corporation; strategy

1. Introduction

The development of technique and technology had its own particularly prominent epochs, so it all started with the first industrial revolution, dating from the turn of the 18th to the 19th century. It is the time when the peasant population turns into a civil and industrial one. James Watt and his steam engine played a major role in the first industrial revolution. The second started in 1870 and lasted roughly until 1914 and was marked by the rapid development of industry, especially steel and oil. In addition, there was a strong development of electrical engineering and mechanical engineering. The third industrial revolution began in the 1980s of the last century. It was marked by personal computers, the Internet and the explosive development of information and communication technology.

The present time is precisely the era of the fourth industrial revolution, and its main determinants are the following [10]:

- Robotics,
- Autonomous vehicles
- Internet of things,
- 3D printing
- Nanotechnologies,
- Quantum computers,
- Artificial Intelligence.

In the desire to replace human physical work, the development of machines that perform a series of mechanical actions, i.e. robots, is expected to be the basis for the largest number of industries by 2025 [4]. The Internet of Things is a new technology that networks and connects devices. With that step, networked devices and sensors enter our lives, becoming part of the Internet of Things. Today, every serious car company already owns or is developing new models and has in its program the creation and development of an autonomous model, i.e. a car without a driver. Printing, printing in three dimensions is a new method of quickly creating the necessary prototypes of various objects and devices, by which objects in space are created directly from the 3D CAD program. Three-dimensional printing is particularly valuable in many fields, such as: design, informatics, electronics/electrical engineering [10]. Nanotechnology, design, preparation, characterization and application of materials, devices and systems that are functional organized in the range of sizes 1 to 100 nm and characterized by phenomena and properties that occur only at these dimensions. Nanomaterials exhibit different properties compared to macro materials, which enables them to be used in various new applications. One or more properties can be different, individual particles can have multiple functionality, and sometimes the properties can be precisely regulated by size.[15]. Quantum computing in its simplest sense refers to technology that makes newly made computers easier adoption of transformational techniques to provide better results. This particular computation uses certain principles of quantum physics to revolutionize the entire process in which computing devices are operated today. Quantum computers can solve complex problems much faster than classical computers and have potential applications in cryptography and optimization.[16]. Artificial intelligence (UI, according to the English acronym AI, from Artificial Intelligence), is a part of computer science (informatics) that deals with the development of the ability of computers to perform tasks that require some form of intelligence, i.e. being able to navigate new situations, learn new concepts, draw conclusions, understand natural language, recognize scenes, etc. [10] The goal of this work is precisely research into the possibility of applying artificial intelligence in concrete economic practice. Corporate communication was chosen as the object of research, the goal of which is to make strategic decisions. Accordingly, the essential elements of strategic decisions will be considered. It is hypothesized that artificial intelligence can more quickly and reliably determine the essential elements or components of a business strategy. In this sense, the possibilities of artificial intelligence will be explored and a model of corporate communication and creation of strategic decisions will be established with the support of artificial intelligence. As part of the discussion and conclusion, the researched material as well as the communication model will be considered, from which the help of artificial intelligence in making strategic decisions in large corporations will be determined, thus confirming the hypothesis and the credibility of the research goal.

This is an intriguing topic that in itself opens up new spaces and goals for research into the use of artificial intelligence in various areas of the economy and all other social activities.

2. Corporate communication and strategic decisions

Artificial intelligence is already in use today in various economic activities. In this case, the possibility of its application as part of the information system of a large company or corporation is discussed. By its very definition, a corporation is a capital company that obtains funds for establishment and operations by issuing shares. It is a more developed form of a capitalist enterprise, and it is one of the forms of a commercial company. Corporations have eliminated almost all the risks that other legal forms of companies have because they managed to eliminate and find almost all the weaknesses that could endanger them or put them at risk, and in the case of bankruptcy of the corporation, the owner of the shares loses only what he invested in the shares. [12]. In addition, corporations have particularly developed in the USA and Great Britain, where the term refers to various types of organizations that have legal personality. In the US, a corporation is denoted by the suffix Inc. (abbreviated from incorporated), and in Great Britain with the addition of Ltd. (abbreviation of limited: (with limited guarantee). [15]. In complex companies or corporations there is a complex system of communication that includes several organizational forms and levels. The goal of this and any other communication can be itself, another person, or other entity such as a group, organization, or society. Similar to many related sciences, there are dozens, even hundreds, of different definitions of communication, and there is still no single one that is accepted by the majority of scholars working in this scientific field .[2].

Corporate communication is also a kind of orchestration of all instruments in the field of organizational identity (communication, symbols and behaviour of organization members) in such an attractive and realistic way that it creates or reflects a positive reputation among groups with which the organization is in a relationship of mutual dependence. The result is a great competitive advantage for the organization."[11].

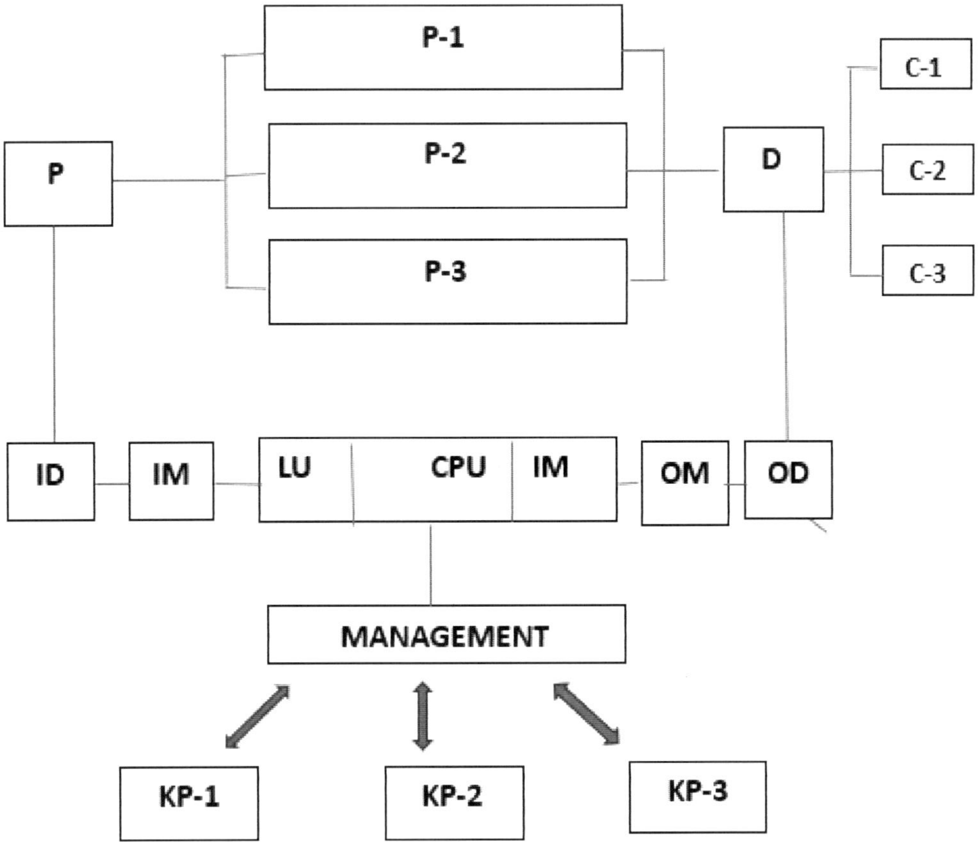

Fig. 1. The communication process of managing a corporation of three types of production

Legend: P-Sales, P1-P3-Production Phases, D-Distribution, C1-C3-Customers, ID-Input Device, IM-Input Module, LU-Logic Unit, CPU-Central Processing Unit, IM-Internal Memory, OM -output module, OD-output device, KP1-KP3-cooperative dislocated drives

In a practical sense, Figure 1 shows an example of a corporation that has three types of production at its headquarters. The picture shows the corporate management and communication process from the procurement of inputs to the distribution of products to customers. Complete control of the production process, system maintenance and cost process is enabled by one of the well-known systems that can be used to manage a complex production or service process, the programmable logic controller known as PLC (Programmable logic controller). According to the cited source, a programmable logic controller is defined as "A digital electronic device that uses programmable memory to store commands that require the execution of specific functions, such as logic functions, counting, timing, calculation, in order to control various types of devices and processes through digital and analogue input-output modules".[6]. Programmable logic controllers are industrial computers whose hardware and software elements are adapted to work in industrial conditions, and which can be easily programmed and incorporated into existing industrial but and business or organizational systems.

The presented technical approach can be connected using information technology to the management system, especially with costs, where each operation can be observed separately. In this way, cost savings can be investigated and accepted. That is why it is necessary to analyse and establish certain laws regarding the movement of yields, that is, income and costs in certain stages of production.

In this case, the subject of consideration will be focused on communication in order to make decisions with special emphasis on strategic decisions. The decisions themselves should be the result of a choice between several alternatives, and strategic decisions are oriented to the long-term business of the company and are characterized by three features [1]:

1. they are rare, usually without precedent
2. they are important, they require considerable resources and human efforts
3. they are directive, they create a basis for other decisions and actions in the company.

The fundamental responsibility of managers at the strategic level is in setting goals, defining the strategy for their achievement, continuous monitoring and analysis of the company's environment, and making strategic decisions, that is, decisions that affect the company as a whole [1]. In addition, effective managers at the strategic level of management are people who see the company as a whole, who can balance current business needs in relation to future ones and make up-to-date decisions based on that [3].

In order to emphasize the importance of strategic decision-making, the following explanations differentiate strategic from other types [9]:

1. Strategic decisions are extensive, risky and have significant long-term effects, require large amounts of resources and have long-term consequences for the company as a whole.
2. They represent a bridge between the designed and realized strategy
3. They eliminate boundaries between functions because, for example, strategic decisions on capital investment require the integration of finance and strategy.

The aforementioned facts that arise from corporate communication in order to make strategic decisions are the basis for the use of artificial intelligence in that process. Making specific strategic decisions is preceded by a thought-organizational process.

The process of making strategic decisions in a complex corporation begins with a vision (in some cases also a mission), continues with the definition of strategic goals and the determination of the plan as the final document of strategic decision-making. This is followed by the process of strategic management in order to realize the business strategy. Strategic management implies the realization of goals or plans and constant and periodic control. At the end of the strategic period, it is necessary to perform a comparative analysis in order to determine the success of the set strategy. [6]

3. Characteristics of artificial intelligence

Artificial intelligence (AI) is the most important general-purpose technology of our times. [17] During the development of society, first physical jobs but also certain routine mental actions were increasingly performed by machines. In this sense, artificial intelligence is increasingly at the origin of occupations today. The term artificial intelligence refers to those activities (algorithms, programs, etc.) that try to imitate the natural intelligence of humans. The combination of robots and artificial intelligence leads to the development of autonomous systems, which is a trend today. Autonomous systems are developed in a targeted way so that they can surpass humans in some properties (physical strength, memory capacity, calculation speed, parallel development of several controlled actions, etc.). The following three types of artificial intelligence are known [7]:

- Technical,
- Biotechnical,
- Biological.

The basic elements of technical artificial intelligence are: artificial neural networks, fuzzy logic systems, genetic or evolutionary algorithms, and expert systems. The term artificial neural network means a complex system, composed of elements (artificial neurons), which interact in a certain way, with each other and with the environment of the system (artificial neural network weights), so that they build a functional whole [7]. The advantages of neural computers, based on artificial neural networks compared to classical (serial) computers, are manifested in massive parallelism in information processing, then in adaptability and in the ability to learn [7]. Another technical approach is a system of indirect logic that works in linguistic form: if-then. For example: if the room is cold, then open the radiator valve. In this way, it is possible to linguistically describe the dynamics of a system whose mathematical model we do not know. For this reason, among other things, this method is widely accepted in the field of control systems. This approach has a limitation because the number of behaviour rules (if-then rules) grows exponentially with the increase in the number of input and output variables of the system, and with the number of adopted input and output fuzzy sets[14].

Genetic or evolutionary algorithms, as elements of artificial intelligence, use the principles of Darwin's theory of selection the better, the stronger win, i.e. different evolutionary paradigms. That is why genetic algorithms are suitable for searching (selecting) the best (optimal) solutions. [7]. Expert systems represent a set of computer programs, which are generated in such a way as to imitate a human expert in a field of interest. Newer expert systems contain the already mentioned elements of artificial intelligence, and they themselves belong to this set [4]. It is important to keep in mind that expert systems are not only programmed according to the decision-making mechanism and do not always behave according to expectations, that is, they can be programmed to learn while drawing conclusions[13]. Biotechnical artificial intelligence is being developed as a combination of technical and biological components of artificial intelligence. These are various experiments on connecting the human brain with electronic chips, electro-mechanical sensors and authors, etc. The value of this type of intelligence is reflected in the rehabilitation of people with various impairments (for example, vision) and disabilities (for example, hands, legs, etc.). On the other hand, biotechnical artificial intelligence can be accepted as a transitional phase to the development of biological artificial intelligence. This artificial intelligence relies on genetic engineering [7]. From the previous definitions, i.e. the particularities of artificial intelligence, it is evident that there are already various possibilities for its useful use today. In any case, it should be borne in mind that artificial intelligence, which is a special software, cannot create certain new solutions on its own for the time being, but must have an extremely complex database or memory of different conclusions according to the problem.

In this sense, Figure 3 shows an example of the use of artificial intelligence in a practical activity. The image shows a management process where all information and communications take place through one contact center known as CRM (Customer Relation Management) [8]

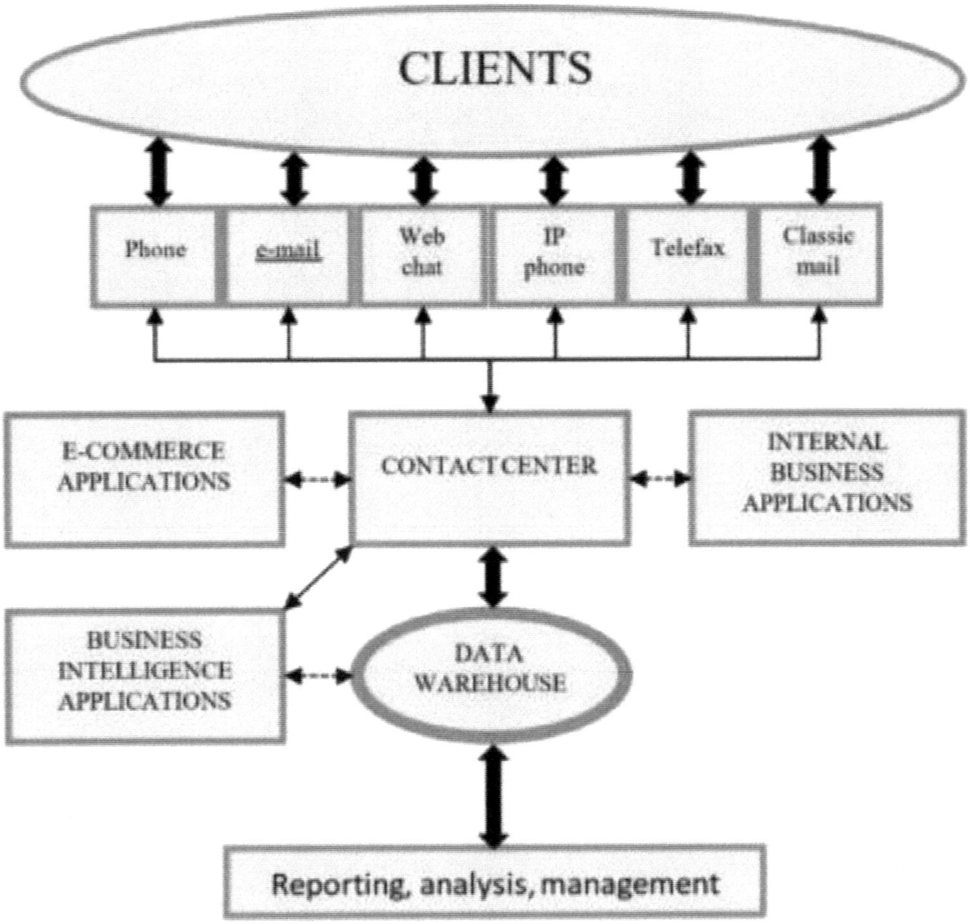

Fig. 2. An example of the use of artificial intelligence [5]

In this case, it is a communication process with clients in the technical maintenance of machines and equipment where the outsourcing maintenance provider is in constant contact with several clients and, if necessary, transmits information about a maintenance problem (facility or equipment). If he doesn't have a solution, he consults a database or artificial intelligence. For complex problems where a quick decision is needed, artificial intelligence can offer a quick solution. The condition for this is a large amount of problems and corresponding solutions stored, which must contain artificial intelligence software. This example can be used in various technical and other organizational management processes and also in making corporate strategic decisions.

4. A model of corporate communication and creation of strategic decisions with the support of artificial intelligence

Given that strategic decisions are of a long-term nature and very important for the direction and intensity of the corporation's development, it is necessary to approach their definition using human resources and information technologies. In addition, during the definition, the resources of the corporation in the headquarters itself, the management and the resources of its dislocated plants should be consulted. This is possible with the help of information technologies, and especially with the use of software that contains elements of artificial intelligence. Figure 4 shows a model of an example of management and communication of a corporation of three types of production, where each type has dislocated cooperative facilities that produce its inputs. In the technical and IT sense, this is possible by combining PLC with CRM as shown in the figure 3.

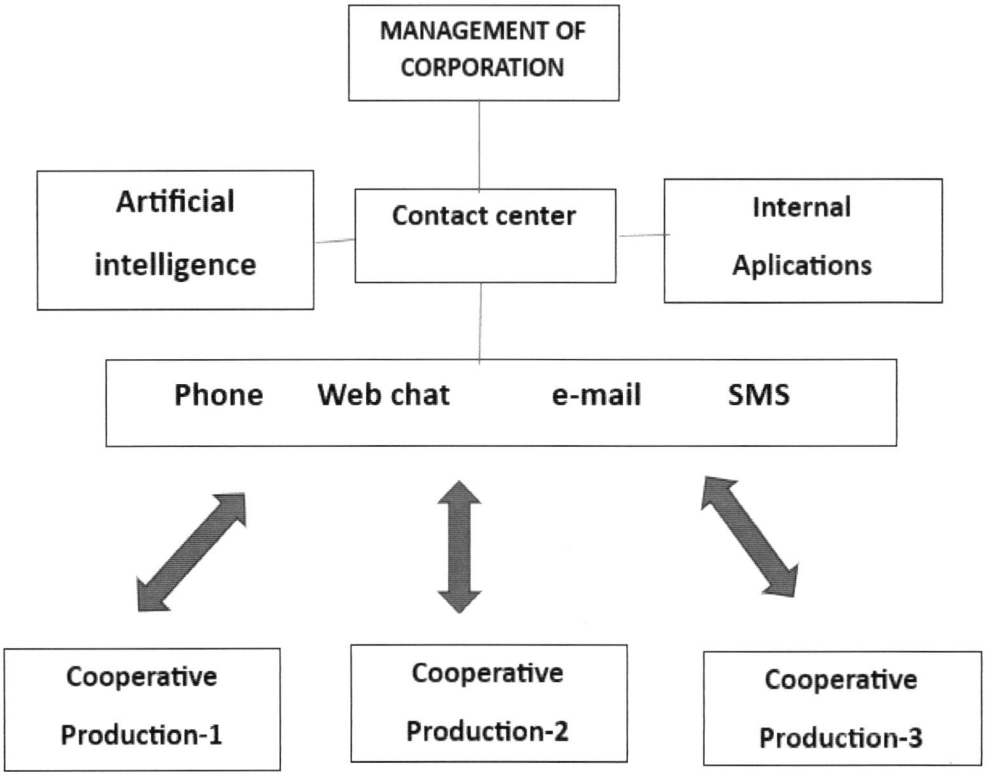

Fig. 3. Communication process of the corporation with the support of artificial intelligence

Figure shows an example of the communication process in this case in order to define strategic decisions at the corporate level. The appropriate human resources of the main administration, as well as personnel from cooperative plants, participate in the process. In this sense, the corporation's computer is connected to the CRM through the contact center. This is made possible by software that contains, in addition to drivers, special software for internal applications that allow connection to dislocated drives. A special role is played by artificial intelligence software, which contains a number of solutions that participants in the creation of strategic decisions can use to create proposals, test solutions and make the final decision. As an effective connection with the dislocated facilities, using the contact center, IT solutions were installed for fast communication through already known IT solutions such as; telephone, Web chat, e-mail and SMS. By using the aforementioned IT technologies and appropriate software, especially artificial intelligence, the management of the corporation with the help of the plant management can create more variants and choose the optimal strategic decision proposal. A special advantage of this kind of communication is the speed of proposing and testing certain decisions, and this is made possible by artificial intelligence.

5. Discussion

The previous considerations show, first of all, the complexity and great responsibility of all participants when making strategic decisions, especially when it comes to large complex companies or corporations. In modern times, information technologies can provide great help, both as support for the production and the complete business process. This enables quick, safe and precise access to relevant information when creating and making strategic decisions. A consideration of the special features of artificial intelligence shows that there are already great possibilities for its use even in making business decisions. There are already certain hardware and software solutions that are used to organize discussions where the participants are dislocated. In this example, it is about the communication of personnel at the headquarters of the corporation with the participation of participants from certain dislocated plants. This enables them to have CRM through the contact center. In support of the participants is artificial intelligence with the help of internal application software that enables the participation of dislocated participants. Considering today's possibilities of artificial intelligence in the application of making corporate strategic decisions, it is possible to increase the efficiency of the process from the previous considerations. This fact alone indicates that the hypothesis of this work as well as the goal of the research have been achieved. In addition, the modernity of the topic creates conditions for continued research. In addition, the development of information technology, especially artificial intelligence, will open up new opportunities for its use in various areas of the economy and other social activities.

6. Conclusion

Based on the research and consideration of the previous content, several conclusions are visible. First of all, it is an absolute justification for researching this very topic. In addition, the target was hit as well as the hypothesis set. Considering the topicality of a topic such as artificial intelligence, research into the possibility of its application in the economy is certainly one of the priorities. In this case, it is an application in making strategic decisions in a complex economic system such as a corporation. Actuality stems from the fact that strategic decisions determine the multi-year direction and intensity of the corporation's development. That is why, when considering and making strategic decisions, all the most important personnel and resources must be included in the decision-making process. This particularly applies to information technology, and in this case it is artificial intelligence. The paper first specified the purpose of using artificial intelligence, which is the process of making strategic decisions. By researching the experience, peculiarities and possibilities of artificial intelligence, it was determined that for now it is a great support for the participants who create and make strategic decisions. After that, a functioning model was set up where IT technologies create a fast and secure connection between decision-making participants. In particular, artificial intelligence can provide support in the creation and testing of individual proposals, as well as in the optimization of several possible combinations for individual decisions. The effectiveness of this process depends to a large extent on the database and solutions and on the perfection of the artificial intelligence software. In any case, the consideration of corporate decisions as an object and the uniqueness, possibilities of artificial intelligence and by defining the process and model of its functioning, showed the goal of the work and confirmed the hypothesis. With regard to the development of artificial intelligence, there is an opportunity to continue research. This applies to many activities in the economy and all other activities.

7. References

[1] Buble, M. (1978) Strategic Management, Faculty of Economics Split, p. 125.
[2] Čerepinko, D (2012) Information, Communicology: a brief overview of the most important theories, concepts and principles, University of Varaždin
[3] Drucker, P. F(1974) Management: Tasks, Responsibilities, Practices, Harper-Collins, New York, USA
[4] Hendler, J (1992).: A I into the 21st Century, IEEE Expert: Intell. Systems and Their Applications, vol. 7, no. 6, pp. 6-9.
[5] Kraljević, D., Lacković, K., Šojo, R. (2020) The Information – Communication Process a Business with Outsourcing for the Maintenance of Complex Technical System, Technical Journal, Vol 14, N. 2, p. 194-200.
[6] Lacković, Z. (2017): Lean-construction, Alberta, Osijek. p. 121.
[7] Novaković, B. (2015): Artificial intelligence, https://www.researchga.tenet/8403653 (excluded 07/02/2023)
[8] Panian, Ž. (2003). Relationships with clients in e-Business, Sinergija, Zagreb, p.115.
[9] Papadakis, M. V. et al.(1998) Strategic decision-making processes: the role of management and context, Strategic Management Journal, 19, p. 115-147.
[10] Prister, V., (2019), Artificial intelligence, Media Culture and Public Relations, Vol. 10, N.1, pp. 67-72.
[11] Tench, R., Yeomans L. (2009), Discovering Public Relations, Croatian Association for Public Relations, Zagreb
[12] Tomić, Z., (2016), Theory and practice of public relations, Synopsis, Zagreb, p. 329.
[13] Warwick K. (2012)Artificial intelligence: the basics, Routledge, p.12.
[14] Zadeh, L. A.(1965): Fuzzy Sets, Inform Control 8, p. 338-352
[15] Croatian encyclopedia
[16] Https://www.europarl.europa.eu/news/hr/headlines sosyeti Artificial intelligence (03.06.2023)
[17] Tekic, Z, Cosic,I., Katalinic, B.. (2019) Manufacturing and the rise of artificial intelligence, Annals of DAAAM & Proceedings, p. 192.

DOI: 10.2507/34th.daaam.proceedings.003

ON THE EFFECT OF ADVANCED PVD COATING ON PERFORMANCE OF CUTTING DIES IN AUSTENITIC STEEL

Miroslav Piska & Kristyna Holkova

This Publication has to be referred as: Piska, M[iroslav] & Holkova, K[ristyna] (2023). On the Effect of Advanced PVD Coating on Performance of Cutting Dies in Austenitic Steel, Proceedings of the 34th DAAAM International Symposium, pp.0016-0025, B. Katalinic (Ed.), Published by DAAAM International, ISBN 978-3-902734-41-9, ISSN 1726-9679, Vienna, Austria
DOI: 10.2507/34th.daaam.proceedings.003

Abstract

Effective machining of austenitic steels has been a long-term challenge due to the unique combination of mechanical properties, strain hardening, poor heat conductivity, and tendency to build up-edge formation of the steel when cutting. The round HSS cutting dies M10x1,5-6g, uncoated and PVD-coated with sputtering technology - TINALOX® GOLD, ALOX SN^2, and HIPIMs technology - FerroCon®, FerroCon ®+Ti,Al(C,N), three samples per each variant, produced in the company CemeCon Ivancice, Czech Republic. All PVD-coated cutting dies showed ~~a~~ very good cutting performance compared to the uncoated HSS dies, prolonging the tool lives 4-9 times overall. The longest effective time of chip formation for TINALOX® GOLD was 43-45 minutes. The higher recorded temperatures of chips (107-111°C) resulted in better formability of the material, compared to 104-105°C for uncoated HSS tools. The PVD coatings worked like ceramic insulators, preventing intensive heat flux into the tool, and enhancing ~~so~~ the plastic deformation of the austenitic steel. The best results for the round HSS cutting die were achieved repeatedly with the sputtering technology - TINALOX® GOLD, which provided excellent tribology between the tool, chip, and workpiece and high-quality threads.

Keywords: austenite; threads; PVD; wear; quality;

1. Introduction

Thread production is a widely used industrial operation that can be utilized in many ways for producing internal or external threads [1],[2],[3],[4]. Internal threads made by tapping are easier to test due to better access to the workpiece and dense hole arrangement. However, there is worse access for direct measurement of surface quality inside, so a precise splitting of samples is needed. There are many trends in the optimization of the technology [5],[6], nevertheless, the prevailing cutting materials in practice are HSS, protected with TiN, Ti(C,N) or Ti(Al,N) coatings made by physical vapor technologies (PVD) [7],[8]. The reason for HSS stems also in higher cutting speeds needed for cemented carbides that generate also high feed speeds, which can be a hazardous operation. The success in the coating of cutting flutes with hard and resistant layers depends not only on many physical and technological parameters of the deposition itself but also on the pre-processing of the functional tool surfaces [9],[10],[11],[12],[13]. The physical vapor deposition techniques, such as arc evaporation or magnetron sputtering, transport the atomic material according to the bias voltage and electromagnetic field in curved paths from the source to the substrate.

It results in irregular crystallization of the coating layer on the substrate, especially at the tooltip, and a breakage of the tip can occur easily. On the other hand, a shadowing effect results in the growth of an inhomogeneous film, so apart from thinner layers, poor adhesion can be found. This effect can be suppressed effectively by high-power impulse magnetron sputtering and a higher ionization rate of deposition material [14],[15]. The incoming particles have higher arrival energy and can align themselves in the shape of the substrate. By the HiPIMS discharge, a plasma density of up to 10^{19} m^{-3} can be achieved, resulting in the ionization degree of the target material up to 90 % [16]. The presence of energetic particles in the flux of coating material not only enables the coating of substrates placed perpendicularly to the target [17] but also effectively changes the morphology and associated properties of the coating. The coatings deposited by HiPIMS were reported to have a smooth, droplet-free surface [18], increased hardness [19], density [20], and adhesion [21]. In addition to that, HiPIMS also allows better control of thin film growth by phase tailoring [17], selective substrate biasing [22], or guiding the deposition material to the desired area of the substrate [23].

Furthermore, the optimal radius of a cutting edge at the recommended mesometric scale (in the order of magnitude ranging from 1 to 100 μm) [9] and it is difficult to reach also. These technologies belong to the know-how of many companies dealing with coating because effective coating depends on the surface quality, exact radius of cutting edge, etc. The standard methods for measurement of coating thickness by the so-called *kalotest* are replaced by more sophisticated measurements using X-Ray spectrometric methods [24]. The wear and tool life of tools for the production of threads can be studied from several aspects directly [25] or indirectly via torque analyses [26]. The torque data is possibly the most relevant signal for tool wear of tap tools according to the concept of Industry 4.0.

Precise machining of austenitic steels [27],[28],[29] is associated with many problems and difficulties due to the mechanical properties of the material. High strain hardening and low thermal conductivity can be mentioned as ~~the~~ typical problems resulting in long chips, high cutting temperatures, and a tendency to produce built-up edges. Testing of external threads made by a cutting die seems to be problematic in general, because of the limited length of tested samples and costly experiments. PVD coating seems to be problematic regarding the accessibility in the cutting die hole and grooves, but the worst problem stems from the complicated deburring of cutting edges needed for coating. Our previous industrial research has shown that the important reasons causing premature edge breakages and the production of wrongdoings are as follows:

- wear of tools that cannot be easily quantified or predicted safely in workshop conditions,
- non-standard or inferior quality cutting dies,
- non-convenient cutting speed,
- a mismatch of machine feed can cause the tool to break in tension or compression,
- a misalignment of the cutting die and workpiece axes,
- a clogging with chips due to poor cutting performance and poor chip evacuation,
- a poor cooling or lubricating (a use of improper cutting fluid or not powerful cooling flow).

Many of the problems can be detected with force sensors or energy consumption because they reflect the forces and cutting conditions [30], however, the number of applications ensuring the top quality of threads without any flaws is booming. The integrity imperfections can be harbors for infections, places for contaminations, or fatigue crack nucleation. The basic aim of this work is to find the best cutting material for excellent thread production made by round cutting dies.

2. Experimental Works

The material of the workpiece was austenitic stainless steel, CSN 10088-1 1.4301 (X5CrNi 18-10), with dimensional and shape deviation tolerances according to EN 9445– Table 1,2. The material is the most used stainless steel, referred to by the consumer names food-grade or 18-10. It is used in the food, pharmaceutical, and cosmetic industries, as well as in facade and residential architecture. The blank rods ϕ10h6/6000 mm were cut and machined to the dimensions according to Fig.1. The fastened workpieces were prevented from unwanted rotation with a hardened cross pin ϕ3 mm and mounted to a 3-jaw chuck in the machine spindle VR5A (Kovosvit, Sezimovo Usti) controlled with the Sinumerik 802D. The cutting dies were clamped to a standard holder, which was fastened to a 3-jaws chuck and fixed with screws on the top of the dynamometer KISTLER 9272. The radial alignment at the end was below 0.05 mm. The dynamometer set was placed into the CNC drill press– Fig. 2. Kistler dynamometer 9272, charge amplifiers 9011, and the Dynoware program for force and torque analyses of the sample loading were used. The sampling rate of 60kHz, the low-pass filter, and the long-time constant were set for all data acquisition. A special CNC program was written for the automatic control of the cutting operations.

Chemical composition (weight %)				
C	Si	Mn	P	Ni
≤0.08	≤0.10	≤2.0	≤0.045%	≤8.0-10.5
Cr	S	Fe		
18.00-20.00	≤0.030%	balance		

Table 1. Chemical composition of the tested material

Yield point R_e [MPa]	Tensile strength R_m [MPa]	Young modullus [GPa]
240 - 250	520 – 640	190

Table 2. Mechanical properties of the tested material

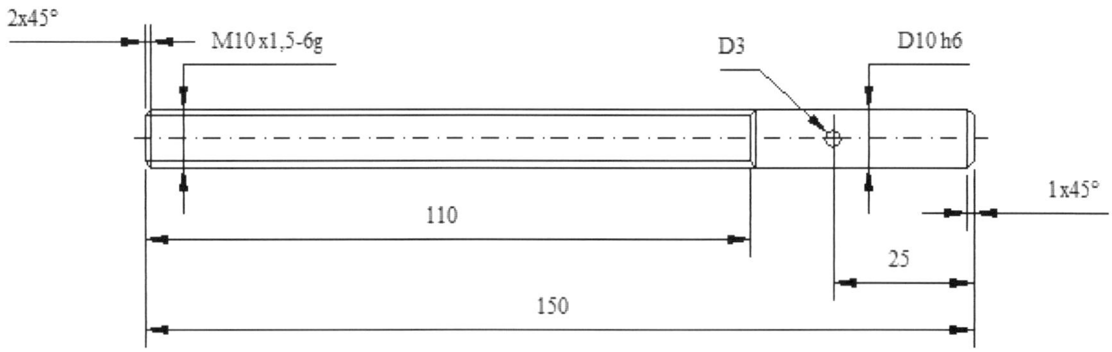

Fig. 1. Sketch of the specimen

a) b)

Fig. 2. Experimental set-up: (a) Kistler 9272 set-up and the machine (in the background); (b) a detail of the thread-cutting operation

The round cutting dies SE-B-gel-VA (Emuge Franken, Germany), article number D0103500.0100, constructional design DIN EN 22568, geometry VA, right-hand sense of rotation, lapped surface quality made from cutting material HSSE were used. Chamfer length was 2 threads, so 8 cutting edges effectively cut the specimen. Round HSSE die is recommended for universal application in stainless steel with excellent quality on the best CNC machines. Nevertheless, the cutting edges were not burr-free, so the ultrasonic polishing machine Piggy Sonic Mark 3 (NOVAPAX CZ s.r.o., Prague, Czech Republic) and diamond micro-grinding tools had to be used before the PVD coating - Fig. 3.

The cutting dies were tested as uncoated, but also PVD coated with unit CC800®/9ML Flex HiPIMS (CemeCon, Wurselen, Germany) working in the standard sputtering technology as CC800® (Tinalox Gold®, Alox SN^2) and FerroCon® and FerroCon® + TiAl(C,N)® working in the HiPIMS regime – Table 3. Each combination was tested three times. The PVD coating was made by the company CemeCon s.r.o., Ivančice, Czech Republic. The coating unit has six cathodes equipped with shutters. Two cathodes with pure Ti targets (99.999%) were used in DCMS mode for etching and four Ti targets with embedded Al slugs were powered by HiPIMS to provide the coating material.

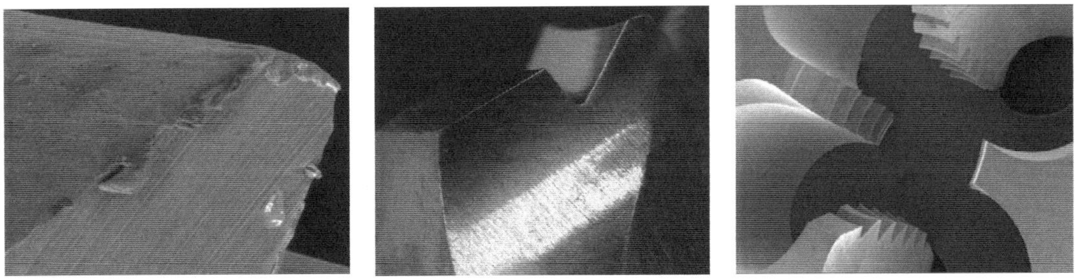

a) b) c)

Fig. 3. Deburring of cutting edges of round dies: a) before polishing, b) after polishing, c) after polishing and coating

The die was placed at the central substrate table, and the axis of the die was oriented perpendicularly to the target surface. Prior to the coating, etching with Ti ions was performed. The average power on the cathodes was 11.95 kW. A bias voltage of –59 V was applied to the substrate table. A mixture of nitrogen, argon, and krypton gases was used, with flow rates of 240, 311 and 213 ml.min^{-1}, respectively. The deposition rate was 39 nm.min^{-1} approximately. The scanning electron microscope MIRA3 (TESCAN, Brno, Czech Republic) was used for analyses of the secondary and backscattered electrons (SE, BSE), energy-dispersive X-ray spectroscopy, and measurement of the real thickness of the coating t – see Fig. 4 - according to equation (1):

$$t = \sqrt{R^2 - (\frac{C1}{2})^2} - \sqrt{R^2 - (\frac{C2}{2})^2}, \quad (1)$$

where R is the radius of the polishing ball, C1 represents the outer diameter, and C2 is the inner diameter of a dent in the surface.

Coatings	**Structure**	**Microhardness HV 0.05**	**Maximal working temperature [°C]**	**Thickness [µm]**
Tinalox Gold®	monolayer, $Ti(Al_xN_{1-x})$ supernitride	2300	600	2.8-3.0
Alox SN^2	multilayer, nanocomposite	3500	1100	2.7-2.9
FerroCon®	nanocomposite	3800	1100	2.8-2.9
FerroCon® + Ti,Al(C,N)	multilayer, nanocomposite	3800	1100-1150	2.9-3.0

Table 3. An overview of the coatings and their specifications (some details presents a company know-how)

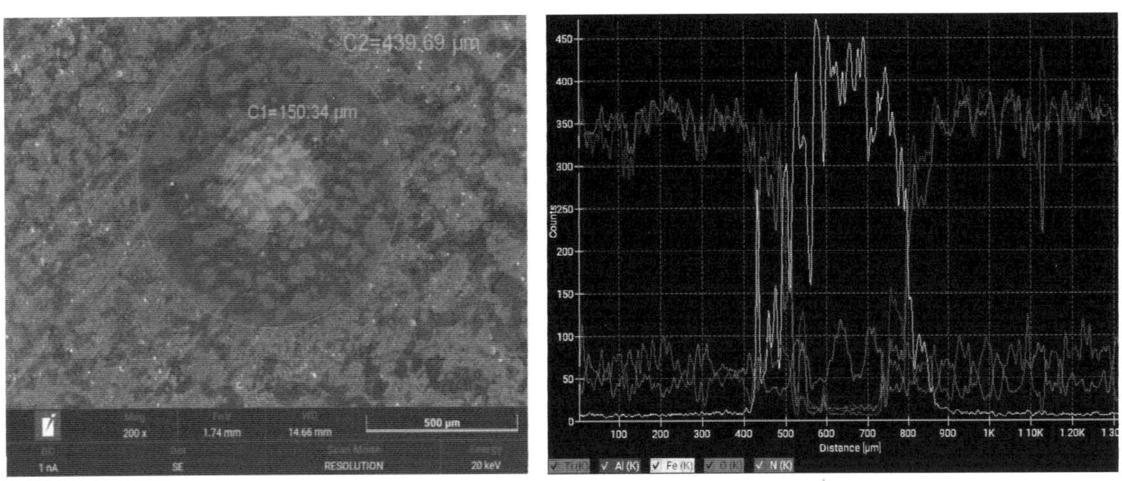

a) b)

Fig. 4. The principle of coating thickness measurements – the kalotest (a) and the linear EDS with the distribution of elements across the section, corresponding with the diameter of the ball and thickness of the layers

Cutting speeds 2.8-3.9-5.3-7.0-9.4 m/min were set in the tests, but the basic speed was 2.8 m/min. The cutting oil FUCHS ECOCUT 532 LE was used for the lubrication of machined surfaces. The ECOCUT 532 LE is a high-performance cutting oil that is chlorine-free, for machining difficult (stainless) materials and very difficult-to-machine materials (density at 15 °C - 0.94 g/ml, flashpoint 210 °C, kinetic viscosity at 40 °C 32 mm^2/s). The oil was applied to the tool with a brush during all cutting and chip production. The temperature of the cutting was recorded by the thermo-camera FLIR EP XT. The instant inspection of the threads was conducted by the caliper M10x1,5-6g VÖLKEL (MT Nastroje, Moravska Trebova, Czech Republic) and by the micrometric caliper INSIZE Czech, Ivancice). The shape deviation and quality parameters were checked by the Bruker Alicona IF G5, Graz, Austria, and by the optical microscope 3D SMART (HotAir, Ostrava, Czech Republic).

3. Results

The selected time series of the torque moments for dies with different coatings can be seen in Fig. 5. All coatings performed very well, but they differed in the stability of the torque moments reflecting the chip production.

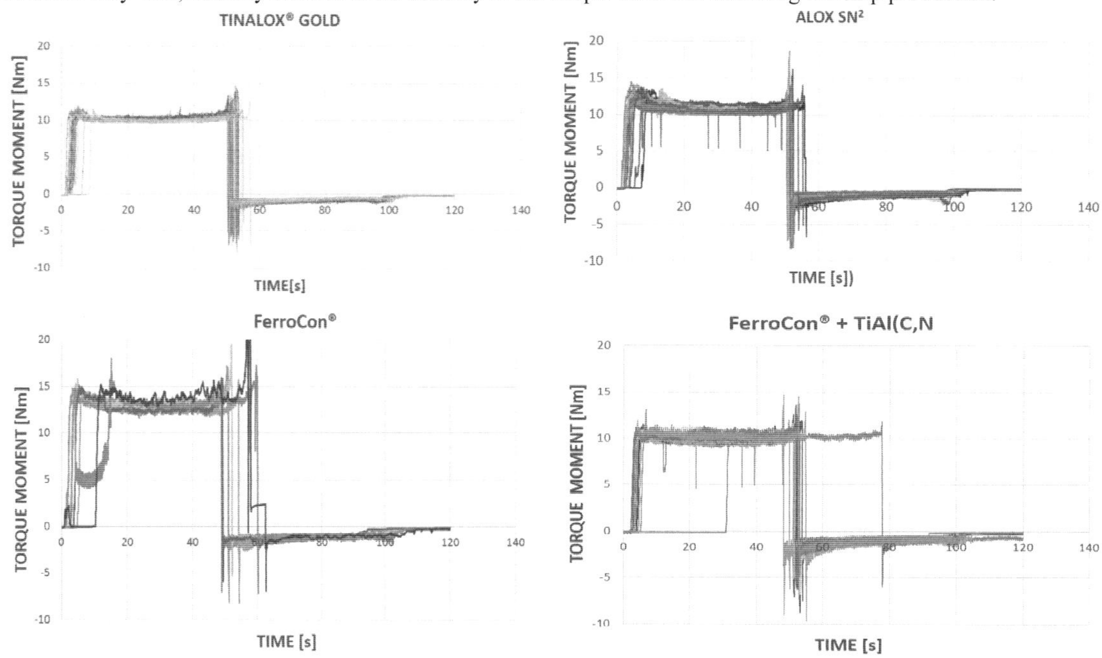

Fig. 5. The time series of torque moments for various PVD coatings

The TINALOX® GOLD and ALOX SN² exhibited excellent performance even at increased cutting speeds – Fig. 6, where no rise in torque moment was measured.

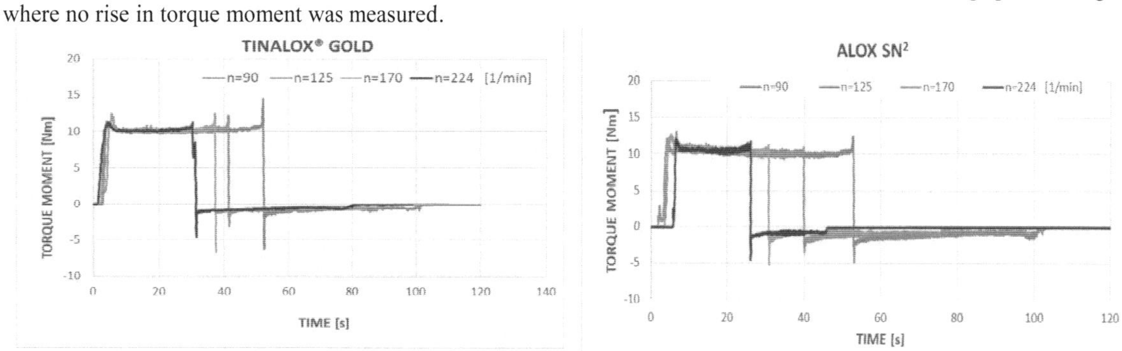

Fig. 6. The torque courses for selected PVD coatings - the effect of cutting speed on loading of the selected tools

The analysis of the cutting die design confirmed an irregular undeformed chip cross-section, but the specific cutting force acting on the cutting edges [30] was approximately stabilized – Fig. 7. It confirms a very good cutting geometry from the producer, because no cutting edge is too overloaded.

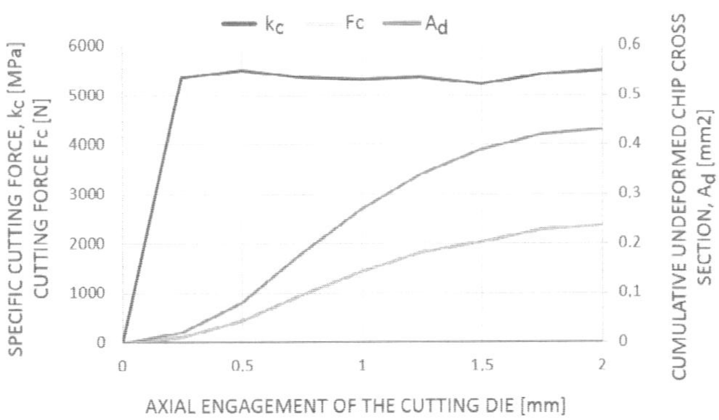

Fig. 7. A detailed analysis of the undeformed chip cross-section Ad, torque moment, cutting force Fc and specific cutting force kc acting on the cutting edges at the beginning of cutting (a selected pass of the TINALOX ® GOLD).

The thermal fields showed the following ranges of maximal temperatures for the cutting materials (Fig. 8):

- HSS: 104-105°C
- Tinalox Gold®: 107-111°C
- Alox SN^2 : 114-117°C
- FerroCon® + TiAl(C,N)® : 118-127°C
- FerroCon® : 128-130°C.

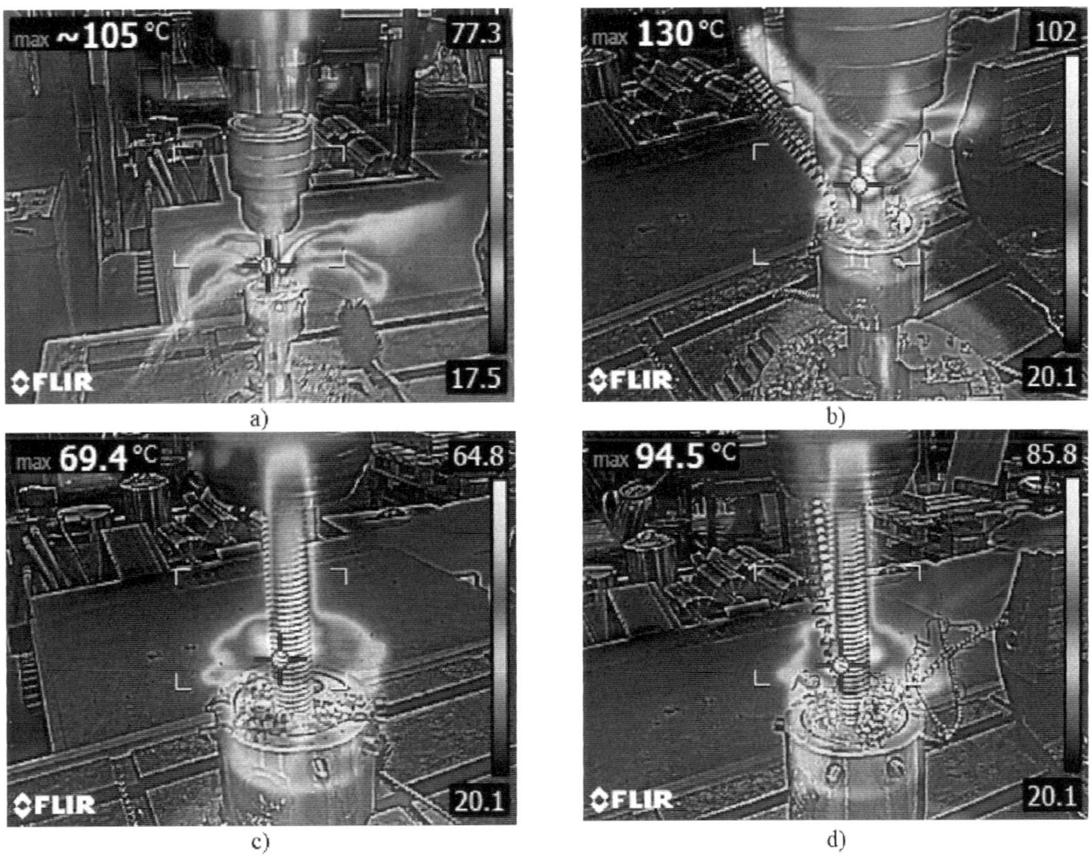

Fig. 8. Distribution of temperatures when cutting threads with a round die: a) uncoated HSS, b) HSS coated with FeroCon®, c) the workpiece after cutting by uncoated HSS, d) the workpiece after cutting by HSS coated with FeroCon®.

The change in surface quality could be seen visually first because of the change in the light contrast– Fig. 9.

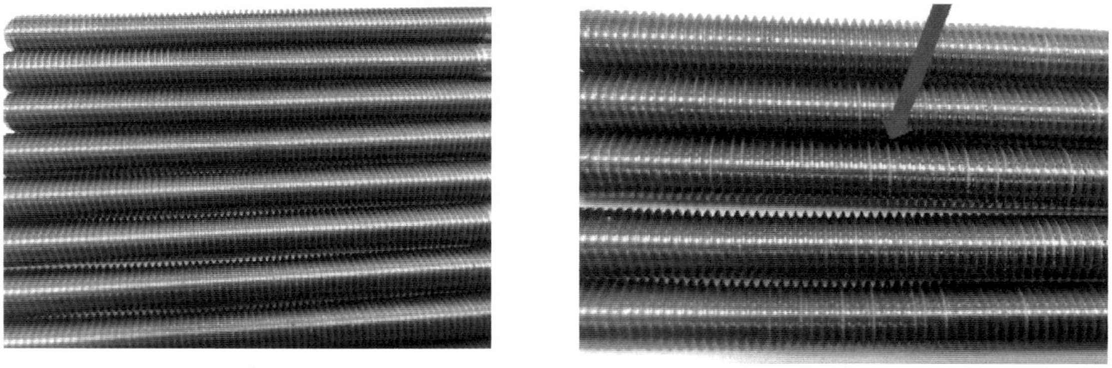

a) b)

Fig. 9. Selected samples: a) excellent surface quality of tested samples, b) first appearance of defects on the surface

Analysis of surface morphology by Alicona Infinity Focus G5 revealed many deviations from the right shape and dimensions of the threads – see Fig. 8,9. Most of them were initiated by the built-up edge, chipping of the edge, and a combination of the phenomena. However, excellent profiles have been found for the majority of the samples (except for uncoated HSS), and chip removal was running smoothly – Fig. 10. According to the quality of the threads, stability of the torque moment, and some other wear effects, the total cutting performance can be seen in Fig. 11.

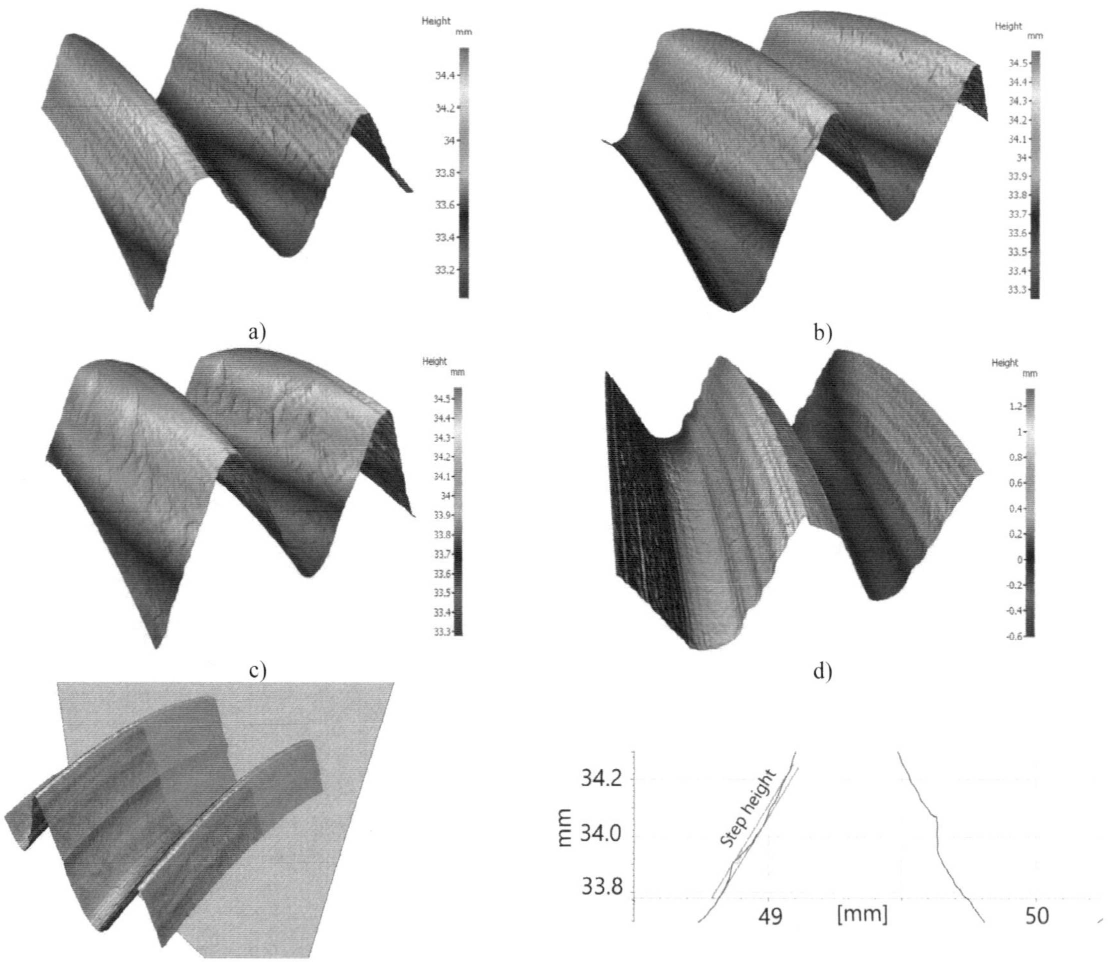

a) b) c) d) e) f)

Fig. 10. An overview of typical results a) the correct design of the thread, b) incorrect crest, c) side defects, d) completely affected all produced surfaces, e) a cross-section of the threads, f) evaluation of the shape discrepancies in the thread design.

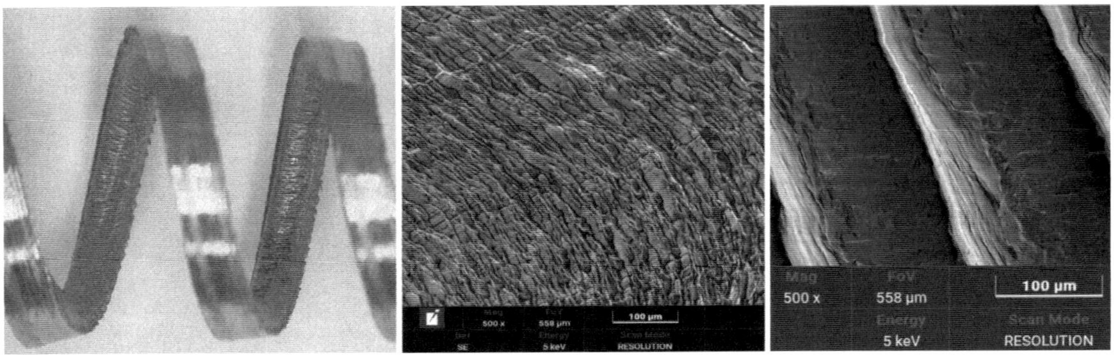

a) b) c)

Fig. 11. Chip produced by the TINALOX® GOLD coating: a) overview of the chip, b) detail from the chip, c) a magnified detail of the chip – a production of fine segments, with the convenient plastic flow of the austenite

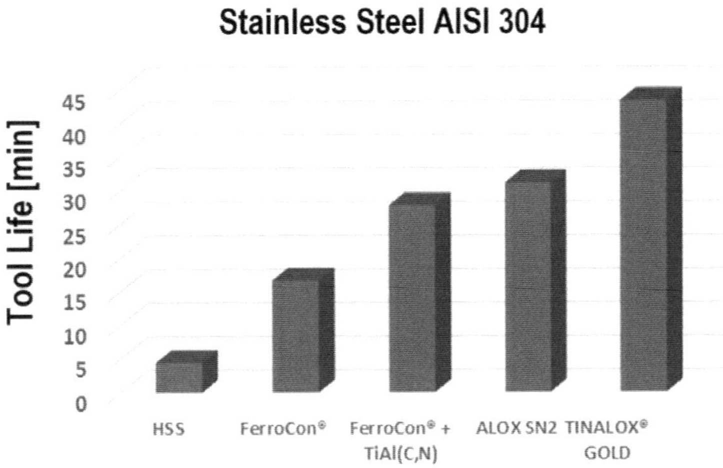

Fig. 12. The average tool life of the tested cutting dies. The integrated criterion of the wear was applied – a deterioration of the surface quality, irregular torque moment when cutting, and edge chipping.

4. Discussion

The experimental results confirmed the previous results in phenomena that could be expected in the technology of thread production in austenitic steel [27-29]. However, the standard sputtering technology is exhibited to prevail due to better tribology of the tested coatings against steel in the test conditions where the hardest coating is more sensitive to edge chipping and production of built-up edges, affecting the design of tools and produced threads. The higher temperatures of chips and workpiece surface seduce to exclude the lubrications, however, such tests ended with catastrophic failure and fracture of tools.

5. Conclusions

The four hard coatings have been proven to be very effective in thread cutting in austenitic steel AISI 304, all of them much better compared to the uncoated HSS round dies. The coatings provided a high wear resistance and prevented the cutting edges from production of built-up edges, and cold welding of the chip to the cutting tool. The longest effective time of chip formation for TINALOX® GOLD was 43-45 minutes. The higher recorded temperatures of chips (107-111°C) resulted in better formability of the material, compared to 104-105°C for uncoated HSS tools.

The PVD coatings worked like ceramic insulators, preventing intensive heat flux into the tool, and enhancing ~~so~~ the plastic deformation of the austenitic steel. The higher working temperatures resulted in a more convenient plastic flow of workpiece material and higher quality of the cut surfaces. The production of built-up edges is the worst problem in austenitic steel because it affects the thread quality and integrity. The best results with TINALOX® GOLD and ALOX SN^2 reflect good tribological properties. The phenomenon of BUE can be detected by the torque measurements, where sudden drops or rises in the loading reflect the problem, and the production of defective threads follows.

Anyway, to reach such good results with PVD coatings for threading in austenite steel means that the dies must be free of any burrs after grinding or polishing. These results are important for the next improvement of cutting die production – the PVD technology can significantly increase the cutting performance of the tool, but an effective deburring of the acting edges is needed. Ultrasonic polishing with diamond tools proved to be a very effective way for it, but it should be automatized or the CNC grinding machines should be able to do it. Perfect integrity of the tool surface and advanced PVD coating prevent built-up production and enhance the quality of threads, tool life, and economy of the production. The best results for TINALOX® GOLD will be tested also for tool-supportive guides (e.g. reaming heads) and work materials suitable for welding.

6. Acknowledgment

This research was funded by the Ministry of Industry and Trade in the Czech Republic. PID: grant FV40313, Application of the New Surface Treatment Technologies in Metal Packaging Industry and by the Brno University of Technology, Faculty of Mechanical Engineering, Specific research „Modern technologies for processing advanced materials used for interdisciplinary applications", FSI-S-22-7957. A special thanks belong to the company CemeCon, s.r.o., Ivančice for their kind support of PVD technology.

7. References

[1] Tlusty, J. (1999) Manufacturing Process and Equipment. 1st edition. Prentice Hall, 928 pp. ISBN 10-0201498650.
[2] Shaw, M.C. (2005) Metal Cutting Principles. Oxford University Press, 2nd ed., pp. 651, ISBN 0-19-514206-3
[3] Black, B.J. (2010) Workshop Processes, Practices and Materials. 4th ed., Elsevier Ltd, Oxford, UK.
[4] Patel, H.J.; Patel, B.P.; Patel, S.M. (2012) A review on thread tapping operation and parametric study. International Journal of Engineering Research and Applications (IJERA), 2(3): 109–113
[5] Fromentin, G., Poulachon, G., Moisan, A. (2002). Thread forming tapping of alloyed steel, ICME Proceedings, Naples, Italy, 115-118.
[6] Píška, M.; Sliwkova, P. (2015) A study of cutting and forming threads with coated HSS taps. Journal of Machine Engineering, 15(3): 65–74
[7] Bouzakis, K.-D.; Michailidis N.; Skordaris, G.; Bouzakis, E.; Biermann, D.; M'Saoubi, R. (2012) Cutting with coated tools: Coating technologies, characterization methods and performance optimization. CIRP Annals – Manufacturing Technology, 61(2): 703–723. doi: 10.1016/j.cirp.2012.05.00
[8] Inspektor, A. Salvador, P.A. (2014) Architecture of PVD coatings for metalcutting applications: A review. Surface & Coatings Technology 257138–153
[9] Rech, J. (2006) Influence of cutting edge preparation on the wear resistance in high speed dry gear hobbing. In Wear. Vol. 261, Issues 5–6, pp. 505-512.
[10] Dana, M.; Zetek, M. & Schorník, V. (2016). Effect of Cutting Edge Geometry on Cutting Tool Life when Drilling Inconel 718, Proceedings of the 26th DAM International Symposium, pp.0709-0714, B. Katalinic (Ed.), Published by DAAAM International, ISBN 978-3-902734-07-5, ISSN 1726-9679, Vienna, Austria DOI: 10.2507/26th.daaam.proceedings.
[11] Vopat, T.; Kuruc, M.; Simna, V.; Zaujec, R. & Peterka, J. (2017). Cutting Edge Microgeometry and Preparation Methods, Proceedings of the 28th DAAAM International Symposium, pp.0384-0391, B. Katalinic (Ed.), Published by DAAAM International, ISBN 978-3-902734-11-2, ISSN 1726-9679, Vienna, Austria DOI: 10.2507/28th.daaam.proceedings.
[12] Adoberg, E.; Podgurski, V.; Peetsalu, P.; Lind, L.; Mikli, V.; Hvizdos, P.; Kulu, P. (2012) The effect of surface pre-treatment and coating post-treatment to the properties of TiN coatings. Estonian Journal of Engineering, 18(3): 185–192. doi: 10.3176/eng.2012.3.03
[13] Vozar, M.; Patoprsty, B; Vopat, T. & Peterka, J. (2019). Overview of Methods of Cutting Edge Preparation, Chapter 20 in DAAAM International Scientific Book 2019, pp.251-264, B. Katalinic (Ed.), Published by DAAAM International, ISBN 978-3-902734-24-2, ISSN 1726-9687, Vienna, Austria DOI: 10.2507/daaam.scibook.2019
[14] Bräuer, G., Szyszka, B., Vergöhl, M., Bandorf, R. (2010). Magnetron sputtering - Milestones of 30 years. Vacuum, 84(12), pp. 1354–1359
[15] Anders, A. (2017). Tutorial: Reactive high power impulse magnetron sputtering (R-HiPIMS). Journal of Applied Physics, 21(17), p. 171101
[16] Bohlmark, J., Alami, J., Christou, C., Ehiasarian, A. P. and Helmersson, U. (2005). Ionization of sputtered metals in high power pulsed magnetron sputtering. Journal of Vacuum Science & Technology, 2005, Vol.23, No.1, pp 18–22, ISSN 0734-2101
[17] Alami, J., Eklund, P., Andersson, J. M., Lattemann, M., Wallin, E., Bohlmark, J., Persson, P., Helmersson, U. (2007). Phase tailoring of Ta thin films by highly ionized pulsed magnetron sputtering. Thin Solid Films, 2007, Vol.515, No.7–8, pp 3434–3438, ISSN 00406090
[18] Sarakinos, K., Alami, J., Wuttig, M. (2007). Process characteristics and film properties upon growth of TiO x films by high power pulsed magnetron sputtering. Journal of Physics D: Applied Physics, 2007, Vol.40, No.7, pp 2108–2114, ISSN 0022-372

[19] Engwall, A. M., Shin, S. J., Bae, J., Wang, Y. M. (2019). Enhanced properties of tungsten films by high-power impulse magnetron sputtering. Surface and Coatings Technology, 2019, Vol.363, pp 191–197, ISSN 02578972

[20] Samuelsson, M., Lundin, D., Jensen, J., Raadu, M. A., Gudmundsson, J. T., Helmersson, U. (2010). On the film density using high power impulse magnetron sputtering. Surface and Coatings Technology, 2010, Vol.205, No.2, pp 591–596, ISSN 02578972

[21] Ehiasarian, A.P., Wen, J. G., Petrov, I. (2007). Interface microstructure engineering by high power impulse magnetron sputtering for the enhancement of adhesion. Journal of Applied Physics, 2007, Vol.101, No.5, ISSN 0021-8979

[22] Greczynski, G., Lu, J., Jensen, J., Bolz, S., Kolker, W., Schiffers, Ch., Lemmer, O., Greene, J. E., Hultman, L. (2014) A review of metal-ion-flux-controlled growth of metastable TiAlN by HIPIMS/DCMS co-sputtering. Surface and Coatings Technology, 2014, Vol.257, pp 15–25, ISSN 02578972

[23] Bohlmark, J., Östbye, M., Lattemann, M., Ljungcrantz, H., Rossell, T., Helmersson, U. (2006) Guiding the deposition flux in an ionized magnetron discharge. Thin Solid Films, 2006, Vol.515, No.4, pp 1928–1931, ISSN 00406090

[24] ISO 3497. (1990) Measurement of Coating Thickness of Metallic Coatings, X-Ray Spectrometric Methods. Available online: https://www.iso.org/standard/8850.html (accessed on 12 May 2022)

[25] Landeta, J.F.; Valdivielso, A.F.; López de Lacalle, L.N.; Girot, F.; Pérez Pérez, J.M. (2015) Wear of form taps in threading of steel cold forged parts. Journal of Manufacturing Science and Engineering, 137(3): 031002-031002-11. doi: 10.1115/1.4029652

[26] Korhonen, H. , Koistinen, A. & Lappalainen, R. (2018): Improvements in the thread cutting torque for a 6082-T6 aluminum-based alloy with tapping tools utilizing diamond coating, Machining Science and Technology, DOI: 10.1080/10910344.2017.1402930

[27] Reiter, A.E., Brunner, B., Ante, M., Rechberger, J. (2006). Investigation of several PVD coatings for blind hole tapping in austenitic stainless steel. Surface and Coatings Technology, 200(18-19), pp. 5532–5541. https://doi.org/10.1016/j.surfcoat.2005.07.100

[28] Píška, M.; Polzer, A. (2012) O n the advanced PVD coatings for threading in austenitic steel. Proceedings of the 23rd DAAAM International Symposium on Intelligent Manufacturing and Automation, October 24–27, Zadar, Croatia. 23(1): 0831–0834.

[29] Píška, M.; Sliwkova. (2014). On The Cutting Performance of Coated HSS Taps when Machining of Austenitic. Technological Engineering. Volume XI, number 1/2014 ISSN 1336 – 5967, pp. 28-32. DOI: 10.2478/teen-2014-0006

[30] ISO 3002-4 (Basic Quantities in Cutting and Grinding - Part 4: Forces, energy, power).

DOI: 10.2507/34th.daaam.proceedings.004

MODELING OF THE PERIODIC LATTICE STRUCTURE FOR EXPERIMENTAL, NUMERIC AND VIBRATION ANALYSIS

Davorka Šaravanja, Mario Soldo

This Publication has to be referred as: Saravanja, D[avorka] & Soldo, M[ario] (2023). Modeling of the Periodic Lattice Structure for Experimental, Numeric and Vibration Analysis, Proceedings of the 34th DAAAM International Symposium, pp.0026-0034, B. Katalinic (Ed.), Published by DAAAM International, ISBN 978-3-902734-41-9, ISSN 1726-9679, Vienna, Austria
DOI: 10.2507/34th.daaam.proceedings.004

Abstract

The paper describes the compression testing of 3D printed plastic lattice structures, the categorization of the selected material by modeling and 3D printing of plastic samples and their testing on a universal machine with a tensile test. Modeling of lattice structures was performed for experimental and numerical analysis as well as additional vibration analysis to optimize the isolation and damping properties of the lattice structure. Numerical analysis using the ANSYS program compares the results obtained by the compression test and the finite element method. Experimental and numerical analysis showed that the shape, i.e. the geometry of the unit cell affects the value of the maximum force, and the vibration analysis showed that in additive manufacturing, the orientation has no significant effect on the vibration parameters except for the modulus of elasticity and that the increase in damping can be achieved by adding external components to the lattice structure in order to improve its ability to absorb vibration energy, such as highly elastic materials, special types of dampers or completely new elements. In the conclusion, the future phases of the research are defined.

Keywords: lattice structure; tensile and compressive test; analysis; vibration damping; isolation.

1. Classification and application of lattice structures in engineering

Functionally periodic lattice structures are defined as networks of connected solid struts or plates that form the edges and faces of the cell. They are divided into solid two-dimensional (2D) cellular materials (examples are honeycombs) and three-dimensional (3D) cellular structures, which are further divided into 3D open-cell foams and 3D closed-cell foams. They can be stochastic cellular structures (foams) and non-stochastic cellular structures (periodic cellular structures; lattices) as shown in Fig. 1.

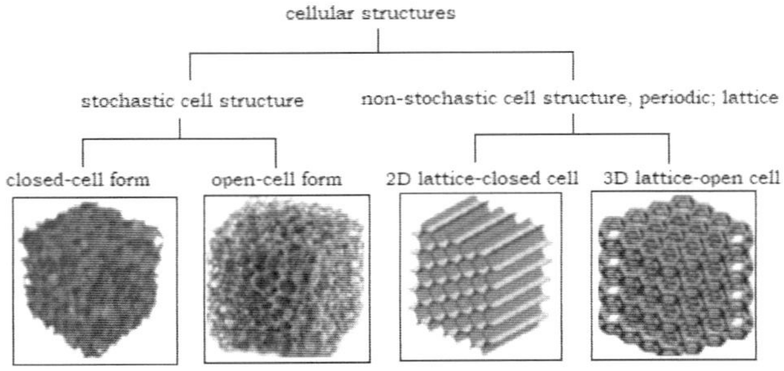

Fig. 1. Classification of cellular structures [1]

Stochastic structures are further divided into foams with open cells and foams with closed cells, while non-stochastic cellular structures or lattices are divided into two-dimensional lattices (with closed cells) and three-dimensional lattices (with open cells). Lattice structures are also defined as topologically determined, three-dimensional open-cell structures composed of one or more repeating unit cells. These cells are defined by certain dimensions and connected by component supporting elements (struts) which are connected by specific nodes [2] [3]. Lattice structures are divided here into 2D and 3D structures, where 2D lattice structures are considered two-dimensional shapes extruded into the third dimension. It is also stated that 3D lattice structures are often found in the literature under the name of strut-based lattices (Fig. 2 left), while there are also triply periodic minimal surfaces (TPMS) (Fig. 2 right) .

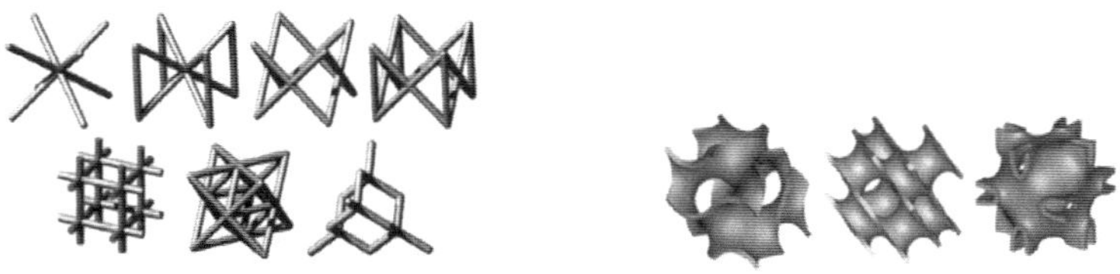

Fig. 2. Structures based on struts; triply periodic minimal surfaces [2]

There are also several more recent divisions of lattice structures: an example is lattice structures divided depending on the arrangement of the unit cell into random and periodic lattice structures. A special subtype of lattice structures is the increasingly common graded (gradient) lattice structure, (Fig. 3).

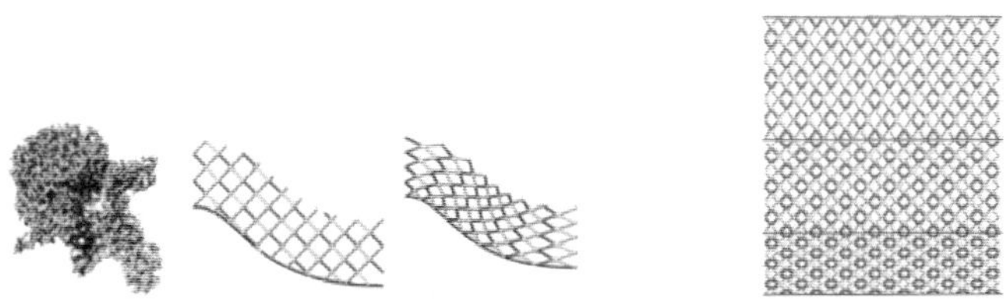

Fig. 3. Random and periodic lattice structures; graded periodic lattice structure [4]

Functionally graded lattice structures are a unique type of lattice structures where the density of the structure is optimally distributed, thus improving the performance of the lattice structure in contrast to uniformly distributed density [5]. Lattice structures have outstanding properties such as lightness, high specific strength and stiffness, energy absorption, high damping and high vibration reduction. Due to their excellent properties, lattice structures have found wide application in the aviation industry, automotive industry, bioengineering as well as other industrial fields. For example, in the field of aviation, not only are excellent mechanical properties expected anymore, but also reduced structural weight. Lattice structures have both biocompatibility and high strength, which can be incorporated into human tissue and bones for the purpose of replacing diseased organs. Lattice structures are particularly widely used in medicine due to their flexible mechanical properties and structural characteristics that can meet specific requirements [4]. The introduction of additive manufacturing in the production of complex cellular structures enabled their wide application in engineering, especially in aircraft structures, automotive parts and engineering medical products. Aircraft components must be strong, but also lightweight, and are usually made of various alloys and ceramics that are very expensive and hardened at ultra-high temperatures. Due to their good strength-to-weight ratio, lattice structures are used in the aerospace industry to improve performance-to-weight ratios that can increase flight efficiency.

2. Previous research on periodic lattice structures

Periodic lattice structures as cellular solids have highly efficient energy absorption properties and are suitable for further improvement by designing their fundamental structures. Previous research and studies have classified different groups of trusses based on their topology and which have shown the way to a new strategy to increase their ratio of energy absorption to weight under compression, but also the possibility of optimizing vibration isolation and vibration transmissibility (Fig. 4.). Experimental compression and vibration excitation test results and numerical data obtained from finite element analysis show that a uniform design with an even distribution of relative density provides the highest initial stiffness among all 3D printed lattice architectures [6]. However, a graded design with a rational variation of relative density can significantly increase the stiffness and energy absorption capacity of trusses that are exposed to high compressive stresses as well as vibrational excitation forces. Specific structures in which the relative density of cellular cells varies normally to the direction of the external pressure force can increase the stiffness and energy absorption capacity of cellular solids from 60% to 110% [1]. The possibilities of designing unicellular lattice structures that can combine the properties of light weight and energy absorption by rational variation of porosity within the cellular architecture have also given excellent results after research.

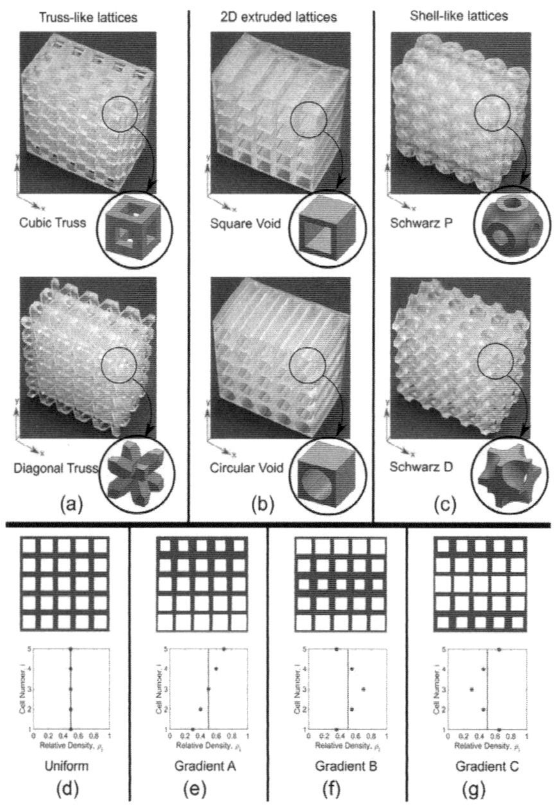

Fig. 4. Different categories of lattice structures with periodic microarchitectures [1].

3. Characterization of the material

PLA (polylactide) material was chosen for 3D printing of plastic test tubes for research purposes due to its good mechanical properties, relatively low price, but also because they have not been sufficiently researched. PLA is the optimal choice for prototypes that have special requirements and is by far the most popular material used in FDM/FFF (Fused Deposition Modeling / Fused Filament Fabrication) 3D printing. Since the material properties of the raw material change during the printing process due to a series of parameters (extruder temperature, substrate temperature, air humidity, print speed, age of the material, etc.), it is necessary to strictly adhere to the same parameters in order to be able to achieve reproducibility of the results. In the program SolidWorks 2020, a test tube was modeled for the characterization of PLA material according to the ISO 527-2 standard (Fig. 5.), and printed on a Zortrax M200 Plus printer in .stl format, using the appropriate program Z-Suite (Fig. 6.). Zortrax M200 Plus is a professional FFF printer that is designed for maximum reliability and continuous operation for a long period of time. The working volume of this printer is 200x200x180 mm, and when printing it is important that the model can fit into such a volume.

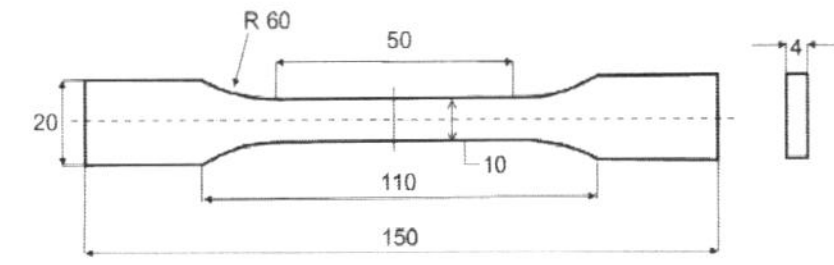

Fig. 5. Standard test sample dimensions according to the ISO 527-2 standard

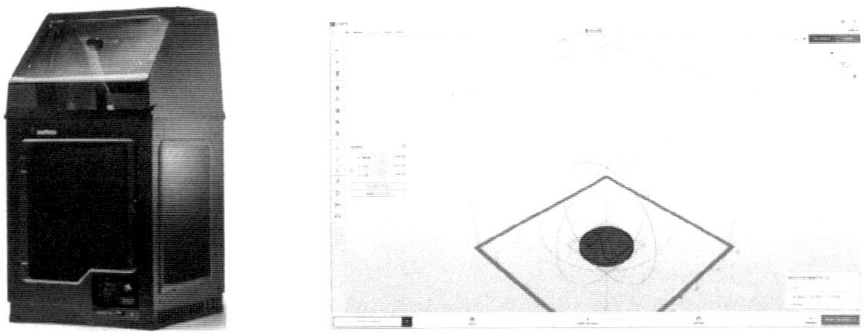

Fig. 6. Zortrax M200 Plus printer; The working environment of the Z-Suite program

In order to enable test reproducibility and determine valid results for subsequent simulation, 5 identical samples (Fig. 7.) are selected for printing and subjected to tensile testing on a universal testing machine with highlighted specifications and the Trapezium X program for test test settings, Fig. 8.

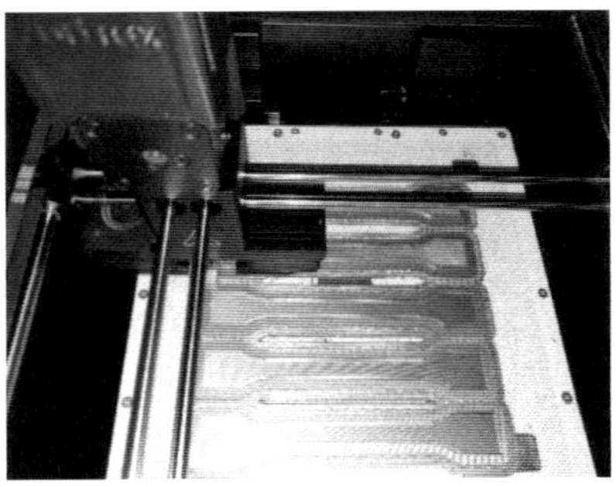

Fig. 7. The process of printing test samples for testing material properties

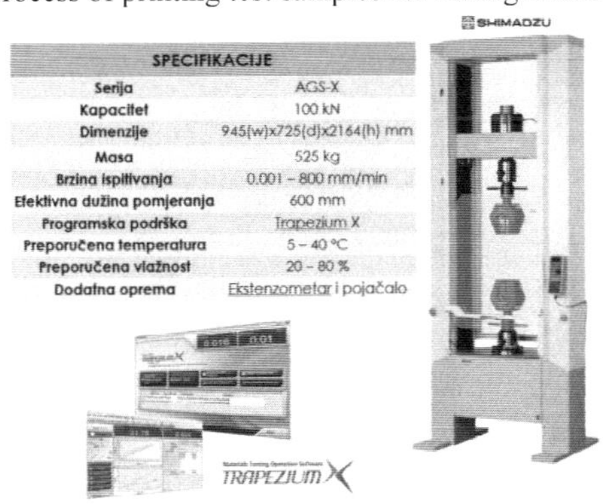

Fig. 8. Universal compressive, tensile, and two- or three-toint tending testing machine with specifications highlighted and Trapezium X test setup program

For the tensile test of the samples, the test parameters of all 5 printed samples were defined and the average of the results was requested. An initial speed of 1 mm/min was selected until a deformation of 0.5% was reached. After reaching a deformation of 0.5%, the speed of the test was 20 mm/min until the test tube broke, which ended the test. After testing all 5 samples, all the necessary data are read from the graphs, which are used for further numerical analysis of lattice structures using the finite element method. The average maximum force was 1,682 N, the modulus of elasticity was 2,157 MPa (between 0.05% and 0.25% deformation), while the average deformation at the break was 5.7%. (Fig. 9.)

Product Name	PLA prosjek epruveta 1-5	Operator	Mario Soldo
Standard	ISO 527-2	Test Date	28/06/2023
Temperature	24° C	Testing Machine	AGS-X
Test Type	Tensile	Speed	1mm/min
Shape	Plate		

Name	Break Stress	Break Strain	Max Force	Elastic
Parameters	Sensitivity: 10	Sensitivity: 10	Calc. at Entire Areas	Stroke Strain 0.05 - 0.25 %
Unit	MPa	%	N	MPa
S1	39.29	5.8	1680	2193
S2	38.90	6.4	1708	2162
S3	37.84	5.1	1647	2136
S4	40.27	4.9	1713	2164
S5	38.47	6.3	1664	2130
Average	38.95	5.7	1682	2157
Standard Deviation	0.91144	0.68191	28.2365	25.1992
Range	2.43000	1.50000	66.0000	63.0000

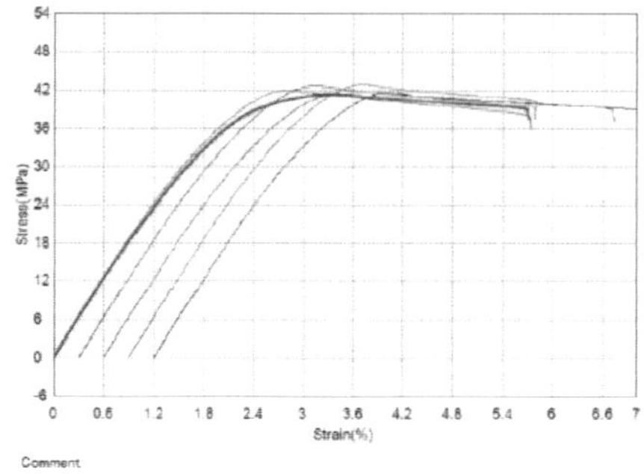

Fig. 9. Tensile test settings for testing plastic test samples; tensile test results of 5 PLA test samples for the purpose of material characterization

4. Experimental and numerical analysis

There are several software packages for static and dynamic structural numerical simulation, which also include the function of modal analysis of the lattice structure to external vibration excitation, e.g. CATIA, SolidWorks Simulation, ANSYS Workbench, ABAQUS, nTop for control and fine tuning, optimization of the lattice structure with variable thickness and smooth transitions [7]. The most important input parameters for the execution of the simulation are boundary conditions, material addition, coefficient of friction, joining of samples, test speed, mass change, amount of excitation force during vibration testing and, of course, simulation results. It must be noted here that pattern linking is difficult to perform due to the complexity of the functionally graded pattern lattice, and in this case pattern symmetry can be a valuable tool when performing a simulation of the functionally graded lattice structure. The goal of using tools and different simulation methods during experimental testing is to reduce weight, and this is achieved by: choosing materials, consolidating individual cells and their individual parts, and optimizing topology.

4.1. Modeling and printing of lattice structure

To test the behavior of periodic lattice structures, the Simple Cubic sample of unit cells was chosen (because of its simplicity of geometry and accuracy of printing), which is multiplied into a cubic volume (dimensions 50x50x50 mm) using the newly designed Grasshopper algorithm in the program Rhinoceros. The volume of the unit cell is 10x10x10 mm. Multiplication is selected in 3 directions by the same repeating principle (5x5x5), Fig. 10. [8]

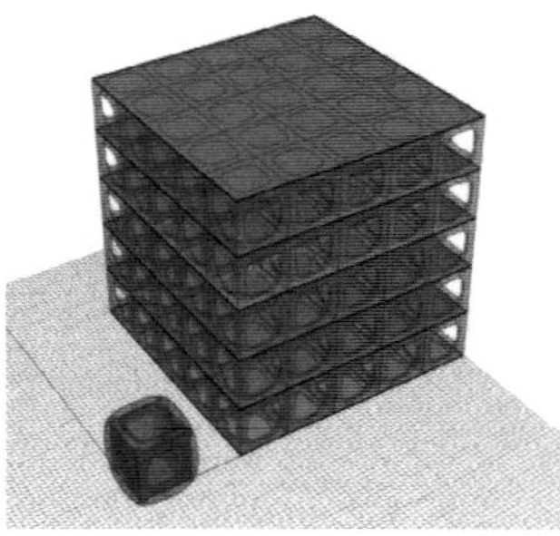

Fig. 10. The final model of the Simple Cubic lattice structure

For the generation process, the Grasshopper algorithm was used to construct samples for compression testing in order to determine the mechanical properties of PLA homogeneous lattice structures obtained by FDM printing. To achieve a relative density of the lattice structure of 0.2, a relative density calculation plugin was inserted and a strut radius of 1.43 mm was selected using the MultiPipe command. After modeling, the model is saved in .step format for loading in the Z-Suite program to generate .stl files that are suitable for loading on a Zortrax 3D printer.

4.2. Compression test of lattice structure

The compression test of the Simple Cubic periodic lattice structure was performed on a Shimatzu universal machine with compression plates in place [9]. The test settings were: test speed 5 mm/min until a traverse movement of 38 mm is achieved; the height or the distance between the plates is 51 mm; the total deformation of the lattice structure is 37 mm (74 %), because by then the lattice structure would have been completely damaged and the phase of thickening of the material would have begun.

Out of 6 printed lattice structures, 4 samples with the best print and the closest mass compared to the original model from the Rhinoceros program were selected for testing. The other 2 samples were discarded from further testing due to printing errors (the error is in the discontinuously connected struts between the unit cells. The layer-by-layer collapse of the lattice structure is shown in Fig. 11.

Fig. 11. Collapse of the lattice structure with respect to the increase in the percentage of deformation

The results showed that the maximum force occurs at an average of 1.2 mm of deformation (1 mm is deducted from the results due to idle travel of the traverse) and is 13,044 N (Fig. 12).

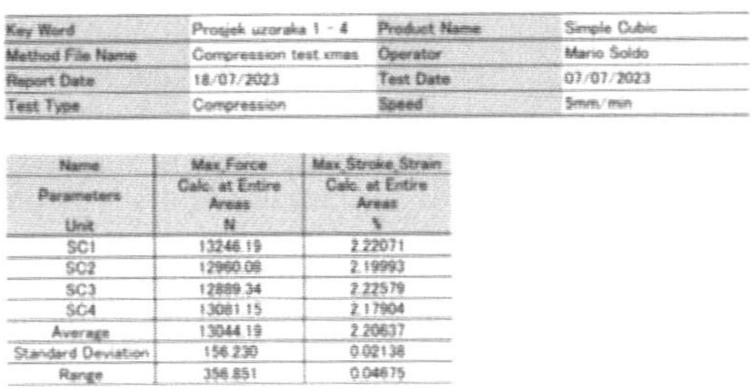

Key Word	Prosjek uzoraka 1 - 4	Product Name	Simple Cubic
Method File Name	Compression test.xmas	Operator	Mario Soldo
Report Date	18/07/2023	Test Date	07/07/2023
Test Type	Compression	Speed	5mm/min

Name	Max_Force	Max_Stroke_Strain
Parameters	Calc. at Entire Areas	Calc. at Entire Areas
Unit	N	%
SC1	13246.19	2.22071
SC2	12960.08	2.19993
SC3	12889.34	2.22579
SC4	13081.15	2.17904
Average	13044.19	2.20637
Standard Deviation	156.230	0.02138
Range	356.851	0.04675

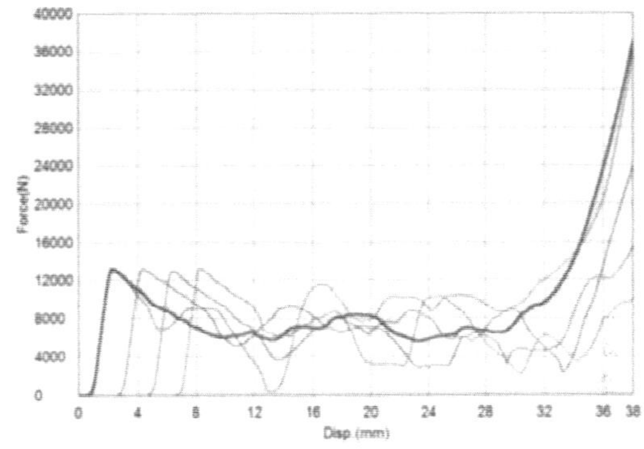

Fig. 12. Graphs and average of 4 compression test samples of Simple Cubic periodic lattice structure

4.3. Numerical analysis of the lattice structure model

After the experimental analysis, a numerical analysis was performed in the ANSYS program. By entering data in the ANSYS program for numerical analysis using the finite element method, a connection between experimental and numerical results was established. In Engineering Data, the properties of the PLA material according to the tensile test on the tear machine were entered. The difference between the standard PLA material in ANSYS and the selected one is in the Poisson's ratio (previously 0.39) which is set to 0.33 according to scientific papers that investigated the PLA material. Another difference is in Young's modulus of elasticity, which was determined experimentally and was 2,157 MPa. These two values have the greatest influence on the final results of the numerical analysis. In the simulation, boundary conditions were taken: a deformation of 1.2 mm on the upper surface of the sample and a fixed lower surface of the sample Fig. 13.

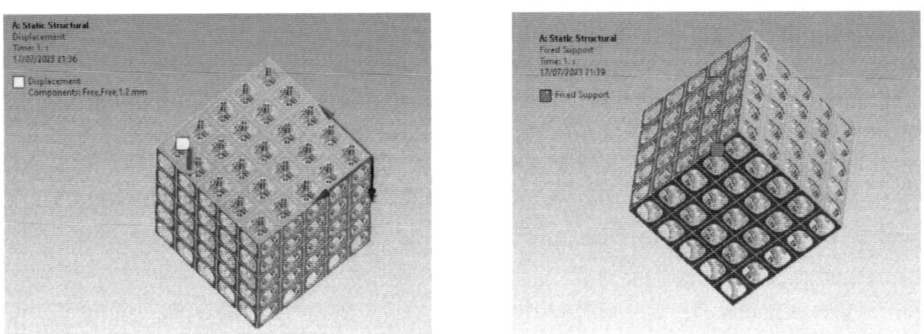

Fig. 13. Simulation boundary conditions: deformation of 1.2 mm on the upper surface of the sample; fixed bottom surface of the sample

The network is generated according to standard settings, and as a result of the simulation, the value of the force that should be applied to the upper surface is selected. In this way, actual compression tests on samples are presented. Numerical analysis concluded that the maximum force (required to deform the Simple Cubic lattice structure by 1.2 mm) is 12,225N. Such simulation settings are accepted as valid for further research since the error of experimental and numerical analysis is within 10%.

Experimental and numerical analysis showed that the periodic lattice structure with reduced mass gives higher values of the required force for breaking. Also, it is possible to choose a different shape of the unit cell and achieve the same relative density with the algorithm. Experimental and numerical analysis thus show whether the shape - that is, the geometry of the unit cell affects the value of the maximum force.

5. Vibration analysis of lattice structure

The effect of mechanical vibrations is very harmful and leads to a reduction in the life of machine components due to the strength of the vibrations. Isolation is of key importance in reducing and controlling the height of vibration amplitudes. Due to the large influence on the lattice structures and because of the non-uniformity in the structure itself, it is difficult to predict the absorption of vibration energy and the behavior of oscillations, as well as all other mechanical properties of the lattice structures. Various cellular materials have been proposed as good substitutes for providing essential isolation of machinery in order to control vibrations and reduce their effect through energy absorption. The mechanical properties and vibrational behavior of such cell structures can be investigated numerically using the finite element method and experimentally. The assumption is that the unit cell of the structure, which is almost closed in all directions, will have a limited damping capacity. However, the advent of additive manufacturing opens up more opportunities for the development of structures with unique combinations of open, uniform and periodically distributed unit cells, i.e. lattice structures which find wide application in everyday life due to their good ability to dampen oscillation waves and control vibration behavior together with other mechanical properties. Periodic and lattice cell structures have improved vibrational properties compared to solid materials.

As the vibration wave passes through the structure, it is damped based on the topology and geometry of the structure, and the interconnected amplitudes are reduced. By considering design, stiffness and damping properties with proper selection of material type topology and physical parameters, lattice structures can be controlled to achieve the required vibration damping. Fig. 14 shows the vibration damping and isolation through the lattice structure. In this case, the lattice structure can serve as a combination of spring and shock absorber.

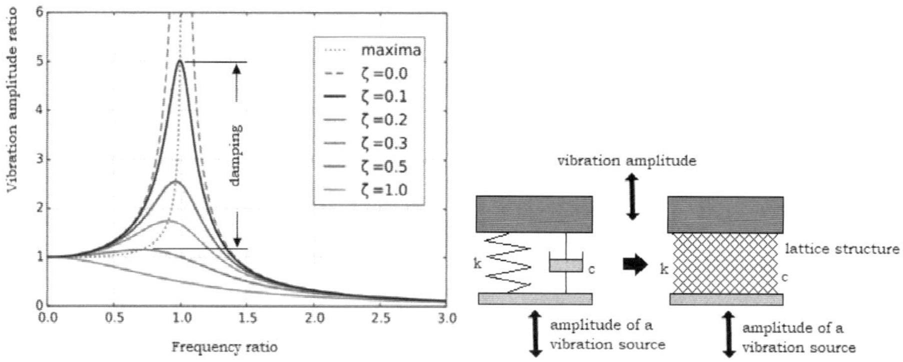

Fig. 14. Damping of vibrations and isolation through a lattice structure

Additive manufacturing offers faster production of complex geometries and savings on wasted material. Lattice structures are preferred over porous materials because they show better application as isolators.

The theory of vibration isolation is based on the excitation force, which is expressed by equation (1):

$$F(t)=F_O\sin(\omega t) \tag{1}$$

where F_O is the excitation force amplitude, ω is the excitation frequency.

Vibration isolation occurs when the ratio of excitation frequency to natural frequency is $r>\sqrt{2}$. Also, isolation is dependent on damping, i.e. isolation is better when damping is less (see picture above). Therefore, to control and adjust the vibrations of any machine, it is essential to use two main parameters: natural frequency and damping ratio. The natural frequency of an undamped system is denoted by ω_n and is measured in Hertz (Hz) or the number of oscillations per second. The damped natural frequency ω_d is the natural frequency of the damped system that determines how quickly the vibrations decrease between two successive bounces and is defined by equation (2):

$$\omega_d = \sqrt{1-\zeta^2}\omega_n \tag{2}$$

where ζ – damping ratio.

Therefore, damping causes the dissipation of the energy of the vibrating system in order to protect it from oscillations and in a certain period of time bring it to stability based on the intensity of the damping coefficient. If the damping ratio is 0 ($\zeta=0$) the machine is not damped, if it is less than 1 ($\zeta<1$) it means that the vibration system is in insufficiently damped operation mode; it is in the critical state when the damping ratio is 1 ($\zeta=1$), and the overdamped mode occurs when the damping ratio is greater than 1 ($\zeta>1$). This means that higher values of the damping ratio indicate that the vibration decreases faster. In addition to the damping ratio, the frequency ratio can show whether the isolator is working properly by checking whether the amount $r>\sqrt{2}$ [10].

The transmission of vibrations is the most common factor on the basis of which the isolation of the vibration system can be evaluated. It refers to the reduction of oscillating force or motion transmitted through the vibration system equipment. In the case of the excitation force, the transmissibility represents the ratio of the amplitude of the force that reached the foundation to the corresponding amplitude of the force that was induced inside the equipment. Displacement transmissibility (T) is considered the main parameter for vibration isolation and is defined by equation (3):

$$T = \sqrt{\frac{1+(2\zeta r)^2}{(1-r^2)+(2\zeta r)^2}} \qquad (3)$$

where r – frequency ratio [10].

Transmissibility can also be defined as the ratio of the amplitude of the motion that is transmitted to the equipment and the amplitude that was applied in the case of excitation of the motion [11]. Thus, minimal transmissibility is a sign of a good insulator, e.g. only a small value of force or motion can be transmitted through an insulator of minimal transmissibility and vice versa. With this aim, the modeling and management of lattice structures under vibration excitation is one of the best methods for analyzing optimal vibration isolation. Fig. 15 shows the load setting of the vibration system or, in this case, illustrates the vibrations in the vertical direction of the load.

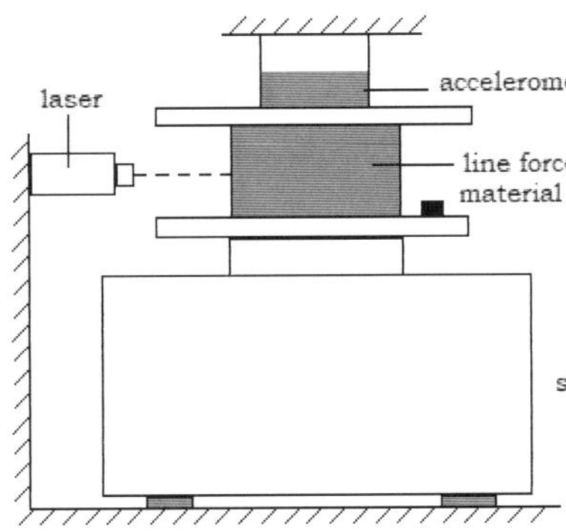

Fig. 15. Device for vibration excitation with appropriate load-vibration in the vertical direction

In this paper, the vibration analysis foresees the examination of the lattice structure under vibration excitation using a vibroplatform with a signal generator, changing different vibration modes of operation that include the own frequency and the isolation frequency of each sample, and it was determined how lattice structures can be used for vibration damping, and by using larger unit cells with with a small volume fraction, the natural frequency can be reduced [12]. Also, the influence of additive manufacturing was determined, which can be an important indicator of damping because the natural frequency, modulus of elasticity and density were compared to production values.

6. Conclusion

In this paper, an experimental and numerical analysis was performed and guidelines were given for the vibrational analysis of the Simple Cubic periodic lattice structure obtained by extruding PLA material using the FFF method on the Zortrax M200 Plus printer. Characterization of PLA material was performed on test samples (according to ISO 527-2 standard) obtained according to clearly defined parameters of the print on a Shimadzu AGS-X tear machine and the value of the elastic modulus of 2 157 MPa was obtained in the range of 0.05% to 0.25% deformation. This value was subsequently used for numerical analysis. The next step was to generate a Simple Cubic periodic lattice structure in the Rhinoceros program using the Grasshopper program according to the previously defined algorithm. The models were saved in .stl using the Z-Suite program. The samples were printed according to the previously defined print parameters according to which the test samples were also printed. A compression experiment at a set speed of 5mm/min at 74% deformation determined the maximum force of 13,045 N that was required for the first initial fracture of the sample. It was concluded that this force appears at 1.2 mm of deformation.

After the experimental analysis, a numerical analysis was performed in the ANSYS program. The previously determined parameter of the modulus of elasticity (from testing the test tubes in the tensile experiment) of 2 157 MPa was entered, and Poisson's ratio of 0.33 was added. A deformation of 1.2 mm acts on the upper surface of the model, while the lower surface of the model is fixed. Numerical analysis concluded that the maximum force (required to deform the Simple Cubic lattice structure by 1.2 mm of deformation) is 12,225N. Such simulation settings are accepted as valid for further research since the error of experimental and numerical analysis is within 10%.

Future research that will be based on the conclusions from this paper may refer to changing the diameter of the supports of the Simple Cubic lattice structure in order to reduce the relative density. Experimental and numerical analysis will show whether such a periodic lattice structure with reduced mass can obtain higher values of the required breaking force. Also, it is possible to choose a different shape of the unit cell and achieve the same relative density with the algorithm. In this way, experimental and numerical analysis would show whether the shape - that is, the geometry of the unit cell affects the value of the maximum force. It is also possible to work on changing the multiplication of the sample within the given volume by keeping the mass or relative density and comparing the obtained results. Finally, future research can refer to the incorporation of this algorithm into production processes for the purpose of reducing the total weight of the product and generally improving the mechanical properties of the final product.

After these analyses, a vibration analysis is provided, which includes the modeling and management of lattice structures under vibration excitation as one of the best methods for the analysis of optimal isolation from vibrations. After the vibration analysis, it was concluded that in additive manufacturing, the orientation has no significant effect on the vibration parameters except for the modulus of elasticity. An increase in damping could be achieved by adding external components to the lattice structure to improve its ability to absorb vibration energy, such as highly elastic materials, special types of dampers or completely new elements. Also, the damping properties can be improved based on the manipulation of the 3D printing manufacturing settings. It has not been determined how the topology of the structure affects the damping properties. Therefore, it is recommended that further research be carried out on how to improve the damping effect of graded lattice structures by testing different lattice topologies and changing geometric parameters such as the shape of the cross-sectional surface.

7. References

[1] Niknam H., A.H. Akbarzadeh. (2020). Graded lattice structures: Simultaneous enhancement in stiffnessand energy absorption, Available from: https://www.researchgate.net/publication/347131181 Accessed: 2023-07-10

[2] Maconachie, T. et al. (2019). SLM lattice structures: Properties, performance, applications and challenges, Mater. Des., vol. 183, p. 108137, 2019, DOI: 10.1016/j.matdes.2019.108137.

[3]. Hanzl, P.; Zetková, I.; Daňa, M. (2017). A Comparison of Lattice Structures in Metal Additive Manufacturing, https://www.daaam.info/Downloads/Pdfs/proceedings/proceedings_2017/067.pdf

[4] Pan, C.; Han, Y.; Lu, J. (2020). Design and optimization of lattice structures: A review, Appl. Sci., vol. 10, no. 18, pp. 1–36, 2020, DOI: 10.3390/APP10186374. Available from: https://www.researchgate.net/publication/347574572 Accessed: 2023-07-12

[5] Nguyen, C.H.P.; Kim, Y.; Choi, Y. (2019). Design for Additive Manufacturing of Functionally Graded Lattice Structures: A Design Method with Process Induced Anisotropy Consideration, Int. J. Precis. Eng. Manuf. - Green Technol., vol. 8, no. 1, pp. 29–45, 2021, DOI: 10.1007/s40684-019-00173-7.

[6] Monkova, K. et al. (2021). Mechanical Vibration Damping and Compression Properties of a Lattice Structure, Available from: https://www.researchgate.net/publication/350182272 Accessed: 2023-07-10

[7] https://www.ntop.com/software/capabilities/lattice-structures/ (2023). Accessed on: 2023-08-01

[8] Almesmari A., Sheikh-Ahmad J., Jarrar F., Bojanampati S. (2023). Optimizing the specific mechanical properties of lattice structures fabricated by material extrusion additive manufacturing. J. Mater. Res. Technol., vol. 22, pp. 1821–1838. DOI: 10.1016/j.jmrt.2022.12.024, Available from: https://www.researchgate.net/publication/366220596 Accessed: 2023-07-11

[9] Yang J. et al. (2022). Compressive properties of bidirectionally graded lattice structures. Mater. Des., vol. 218, p. 110683. DOI: 10.1016/j.matdes.2022.110683, Available from: https://www.researchgate.net/publication/360445653 Accessed: 2023-07-12

[10] De Silva, C.W. (2007). Vibration: Fundamentals and Practice, CRC Press Taylor&Francis Group, ISBN: 0-8493-1987-0, Boca Raton, London, New York

[11] Šaravanja, D.; Petković, D. (2010). Vibrating Diagnosis-Theory and Practice, University of Mostar, University of Zenica, ISBN: 978-9958-9263-1-0, Mostar-Zenica

[12] Elmadih, W. (2019). Additively manufactured lattice structures for vibration attenuation, Dissertation, University of Nottingham, Available from: https://www.nottingham.ac.uk/research/groups/advanced-manufacturing-technology-research-group/documents/manufacturing-metrology-team/waielelmadih.pdf Accessed: 2023-07-12

DOI: 10.2507/34th.daaam.proceedings.005

APPROACHES TO IMPLEMENTATION OF THE 3D LASER SCANNING TO ENSURE THE SEARCH AND DOCKING OF AUTONOMOUS UNDERWATER ROBOTS WITH A WALKING BASE

Viatcheslav Yastrebov, Valentin Pryanichnikov,
Vadim Chernyshev & Nikolay Sharonov

This Publication has to be referred as: Yastrebov, V[yacheslav]; Pryanichnikov, V[alentin]; Chernyshev, V[adim] & Sharonov, N[ikolay] (2023). Approaches to Implementation of the 3D Laser Scanning to Ensure the Search and Docking of Autonomous Underwater Robots with a Walking Base, Proceedings of the 34th DAAAM International Symposium, pp.0035-0039, B. Katalinic (Ed.), Published by DAAAM International, ISBN 978-3-902734-41-9, ISSN 1726-9679, Vienna, Austria
DOI: 10.2507/34th.daaam.proceedings.005

Abstract

Underwater operations, such as docking with the base module to get/adjust the instructions, as well as recharging the onboard power batteries, are increasingly being robotized in order to solve safety problems and reduce costs. Robotic systems typically rely on vision sensors to perform fundamental tasks such as navigation and object recognition. In particular, active optical 3D scanners are implemented to solve a number of specific tasks related to obtaining images underwater. However, their capabilities are limited by a range of several meters. The paper considers an approach, that allows increasing of an operational range for the seabed walking robot. The above, in turn, acts as the base for the autonomous floating robots, carrying sensor devices.

Keywords: intelligent 3D-sensing laser system; underwater walking robot with docking autonomous satellites.

1. Introduction, Problem statement

When using underwater vehicles, a significant part of the time is spent on the operations of lifting and lowering them to recharge and remove information. The paper considers a technology involving the use of an underwater walking base equipped with a docking station for receiving and servicing several smaller underwater robotic satellites carrying a sensor suspension. In such a scheme, an essential task is to automate the search by satellites for a docking station from sufficiently significant distances of about 50 meters. This significantly extends the operational time in such a robotic complex. An autonomous underwater vehicle should automatically (without communication with the main carrier) recognize a docking node having a pre-known geometric 3D shape with a dimensions of 1-2m. The shape for recognition is made of pipes (round or rectangular in cross-section, well reflecting both light and acoustic radiation. It is advisable to visualize and recognize this object from a distance of 20-30 meters and at about the same depth in possibly turbid water in automatic mode based on the received image and, in the future, to dock to it also automatically. It is highly desirable that the docking node remains in passive mode and does not emit anything continuously or in response to a request.

The mooring process is controlled, respectively, by the same vehicle in automatic mode. The range of action of the sonar forming the image along the beam is 30-8 m. At a closer distance <8 m, it is planned to use optical means, on which we will focus below (Fig.1)

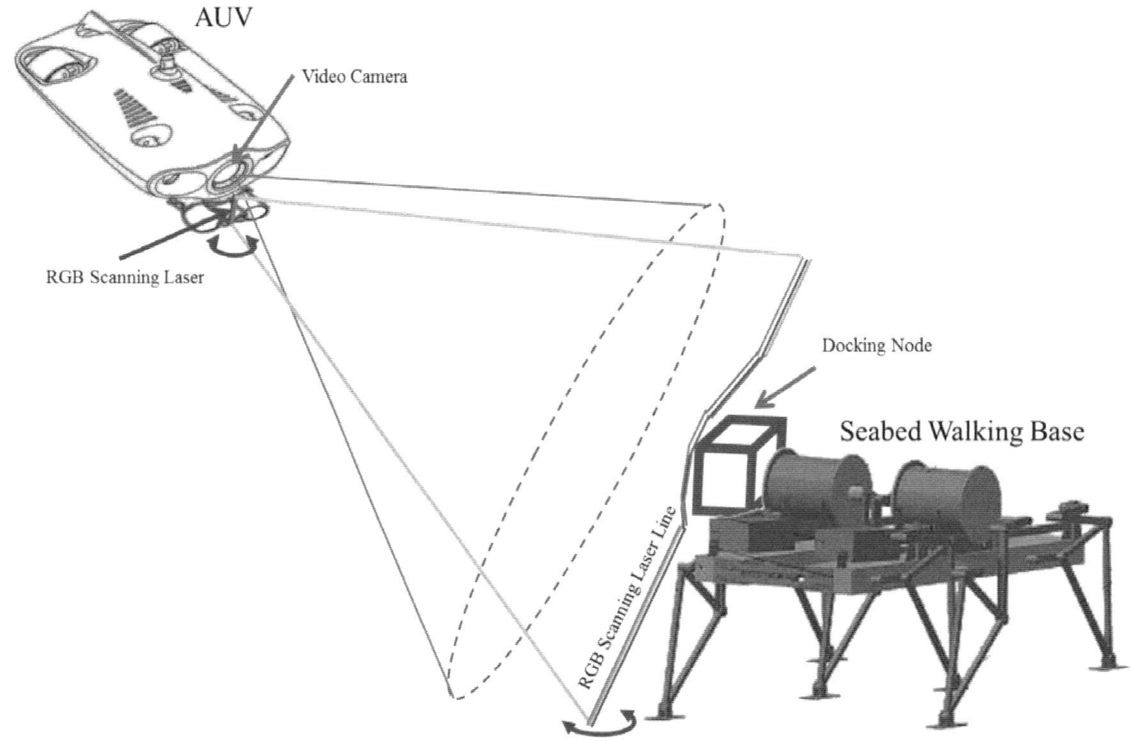

Fig. 1. AUV Docking by means of 3D RGB Underwater Laser Scanner

2. Optical 3D sensors – State of the Art

Optical 3D sensors can be categorized as active or passive. Active optical 3D scanners are more commonly used due to the domain-specific challenges of underwater imaging. An underwater sensor is said to use active (or structured) light when it projects light patterns onto the scene in a controlled way. These patterns can be a point, a line or more complex shapes. In active techniques, the information given by the structure of the pattern is key to reconstruct the scene in 3D. Active techniques determine the 3D position of the points in the environment either by time of flight ToF or by triangulation principles [1]. Passive lighting relies solely in ambient light to illuminate the scene, although artificial diffuse light may be used in dark environments. Passive techniques in underwater environments typically use stereo vision or structure from motion (SfM). Active light triangulation sensors find the 3D position of a point in the scene by combining geometrical information from the light emitter and the receiver. This way, the position in space of the scanned point coincides with the intersection of two light rays: the one sent by the projector and the one going from the camera focal point through the illuminated pixel. The main difference in performance between ToF and triangulation sensors concerns the scan range and depth resolution. The depth resolution of a ToF scanner depends on the resolution of the time or phase measurement but not on the scan distance, unlike for triangulation scanners.

3. Classification of 3D scanning techniques

The search for a solution led us to the most priority methods, which are:

- Active techniques that do not perform beam steering are of most priority: LLS - laser line scanner & DOE - diffractive optical element;
- Laser steering technologies are classified according to whether they involve mechanical elements or not;
- Scanning technologies can also be classified in raster and random-access scanning. Raster scanners (such as polygon mirrors or MEMS in resonant mode) need to steer the flying spot or line through the entire FoV (Field of View) before they can start to scan again.
- Random-access scanners can dynamically modify the scanned Field of View (FoV) in order to increase spatial resolution or decrease acquisition time, which makes them more flexible.

4. Searching for an approachable to implement Active Light Projection technologies

Active optical 3D scanners project light in a known direction and/or at a known instant in time, which gives essential information for the 3D scene reconstruction. Their main advantage in underwater applications is that they help provide a more homogeneous, denser point cloud, less dependent on the scene's texture than passive methods such as stereo vision. Scanning technologies can also be classified in raster and random-access scanning. Raster scanners (such as polygon mirrors or MEMS in resonant mode) need to steer the flying spot or line through the entire FoV before they can start to scan again. Random-access scanners can dynamically modify the scanned FoV in order to increase spatial resolution or decrease acquisition time, which makes them more flexible. Usually, the main performance criteria of laser scanners are scanning speed, FoV, resolution, and accuracy. Let's consider the two light projection techniques that do not use any type of beam-steering mechanism.

1. The whole scene can be illuminated at once using a homogeneizer diffractive optical element (DOE) (see Figure). This technique needs a 2D ToF sensor [4] in order to resolve the 3D position of the scanned scene [1]. This is considered an active method because, even though the light direction is not actively controlled, its time structure provides information for the 3D reconstruction.

2. A variation of no beam steering could be a special diffractive optical element (DOE) to project a pattern of lines and resolve their 3D position by triangulation. Acquiring the whole scene at once has the advantage of being robust against the high dynamics of the scanned scene and against the sensor movement. However, it usually comes at the cost of reducing lateral resolution [2], [3].

3. A fixed laser line scanner (also called profiler) can be used. This setup usually consists of a laser line module and a camera. The relative position of the projected laser plane with respect to the camera is always the same. Even though the laser light is not swept across the scene, it is considered an active technique because the plane equation of the light is essential to reconstruct the 3D points. This 3D reconstruction is usually done by triangulation. This configuration makes these scanners simpler to build and calibrate, but they always need to be attached to a moving platform, usually either an UUV or a rotating tripod. Therefore, the accuracy of the final reconstruction depends greatly on the accuracy of the pose of the platform.

4. Mechanical beam steering is achieved by moving an object, usually a mirror, in a controlled way. A laser beam can also be steered by moving the whole sensor. However, the smaller the inertia of the moving part, the faster the scanning can take place. Also, if the moving element is surrounded by air, the friction forces are smaller than in water. Hence, the moving part of an underwater scanner is usually placed inside a sealed housing. In all mirror systems, the reflection angle is twice the mirror tilting angle. Consequently, mechanical scanners can achieve high deflection angles more easily. Moreover, mechanical systems are usually suitable for a wider range of laser wavelengths and since the mirror's surface generally has very broadband reflectance.

5. Search conclusion

From the point of quick prototyping and testing implementation, operability and efficiency, we came to the conclusion: mechanical laser scanners provide higher angular resolution and maximum deflection angle, while non-mechanical devices provide much higher speeds.

6. Experimental Setup for LLS - laser line scanner

Our experimental RGB laser module produces a bright white spot, which is then going through the diffraction laser line lens. The above converts the laser spot to a laser line polarization. Laser line lens is mounted inside the hollow shaft of the rotating gear motor, angular speed of the rotation is controlled remotely. Resulting laser line is directed to the round mirror attached to the shaft of the servomotor, controlled from the Arduino-UNO microprocessor Underwater video camera, mounted rigid at the same meridian as the mirror, is capturing the distortion of the scanning laser line reflection (or scattering) and the software (being developed) for reconstruction of the 3D image of the scanned object using triangulation algorithm. Depending on the water turbidity, few tests of laser line reflection are automatically conducted before the final scan. Test includes variation of laser color and intensity and angular speed of the scanning mirror (Fig.2).

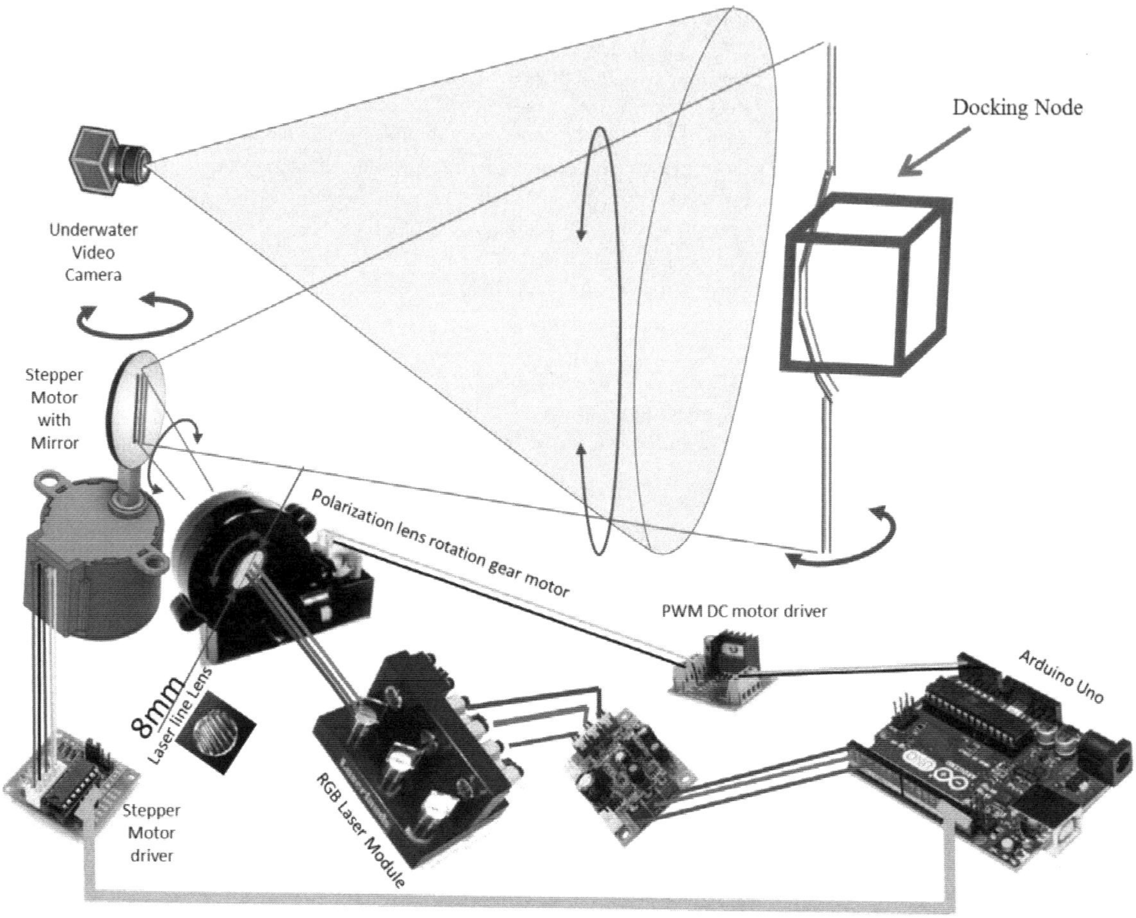

Fig. 2. Experimental Setup for LLS - laser line scanner

7. Conclusion

The already carried out experiments confirmed the feasibility of this approach for solving the problems of docking a small underwater vehicle with a base station, while the need for matching the polling frequencies of the sensors of the control cycle by extrapolating and interpolating sensor readings to ensure real time and minimize calculations using microprocessors was noted. The solutions also rely on our earlier developments [4], [5], [6], [7]. An effective solution has been found for the problem of finding a satellite docking station in the developed robotic complex, which allows significantly expanding the operational capabilities of the system. In the future, it is planned to switch from an experimental installation to the implementation of the system directly on a walking machine with the appropriate software refinement. The study was partially funded by the grant from the Russian Science Foundation № 22-29-01589.

8. References

[1] Miguel Castillón, Albert Palomer, Josep Forest, and Pere Ridao. (2019) State of the Art of Underwater Active Optical 3D Scanners. Sensors (Basel). 2019 Dec; 19(23): 5161. Published online 2019 Nov 25. doi: 10.3390/s19235161.Available from: https://www.ncbi.nlm.nih.gov/pmc/articles/PMC6928952/ Accessed: 2023, September 28.

[2] Himri, Rida, Gracias. (2019) 3D Object Recognition Based on Point Clouds in Underwater Environment with Global Descriptors: A Survey. Sensors (Basel). 2019 Oct 14;19(20). pii: E4451. doi: 10.3390/s19204451. Available from: https://www.ncbi.nlm.nih.gov/pubmed/31615081 Accessed: 2023, September 20

[3] Small Form-factor Structured Light Pattern Generator Reference Design for Portable 3D Scanners. (2015) TIDA-080001 Texas Instruments. http://www.ti.com/tool/TIDA-080001#descriptionArea Understanding laser-based 3D triangulation methods. Available from: https://www.vision-systems.com/factory/article/16738248/understanding-laserbased-3d-triangulation-methods Accessed: 2023, September 28

[4] Pryanichnikov V.E., Chernyshev V.V., Arykancev V.V., Aryskin A.A, Ksenzenko A.YA., Petrakov M.S., Travushkin A.S., Eprikov S R. (2019) Supervizornoe upravlenie shagayushchej podvodnoj bazoj s avtonomnymi robotami-

sputnikami. Intellektual'nye adaptivnye roboty // Informacionno-izmeritel'nye i upravlyayushchie sistemy. 2019. T. 16. № 1–2. S. 49–56. ISSN 2070-0814. http://radiotec.ru/ru/journal/Information-measuring_and_Control_Systems/number/2015-7

[5] Pryanichnikov V.E. (2023) Razrabotka distancionnyh sensorno-upravlyayushchih sistem, ispol'zuyushchih svyortki izobrazhenij / Zapis' i vosproizvedenie ob"yomnyh izobrazhenij v kinematografe, nauke, obrazovanii, media i v drugih oblastyah: XV Mezhdunarodnaya nauchno-prakticheskaya konferenciya, Moskva, 3–5 aprelya 2023 g.: Materialy i doklady // pod obshchej redakciej O. N. Raeva. — Moskva : IPP «KUNA», 2023. — 347 s., S.87-101, ISBN 978-5-98547-144-1, https://www.rsuh.ru/education/minot/conf_3d-2023.php Accessed: 2023, September 20

[6] Davydov D.V., Eprikov S.R., Kirsanov K.B., Pryanichnikov V.E. (2017). Service Robots Integrating Software and Remote Reprogramming, Proceedings of the 28th DAAAM International Symposium, pp.1234-1240, B. Katalinic (Ed.), Published by DAAAM International, ISBN 978-3-902734-11-2, ISSN 1726-9679, Vienna, Austria DOI: 10.2507/28th.daaam.proceedings.172. Accessed: 2023, September 20

[7] Chernyshev V. V., Pryanichnikov V. E., Arykantsev V. V. (2021) Algorithms of Self-control of the Underwater Walking Device According to Information on a Collision of Feet with an Unorganized Support Surface. In: Yuschenko A. (Eds) Modern Problems of Robotics. MPoR2020. Communications in Computer and Information Science, vol 1426. Springer, Cham. DOI https://doi.org/10.1007/978-3-030-88458-1_14, Accessed: 2023, October 10, Print ISBN978-3-030-88457-4 Online ISBN978-3-030-88458-1.

DOI: 10.2507/34th.daaam.proceedings.006

DESIGN AND OPTIMIZATION OF THE WELDING ROBOT SYSTEM FOR ALUMINIUM FRAMES USING SIMULATION SOFTWARE

Arbnor Pajaziti, Kemajl Stuja, Afrim Gjelaj & Leicht Calvin

This Publication has to be referred as: Pajaziti, A[rbnor]; Stuja, K[emajl]; Gjelaj, A[frim] & Calvin, L[eicht] (2023). Design and Optimization of the Welding Robot System for Aluminium Frames using Simulation Software, Proceedings of the 34th DAAAM International Symposium, pp.0040-0049, B. Katalinic (Ed.), Published by DAAAM International, ISBN 978-3-902734-41-9, ISSN 1726-9679, Vienna, Austria
DOI: 10.2507/34th.daaam.proceedings.006

Abstract

The aim of this study was to evaluate Welding Robot Systems (WRS) for aluminium parts in terms of energy efficiency, cycle time and workspace, automation process and cost. The study also considers the reachability, accessibility and positioning accuracy requirements for automated robotic welding to ensure the ability to weld any part of the product. The most recent methods and technologies such as seam tracking, simulation software, digital twin and virtual reality are used in this work to determine compliance with the concept of intelligent manufacturing. Using the methodical approach, different layouts of the WRS are designed to meet different industry requirements and to handle different configurations for different products. A robot simulation tool has supported all design steps, evaluating different scenarios for different system configurations. The virtual reality capabilities of the simulation tool provided the ability to check the movement of the robots through fixtures and parts in "near real" dimensions. Intelligent system software would be used to ensure the accepted quality of the parts and to automate various activities. The study also considered intelligent welding systems, as they have shown great potential in experimental settings and there is growing interest in their application in industry.

Keywords: Robot System; Simulation; Intelligent Welding.

1. Introduction

Welding allows the components to be permanently joined together. This joining allows the components to withstand the mechanical stresses throughout their life cycle. As this process is accompanied by various side effects such as exhaust fumes, heat, etc., which would have health effects on humans, welding is carried out with the robot systems. To get an idea of how large the share of welding applications is in the industry, Figure 1 is shown. According to "World in Robotics 2022" [1], it has the second biggest installations of industrial robots.

Despite great advances in processes such as bonding, clinching, etc., welding remains the only alternative due to the mechanical properties of the welded parts. Robotic systems are therefore also being developed and optimized for a range of products. Some robot manufacturers even offer complete solutions that include welding equipment, safety components, exhaust systems, etc. This significantly reduces engineering, assembly and set-up costs, whereby the robot vendor then takes responsibility for the safety of the robot system.

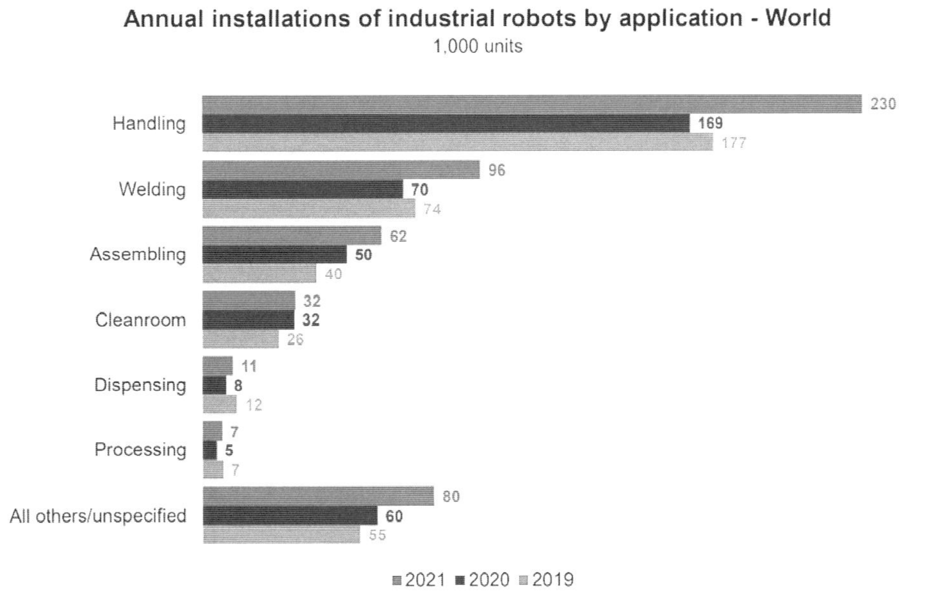

Fig. 1. Annual installations of industrial robots by application [3]

Most used technology for welding is MIG/MAG (Metal Inert Gas /Metal Active Gas). MIG/MAG welding is a widely used manufacturing process to produce components with high strength properties. This process involves permanently joining two or more components through the application of thermal energy [2]. In MIG/MAG welding, a filler metal or welding wire ignites the arc when it touches the component. To protect the arc from the reactive oxygen in the environment, a so-called shielding gas also flows through the gas nozzle [3]. This displaces the oxygen during welding and thus prevents oxidation of the arc and the weld pool [3].

In recent years, various tools/systems have been developed to enable weldability when joining components made of different materials on the one hand and to increase energy efficiency on the other. Intelligent Manufacturing (IM) and Smart Manufacturing (SM) are terms that refer to the use of technologies such as Artificial Intelligence (AI), and Machine Learning (ML) in the manufacturing industry. Both terms refer to the integration of intelligent systems and technologies into Manufacturing Processes (MP) to improve efficiency and enhance product quality [4]. The main difference between IM and SM is that IM is based on the use of AI and ML to automate and optimise MP. Smart manufacturing, on the other hand, refers to the use of Internet-of-Things (IoT) technologies and other digital solutions to improve and optimise MP. In general, it can be said that IM is a subset of SM, as IM builds on SM technologies. However, there is also overlap between the two areas and the boundaries between them are not always clearly defined. Due to the digitalisation of the industry, digitalisation is also becoming increasingly important in the field of welding robots. This can be seen in Fig. 2.2, which shows the number of papers published on SM and IM over the years [5]. From 2015 onwards there has been an increase in scientific papers on these topics. For this purpose, numerous simulation tools and Power Packs such as "Arc Welding Power Pac for RobotStudio" have also been developed. In this paper, the state of the art of software for MIG/MAG welding is analyzed. The most recent software from the main robot manufacturers and software vendors is examined.

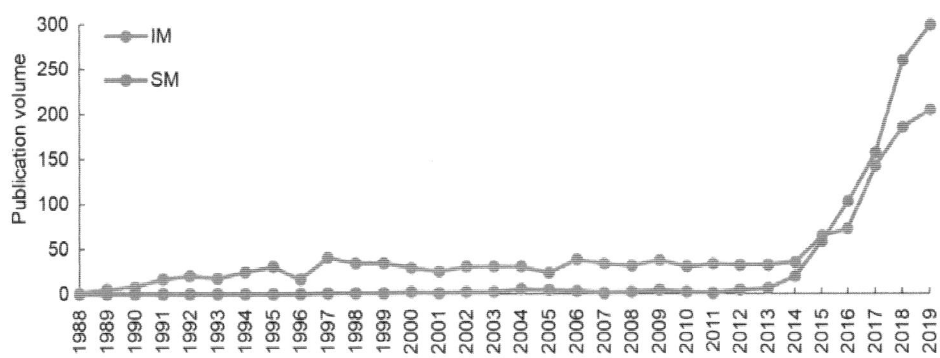

Fig. 2. Publication volume on SM and IM during 1988-2019 [5]

2. Problem description

The rapid advancement of technology has greatly impacted the efficiency and effectiveness of various industries, including the market for MIG/MAG welding robots. As a key economic sector, it is important to understand the current state of the art in terms of software and simulation software used in this market.

While the main manufacturers of these products offer similar features, there may be a lack of representation from pure software manufacturers, and it is unclear which features truly represent the current state of the art. In order to stay competitive and meet the demands of an ever-evolving industry, it is necessary to examine the latest features and assess their alignment with the concept of intelligent manufacturing.

Welding Robotic System (WRS) has been proposed based on the requirements from the industry. An industrial robot with 6 axes should work in parallel with 2 (two) working tables, so the working space of the industrial robot should cover the area of 350°-360° for each table of the WRS. Welding gun cleaner should be able to be able to match with welding robot. Manual loading and unloading of product. One time fixing of the product onto the fixture (see Figure 3).

An Industrial robot has to reach the entire working space of the structure to be welded.

Workpiece material: aluminium, wall thickness: 2,5 mm – 15 mm; Max. Frame Size: Length: 1500x750 mm, and Min. Frame Size: Length: 1200x750 mm. Product Tolerance: ±0.05 mm. No gap between welded profiles. The dimensions of the profiles are mainly for the plates for building.

The cylinders or fast clamps or other solution need to be used to fix the position of the product with the lever principle to ensure the positioning requirements of automatic welding of the robot, so as to ensure that it can be welded to each part of the product. According to the requirements of production efficiency, two sets of one axis positioner need to be designed, one for loading and unloading of the workpiece, one for welding.

In addition to robotics, the accompanied software can be used to easily program industrial robot and software usage. If possible to add the instrument Point to Point recording of the material to be welded.

The working tables should be easily rotated in order continue with welding on opposite side.

Special protective mechanism to prevent welding spatter or dust in the air which will cause bad impact of precision parts. In order to adapt to the different conditions of different requirements, be able to do different configuration for industry.

The anti-collision device needs to been considered to feedback the collision information in time and protect the key parts of the robot. Additional request from the industry is related to the robot working software for welding.

One of the main requirements regarding the robotic arm for welding is its connection with various working programs such as Autodesk Inventor, SolidWorks, etc.

Why this advantage would be more acceptable we will give the following explanation:

If this data were to be transferred to the RWS, it would not be necessary to compile a robot programming for different details each time, but any robot programming information would be obtained from the working program, and the only action that would be done would be to centre the robot on the first point of the weld and everything would be easier.

TIG welding, MIG welding can be used for Aluminium welding. Simulation software needs to be used in the MIG/MAG welding robot market, and to outline the Intelligent Welding Systems (IWS) developing process.

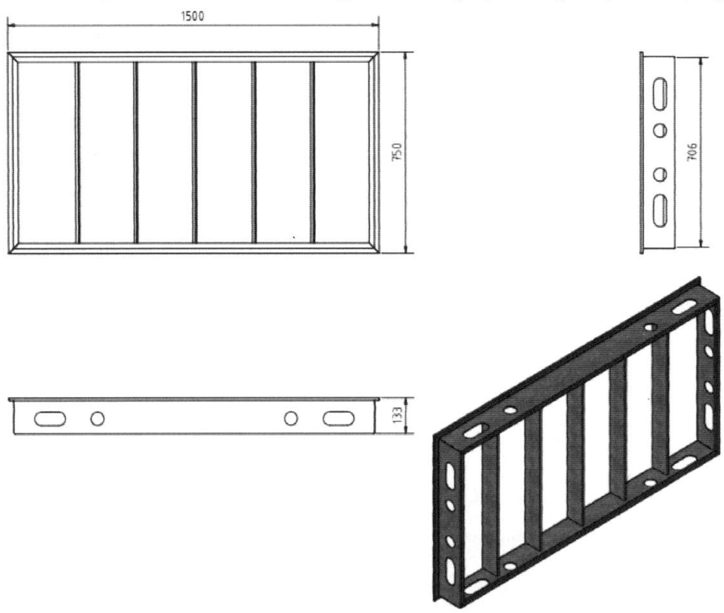

Fig. 3. Dimensions in mm of the frame to be welded

3. Materials und Methods- Design of mechanical system of the intelligent robot welding systems

As described in the problem statement above, in this study case, aluminium frames in various sizes are to be welded. The customer's list of requirements is very long and part of it is shown in Table 1. From this list, the requirements must be fulfilled, whereby there are also customer wishes and customer recommendations which are also fulfilled if possible. In other words, the designed robotic cell must fulfil multi-criteria (technical, economic, safety, environmental, etc.). On the other hand, there are many companies (such as ABB) that offer standardized solutions for welding applications. In most cases, these solutions are also configurable, which drastically reduces engineering costs. But, in most cases the customer is looking for an offer that is more cost effective (in most cases customized) than standardized solutions. In addition, there are so many component manufacturers competing for the same robot system function carrier. This makes the task very difficult to solve, especially with the limited time to solve the problem/provide the solution to the customer. In this case, students (education) and system integrators - small and medium sized companies (industry) - become uncertain and sometimes overwhelmed. To remedy this situation, this paper describes a method based on VDI-2221 (Verein Deutscher Ingenieure -2221). This method has already been explained in the papers [6], so the steps are not explained explicitly.

3.1. Customer Requirements - the first stage of the methodical approach-for welding robot system

The requirements list is the first stage of the design process. In this stage, the system integrators/students create a list of customer requirements in terms of the desired functionality of the robot applications. It is very important that all requirements are included in the table. Later, it is very easy to check that the requirements have been met. This means that checking row by row, rather than reading the whole project folder, saves a lot of time. At the same time, this provides more transparency between the customer and the system integrator.

Here are the requirements that are important in the case study:

- The parts are aluminium frames with different dimensions and geometric complexity (see Figure 3). The maximum dimensions are 1500mm long x 600mm wide x 200mm high.
- There are no customer fixtures. This means that the fixture must be designed.
- The frame is to be welded using a conventional process such as inert gas (MIG) or tungsten arc (TIG) welding. There is also no customer specification for the welding process. This means that there are a number of different processes to evaluate in order to select the best and most cost effective. This will depend on the material, material thickness and mechanical requirements.
- Loading and unloading can be done manually (by the operator), which has a direct impact on safety, but also requires cycle time analysis check- using simulation software. Cycle Time should be max: <4 Minutes/ Frames.
- In this case the material flow is not continuous. There are also interruptions while the operator loads and unloads the parts and restarts the machine.
- Safety requirements should be covered by current standards and machine guidelines.
- Safety requirements: the cost of the project must not exceed the budget. In these Case < 350.000 Euro.

Requirements			**Demands**	**Wishes**
Technical Req.				
	Workpiece	Weight	10kg	15kg
		Content	Aluminium Frames	
		Dimension	Max 1.500mm x 600mm x 200 mm	
		Cycle Time	<4 Minutes/ Frames	3
		Loading/Unloading	Manual by operator	
	Fixture	Weight	Manual Loading	Semiautomatic Loading
	Processing	Type	MIG/TIG	
		Material Workflow	Discontinuously	Continuously
Safety Req.				
	Mechanical		National guidelines/ ISO	
	Electrical		National guidelines/ ISO	
Secure Req.	**Software**	Authorization	OPC UA	
	Hardware			
Cost Req.			<350.000€	

Table 1. Short list of customer requirements for welding of aluminium Frames.

3.2. Conceptual design – searching for suitable solutions of welding robot system.

The conceptual design stage is a process of generating different proposals (Variants or Solution candidates) that can be used to build the welding robot system. Usually there are complex system with lot of components. To reduce the complexity, the whole system (based on VDI- approach) can be divided into subsystems with less complexity (see the Table 2). That is the first step in this stage. The left column of the table shows the Subsystems/subfunctions. If the subsystem is still complex, it can be divided into smaller function carriers, e.g., "device for positioning of parts" and so on.

The second step in this stage is finding solutions for each subsystem. More than one solution (sub-solution 1, 2 … n) should be found for each subsystem. For a better understanding or management of the sub-solutions, they can be placed in a box (called Zwicky "morphological box"). This approach stimulates creativity to find new solutions.

The third step in this stage of methodical approach is combining of sub-solutions to find the whole solution of the system. Another advantage of the morphological box is the creation of different solutions (solution candidates) or variants of the system by combining (mapping) different sub-solutions (if technically feasible). After combining/mapping the possible solutions, the possible candidate concepts are selected for evaluation. However, during the selection process of variants should take into account all customer requirements, compliance with security guidelines and technical/economic requirements. This requires a great deal of experience on the part of the system integrator [6].

	Sub-Solutions for the Sub-Functions			
Sub-Functions of robot system	**Sub-Solution 1**	**Sub-Solution 2**	**....**	**Sub-Solution n**
Handling/TCP Moving/ Robot	6 axis Welding Robot	6 axis universal Robot	....	Cartesian Robot, Hyper redundant robot [11]
Welding System	MIG	TIG	....	Laser Welding
Robot Moving	Stationary	Linear axis		Linear axis hanging
End-Effectors	22-degree Weld-Gun	45-degree Weldgun		Tandem Weld gun
Parts Positioning	Linear Axis	2 Axis	....	3 Axis
	Two Stations/ two Devices	Two Stations / one device		
Part inspection/scanning	Tactile sensor	Using IR Sensor	....	Vision systems
Exhaust/extractor system	Localy/partial	full		
Safety components	Safety Gates	Light Grids	....	Optical monitoring
....	....	....	.....	....
User Authorization	OPC UA	Windows embedded IOT	....	Linux
Solution Candidate (SC) - Variants	SC1	SC2		SCn

Table 2. Simplified "morphological box" (Zwicky box) for creating of solutions -variants.

Let's look at the selection of the first candidate (SC1) in the table.

- The main component of the system is the robot (manipulator). There are different kinematics that can be used to perform the welding task, but nowadays there are special robots for special tasks. You can use a welding robot instead of a universal robot. The advantage of the welding robot is that the media (gas, filler material, etc.) go directly into the holes of the wrist. This means that there is no noise during welding. The disadvantage is that this type of robot can only be used for welding or in a water jet system. For other applications, such as palletising, this robot kinematics is unsuitable. A Cartesian robot cannot be used in this case study.
- There are also different welding processes (first column, row 2). Possible candidates are MIG, TIG, laser, etc.
- For the Robot moving subsystem (Function), the question is whether the robot is stationary or in a linear motion axis. This would make sense for long parts or when operating several stations in a row.
- The next sub-function deals with the torch. It is obvious that a torch will be used. For simple parts or good accessibility, a torch with a small bend (22 degrees) will be sufficient. For complicated parts and tight spaces, for example, 45 or 60 degrees should be used.
- Through the other partial solutions considered, different variants can ultimately be created. The following illustrations show different layouts. This means that different options can be considered. The best solution will depend on the requirements and evaluation criteria.

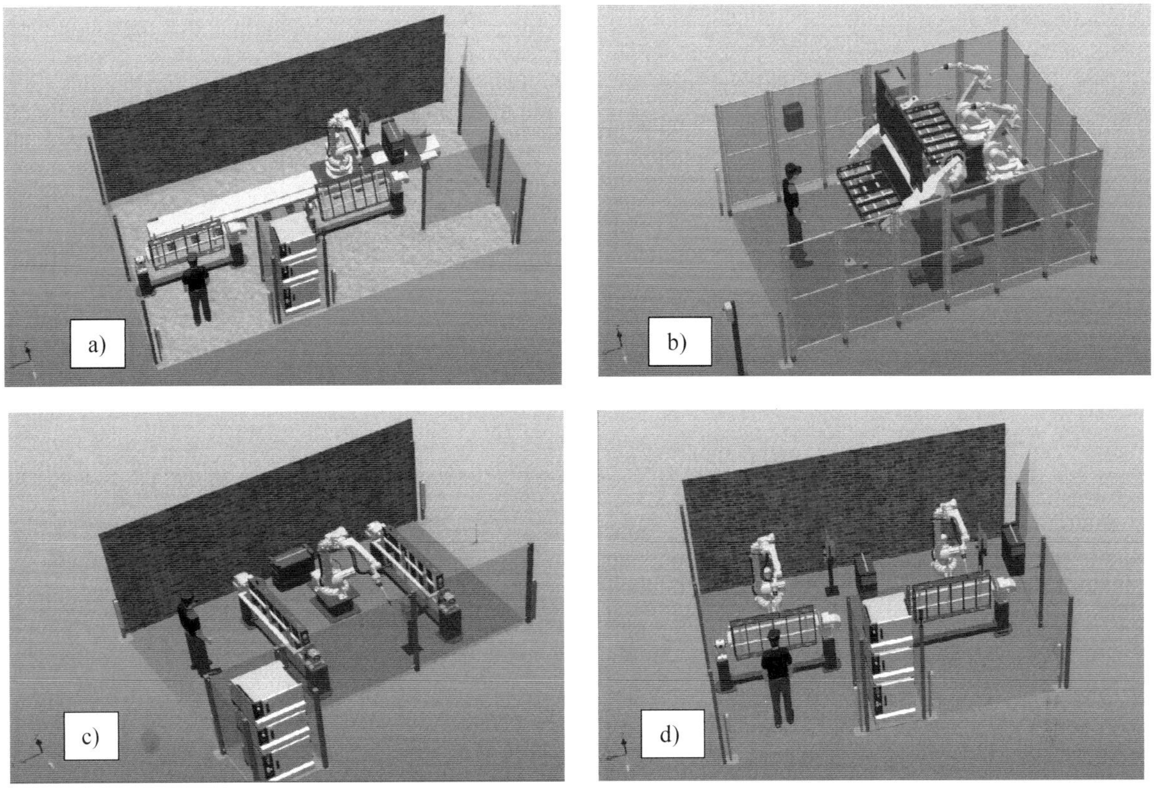

Fig. 4. Different Variants (Solution Candidates) for WRS: a) Robot in linear moving axes, b) Two Robots and K-Positioner, c) One robot and two L-Positioner and d) Two Robot/two L-Positioners

As can be seen in the Figure 4, different combinations of subsystems are possible, resulting in different solutions for the overall robotic welding system. The next step is to evaluate the generated layouts according to predefined criteria and customer requirements. The next figure shows the evaluation criteria for this case. If all the criteria have the same weighting coefficient, this looks like a regular polygon. If the criteria have different weighting coefficients, then this would be seen as a non-regular polygon.

In this case, the solutions can be scored between 0 (inappropriate solution) and 1 (ideal solution). Where 0.75 would all be favourable. Ideal solutions are very rare when multiple criteria are used (see Figure 5).

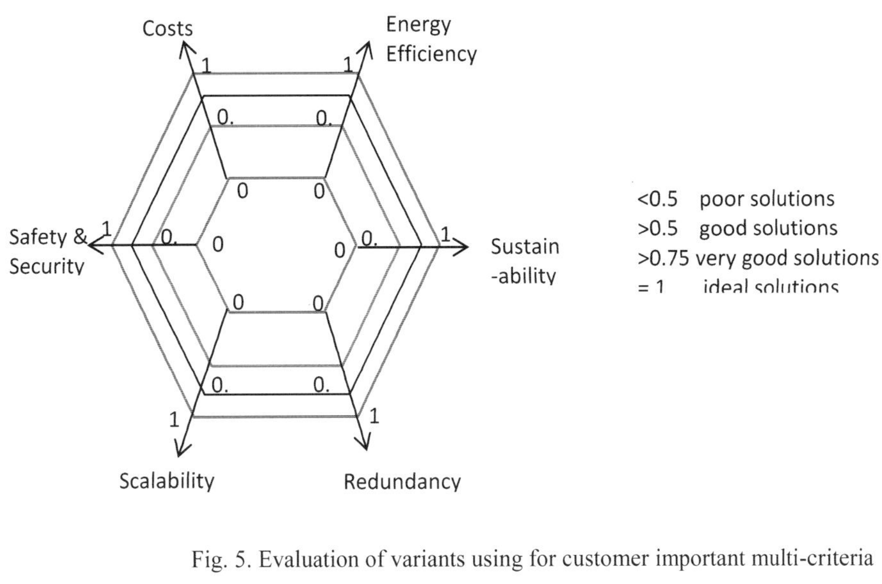

Fig. 5. Evaluation of variants using for customer important multi-criteria

3.3. Preliminary design

The third phase of the design process according to the VDI is the preliminary design. All the necessary components (parts, fixtures, etc.) must be designed in CAD. As the parts interact dynamically with the robot, it is also necessary to simulate the system. Simulation can be used, for example, to determine whether the fixture is suitable. To answer the question: is the welding process even possible without collision with the fixture ore any clamps? If cycle time analysis is also important, the robot manufacturer's simulation tools must be used more often. Simulation can also be used to answer questions about reachability, accessibility and optimisation of the overall cycle. With the help of simulation, it is possible to determine why a D-positioner (which is expensive Figure 6 below) should be preferred to a K-positioner (Figure 4 b above). The D-positioner has an extra axis and allows better positioning of the workpiece. With the K-table, in this example, you would also have to weld overhead or use an inappropriate welding torch angle.

3.4. Final design-layout

The final stage of the design process is to finalise and optimise the best solution that is the result of the previous stage. While CAD is used to optimise the size and arrangement of the components in the robot cell, simulation is used to check and optimise the various movements of the robot and the dynamics of the components in the robot cell. In addition, activities related to the overall material flow and process need to be optimised. It is important to understand that simulation is only an off-line method for optimisation, whereas process monitoring is an on-line tool for process monitoring and optimisation. But the advantage of simulation is that the robot's movements can be incorporated into the robot's programming and in that way the structure of the robot program can be created. The figure below shows the final design. All the components of the systems can be accommodated in a compact area of 6000 mm by 3000 mm. Here are the main subsystems:

- The manipulator is an ABB welding robot 2600ID (integrated dress pack), payload 8kg, reach 200cm.
- The workpiece positioner is an ABB IRBP type D (with five axes (Station A- 2 axes, Station B- 2 axes and Interchange Station 1 axis, only for changing the station (A or B), i.e., not for synchronised movement with the manipulator).
- Binzel ID 45-degree welding torch for better access to narrow parts to be welded.
- Torq nozzle cleaner, welding wire cutter and automatic tool centre measurement.
- Robot and welding controller.
- All safety fences and light curtains (for fast part change, door safety and all safety components according to the risk analysis [7]. On entering the warning zone, the manipulator/positioner moves at a reduced speed of max: 250 mm/second.
- Workpiece Fixtures.
- Operator panel.

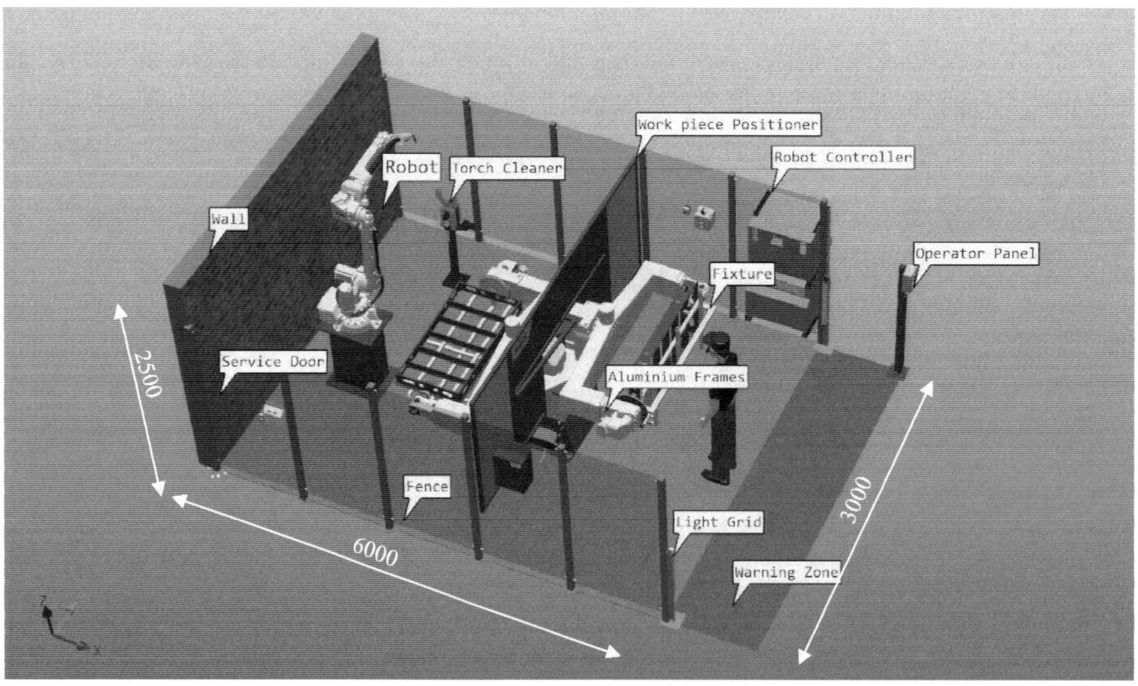

Fig. 6. Layout and simple safety concept for WRS

4. Intelligent welding system for control of the weld process -design of the information system

The IWS uses advanced computer algorithms and sensors to provide automated welding control. This technology improves welding accuracy and efficiency. It can be used in a wide range of industries, including construction, automotive and shipbuilding. A comparison between traditional welding systems and intelligent welding systems is shown in Figure 7. Key technologies such as powerful computers, IoT, big data and AI are required to enable intelligent welding systems. These will help reduce welding defects and improve the quality of welded parts. At the same time, these systems can also reduce the need for skilled human resources. This enables higher levels of automation and cost savings.

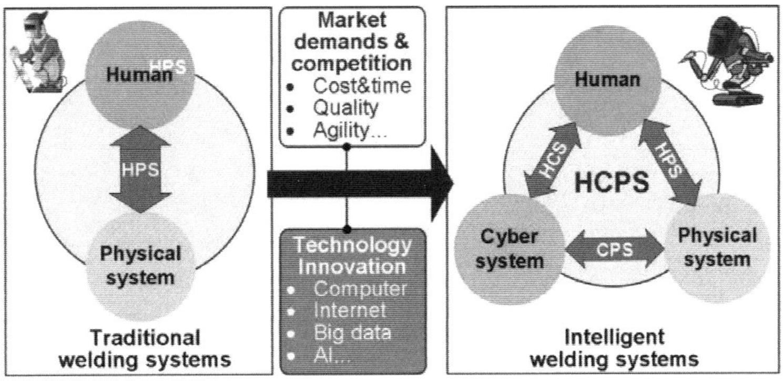

Fig. 7. Welding systems: the evolution from HPS to HCPS [4]

4.1. Seam Tracking

The general purpose of seam tracking is to automate the welding process without having to manually adjust the welding path beforehand. There are several possible approaches for tracking the seam, which are listed in Table 3.

Sensors	Advantages
Arc sensor	Cheap compared to vision-based sensors
Vision sensor	Can detect weld seam position and weld groove dimensions
Laser vision sensor	Error values in seam process is minimum
Ultrasonic sensor	Weld quality can also be inspected
Electromagnetic sensor	Not affected by arc light, fumes
Composite sensor	Advantages of both arc sensor and vision sensor can be obtained
Tactile sensor	Low cost compared to other
Touch sensing	The offset of workpiece can be recognized
Weld pool feature observation	Weld pool characteristics and geometry can be observed on the spot

Table 3. Listing of sensors with their Advantage [8]

Laser vision and optical sensors are most used for seam tracking. These intelligent sensors are very accurate and can send data corrections to the robot and welding controller in real time. Each robot target is declared as a persistent variable in the robot program. The data received from the sensors is used by the robot controller to modify the persistent variables in the program being executed. At the same time, the welding parameters are adjusted using an intelligent neural network or similar algorithm. Most of these algorithms are already part of the Fronius or Lincoln welding system. Based on the paper from Amruta Et al. [8], neural networks and the fuzzy approach are used for the welding parameter controller.

4.2. Digital Twin- virtual reality

Virtual reality is an ever more rapidly advancing technology. In the software "RobotStudio" supplied by ABB [9], a function for virtual reality is offered directly. For the practical test, the Meta Quest 2 is used with a link cable [10]. A welding cell is here the simulation model. This example is intended to highlight the advantage of virtual reality. While in the real process it is not possible to stand directly next to the welding process, this is possible with virtual reality. The VR headset was connected to the computer and the "Virtual Reality" button was pressed. Because Steam and Steam VR were installed on the computer, the emulation was automatically started via Steam VR. In the simulation, the controls were explained right away; they are intuitive and feel very natural. The simulation can be started via a play button. However, due to version differences, the entire simulation could not be run.

4.3. Interaction of the subsystems

Virtual Reality (VR), problems such as collisions and reachability can be easily identified in parallel simulations. Without this virtual verification, these problems would only be discovered directly on the robot, resulting in an extended production stop. After the program is loaded onto the robot, final adjustments are made, in the sense of online programming, to create a high-quality weld seam. Production can then be restarted. Throughout the entire product processing process, all features of the IWS are running in the background. This means that if there are deviations during production, they automatically generate a correction. Therefore, there is a continuous and permanent exchange and comparison of data throughout the entire production process. These are represented in real time in the simulation program. This allows for on-the-fly corrections to be made. By using VR during offline programming and using IWS features such as seam tracking or digital twin, the line between offline and online programming is becoming increasingly blurred (see Figure 8) [5].

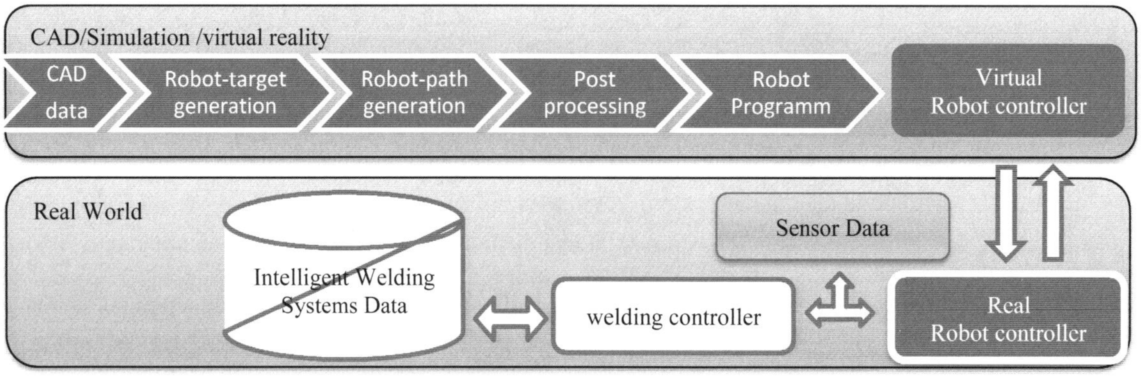

Fig. 8. Offline-Programing with IWS combined

5. Results and discussion

The design of technical systems according to the VDI 2221 guideline not only helps to solve complex tasks, but also to find the best solution to the problem. This means that the robot welding system is already optimally designed. The next step is to develop a material flow strategy that utilises the synergies to ensure the energy efficiency of the system. Although most of the energy is consumed by the welding process itself to melt the materials, the robot's movements should also be optimised for energy efficiency - ISO 50001:2018 [11]. The simulation tool RobotStudio offers the possibility to record energy measurements of the virtual robot (only the robot's movements) in the digital twin [12]. An example is shown in the figure below.

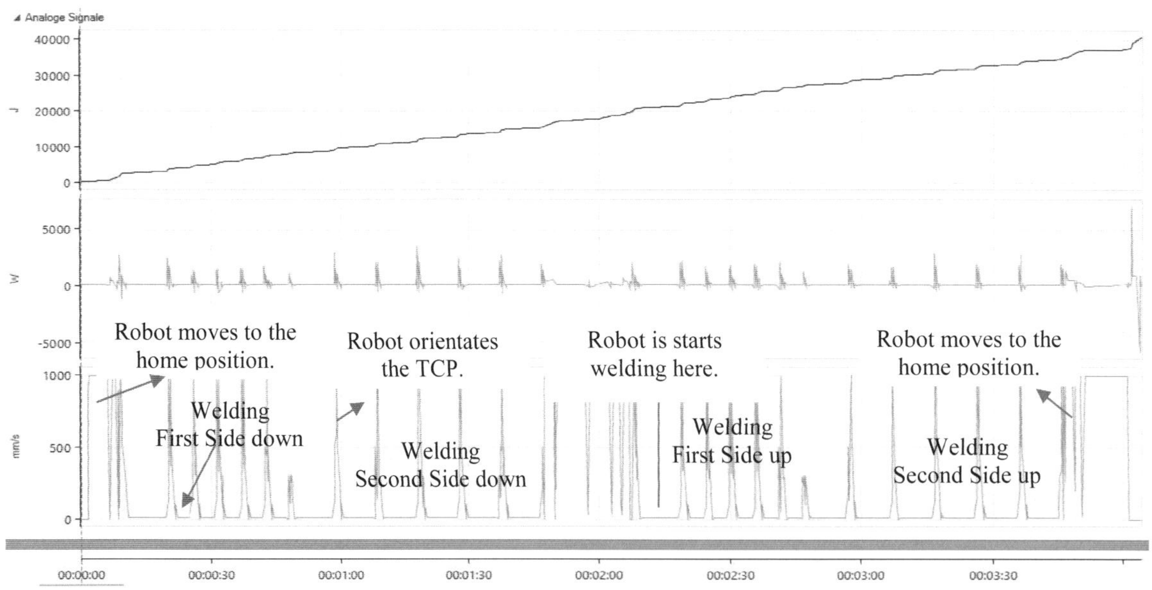

Fig. 9. Energy profile for double-sided welding strategy for aluminium frames

For each scenario it is possible to generate these profiles and thus select the best sequence scenario. For example, in the Figure 9, data are collected for the Tool Centre Point (TCP) movement in mm/sec relative to the work object coordinate systems, the work done for the manipulator only in watts and the energy consumption of the manipulator in joule (J).

All this is done offline in the virtual environment before the plant is brought online. From here, the intelligent welding process takes over, correcting the robot's movements and monitoring and controlling the welding process parameters. For a good weld, it is important to know how the parameters relate to the weld [13]. An artificial neural network has to be developed and then used to predict the hardness and tensile strength of the weld. The prediction is based on the input parameters of gas pressure, welding current and speed. Furthermore, for defect detection a fuzzy logic interference system can be used.

6. Summary, Conclusion and outlook

In conclusion, the current state of the art in welding technology does not prominently feature advanced algorithms for parameter optimization and defect detection, as evidenced by the limited availability of public information. However, recent advancements have shown promising initial steps towards the automation of these processes through the utilization of neural networks. The application of machine learning techniques in welding is still in its early stages, and it is evident that further research and development are required before these approaches can be effectively deployed in industrial applications.

The ongoing efforts to incorporate machine learning into welding systems present a significant opportunity for advancing the field. By leveraging the capabilities of neural networks and other advanced algorithms, it becomes possible to optimize welding parameters and enhance defect detection with greater precision and efficiency. The potential benefits of such automation include increased productivity, improved weld quality, and reduced reliance on manual adjustments and human inspection. However, it is essential to acknowledge that the integration of advanced algorithms into industrial welding applications will not happen overnight. The race for machine learning in welding has just begun, and substantial time and resources will be required to overcome the existing challenges and refine these techniques. Factors such as data availability, algorithm robustness, and computational requirements need to be carefully addressed to ensure reliable and practical implementation.

Looking ahead, the future of intelligent welding systems lies in the continued exploration and advancement of machine learning approaches. As research progresses and knowledge deepens, we can anticipate the development of more sophisticated algorithms tailored specifically to welding processes. These algorithms will not only automate parameter optimization and defect detection but also enable real-time decision-making, adaptive control, and seamless integration within existing welding infrastructure.

7. References

[1] Christopher, M. (2022). Industrial Robots, World Robotics, VDMA Services GmbH, Frankfurt am Main

[2] Donhauser, V. (2017). Entwicklung von technischen Entscheidungsgrundlagen zur Implementierung automatisierter Schweißtechnologien. Merseburg

[3] https://www.fronius.com/de/schweisstechnik/schweisswelten/mig-mag-schweissen. (2022). Fronius, Accessed on 5.10.2022

[4] Baicun, Wang, et al. (2020). Intelligent welding system technologies: State-of-the-art review and perspectives, 56, s.l.: Journal of Manufacturing Systems, pp. 373-391. 0278-6125

[5] Baicun, Wang, et al. (2022). Smart Manufacturing and Intelligent Manufacturing: A Comparative Review. 6, Bd. 7. 2095-8099

[6] Stuja, K.; Poszvek, G.; Walter, W. & Markl, E. (2018). Integrated Method for The Design and Evaluation of Safety & Secure Manufacturing Systems, Proceedings of the 29th DAAAM International Symposium, pp.157-163, B. Katalinic (Ed.), Published by DAAAM International Vienna, ISBN 978-3-902734-20-4, ISSN 1726-9679, Vienna

[7] ISO 50001. (2018). Energy management systems - Requirements with guidance for use (ISO 50001:2018)

[8] Amruta, Rout, B.B.V.L, Deepak und B.B, Biswal. (2019). Advances in weld seam tracking techniques for robotic welding: A review, Bd. 56. 0736-5845

[9] https://new.abb.com/products/robotics/robotstudio. ABB AG. Accessed on 21.08.2021

[10] https://www.meta.com/at/quest/accessories/link-cable/. (2022). Meta. Accessed on 20.12.2022

[11] Dominik Diek, Kemajl Stuja & Mohamed Aburaia (2022). Design and development of a flexible, cable-controlled and hyper redundant robot with Strength-Adjustable ball joints, Proceedings of the 33rd DAAAM International Symposium, pp.0024-0028, B. Katalinic (Ed.), Published by DAAAM

[12] Fuller, Aidan, Fan, Zhong and Day, Charles. (2022). Digital Twin: Enabling Technologies, Challenges and Open Research. Accessed on 15.11.2022

[13] Rishikesh, Mahadevan R, et al. (2021). Intelligent welding by using machine learning techniques. January 2021. 46

DOI: 10.2507/34th.daaam.proceedings.007

A MODEL OF AN INTELLIGENT AUTOMATION SYSTEM FOR MONITORING OF SENSOR SIGNALS WITH A NEURAL NETWORK IMPLEMENTATION

Irina Topalova* & Pavlinka Radoyska

This Publication has to be referred as: Topalova, I[rina] & Radoyska, P[avlinka] (2023). A Model of an Intelligent Automation System for Monitoring of Sensor Signals with a Neural Network Implementation, Proceedings of the 34th DAAAM International Symposium, pp.0050-0055, B. Katalinic (Ed.), Published by DAAAM International, ISBN 978-3-902734-41-9, ISSN 1726-9679, Vienna, Austria
DOI: 10.2507/34th.daaam.proceedings.007

Abstract

The automation systems today combine the capabilities of AI to process large databases in real-time work, aiming to predict equipment or machine failures. Essential to the reliable and efficient operation of automated systems is the application of AI to monitor their current state. Tracking the status of the sensors in each cycle of machine operation through neural networks would provide an adaptive reflection of faulty or correct behaviour of the automated system. The present study presents a model of an intelligent automation system for monitoring the sensor signals with a neural network implementation. An algorithm for working in two basic modes of a programmable logic controller in this integration is proposed. The neural network is trained with a large number of combinations of sensory signals, corresponding to states of correct behaviour and system faults. Depending on the classification accuracy or currently occurring wrong sensor signals, a retraining method is developed for both modes of operation. The main purpose of the research is to show the effectiveness of the method for classifying the sensor behaviour, in terms of dynamic reaction of the system. The obtained results are discussed and a proposal is made for further development of the research.

Keywords: Automated control; Sensor signals; Neural network; Classification; Programmable logic controller.

1. Introduction

AI (artificial intelligence) automation systems provide real-time integration of PLCs (Programmable Logic Controller) and AI data processing capabilities. Gradually, automation will begin to replace human cognitive abilities, providing greater speed, accuracy and consistency of operations. Implementing machine learning and artificial intelligence is a key stone of contemporary innovations in production automation as discussed in [1]. Several forward-thinking organizations and institutions are paying attention to convergent automation and cognitive technologies by investing in their development of such systems [2]. As manufacturers begin to embed intelligence into automation systems such as PLCs, SCADA (Supervisory Control and Data Acquisition), etc., this new wave of cognition is now a fact of in industrial automation. This evolution of PLCs makes and evolves them as part of a robust, flexible and supportive AI system that exhibits intelligent capabilities.

The future is open to machines that can operate on their own with little manual intervention, have computing capabilities on demand, and can autonomously optimize the resources and processes involved in automated production. The large number of PLC manufacturers that apply unified and standardized methods and integrated tools to ensure the functionality of controllers, as well as their integration with neural networks in hardware or cloud-based platforms, have allowed the application of AI at different levels in the hierarchical structure of control systems - field level, control level and management level [3] [4]. Tracking the status of the sensors in each cycle of machine operation through neural networks would provide an adaptive reflection of faulty or correct behaviour of the automated system. The present study presents a model of an intelligent automation system for monitoring the sensor signals with a MLP (Multi-Layered-Perceptron) neural network implementation. The objective of the study is to show how the neural network can be trained to recognize combinations of sensory signals, corresponding to states of correct behavior and system faults, as well as to dynamically retrain itself with currently emerging new states. An algorithm for working in two basic modes (off-line and on-line) of a programmable logic controller in this integration is proposed. The neural network is trained with a large number of combinations of sensory signals, corresponding to states of correct behaviour and system faults. Depending on the classification accuracy or currently occurring wrong sensor signals, a retraining method is developed for both modes of operation. The main purpose of the research is to show the effectiveness of the method for classifying the sensor behaviour, in terms of dynamic reaction of the system. The obtained results are discussed and a proposal is made for further development of the research.

2. State of the Art

In the field of automation, more and more manufacturers of control equipment are developing the capabilities of AI to work together with PLCs as the main devices in the control systems. With the introduction of the S7-1500 TM neural processor (NPU) module for the Simatic S7-1500 controller and ET 200MP I/O system, Siemens has integrated AI into the controller itself. The S7-1500 TM NPU module is suitable for use at the field level in the control structure, as well as wherever reliable, fast, deterministic and real-time solutions are required [5]. The S7-1500 TM NPU module works using a trained neural network and stored knowledge on an SD card. Users can connect Gigabit Ethernet and USB 3.1 compatible sensors, such as cameras and microphones, to the module's integrated interfaces. CPU data transmitted from the backplane can also be used as input data. The processing results are then evaluated in the processor program [6]. The TM NPU transfers the result of the processing operation to the CPU via the backplane bus. The CPU then allows further evaluation to reflect the results in the user program. But the typical area of application is a visual quality check in production plants, with an implemented application shown in [6]. But no examples have been published for implementations with digital sensory input signals and transferring the results from the neural module to the CPU via the backplane bus. The Allen-Bradley Control Logix and Compact Logix controllers support machine learning and artificial intelligence only through the integration of third-party software tools. They implement cloud based Neural network technologies, but in the field of energetics [7]. The AC500 PLC series of ABB, features an integrated AI controller that can be programmed using MATLAB and Simulink3, but with no examples with AI applications and integration through PROFINET bus, protocol that efficiently manages Inputs and Outputs exchange between the controller and I/O Devices in real-time operation [8] [9]. The iQ-R Series PLCs of Mitsubishi Electric, support machine learning and artificial intelligence through the integration of third-party software tools. The iQ-F series is an all-in-one programmable controller that can be used for stand-alone applications or networked system applications, but the proposed applications are addressed only for Data and statistical analysis [10]. The Modicon M580 controller of Schneider Electric supports machine learning and artificial intelligence through the integration of third-party software tools. This feature allows the controller to perform mostly advanced analytics and predictive maintenance [11]. The authors Marugán, A.P. & Márquez, F.P.G. present a method based on neural networks for a dynamic generation of a control strategy, but not based on input sensor signals. It suggests that the thresholds used for generating alarms can vary and, therefore, the control of the wind turbine will be adapted to each specific wind turbine [12]. Wang, W., Harrou, F. § Bouyeddou, B. propose an application of a neural network for recognition of various heterogeneous cyberattacks, also working at the SCADA level [11]. It is reasonable due to the fact, that these attacks are carried out at the management level. Both methods given in [11] and [12] work on SCADA management level, which does not imply efficient operation in terms of speed in the exchange of data between the PLC, the neural module and the input-output data.

In this study a model of an intelligent automation system for monitoring of sensor signals with a MLP neural network implementation is presented. It shows how the neural network can be trained to recognize combinations of sensory signals, corresponding to states of correct behavior and system faults, as well as to dynamically retrain itself with currently emerging new sensory states. The emphasis is on the training and retraining of the MLP network in the two main modes off-line and on-line. Depending on the classification accuracy or currently occurring wrong sensor signals, a retraining method is developed for both modes of operation. The main purpose of the research is to show the effectiveness of the method for classifying the sensor behaviour, in terms of dynamic reaction of the system. In this way, the dynamics in the changes of input combinations will be correctly classified by the MLP NN, especially after long-term operation of the automatic complex, when the majority of sensory combinations unforeseen in the training data would occur. A number of input training data representing real sensor signals were used in the experiment. The obtained results show very good adaptability and accurate classification of the neural network when tested with a large sample of dynamically changing input data.

3. Description of the proposed conceptual system model

The proposed conceptual system model is presented in Figure 1. The data of connected sensors and the data from the user program of the CPU is processed with a high data throughput/high data rate through neural network in the MLP NN. The sensors are connected via Gigabit Ethernet and USB 3.1 to the integrated I/O (Input/Output) interfaces of the module. The MLP NN transfers the result of the corresponding processing operation to the CPU via the backplane and PROFINET bus. The CPU then allows further evaluation to take place in the user program and the PLC outputs to the actors. Displaying the current results on the HMI (Human Machine Interface) device. After the model is trained, it is exported in a format that is compatible with the MLP NN module. The trained neural system is exported/loaded to the SD card of the module. The system works in two base modes - off-line and on-line mode. The following is an exposition of the proposed functional algorithm.

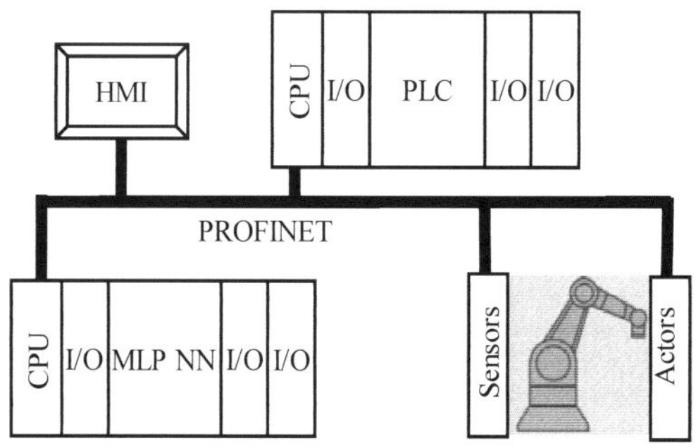

Fig. 1. Conceptual system model

3.1. Proposed functional algorithm

Off-line mode:

1. Train the MLP with a large sample of combined sensor data for both classes - "normal values" and "incorrect values";
2. Test the MLP NN, submitting large sample of combined sensor data of both classes, that did not participate in the training phase. If the classification accuracy is under 80%, optimize the MLP NN parameters: increase the number of neurons in the hidden layer, reduce the preset MSE (Mean Square Error) training error;
3. Add the misclassified sensor combinations to the training sample, assigning the correct class and train the MLP NN again;
4. Download of the obtained matrix of weight coefficients (the trained knowledge base) in the SD card of the NN hardware module.

On-line mode:

1. In real-time operation, the current combination of sensor values is fed within each cycle (current state of the automaton complex) to the inputs of the trained MLP NN and to the PLC.

2/a If any of the sensor values are out of range in the current program input combination and the MLP NN does not correctly classify it as the "incorrect values" class,

2/b or all the sensor values are within the allowable range in the current program input combination, but the MLP NN does not correctly classify it as class "normal values".

In this case, the current misclassified input sensor combination is added to the training sample of sensor combinations as the right corresponding class, and the off-line training mode is set again.

3. Update the obtained matrix of weight coefficients (the trained knowledge base) in the SD card of the NN hardware module and continue in on-line mode.

In this way, the dynamics in the changes of input combinations will be correctly classified by the MLP NN, especially after long-term operation of the automatic complex, when the majority of sensory combinations unforeseen in the training data would occur.

4. Experimental data and results

To conduct the experiment, water pump sensor data, freely available in the open platform for datasets *Kaggle* [13], were used. The training file of the neural network includes the values of 49 sensors with levels corresponding to normal operation of the sensors, as well as those outside their operating range that occurred during real emergency situations. Included are 300 combinations of actual „normal values" during operation and 300 combinations of „incorrect values". These values are grouped in the training file into two classes of sensor signal combinations, the class "normal values" and class "incorrect values". Figure 2 represents the first 300 correct and the next incorrect combinations of sensor values for three of the sensors.

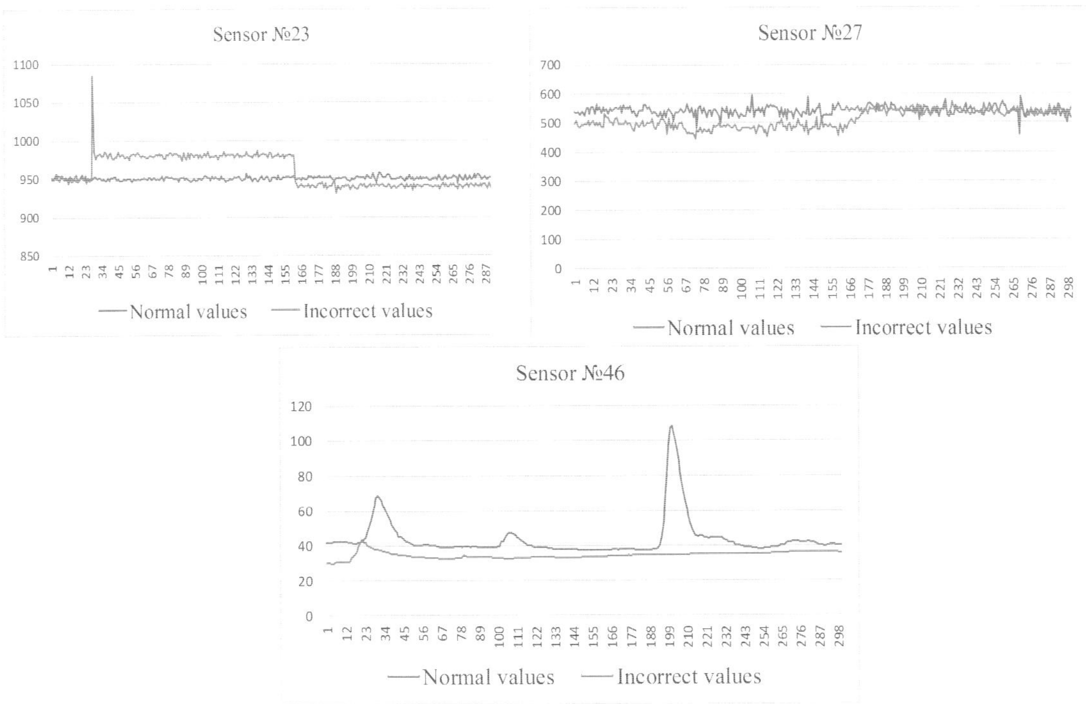

Fig. 2. 300 trained sensor values for classes "Normal values" and "Incorrect values" for three of the sensors

The trained neural network is of the MLP type with a 49-20-2 structure (49 input neurons, 20 neurons in the hidden layer and two neurons in the output layer - corresponding to the two defined classes of sensory signals - "normal values" and "incorrect values"). The MLP structure was trained until reaching 0.1% MSE in the output layer. This was achieved after 12354 iterations. A tangent hyperbolic activation function was applied to all three layers of the MLP structure and a standard Backpropagation training algorithm was applied.

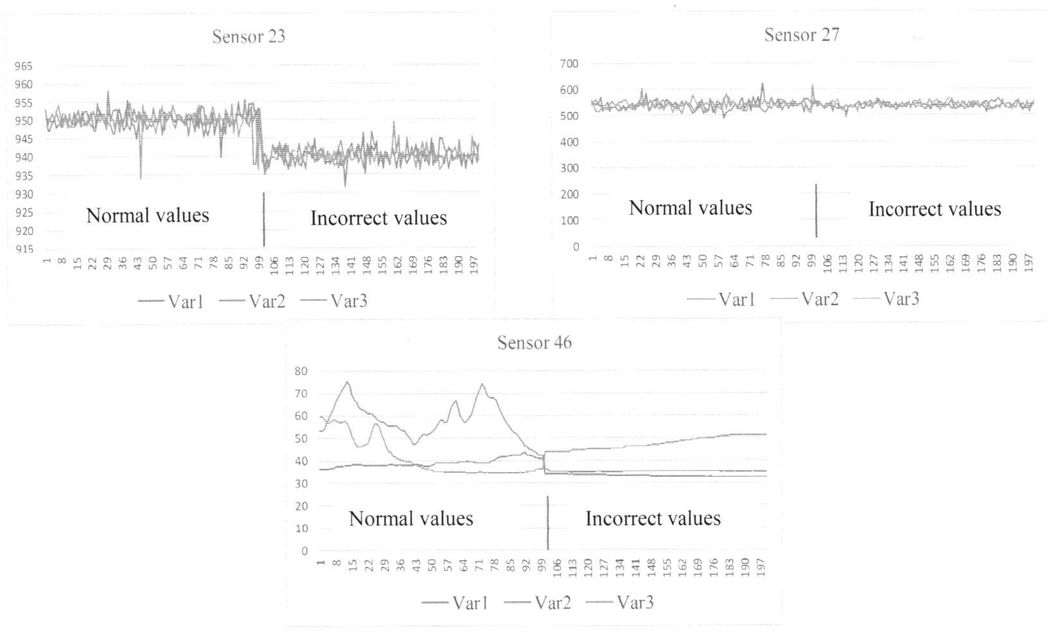

Fig. 3. 100 different instances for each of the three variants in the test phase

In the test phase, when performing steps 2 and 3 in the off-line mode and steps 2a and 2b in the on-line mode of the algorithm, three variants of combinations of sensor signals for both classes, which did not participate in the training sample, were applied. For each of the 49 sensors, 100 different instances were selected for each of the three variants for class "Normal values" and the next 100 - for class "Incorrect values". They are represented in Figure 3.

4.1. Test results

The next section presents the results of testing the trained MLP NN before and after retraining in steps 2 and 3 in off-line mode and steps 2a and 2b in on-line mode. The graphs show the ideal values set in the training phase for the two classes "Normal values" and "Incorrect values", as well as the values obtained in the testing phase. Graphs for one of the three variants tested are presented in Figure 4. The effect of retraining is visible, whereby there are no misclassified input combinations in the "after retrained" plots. The two misclassified "normal values" sensor combinations in output 1 and the three misclassified "incorrect values" sensor combinations in output 2 are recognized correctly after retraining the neural network.

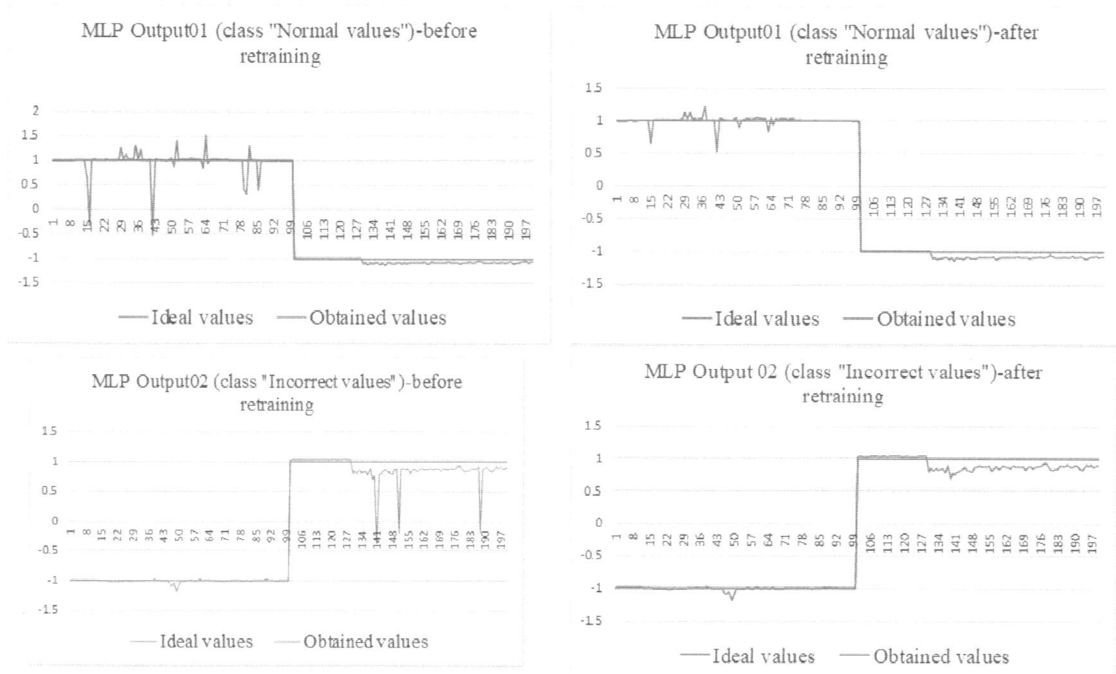

Fig. 4. MLP NN outputs in test phase before and after retraining

4.2. Classification results

The results of the classification before and after the retraining of the MLP NN are presented in Table 1. The classification accuracy was calculated as the number of incorrectly classified input sensor combinations in on-line mode/the total number of tested combinations in each of the three sample variants. After the retraining proposed in steps 2a and 2b of the algorithm, the neural network is dynamically retrained. In this way, the dynamics in the changes of input combinations will be correctly classified by the MLP NN , especially after long-term operation of the automatic complex, when the majority of sensory combinations unforeseen in the training data would occur.

MLP NN - Classification accuracy [%]					
Variant1		Variant2		Variant3	
Before retraining	After retraining	Before retraining	After retraining	Before retraining	After retraining
95%	100%	91%	100%	87%	100%

Table 1. Presentation of classification accuracy

5. Conclusion

The proposed conceptual model and algorithm for dynamic retraining of the neural network is tailored to the specifics of data exchange in an automated PLC control system and operation of an integrated hardware module with neural networks. The main benefits and contributions of this study are as follows:

1. The application of deep learning and NN retraining enables adaptive self-tuning of the control system at the field communication level;
2. By statistically accumulating and storing sensor combinations corresponding to incorrect sensor values or wrong sensor combinations, it would be more difficult to determine wrong combinations by searching the accumulated database. It is much more efficient to update the database by retraining, as shown in the proposed algorithm;
3. The dynamics in the changes of input combinations will be correctly classified by the MLP NN, especially after long-term operation of the automatic complex, when the majority of sensory combinations, unforeseen in the training data would occur;
4. If the art of the sensor signals need to be changed, retraining can be easily performed without significantly changing the algorithm itself. The presented results and the achieved high classification accuracy after retraining for all three investigated variants show a good generalization of the method;
5. The proposed method can be easily integrated in different real automated systems, supporting standard communication protocols and data exchange, between the neural module and PLC.

As a future development and continuation of the research, extended testing of the method with a large number of different sensor combinations in learning differente structures of the MLP NN, aiming to minimize the number of neurons in the inner hidden layer, can be indicated. This would speed up the real-time operation of the neural network in the hardware module. Additional implementation and testing of the model in the Siemens S7-1500 TM NPU modules is also pending.

6. References

[1] Tekic, Z; Cosic, I & Katalinic, B. (2019). Manufacturing and the Rise of Artificial Intelligence: Innovation Challenges, Proceedings of the 30th DAAAM International Symposium, pp.0192-0196, B. Katalinic (Ed.), Published by DAAAM International, ISBN 978-3-902734-22-8, ISSN 1726-9679, Vienna, Austria, DOI: 10.2507/30th.daaam.proceedings.025.

[2] Sundaram, K. & Natarajan, N., (2018). Artificial Intelligence in the Shop Floor (Envisioning the Future of Intelligent Automation and its Impact on Manufacturing), Frost & Sullivan White Paper, pp.3-10.

[3] Neil, St., (2019), Siemens Adds Artificial Intelligence to Control Logic, Sept. 26, 2019.

[4] Topalova, I. & Staeva, Y. (2022). Classification of Two-dimensional Mechanical Parts Using a Convolutional Neural Network, Proceedings of the 33rd DAAAM International Symposium, pp.0042-0049, B. Katalinic (Ed.), Published by DAAAM International, ISBN 978-3-902734-36-5, ISSN 1726-9679, Vienna, Austria, DOI: 10.2507/33rd.daaam.proceedings.007.

[5] MANUAL SIMATIC 6ES7556-1AA00-0AB0, S7-1500 / ET 200MP, Edition 07/2023 A5E46384784-AC, Copyright © Siemens AG 2019 - 2023.

[6] Siemens Industrial Portal, (2023), Getting Started with the Siemens Industrial AI Portfolio,(2023), Entry ID: 109781466, Entry date: 09/29/2023.

[7] Shell Global, (2021), Shell announces new AI-Powered applications through Open AI Energy Initiative, (2021), 9 Nov 2021.

[8] ABB Portal, (2023), PLC Application Examples for Automation Builder, Application Example AC500 HVAC Application Library (V1.0.3) package.

[9] Xiu, Ji, (2019). PROFINET in Practice: Installation, Maintenance, Design and System Engineering. Independently published. ISBN 978-1-07707-721-8.

[10] Modicon M580 Hardware, (2023), Reference Manual, Original instructions, EIO0000001578.13, 02/2023.

[11] Marugán, A.P. & Márquez, F.P.G. (2018). SCADA and Artificial Neural Networks for Maintenance Management. In: Xu, J., Gen, M., Hajiyev, A., Cooke, F. (eds) Proceedings of the Eleventh International Conference on Management Science and Engineering Management. ICMSEM 2017. Lecture Notes on Multidisciplinary Industrial Engineering. Springer, Cham., doi.org/10.1007/978-3-319-59280-0_75.

[12] Wang, W., Harrou, F., Bouyeddou, B. et al., (2022), A stacked deep learning approach to cyber-attacks detection in industrial systems: application to power system and gas pipeline systems. Cluster Comput 25, 561–578 (2022), doi.org/10.1007/s10586-021-03426-w.

[13] The largest AI & ML community, kaggle.com/datasets/anseldsouza/water-pump-sensor-data, (2018).

DOI: 10.2507/34th.daaam.proceedings.008

INFERENCING – NEXT STEP OBSERVATION

Zeljko Seremet & Kresimir Rakic

This Publication has to be referred as: Seremet, Z[eljko] & Rakic, K[resimir] (2023). Inferencing – Next Step Observation, Proceedings of the 34th DAAAM International Symposium, pp.0056-0061, B. Katalinic (Ed.), Published by DAAAM International, ISBN 978-3-902734-41-9, ISSN 1726-9679, Vienna, Austria
DOI: 10.2507/34th.daaam.proceedings.008

Abstract

A production software system is observable to the extent that new internal system states can be understood without making arbitrary guesses, predicting those failure modes in advance, or shipping new code to understand that state. In this way, the concept of observation of management theory is extended to the field of software engineering. But it is with modern distributed systems that the scalpel and searchlight afforded by observability tooling becomes absolutely nonnegotiable. In distributed systems, the ratio of somewhat predictable failure modes to novel and never-before-seen failure modes is heavily weighted toward the bizarre and unpredictable. Those unpredictable failure modes happen so commonly, and repeat rarely enough, that they outpace the ability for most teams to set up appropriate and relevant enough monitoring dashboards to easily show that state to the engineering teams responsible for ensuring the continuous uptime, reliability, and acceptable performance of their production applications. In article explored monitoring and observability and how they differ. It looked at how observability is poorly defined today with loose boundaries, which results in uncontrolled data, tool, and spend sprawl. Meanwhile, the latest progress in AI could resolve some of the observability problems we have today with a new class of Inferencing solutions based on AI.

Keywords: observability; AI; software; engineering; montoring

1. Introduction

In the software development industry, the topic of observation has generated a lot of interest and is often on hot new topic lists. Considering its applicability in the field of SRE, as a narrow subfield of software engineering, it is used when debugging at all levels of the developed system. But as things seem to inevitably go when a new hot topic sees a spike in adoption interest, complex ideas become too ripe for misunderstanding without a deeper look at the many nuances contained within a simple topic label. The aim of the article is to clarify the area of observability in detail. Beginning with the mathematical origins of the term "observability" to examining how software development professionals have adapted it to describe the characteristics of production software systems. [1] [2] [3] [4].

Furthermore, explains what "observability" means, how to determine if a software system is observable, why observability is necessary, and how observability is used to find problems in ways that are not possible with other approaches. It also looks at why observability customization is needed for use in production software systems, and the underlying problem that drives the development of the next phase of inferencing.

2. The Mathematical Description of Observability

The term "observability" was coined by engineer Rudolf E. Kálmán in 1960 [5]. It has since grown to mean many different things in different communities. Let's explore the landscape before turning to our own definition for modern software systems. In his 1960 paper, Kálmán introduced a characterization he called observability to describe mathematical control systems. In control theory, observability is defined as a measure of how well internal states of a system can be inferred from knowledge of its external outputs.

This definition of observability studies observability and controllability as mathematical duals, along with sensors, linear algebra equations, and formal methods. This traditional definition of observability is the realm of mechanical engineers and those who manage physical systems with a specific end state in mind [2]. In a textbook focused on mathematics and process engineering, observability has a formal meaning in traditional systems engineering terminology. However, when this same concept is adapted for use with narrower virtual software systems, it opens up a radically different way of interacting with and understanding written code.

3. Applying Observability to Software Systems

Kálmán's definition of observability can be applied to modern software systems. When adapting the concept of visibility to software, additional considerations specific to the domain of software engineering must also be considered. For a software application to have visibility, it must be able to do the following:

- Understand the inner workings of your application
- Understand any system state the application may have entered, even a new one that has never been seen before and could not have been predicted
- Understand the internal workings and state of the system solely by observation and testing using external tools
- Understanding the internal state without sending any new custom code to handle it (as this implies necessary prior knowledge to explain)

A good litmus test to determine if these conditions are true is to ask yourself the following questions:

- Can open questions about the inner workings of the applications be continuously answered to explain any anomalies without reaching investigative dead ends (ie the problem may be in a particular group of things but cannot be further disaggregated to confirm)?
- Can one understand what a particular user of the software might experience at any given time?
- Can any cross-section of system performance be quickly seen, from top-level summary views to individual and exact user requests that may be contributing to slowness (and anywhere in between)?
- Can arbitrary groups of user requests be compared in ways that allow one to correctly identify which attributes are commonly shared by all users experiencing unexpected behavior in the application?
- Once suspicious attributes are found within a single user request, can all user requests be searched to identify similar patterns of behavior to confirm or rule out suspicions?
- Can I identify which system user generates the most load (and therefore slows down the application the most), as well as the 2nd, 3rd or 100th user that generates the most load?
- Can it be identified which of those users generating the most load have only recently started to affect performance?
- If the 142nd slowest user complained about performance speed, can their requests be isolated to understand why exactly things are slow for that particular user?
- If users complain about timeouts, but graphs show that 99, 99.9, even 99.99 percent of requests are fast, can the hidden timeouts be found?
- Can questions like the previous ones be answered without first having to anticipate that one day they might have to change them (and therefore set up special monitors in advance to collect the necessary data)?
- Can such application questions be answered even if this problem has never been seen or debugged before?
- Can questions like the previous one be answered quickly, so that a new question can be iteratively asked, and then another, and another, until the real source of the problem is reached, without losing the train of thought (which usually means getting the answer within seconds instead of minutes)?
- Can questions like the above be answered even if the problem has never happened before?
- Can any fault in the system be quickly (within minutes) isolated, no matter how complex, deeply buried, or hidden within the stack?

Meeting all of the previous criteria is a high bar for many software engineering organizations to clear. If that tape can be removed, it will no doubt come to a conclusion as to why observation has become such a popular topic for software engineering teams. Put simply, the definition of "observability" for software systems is a measure of how well one can understand and explain any state the system may get into, no matter how novel or bizarre. One must be able to comparatively debug that bizarre or novel state across all dimensions of system state data, and combinations of dimensions, in an ad hoc iterative investigation, without being required to define or predict those debugging needs in advance.

If any can be understood any bizarre or novel state without needing to ship new code, the ability to observability is achieved. Adapting the traditional concept of observability for software systems in this way is a unique approach with additional nuances worth exploring. For modern software systems, observability is not about the data types or inputs, nor is it about mathematical equations. It is about how people interact with and try to understand their complex systems.

Therefore, observability requires recognizing the interaction between both people and technology to understand how those complex systems work together. If this definition is accepted, many additional questions emerge that demand answers:

- How does one gather that data and assemble it for inspection?
- What are the technical requirements for processing that data?
- What team capabilities are necessary to benefit from that data?

The application of observability to software systems has much in common with its control theory roots. However, it is far less mathematical and much more practical. In part, that's because software engineering is a much younger and more rapidly evolving discipline than its more mature mechanical engineering predecessor. Production software systems are much less subject to formal proofs. That lack of rigor is, in part, a betrayal from the scars we, as an industry, have earned through operating the software code we write in production.

As engineers attempting to understand how to bridge the gap between theoretical practices encoded in clinical tests and the impacts of what happens when the code runs at scale, there was not looking for a new term, definition, or functionality to describe how it got there. It was the circumstances of managing systems and teams that led to the evolving practices away from concepts, like monitoring, that simply no longer worked. The industry must move beyond the current gaps in tools and terminology to overcome the pain and suffering caused by outages and the lack of more proactive solutions [2] [4].

Observability is the solution to that gap. Complex production software systems are a mess for a variety of both technical and social reasons. Therefore, both social and technical solutions are needed to get out of this hole. Observability alone is not the entire solution to all software engineering problems. But it helps to see clearly what's happening in all the obscure corners of the software, where you are otherwise typically stumbling around in the dark and trying to understand things.

4. Why Observability Matters Now

The traditional approach of using metrics and monitoring of software to understand what it's doing falls drastically short. This approach is fundamentally reactive. It may have served the industry well in the past, but modern systems demand a better methodology.

For the past two or three decades, the space between hardware and its human operators has been regulated by a set of tools and conventions most call "monitoring." Practitioners have, by and large, inherited this set of tools and conventions and accepted it as the best approach for understanding that squishy virtual space between the physical and their code. And they have accepted this approach despite the knowledge that, in many cases, its inherent limitations have taken them hostage late into many sleepless nights of troubleshooting. Yet, they still grant it feelings of trust, and maybe even affection, because that captor is the best they have [3].

With monitoring, software developers can't fully see their systems. They squint at the systems. They try, in vain, to size them up and to predict all the myriad ways they could possibly fail. Then they watch—they monitor—for those known failure modes. They set performance thresholds and arbitrarily pronounce them "good" or "bad." They deploy a small robot army to check and recheck those thresholds on their behalf. They collect their findings into dashboards. They then organize themselves around those robots into teams, rotations, and escalations. When those robots tell them performance is bad, they alert themselves. Then, over time, they tend to those arbitrary thresholds like gardeners: pruning, tweaking, and fussing over the noisy signals they grow [2].

5. Is This Really the Best Way?

For decades, that's how developers and operators have done it. Monitoring has been the de facto approach for so long that they tend to think of it as the only way of understanding their systems, instead of just one way. Monitoring is such a default practice that it has become mostly invisible. As an industry, we generally don't question whether we should do it, but how.

The practice of monitoring is grounded in many unspoken assumptions about systems. But as systems continue to evolve—as they become more abstract and more complex, and as their underlying components begin to matter less and less—those assumptions become less true. As developers and operators continue to adopt modern approaches to deploying software systems (SaaS dependencies, container orchestration platforms, distributed systems, etc.), the cracks in those assumptions become more evident.

More people, therefore, are finding themselves slamming into the wall of inherent limitations and realizing that monitoring approaches simply do not work for the new modern world. Traditional monitoring practices are catastrophically ineffective for understanding systems. The assumptions of metrics and monitoring are now falling short. To understand why they fail, it helps to examine their history and intended context [2] [6].

6. Why Are Metrics and Monitoring Not Enough?

In 1988, by way of the Simple Network Management Protocol (SNMPv1 as defined in RFC 1157), the foundational substrate of monitoring was born: the metric. A metric is a single number, with tags optionally appended for grouping and searching those numbers. Metrics are, by their very nature, disposable and cheap. They have a predictable storage footprint. They're easy to aggregate along regular time-series buckets. And, thus, the metric became the base unit for a generation or two of telemetry—the data collected from remote endpoints for automatic transmission to monitoring systems.

Many sophisticated apparatuses have been built a top the metric: time-series databases (TSDBs), statistical analyses, graphing libraries, fancy dashboards, on-call rotations, ops teams, escalation policies, and a plethora of ways to digest and respond to what a small army of robots is telling. But an upper bound exists to the complexity of the systems you can understand with metrics and monitoring tools. And once you cross that boundary, the change is abrupt. What worked well enough last month simply does not work anymore. It starts back to low-level commands like strace, tcpdump, and hundreds of print statements to answer questions about how the system is behaving daily [2].

It's hard to calculate exactly when that tipping point will be reached. Eventually, the sheer number of possible states the system could get itself into will outstrip the team's ability to pattern-match based on prior outages. Too many brand-new, novel states are needing to be understood constantly. The team can no longer guess which dashboards should be created to display the innumerable failure modes. Monitoring and metrics-based tools were built with certain assumptions about architecture and organization, assumptions that served in practice as a cap on complexity. These assumptions are usually invisible until they are exceeding them, at which point they cease to be hidden and become the bane of the ability to understand what's happening. Some of these assumptions might be as follows:

- The application is a monolith.
- There is one stateful data store ("the database"), that is running.
- Many low-level system metrics are available (e.g., resident memory, CPU load average).
- The application runs on containers, virtual machines (VMs), or bare metal
- System metrics and instrumentation metrics are the primary source of information for debugging code.
- A static and long-running set of nodes, containers, or hosts to monitor.
- Engineers examine systems for problems only after problems occur.
- Dashboards and telemetry exist to serve the needs of operations engineers.
- Monitoring examines "black-box" applications in much the same way as local applications.
- The focus of monitoring is uptime and failure prevention.
- Examination of correlation occurs across a limited (or small) number of dimensions.

When compared to the reality of modern systems, it becomes clear that traditional monitoring approaches fall short in several ways. The reality of modern systems is as follows:

- The application has many services.
- There is polyglot persistence (i.e., many databases and storage systems).
- Infrastructure is extremely dynamic, with capacity flicking in and out of exisence elastically.
- Many far-flung and loosely coupled services are managed.
- Engineers actively check to see how changes to production code behave, to catch tiny issues early, before they create user impact.
- Automatic instrumentation is insufficient for understanding what is happening in complex systems.
- Software engineers own their own code in production and are incentivized to proactively instrument their code and inspect the performance of new changes as they're deployed.
- The focus of reliability is on how to tolerate constant and continuous degradation, while building resiliency to user-impacting failures by utilizing constructs like error budget, quality of service, and user experience.
- Examination of correlation occurs across a virtually unlimited number of dimensions.

The last point is important, because it describes the breakdown that occurs between the limits of correlated knowledge that one human can be reasonably expected to think about and the reality of modern system architectures. So many possible dimensions are involved in discovering the underlying correlations behind performance issues that no human brain, and in fact no schema, can possibly contain them. With observability, comparing high-dimensionality and high-cardinality data becomes a critical component of being able to discover otherwise hidden issues buried in complex system architectures [2] [3] [4].

7. The problem with observability

The fundamental problem with observability is — not knowing what information might be needed beforehand. The data required depends on the issue. The nature of production errors is that they are unexpected and long-tail: if they could've been foreseen, they'd have been fixed already.

This is what makes observability fuzzy: there's no clear scope around what and how much to capture. So, observability became "any data that could potentially help us understand what is happening". Today, the best way to describe observability as it is implemented is — "Everything outside of metrics, plus metrics."

Fig. 1. Observability in practice [7]

A perfectly observable system would record everything that happens in production, with no data gaps. Thankfully, that is impractical and prohibitively expensive, and 99% of the data would be irrelevant anyway, so an average observability platform needs to make complex choices on what and how much telemetry data to capture. Different vendors view this differently, and depending on who you ask, observability seems slightly different.

7.1. Commonly cited descriptions of observability are unhelpful

Common articulations of observability, like "observability is being able to observe internal states of a system through its external outputs", are vague and do not give a a clear indication of what it is, nor guide in deciding whether there is sufficient observability for the purpose. In addition, most of the commonly cited markers that purport to distinguish observability from monitoring are also vague, if not outright misleading [2] [4].

8. Inferencing — the next stage after Observability?

To truly understand what the next generation would look like, it starts with the underlying goal of all tools, such as OpenTelemetry, AppDynamics, Datadog, Dynatrace, Grafana, Grafana, and so on. This is to keep production systems healthy and running as expected and, if anything goes wrong, to allow you to quickly understand why and resolve the issue. That are three distinct levels in how tools can support:

- Level 1: "Tell me when something is off in my system" — monitoring
- Level 2: "Tell me why something is off (and how to fix it)" — let's call this inferencing
- Level 3: "Fix it yourself and tell me what you did" — auto-remediation

Traditional monitoring tools do Level 1 reasonably well and help us detect issues. We have not yet reached Level 2 where a system can automatically tell us why something is breaking [7]. So, we introduced a set of tools called observability that sit somewhere between Level 1 and Level 2, to "*help understand why something is breaking*" by giving us more data [8].

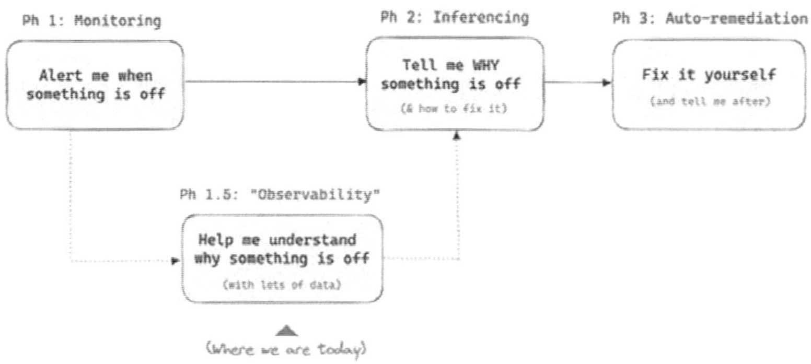

Fig. 2. Evolution of production troubleshooting systems [7]

9. Inferencing — Observability plus AI

Over the last years, rapid advancements in Generative AI have impacted nearly every industry. Likewise, in computer science, the next step after observability is inference — where the platform can reasonably explain why an error occurred, so we can fix it. This becomes possible now in 2023 with the rapid evolution of AI models over the last few months.

Imagine a solution that:

- Automatically surfaces just the errors that need immediate developer attention.
- Tells the developer exactly what is causing the issue and where the issue is: this pod, this server, this code path, this line of code, for this type of request.
- Guides the developer on how to fix it.
- Uses the developer's actual actions to improve its recommendations continuously.

There is some early activity in this space, but this is an open space yet and we can expect several new companies to emerge here over the next couple of years.

10. Conclusion

A production software system is observable to the extent that new internal system states can be understood without making arbitrary guesses, predicting those failure modes in advance, or shipping new code to understand that state. In this way, the concept of observation of managment theory is extended to the field of software engineering. In article explored monitoring and observability and how they differ. It looked at how observability is poorly defined today with loose boundaries, which results in uncontrolled data, tool, and spend sprawl. Meanwhile, the latest advances in artificial intelligence could solve some of the observability issues (eg why something is off and how to fix it) with a new class of Inferencing solutions based on AI. Future work will investigate the shortcomings of Inferencing solutions based on AI on real applications.

11. References

[1] Seremet, Z., & Rakic, K. (2022). Platform Engineering and Site Reliability Engineering: The Path to DevOps Success, DAAAM International Scientific Book 2022, Vol. 21, ISSN 1726-9687, ISBN: 978-3-902734-34-1, Editor: B. Katalinic hard cover, Publisher DAAAM International Vienna, Vienna

[2] Majors, C., Fong-Jones, L. & Miranda, G. (2022). Observability Engineering, Published by O'Reilly Media

[3] Russ, M. (2019). Chaos Engineering Observability, first ed., O'Reilly Media, Inc

[4] Sridharan, C. (2018). Distributed Systems Observability, first ed., O'Reilly Media, Inc

[5] Kálmán, R. E. (1960). On the General Theory of Control Systems, IFAC Proceedings Volumes 1, no. 1: 491–502.

[6] Grobmann, M. & Klug. C. (2017). Monitoring container services at the network edge, in: 29th International Teletraffic Congress, ITC 29, ITC Press, pp. 130–133.

[7] Samyukktha, T. (2022). Inferencing: The AI-Led Future of Observability?, Available from: https://dzone.com/articles/inferencing-the-ai-led-future-of-observability Accessed: 2022-09-20

[8] Sheldon, R. (2023). Top observability tools for 2023, Available from: https://www.techtarget.com/searchitoperations/tip/Top-observability-tools, Accessed: 2023-09-20

DOI: 10.2507/34th.daaam.proceedings.009

CHOOSING AN OPTIMAL STRUCTURAL VARIANT OF A BASIC SIZE FOR A SIZE RANGE OF MODULES

Velizar Zaharinov, Ivo Malakov & Hasan Hasansabri

This Publication has to be referred as: Zaharinov, V[elizar]; Malakov, I[vo] & Hasansabri, H[asan] (2023). Choosing an Optimal Structural Variant of a Basic Size for a Size Range of Modules, Proceedings of the 34th DAAAM International Symposium, pp.0062-0071, B. Katalinic (Ed.), Published by DAAAM International, ISBN 978-3-902734-41-9, ISSN 1726-9679, Vienna, Austria
DOI: 10.2507/34th.daaam.proceedings.009

Abstract

In the present article the solution of the problem for choosing an optimal structural variant of a basic size of a size range of modules for driving automatic sliding doors is considered. For this purpose, the necessary parameters of the movement of the doors and their dimensions and mass are analysed. A functional structure of the module is built and alternative device variants are developed for the implementation of each function. The set of possible structural variants is constructed with compatibility between devices in mind. The problem of choosing the optimal structural variant is formulated. Based on an analysis of the requirements for the module, criteria for evaluating the structural variants and constraints are selected. A mathematical model of the problem for choosing an optimal structural variant is built. By means of algorithmic and software support, the formulated multi-criteria optimization problem is solved both with equal objective functions and also with different priority of some of them, with the aim of enriching the possibilities of the decision maker to find Pareto optimal solutions that best meet its requirements and limitations.

Keywords: design; optimization; size ranges; automatic doors; linear module.

1. Introduction

Automated sliding doors belong to automatic doors, which are a section of architectural systems with a large application in construction. They are used in malls, shops, offices, hotels, etc. as in practice there is hardly any public building without such type of doors installed. Apart from public buildings, automatic doors are also used in private homes, usually when it comes to large opening areas and when using heavy wings with relatively large overall dimensions.

A number of automated sliding door drive systems are available on the market. Some of the leading manufacturers are Dormakaba [1], Alumil [2], GEZE [3], etc. Dormakaba offers two types of drives for sliding doors – with a rotary electric motor and belt drive [4] and with a linear electric motor [5]. Alumil uses in its system for automated sliding doors a module with an electric drive and a toothed belt [6], converting the rotary motion of the electric motor into a linear one. Geze use, like Alumil, an electrically driven linear motion module with a toothed belt to convert rotation to translation [3]. The leading manufacturers offer a variety of typologies and functions, but the drive system, in general, is based on one main type. It is a linear motion module that is sized to drive the heaviest wings of the respective system.

Thus, for assignments that require smaller wings, it is necessary to equip the automatic sliding door with an oversized drive, which leads to higher operating costs. The solution to this problem is the development of a size range [7], [8], [9], [10] from automatic sliding door drive modules. The first step in the development of the size range is the determination of the structure of the base variant (the main size) [8], [9], [11], [12], i.e., the structure that will be used to build the modules of the size range. The selection of the structure of the base variant is a stage that is fundamental to the design of a size range. This is so because the technical and economic characteristics of the basic size determine the properties of the entire range (family) of sizes [8]. The results of this stage are of particular importance for the development of the size ranges, since the errors made here are multiplied in the other elements of the range. Therefore, the choosing of an optimal structural variant of a module for driving automatic sliding doors is a prerequisite for the effective development of a size range of modules. The purpose of this article is to present the choosing of an optimal structural variant of a basic size, which will be the basis for the construction of a size range of modules for driving automatic sliding doors.

2. Development of the set of possible structural variants

2.1. Selection of typology and requirements for the basic size

Automated sliding doors are offered in the form of different typologies. By the term "typology" is meant the arrangement and the way the wings move. In Fig. 1 are shown examples of different typologies.

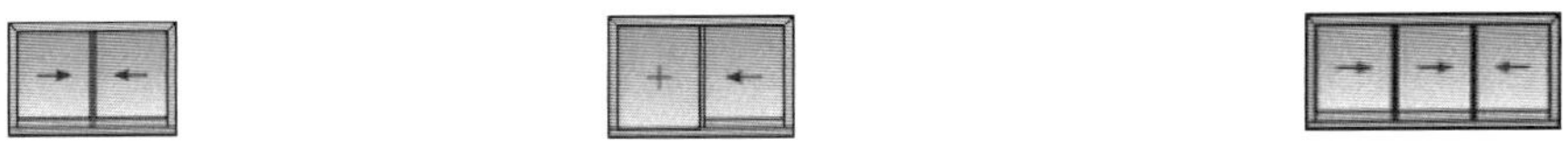

(a) (b) (c)

Fig. 1. Example of different typologies of sliding doors [2]

Due to the different requirements related to installation, mass of the driven wings, frame geometry, etc. with the different typologies, it is necessary to choose a specific one for which to develop a drive. The typology for which the size range is going to be developed and the corresponding drive unit for the automatic door wings is shown in Fig. 1(a). This typology is widespread and is used both for entrance doors and for interior doors. It is characterized by the presence of two movable wings that open towards the centre of the door opening. The wings are guided by steel rollers rolling on a rail in the lower part of the frame. In its upper part, the wing is supported by the geometry of the frame, which functionally prevents the wing from overturning and assists in guiding. After market research, it was found that the dimensions of the wings for automatic sliding doors vary in length from 600 mm to 3000 mm, and in height from 1800 mm to 2800 mm. The mass of one wing can vary widely – from 20 kg to 700 kg [1], [2], [3]. The requirements for the basic size are shown in Table 1, and the specifications - in Table 2. For the requirements, a priority for implementation is indicated, with 10 indicating the highest priority ("demand" type requirement), and values between 9 and 1 varying in the degree of priority for implementation ("wish" type requirements). From the specification table it can be seen that the basic size will be developed taking into account the parameters of the wings with the largest dimensions and mass in the ranges indicated above. This decision was made on the basis of the consideration that the devices selected to fulfil the functions of the drive must be able to drive even the heaviest doors available on the market.

2.2. Development of a functional model of the basic size

In Fig. 2 is shown the overall function of the drive module. It is the positioning of the sliding door wing when control signals are given by the user. In order to be able to perform the overall function, the drive module must be provided with a certain set of inputs (Fig. 2 and Table 3): wing – this is the wing of the sliding door, suspended on a guide system; supporting structure – this is the mechanical system to which the module will be connected; • Power supply – source/sources of energy powering the motor of the module; power supply – source/sources of energy powering the motor of the module; motor control signals – signals from the control system to the motor to execute the commands given by the user. The inputs of the system are converted by the overall function into a certain set of outputs (Fig. 2 and Table 3): positioned wing – the position of the wing changes according to the user's assignment; noise – the level of noise produced by the module must be taken into account in view of its possible application in residential areas; signals for completed movement – signals from position sensors to the control system, confirming completed positioning (reach to position).

№	Type	Requirement	Priority
1	D	Integration into the profiles that make up the frame structure	10
2	W	Low price	5
3	W	Low energy costs	6
4	W	Low noise	8
5	W	Easy installation	7

Table 1. Basic size requirements

№	Requirement/Metric	Value	Dimension
1	Mass of the wing being moved	700	kg
2	Stroke	2,9	m
3	Linear speed	0,1	m/s

Table 2. Basic size specifications

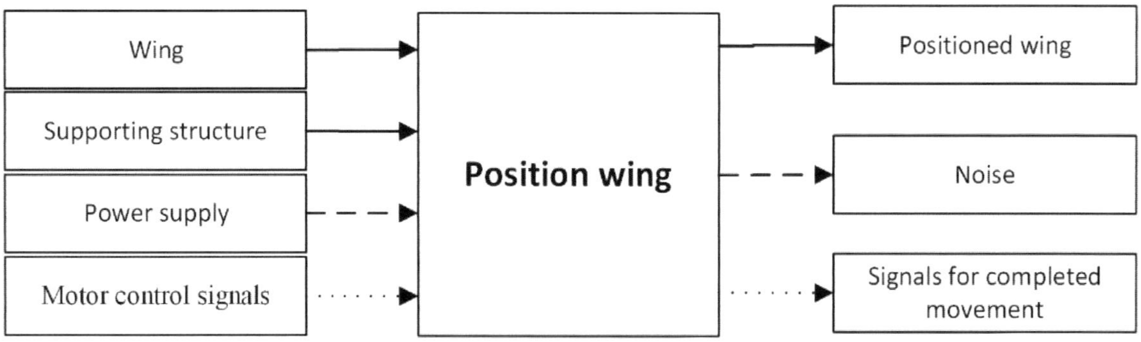

Fig. 2. General function of the basic size

Inputs			Outputs		
Material	*Energy*	*Information*	*Material*	*Energy*	*Information*
Wing	Power supply	Motor control signals	Positioned wing	Noise	Signals for completed movement
Supporting structure					

Table 3. Inputs and outputs for the drive module classified into three groups

The process of converting the inputs of the overall function into its outputs is explained by the list of functions (Table 4) and the tree of functions (Fig. 3). The function tree was obtained after analysing the components included in the structural composition of existing modules.

Designation	Function	Explanation
F1	Drive	Converting supply energy into useful mechanical motion
F2	Transmit motion	Realization of transmission - direct or with transmission ratio
F3	Connect	Connecting the elements of the system to be driven to the drive
F4	Convert motion	Converting motion from one form to another (rotation to translation)
F5	Provide position information	End position detection (wing open/closed)

Table 4. Functions list

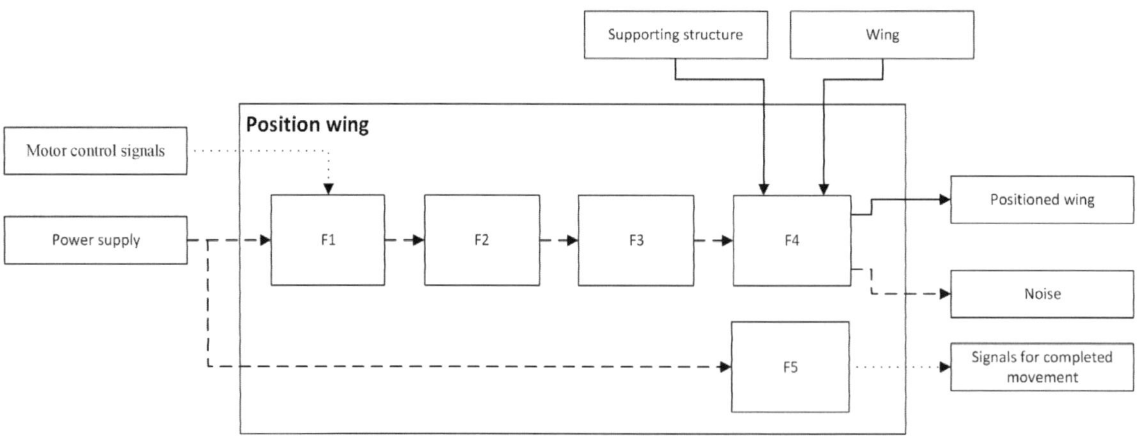

Fig. 3. Functional structure

2.3. Development of alternative variants of devices for the implementation of functions

For each specific function of the designed module (basic size), possible implementation variants are developed. The developed variants for F1 to F5 are summarized in Table 5. The criterion for inclusion is quality of implementing the function [13]

Function	Variant 1	Variant 2	Variant 3	Variant 4	Variant 5	Variant 6	Variant 7
F1	Stepper motor for direct drive	Hybrid servo motor for direct drive	AC servo motor for direct drive	Stepper motor for transmission drive	Hybrid servo motor for transmission drive	AC servo motor for transmission drive	Linear electrical motor
F2	Worm gear	Planetary gear	Bevel gear				
F3	Metal bellow coupling	Self-aligning coupling	Elastic coupling				
F4	Toothed belt drive	Chain drive	Gear rack and pinion drive				
F5	Limit switch	Proximity sensor	Reed switch	Hall effect sensor	Light proximity sensor		

Table 5. Alternative solutions

2.4. Network model of the set of possible structural variants

The set of possible structural variants X that fulfil the overall function of the module is represented in the form of a network model in Fig. 4. The vertices in the model are arranged in 5 columns according to the functions of the basic size. The possibilities of combining the devices x_n^l into structures x performing the overall function of the drive system are shown by arrows. Each path connecting the beginning (S) and the end (E) of the network model represents a possible structural variant. The network model shown in Fig. 4 describes the possible combinations of devices and their compatibility. In the developed variants, devices x_1^1, x_1^2, x_1^3, and x_1^7 are polyfunctional (can perform more than one function). Devices x_1^1, x_1^2, and x_1^3 can perform functions F1 and F2, and device x_1^7 can perform functions from F1 to F4 inclusive.

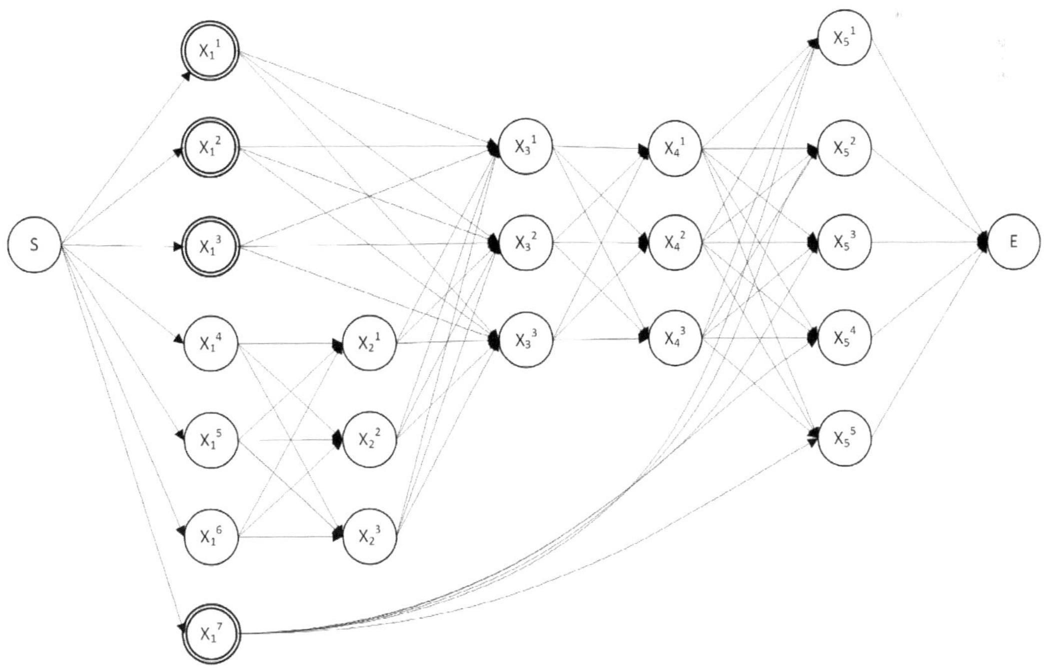

Fig. 4. Network model of the set of possible structural variants

3. Choosing of optimization criteria and their measurement

The choosing of criteria is according to the requirements for the basic size formulated in Table 1. Four criteria are chosen for the evaluation of alternative structural variants: production costs of the developed module in euros, efficiency factor, occupied space in cubic centimetres, and service and repair time. Therefore, the considered problem is a multi-criteria optimization problem with compatibility constraints between individual devices. The compatibility constraint originates from the presence of the polyfunctional devices x_1^1, x_1^2, x_1^3, and x_1^7.

Three of the criteria require a minimum value for the developed alternative structural variants, and one (efficiency factor) - a maximum. The fourth criterion is assessed (measured) by pairwise comparisons of the developed variant. The assessment was carried out by experts according to the method of Voichinskiy and Jansson [14]. The valued criteria (transformed into objective functions) are shown in Table 6.

Function	Variant 1	Variant 2	Variant 3	Variant 4	Variant 5	Variant 6	Variant 7
F1	320,04 -0,431 1183 0,046	343,20 -0,431 1331 0,049	1154,36 -0,186 2805 0,054	70,44 -0,431 325 0,052	315,68 -0,431 323 0,040	703,62 -0,186 800 0,044	11151,00 -0,163 1800 0,044
F2	684,46 -0,357 256 0,071	673,72 -0,020 94 0,075	1697,86 -0,041 3927 0,068				
F3	64,92 -0,020 76 0,045	225,16 -0,020 46 0,047	27,36 -0,010 25 0,050				
F4	1003,53 -0,051 2850 0,083	354,87 -0,073 3660 0,087	1468,11 -0,051 5400 0,079				
F5	12,80 -0,010 2 0,067	24,19 -0,010 20 0,054	6,84 -0,010 2 0,047	14,76 -0,010 6 0,052	23,00 -0,010 3 0,049		

Table 6. Values of the criteria for the variant solutions of the functions

4. Building the mathematical model of the problem

The solving of the following problem is assigned: Given a set of possible structural variants of the basic size, determine the optimal structural variant so that:

$$\begin{aligned} min\, C(x) &= \sum_{n=1}^{5} C(x_n^l), min\, P(x) = \sum_{n=1}^{5} P(x_n^l), \\ min\, V(x) &= \sum_{n=1}^{5} V(x_n^l), min\, T(x) = \sum_{n=1}^{5} T(x_n^l) \end{aligned} \tag{1}$$

where: $x \in X$; $n = 1 \div 5$; $|L_1| = 7$, $|L_2| = 3$, $|L_3| = 3$, $|L_4| = 3$; $|L_5| = 5$; $C(x)$ – production costs, EUR; $P(x)$ – efficiency; $V(x)$ – occupied space, cm^3; $T(x)$ – service and repair time. With the mathematical model built in this way, the value of the relevant objective function for a given possible structural variant of the system is defined as the sum of the technical and economic characteristics of the devices included in this structural variant. Mathematically, to obtain the value of the objective function for the second criterion – efficiency, it is necessary to multiply the efficiency coefficients of the devices making up the variant. Since model (1) uses summation, a logarithm is applied to the actual efficiency values [15] for the devices to obtain the values of the second objective function. The logarithmization was performed with the Napier's constant as the base (the so-called natural logarithm).

5. Solving the problem

Due to the presence of polyfunctional devices, it is necessary to first decompose the set of structural variants of the basic size into subsets where there are no constraints on compatibility. For this purpose, the PolyOptimizer [16] optimization software environment is used.

The set of possible structural variants of the drive system is decomposed into three subsets, the elements of which are shown in Table 7.

Subset	X_1	X_2	X_3	X_4	X_5	Number of variants
1	$x_1^1;\ x_1^2;\ x_1^3$		$x_3^1;\ x_3^2;\ x_3^3$	$x_4^1;\ x_4^2;\ x_4^3$	$x_5^1;\ x_5^2;\ x_5^3;\ x_5^4;\ x_5^5$	135
2	$x_1^4;\ x_1^5;\ x_1^6$	$x_2^1;\ x_2^2;\ x_2^3$	$x_3^1;\ x_3^2;\ x_3^3$	$x_4^1;\ x_4^2;\ x_4^3$	$x_5^1;\ x_5^2;\ x_5^3;\ x_5^4;\ x_5^5$	405
3	x_1^7				$x_5^1;\ x_5^2;\ x_5^3;\ x_5^4;\ x_5^5$	5

Table 7. Elements of structural variant subsets

The total number of possible structural variants of the designed system is 545. To determine the upper and lower bounds of the objective functions, 24 single-criteria optimization problems are solved, and to find a solution to the multi-criteria problem, it is necessary to solve 3 multi-criteria optimization problems (one for each subset). The considered problem is a multi-criteria optimization problem with compatibility constraints between devices. It is solved using the PolyOptimizer dialog system.

5.1. Solution under equal priority of the objective functions

The compromise solution of the problem is found for objective functions of equal importance. Table 8 shows the values of the objective functions for the variant, as well as the devices entering the structure of the module. In Fig. 4 is shown the solution from PolyOptimizer.

Variant x_1^* (Table 8) is a timing belt mechanism driven by a stepper motor without a transmission mechanism (direct drive). The motor is connected to the mechanism by a self-adjusting clutch, and the end positions of the wing are established by a reed contact. When analysing the proposed solution, the following conclusions can be drawn: production costs have the smallest deviation from their optimum, and the largest - service and repair time; the compromise variant has significant deviations from the optimum of the objective functions; the optimum is found in the first subset. In Fig. 5, the devices that make up the structural variant are marked in green, and the values of the elementary devices that make up the subset (after decomposition) are written in red.

Structural variant x_1^*				
$x_1^* = \{x_1^1; x_3^1; x_4^1; x_5^3\}$				
№	Objective function	Value	Deviation from the optimum for the objective function	Upper and lower boundary
1	$C(x)$, EUR	1395,76	$w_1 = 0{,}0656$	$709{,}11 \le C(x) \le 11175{,}19$
2	$P(x)$	-0,512	$w_2 = 0{,}4722$	$-0{,}891 \le E(x) \le -0{,}173$
3	$V(x)$, cm³	4111	$w_3 = 0{,}2742$	$1802 \le V(x) \le 10\ 223$
4	$T(x)$	0,221	$w_4 = 0{,}5417$	$0{,}091 \le T(x) \le 0{,}331$

Table 8. Structural variant with equal priority of the objective functions

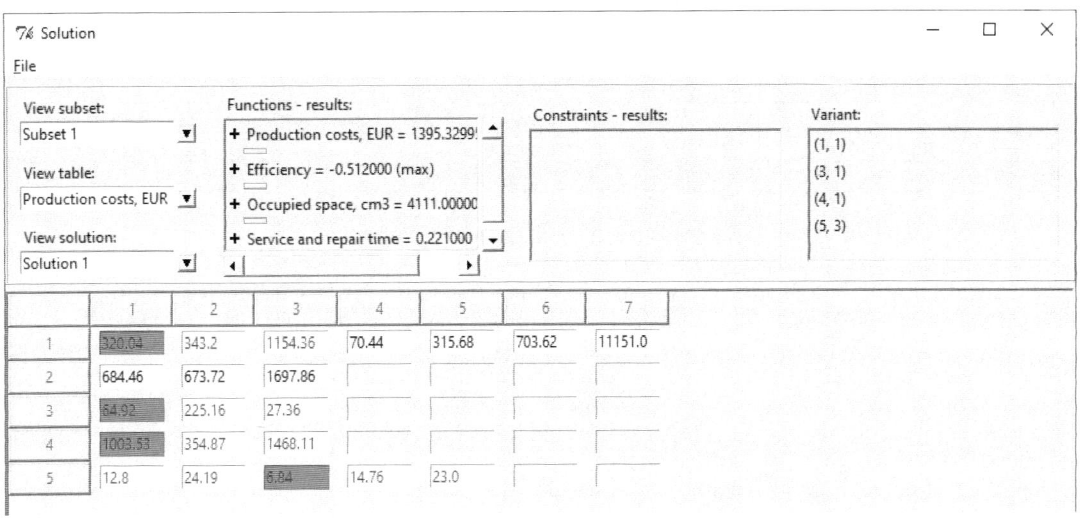

Fig. 5. Solution x_1^*

5.2. Solution when setting a different priority for criteria

If the solutions obtained do not satisfy the decision maker (DM), it can continue to search for other solutions by changing the priority of the criteria. The investigation of the problem continues with the introduction of different weighting coefficients.

Due to the significant deviations of the objective functions from their optima, the problem offers good opportunities for researching alternative solutions giving priority to individual objective functions. Initially, the problem is solved with a higher priority for production costs. Therefore, the following problem is set: Determine a weight vector that assigns a higher priority to the production cost criterion $C(x)$ of the module.

Saaty's [17] method is utilized for prioritizing production costs. In Fig. 6 is shown the user interface of the method in the PolyOptimizer dialog system, through which the parameter values are entered. The binary comparisons for the particular problem are six. Production costs are prioritized over efficiency, space occupied, and service and repair time. DM has set these parameters through the sliders (Fig. 6). The figure shows that equal importance is set for the remaining criteria by placing the sliders midway between the comparison pairs. After calculating the weighting coefficients, the problem can be solved with the thus entered information about the priority of the criteria. The values obtained for this solution are shown summarized in Table 9. The calculated priority vector is $p_1 = \{0{,}7500; 0{,}0833; 0{,}0833; 0{,}0833\}$.

The differences between this structural variant and variants x_1^* are in one device – performing the function of motion conversion. Manufacturing costs improve by 6% relative to x_1^*. All other objective functions worsen – efficiency by 3%, space occupied by 10%, and service and repair time by 2%. If the proposed variant satisfies the DM, then it is selected for the solution of the problem. Otherwise, a new priority is set. Using the PolyOptimizer dialog system, the situation where higher priority is given to efficiency was investigated.

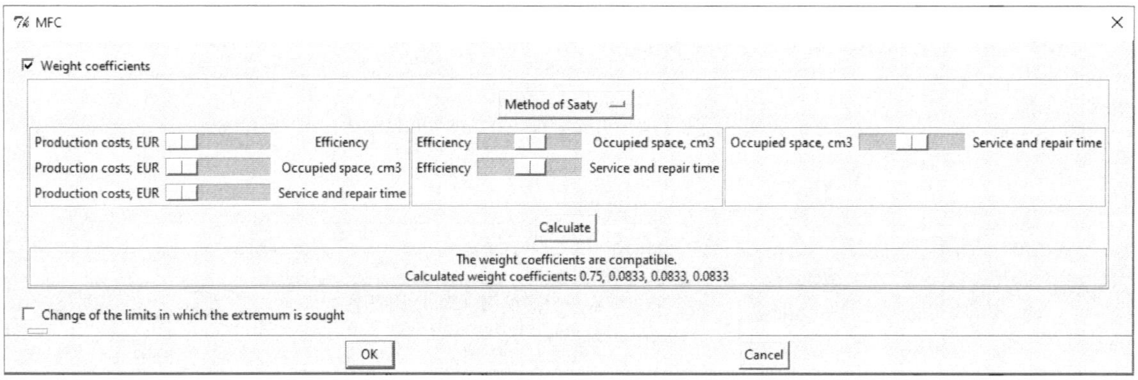

Fig. 6. Prioritize production costs

Structural variant x_2^*				
$x_2^* = \{x_1^1; x_3^1; x_4^2; x_5^3\}$				
№	Objective function	Value	Deviation from the optimum for the objective function	Upper and lower boundary
1	$C(x)$, EUR	746,67	$w_1 = 0{,}0036$	$709{,}11 \leq C(x) \leq 11175{,}19$
2	$P(x)$	-0,534	$w_2 = 0{,}5028$	$-0{,}891 \leq E(x) \leq -0{,}173$
3	$V(x)$, cm3	4921	$w_3 = 0{,}3704$	$1802 \leq V(x) \leq 10\ 223$
4	$T(x)$	0,225	$w_4 = 0{,}5583$	$0{,}091 \leq T(x) \leq 0{,}331$

Table 9. Structural variant with priority of production costs

When prioritizing using Saaty's method, a solution x_3^* (Table 10) is obtained, which is significantly different from x_1^*. A linear electric motor was chosen for it. It is a polyfunctional device and performs all the functions of the system without one. The choice worsens the production costs to a large extent – by 93%, but the remaining objective functions achieve their optima. The calculated priority vector is $p_2 = \{0{,}0833; 0{,}7500; 0{,}0833; 0{,}0833\}$.

It is found that when prioritizing the remaining objective functions, by Saaty's method (analogous to Fig. 6), the solutions coincide with solution x_3^*. This is because given the priority of one objective function and all else being equal, Saaty's method assigns a large weighting factor to the priority objective function. Therefore, to determine the priority of the occupied space, the method of Voichinskiy and Jansson [14] is used (Fig. 7). The solution found is shown in Table. 11.

Structural variant x_3^*				
$x_3^* = \{x_1^7; x_5^3\}$				
№	Objective function	Value	Deviation from the optimum for the objective function	Upper and lower boundary
1	$C(x)$, EUR	11157,84	$w_1 = 0{,}9983$	$709{,}11 \leq C(x) \leq 11175{,}19$
2	$P(x)$	-0,173	$w_2 = 0{,}0000$	$-0{,}891 \leq E(x) \leq -0{,}173$
3	$V(x)$, cm3	1802	$w_3 = 0{,}0000$	$1802 \leq V(x) \leq 10\ 223$
4	$T(x)$	0,091	$w_4 = 0{,}0000$	$0{,}091 \leq T(x) \leq 0{,}331$

Table 10. Structural variant with priority of efficiency

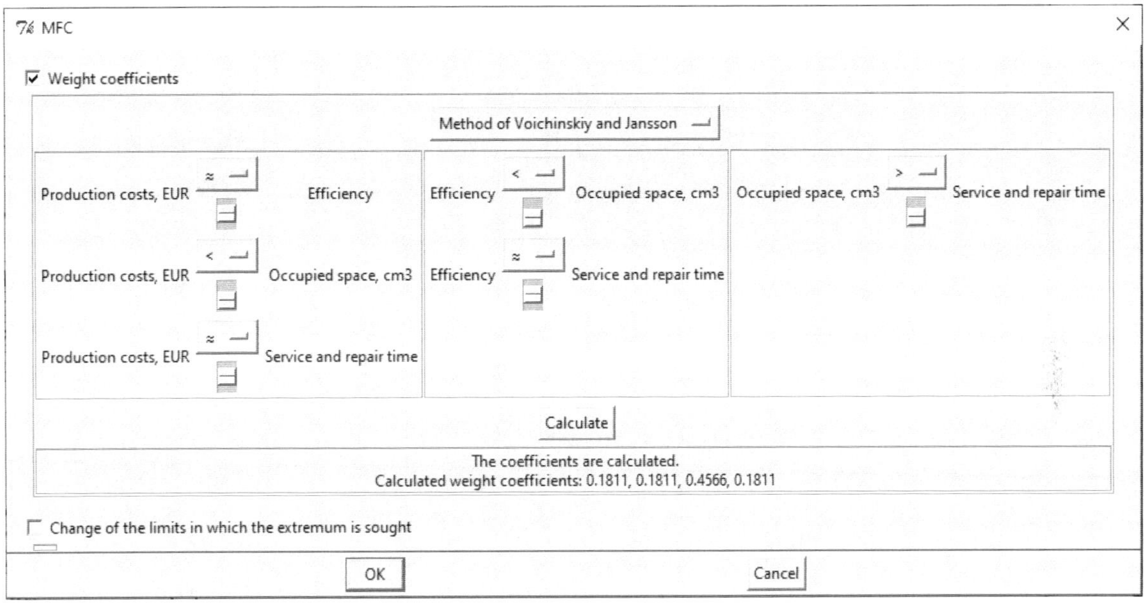

Fig. 7. Voichinskiy and Jansson prioritization of the occupied space objective function

Structural variant x_4^*				
$x_4^* = \{x_1^1; x_3^3; x_4^1; x_5^1\}$				
№	Objective function	Value	Deviation from the optimum for the objective function	Upper and lower boundary
1	$C(x)$, EUR	1363,73	$w_1 = 0{,}0626$	$709{,}11 \leq C(x) \leq 11175{,}19$
2	$P(x)$	-0,502	$w_2 = 0{,}4582$	$-0{,}891 \leq E(x) \leq -0{,}173$
3	$V(x)$, cm3	4060	$w_3 = 0{,}2681$	$1802 \leq V(x) \leq 10\ 223$
4	$T(x)$	0,246	$w_4 = 0{,}6458$	$0{,}091 \leq T(x) \leq 0{,}331$

Table 11. Structural variant with priority of the occupied space

With this solution, the occupied space is improved by 1%, production costs and efficiency are also improved by 0,3% and 1% respectively. Service and repair time is worsened by 10%.

In Fig. 8, the four solutions found are shown in summary. Graphical representation is made through a radar diagram, on which each axis represents an evaluation criterion. The different coloured rectangles plotted on the diagram are the four solutions. The values on the axes are the percentage deviations of the solutions found for each criterion. The ideal solution that simultaneously achieves an optimum by all criteria is the point of intersection of the axes in the centre of the diagram. The graph can be likened to a target – the more concentrated the variant's rectangle is around the zero point, the closer it is to the ideal solution. The graph shows that the compromise solution x_1^* is the most concentrated, with solutions x_2^* and x_4^* also close to it. Solution x_3^* is highly directed towards the optimum of production costs and is therefore visible on the graph as a vertical line pointing to the objective function. On the basis of the solutions found and their graphic analysis, a decision was made by the DM, that the basic size will be based on variant x_1^* (Fig. 9).

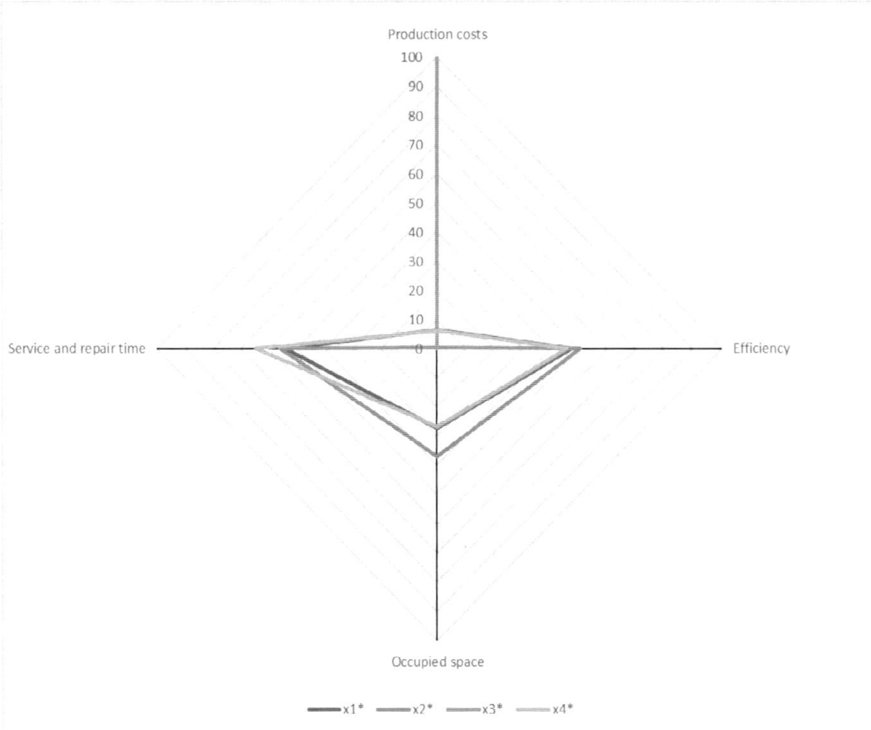

Fig. 8. Graphical interpretation of the solutions found

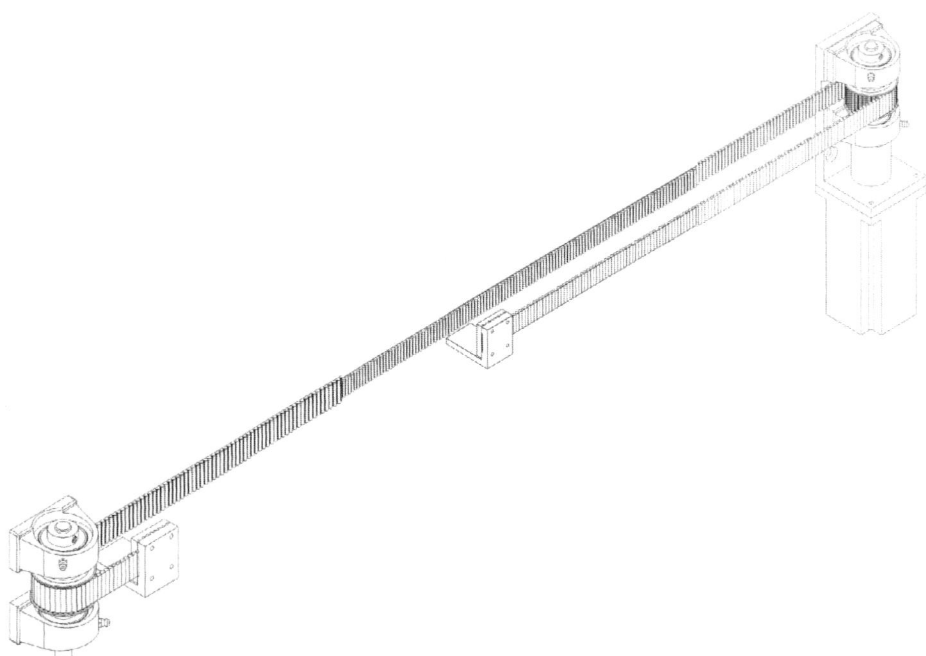

Fig. 9. Variant x_1^* drive elements

6. Conclusion

The following results have been achieved in the present development:

- A sliding door typology has been chosen for which to develop a size range of drive modules;

- After researching the market, the requirements and specifications for the developed basic size of the range are formulated;
- The functional structure of the basic size is built, where its overall function with inputs and outputs and the function tree are formulated;
- Solution variants are developed for each function of the functional structure;
- The set of possible variants is determined by means of a directed graph and the definition of polyfunctional devices and compatibility;
- A presentation and analysis of the problem of choosing the optimal structural variant is made;
- Criteria for evaluation of the developed variants are selected;
- A significant amount of data is collected and the evaluation criteria are valued – converted into objective functions;
- A mathematical model of the problem is built;
- All problem modelling data is entered into a dialog system for multi-criteria optimization, with the help of which to facilitate analysis and finding solutions;
- The set problem of choosing an optimal structural variant of a basic size of a size range is solved under different decision-making conditions and the solutions found are analysed;
- A variant is selected to be the basis for the sizes' structure in a size range of modules for the automatic drive of sliding doors.

In addition to the achieved final goal – the choice of the optimal variant of the basic size, the present development is also an illustrative example that can be generalized to practically any type of technical system. The only requirement is the use of a systematic approach to design and functional thinking (functional representation of the structure of the product being developed). The object of future development is the construction of a size range that integrates the solution chosen in the present work as the basic size. This will also require the development of tools that will support the process of solving the problem and will expand the capabilities of the dialog system used.

7. Acknowledgments

The results in the present research are obtained under project No. 231ПД0028-06 financed by the grant for scientific research at Technical University - Sofia.

8. References

[1] Dormakaba. (2022). Meeting all your access needs. Dormakaba.
[2] Alumil (2023). SMARTIA M630 PHOS, Technical Brochure
[3] GEZE (2023). Automatic Sliding Door, Product Brochure
[4] Dormakaba (2021). ES 200 T Telescopic sliding door operator, Technical Brochure
[5] Dormakaba (2021). CS 80 Magneo, Technical Brochure
[6] Alumil (2017). SUPREME S650 e-MOTION, Technical Brochure
[7] Simpson, T., Maier, J.R.A. & Mistree, F. (2001). Product platform design: Method and application. Research in Engineering Design 13(1), pp. 2–22
[8] Pahl, G. & Beitz, W. (2007). Konstruktionslehre. Methoden und Anwendung. Berlin/Heidelberg, Springer-Verlag
[9] Mueller, D. (2011). A cost calculation model for the optimal design of size ranges. Journal of Engineering Design. Vol. 22, Issue 7, pp. 467-485
[10] Lotz, J. (2018). Beherrschung von Unsicherheit in der Baureihenentwicklung. Dissertation, Darmstadt
[11] Lotz, J., Freund, T, Würtenberger, J. & Kloberdanz H. (2015). Principles to develop size ranges of products with ergonomic requirements, using a robust design approach. Proceedings of the 6th International Conference on Applied Human Factors and Ergonomics (AHFE 2015) and the Affiliated Conferences, AHFE 2015, pp. 1250-1257
[12] Pahl, G., & Beitz, W. (1974) Baureihenentwicklung; Konstruktion 26
[13] Kaisheva, D.; Anchev, A.; Valkov, S.; Dunchev, V.; Kotlarski, G.; Stoyanov, B.; Ormanova, M.; Atanasova, M.; Petrov, P. (2022) Influence of Beam Power on Structures and Mechanical Characteristics of Electron-Beam-Welded Joints of Copper and Stainless Steel. Metals 2022, 12, 737. https://doi.org/10.3390/met12050737 - IF2.351 / IF2.695-01.202
[14] Voichinskiy, A., Jansson, E. (1988) Manufacturability of products in instrumentation. Leningrad, Mashinostroenie
[15] Dobreva, A. & Pavlov, P. (2021). Energy Efficiency of Worm Gear Drives, Proceedings of the 32nd DAAAM International Symposium, pp.0110-0116, B. Katalinic (Ed.), Published by DAAAM International, ISBN 978-3-902734-33-4, ISSN 1726-9679, Vienna, Austria DOI: 10.2507/32nd.daaam.proceedings.016
[16] Malakov, I. K. & Zaharinov, V. V. (2012) Interactive software system for multicriteria choosing of the structural variant of complex technical systems, 23rd DAAAM International Symposium on Intelligent Manufacturing and Automation 2012; pp.199-204
[17] Saaty, T. (1994) Fundamentals of decision making and priority theory with the AHP. RWS Publications:Pittsurgh

DOI: 10.2507/34th.daaam.proceedings.010

TECHNICAL AND TECHNOLOGICAL CULTURE – ENGINEERS POINT OF VIEW

Ljubica Popović, Damir Purković, Nemanja Sremčev & Ilija Ćosić

This Publication has to be referred as: Popovic, L[jubica]; Purkovic, D[amir]; Sremcev, N[emanja] & Cosic, I[lija] (2023). Technical and Technological Culture – Engineers Point of View, Proceedings of the 34th DAAAM International Symposium, pp.0072-0076, B. Katalinic (Ed.), Published by DAAAM International, ISBN 978-3-902734-41-9, ISSN 1726-9679, Vienna, Austria

DOI: 10.2507/34th.daaam.proceedings.010

Abstract

Technical and technological culture remained in the shadow of innovation culture. If we want to respond to the demands of future technological development, it is important to reinvest in the concept of Technical and Technological Culture and engineers, as its bearers, have to ask themselves about the impact of technique and technology on society and the world we live in. This paper will present the results of the examination of attitudes and beliefs about the development of Technical and Technological Culture and its connection with socio-demographic factors. Respondents in this research are engineers of technical sciences: professors, students and employees in the economy/IT industry/NGO. For this research, was created an instrument, upgraded to the already existing instrument for examining Technical and Technological Culture. The results of this examination will indicate the attitudes and beliefs of engineers about the development of technical and technological culture and the connection of socio-demographic factors with its development.

Keywords: Technical and technological culture; engineers; attitudes; socio-demographic factors.

1. Introduction

There was not much research in the past indicating the importance of Technical and Technological Culture. It remained in the shadow of innovation culture. The introduction of technological innovations into everyday life, the readiness to recognize new solutions, and the acquisition of abilities to exploit technological products are related to personal, organizational and local culture, and should be analyzed as a separate category – Technical and Technological Culture [1]. Technological changes and the introduction of new technological concepts lead to the emergence of entirely new needs in terms of behavior and competencies of those involved in production processes [2]. The subject of this pilot research is the attitudes and beliefs of engineers in technical sciences regarding the development of Technical and Technological Culture and to determine socio-demographic factors that may influence its development.

2. Review of literature

According to Pochanek's definition, [3] Technical (Technological) Culture is the entirety of social achievements in the field of technical sciences and their application, as well as all the knowledge and skills needed to understand achievements, use them correctly, transfer them to the younger generation, and create new values in this field.

Wyrwicka [1] has analised a literature review, and delivered an insight into the research's views of technological culture. According to Wyrwicka [1] Technical and Technological Culture is a system of permanent attitudes and competencies of people that enable the proper use of existing technology products to change the pattern of quality of life. It is expressed in relatively enduring and positive attitudes toward technology phenomena (design and production), in possessing knowledge, and above all, in ethical behavior and attitudes of people in various technological situations [1]. Technical and Technological Culture is manifested in relatively long-term activities and good ethical attitudes that enable the correct application of existing technologies and the development of new ethical solutions to improve the efficiency of life cooperation processes [4].

Chalas states [5] that Technical and Technological Culture must recognize the fundamental value of life, as modern humans can achieve much in the field of biological engineering and the study of life, thereby improving and enhancing the quality of human life, while on the other hand, they can use technology to destroy human lives. Technical and Technological Culture is a logical, artistic and positive attitude toward the use of technology to improve the quality of economic, social and spiritual daily life by the level of technical and technological progress [6].

Modern technologies today are not only the product of men but also represent a precondition for their existence. They continuously change humans and their relationship with the environment [7]. Knowledge and awareness are important human characteristics that define and distinguish them from the world around them [8]. The development of Technical (Technological) Culture accompanied the development of education enriched with new disciplines, leading to the establishment of universities in the 12th and 13th centuries [8]. Thanks to universities, Europe began to take a leading role in the world of science and technology [8].

Innovations drive many societies, but they are often thought of only in terms of solutions, rarely in terms of the relationship that humans maintain with technical objects and/or the object's relationship with the society of which it is a part [7]. Technical and Technological Culture has remained in the shadow of innovation culture, although it is a key factor for innovation with awareness [7] and could be a prerequisite for the reliability of work processes and support for maintaining the standard of quality [1].

If we want to respond to the demands of future technological development, it is important to reinvest in the concept of Technical and Technological Culture and for engineers, as bearers of Technical and Technological Culture, to ask themselves what impact techniques and technologies have on society and the world in which we live.

3. Research and results

3.1. Instrument development

In the framework of this pilot research, we constructed a questionnaire to examine attitudes and beliefs about the development of Technical and Technological Culture, which was created based on a list of questions related to the perception of the development of Technical and Technological Culture (list of survey questions by Wyrwicka [1]), which applied in her research [1], and for which we obtained the author's consent. This created questionnaire contain statements about Technical and Technological Culture, and it is located in the second and third part of our instrument. In these parts we used a 5-point Likert scale. The first part of the questionnaire consists of the socio-demographic characteristics of the participants: their gender, age, education, professional status, years of experience, size and ownership of institution, and material status.

3.2. Research results and discussion

Socio-demographic characteristics of the sample: The participants in this study were engineers from technical faculties, including professors (39,13%), students (10,87%), and employees in industry/ IT sector/ non-governmental organizations (50%). Gender of the total number of participants: 58,7% were male, and 41,3% were female. Age of the participants: under 20 years (0%); between 20 and 24 years (4,35%); between 25 and 29 years (6,52%); between 30 and 34 years (17,39%); between 35 and 39 years (15,22%); between 40 and 49 years (45,65%); between 50 and 59 years (6,52%); 60 and more years (4,35%). Level of education: Undergraduate Academic Studies – Bachelor (UAS) =17,39%; Master Academic Studies (MAS) =**36,96%;** Doctoral Academic Studies (DAS) =**39,13%;** others 6,52%. Years of work experience: 1 year (4,35%); from 1 to 5 years (19,56%), from 6 to 10 years (13,04%), from 11 to 15 years **(23,91%)**, from 16 to 20 years (21,74%), from 21 to 25 years (8,7%), from 26 to 30 years (2,17%), from 31 to 35 years (4,35%), 35 years and more (2,17%). The size of the institution: micro:1-10 employees (2,17%), small: 11-50 employees (17,39%), medium: 51-250 employees (17,39%), large: 250 and more employees (**63,04%**). Ownership structure of the institution: domestic-public=**58,69%**; domestic-private= 21,74%; foreign ownership=15,22%. Material status: very satisfied are 17,39% of participants, satisfied are 50% of participants, neither satisfied nor dissatisfied are 28,26% of participants, and dissatisfied are 4,35% of participants.

The participants expressed the following attitudes and beliefs about Technical and Technological Culture: 65,22% of participants understand the meaning of the term Technical and Technological Culture, 6,52% don't understand the meaning, and 28,26% are uncertain about the understanding of the meaning of the term of Technical and Technological Culture.

Pearson's correlation coefficient indicates a positive and relatively weak association between understanding of the term of Technical and Technological Culture and: age (r=0,433); years of work experience (r=0,365); ownership structure of the institution (r=0,272); and professional status (r=0,181). The correlation between the understanding of the term Technical and Technological Culture and the level of education is positive and insignificant (r=0.016). The correlation between the understanding of the term Technical and Technological Culture and the size of the institution is negative and insignificant (r= -0,043).

In the second part of the questionnaire, respondents stated their attitudes and beliefs regarding Technical and Technological Culture (understanding, education, promotion, legal regulation etc.). The results indicated the following:

- Participants understand the meaning of the term Technical and Technological Culture (65,22%).
- They emphasize the importance of educating young engineers about Technical and Technological Culture (84,78%).
- They believe that it is important to promote Technical and Technological Culture to the general public (78,26%) and especially among scientists and engineers (82,61%).
- They also believe that this topic is insufficiently represented in the media (32,61%) or they are uncertain about that (54,35%).
- Regarding the legal regulation of Technical and Technological Culture participants believe that it does not exist in their country (36,96) or they are uncertain about that (52,17%).
- That Technical and Technological Culture should be regulated by legal regulations believes 54,35% of participants.

The third part of the questionnaire referred to views on the meaning of Technical and Technological Culture. In addition to the statements made by Wyrwicka in her list of survey questions, we created the new statements, that our participants stated as more significant. They point out the most important:

- The awareness of sustainable development (93,47%);
- The impact on environmental protection (89,13%);
- Technical and technological literacy of young people (86,96%);
- Using renewable energy sources (84,78%);
- The development of positive attitudes of young people towards Technical and Technological Culture (84,78%);
- Developing a responsible attitude towards work, tools and own environment (82,62%);
- The social impact of modern technologies (69,56%) (Figure 1.) .

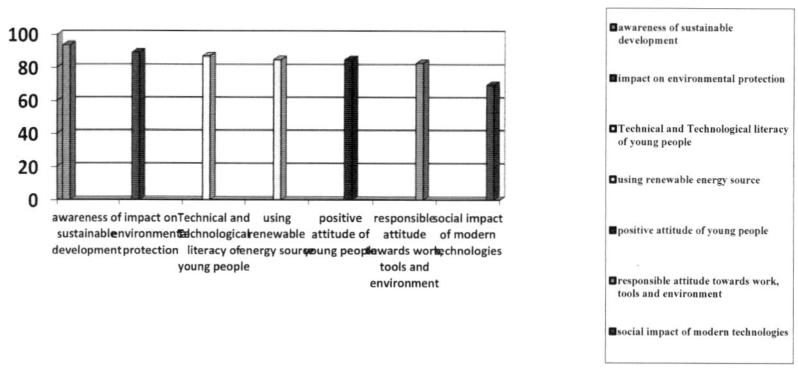

Fig. 1. Expressed point of view on the meaning of Technical and Technological Culture

The statements we used (from the list of survey questions by Wyrwicka) also indicated:

- Obeying norms – 76,09%;
- Establishing procedures – 73,91%;
- Standardization – 76,09%;
- Unified approaches to technical problems -67,39%;
- Prevention (anticipation of future requirements) -69,56%;
- Thriftiness – 63,04%;
- Efficiency – 73,91%;
- Professional solidarity and cooperation – 73,91%;
- Reliability – 73,91%;
- Discipline – 69,56%;
- Involvement in work – 73,91%.

Research conducted by Wyrwicka [1] identified statements that participants perceived as part of Technical-Technological Culture:

- Obeying norms – 72,4% of responses
- Establishing procedures – 62,6% of responses
- Standardization – 50,2% of responses
- Unified approaches to technical problems -37,5% of responses
- Prevention (anticipation of future requirements) -23% of responses
- Thriftiness – 30% of responses
- Efficiency – 50,6% of responses
- Professional solidarity and cooperation – 35,4% of responses
- Reliability – 45,5% of responses
- Discipline – 35,2% of responses
- Involvement in work – 39,4% of responses.

Based on the analysis of the obtained data, the author of this research [1] suggests that Technical and Technological Culture will be more important in future business environments and should be nurtured in every organization through organizational learning and continuous improvement. If we compare the results obtained in this research and the results obtained by Wirwicka [1], we can conclude that for both group of participants (more than 50%) the priority is: Obeying norms, Standardization, Establishing procedures, and Efficiency (Table 1.).

Statements	% of participants	
	In our research	From research by Wyrwicka
Obeying norms	**76,09%**	**72,4%**
Standardization	**76,09%**	**50,2%**
Establishing procedures	**73,91%**	**62,6%**
Efficiency	**73,91%**	**50,6%**
Reliability	**73,91%**	**45,5%**
Involvement in work	**73,91%**	**39,4%**
Professional solidarity and cooperation	73,91%	35,4%
Prevention (anticipation of future requirements)	69,56%	23%
Discipline	69,56%	35,2%
Unified approaches to technical problems	67,39%	37,5%
Thriftiness	63,04%	30%

Table 1. Expressed comparing results from a list of survey questions by Wyrwicka

The results of the research [9] indicate that introducing content to form Technical and Technological Culture among students, as future engineers, allows for an increase in the level of development of indicators of Technical and Technological Culture. It also provides integration of special, professional and technological training that contributes to the formation of a specific technological view of the world and mastery of Technical and Technological Culture [9]. The results of the research we conducted support these claims and indicate the importance of educating young engineers about Technical and Technological Culture, which implies the introduction of Technical and Technological Culture in the content of the education in Technical faculties. They also indicate the importance of promoting and affirming Technical and Technological Culture in the media and between the general public and among experts and engineers, and the importance of introducing legal regulations for Technical and Technological Culture.

The results of the conducted research indicated the importance of Technical and Technological Culture [10] and determined that Technical and Technological Culture significantly moderates the relationship between technological acquisition and innovative investment and moderates the relationship between technological acquisition and attitudes toward digital innovations. We believe that Technical and Technological Culture has been unfairly overshadowed by innovation and that our research will encourage further research into its impacts.

4. Conclusion

From the Middle Ages to the present day, there has been an awareness that liberal and mechanical arts should not be separated but should complement each other, that they are interdependent, that knowledge enlightens man and work dignifies him, and that only their balanced development can improve humanity and mankind [8]. This points to the need for engineers and experts to develop a more comprehensive and humane approach to their knowledge and work, which we consider here under the concept of Technical and Technological culture. The results of this research have provided insights into the attitudes and beliefs of engineers in technical faculties regarding the development of a Technical and Technological culture. The research shows that engineers largely understand the concept of Technical and Technological culture, that they most often identify it with sustainable development and environmental protection, and that it is still related to the person's experience, company ownership structure, and professional status.

The results show the importance of educating young engineers about Technical and Technological Culture. In addition, the results point out weaknesses that can be improved in institutions, but also strengths that can be considered in the development of Technical and Technological Culture, such as the importance of introducing legal regulations for Technical and Technological Culture. Furthermore, one of the implications of this research could be an incentive for a proactive policy of transforming national culture through changes in education, such as introducing the subject of Technical and Technological Culture into the engineering education program, developing responsible engineering and promoting Technical and Technological Culture in the professional and wider public.

The instrument validated through this research will serve for further research on influential factors in the development of Technical and Technological Culture. Future research could include not only technical, but also technological faculties, secondary and primary schools and representatives of companies and institutions that want to contribute to the promotion and improvement of the development of Technical and Technological Culture.

5. Acknowledgments

This research (paper) has been supported by the Ministry of Science, Technological Development and Innovation through project no. 451-03-47/2023-01/200156 “Innovative scientific and artistic research from the FTS (activity) domain”.

6. References

[1] Wyrwicka, M.K. (2011). Technological Culture and Its Interpretation: Research Results in Poland, Human Factors and Ergonomics in Manufacturing and Service Industries, 21(2), pp.178-187. doi: 10.1002/hfm.20254

[2] Wyrwicka, M.K., Chuda, A. (2019). The Diagnosis of Organizational Culture as a Change’s Factor in the Context Application of Design Thinking, LogForum, 15(2), pp.279-290.

[3] Pochanke, H. (1985). Dydaktyka techniki: praca zbiorowa, Państwowe Wydaw. Didactics of technology: collective work, National Publishing House, Naukowe, ISBN: 8301062223, 9788301062224.

[4] Leidner, D., Kayworth, T. (2006). Review: A Review of Culture in Information Systems Research: Toward a Theory of Information Technology Culture Conflict. MIS Q.30, pp.357-399. doi:10.2307/25148735

[5] Chalas, K. (2014). Technical culture and human axjosphere, Advances in Science and Technology Research Journal Vol. 8, pp.107–110.

[6] Rosenzweig, P.M. (2022). No Title. Acad. Manag. Rev. 18, pp.370-374. doi:10.2307/258765

[7] Didier, J. (2018). Technical Culture and Innovation Culture: Reconciling Through Design. In: Chouteau, M., Forest, J., Nguyen, C. (eds), Science, Technology and Innovation Culture, Vol. 3., Chapter 7., pp.117-137. London, UK: ISTE Ltd. doi:10.1002/9781119549666.ch7

[8] Krasic, S. (2013). Humanistic and Technical Education in Antiquity and the Middle Ages, Chapter 01 in DAAAM International Scientific Book 2013, pp.001-052, B. Katalinic & Z. Tekic (Eds), Published by DAAAM International, ISBN 978-3-901509-94-0, ISSN 1726-9687, Vienna, Austria DOI: 10.2507/daaam.scibook.2013.01

[9] Rubtsova, E., Bogolyubova, I., Starodubtseva, G., Lyubaya, S. (2019) Technological Culture of Future Engineers in the Context of Modern Socio-Economic Development of the Society, IOP Conference Series Earth and Environmental Science, 315(2):022002 doi: 10.1088/1755-1315/315/2/022002

[10] Weiyu, Z., Othman, A., Guli, T. (2022). Role of Technological Acquisition and R&D Expenditure in Innovative Investment Front Psychol. Apr. 5; 13:855295 doi:10.3389/fpsyg.2022.855295 PMID: 35478744; PMCID: PMC 9037329

DOI: 10.2507/34th.daaam.proceedings.011

ANALYSIS OF PASSWORD SECURITY POLICIES AND THEIR IMPLICATIONS ON REAL-LIFE SECURITY

Jasmin Redzepagic, Vedran Dakic, Josip Stanesic & Matej Basic

This Publication has to be referred as: Redzepagic, J[asmin]; Dakic, V[edran]; Stanesic, J[osip] & Basic, M[atej] (2023). Analysis of Password Security Policies and Their Implications on Real-Life Security, Proceedings of the 34th DAAAM International Symposium, pp.0077-0081, B. Katalinic (Ed.), Published by DAAAM International, ISBN 978-3-902734-41-9, ISSN 1726-9679, Vienna, Austria
DOI: 10.2507/34th.daaam.proceedings.011

Abstract

This paper examines current password security guidelines and best practices, as well as their effectiveness in preventing unauthorized access and data breaches. The research also explores the impact of these policies on user behaviour and the potential trade-offs between security and user experience. The focus is especially on password length and mandatory change requirements since the best practices conflict across different standards. The findings of this study have important implications for organizations in developing and implementing password security policies that effectively balance security and usability. The research will compare different criteria on what defines a "good" password and by applying different password analysis methods establish what objectively should be used in real-world scenarios.

Keywords: password security; password complexity; end-user security; security guidelines; security policy.

1. Introduction

The interconnected world we operate in today is based on the idea that any user can be identified by a standardized process, and we call that process authentication. Sometimes we also use the term identification for the same process, but authentication is more precise. Authentication requires proof of a user's identity while identification just requires identifying the user. The main difference is that by providing a username, the user is identified. By successfully using a password or some other form of authentication, the user is authenticated. There are different ways the user can be authenticated. Usually, we are talking either about something that the user knows, for example, the password or a PIN, something that the user has, for example, a token or a list of one-time passwords, or something that the user is, for example, his voice print fingerprint, or his retina scan.

The biggest problem is that, in most user scenarios that we are encountering today, things that users have or features that users have are unable to be verified because the systems themselves are integrated into the cloud. The ability of the system to provide the user with a safe way of communicating the information needed to verify (for example) a token is either complicated or completely impossible. This applies to physical tokens, a solution to this exists in the form of digital tokens that provide one-time passwords, but they are just slowly gaining traction among users.

This leaves us with the password as the primary source of authentication for the user. Most of the systems today combine passwords with some other forms of identification usually a one-time password or some other generated identification token and we call that two-factor identification but the password as such remains one of the most important ways of identifying a user. On the other hand, passwords are often considered one of the least secure ways of authenticating users so their complexity and resistance to cracking is a big problem. To make matters worse, password complexity is a heavily debated subject since a standard benchmark is not agreed on [1].

2. Passwords

From a security perspective, this means that creating and maintaining a healthy password policy is something that is one of the most important things when it comes to creating a secure environment. The problem that this poses for the administrators is that the passwords themselves are heavily dependent on the users and their ability to both create working passwords and safely keep them secure. For a long time, the main strategy for creating a secure environment depended on a clear-cut password policy. This usually involves a couple of things: creating a set of requirements for the password length complexity history and age, creating an environment that will ensure that users are adhering to this policy, and educating users so that the policy is actively enforced.

Before defining the best practices for passwords, we need to define the metrics. When we are talking about passwords, we are usually talking about the following characteristics a password can have.

- Length: A minimum length requirement for passwords, e.g., at least 8 characters. This metric is usually taken as the most important one. A reason for this is that the most common way of guessing a user's password by brute forcing all the combinations is something that is done quickly if a password is short.
- Complexity: A requirement for a mix of distinctive character types, such as upper and lowercase letters, numbers, and special characters. This requirement for a password also comes from the method of brute forcing the passwords. Since we are dealing with permutations and combinations of different characters increasing the character set is by itself a terrific way to slow down the brute forcing process. Also, we must prohibit the user from using known words available in the wordlists.[3][4]
- History: Prohibiting the reuse of previously used passwords. Since users sometimes inadvertently give away or lose their old passwords, it is usually a clever idea to make sure that all the passwords are not able to be reused. Password reuse is extremely common among users, following figure shows percentages of reused passwords from an analysis done by Google in 2018 [2]. There is no reason to believe percentages have changed much since providers can not compare user passwords and warn them of the reuse.

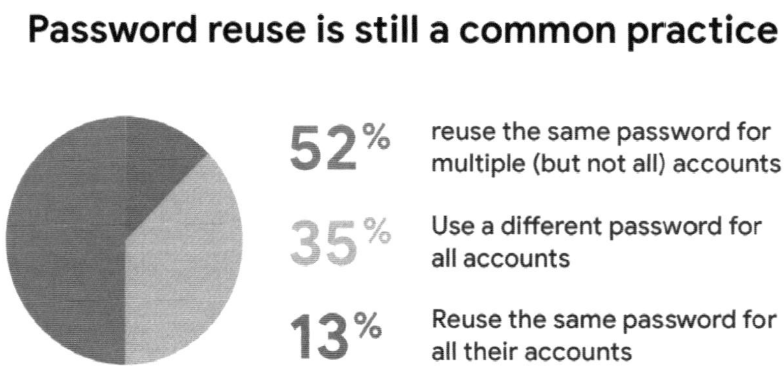

Fig. 1. Password reuse practices in US

- Age: Maximum password age, after which the password must be changed. This requirement is something that is a hot topic these days and will be mentioned a little bit later in this paper.
- Blacklist: A list of prohibited passwords, such as "password" or "123456". By using such a list one can make sure that the users avoid passwords that are easily guessed especially the ones that are available on the Internet on the so-called world lists.[3]
- Two-factor authentication: An additional step of verification, such as a code sent to a phone, in addition to a password. The reason why two-factor identification is important is that with the availability of mobile phones, it is relatively easy to make sure that the password is not the only thing standing between the adversary and the system he is trying to log in to.

Let's use these as they are the most used metrics when defining granular password policies.

2.1. Password retrieval methods

A major point about password storage and retrieval is to take two things into account. First is to never store the password in its plain text form. If a password is stored in plain text form, there is no point in trying to enforce any policy since the only thing that the adversary needs to do is to get to the password database and use the passwords as they are. So, we presume that we are dealing with passwords that have been stored using one of the secure hashing functions which means that they require some time to decipher either by using one of the methods that use brute force functions or by using so-called rainbow tables or words password guessing. [4]

We are not going to deal with password guessing used by dictionary-based attacks because we will presume that the password policy already has dictionary checks implemented so that the passwords that the users can create cannot be in the dictionary. Another thing we are going to disregard is using rainbow tables. Rainbow tables basically mean that for some of the encryption or hashing methods, somebody has already calculated all the possible hashes and that there is an available lookup table for an adversary to directly decipher the password for a given hash. Rainbow tables exist for some of the ciphers and for passwords of limited length. Both these problems can be solved by using a well-known good cipher and limiting the minimum password length.

This leaves us with brute-force attacks, which are a huge field of research. If we need to separate these attacks into different categories, we can talk about using either CPUs or GPUs to decipher passwords or we can deal with dedicated hardware designed specifically to decipher a particular cipher method. Either way, we are dealing with methods that require guessing passwords by creating a hash out of our presumed password and then comparing this hash to a hash of a password that we are trying to guess. This of course requires time and what we are trying to accomplish is to create passwords that take so much time to crack so that the process itself makes no sense to the adversary. [5]

For the last couple of decades, the usual thinking was that passwords should be as complicated as possible because what we are trying to do is to create a large set of characters and in that way a huge initial set of possible passwords so that the password guessing becomes more complex.

2.2. Password policies

So almost by definition, all the different password policies define passwords as having to have all the special characters, upper- and lower-case letters, numbers, and all the other possible characters that exist. Usually, the password policy also defines at least 6 to 8 characters as a minimum length for a password, and the user is usually required to change this password regularly. Depending on the requirements usual time interval for this mandatory change is between 30 and 180 days.

Although minimum length and complexity requirements are defined in standards mandatory change is not defined. Mandatory change has been so deeply ingrained in the security policies of companies that even administrators believe that this in fact is part of some standard and enforce it.

Combining these different policy requirements into something that the user can or must do usually means that the user is going to create a short, memorable password. Passwords are then going to be changed slightly whenever mandatory change comes into effect but since the password is going to contain many different characters there is a huge change that the password itself is going to be written down somewhere. This means that users are complying with a policy but at the same time are creating the same password environment that the policy itself was designed to avoid.

2.3. Policy shifts

It is not unusual then that different companies including some of the largest ones like Microsoft have decided to change password policies in a way to stimulate users to create a password that is long and is not complicated in a way we may expect. Instead of relying on complexity, they require the users to create a password that has an exceptionally large length. This makes passwords much more complicated to brute force. But we have another problem.

Almost all the systems come with a predefined maximum length of the password which is now considered a security risk. When we consider the entropy of a password, more precisely the amount of time required to brute force a particular password based on its length it is it is much better to have a long password based on dictionary words than a short password based on special characters.

Also, since we are depending on the password to be extremely difficult to guess it is advisable to eliminate the need for the user to change his passwords from time to time since this almost guarantees that the user is going to create a lot of passwords that are going to be the same or similar.

New recommendations, therefore, are for the users to choose not a password as a word that is going to be used but switch to a "passphrase". A phrase means using multiple words that are on the one hand easy to remember for that user but are hard to brute force since the phrase can contain tens if not hundreds of characters. For example, for a given hashing function and eight 8-character passwords taken from a set of 64 different characters is going to be hundreds of orders of magnitude easier to guess than a simple phrase that uses only lowercase letters but is longer than 32 characters.

Number of characters in password	Numbers Only	Lowercase Letters	Lowercase Upper & Letters	Numbers, Upper & Lowercase Letters	Numbers, Upper & Lowercase Letters, Symbols
4	Instantly	Instantly	Instantly	Instantly	Instantly
5	Instantly	Instantly	Instantly	Instantly	Instantly
6	Instantly	Instantly	Instantly	Instantly	4 Seconds
7	Instantly	Instantly	22 Seconds	42 Seconds	6 Minutes
8	Instantly	3 Seconds	19 Minutes	48 Minutes	7 Hours
9	Instantly	1 Minutes	11 Hours	2 Days	2 Weeks
10	Instantly	1 Hours	4 Weeks	6 Months	5 Years
11	Instantly	23 Hours	4 Years	38 Years	356 Years
12	25 Seconds	3 Weeks	289 Years	2K Years	30K Years
13	3 Minutes	11 Months	16K Years	91K Years	2M Years
14	36 Minutes	49 Years	827K Years	9M Years	187M Years
15	5 Hours	890 Years	47M Years	613M Years	14Bn Years
16	2 Days	23K Years	2Bn Years	26Bn Years	1Tn Years
17	3 Weeks	812K Years	539.72M Years	2Tn Years	95Tn Years
18	10 Months	22M Years	7.23Bn Years	96Tn Years	6Qn Years

Table 1. The time it takes to crack passwords using publicly available technology in 2023

This runs a little bit counterintuitive to what a typical administrator right now thinks about password policies since the requirement to have a large set of possible characters is something that a lot of administrators are used to. Having them recommend to their users that they should use a simple but long phrase instead of a shorter but more complicated password is something that a lot of administrators are not inclined to do.

The main reason for this is simply because they don't understand the statistical and mathematical consequences of trying to guess passwords that are longer in length. In Table 1 we are providing a rough idea of how the length of a password correlates to the time it takes to crack the password using publicly available technology [6]. Note that 8-character passwords are vulnerable to brute force attacks in a short time frame but are still recommended as viable password lengths. Also, some of their systems may not be capable of handling longer passwords since a lot of them have been designed with fixed maximum password length and this is something that requires not only a change of policy but also a change in software. That is a huge problem for a lot of legacy systems.

3. Findings, recommendations, and future research

There are more than a few takeaways from the methodology that we're using in our paper. Companies try to enforce password policies to their employees by issuing a blanket policy that "everyone must adhere to", without spending any time considering what that means in practice. Even forgetting the fact that this can lead to employee alienation, it's clear as day that people will always be people and either try to circumvent these rules or try to make their lives easier, in the form of writing down their passwords on post-it notes or in plain-text format in their mobile phones. We can avoid most of these problems by implementing a few recommendations, such as:

- Password length: Any maximum length requirement for the password should be completely avoided. The minimum length for the password should be kept to at least 16 if not more characters to stimulate the user to use a phrase not a single word or a combination of characters. This should be explained to the user.
- Mandatory password changes: Any requirement to periodically change your password should be avoided because it creates an artificial sense that the user is going to change the passwords and make them unique every time. This requirement means that users are just simply reusing old passwords and the risk of this password becoming available due to sheer negligence is almost guaranteed since the user is often going to write the password down.
- User training: Training is something that should be focused on trying to explain to the users that passwords themselves are not something that should be complicated but something that should be long and simple to remember. Users need to understand the reasoning behind this since they spent almost all their working lives being told otherwise.

A good example is to let the users choose a phrase from a book or part of the lyrics of a familiar song. This is going to make for a much better password than a nonsensical set of different characters and if this is presented to a user in such a way that the user understands the idea behind it users are going to be much more inclined to create a good phrase and keep it secret.

These requirements don't even take into consideration the fundamental truth of IT security – managing security in our computing environment is a collection of continuous activities [7]. By prescribing password policies, complexities, and requirements we are just fighting one of the security problems, not thinking about security. This does not only apply to regular users – it applies to everyone, including system engineers and application developers, as it can be difficult having them conform to different security challenges [8].

Our recommendations are based on the premise that the current technologies for guessing or cracking are going to grow linearly and that there are not going to be any breakthroughs in password-guessing technology. The advances in quantum computing unfortunately mean that some of the methods that we are talking about are going to completely change and current research does not point to a solution to this problem. And, even more to the point, companies nowadays face several types of cyber-crime-related attacks [9],[10]. We just do not want to give cyber-criminals more fuel to do even more damage.

At the same time, this is the single most important limitation in this paper. Focusing on passwords is important, but trying to predict new technologies that may make password guessing even more efficient is almost impossible. Further improvement in security should be first switching to passphrases but in the long run, only technologies that completely avoid using user-generated passwords will enable us to maintain secure authentication in the longer run.

Our future research could include topics like SSO (Single Sign-On) and security from the context perspective. For example, it's wrong to consider senior IT engineers who deal with cloud services every day and a novice IT administrator who only has limited experience with security policies and security as the same. Context plays a role, as does company culture, working environment, and systems/applications used. Single Sign-On systems fundamentally change the way that we approach security. Having in mind that external authentication systems like Azure Active Directory and a host of other SAML/OAuth-based systems have support for MFA (Multi-Factor Authentication), there will be fundamental changes in the way we approach user authentication in the future. Some of these systems will make authentication easier, but the question remains – at what cost?

4. Conclusion

Passwords are still the primary way of protecting most of the low and medium-security data and due to their importance for the average user, some changes must happen when it comes to the way passwords are generated and used. Password policies should be changed to reflect the realities of the brute force hash guessing methods. As methods keep getting quicker and more sophisticated passwords must be exchanged for passphrases, reflecting the need to get them as long as possible. Users need to be educated on why it is important to have long passwords and keep them safe. The requirement to change passwords regularly should be avoided, using long passphrases will mean that users will unavoidably write the phrases down and this will void all the ideas from this paper. As a result, we wholeheartedly agree that the strength and frequency of passwords (changes) are the wrong focus. We can be quite a bit more user-friendly with long passwords based on user's preferences, which will make other security issues smaller – for example, fewer passwords will leak via paper or digital trail. Users' education should also involve incentives to use MFA and other authentication technologies that will enable them to move away from using only passwords or if possible from using passwords at all.

5. References

[1] Carnavalet, X. & Mannan, M. (2015). A Large-Scale Evaluation of High-Impact Password Strength Meters, ACM Transition on Information and System Security, Vol. 18, Issue 1, Article No. 1, ISSN 1094-9224, DOI: 10.1145/2739044

[2] https://services.google.com/fh/files/blogs/google_security_infographic.pdf, (2019). Google Online security survey, Accessed on: 20-09-2023

[3] Houshmand, S.; Aggarwal, S. & Flood, R. (2015). Next Gen PCFG Password Cracking, IEEE Transactions on Information Forensics and Security, Vol. 10, No. 8, pp. 1776-1791, ISSN 1556-6021, DOI: 10.1109/TIFS.2015.2428671

[4] Tatli, E. I. (2015). Cracking more password hashes with patterns, IEEE Transactions on Information Forensics and Security, Vol. 10, No. 8, pp. 1656-1665, ISSN 1556-6021, DOI: 10.1109/TIFS.2015.2422259

[5] Binnie, C. (2016). Password Cracking with Hashcat, In: Linux Server Security: Hack and Defend, Shimonski, R., Tulton, A.O. (Ed), 99-111, Wiley, 978-1-119-27765-1, ISBN: 978-1-119-28309-6, United States of America

[6] https://www.homesecurityheroes.com/ai-password-cracking/?password=ufskE34%24%240, (2023). An AI Just Cracked Your Password, Accessed on: 20-09-2023

[7] Dakic, V.; Jakobovic, K. & Zgrablic, L. (2022). Linux security in physical, virtual and cloud environments, Proceedings of the 33rd DAAAM International Symposium, ISSN 1726-9679, ISSN 1726-9679, ISBN 978-3-902734-36-5, DOI: 10.2507/33rd.daaam.proceedings.021

[8] Dakic, V.; Redzepagic, J. & Basic, M. (2022). CI/CD toolset security, Proceedings of the 32nd DAAAM International Symposium, ISSN 1726-9679, ISBN 978-3-902734-36-5, DOI: 10.2507/33rd.daaam.proceedings.022

[9] Moric, Z.; Redzepagic, J. & Gatti, F. (2021). Enterprise tools for data forensics, Proceedings of the 32nd DAAAM International Symposium, ISSN 1726-9679, ISBN 978-3-902734-36-5, DOI: 10.2507/32nd.daaam.proceedings.014

[10] Inglesant, P. & Sasse, M.A. (2010). The True Cost of Unusable Password Policies: Password Use in the Wild, Proceedings of the SIGCHI Conference on Human Factors in Computing Systems Computing Systems, pp. 383–392, ISBN: 978-1-60558-929-9, DOI: 10.1145/1753326.1753384

DOI: 10.2507/34th.daaam.proceedings.012

PREVENTION OF DNS AMPLIFICATION ATTACKS

Josip Stanešić, Zlatan Morić, Vedran Dakić & Matej Bašić

This Publication has to be referred as: Stanesic, J[osip]; Moric, Z[latan]; Dakic, V[edran] & Basic, M[atej] (2023). Prevention of DNS Amplification Attacks, Proceedings of the 34th DAAAM International Symposium, pp.0082-0087, B. Katalinic (Ed.), Published by DAAAM International, ISBN 978-3-902734-41-9, ISSN 1726-9679, Vienna, Austria DOI: 10.2507/34th.daaam.proceedings.012

Abstract

The Domain Name System (DNS) is a critical internet infrastructure component, that translates domain names to IP addresses. This study examines the persistent threat of DNS amplification attacks, which exploit certain DNS servers to magnify query responses, causing network congestion. Despite existing mitigations like Response Rate Limiting (RRL) and DNS Security Extensions (DNSSEC), these attacks remain prevalent. Our comprehensive analysis of over 1.7 million IP addresses reveals that approximately 14.77% of Internet DNS servers support recursive queries without access controls. Notably, vulnerable servers are distributed globally, with Asia, Africa, and South America showing the highest vulnerability rates. This research underscores the urgency of enhancing DNS server security. Recommendations include disabling recursion or implementing strict access controls, deploying rate-limiting measures, and restricting "ANY" queries to mitigate DNS amplification attacks. Collaboration between regulatory bodies and network operators is crucial, especially for government infrastructure. In conclusion, this study provides crucial insights into the state of public DNS servers, their vulnerabilities, and the ongoing threat of DNS amplification attacks. As the internet evolves, vigilance and proactive measures are essential to protect DNS service integrity and availability.

Keywords: DNS; DNS security; Amplification attack; Recursive DNS; DNS RRL.

1. Introduction

Domain Name System (DNS) is a distributed naming system. It enables looking up values of resource records that are distributed in different zones. The most common use case for DNS is translating hostnames to IP addresses. A mapping of a hostname to an IP address is a resource record of type A. There are different record types. Some of the most common types of resource records are A, SOA, AAAA, PTR, MX, NS, and CNAME. and AAAA records are used to translate hostname into IPv4 and IPv6 records respectively. PTR records are used to translate IP addresses into hostnames. CNAME records server as aliases for A records. SOA records contain information about the DNS zone and NS records identify DNS servers authoritative for the zone.

Hostname together with the zone name represents a Fully Qualified Domain Name (FQDN), for example, the host webserver in DNS zone example.com would have FQDN of webserver.example.com. Zones themselves are hierarchically organized into an inverted tree structure with the . or root zone at its start, top-level domains like .com. one level beneath it, and domains like example.com. beneath TLDs as seen in Fig. 1. Domain hierarchy. Root zone has information on which servers are responsible for TLDs, and TLDs hold information on which servers are responsible for domains underneath them.

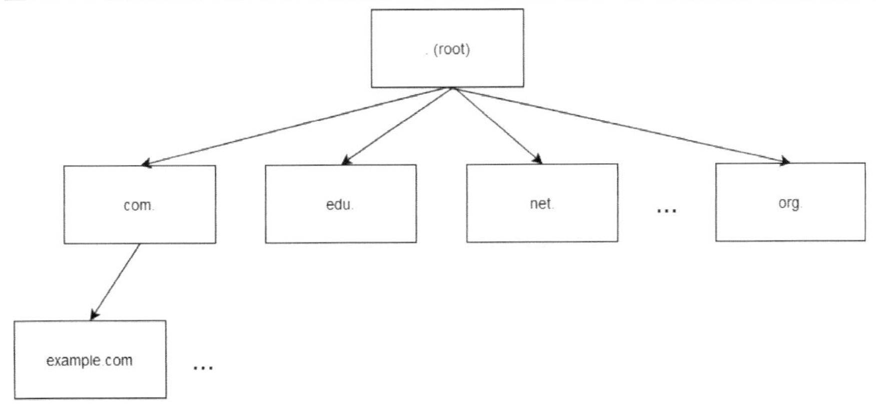

Fig. 1. Domain hierarchy

The client makes a DNS query each time it wants to access a host by referencing its hostname. Resolver is a program that retrieves answers from DNS servers in response to client queries. Queries themselves can be iterative or recursive. Iterative query means that the server if it is not authoritative for a DNS zone will refer the client to another server and the client must contact the next server. Recursive query means that the DNS server upon getting a request for a zone it is not authoritative for, will query further nameservers to obtain it and return it to the client. This enables more efficient caching of DNS responses. A DNS server can be authoritative for one or more zones, and it is considered authoritative if it contains zone data (resource records) about a specific zone. Also, it needs to be referenced as authoritative by the DNS server one step higher in the hierarchy. A recursive server is a server which will respond to recursive queries.

Having a recursive server inside a network is useful as it will lower the number of queries done due to the caching of responses. DNS servers themselves cache query results for efficiency reasons. In a scenario in which there is nothing cached and there is no recursive DNS server, a client will contact its configured DNS server for a record. If the DNS server is not authoritative for that zone it will forward its request to a root server. The root server will point to the authoritative TLD server. TLD server will point to the authoritative server for that zone, which should return an answer about the requested record. The process would be repeated if another client requires information about the same host. By having a recursive server, the second client would get a cached response without having the whole query process repeated.

The problem with DNS amplification attacks with public DNS servers is that attackers can spoof their source IP addresses to the IP of their target, request records with large responses, and exhaust the bandwidth of their target with responses. Security of public DNS servers is hard to improve as their ownership and the possibility to remediate the configuration is distributed among many entities across the globe.

In scientific literature, DNS amplification attacks were covered for the first time in 1999 [1]. Even 24 years later, solutions recommended then are still not applied, leaving the Internet vulnerable to DDoS attacks. DNS was the protocol used in 24% of UDP reflected amplification attacks observed from April 1, 2021, to March 31, 2022, incoming and outgoing from Azure [2]. This paper aims to investigate the prevalent issue of DNS amplification attacks, which exploit specific DNS server vulnerabilities to magnify query responses. The central questions guiding this research are:

- What is the extent of vulnerability among public DNS servers to DNS amplification attacks, and how prevalent is support for recursive queries without access control?
- Which geographic regions are most affected by these vulnerabilities, and what factors contribute to their distribution?
- What effective mitigation strategies can be recommended to fortify DNS servers against DNS amplification attacks and enhance internet security?

2. Related works

There have been multiple works regarding DNS protocol vulnerabilities, attacks on DNS protocol and DNS servers, improving the security of DNS, and mitigation techniques for the mentioned attacks. DNS was first introduced by RFC 882 in 1983 [3]. 14 years later DNSSEC was introduced in RFC 2065 [4] to add mechanisms that will assure data integrity and authentication by adding cryptographic signatures to DNS records. Unfortunately, this did not make DNS fully secure and immune to malicious usage.

The paper "Motivation for Behaviour-Based DNS Security: A Taxonomy of DNS-Related Internet Threats" by Nikolaos Chatzis from 2007. discusses various DNS-related attacks and categorizes them into three main categories. Name server vulnerabilities in which attackers are exploiting weaknesses in name server implementations to obtain unauthorized control or perform reconfiguration, and corruption of zone files. These types of attacks are achieved by exploiting software vulnerabilities or misconfiguration. Authenticity and Integrity Attacks in which attacks to authenticity and integrity are done to misdirect users to fraudulent sites or to perform DDoS attacks.

These types of attacks include domain hijacking, domain theft, cache poisoning, man-in-the-middle attacks, attacks against non-existent names, and reflection attacks using DNS servers (DNS amplification). Consumption attacks can be further split into two more categories: exhausting resources from a single name server or exploiting DNS servers to attack other Internet hosts. Other Internet hosts can be attacked with reflection and amplification attacks. As DNS records can be quite long attackers can use them to deliver malware. Another misuse of DNS is by creating DNS tunnels to avoid firewalls. The paper also emphasizes that while name server vulnerabilities and authenticity/integrity attacks are being addressed, consumption attacks require additional research work [5].

Amplification attacks are effective because of the high amplification rate. "By combining different response types, the amplification effect can reach up to a factor higher than 60. If, for example, the response consists of a 122-byte A type response, a 4000-byte TXT response, and a 222-byte SOA response, the total response consists of 4320 bytes. This yields an amplification factor of 73. Other amplifications are possible depending on the query size and the experienced packet distributions in an actual attack. Due to networking limits, traffic collisions, and other factors, the effective rate of an attack will be significantly smaller than the amplification's theoretical upper limit." [6] One of the most famous DNS amplification attacks is the 2013 attack on SpamHous. Attackers were requesting ripe.net zone files from open DNS resolvers and spoofed IP addresses CloudFlare issued for SpamHous. Open resolvers were responding with a complete DNS zone file creating 75 Gbps of attack traffic. Attackers were issuing requests approximately 36 bytes long and getting responses of approximately 3000 bytes, resulting in almost a 100x amplification factor [7].

CISA describes the DNS Amplification attack as a popular form of a distributed denial of service (DDoS) attack that relies on publicly accessible DNS servers that support recursion to overwhelm the target with DNS response traffic. Attackers will make a DNS request while spoofing the source IP to be that of the target's with a small request which will result in a very large response. Mitigation techniques recommended are to disable recursion on publicly accessible DNS servers or to allow it only from authorized clients. In case the attack does not rely on recursive DNS servers' response rate limiting (RRL) should be implemented. Authoritative and recursive servers should run on different systems, with response rate limiting implemented on authoritative servers and access control lists implemented on recursive servers. Another mitigation can be implemented by Internet Service Providers by implementing controls to filter network traffic on their network to reject packets with source addresses not reachable via the actual packet's path [8]. Based on the results of the Spoofer project conducted by the Centre for Applied Internet Data Analysis (CAIDA) around 20% of Autonomous Systems do not block spoofing [9].

Bind introduced RRL support with version 9.9.4 in 2013 [10] and Windows Server 2016 in 2015 [11]. In a 2020 paper, "The Impacts of DNS Protocol Security Weaknesses" the authors state that "DNS is a highly sensitive part of every ICT system. If attackers can take control of DNS, it would give them unlimited possibilities to abuse the organization in different aspects. DNS is a key component in the concept of multilayered security." [12] In comparison to [5]. As opposed to 13 years ago attackers are still using similar attacks and combining them with other attacks to achieve their goals. One example of that is a DNS rebinding attack in which attackers trick the victim's browser into making requests to local network devices. In a 2019 quantitative study [13] researchers found that about 16% of authoritative DNS servers employ some sort of rate limiting.

Waseda University has conducted a large-scale survey into the adoption of various DNS security mechanisms - DNSSEC, DNS Cookies, CAA, SPF, DMARC, MTA-STS, DANE, and TLSRPT and in doing so identified what effects adoption rates and reported its results in a blog post [14]. They discovered that DNS security that is easier or cheaper to deploy is more widely adopted and that root servers and top-level domains are leading adopters of DNSSEC and DNS Cookies.

One of the most important guidelines for securing DNS is NIST's Secure Domain Name System (DNS) Deployment Guide [15] which should be followed by all administrators of private and public DNS servers. Due to the distributed nature of DNS, all entities that host DNS servers should take part in protecting such an important system and making sure it is not misused. It does not mention ANY records, but Cloudflare recommends their removal and disputes their usefulness in real-life scenarios [16].

In a 2015 paper researchers reported statistics on DNS servers after tracking them for 13 months starting on January 31st, 2014. During that time, DNS resolvers dropped from 26.8 to 17.8 million servers. Also, by comparing responses they found out that more than 3 million servers redirect specific domain names to landing pages for censorship [17]. In a 2019 paper researchers found that there are around 3 million open resolvers on the Internet, and of those 110k provide incorrect IP addresses as a DNS response with more than 26k servers returning IPs reported as malicious. [18]

Attackers can use information stored inside DNS records with other information related to a particular entity to perform OSINT investigations as demonstrated in "Croatian Bank Security Analysis by Publicly Available Data" [19].

DNS did not lose its importance with IPv6 or move from self-hosted to cloud and hybrid infrastructure. DNS as a service, operated by cloud providers offers easier protection from DDoS attacks, higher availability, and moving the responsibility of securing DNS servers to cloud providers [20]. Cloud vendors due to the centralization of resources have higher budgets and easily available infrastructure distributed throughout the globe. Although by centralizing where the DNS zones are hosted a potential compromise can be more damaging. Applications designed for the cloud still reference other hosts and services through their DNS names.

DNS in Internet of Things (IoT) environments offers to solve multiple use cases. It can help with uniquely identifying devices, autoconfiguration, security, and interoperability. One of the challenges is that IoT devices are designed with constricted capabilities. [21].

In 2015 papers authors stated that "The concept of IoT represents the evolution of the Internet and its appliance is continuously growing. According to estimates, through this concept, 50 billion devices will be connected by 2020 which places heavy demands and challenges in maintaining the required safety level of such an environment." [22] As the growth of IoT added a huge number of devices with often low levels of security, attackers are using them to create botnets for malicious purposes. One of the biggest DDoS attacks was done by the Mirai botnet against the Dyn DNS service provider by using several hundred thousand devices. For approximately two hours, the attack on Dyn made multiple internet platforms inaccessible including PayPal, Twitter, Reddit, Sony, Amazon, Netflix, Spotify, Pinterest, SoundCloud, Squarespace, and several major news websites. Due to the attack on Dyn and its business interruption, it lost 8% of its customer base [23].

3. Data Collection

To find all DNS servers on the Internet masscan [24] was used to find all IP addresses listening on port 53 UDP and TCP. Afterwards, to get valid DNS recursive DNS servers, all of them were queried for A record for google.com. Those that provided a valid DNS response were later tested for recursion. A response was considered valid if it was returned in 3 seconds. IPs were related to a country by using the MaxMind GeoLite2 database.

Initially, we wanted to test servers for zone transfer, recursion, DNSSec, RRL, DNS Socket Pool utilization, and DNS Cache locking, but we were unable to do so without deducing zones which the DNS server is hosting from its IP address. Performing reverse lookup was not a reliable method as DNS servers can host multiple zones and the FQDN of the DNS server does not need to match the zone it is hosting. Zone transfer and DNSSec require knowing the zone name, DNS socket pool utilization requires tracking which port does DNS server uses to issue queries, and checking DNS cache locking requires effectively attempting to do DNS cache poison on every DNS server. Testing RRL was not conducted as DNS servers can be behind a load balancer or have different RRL based on the domain as noted in [13]. Another limitation of our data collection is that we did not track changes to these settings over a longer period. As opposed to [25] our focus was not analysing data collected at Internet eXchange Points but collecting data from servers like it was done in [17].

4. Data analysis and results

Examining the entirety of public IPv4 addresses culminated in the identification of a substantial dataset comprising 1,740,663 distinct IP addresses, each characterized by the presence of an open port 53—an indicative marker of potential involvement with the Domain Name System (DNS), an integral facet of internet functionality. IP addresses in the resulting data set were further analysed by assessing the responses IP addresses returned for DNS queries to identify valid DNS server installations within this subset, with a focus on ascertaining the subset capable of facilitating recursive query resolution.

Parsing the responses yielded a compelling insight: within the subgroup of IP addresses harbouring an open port 53, 957,806 addresses exhibited a coexistence of DNS server installations. DNS installations were identified by receiving a valid DNS response. Valid DNS servers were further tested for the presence of the Recursion Available (RA) flag within their responsive DNS messages, signifying their potential for engagement in the mechanics of recursive query resolution.

A total of 141,443 servers emerged, each endowed with the capacity to undertake recursive query resolution. This means that of all DNS servers on the Internet, 14.77% support recursive resolution without any access control. It is theoretically possible that an access control exists albeit very wide. However, our research was not confined to empirical observation alone; rather, it underscored a formidable concern, thereby enhancing the gravity of our findings. The spectre of DNS amplification attacks underscores the necessity to fortify these servers against vulnerabilities susceptible to malicious exploitation.

Transitioning from empirical revelations to a broader contextualization, we shift our gaze to the geographical distribution of these vulnerabilities. Our inquiry led us to discern that the top 20 countries harbouring vulnerable servers are China (21,393), the United States (20,661), South Korea (10,147), Bangladesh (8,131), Indonesia (7,853), Russia (7,733), Brazil (5,978), India (5,276), France (3,460), Ukraine (2,573), Japan (2,485), Germany (2,051), Argentina (1,961), Poland (1,761), Turkey (1,756), Colombia (1,750), South Africa (1,668), Taiwan (1,572), United Kingdom (1,537), and Canada (1,468). We observed a correlation between the number of DNS servers and the prevalence of vulnerable instances. Notably, countries with a higher number of DNS servers exhibited a proportionate increase in vulnerable instances. We found that the correlation between the number of DNS servers and the number of ones that accept recursive queries is 0.788955.

However, our investigation did not cease at a mere enumeration of vulnerable instances; it extended to a nuanced exploration of the correlation between the number of DNS servers and the percentage of instances supporting recursion. Intriguingly, a negative correlation emerged (-0.127182), emphasizing that an abundance of DNS servers does not inherently correlate with an equal proportion of instances supporting recursion. This observation underscores the complexity of the underlying dynamics.

To circumvent potential distortions introduced by countries with minimal DNS servers, we analysed by excluding nations with fewer than 100 DNS servers. This refined analysis revealed a distinct hierarchy of vulnerability.

Bangladesh (91.68%), Yemen (89.03%), Dominican Republic (85.48%), Colombia (74.98%), Palestine (72.62%), South Korea (70.63%), Guatemala (70.04%), Lebanon (69.23%), Kenya (67.26%), Cambodia (65.78%), Philippines (65.17%), Pakistan (62.62%), Honduras (61.87%), Libya (61.82%), Nicaragua (60.74%), Indonesia (59.89%), Albania (58.67%), Bolivia (57.91%), Venezuela (57.38%), and Mongolia (57.07%) were identified as the countries with the highest percentages of vulnerable servers. Seeking to distil broader patterns, we aggregated countries into continents.

Continent	DNS server instances	Recursive DNS servers	Percent
North America	344928	24332	7.05%
Asia	286221	77224	26.98%
Europe	267283	22963	8.59%
South America	33014	11793	35.72%
Oceania	13156	1502	11.42%
Africa	8571	3216	37.52%

Table 1. DNS and recursion across continents

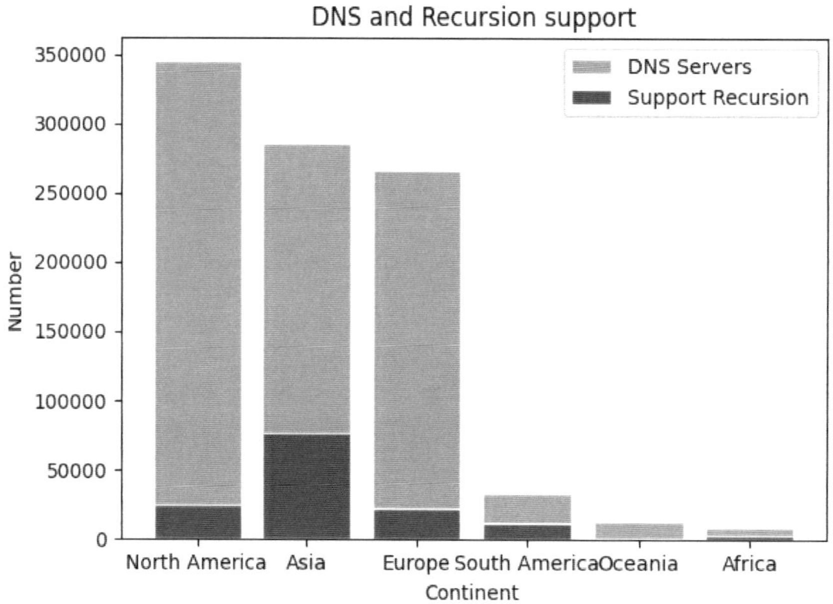

Fig. 2. DNS and recursion support

Africa, South America, and Asia have the highest number of vulnerable servers with more than 20% of them being recursive as seen in Table 1 – DNS and recursion across continents, and Fig. 2. DNS and recursion support. This continental-level analysis accentuates the necessity for comprehensive security measures and vigilance, particularly within previously mentioned regions. On the other side, Europe, and North America with many DNS servers have a small percentage of vulnerable DNS servers. MaxMind GeoLite2 database used was not able to correlate 3417 IP addresses with a country/continent.

For further research, there are multiple directions of whom all can contribute to the ongoing efforts to strengthen the security of DNS infrastructure and protect against DNS amplification attacks, which continue to pose a significant threat to the stability and availability of the Internet. One of the possible directions is tracking how the number of public and recursive DNS servers change over time and how does increasing number of IoT devices will affect the amount of valid and malicious DNS traffic. Another direction would be tracking the implementation of global DNS security mechanisms and correlating to multiple other factors like changes in attack methodology used by malicious actors, the correlation between implementation of security mechanisms and industry and government regulations of IT infrastructure, and the level of awareness and education of DNS server administrators.

5. Conclusion

High amplification rates created by queries with large responses provide an easy way for attackers to make their attacks more efficient and destructive. The growth of IoT represents a challenge for defence against DDoS attacks as number of devices connected to the Internet rapidly increases.

Based on the results of our study we identified that the number of DNS servers on the Internet is decreasing as compared to the results of previous studies. Among the remaining DNS servers, 14.77% of them is recursive which can be abused by malicious actor and should be properly secured. The number of DNS servers that support recursion does not correlate directly with the number of DNS servers but depends on other factors as well. The highest number of recursive DNS servers was found in Asia.

Responsibility for the implementation of security measures lies with operators of specific DNS servers. In case when those servers are part of government infrastructure regulatory bodies should enforce adherence to best practices. Operators of DNS servers should completely disable recursion or restrict it by ACLs. If required recursive DNS servers can be deployed only for internal clients. Deploying rate limiting will also decrease the efficiency of amplification attacks as the total number of queries will decrease. Blocking or restricting ANY queries will limit the size of records attackers can request. NIST Secure DNS deployment guide offers useful guidelines to secure DNS servers but due to its age, it is ready for a new revision. For future work, it would be useful to measure the implementation of DNS security measures over time and track how the attacker's methodology changes.

6. References

[1] https://web.archive.org/web/20070203060758/http://www.ciac.org/ciac/bulletins/j-063.shtml (1999). J-063: Domain Name System (DNS) Denial of Service (DoS) Attacks, Computer Incident Advisory Capability, Accessed on: 2023-06-02

[2] https://www.microsoft.com/en-us/security/blog/2022/05/23/anatomy-of-ddos-amplification-attacks (2022) Azure Network Security Team, Anatomy of a DDoS amplification attack, Accessed on: 2023-06-01

[3] Mockapetris, P. (1983), Domain Names - Concepts and Facilities, IIS

[4] https://datatracker.ietf.org/doc/html/rfc2065 (1997). Domain Name System Security Extensions, Internet Requests for Comments

[5] Chatzis, N. (2007), Motivation for Behaviour-Based DNS Security: A Taxonomy of DNS-Related Internet Threats in The International Conference on Emerging Security Information, Systems, and Technologies (SECUREWARE 2007), Valencia, Spain

[6] Anagnostopoulos, M.; Kambourakis, G.; Kopanos, P.; Louloudakis G. & Gritzalis, S. (2013), DNS amplification attacks, Computers & Security, doi:10.1016/j.cose.2013.10.001

[7] M. Prince (2013). The DDoS That Knocked Spamhaus Offline (And How We Mitigated It)

[8] https://www.cisa.gov/news-events/alerts/2013/03/29/dns-amplification-attacks (2013). DNS Amplification Attacks, Cybersecurity and Infrastructure Security Agency

[9] https://spoofer.caida.org/summary.php (2015). State of IP Spoofing, Center for Applied Internet Data Analysis

[10] Behjat A. (2013), BIND 9.9.4 Released, Internet Systems Consortium

[11] https://learn.microsoft.com/en-us/archive/blogs/teamdhcp/response-rate-limiting-in-windows-dns-server (2015). Response Rate Limiting in Windows DNS Server, Microsoft

[12] Dimitrov, W. & Panayotova, G. (2020). The Impacts of DNS Protocol Security Weaknesses

[13] Deccio, C.; Argueta, D. & Demke, J. (2019). A Quantitative Study of the Deployment of DNS Rate Limiting, IEEE, 2019, pp. 442-447.

[14] https://blog.apnic.net/2021/11/26/adoption-of-dns-security-mechanisms-related-to-ease-of-use-cost (2021). Adoption of DNS security mechanisms related to ease-of-use, cost, APNIC, Accessed on: 2023-05-22

[15] Chandramouli, R. & Rose, S. (2013). Secure Domain Name System (DNS) Deployment Guide

[16] Majkowski, M. & Guðmundsson Ó. (2015), Deprecating the DNS ANY meta-query type

[17] Kührer, M.; Hupperich, T.; Bushart, J.; Rossow, C. & Holz, T. (2015). Going wild: Large-scale classification of open DNS resolvers, 15th ACM Internet Measurement Conference (IMC)

[18] Park, J.; Mohaisen, M & Mohaisen A. (2019). Investigating DNS Manipulation by Open DNS Resolvers, CoNEXT '19 Companion: Proceedings of the 15th International Conference on emerging Networking EXperiments and Technologies, doi:10.1145/3360468.3368172

[19] Matvej, E.; Morić, Z. & Papić, S. (2020). Croatian Bank Security Analysis by Publicly Available Data in Proceedings of the 31st International DAAAM Symposium

[20] Aishwarya, C.; Sannidhan, M. S. & Rajendran, B. (2014). DNS Security: Need and Role in the Context of Cloud Computing in 3rd International Conference on Eco-friendly Computing and Communication Systems, Mangalore, India

[21] Ayoub, I.; Balakrichenan, S.; Khawam, K. & Ampeau, B. (2023). DNS for IoT: A Survey, Sensors

[22] Cvitić, I.; Vujić, M. & Husnjak, S. (2016). Classification of Security Risks in the IoT Environment, in Proceedings of the 26th DAAAM International Symposium

[23] Young, K. (2022). Cyber Case Study: The Mirai DDoS Attack on Dyn

[24] Graham, R. D. (2021), Masscan, Github, https://github.com/robertdavidgraham/masscan, Accessed on: 2023-04-04

[25] Nawrocki, M.; Jonker, M.; Schmidt, T. C. & Wählisch, M. (2021). The Far Side of DNS Amplification: Tracing the DDoS Attack Ecosystem from the Internet Core, In Proceedings of the 21st ACM Internet Measurement Conference (pp. 419-434), doi: 10.48550/arXiv.2109.01104 ecosystem from the internet core..

DOI: 10.2507/34th.daaam.proceedings.013

DENIAL OF SERVICE ATTACKS USING THE EXAMPLE OF CROATIAN HOSTERS

Kristijan Dizdar, Zlatan Morić, Vedran Dakić & Matej Bašić

This Publication has to be referred as: Dizdar, K[ristijan]; Moric, Z[latan]; Dakic, V[edran] & Basic, M[atej] (2023). Denial of Service Attacks Using the Example of Croatian Hosters, Proceedings of the 34th DAAAM International Symposium, pp.0088-0095, B. Katalinic (Ed.), Published by DAAAM International, ISBN 978-3-902734-41-9, ISSN 1726-9679, Vienna, Austria
DOI: 10.2507/34th.daaam.proceedings.013

Abstract

This research delves into Distributed Denial of Service (DDoS) attacks, investigating their types, tactics, and mitigation. DDoS attacks are categorized into volumetric, TCP-based, and application layer attacks, revealing their mechanics and accessibility through stressors and booters. Additionally, the study explores attack methods, such as UDP flood and amplification attacks, emphasizing their impact and challenges. It highlights the need for proactive cybersecurity measures. The research also discusses booters and stressors, tools used by cybercriminals, and mitigation techniques involving scrubbers, firewalls, and load balancers. In conclusion, this research contributes to understanding the evolving DDoS threat landscape, emphasizing the importance of advanced detection and mitigation. Collaborative defence strategies and economic implications are also discussed, promoting network security in an interconnected world.

Keywords: Denial-of-Service; Distributed Denial-of-Service; penetration testing; cyber-attacks.

1. Introduction

The study under scrutiny immerses itself in the complex realm of Distributed Denial of Service (DDoS) attacks [1], a profound concern within the domain of cybersecurity. With a multifaceted focus, the primary objective of this research is to provide a comprehensive examination of DDoS attacks, shedding light on their diverse manifestations, repercussions, and countermeasures. By doing so, it aims to make a significant contribution to the understanding and mitigation of these disruptive cyber threats, which continue to pose substantial challenges to network security.

In pursuit of its overarching goal, this study tackles a series of fundamental questions. It delves into the different types of DDoS attacks, dissecting how each exploits specific vulnerabilities inherent in network protocols. It proceeds to unravel the tactics employed by malicious actors in launching DDoS attacks and examines the profound impact of these tactics on targeted systems. Furthermore, the study scrutinizes the existing mitigation techniques against DDoS attacks, carefully assessing their limitations and challenges. Ultimately, it strives to offer valuable recommendations to fortify cybersecurity measures in the face of ever-evolving DDoS threats. To provide a solid foundation for its exploration, this study makes extensive reference to seminal works within the field. It notably draws upon key references such as server's response to UDP packets [2], and an in-depth analysis of NTP amplification attacks [3]. Additionally, the study casts its spotlight on DNS amplification attacks [4].

In this endeavor, the study categorizes DDoS attacks into three primary classifications, namely volumetric-based attacks, TCP-based attacks, and application layer attacks[5]. It unearths the underlying mechanics of these attacks and their impact on network infrastructures. Furthermore, it underscores the disconcerting reality that DDoS attacks have become commoditized, available as services categorized as stressors and booters, showcasing their widespread accessibility and potency. This research sets the stage for a comprehensive exploration of DDoS threats and countermeasures, signifying their profound significance in the ever-evolving landscape of cybersecurity.

2. Types of DDoS attacks

Although quite similar, not all DDoS attacks are the same. The main differentiating factor between types of DDoS attacks is the protocol that is being exploited. One of the more popular types of DDoS attacks is a UDP flood. By design, the User Datagram Protocol is session less and connectionless. This means that for the application or service to work over UDP, the server and client do not need to establish a connection for the service to work. One example of such a service would be DNS, which primarily uses UDP port 53. One design pattern of UDP that can be easily exploited to carry out DDoS attacks is the algorithm that each network device goes through when receiving a UDP packet. Under normal conditions, when a server receives a UDP packet, there are two steps, according to conditions[6]:

- Server checks if there is a service running on a particular port, in a listening state for new incoming requests.
- If the first condition is not fulfilled, the server will return ICMP (or ping) a packet back to the sender, informing him that the destination is unreachable.

With every packet sent, the server needs to use its resources to process requests. Every UDP packet has the IP address of a source machine. When the attack is launched from a well–organized botnet, attackers will usually spoof the originating IP address so that the returning traffic does not affect the botnet. As there are no checks such as when using TCP, the server would usually return a response to the source IP address. The way this is leveraged in a DDoS attack is that an attacker would use an entire botnet of infected servers that would then send UDP packets with a spoofed IP address towards the victim. The victim would then try to process these requests and send out ICMP error packets to the spoofed IP sent in the initial request header.

Most modern systems implement some type of rate limit, but assuming the length of each packet is 1500 bytes, the compute and network resources on the victim machine are overwhelmed for the entire duration of the attack. Using this method, the attacker first overwhelms the victim's network bandwidth and later saturates the server resources. In an example of a DDoS attack including 20 20,000 devices, which is nothing out of the ordinary, that send 10 packets per second, the target server needs to process 200 thousand requests per second. Even the most efficient computing resources today are not able to process this amount of sustained traffic.

In UDP-based attacks, a single server could, in theory, generate gigabits per second of traffic. However, this way, the attacker's IP addresses could be tracked. For this reason, attackers usually use methods of amplification to not only amplify the amount of traffic being sent to the victim but also create a scenario in which the original attacker's IP address is not visible from the victim's machine. The most popular methods of amplification are DNS amplification and NTP amplification.

Network Time Protocol is a service used to synchronize time on different devices. NTP has been a part of every network for decades and is one of the oldest Internet protocols still in use. One specific command, monlist, sent to an NTP server is the dominant method of executing an NTP amplification attack. Monlist was designed as a control message command that prints traffic counts collected and maintained by the monitoring facility of an NTP daemon. The monitoring facility of an NTP daemon collects and maintains data of the Most Recently Used hosts or clients that have communicated with the NTP daemon. Monlist command can display a maximum list of 600 entries. As the initial NTP request sent from the client to the server is relatively small, in the case where the monlist command returns a list of 600 servers, the response that the server sends will be 206 times larger than the initial request.[3]

DNS amplification is another variation of amplified DDoS attacks. Domain Name System is the most used protocol on the internet and uses both UDP and TCP port 53. One thing that makes DNS a great protocol for DDoS attacks is the fact that a DNS response packet is significantly larger than a DNS query packet. The method of exploiting DNS for a DDoS attack is the same as with NTP. The attacker sends many DNS queries with a spoofed IP address that points to the victim to several DNS resolvers that then start sending responses to the victim. The average size of a DNS request is 60 bytes. The average size of a DNS response is 3 kilobytes or 3200 bytes, which makes the amplification factor to be around 50x, or even 100x in rare cases.[7]

UDP flood attacks are volumetric-based attacks. The goal of these attacks is to exploit a well-known protocol that most devices use to saturate the network connection on the victim side. Most of the time, completely preventing these types of attacks is not possible, but certain remediation methods are available that can help in scenarios when their infrastructure is a target of a DDoS attack.

Alongside UDP flood attacks, several attacks are exploiting various protocols that use TCP. Transmission Control Protocol, unlike UDP, requires a connection to be established between the client and a server for the communication to be successful. The way the server and a client establish this connection is by performing a 3–way handshake, a 3–step procedure that is used to prepare both sides.

The three-step procedure is as follows:

- Client sends a SYN packet to initiate a connection.
- Server returns a SYN/ACK packet as a response.
- Client sends an ACK packet.

Once this procedure is completed, the TCP connection will be open, making the client available to send and receive data. Attackers can use SYN flood attacks by exploiting this design. An SYN flood attack, also known as a "half-open" attack, is a type of DDoS attack targeting all systems, with services using TCP protocol such as web servers, mail servers, etc.[8]. The way a SYN flood attack is implemented is by exploiting the last step in a 3–way handshake. The attacker will send many SYN packets to the target server, again usually by using a botnet of infected devices. In general, there are 3 main vectors in which a SYN flood attack can be done:

- Direct attack – In this case, attackers do not spoof their IP address, but rather use various routing and filtering rules on their side so that their machine cannot respond to SYN/ACK packets.
- Spoofed attack – As is the case with most UDP flood attacks, in spoofed SYN flood attacks, attackers will also spoof their IP address to me the process of tracing the traffic back to them more difficult.
- Distributed attack – the most common type of a SYN flood attack in which an entire botnet of infected devices attacks the same target, making it almost impossible to trace traffic back to the original attacker.

The reason SYN flood attacks work is that the server will allocate resources upon sending a SYN/ACK packet back to the client. One of the most popular web servers today, Apache, schedules a new thread for each new incoming connection. Although this design is great for normal operations, it can be easily exploited by opening a bunch of connections to the server and never closing them. By doing this, we can schedule a bunch of threads and exhaust memory on the target server without ever compromising our Internet connection. That very reason is the main reason this method of implementing a DDoS attack is the most popular one. As it is not volumetric, it does not require a lot of traffic being sent to the target server which means it can be done using a smaller number of infected devices.

The final category of DDoS attacks we will cover are application layer or Layer 7 attacks. Alongside volumetric and protocol–based attacks, another popular method for implementing a DDoS attack is an HTTP flood attack. Hypertext Transfer Protocol is used for exchanging data between a server and a client, mostly used for serving websites on the server side and rendering websites on the client side. [6]

HTTP flood attack is a specific type of attack where attackers send requests as legitimate users but with a difference in the amount of potentially open connections. HTTP protocol defines different types of requests, with the 2 most used being GET and POST. A GET request is usually used to request something from the server, while a POST request is used to send some data from the client to the server. By exploiting a GET request, an attacker can perform a GET flood. A GET request is what a typical web browser sends to a web server to fetch web content. In this case, attackers impersonate real clients. The key to this attack is that it requests legitimate content served on the web server. Additional configuration pieces that make this attack even more complicated to remediate are the following:

- Using residential IP addresses – modern DDoS attacks are performed by leveraging large botnets of infected devices. These botnets are comprised of IoT devices that are notorious for having subpar security standards. Given that IoT devices are connected to home networks, the originating IP address that the victim will see will be the public IP address that most of the time is not static
- User agents – By configuring User agent fields in HTTP requests, an attacker can impersonate a web browser client, such as Google Chrome, even though the requests might originate from an IoT device.
- Three–way handshake – connections must be open between 2 sides during the entire time of the attack.

On the other hand, a POST flood attack exploits the design of a POST request. As already mentioned, POST requests are used to send data from the client machine to the server, which means that to exploit this request, the attacker must keep the connection alive from the client side and try to exhaust server resources. The idea behind POST flood attacks is to send a small amount of data over a period just to keep the connection open on the server side.

Aside from classic GET and POST flood attacks, a more modern approach is something named Slow Client Attacks. Slow Client Attacks behave like classic GET and POST flood attacks, but the patterns and execution times are different. Slow client attacks are assigned to send specific requests to make the traffic flow slower. This is done because of the design of the protocol itself – the server must keep the connection open until a signal is sent. Denial of service happens when a web server has too many open connections and cannot open new ones for legitimate users. The most common implementations of slow HTTP attacks are:

- SlowLoris[9]
- Slow HTTP Post[10]
- Slow Read attack[11]

All these categories are known as "low and slow" attacks. They have a mechanism in common – sending a small amount of traffic and maintaining slow connections. Unlike traditional attack types, they do not require a lot of bandwidth volume to take down an average web server[12].

Some of the techniques for performing DDoS attacks may seem complicated. In reality, a DDoS attack can be bought as a service online. These services are separated into 2 categories – stressers and booters.

3. Booters and stressors

Booters and stressers are often offered by hackers, or cybercriminals, who make botnets using cloud virtual machines and infected devices all over the world. All stressors have the same methodology in common, using proxy servers and various other techniques for spoofing IP addresses of attack machines to protect their botnet and make it more difficult to trace the attack back to them.[13]

The way These services are usually presented as a toolset used to stress test your infrastructure to make sure that you could handle a DDoS attack if you were ever a victim of one. Both terms have the same meaning in practice. On the front end, hackers usually offer a web page where you can sign up for their services. These web pages are often hosted in non–EU countries to avoid being banned from the Internet. In most countries, it is illegal to create and sell such services. Today, you can buy a subscription and start performing a DDoS attack after just a few minutes of searching the web. Once you land on one of these pages you are immediately presented with the price and the payment method. Prices vary from one provider to the other, and based on their reputation, the prices can be as low as 20 USD or go up to as high as 1000 USD for a single attack.

4. Tests and results

Tests were performed on 3 independent hosting providers available in Croatia. Tests were carried out using "DDoS for hire" services explained earlier in the paper. Tests were announced upfront, and written consent was received from all 3 hosting providers. Results are separated into 3 tables, one for each hosting provider. Test results for the first provider are in the table below:

#	Attack Vector	Target	Rate	Result
1	Advanced HTTPS Flood (without protection)	Website	90K R/S	FAIL
2	Advanced HTTPS Flood (with scrubber on)	Website	90K R/S	FAIL
3	VSE Flood @ 443 port – with protection disabled	Website	120KPS	Partial pass
4	VSE Flood @ 443 port – with protection enabled	Website	120KPS	Partial pass
5	SYN Flood @ 443 port – with protection	Website	1M PPS	FAIL
6	SYN Flood @ 443 port – without protection	Website	1M PPS	FAIL
7	UDP-MIX - against 22 ports – protected	Host	20Gbps	FAIL
8	UDP-MIX against 22 ports - unprotected	Host	20Gbps	FAIL
9	HTTP get flood – without protection	Website	80K R/S	FAIL
10	HTTP get flood – with protection	Website	80K R/S	FAIL
11	HTTP raw get – with protection	Website	50K R/S	FAIL
12	HTTP raw get – without protection	Website	50K R/S	FAIL
13	HTTP null – with protection	Website	50K R/S	Pass
14	HTTP null – without protection	Website	50K R/S	Pass

Table 1. First provider results

Result table for the second hosting provider:

#	Attack Vector	Target	Rate	Result
1	Advanced HTTPS Flood (without protection)	Website	90K R/S	FAIL
2	Advanced HTTPS Flood (with scrubber on)	Website	90K R/S	FAIL
5	SYN Flood @ 443 port – with protection	Website	1M PPS	PASS
6	SYN Flood @ 443 port – without protection	Website	1M PPS	FAIL
7	UDP-MIX - against 22 ports – protected	Host	20Gbps	PASS
8	UDP-MIX against 22 ports - unprotected	Host	20Gbps	FAIL
9	HTTP get flood – without protection	Website	80K R/S	PASS
10	HTTP get flood – with protection	Website	80K R/S	PASS
11	HTTP raw get – with protection	Website	50K R/S	FAIL
12	HTTP raw get – without protection	Website	50K R/S	FAIL
13	DNS-AMP against 53 port - protected	Host	300K R/S	PASS
14	DNS-AMP against 53 port - unprotected	Host	300K R/S	FAIL

Table 2. Second provider results

Result table for the third hosting provider:

#	Attack Vector	Target	Rate	Result
1	Advanced HTTPS Flood – protected subnet #	Website	90K R/S	FAIL
2	Advanced HTTPS Flood – protected subnet #2	Website	90K R/S	FAIL
3	VSE Flood @ 443 port – protected subnet #1	Website	120KPS	PASS
4	VSE Flood @ 443 port – protected subnet #2	Website	120KPS	PASS
5	SYN Flood @ 443 port – protected subnet #1	Website	1M PPS	PASS
6	SYN Flood @ 443 port – protected subnet #2	Website	1M PPS	PASS
7	UDP-MIX - against 22 port – protected subnet #1	Host	20Gbps	PASS
8	UDP-MIX against 22 port–protected subnet #2	Host	20Gbps	PASS
9	HTTP get flood – protected subnet #1	Website	80K R/S	FAIL
10	HTTP get flood – protected subnet #2	Website	80K R/S	FAIL
11	HTTP raw get – protected subnet #1	Website	50K R/S	Partial pass
12	HTTP raw get – protected subnet #2	Website	50K R/S	FAIL
13	HTTP null – protected subnet #1	Website	50K R/S	Pass
14	HTTP null – protected subnet #2	Website	50K R/S	Pass

Table 3. Third provider results

The DDoS testing was a basic analysis that included seven attack vectors. The network attack (UDP flood and SYN flood) was not blocked. While protection was on, it required about five minutes of detection time in terms of UDP flood, only to lower packet loss. Services behind the host were not available during that time. Only one vector had a partial pass, and that was the VSE vector. Five vectors were against the website itself, and none of them were detected by DDoS scrubbing services, as the results were the same.

5. Methods of remediation

Hosting your applications on your infrastructure has many advantages, but handling DDoS attacks surely is not one of them. Apart from the already mentioned method of adding more devices, one common method companies use to remediate this threat is using one of the online services that claim to be able to filter and handle traffic for you, often called scrubbers. One of the most popular companies that provide such services is CloudFlare[14]. CloudFlare offers a wide range of services, all designed to make your infrastructure and your services more resilient and less vulnerable to DDoS attacks.

Based on the types of attacks mentioned earlier, we define different methods for mitigation and prevention of these attacks. Prevention and mitigation solutions for amplified attacks are limited. The main reason behind this is the amount of traffic generated from these attacks, where the surrounding networks usually have problems with processing this type of traffic. Usually, ISPs blackhole all that traffic, but the effects for the site owner are practically the same since, while an attack is ongoing, the server receives packets the whole time.[15]. On the other hand, many scripts available have recognizable patterns that can be easily blocked on edge routers. Along with filtering by pattern, some of the techniques used can be to limit the NTP packet ratio or verify the source IP address.

When it comes to TCP-based attacks, there are more advanced methods of mitigation, both on the server itself and once designed to be placed "in front" of the server. Firewall filtering is a method that mitigates a large portion of SYN flood attacks. Many modern firewalls come with predefined patterns for detecting SYN attacks. They can fully prevent or limit the surface of incoming attacks. A popular example of such systems would be a FortiGate firewall which has a threshold for TCP SYN flood. In general, it intercepts TCP connections before they come to the server itself, and a numeric limit of incoming connections can be implemented along with a waiting time. All stages of a 3–way handshake will be carried out between the client and a firewall, after which the firewall itself makes a "replay" to the original destination. If a threshold is reached, all further packets will be blocked, but with the exception that SYN packets will be allowed if clients send another SYN packet – simply put if the client performs a retry, a full connection will be opened. This is the main benefit of using a firewall as a separate appliance. Something similar could be implemented on the firewall on our server itself on the OS level, but having a firewall on the same machine as the web server means that the same server must process both real and malicious network traffic which could highly increase compute resource consumption on the target server.

The second method we can use to mitigate against SYN flood is half–open connection recycling. Half–open connection recycling means that time for leaving half–open connections can be lowered along with cleaning a certain number of recent connections, but this approach does not have any effects if the attack has a higher volume. As for HTTP flood prevention techniques, scrubbing services still barely manage to mitigate attacks, except for a few industry leading CDN services with additional protection layers in the background. Because of that, many hosting providers decide to invest money and effort in on–premise solutions, which are usually open–source and easier to maintain and configure. Mitigation techniques include:

- Using a CDN
- Web application firewalls
- Web server hardening
- Proxy as a mitigation mechanism

Web application firewalls are usually software suites installed on an endpoint, and they reside between a web server and clients connecting to it. The primary goal of WAFs is to monitor and filter traffic between the two sides. If a condition is fulfilled in terms of server resources, a WAF can be an efficient tool in mitigating not only DDoS attacks but other types of malicious traffic such as SQL injections, cross–site scripting, etc. In terms of mitigating an attack, a firewall is effective against low–volume attacks, but strong attacks reach the server. A WAF works like any other firewall, it has a set of policies protecting against vulnerabilities. As a DDoS attack arrives, a WAF is capable of rate–limiting addresses, blocking them from further attacks. However, attackers often use proxies to hide IP addresses, which can leave a server struggling to block all IP addresses in a timely manner.

Web application firewalls operate in two ways:
- Blacklist – usually by IP ranges pulled from known vendors like *SpamHaus* or by pattern–specific options.
- Whitelist – only specific IP ranges can access the server, or only specific browser agents or specific patterns should be fulfilled to open a webpage.

The cheapest mitigation technique that can be implemented on a customer's premises is to utilize additional publicly available modules, along with web server packages, nginx, and Apache. *Mod_evasive* module is the most popular solution among shared hosting providers. This addon monitors connections for DDoS attacks and brute force attacks. This works similarly to WAFs, and its main feature is that in case of excessive connections to the resource the IP gets banned for a period. This module works only if the client experiences GET and POST flood attacks with higher volume, while low – and – slow attacks go undetected.

Another open–source solution that could be utilized is *HAProxy*.[16] *HAProxy* is a high-availability load balancer and reverse proxy for TCP and HTTP applications. This tool balances traffic to two or more servers in the backend after health checks are performed. *HAProxy* is used not only for web applications but it can be used for all TCP–based applications, such as SQL databases. Companies would usually set up *HAProxy* with an external IP address. From the Internet, the application hosted on our server is available over ports 80 and 443. The request first ends up on the load balancer, behind which multiple backend servers are up and running.

6. Further recommendations for mitigation

DDoS attacks are primarily used for disrupting services and causing business damage through every second of downtime with a huge load of illegitimate traffic. As a countermeasure to these attacks, companies need to utilize a solution that will amortize the effects of received attacks. DDoS mitigation is a process where a protection service of any kind, such as a scrubbing service, firewall, or appliance, protects resources from DDoS attacks, thus protecting resources from being unable to process legitimate requests. As DDoS attacks are developing with new methods and growing each day in terms of their strength, mitigation services are improving at the same pace.

Mitigation can be achieved with three models:
- On-premises
- Cloud-based
- Hybrid environment

There are pros and cons of using various cloud-based and/or hybrid solutions. However, an additional problem is the public or hybrid cloud computing model, where a single physical server can contain multiple virtual machines[17] that could suffer noisy-neighbour problems (influence each other's performance) or have availability problems. This is why these types of simulations are a powerful tool used in testing organizations' vulnerability and exposure to cyberattacks[18]. Furthermore, there's a problem of digital footprint, which is what attackers will try to use to extract useful information to be even more malignant when initiating cyberattacks[19]. The same recommendations that we have for employees of any company apply to our scenarios here, as well – the digital footprint of a service provider needs to be tightly managed and as small as possible.

The mitigation process is divided into four steps: detection, diversion, filtering, and analysis. The first phase is detection, where the environment should be aware of incoming traffic if the service and enterprise using it expect effective mitigation. The primary goal of DDoS is to flood the destination with a lot of traffic, In the detection phase a particular service should detect patterns of incoming attacks and if there is a match, block the traffic. Patterns must be prioritized instead of source IP addresses, as botnets consist of numerous machines with different source IP addresses. The reason behind this is that attacks cannot be blocked by banning IP addresses, as there is a possibility that behind these addresses are legitimate users, and such users would not be able to access the content, especially if addresses are CGNAT–ed.

The process of detection consists of receiving and identifying anomalies in network traffic reaching the resource. To be successful in resolving anomalies promptly, owners or providers should decide how they will face these attacks by blackholing, enabling a BGP speaker, or using a scrubber service. The next phases are interpreted differently by various DDoS protection services. After detecting a DDoS attack, a protection service should be able to drop malicious traffic incoming from a botnet, while legitimate traffic should reach the destination. This is implemented by using routing schemes, where traffic patterns are identified and dropped without blocking the IP address itself, because of possible blocking of further legitimate traffic from the same IP.

Building upon the insights gained from this research, several promising avenues for future investigations emerge. Firstly, researchers should delve deeper into the behavioural analysis of DDoS attacks, exploring advanced machine learning and AI-based techniques to enhance real-time detection and mitigation. Additionally, there is a need to focus on zero-day DDoS attacks, understanding how attackers exploit previously undocumented vulnerabilities in network protocols, and developing proactive defences against such emerging threats.

Furthermore, the intersection of blockchain technology and DDoS mitigation presents an intriguing research area, with potential for decentralized and resilient solutions. As the Internet of Things (IoT) expands, it becomes crucial to explore the vulnerabilities and threats posed by IoT devices in the context of DDoS attacks and develop strategies for securing IoT networks. Additionally, researchers can delve into the impact of quantum computing on DDoS defence, investigating post-quantum cryptography techniques to secure network communication. A comprehensive study of the global DDoS threat landscape, region-specific trends, and emerging patterns can inform adaptive mitigation strategies. Research into human factors, cybersecurity awareness, and incident response training is essential, as is the improvement of DDoS attack attribution methods. Finally, exploring the economic and policy implications of DDoS attacks and collaborative defence strategies that involve information sharing among organizations and service providers are also promising areas for future inquiry.

7. Conclusion

Nowadays many issues cannot be easily addressed. The number of botnets is growing daily, and these threats consist of devices that people did not even expect to cause issues, meaning IoT is the biggest problem. In this paper, we have used a DDoS-for-hire service, consisting of dedicated servers just for that purpose and a small number of infected devices. DDoS-for-hire services are powerful these days and can be made with minimal effort and investment, thanks to revealing methods and a constantly growing community on public repositories, such as GitHub, along with so-called "underground" forums, which can be accessed from a single Google search.

We have selected three independent "DDoS-for-hire" services, popular as "stressors", which can be found on the same resources, meaning that they are practically publicly advertised with one "security" measure: they can only be paid via cryptocurrencies, as they consider it the safest way to hide their identity. These tools were selected by "word of mouth", as sellers often scam people, selling services as the "most powerful" which in practice cannot reach the target with the required strength. The primary goal of this work was to carry out a stress test on several local hosting companies. All of them had subscriptions to scrubbing services, claiming that they could filter the most known attack types.

Testing results:

- None of the providers passed all tests
- One provider failed to mitigate Layer 3 and 4 attacks, even with the scrubber enabled
- Layer 7 attacks were not filtered in most tests

In the analysis, the author did not have full insight into the other side of the test (incoming bandwidth, incoming requests). The only true result that can be seen and is realistic for a reader is load time from the browser, tested from two independent carriers (home line subscription and mobile carrier).
Further, after testing, several mitigation techniques were analysed to find a nearly fully working solution. All of them have a single point of failure, except CloudFlare at this point, as it combines all possible mitigation methods with proprietary technologies (like Under Attack Mode).

Overall, these results are unexpected, as the services they are using are known worldwide, without any recently reported issues. In the appendices, several recommendations should be tried to address these issues. On the other side, there were no recent attacks on hosting providers in Croatia with significant impacts, but this work showed that almost all of them should consider changing their approaches to defending against DDoS attacks. Unfortunately, in most cases, prevention techniques are not cost-effective, especially for the Croatian market. Small businesses will still use these providers as they are placed locally and the simplicity of using these has more advantages over security, but the market is unfortunately not ready for bigger-scale projects and mission-critical web pages, which are already behind CDN providers like CloudFlare or Akamai.

8. References

[1] Osterweil, E.; Stavrou, A. & Zhang, L. (2020). 21 Years of Distributed Denial-of Service: Current State of Affairs, Computer (Long. Beach. Calif)., vol. 53, no. 7, pp. 88–92.
[2] Thomas, D. R. ;Clayton, R. & Beresford, A. R. (2017). 1000 days of UDP amplification DDoS attacks, 2017 APWG Symposium on Electronic Crime Research (eCrime), pp. 79–84.
[3] Rudman, L. & Irwin, B. (2015). Characterization and analysis of NTP amplification based DDoS attacks, 2015 Information Security for South Africa (ISSA), pp. 1–5.
[4] Gupta, V.; Kochar, A.; Saharan, S. & Kulshrestha, R. (2019). DNS Amplification Based DDoS Attacks in SDN Environment: Detection and Mitigation, 2019 IEEE 4th International Conference on Computer and Communication Systems (ICCCS), pp. 473–478.
[5] Douligeris, C. & Mitrokotsa, A. (2003) 'DDoS attacks and defense mechanisms: a classification', Proceedings of the 3rd IEEE International Symposium on Signal Processing and Information Technology (IEEE Cat. No.03EX795), pp. 190–193.
[6] Bijalwan, A.; Wazid, M.; Pilli, E. S. & Joshi, R.C. (2015). Forensics of Random-UDP Flooding Attacks, J. Networks, vol. 10.
[7] https://www.cloudflare.com/learning/ddos/dns-amplification-ddos-attack (2022). DNS amplification DDoS attack | Cloudflare. Accessed: 12-Feb-2023.
[8] Yuan, D. & Zhong, J. (2008). A lab implementation of SYN flood attack and defense, Proceedings of the 9th ACM SIGITE conference on Information technology education, pp. 57–58.
[9] Shorey, T.; Subbaiah, D.; Goyal, A.; Sakxena, A. & Mishra, A. K. (2018). Performance Comparison and Analysis of Slowloris, GoldenEye and Xerxes DDoS Attack Tools', 2018 International Conference on Advances in Computing, Communications and Informatics (ICACCI), pp. 318–322.
[10] Calvert, C.; Kemp, C.; Khoshgoftaar, T.M. & Najafabadi, M.M. (2019). Detecting slow http post dos attacks using netflow features, Proc. 32nd Int. Florida Artif. Intell. Res. Soc. Conf. FLAIRS 2019, pp. 387–390.
[11] Tayama, S. & Tanaka, H. (2017). Analysis of Effectiveness of Slow Read DoS Attack and Influence of Communication Environment, 2017 Fifth International Symposium on Computing and Networking (CANDAR), pp. 510–515.
[12] https://www.cloudflare.com/learning/ddos/ddos-attack-tools/slowloris (2022). Slowloris DDoS attack | Cloudflare, Accessed: 12-Feb-2023.
[13] Haria, S. (2019). The growth of the hide and seek botnet', Netw. Secur., vol. 2019, no. 3, pp. 14–17.
[14] Blankenship, J. (2017). The Forrester Wave: DDoS Mitigation Solutions, p. 16.
[15] Kushwah, G.S. & Ali, S.T. (2017). Detecting DDoS attacks in cloud computing using ANN and black hole optimization, 2017 2nd International Conference on Telecommunication and Networks (TEL-NET), pp. 1–5.
[16] Cruz, J.E.C. & Goyzueta, I.C.A.R. (2017). Design of a high availability system with HAProxy and domain name service for web services, 2017 IEEE XXIV International Conference on Electronics, Electrical Engineering and Computing (INTERCON), pp. 1–4.
[17] Cvitić, I.; Vujic, M. & Husnjak, S. (2016). Classification of Security Risks in the IoT Environment, 26th DAAAM International Symposium, Vienna, Austria, pp. 0731–0740.
[18] Kafol, C. & Bregar, A. (2017). Cyber Security – Building a Sustainable Protection, DAAAM International Scientific Book pp. 081–090.
[19] Matvej, E.; Moric, Z. & Papic, S. (2020). Croatian Bank Security Analysis by Publicly Available Data, Proceedings of the 31st International DAAAM Symposium, pp. 0184–0188.

DOI: 10.2507/34th.daaam.proceedings.014

COMPARATIVE ANALYSIS OF IBM QRADAR AND WAZUH FOR SECURITY INFORMATION AND EVENT MANAGEMENT

Dario Šuškalo, Zlatan Morić, Jasmin Redžepagić & Damir Regvart

This Publication has to be referred as: Suskalo, D[ario]; Moric, Z[latan]; Redzepagic, J[asmin] & Regvart, D[amir] (2023). Comparative Analysis of IBM Qradar and Wazuh for Security Information and Event Management, Proceedings of the 34th DAAAM International Symposium, pp.0096-0102, B. Katalinic (Ed.), Published by DAAAM International, ISBN 978-3-902734-41-9, ISSN 1726-9679, Vienna, Austria
DOI: 10.2507/34th.daaam.proceedings.014

Abstract

The objective of this paper is to conduct a comparative analysis between two prominent SIEM tools: the commercial IBM QRadar SIEM and Wazuh, an open-source security solution. The chosen focal point was to assess the extent to which these tools can meet the security requirements within a typical IT infrastructure. The selection of this paper aimed to evaluate the efficacy of these tools in addressing the security needs prevalent in contemporary IT setups. The practical phase of the research was built around carefully curated scenarios that mirror the daily security incidents encountered by businesses. This deliberate choice was underpinned by the significance of understanding how QRadar and Wazuh interpret and respond to such events.

Keywords: SIEM; QRadar; Wazuh; event management.

1. Introduction

Systems for managing security information and events in the information system, or SIEM (Security Information and Event Management) systems are created to serve as a central place where all devices in the computer system forward their log records. SIEM systems [1][2] enable real-time monitoring, analysis and correlation of security events in the IT system, thus giving computer security experts better visibility into what was really happening within the system during and after a security incident.

This paper describes the working principle of the SIEM system in general, and then analyzes two different solutions - the commercial IBM QRadar system [3][4] and the open source Wazuh system [5] [6]. In the realm of commercial software, the manufacturer's access to the solution's underlying code is typically restricted to the development engineers who created the system. This proprietary nature renders the solution the intellectual property of the company, and end-users are required to procure licenses for utilization. Conversely, open-source solutions embody a contrasting ethos. They are freely accessible to all interested users, as the developers who conceptualized such solutions opt to share the source code for examination, modification, enhancement, and even redistribution. The main point of this analysis lies in the meticulous evaluation of both products, discerning the respective merits and drawbacks inherent in each solution.

The investigation seeks to ascertain whether these systems hold the capacity to effectively cater to the requisites of a conventional IT environment. Beyond this, the research scrutinizes whether one solution exhibits functionalities absent in the other, and whether these specific attributes carry critical implications for bolstering system security. At the heart of this examination is a paramount consideration: the financial rationale behind investing in QRadar.

In essence, this study operates as a discerning compass navigating through the intricate landscape of SIEM solutions. By assessing the tangible strengths and limitations of IBM QRadar and Wazuh, it aims to unravel not only the viability of these systems within a typical IT framework but also the delicate equilibrium between investment and returns in the context of QRadar. Through this exploration, a more comprehensive understanding of these solutions and their applicability in diverse scenarios emerges, aiding decision-makers and IT professionals in shaping their cybersecurity strategies.

2. Background and Theoretical Framework

SIEM systems have emerged as indispensable components of modern cybersecurity strategies. These systems play a pivotal role in aggregating, correlating, and analyzing vast volumes of security data generated by various network and system components. Their overarching goal is to provide organizations with the ability to identify and respond to potential threats in real-time, enhancing the overall cybersecurity approach. A SIEM system combines security information management (SIM) and security event management (SEM) functions. SIM involves the collection, normalization, and storage of security-related data, while SEM revolves around real-time event correlation and threat detection. Companies often have problems with the security of their IT systems because they do not have complete visibility of what is happening within their network.

This happens for several reasons:

- There is no system for detecting anomalies or abnormal activities;
- Lack of insight into events on end users computers;
- Too many different tracking tools that are not integrated;
- Oversaturation with the amount of log entries;
- Higher cost of security management and maintenance;
- Inefficient enforcement of compliance policies within the system;
- Lack of resources or employee skills to interpret security events.

2.1. QRadar: Commercial model

QRadar is a robust and commercially available SIEM solution developed by IBM. The solution offers advanced threat detection and incident response capabilities, utilizing machine learning and behavioural analytics to identify deviations from normal behaviour. The solution's commercial nature entails proprietary software and a licensing model, allowing IBM to tightly control access to its source code. The system has a modular architecture, so that it would enable the visibility of the IT infrastructure, and it is scalable to meet the needs of the end users.

The components of QRadar include:

- Event and Flow Processors - responsible for collecting raw event and flow data from various sources, including network devices, servers, applications, and endpoints
- Data Nodes - store the processed and normalized data, usually distributed across the network to ensure high availability and reduced latency
- Event and Flow Collectors - collects the events from intermediate devices via SPAN (Switchport Analyzer) port to Event and Flow Processors
- QRadar Console - user interface for management and event monitoring
- QRadar Risk and vulnerability Manager - assistant for visualization of network topology and identifying vulnerabilities and misconfigurations.
- App Framework- allows integration of third-party applications and extensions.

Its distributed nature, advanced analytics, and customizable features make it a powerful solution for organizations seeking robust SIEM capabilities.

2.2 Wazuh: Open-Source Model

Wazuh represents an open-source approach to SIEM. It is built on the Elastic Stack and is tailored for extensibility and customization. Wazuh not only gathers and analyses security data but also provides intrusion detection, vulnerability detection, and compliance monitoring. The open-source nature of Wazuh encourages community participation, enabling collaborative development, and transparent code accessibility.

Some of the key advantages of the system are:

- Security analytics – the system collects, indexes and analyses security data that can be used to detect threats, unusual behaviour and intrusions into the organization's IT infrastructure.
- Intrusion Detection – The Wazuh system uses agents on the systems it monitors to detect malware and suspicious anomalies. The agent recognizes hidden files, processes, unregistered network connections, as well as unusual responses to system calls. The server component of the Wazuh system uses signature-based detection to detect intrusions into network nodes by analysing the collected log records.
- Analytics of log records – agents read the log records of operating systems and applications installed on a network node and forward them to a central manager where the records are analysed and stored. The system, using defined rules, enables system managers to detect problems more easily in applications or configurations, attempts to launch harmful operations, and violations of security policies.
- File integrity monitoring – The Wazuh system monitors the file system on a network node. Tracks changes to file permissions and ownership, as well as the users who made those changes. It is possible to combine file integrity data with threat data to identify compromised network nodes and security vulnerabilities.
- Vulnerability detection – Wazuh agents collect information about the software on the monitored nodes and send it to the Wazuh server where a comparison is made with the database of known CVE (Common Vulnerabilities and Exposures) vulnerabilities in order to identify vulnerabilities.
- Configuration assessment - by monitoring the configuration of systems and applications, compliance with defined security policies, rules or standards is ensured. Agents periodically scan monitored nodes for security risks within applications. If the Wazuh system detects deficiencies in the configuration, it sends recommendations on how to correct them.
- Incident Response – Wazuh provides proactive actions that need to be taken to combat detected threats, such as blocking access to threats. The system can also be used for computer forensics and incident response by remotely executing queries or commands and identifying indicators of compromise – IOCs.
- Regulatory compliance – the system provides measures to meet regulations, industry requirements and laws. Also, scalability and cross-platform support ensure compliance standards are met for organizations. Financial institutions and payment processing companies rely on Wazuh to comply with the regulations defined in the PCI DSS (abbr. Payment Card Industry Data Security Standard) standard, as well as additional supported regulations, such as GDPR (abbr. General Data Protection Regulation), NIST 800-53 (abbr. Security and Privacy Controls for Information Systems and Organizations) and HIPAA (abbr. The Health Insurance Portability and Accountability Act).
- Security monitoring of cloud services - using integration modules, security data from cloud services, such as Amazon AWS, Microsoft Azure or Google Cloud, is collected at the API level. It evaluates the system configuration, detects flaws and offers suggestions for improvement.
- Container security – the system provides visibility at the server level, as well as at the level of individual containers. Support for integration with Docker containers enables monitoring of network and storage configuration, as well as of the Docker container image itself.

3. Tools scenarios and analysis

In the experimental part of the work, a test environment consisting of several different virtual computers, as well as a firewall connecting the computers to the network, was set up. QRadar and Wazuh SIEM systems are installed in the working environment, inside LAN part of the network, which collect log records from all virtual devices, and thus monitor security events in the network. PC with Kali Linux OS presents the computer of a malicious user who is trying to hack into the system as seen in figure 1. The CI/CD toolset [7] was used inside test scenario.

Thorough documentation of IT processes is essential [8], as it serves as a critical asset in forensic investigations. These procedures play a crucial role in maintaining the investigation's integrity by providing a clear and unambiguous record of the steps taken during the investigative process. Documentation ensures that all actions, from the initial detection of an incident to the collection of digital evidence, are transparent, accountable, and in compliance with established protocols and legal requirements. Moreover, it aids in preserving the chain of custody, allowing investigators to demonstrate that evidence has not been tampered with or compromised. In addition to its significance in legal proceedings, comprehensive documentation is invaluable for knowledge transfer within an organization, enabling staff to learn from past incidents and continuously improve their incident response capabilities. Therefore, a well-documented IT process not only strengthens the credibility of forensic investigations but also contributes to the overall resilience and security posture of an organization.

3.1. Test-case Scenarios

Several scenarios were selected and the behaviour of both SIEM systems was observed, whether security events and threats are interpreted in accordance with the expected results [9].

The following scenarios were selected:

a) User activity
- User login to the system
- Attempt to change the password
- Changes to user accounts

b) Attacks with successive attempts (*brute-force*) on the remote access service
c) Attacks with successive attempts on the Outlook Web Access service
d) Apache web service scanner
e) Malicious attempts to access the network firewall

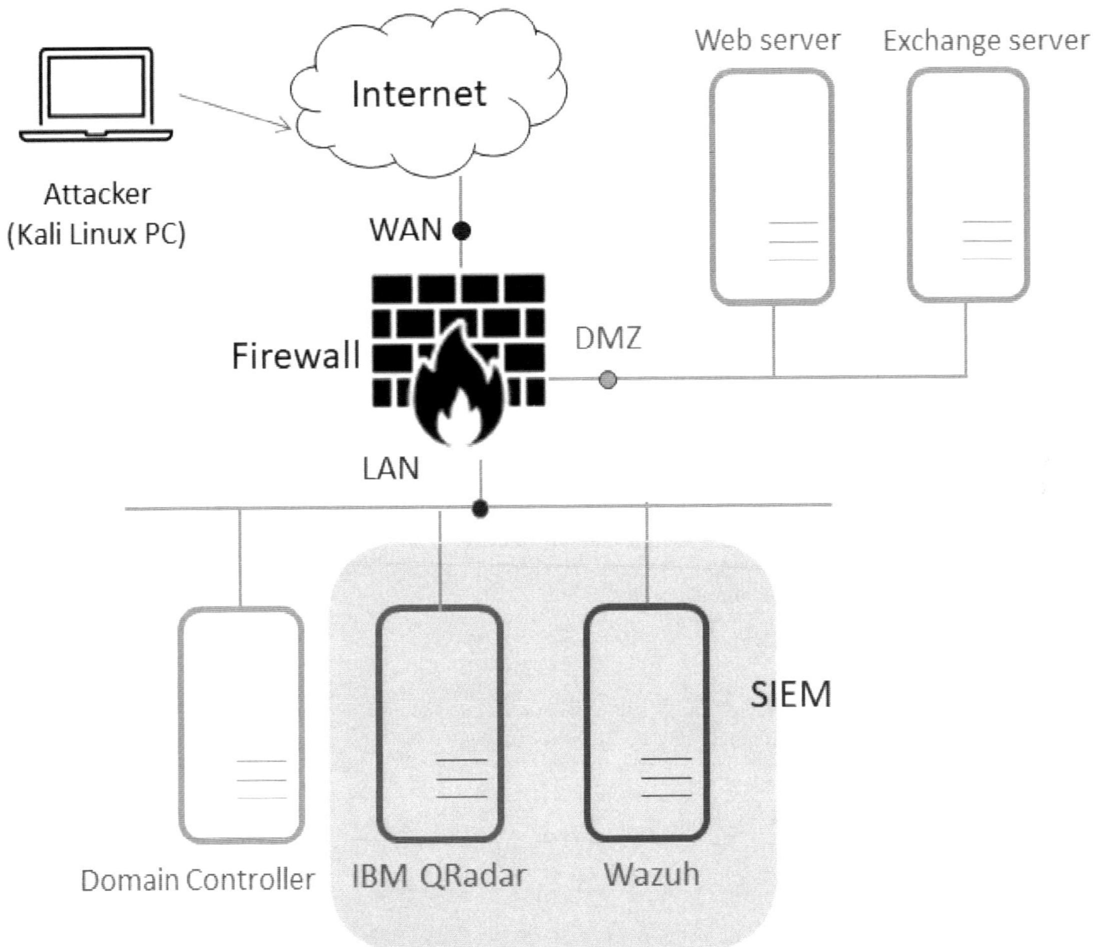

Fig 1. Test scenario topology

3.2 Comparative Analysis

The methodological approach enables to identify which solution excels in specific areas but also discern the nuanced differences that might be vital for organizations with unique security needs [10] [11]. Furthermore, it fosters an objective and evidence-based decision-making process, empowering organizations to select the SIEM solution that aligns most closely with their security goals and operational requirements. In a rapidly evolving cybersecurity landscape, where threats continually evolve, a rigorous comparative analysis is an indispensable tool for organizations striving to safeguard their digital assets effectively. The results of the test scenario are shown in table1.

Scalability is a hallmark of IBM QRadar. It is well-suited for large enterprises dealing with substantial data volumes. The architecture supports distribution, making it an optimal choice for organizations with complex and expansive IT environments. Wazuh also offers scalability, but it may necessitate more manual configuration in comparison to QRadar. This flexibility makes it adaptable to a range of organization sizes, from small to medium enterprises to larger institutions. While feature rich, QRadar can present a steeper learning curve. Efficient utilization often requires training and familiarity with its comprehensive interface. Wazuh distinguishes itself with a more straightforward setup and user interface. This simplicity can be attractive to organizations seeking rapid deployment and user-friendly management.

Scenario	QRadar analysis	Wazuh analysis	Conclusions
User Activity	The system has successfully detected: • Successful login • Unsuccessful login • Password change • Changes over user accounts		Using the machine learning component, QRadar intelligently groups individual events into chained violations. For example, if a user commits multiple unsuccessful reports, then makes a successful login, the system links these events to a violation that a system administrator can investigate. Wazuh does not have a machine learning component but is capable of detecting multiple low-level events and generating a higher-level event.
Attacks with successive attempts on remote access service	The system successfully detected successive unsuccessful attempts to log in to an SSL VPN service, and generated chained violations.	The system has successfully detected successive unsuccessful attempts to sign in to an SSL VPN service.	QRadar generated a chained foul. Wazuh generated an event of a significant level, but it was necessary to search the diary entries by key fields.
Attack on Outlook Web Access service	The system successfully detected an attack on the OWA service, created a chained violation from three different types of journal entries	The system successfully detected an attack on the OWA service, generating individual Level 5 events, and Group Events of Level 10.	QRadar has generated a chained breach that shows that the event is significant and requires further investigation. group events, which can bring attention to system managers for investigation.
Apache web service scanner	The system detected the scanner, however it did not create a violation, because the scanner itself tries to detect if there are certain files on the web server.	The system detected the scanner and created individual events, as well as group events of a more significant level.	QRadar did not create a violation, so this event would go unnoticed. Wazuh created a group event of a significant level, so such a security incident would attract the attention of security experts.
Malicious attempts to access a network firewall	The system detected many unsuccessful attempts to access the Fortigate system through the SSH protocol, and created multiple chain violations that would be immediately spotted and investigated by system managers.	The system detected a large amount of individual unsuccessful access attempts and assigned Level 5.	QRadar created multiple chained violations, and system managers would immediately notice unusual activity. Wazuh also recorded a large number of events, but given the events are not grouped, there is a chance that the attack would go unnoticed if it did not last for a long period of time.

Table 1. Comparative Analysis of IBM QRadar and Wazuh

IBM QRadar receives updates to DSMs, protocols, and decoders on a daily base, as IT equipment manufacturers are constantly developing new hardware and software. IBM also has an IBM Security App Exchange store with extensions that are mostly free for users who own a QRadar license.

The applications placed in this store are developed in cooperation with other manufacturers of security equipment, so they are compatible with QRadar and do not require complex installation. The following example is a real-life example: Cisco released Firepower Management Center (abbr. FMC) version 7.0 earlier this year. The commercial IBM QRadar system in use had the Cisco eStreamer extension installed that could read and decode network streams from Cisco FMC up to version 6.5. As the new version was released, at one point the FMC was updated from version 6.5 to 7.0 and after that it was noticed that the eStreamer function does not work, that is, that QRadar no longer receives records of network streams from the FMC. After spending some time researching, the problem was reported to IBM support. It was reported back that they currently do not support eStreamer for FMC 7.0, but that they will soon rework the extension to support it. Functionality was restored within a month. This kind of problem would not be solved in such a short time by the Wazuh developers for the simple reason that it is an open-source system.

Utilizing Wazuh as a SIEM solution offers significant advantages. Wazuh benefits from a passionate and community-driven development team, driven by their enthusiasm and a commitment to contributing to the broader community. A practical example that highlights this communal spirit pertains to the redirection of network logs from a Fortigate network device to Wazuh. During this process, it became evident that the built-in decoders within Wazuh were outdated, causing inaccuracies in log interpretation. Fortunately, the open-source ethos of Wazuh led to a solution. A diligent online search unearthed a decoder crafted by an altruistic individual who voluntarily shared the code with the wider community, without any expectation of payment or reservation. Their sole motivation was to aid the advancement of the Wazuh tool. Another real-world scenario involved redirecting VPN connection logs from virtual Cisco ASAs (network firewalls) to a production Wazuh SIEM. The integrated decoder in this case proved inadequate in interpreting log records from these virtual ASA devices, misinterpreting them as logs from much older Cisco PIX firewalls that have long been out of production. Again, the power of online collaboration came to the rescue as a suitable decoder was sourced from the internet, adept at accurately interpreting records from these virtual network devices, effectively resolving the issue.

4. Conclusion

Both systems successfully cope with the detection of security incidents presented through the previous scenarios. QRadar and Wazuh can be installed as single and distributed appliances and can cover multiple geographic locations at once. They have similar additional functionalities that can expand their role as security tools. If a company has a complex IT infrastructure, covers dozens of locations, has a large budget for procuring security solutions and resources for setting up multiple hardware-demanding devices, often acquires new IT equipment and needs timely support, it should choose IBM QRadar as its SIEM solution. If the company is smaller, has a limited budget and resources, and relies on the open source community to use or even develop its own solutions, then it is definitely recommended to implement Wazuh SIEM as a security solution. And there is always a compromise solution: using an IBM QRadar system with some basic license that can cover critical infrastructure that needs timely support, while for less essential systems, Wazuh or another open source SIEM can be used.

- IBM QRadar excels in terms of comprehensive security features and scalability, making it suitable for large enterprises with robust security requirements.
- Wazuh offers a cost-effective open-source solution with a user-friendly interface, making it accessible to a wide range of organizations.

The choice between IBM QRadar and Wazuh hinges on an organization's specific security needs, budget considerations, and the level of customization and support required [12]. This comparative analysis aims to provide a foundation for informed decision-making in selecting the most appropriate SIEM solution.

In future works, other SIEM solutions, whether they are commercial or community-driven, can also be used for comparison using the user scenarios and methodology described in this study. This approach can open the possibility of creating a broader comparative framework, allowing organizations to evaluate a more extensive array of SIEM options available in the market. By applying the same user scenarios and rigorous methodology, organizations can make more informed decisions when selecting the SIEM solution that best aligns with their unique security requirements and constraints. Furthermore, this expansion of the comparative analysis can contribute to the ongoing advancement of the field of cybersecurity, fostering a deeper understanding of how various SIEM solutions perform in real-world scenarios and promoting the development of more effective and resilient security strategies.

5. References

[1] Gustavo, G.G.; González-Zarzosa, S. & Diaz, R. (2021). Security Information and Event Management (SIEM): Analysis, Trends, and Usage in Critical Infrastructures, Sensors 21, no. 14: 4759. https://doi.org/10.3390/s21144759

[2] Miller, D.R.; Harris, S.; Harper, A.; VanDyke, S. & Blask, C. (2010). Security information and event management (SIEM) implementation. McGraw Hill Professional., Default Book Series. October 2010, DOI: 10.1036/9780071701082

[3] https://www.IBM.com/community/qradar/wp-content/uploads/sites/5/2020/11/b_qradar_community_edition_7.3.3GA_v1.0.pdf (2021). IBM QRadar Community Edition 7.3.3, Access: 2023-03.

[4] https://www.IBM.com/docs/en/qsip/7.3.3?topic=SS42VS_7.3.3/com.IBM.qradar.doc/c_qradar_pdfs.html (2021). IBM QRadar SIEM V7.3.3 documentation, Access: 2023-03

[5] https://documentation.wazuh.com/current/getting-started/components/wazuh-agent.html (2021). Wazuh agent - Components · Wazuh documentation, Access: 2023-05

[6] https://documentation.wazuh.com/current/getting-started/architecture.html (2022). Architecture - Getting started with Wazuh · Wazuh documentation, Access: 2023-05

[7] Dakić V.; Redžepagić J.; Bašić M. (2022). CI/CD Toolset Security, 33rd DAAAM International Symposium on Intelligent Manufacturing and Automation, Published by DAAAM International, ISBN 978-3-902734-36-5, ISSN 1726-9679, Vienna, Austria, DOI: 10.2507/33rd.daaam.proceedings.022

[8] Moric, Z.; Redzepagic, J. & Gatti, F. (2021). Enterprise Tools for Data Forensics. Annals of DAAAM & Proceedings, 10(2), Published by DAAAM International, ISBN 978-3-902734-33-4, ISSN 1726-9679, Vienna, Austria, DOI: 10.2507/32nd.daaam.proceedings.014

[9] Parn, E.A. & Edwards, D. (2019). Cyber threats confronting the digital built environment: Common data environment vulnerabilities and block chain deterrence, Engineering, Construction and Architectural Management, Vol. 26 No. 2, pp. 245-266., DOI: https://doi.org/10.1108/ECAM-03-2018-0101

[10] Ghafir, I.; Prenosil, V.; Svoboda, J.; & Hammoudeh, M. (2016). A survey on network security monitoring systems. In 2016 IEEE 4th International Conference on Future Internet of Things and Cloud Workshops (FiCloudW) (pp. 77-82). IEEE.

[11] Ndatinya, V.; Xiao, Z.; Manepalli, V. R.; Meng, K. & Xiao, Y. (2015). Network forensics analysis using Wireshark. International Journal of Security and Networks, 10(2), 91-106.

[12] Demchenko Y.; Turkmen F.; Slawik M. & Laat C.D. (2017). Defining Intercloud Security Framework and Architecture Components for Multi-cloud Data Intensive Applications, 2017 17th IEEE/ACM International Symposium on Cluster, Cloud and Grid Computing (CCGRID), Madrid, Spain, 2017, pp. 945-952, DOI: 10.1109/CCGRID.2017.144.

DOI: 10.2507/34th.daaam.proceedings.015

DETECTING SECURITY VULNERABILITIES ON INTERNET-CONNECTED DEVICES

Ema Lovric, Zlatan Moric, Jasmin Redzepagic & Damir Regvart

This Publication has to be referred as: Lovric, E[ma]; Moric, Z[latan]; Redzepagic, J[asmin] & Regvart, D[amir] (2023). Detecting Security Vulnerabilities on Internet-connected Devices, Proceedings of the 34th DAAAM International Symposium, pp.0103-0110, B. Katalinic (Ed.), Published by DAAAM International, ISBN 978-3-902734-41-9, ISSN 1726-9679, Vienna, Austria
DOI: 10.2507/34th.daaam.proceedings.015

Abstract

With the development of technology and the growth of the number of devices connected to the Internet, cyber security is decreasing, and it has become a real challenge to secure systems and protect personal data. OSINT intelligence collects data from all publicly available and open-source databases, which makes enormous amounts of data available to everyone, including cybersecurity experts and attackers. Using the OSINT tool Shodan, this paper shows the vulnerabilities of certain European Union countries' protocols, devices, and services. Except for the analysis, the paper compares the obtained results to show how much certain countries invest in cyber security. This paper aims to increase awareness about cyber security flaws so that users can detect and protect them.

Keywords: Shodan, security investment; security problems; vulnerability scan

1. Introduction

One of the biggest challenges of today's time, accompanied by the rapid development of technology, is to protect private information and secure it against data theft. Users leave traces in the form of confidential data by using the Internet daily for any purpose. The amount of personal data on the Internet is enormous and entails specific security threats. When accessing certain websites and using social networks, blogs, and forums, data on the Internet is remembered and linked to individual persons, organizations, and devices. Accordingly, anyone can find much information about each individual or organization using a particular type of intelligence – OSINT (open source intelligence).

OSINT is a type of intelligence that collects information from publicly available data sources. With the growth of the number of users on the Internet and the amount of information available, several tools have developed that automatically collect information from various sources and make connections between individual entities. One of the OSINT tools is Shodan [1], which scans the IP addresses of devices and all available data by searching devices on the Internet. This paper presents the concept of OSINT and the exposure of information, i.e., the vulnerability of individual protocols, types of devices, and services in the Countries of the European Union. By analyzing the data, a comparison of the obtained results was made, and security recommendations for individual vulnerabilities were proposed. The paper aims to highlight how much respective countries invest in the security of devices exposed to the Internet and draw attention to possible shortcomings.

2. What is OSINT

OSINT is a type of intelligence that collects information and data from free and publicly available data sources. The term OSINT appeared in the 1980s and referred to conventional ways of gathering information such as newspapers, television, radio, or business reports. Information is collected from publicly available and open-source data sources, such as blogs, forums, or social networks. With the development of social networks and their popularization, an intelligence oriented explicitly to social networks – SOCINT began to develop.

With the development of the Internet and its capabilities, the way of collecting data has also developed. Information is often collected using search engines like Google, Bing, Yahoo, and others. In addition to classic search, information is collected through the deep and dark web. Deep web represents every page not indexed on web browsers, i.e., every website to which it is necessary to log in with user data. The dark web allows for secure search and implies every page whose access requires Tor protocol or any other way of protecting anonymity.

2.1. Tools used in OSINT

Maltego [2] is an OSINT tool for collecting data from publicly available data sources. The advantage of Maltego is that for each input, such as an e-mail address, name, IP address, organization, or website, it searches various databases to collect as much information as possible about the given information. With the information collected, the tool processes the data creates links between parts of collected data, and visually displays these links. All complex links of individual entities are presented, which significantly facilitates users to understand the information searched. The Maltego tool is one of the most commonly used tools for penetration testing in the reconnaissance phase where penetration testers seek to gather as much information as possible about the system they intend to attack.

SpiderFoot [3] is another OSINT tool that automatically collects information from various data sources. This open-source tool uses more than 100 publicly available sources to collect and connect information in one place. The advantage of this tool is the excellent documentation provided, which allows users to install the tool and run it easily and allows them to understand the tool itself, its processes, and its properties, and is available on all operating systems. SpiderFoot tool, like other OSINT tools, collects and processes data on domains, IP addresses, DNS records, networks, services, and many other information. When penetrating testing, SpiderFoot dramatically facilitates the collection of information and its analysis.

Metagoofil [4] is an OSINT information-gathering tool that collects metadata from publicly available documents, such as PDF, DOC, XLS, and PPT documents. Metagoofil is a free and open-source tool. To collect publicly available information, Metagoofil first searches available documents through Google then stores them locally and uses Hachoirh and PdfMiner tools to extract metadata from those documents. Like other tools, Metagoofil collects certain information about the default entity and associates it with specific user names, software versions, services, and devices.

3. Shodan

Shodan.io is a search engine that uses a database of all publicly available IP addresses and collects data on all devices connected to the Internet. In addition to searching the database of devices connected to the Internet, Shodan offers several additional tools that participate in better quality data processing. Shodan as a tool allows users to search for not only "normal computers" but also the Internet of Things devices, specialized devices connected to the Internet usually to serve a single purpose. When executing a Shodan search query, only the caption from the detected service is retrieved primarily, the metadata is not searched [5].

Banner grabbing is a method used to obtain information about a device, i.e. information about services, type, and version of software, and operating system, by searching for devices connected to the Internet and scanning for open ports such as HTTP (80), FTP (21) and SMB (445).

Banner grabbing uses tools to establish a connection to the device, such as telnet, nmap, wget, and curl. When a connection to the device is established, a request is sent to the device to which the remote device responds with *a banner* message. This message can then be used to gain additional info on the probed device.

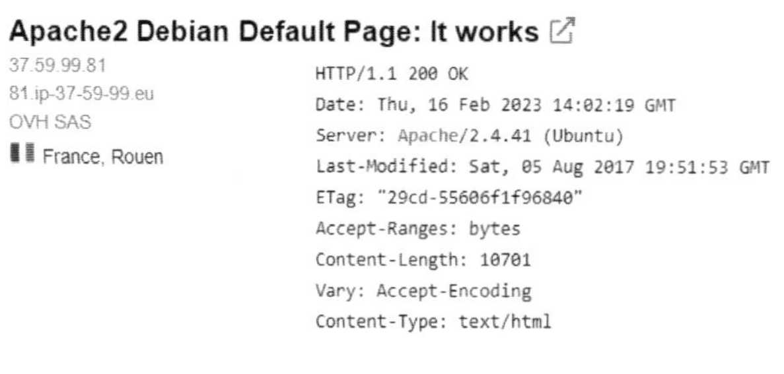

Fig. 1. Banner message

For example, a *search query* "Apache" was performed, which retrieves the results about Apache server users. In the above example in figure 1., a *banner* message is visible from which you can read information about the operating system and the latest changes on the website. To view the metadata of devices and services, the search query is not enough, but certain Shodan filters are used for this, which provide more specific search results. All filters are written as *"filter name": "value"*. The table below shows some of the most common filters, which are later combined to perform more complex searches.

Filter Name	**Description**	**example**
Country	2-letter country code	country: HR
City	Name of the city	City: Zagreb
axis	Operating system	axis: windows
Org	The organization assigned the netblock	org: google
Port	Port number for the service	port:80

Table 1. Example shodan filter

If these are values consisting of two or more words, then those words are enclosed in quotation marks: City: "Novi Vinodolski". Any filter with a label before the filter's name indicates that the specified value is ignored, that is, we want to exclude it from the search. For example, to show all IoT devices in Croatia, except those in Zagreb: Country:City: Zagreb

As noted earlier, filters can be combined. For example, if you search for devices/services in Croatia, running on HTTP or HTTPS protocols, ports 80 and 443 will be added to the filter.
Country:HR port:80,443

4. Previous research - overview

In 2017, the study "Vulnerability Scanning of IoT Devices in Jordan Using Shodan" was conducted [6]. Internet of Things and IoT devices are always connected, providing the users with a set of data or some other functionality intended for online use. The problem is that users are not aware of or ignore that these devices are exposed to the Internet and that all their information is publicly available. The work seeks to raise awareness of IoT device vulnerabilities and potential loss of privacy and security and encourage more use of said devices.

The paper brings together IoT device vulnerability research with statistical analysis in May 2017. The study revealed 40849 IoT devices, of which 9.2% have UPnP (Universal Plug and Play) enabled. By enabling UPnP, the device automatically opens ports and becomes visible to other devices in the network. By analyzing some of the most common services on the Internet, it was found that of all devices that have HTTP enabled, 62% of them have successful HTTP connections. 41% of devices using the SMB protocol have authentication disabled and 26% have anonymous logon enabled, leading to the potential to compromise shared data.

The paper states that the use of telnet protocol, which does not encrypt data, is twice as popular or more frequent than SSH, which is safer. By analyzing other services, it was concluded that only 8% of FTP connections are successful and that a screenshot was created for 35% of RDP devices. A search of the device in Jordan found 16 devices exposed to the Ticketbleed vulnerability and 41 devices exposed to the Heartbleed vulnerability.

As a result of the analysis, the authors propose following experts' recommendations, which implies avoiding vulnerable services and disabling unused services. Users are advised to update devices and their services and appropriate device adjustment and authentication.

Another paper: "Vulnerability Analysis of Internet Devices from Indonesia Based on Exposure Data in Shodan" collected information for all networks publicly available in Indonesia using the Shodan tool [7]. The paper aimed to draw the attention of organizations that manage public networks to the exposure of Internet-connected devices in Indonesia and to become aware of these devices' vulnerabilities. Since there is a large number of IP addresses in Indonesia, categorized into 1699 separate networks (each having its own ASN - Autonomous System Number), in this paper, by a special method of grouping, IP addresses are divided into four categories depending on the level of exposure.

For the largest number of networks, 1075 of them do not have any information on Shodan. The low exposure category houses 614 networks and the medium exposure level category contains 9. Only one network is in the category of high level of exposure.

In the research of the work, all networks were scanned using a bash script that automatically scans all subnets for each network. For the 624 public networks about which there is information on Shodan, 289,715 queries were made and data on ports, services, domains, IP addresses, and operating systems were scanned. Of the nearly 12,000 unique ports found, 13.95% of the information found is related to the HTTP protocol, port 80. The collected operating system information shows that 35.87% of devices use Linux 3. x, followed by Windows Server 2012 R2 Standard with a significantly smaller 9.45%.

Further analysis found 98,152 information points about services that are predominantly distributed across the following services: MikroTik bandwidth-test server (33.37%), Apache httpd(16.54%) and OpenSSH (7.78%). For domain queries, the results of two top-level domains are highlighted – out of a total of 3350, 1881 are unique .id domains and 1242 are domain .com. The study concluded with the discovery of 145,543 unique IP addresses and 790 unique organizations.

5. Analysis of protocol vulnerabilities

In this part of the paper, one of the most commonly used protocols was analyzed – FTP, RDP, SMB, Telnet, and SSL. Some of these protocols do not use encryption at all and their use is not advised [8]. In contrast, other protocols are vulnerable only to some parameters and the correct configuration can protect the entire system [9]. Unless stated otherwise, each query contained the following filter for the countries: country: at, be, bg, hr, cy, cz, dk, ee, fi, fr, fr, de, gr, hu, ie, it, lv, lt, lu, mt, nl, pl, pt, ro, sk, si, es, se.

5.1. Results

After the analysis, the state of vulnerability of individual protocols for the countries of the European Union was determined. In this part of the paper, a comparison of the obtained results is done and the relationship of individual protocols with individual vulnerabilities is presented. Also prominent are the states with the best and worst results.

5.1.1 FTP protocol

By analyzing the FTP protocol, all open ports in the European Union countries were first scanned, then a query was made for these ports that have the possibility of anonymous access. The graph shows the relationship between open ports about the vulnerability enabled by anonymous access.

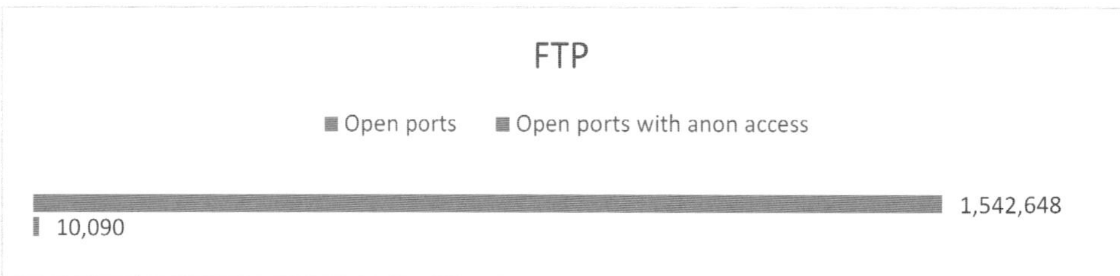

Fig. 2. FTP by port type

By querying the FTP protocol out of a total of 1,542,648 results, 35.76% of the results were found to originate from Germany. By analyzing ftp protocol vulnerabilities, i.e. analyzing ftp protocols with the possibility of anonymous access, it was found that 0.47% of scanned devices are vulnerable in Germany. The results obtained for Germany are twice as high as France's, which is placed second. The fewest scanned devices were detected in Montenegro, with 423 results, or 0.02% of the total results, and two found results of successful anonymous connection, making 0.47% of their devices vulnerable. Croatia is at the bottom of the scale with 3,663 results, of which 0.6% have the possibility of anonymous access. Given that certain countries have a larger population and thus a larger number of devices connected to the Internet is expected, the normalization of the results obtained for the two largest and two smallest results, as well as for Croatia, has been done. The following table shows the normalization of the results found about the number of Internet users to get a more faithful view of the results.

State	Number of Internet users	Open FTP ports	Percentage of open ports relative to the number of Internet users (%)	Open FTP ports with anonymous access capability	Percentage of vulnerable ports relative to the number of Internet users (%)
Germany	79,127,551	551,723	0,6972	2,608	0,0032
France	60,421,689	253,507	0,4195	1,336	0,0022
Cyprus	1,320,400	1,991	0,1508	8	0,0006
Montenegro	547,000	423	0,0773	2	0,0004
Croatia	3,787,838	3,663	0,0967	22	0,0006

Table 2. Normalization of results - FTP protocol

By analyzing the normalized results, it is realized that the number of scanned open ports, as well as the number of vulnerable ports, is still the highest in Germany.

5.1.2 RDP protocol

A search of devices that have port 3389 open for RDP protocol found 453956 results of which as many as 35.74% have a screenshot taken. A graphical representation establishes the relationship between these analyzed results.

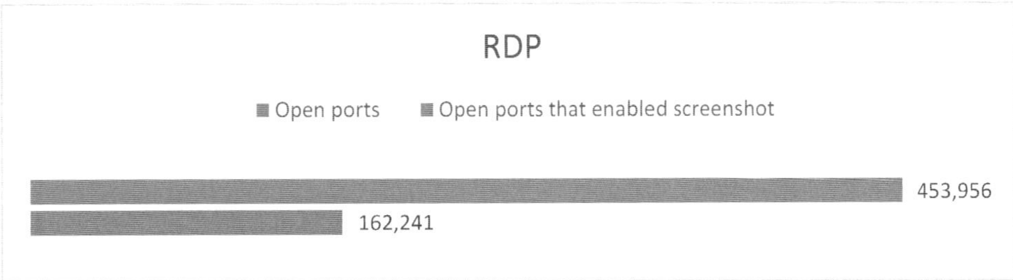

Fig. 3. RDP by ports

Further analysis of the results determined the re-lead of Germany, which contains 34.27% of the total score. In Germany, as in the FTP protocol analysis, twice as many devices were scanned as the country ranked second in the ranking, in this case, the Netherlands. The last country on the table is Montenegro with 220 scanned results, which is 0.04% of the total result. 1,856 results were found for Croatia, which makes up 0.4% of total scanned ports. The following table shows the normalization of the results found about the number of Internet users.

State	Number of Internet users	Open RDP ports	Percentage of open ports relative to the number of Internet users (%)
Germany	79,127,551	155,578	0,1966
Netherlands	16,383,879	79,507	0,4852
Cyprus	1,320,400	478	0,0362
Montenegro	547,000	220	0,0402
Croatia	3,787,838	1,856	0,0489

Table 3. Normalization of results - RDP protocol

By examining the normalized results, it is evident that Germany has a lower percentage of open RDP ports compared to the Netherlands, although in Germany twice as many open RDP ports were scanned. Table 2. also shows the same results obtained in Croatia and Montenegro.

5.1.3 SMB protocol

By analyzing the SMB protocol, queries were made for the total number of open ports, the number of ports using SMBv1, and the number of open ports with disabled authentication. The graph shows the ratio of the results of the above parameters.

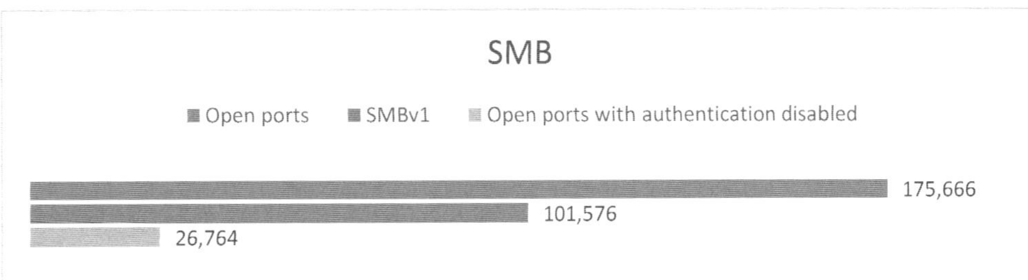

Fig. 4. SMB by protocol type

As in the previous two analyses, scanning the SMB protocol, it was found that Germany has the highest number of results, of the scanned 175 666 of which 51.85% are SMB of the first version and 7.99% of the results with untuned authentication.

Although the most overall results were found in Germany, France, which ranks second with 35,292 results, has the most results in the analysis of disabled authentication and is 63.09% unsafe due to the use of SMBv1. 13.28% of the results obtained in France have disabled authentication on SMB protocols. Montenegro is again in the last place of the rankings with 119 results, of which 50.42% is the first version and only 20.16% with authentication disabled. Croatia has 721 scanned SMB protocols, which ranks 22nd in the overall standings. Of the 721 results listed, as many as 71.56% are SMB of the first version and 33.28% with authentication disabled. The following table 3. shows the normalization of the results found about the number of Internet users.

State	Number of Internet users	Percentage of Open SMB Ports relative to Internet users (%)	Percentage of open SMB ports of the first version compared to Internet users (%)	Percentage of open SMB ports with authentication disabled relative to Internet users (%)
Germany	79,127,551	0,0667	0,0345	0,0053
France	60,421,689	0,0584	0,0368	0,0069
Luxembourg	636,565	0,0232	0,0153	0,0026
Montenegro	547,000	0,0043	0,0109	0,0044
Croatia	3,787,838	0,0190	0,0136	0,0063

Table 4. Normalization of results - SMB protocol

The normalization table noted partially more faithful results of the SMB protocol vulnerability. Compared to internet users in a particular country, Germany has the most scanned SMB ports, but in France, the number of vulnerabilities is greatest. France has the most SMBv1 vulnerabilities and the highest percentage of SMB ports with disabled authentication. Of these countries, Croatia has the highest percentage of vulnerabilities for disabled authentication, while in countries with a smaller number of Internet users, we see lower percentages of vulnerable devices.

For analysis of telnet, ports using OpenSSH were also scanned. Since it is heavily advised to use SSH protocol, the relationship between the obtained results is satisfactory. The use of OpenSSH tools is significantly more frequent than Telnet, which is evident in the following graph:

Fig. 5. Telnet and SSH – comparison by number of ports

By querying devices using Telnet (table 4.), 132,300 results were found. In the first place of the ranking of results by country is Italy, with 21.2% of the total. The last place is occupied by Luxembourg with only 97 results, or 0.07%. In Croatia, 1 556 open telnet ports contained 1.18% of total scanned ports. Given that SSH is a safe alternative to Telnet, the number of results for Telnet ports should be smaller, i.e. the expected share for individual countries should be smaller.

State	Number of Internet users	Percentage of Telnet ports open relative to Internet users (%)	Percentage of open SSH ports relative to Internet users (%)
Italy	54,798,299	0,0512	0,1765
France	60,421,689	0,0360	0,9442
Luxembourg	636,565	0,0152	0,9221
Montenegro	547,000	0,0230	0,0780
Croatia	3,787,838	0,0413	0,1156

Table 5. Normalization of results - Telnet and SSH

The normalized data for scanned Telnet and SSH ports revealed that Italy had the highest proportion of Telnet ports, followed by Croatia. Regarding SSH usage normalized by the number of internet users, France leads the pack. This notably impacts the security of devices connected to the internet nationwide.

By analyzing the SSL protocol, i.e. SSL certificates, a significant number of results were found, namely 2,445,148 expired certificates. A large proportion of these results, namely 814,544 (33.35%) are in Germany. The largest number of certificates issued by users themselves, i.e. *self-signed* certificates for the example.com domain, was also scanned in Germany, where 38.42% of such certificates were found. In Montenegro, 0.03% of expired certificates were scanned, which is a noticeably lower score than all other countries. When analyzing expired SSL certificates, Croatia positioned itself at the bottom of the scale with 0.19% of total searched certificates. For *self-signed* certificates for example.com, 110 results were recognized in Croatia, making 0.07% of the total results.

State	Number of Internet users	Expired certificates relative to Internet users (%)	Self-signed certificates for example.com about Internet users (%)
Germany	79,127,551	1,0294	0,0750
France	60,421,689	0,8141	0,0473
Cyprus	1,320,400	0,4071	0,0042
Montenegro	547,000	0,1592	0,0011
Croatia	3,787,838	0,1275	0,0029

Table 6. Normalization of results – Certificate expiration

With the normalization of the obtained results, a table was presented in which the percentage of expired certificates for Internet users in Germany is significantly higher than in other countries. Behind Germany, there is France with a slightly smaller percentage, but still large enough compared to other states. It is also evident that of the above results in table 5., Croatia has the lowest percentage of expired certificates.

6. Recommendations for protection against attacks

By analyzing device vulnerabilities by certain protocols, it was found that many devices are unprotected and can be compromised. Users do not necessarily have confidential or sensitive information, but it violates the privacy of the data. Given that these are frequently used protocols, several security recommendations for protection against attacks exist.

Earlier in the paper, it was noted that several FTP servers allow anonymous login, so the first step in security protection is to disable anonymous login, i.e. to force authentication on servers. Since the FTP protocol itself is not a secure protocol because it does not encrypt data, it is advisable to use a secure version of the protocol – FTPS (FTP Secure) or SFTP (SSH FTP) [20]. FTPS is an upgraded version of FTP that uses SSL/TLS to encrypt data, and SFTP is an upgraded FTP that runs on an SSH connection. Both protocols transmit encrypted data, but the difference is that FTPS is based on FTP and still uses TCP port 21, unlike SSH-based SFTP which uses TCP port 22. Unlike the FTPS protocol, SFTP does not offer the possibility of anonymous logins to the server.

By allowing the RDP protocol directly from the Internet, users are exposed to potential *BruteForce* attacks where attackers use automated tools to try to decipher the user account's password. The first step in protecting the RDP protocol is to set strong or complex passwords and adjust the system so that only a certain number of incorrectly entered passwords are allowed, after which the user account is locked. Users are advised to update the software regularly as updated software versions may contain patches of potential vulnerabilities. Also, when protecting the RDP protocol, it is necessary to take into account the restriction of access and connections on firewalls for RDP port 3389. The analysis shows that some devices do not have SMB authentication set. As a first security recommendation, authentication is necessary to secure the device. Furthermore, the use of the SMB protocol of the first version is not recommended due to certain security vulnerabilities that are listed earlier in the paper, but it is recommended to use the latest SMB version.

All traffic used by Telnet is not encrypted and not secure, therefore it is advised to avoid using Telentin in all cases. As an alternative to Telnet, a good practice is using the SSH protocol because all traffic, including login information, is encrypted. OpenSSH is a tool that allows users to connect to a remote computer with data encryption and secure transfer between devices.

The SSL protocol is considered outdated, and security experts advise using the TLS protocol, an upgraded version of SSL that ensures secure access to web servers. Given that the work has done an analysis of expired certificates and that a significant number of results have been found, users need to be aware of the importance of using the correct certificates and their renewal. Setting up automatic certificate renewal will avoid some of the vulnerabilities of SSL certificates. The use of certificates created and issued by users themselves, i.e. self-signed certificates, can endanger not only individual users, but also the entire system, so it is advised to avoid using them. When creating a website, users can use free SSL certificates to confirm their identity. If self-signed certificates are used, it is a necessary step to ensure that the attacker cannot reach the root certificate, otherwise, all the authorities to which he issued the certificate would be endangered[10] [11].

7. Conclusion

The development of technologies and the increase in the number of devices connected to the Internet have increased their security exposure. Most private users leave devices connected to the Internet with predefined configuration and predefined access data, which poses significant security threats and challenges against them. Especially when it comes to devices from the Category of Internet of Things whose purpose is to facilitate the daily life of users. but for this to be so, it is necessary to be aware that all these devices are publicly available and that predefined settings cannot be a satisfactory solution.

In parallel with the development of technologies, intelligence has been developed whose primary purpose is to collect data on all available and public devices on the Internet – OSINT. Although OSINT is often used with negative intentions, it is necessary to take advantage of this intelligence's benefits to find vulnerabilities and ensure. One of the OSINT tools useful in searching all devices connected to the Internet is Shodan, a browser that can find devices such as desktop computers to nuclear power plants. The paper presents a Shodan search of all European countries according to FTP, RDP, SMB, Telnet, and SSL protocols and analyzes the results obtained. Given that the number of found results depends on the number of inhabitants of a particular country, i.e. the number of Internet users, the paper normalized prominent data.

Potential future research directions should include development of more comprehensive and user-friendly tools for scanning and auditing of the devices connected to the internet. These tools could extend their scope beyond the protocols discussed in the text to encompass a broader array of services and devices. Exploring the utilization of machine learning and artificial intelligence for the analysis of the extensive data gathered through OSINT tools like Shodan holds significant promise as a research work. Such an approach has the potential to facilitate automated detection of vulnerabilities and mitigation of threats.

8. References

[1] Matherly, J. (2018). The Complete Guide To Shodan: Collect. Analyze, Amazon, ASIN: B01CDIU880, Amazon.com Services LLC

[2] https://www.maltego.com/about-us/, (2023). Maltego Technologies website, Accessed on: 2023-09-05

[3] https://www.spiderfoot.net/, (2023). Spiderfoot subsite, part of Intel471. Accessed on: 2023-09-05

[4] https://github.com/opsdisk/metagoofil, (2023). Kali Linux Fork of Metagoofil Tool Tepository. Accessed on: 2023-09-05

[5] Albataineh, A. & Alsmadi, I. (2019). IoT and the Risk of Internet Exposure: Risk Assessment Using Shodan Queries, 2019 IEEE 20th International Symposium on A World of Wireless, Mobile and Multimedia Networks (WoWMoM), Washington, DC, USA, pp. 1-5, ISBN:978-1-7281-0271-9, DOI: 10.1109/WoWMoM.2019.8792986

[6] Al-Alami, H.; Hadi, A. & Al-Bahadili, H. (2017). Vulnerability Scanning of IoT Devices in Jordan using Shodan, 2017 2nd International Conference on the Applications of Information Technology in Developing Renewable Energy Processes & Systems (IT-DREPS), ISBN: 978-1-5386-1987-2, DOI: 10.1109/IT-DREPS.2017.8277814

[7] Novianto B.; Suryanto, Y. & Ramli, K. (2020). Vulnerability Analysis of Internet Devices from Indonesia Based on Exposure Data in Shodan. Proceedings of International Conference on Science, Technology, Engineering and Industrial Revolution (ICSTEIR 2020), Vol 1115, ISSN 1757-8981, DOI: 10.1088/1757-899X/1115/1/012045

[8] Dakic, V.; Jakobovic, K. & Zgrablic, L (2022). Linux Security in Physical, Virtual, and Cloud Environments, Proceedings of the 33rd DAAAM International Symposium, pp.0151-0160, B. Katalinic (Ed.), Published by DAAAM International, Vol. 33, No.1, ISBN 978-3-902734-36-5, ISSN 1726-9679, DOI: 10.2507/33rd.daaam.proceedings.021, Vienna, Austria

[9] Perakovic, D.; Husnjak, S. & Remenar, V. (2012). Research Of Security Threats In The Use Of Modern Terminal Devices, Annals of DAAAM for 2012 & Proceedings of the 23rd International DAAAM Symposium, Volume 23, No.1, ISSN 2304-1382 ISBN 978-3-901509-91-9, CDROM version, Ed. B. Katalinic, Published by DAAAM International, Vienna, Austria

[10] Babic, M.; Stanojevic, M.; Ostojic, G.; Tegeltija, S. & Stankovski, S. (2022). Industrial Cyber Security Aspects in ICS Applications, Proceedings of the 33rd DAAAM International Symposium, pp.0326-0331, B. Katalinic (Ed.), Published by DAAAM International, ISBN 978-3-902734- 36-5, ISSN 1726-9679, DOI: 10.2507/33rd.daaam.proceedings.045, Vienna, Austria

[11] Reithner, I.; Papa, M.; Lueger, B.; Cato, M.; Hollerer, S. & Seemann, R. (2020). Development and Implementation of a Secure Production Network, Proceedings of the 31st DAAAM International Symposium, pp.0736-0745, B. Katalinic (Ed.), ISBN 978-3-902734-29-7, ISSN 1726-9679, DOI: 10.2507/31st.daaam.proceedings.102, Vienna, Austria

DOI: 10.2507/34th.daaam.proceedings.016

ROBOTIZATION OF IN-ORBIT LARGE-SCALE STRUCTURE ASSEMBLY

Vladislav Kopylov, Igor Shardyko* & Victor Titov

This Publication has to be referred as: Kopylov, V[ladislav]; Shardyko, I[gor] & Titov, V[ictor] (2023). Robotization of In-Orbit Large-scale Structure Assembly, Proceedings of the 34th DAAAM International Symposium, pp.0111-0120, B. Katalinic (Ed.), Published by DAAAM International, ISBN 978-3-902734-41-9, ISSN 1726-9679, Vienna, Austria
DOI: 10.2507/34th.daaam.proceedings.016

Abstract

Large-scale spacecraft structures provide the better part of the critical features of a satellite. The increase in size of the structures leads to a greater power reserve, an increased number of relay channels, better resolution quality, and so on. Nowadays, deployable from folded-state structures have reached the technological limits of their own size. In order to overcome these limits, it is often proposed to assemble space structures directly in space. However, a lot of issues remain that need a feasible and efficient solution. In this article, the key aspects of in-orbit robotic assembly technology are discussed, such as the technology of mechanical connections, robotic motion and operation performance, and the overall layout of the space structure. Then special attention is paid to the mathematical model that is necessary to build the control system, whose basic design is also proposed. The total concept makes it possible to assemble large-scale structures providing oscillation damping relying upon existing spacecraft control system capabilities.

Keywords: robotic assembly; large-scale space structures; spacecraft; solar panels; truss structures.

1. Introduction

In-orbit assembly of large-scale structures gives huge advantages that were clear from the origin of the space industry. Assembly operations have been tested in all EVA missions since the first spacewalk (A. Leonov mounted a TV camera to the spacecraft outer surface). The most significant work was carried out on the space stations. Namely, due to the assembly in space, it has become possible to achieve the unprecedentedly big power reserve of the ISS provided by the ITS-truss with solar panels and radiators installed on it [1]. However, the ITS-truss itself was delivered in the cargo bay of a space shuttle that was already assembled. Nowadays, it is not possible to repeat such an assembly as no means of delivery exist that could provide sufficient payload volume.

The experiments “Sophora”, “Rapana”, and “Strombus”, conducted from 1991 to 1996 on the Mir space station [2], have shown the fundamental possibility of assembling a sufficiently precise and stable truss mast from short segments directly on the orbit. All these experiments assumed the participation of a cosmonaut to perform each connection, which means high labour intensity in the assembly and prevents the application of the technology on unmanned spacecraft, especially on high orbits. Different approaches to robotic assembly are discussed in [3][4][5][6][7][8]. In these research, a number of issues are investigated, such as accurate placement of the assembled part or unit, reliable fixation, structure scalability, and easy handling of parts with robotic grippers.

Besides, another important issue is the coupling (mutual influence) of the robot motion with spacecraft motion, which was investigated in articles and experiments on on-orbit servicing [9][10][11][12][13]. Recently, researchers have also paid attention to the relatively low stiffness of the space-assembled structures, which translates into the need to consider the natural oscillations of the assembled truss when the robot moves [14]. Application of robotics in projects of in-orbit assembly as well as on-orbit servicing is commonplace nowadays, an example can be seen in [15]. In addition to the abovementioned challenges, it is desirable to provide geometry repeatability during the assembly of large-scale structures for two reasons. First, such structures are often used as support for precise instruments like radio-reflectors or arrays of optical sensors. Second, the deviations from the nominal shape of long structures disturb the spacecraft balance and lead to excessive workload for the reorientation system. Most proposals for robotic assembly involve the fixation of connections by locks that necessarily works with backlash. Taking into account that there are a lot of connections in the structure, the total backlash may reach substantial levels. In its entirety, assembly technology comprises the following components:

- Assembly robot
- Connection part
- Standard construction parts
- Assembly tools
- Control algorithms of robotic arms for the task of installing a new part
- Control algorithms for robot motion across the structure.

2. Description of assembly technology components

2.1. Connection part

A simple idea of connection is proposed in [16,17], which is to fuse two concentric rings made of low-melting metal that are located in the grooves of the peg and the sleeve. The connection forms when these rings solidify being fused into one. This ring works as a locking element. We have reworked the connection from [17] in order to increase manufacturability and reduce size. The resulting connection has several advantages in terms of robotic assembly [18]. The connection does not require effort to perform assembly (the resistance force during assembly does not exceed 16 N), and the assembly is done in one motion. The locking element does not contain any moving parts, which provides its high reliability. The connection is formed in a way that provides a specific feature, which is the total absence of play in the axial direction. At the same time, load capacity is very high, especially with respect to small size and weight (1200 N tensile load for 12 g weight). Also, the time to perform the connection is small, and the application of forced ventilation will allow for further shortening the duration several times. The absence of solder and main metal adhesion makes possible multiple assembly and disassembly without the loss of solder. Design features and test results are presented in [19] and omitted here. The structure of the connection is clear from Fig. 1. The spring that the unit contains allows assembly from two initial states: when the outer cone is shifted backwards and when the outer cone is situated, such as after the fixation of the connection. In return, it makes it possible to assemble closed structures like triangles (statically indeterminate).

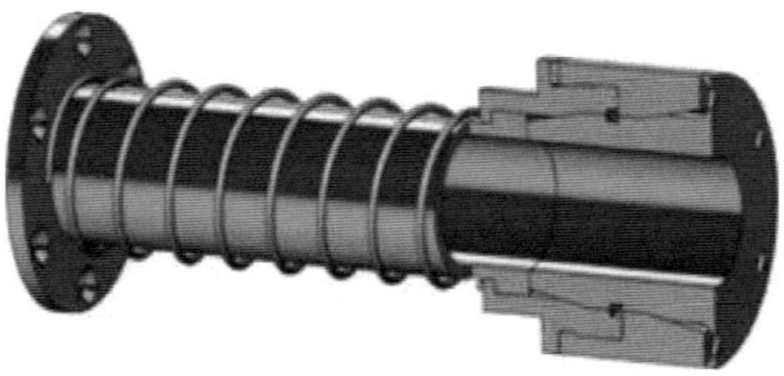

Fig. 1. Standardized connection for robotic assembly

2.2. Standard constuction part

Rods and plates are widespread examples of standard construction parts. The proposed robots capable of motion across the spacecraft structure usually employ two types of attachment to the structure. The first type is represented by specialized attachment sites adapted to the robot gripper (rigging points), while universally applicable structures like handrails belong to the second type. Other types of attachment to the spacecraft structure, like usual rails or even the absence of attachments, can be found in the literature but are scarce, and they have several drawbacks. Handrails are more universal in application but require complicated grippers that reduce the reliability of the assembly robot. Conversely, rigging points, though increasing the weight of the structure, usually keep this increase insignificant, especially for "plate" parts. For rod-type trusses, the most convenient attachment site is a fitting where several rods meet; thus, only 5% to 15% of the fitting mass falls on one rod, depending on the truss structure. In papers [19][20], a method to form carbon panels for solar batteries was proposed that provides high stiffness at a low weight. It is not difficult to modify this method by placing the rigging point fixators into the mold, so the "plate" element can be obtained, that is suitable for the robot to navigate through. The assembly process is shown in Fig. 2.

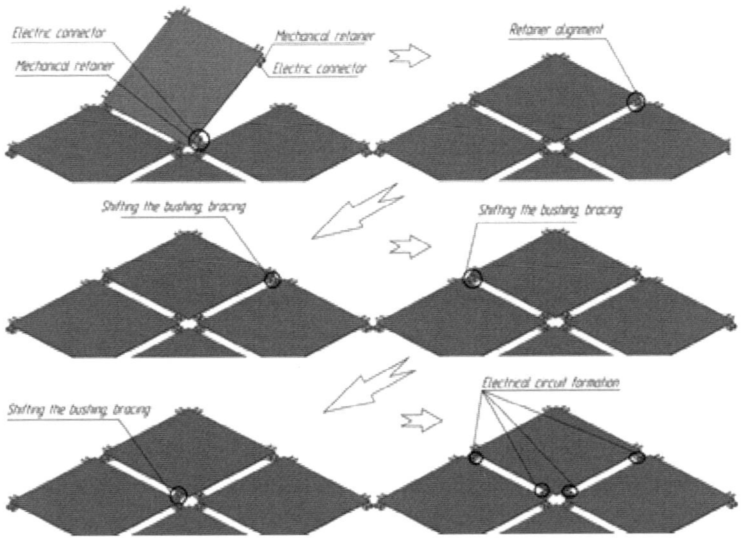

Fig. 2. The assemble process of panel structure

The assembly fitting has two switchable options, which make it possible to install each new assembly part with an arbitrary state of the fitting, either compressed or stretched, while the choice of the option is made by placement of a fixator in the robot gripper. As the assembly fitting allows rotation around its own axis, two fittings can be assembled on one side of the "plate" assembly part. Together with the choice of the fitting option, a two-dimensional structure can be assembled from panels. Truss assembly is performed the same way.

2.3. Assembly tools

The design of the tools for robotic assembly follows the type of connection. As the assembly comprises the melting and solidification of the locking element, the tool must contain a heater and a cooler. The choice from the two connection options is performed before the assembly with the help of a pin actuated by an electromagnet, while the gripper jaws hold the assembly unit. The design of the tool is shown in Fig. 3.

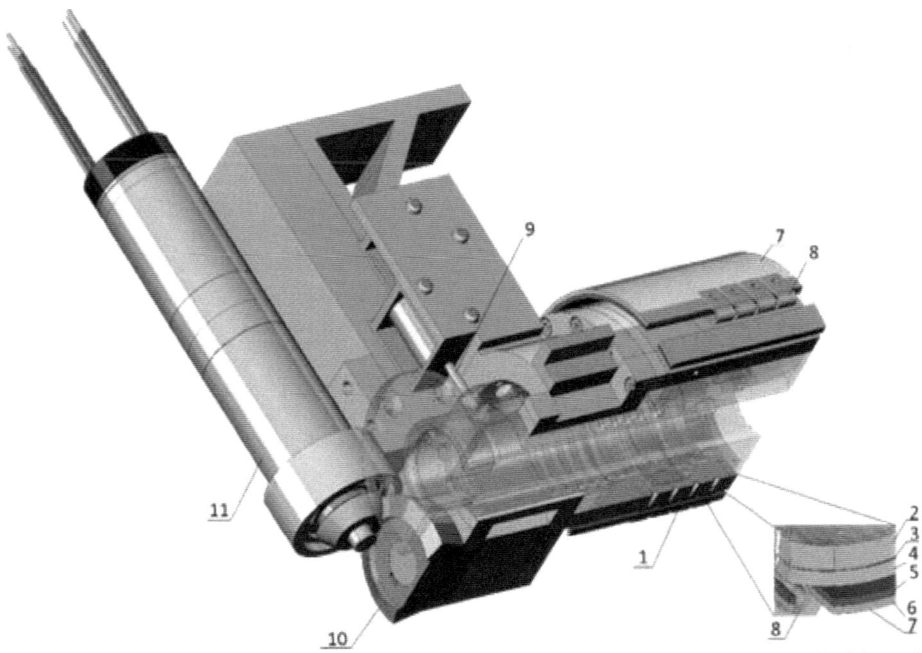

1 – jaw, 2 – load wall, 3 – thin-film heater, 4 – heater clamp, 5 – radiator, 6 – wafer, 7 – MLI, 8 – bimetallic plate, 9 – rod of position choice, 10 – jaws actuator, 11 – electric motor

Fig. 3. General view of the assembly tool

The tool works as follows. Before the assembly starts, the jaws grasp the element to assemble directly at the assembly fitting body, and then the robot performs the preliminary docking. If necessary, the outer bushing of the fitting may be shifted, and the new position is fixed by the rod of position choice. As the fitting is installed in the position for assembly, the heater turns on and the locking rings melt. The end of melting can be deduced from the temperature of the jaws, which only slightly differs from the ring temperature, or by timing, i.e. when the specified time expires. The temperature increase curve has a noticeable break during melting that was captured in the experiment; see Fig. 4. Heating efficiency is increased by the MLI that covers the outer surface of the radiator.

This insulation almost totally blocks the heat outflow from the heating area, excluding only heat bridges between the connection element and the structure being assembled. It is assumed to place a heat-insulation spacer made of asbestos-reinforced laminate or ceramics in these areas. As a result of jaws heating-through in bimetallic plates connecting the radiator and the jaws, stresses increase. The radiator is kept from opening by a magnetic latch that releases at the command of the robot, and the radiator flaps open. The inner side of these flaps is covered by thermo-regulating paint ($As < 0.2$, $\varepsilon > 0.9$), which causes fast cooling of the gripper after the heater turns off. The connection gains strength already when the temperature falls 50 °C below the melting point, therefore, at this moment the gripper can be opened and removed, and further cooling takes place in the structure. When the gripper is sufficiently cooled, bimetallic plates close the radiator flaps back, which are latched by turning on the electromagnet.

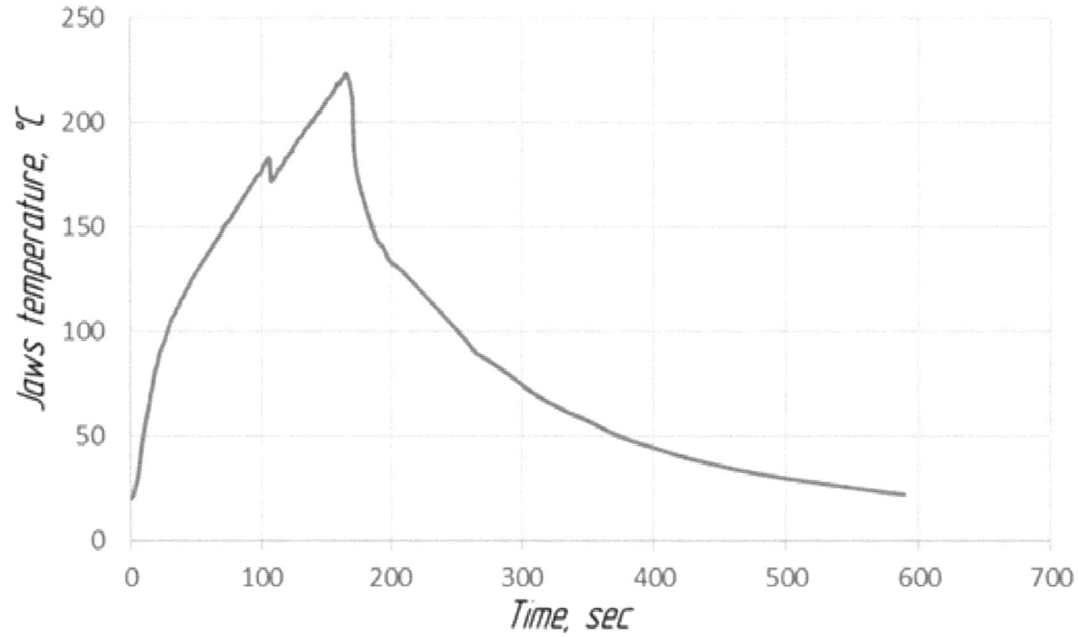

Fig. 4. Temperature of the jaws in the experiment

2.4. Assembly robot

The robot must be capable of performing the following tasks:
- Navigate through the assembled structure
- Carry construction parts, withdrawing the new ones from the warehouse on the servicer spacecraft if necessary
- Perform assembly operations using appropriate tools.

It is also desired that robot should assist in receiving cargo with assembly parts if they are delivered in orbit by a separate launch, i.e. by helping to perform docking of the cargo bay with the servicer spacecraft. The robot is equipped with a rigging tool as an end-effector to conduct locomotion tasks. It seems complicated to fix the robot on the assembled structure with assembly tools because the jaws of the tool are designed to firmly hold the relatively low weight of one assembly part, not the robot itself. However, the rigging tool can be manufactured in a single package with the assembly tool, so the robot does not need to interchange tools at all.

During the assembly process, the robot must accurately place the assembling part while at the same time compensating for the impact from docking. A joint capable of precise tracking of output angle as well as output torque is the most suitable for this task. It is also desired that the physical stiffness of the joint be substantially lower than the stiffness of the objects with which the robot interacts.

The element with low stiffness represents mechanical capacity. Actually, if we introduce a reciprocal to the stiffness quantity, which is compliance p, then the dynamics equation neglecting friction can be written down as

$$m\ddot{x} + \frac{1}{p}x = 0 \tag{1}$$

where x is the generalized coordinate of motion and m is the measure of inertia (mass of moment of inertia).

This equation corresponds to the equation of current in the circuit with capacity p and inductance m. Thus, the increase in compliance has the same effect as the introduction of capacity into an electric circuit, i.e. it provides filtering of low-frequency oscillations, smooths shock impacts, and allows energy storage. The type of joints that contain a compliant element at the output of the gearbox is now known as series elastic actuators (SEA) [21]. The assembly robot should be based on these joints in order to fulfill its purpose.

The joint controller employs measurements from two position sensors: a Hall array on the motor shaft and an optical sensor on the joint output shaft. Due to low backlash as well as high stability of the ratio of harmonic drive, the rotation angle of the gearbox slow shaft is equal with high accuracy to the rotation angle of the motor divided by the gear ratio. Consequently, the measurement error for this angle is equal to the Hall array accuracy divided by the gear ratio as well. By comparing the indirectly measured rotation angle of the gearbox slow shaft with the actual joint output rotation angle and multiplying the difference by the known joint stiffness, one can obtain the external torque applied to the joint. The shape of the elastic element can be designed following different approaches to torsion spring development [22][23][24] [25].

The assembly robot comprises two robotic arms, a base unit, and a vision unit. Each of the arms consists of seven modular joints of three different sizes. A support-bracket with a rigging tool that represents a construction parts magazine is mounted on the base unit. Since construction parts should be equipped with rigging points required for robot motion, it seems most convenient to mount them by rigging point. Before assembly starts, the robot chooses the position to perform it and takes the necessary part from the magazine with the help of the assembly tool.

3. Robotic arm operations during assembly

The vision unit provides lighting for an assembly site as well as the initial orientation of the assembly robot on the basis of the position of tags situated on the construction parts and read by the unit. This operation is necessary because the absolute position error that is accumulated from the docking point of one arm of the robot to the site of the assembly performed by another arm can reach 10 mm, which prevents the mating of lead-in elements of construction parts. By measuring the actual robot pose and the coordinates of the assembly site, the uncertainty can be reduced to 1-2 mm, so it becomes possible to realize preliminary docking followed by alignment along the conical surfaces of the connection parts. The alignment is performed in impedance mode, i.e. the robotic arm intends to minimize the reaction of the support part in the direction orthogonal to the supposed docking axis while moving the tool along that direction, correcting the estimated position of the docking axis based on the reaction force feedback. When the docking is finished, the resistance force starts to grow sharply due to the contact of the peg with the bottom of the conical funnel of the unit. Fig. 5 shows the force portrait measured in the experiment.

The curve in Fig. 5 does not show the growth of force after the assembly has been performed because the experiments were stopped as soon as the funnel bottom was reached; the growth of force then was almost instantaneous and therefore not of interest. The experiments were held in imitation of the impedance mode, i.e. the assembly unit was fixed in a chuck, providing compliant centering and alignment. The chuck was set to move slowly and uniformly in a vertical direction, while the projection of reaction force onto the axis of the fixed part of the connection was logged by a dynamometer. The obtained dependence has three characteristic areas: a gradual increase in reaction force in the process of initial alignment, "a shelf" during the alignment of cylindrical parts, and an abrupt fall when the alignment has been reached. The reaction force in the first area of the curve is defined by the combination of the friction between the peg and the sleeve and the projection of the reaction force of the sloped surface onto the docking axis, and initially the latter part prevails.

The main cause of the constant-force area is the friction that is the result of the peg-sleeve misalignment. This misalignment becomes possible due to axial clearance in the H/h fit of the connection, and the primary cause is the noncentrality of the initial contact. After the moment when a substantial part of the connection is centered by the cylindrical surfaces of the peg and the sleeve, the only resistance force becomes the peg-sleeve friction that is caused by the weak preload of the peg by the compliant alignment system. This preload force is much less than the reactive force of the peg-sleeve elastic deformation during alignment, so the resistance reduces sharply. As the experiment results show, the force portrait can be quite accurately approximated by a piecewise linear function, which is employed by the assembly robot to control the performance of the task.

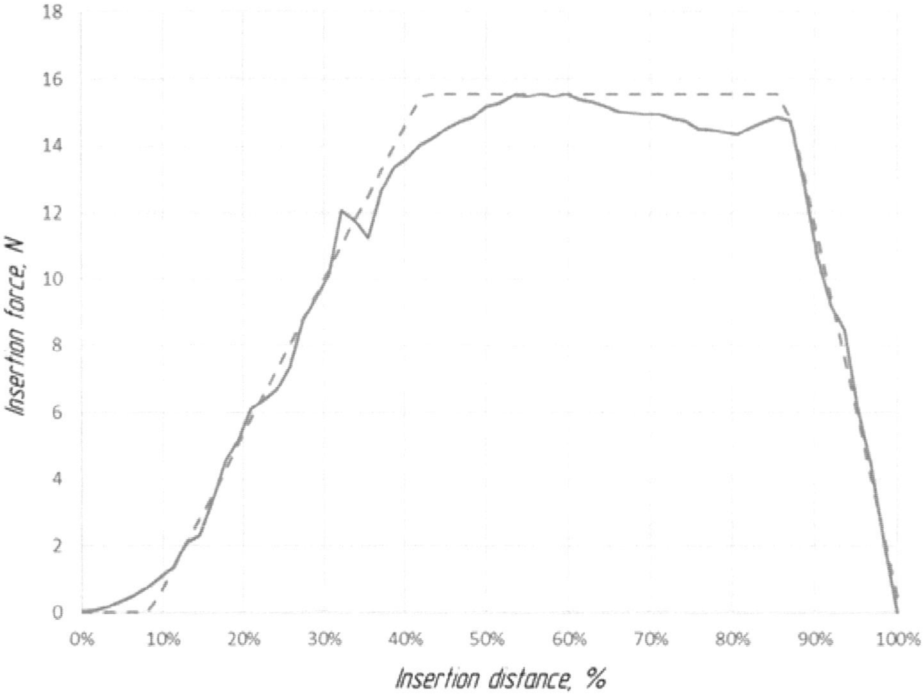

Fig. 5. Contact force during assembly process

The abrupt fall of the resistance force on the last area of the curve becomes possible only in the case of the assembly of a statically determinate structure. In practice, such structures only exist as an intermediate stage of the full assembly process. Installing the closing fitting in a statically indeterminate structure creates forces that are caused by the elastic deformation of the construction part, which results from errors in the installation of the fittings into the structure. These forces depend not only on the peg and sleeve production quality but also on the stiffness of the construction part. The force portrait distorts in this case and depends on the combination of the peg and sleeve tolerance zones. Since it is more convenient to change the shaft tolerance, it is assumed that the sleeve tolerance zone is always H7 (ISO), while the tolerance zone for the cylindrical part of the peg is varied. The resulting force portraits are shown in Fig. 6.

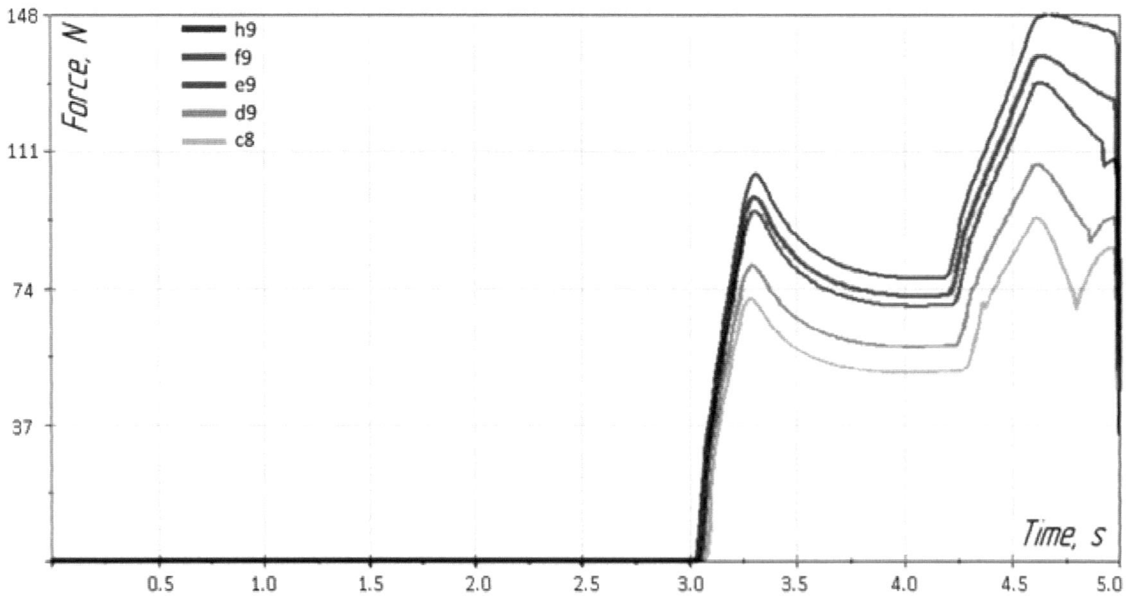

Fig. 6. Contact force during the assembly in a statically indeterminate case

In the statically indeterminate case, the connected parts tend to align not along the common axis but in some random direction that depends on the direction of the accumulated production error. At the same time, when cylindrical parts are mating, the reaction force component caused by the sleeve wall is zero, as in the case of a statically determinate structure, so the overall nature of the load change remains the same on the first two docking stages. However, after the final alignment of the cylindrical parts, the friction force grows rapidly as its cause is not the weak preload (or, in other words, a small transverse force at the robotic tool), but a significantly higher elastic force from the structure deformation. An increase in the guaranteed clearance (see Fig. 6) reduces internal stresses along with the maximal value of the reaction force during docking. This, however, reduces the accuracy of the structure because the uncertainty caused by the random combination of clearances grows. This effect can be lowered if we first assemble the structure on Earth, then mark the parts and assemble them in orbit in the same manner, reproducing the same structure. As a result, the backlash will be zero, and the combination of clearances will be repeated with high accuracy, providing high repeatability if not absolute accuracy. If necessary, the resulting absolute errors can be compensated either mathematically (compensation matrix for orientation) or mechanically by the introduction of a special adjustment platform, whose geometry is again known from the on-ground assembly testing.

4. Control algorithms for robot motion across the structure

Motion of the assembly robot across the assembled structure causes oscillations of this structure and angular momentum variation of certain parts, e.g. the main unit with orientation sensors and the guidance system of the spacecraft. This is associated with mass redistribution and the motion itself. The influence of the change in structure configuration as well as cooperative and uncooperative docking has been considered in many papers [15], [26][27][28][29].

When the assembler moves and construction parts of the structure are carried over, the angular momentum of the whole spacecraft remains constant; however, the angular momentum of certain parts changes, which is due to the variation of the mass distribution across the spacecraft along with the energy introduced into the system by the assembler. The spacecraft is controlled by angular velocity and position, which are determined by instruments located on the base structure, and these quantities do not coincide with those of the inertia ellipsoid of the spacecraft. At the same time, the tidal torque, which is defined by the position of the inertia ellipsoid with respect to orbit, is crucial for stretched spacecraft. As a result, the control system incorrectly compensates for disturbances or even starts such compensation when there are none.

In order to avoid such parasitic compensations, the contribution of the assembler to angular momentum variation should be as low as possible and zero on average. As simulation results show, this condition almost holds for a robot moving human-like when robotic arms move in parallel planes as human legs do. The speed of a distal point (known as the tool center point) for this motion should change in accordance with a trapezoidal law. The estimation of the angular momentum distribution of the subsystem "robot-construction part" for the particular case of motion across the assembled truss is shown in Fig. 7. In simulation, a robot with a total mass of 175 kg and a joint speed limit of 30 °/s was employed. It is clear that variation in angular momentum can be substantial, but the average value is close to zero, which is also the most probable. The deviation of the average value from zero is caused by the inertia variation in the assembly process. The major finding from this estimation is that an assembler robot of about 200 kg is proportionate for a spacecraft of 2000 kg and more because the control system of such a spacecraft is capable of compensating disturbances due to robot locomotion [30].

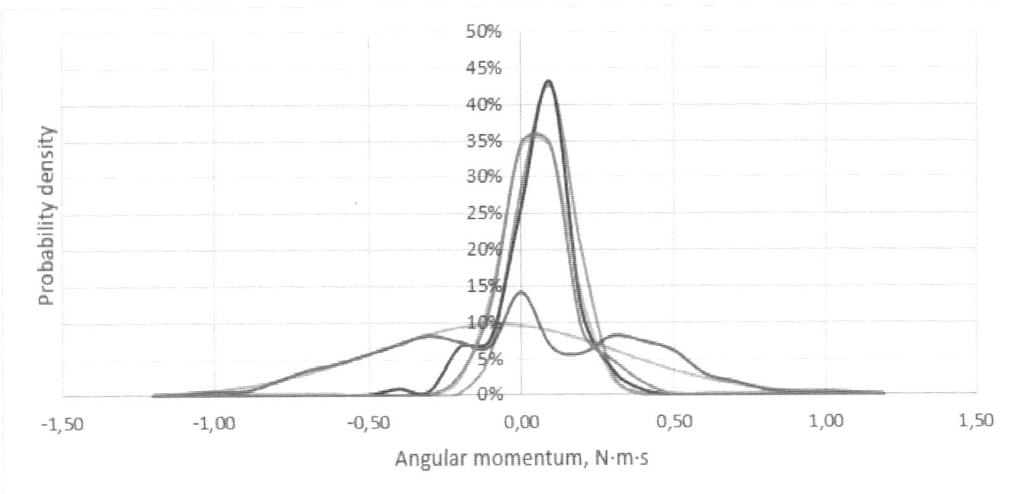

Fig. 7. Angular momentum distribution for locomotion of the assembly robot.
Projections on coordinate axes: OX – red, OY – green, OZ - blue

Locomotion of the robot across the assembled structure involves another issue, which is the unavoidable oscillations of the structure caused by the robot at the time of grasping and releasing the rigging point. Eigenfrequencies of significantly stretched objects like reflectors or solar panels are in the range up to 10 Hz, which means that resonance is possible with the frequencies of the motion control system. Measures directed to avoid this resonance (frequency separation) are ineffective in the case of assembled spacecraft because the eigenfrequency of the structure changes in the process of assembly. Properly coupled performance of the assembler robot and the spacecraft can be guaranteed by damping the arising oscillations through energy dissipation in the motors of the robotic arms. Physically, the damping happens in the course of turning the energy of structure oscillations into the elastic potential energy of the torsion springs in robotic joints, which is further dissipated by activating the generator mode of a brushless electric motor actuating the joint.

Implementation of the latter control mode and its simplification were presented in [31] and [32], respectively. Since the coupling of a robotic arm with an elastic truss is a more complex system than a single loaded elastic joint, it is expected to see some effects in the operation of the system that haven't been observed in [31][32] and are analogous to them. The study of this dynamical system is by itself a topic for an article; however, the overall approach to building a model is rather concise.

The dynamics of the host spacecraft are described by the equation of rotational motion of a rigid body along with an appropriate kinematics relation:

$$J\dot{\omega} + \omega \times (J\omega) + \dot{J}\omega = T_R \tag{2}$$

$$2\dot{\Lambda} = (\Lambda \circ \boldsymbol{\omega}) \tag{3}$$

where J is the spacecraft inertia tensor, ω is the spacecraft angular velocity, and Λ is the quaternion of spacecraft orientation in an inertial system. Rigid-body approximation is justified in this case as variation in the inertia tensor due to oscillations is irrelevant in size and time, resulting in zero contribution taking average over time. Oscillation equations in generalized (modal) coordinates can be written as

$$L_n^T \widehat{M}_{\Sigma_n} L_n \ddot{\xi}_n + L_n^T \widehat{N}_{\Sigma_n} L_n \dot{\xi}_n + L_n^T \widehat{K}_{\Sigma_n} L_n \xi_n = L_n^T \widehat{Q}_n \tag{4}$$

where n is an index of the truss component, ξ_n is a vector of generalized coordinates, L_n is a transformation matrix between independent and dependent variables, $\widehat{M}_{\Sigma_n}$ is the inertia matrix, $\widehat{K}_{\Sigma_n}$ is the stiffness matrix, and $\widehat{N}_{\Sigma_n}$ is the damping matrix defined as:

$$\widehat{N}_{\Sigma_n} = \alpha \widehat{M}_{\Sigma_n} + \beta \widehat{K}_{\Sigma_n} \tag{5}$$

where α, β are scalar Rayleigh damping coefficients for truss material.

The vector of generalized forces Q is defined as follows:

$$\widehat{Q}_n = L_n^T T_{CBn}^T F_n^f \tag{6}$$

where T_{CB} is the transformation matrix for the Craig-Bampton method [33], and F_n^f is the vector of forces in the component.

Matrices of inertia, stiffness, and damping are determined through the direct stiffness method [34]. Equation (6) can be inverted to find the actual force vector at the reference point. To achieve this, trusses are broken into n components when being reduced, and each of them contains a rigging point. The rigging point is then chosen as the interface point for the reduction of stiffness and inertia matrices. Resulting forces are fed to the input of the dynamic model of the robotic arm equipped with the damping controller, which closes the model.

5. Conclusion

In this paper, implementation issues of robotic assembly of space structures have been considered, which include the design of simple and reliable mechanical connections, algorithms of robotic assembly, oscillation damping, and contact force minimization. The appropriate solutions were formulated, in particular, the physics of the development process of the proposed mechanical connection were studied, and the design of both the robot assembler and the assembled structure was manifested. It was concluded that in order to provide adequate performance of the system, it is necessary to analyze the coupled dynamics of the robot and the structure, assuming controllability of the servicer spacecraft and employment of the damping controller for robot joints. It is expected that the application of SEA joints in the robotic arms of the assembler should solve the main problems arising during the assembly of large-scale non-rigid (compliant) structures. The mathematical model of the coupled robot-structure dynamics was obtained and presented, and our future plans involve a simulation study of the system as well as thorough design and validation of the combined control system of the robot and the spacecraft.

6. Acknowledgments

The work was supported by the Ministry of Science and Higher Education of the Russian Federation. State assignment 075-01595-23-01 “Large-sized space structures assembly methods and technologies research” (FNRG-2022-0007, ID 1021060307686-0-2.2.2).

7. References

[1] https://spacecraftearth.com/portfolio/iss-integrated-truss-structure/, (2017). ISS Integrated Truss Structure, Accessed on: 2023-07-18

[2] Trusov, S.N. & Chernyavskiy, A.G. (1996). “Opyt primeneniya splavov s effectom formy pri sooruzhenii krupnogabaritnykh konstruktsyi v otkrytom kosmose”, Journal of Technical Physics, Vol. 66, No. 11 (in Russian)

[3] Doggett, W. (2002). Robotic assembly of truss structures for space systems and future research plans, Proceedings of IEEE Aerospace Conference, 9-16 March 2002, Big Sky, MT, USA

[4] Jenett, B.; Cellucci, D. & Cheung, K.C. (2015). SpRoUTS (Space Robot Universal Truss System): Reversible Robotic Assembly of Deployable Truss Structures of Reconfigurable Length, Proceedings of AIAA Space 2015 Conferences and Exposition, 31 Aug-2 Sep 2015, Pasadena, California

[5] Hoyt, R.P.; Slosad, J.T. & Moser, T.J. (2016). In-Space Manufacturing of Constructable™ Long-Baseline Sensors using the Trusselator™ Technology. Proceedings of AIAA SPACE 2016. 13-16 September 2016, Long Beach, California

[6] Patane, S.C.; Schomer, J.J. & Snyder, M.P. (2018). Design Reference Missions for Archinaut: A Roadmap for In-Space Manufacturing and Assembly. Proceedings of 2018 AIAA SPACE and Astronautics Forum and Exposition. 17-19 September 2018, Orlando, FL

[7] Wong, I.; Chapin, W. & Komendera, E. (2018). Validation of Operations for the In-Space Assembly of a Backbone Truss for a Solar-Electric Propulsion Tug. Proceedings of 2018 AIAA SPACE and Astronautics Forum and Exposition. 17-19 September 2018, Orlando, FL

[8] Schervan, T.A.; Kortmann, M.; Kai-Uwe Shronder, P. & Kreisel, J. (2017). iBOSS Modular Plug & Play - Standardized Building Block Solutions for Future Space Systems Enhancing Capabilities and Flexibility, Design, Architecture and Operations. Proceedings of 68 International Astronautical Congress. Adelaide, Australia, 25-29 September 2017

[9] Shah, S. V.; Sharf, I. & Misra, A. (2013). Reactionless path planning strategies for capture of tumbling objects in space using a dual-arm robotic system. Proceedings of AIAA Guidance, Navigation, and Control (GNC) Conference (p. 4521)

[10] Kasai, T.; Oda, M. & Suzuki, T. (1999). Results of the ETS-7 mission - rendezvous docking and space experiments. Proceedings of 5th International Symposium on Artificial Intelligence, Robotics and Automation in Space (i AIRAS), Noordwijk, Netherlands, 1999

[11] https://www.nasa.gov/pdf/167813main_RP-06-119_05-020-E_DART_Report_Final_Dec_27.pdf, (2006). NESC Review of Demonstration of Autonomous Rendezvous Technology (DART) Mission Mishap Investigation Board Review (MIB). Accessed on: 2023-07-18

[12] Dennehy, C. J. & Carpenter, J. R. (2011). A summary of the rendezvous, proximity operations, docking, and undocking (rpodu) lessons learned from the defense advanced research project agency (darpa) orbital express (oe) demonstration system mission (No. NESC-RP-10-00628).

[13] Henshaw, C. G. (2014). The darpa phoenix spacecraft servicing program: Overview and plans for risk reduction. Proceedings of International Symposium on Artificial Intelligence, Robotics and Automation in Space (i-SAIRAS). European Space Agency.

[14] Yuchen She, Shuang Li, Yufei Liu, Menglong Cao, "In-orbit robotic assembly mission design and planning to construct a large space telescope," J. Astron. Telesc. Instrum. Syst.

[15] Lopota, A.; Dalyaev, I.; Shardyko, I.; Kuznetcova, E. & Belezyakov, I. (2016). Means of robotic support for on-orbit servicing. Proceedings of the 26th DAAAM International Symposium (pp. 0865-0870). 2016.

[16] Pochezhertsev, A. G. & Kopylov V. M. (2020). Universal Docking Assembly Design for Automatic Assembly of Large Untight Structures in Near-Earth Space. Proceedings of 2020 International Conference on Industrial Engineering, Applications and Manufacturing (ICIEAM). – 2020. – pp. 1-6

[17] Kopylov V.M. & Pochezhertsev A.G. (2023). System for robotic assembly of large-sized trusses, including on earth's orbit or moon. Patent RU2790311C2.

[18] Kopylov V. M. & Volnyakov K.A. (2022). Малогабаритные средства адаптации КА для возможности робототехнического обслуживания на орбите. (Small-size Means of Adaptation of Spacecraft for On-orbit Servicing Capability, in Russian). Proceedings of XIV Open Russian Scientific and Technical Conference “Robotics and Artificial Intelligence” (RAI) – 2022. – pp. 37-42

[19] Drondin, A.V.; Zernov, O.D. & Yanchur, S.V. (2017). Manufacturing method of celled honeycomb filler from composite materials. Patent RU2623781C2.

[20] Rebrov, S. G.; Yanchur, S. V.; Drondin, A. V. & Zernov, O. D. (2019). “Developing the concept of solar energy units robotic assembly in orbit”, Aerospace MAI Journal, Vol. 26, No. 1, pp. 201-211

[21] Pratt, G A. & Williamson, M.M. (1995). Series elastic actuators. Proceedings of 1995 IEEE/RSJ International Conference on Intelligent Robots and Systems. Human Robot Interaction and Cooperative Robots. Vol. 1. IEEE, 1995.

[22] Carpino, G.; Accoto, D.; Sergi, F.; Tagliamonte, L.N. & Guglielmelli, E. (2012). "A novel compact torsional spring for series elastic actuators for assistive wearable robots", Journal of Mechanical Design, Vol. 134, No.12

[23] Tagliamonte, N. L.; Sergi, F.; Carpino, G.; Accoto, D. & Guglielmelli, E. (2010). Design of a variable impedance differential actuator for wearable robotics applications. Proceedings of 2010 IEEE/RSJ International Conference on Intelligent Robots and Systems (pp. 2639-2644). IEEE.

[24] Zhou, M. (2021). "A method to determine the topology of custom torsional elastic element for the lightweight rotary series elastic actuator", Journal of Physics: Conference Series, Vol. 1939, No. 1, IOP Publishing

[25] Diftler, M. A.; Mehling, J. S.; Abdallah, M. E.; Radford, N. A.; Bridgwater, L. B.; Sanders, A. M.; ... & Ambrose, R. O. (2011). Robonaut 2-the first humanoid robot in space. Proceedings of 2011 IEEE international conference on robotics and automation (pp. 2178-2183). IEEE.

[26] Rognant, M.; Cumer, C.; Biannic, J. M.; Roa, M. A.; Verhaeghe, A. & Bissonnette, V. (2019). Autonomous assembly of large structures in space: a technology review. Proceedings of EUCASS 2019.

[27] Fukazu, Y.; Hara, N.; Kanamiya, Y. & Sato, D. (2009). Reactionless resolved acceleration control with vibration suppression capability for JEMRMS/SFA. Proceedings of 2008 IEEE International Conference on Robotics and Biomimetics, pp. 1359-1364. IEEE.

[28] Costa, A.; Abdel-Rahman, A.; Jenett, B.; Gershenfeld, N.; Kostitsyna, I. & Cheung, K. (2019). Algorithmic approaches to reconfigurable assembly systems. Proceedings of 2019 IEEE Aerospace Conference, pp. 1-8. IEEE

[29] Martinez-Moritz, J.; Rodriguez, I.; Nottensteiner, K.; Lutze, J.-P.; Lehner, P. & Roa, M. A. (2021). Hybrid Planning System for In-Space Robotic Assembly of Telescopes using Segmented Mirror Tiles. Proceedings of 2021 IEEE Aerospace Conference, pp. 1-16. IEEE.

[30] Mkrtychyan, A. R.; Bashkeev, N. I.; Yakimovskii, D. O.; Akashev, D. I. & Yakovets, O. B. (2015). "Control moment gyroscopes for spacecraft attitude control systems: Current status and prospects", Gyroscopy and Navigation, Vol. 6, No. 3, pp. 236-240.

[31] Albu-Schäffer, A.; Ott, C. & Hirzinger, G. (2007). "A unified passivity-based control framework for position, torque and impedance control of flexible joint robots", The international journal of robotics research, Vol. 26, No.1, pp. 23-39.

[32] Shardyko, I.; Samorodova, M. & Titov, V. (2020). Development of control system for a SEA-joint based on active damping injection. Proceedings of 2020 International Conference on Industrial Engineering, Applications and Manufacturing (ICIEAM), pp. 1-6. IEEE.

[33] Craig Jr, R. R. & Bampton, M. C. (1968). "Coupling of substructures for dynamic analyses", AIAA journal, Vol. 6, No. 7, pp. 1313-1319.

[34] Felippa, C. A. (2004). Introduction to finite element methods, University of Colorado, T. 885.

DOI: 10.2507/34th.daaam.proceedings.017

SMART DESIGN OF CORE COMPONENTS IN ENGINEERING EDUCATION

Vyarka Ronkova & Antoaneta Dobreva*

This Publication has to be referred as: Ronkova, V[yarka] & Dobreva, A[ntoaneta] (2023). Smart Design of Core Components in Engineering Education, Proceedings of the 34th DAAAM International Symposium, pp.0121-0127, B. Katalinic (Ed.), Published by DAAAM International, ISBN 978-3-902734-41-9, ISSN 1726-9679, Vienna, Austria
DOI: 10.2507/34th.daaam.proceedings.017

Abstract

The paper presents an investigation concerning the specific features of academic training in fundamental engineering subjects in the field of Higher education. The main objective of the research presented is to ensure a smart design of core components in engineering education by implementing innovative practices and improvements in two of the main four components of the concept of Education 4.0: advanced educational teaching and learning models and instruments for increasing the competencies of future engineers. The statistical and experimental study has been carried out according to an advanced methodology containing six main stages. The target group of research consists of first-year students from 22 engineering Bachelor degree courses. Based upon the research results, a system of advanced educational models, methods and tools for increasing the competencies of future engineers aiming to increase the training quality in the Higher education has been suggested. Conclusions and recommendations have been made, the authors' vision for future work in the area investigated is presented.

Keywords: Engineering Education; Smart Design of Educational Technologies; Competences; Learning methods.

1. Introduction

Technological progress has a great impact on the industrial sector and other important areas. In the field of education, the application of current and emerging technologies is combined with innovative pedagogical approaches and good practices. Four main components of education have been formulated, so that they can be used as benchmarks for designing new projects in the field of educational innovation: competences, learning methods, information and communication technologies, infrastructure, [1].

During the recent decades, the interest in engineering education and research has grown worldwide. During the same period, however, there is a constant shortage of students in technical bachelor degree courses in higher education institutions. For Bulgaria, this problem is particularly important for the field of mechanical and general engineering. As it is known, adequate training of qualified engineering professionals is possible only in close cooperation between schools, universities and industry. University curricula are expected to meet the needs of enterprises developed in a particular region.

In order to solve such a problem, the authors of [2] initiated a joint project with the participation of three universities in the Baltic region. The project described aims to develop a new approach to preparing pupils and students in choosing the right profession through the joint efforts of universities, schools and enterprises. At the same time, global perspectives and experience indicate that engineering higher institutions will increasingly focus on the development of "skills for effectiveness in a global environment", i.e. they have to provide students with a range of opportunities to work in different counties and to adapt to different cultures making the role of educational technologies even more important and responsible, [3].

2. Theoretical analysis of the special features of the education in fundamental engineering subjects

Certain number of significant studies have investigated the importance of knowledge management concerning engineering higher education outcomes and innovations [4], [5], [6]. The authors of [5] emphasize the extremely important role of creating a sustainable functional and theoretical approach to teaching and learning through the application of a theoretical framework related to knowledge. According to the authors of [7], the correct decision-making in managing the knowledge flow and volume in fundamental engineering subjects involves a critical evaluation of proofs which requires time and efforts.

In [8], the importance of knowledge management, providing the link between universities and engineering practice, has been emphasized. Teaching Engineering Graphics as a fundamental subject creates important creative perceptions according to [9], which is related to introducing to the first year students the traditional ways of creating graphical presentations. The knowledge of geometry is very important for the engineering students' further education, because the understanding of geometry principles is a key feature of human intelligence. In addition, interactive work with 3D-models is important for clarifying spatial representations and relationships, especially if sufficient effort is implemented into combining graphical drawings, models and text descriptions, [9].

Based upon the theoretical analysis of the special features of education and training in fundamental engineering subjects, the research objective has been deduced: smart design of engineering education by applying innovative practices and improvements in two of the four core components of the concept of Education 4.0 (advanced educational teaching and learning models and tools to enhance the competencies of future engineers) has to be ensured.

3. Research methods

In order to fulfil the objective of the research, an improved methodology has been developed. It consists of several stages:

1) Investigating the reasons for an inappropriate selection of a bachelor's major by some of the future students;
2) Specifying the consequences of the inappropriate choice made;
3) Selection of the target group of the research aiming to establish the challenges in the education of students from different secondary upper schools as future engineers;
4) Statistic and experimental research of the preliminary general technical training of future students taking into account the role of curricula in upper secondary schools in achieving their technical training and in the professional orientation of future students;
5) Analysis and discussion of research results;
6) Defining a system of teaching, learning methods and additional activities aiming to improve the quality of education in the higher education system based upon the results of the investigation. The authors' team outlines the following limitations: the methodology will be applied in the field of subjects with design engineering focus.

4. Reasons and consequences related to an inappropriate selection of Bachelor degree course

Future students do not always have a clear idea about the kind of specialists they want to be. Sometimes the selection of a Bachelor degree course when applying to a university is based upon the fashion trends in the surroundings of young people. The wishes of the future students are not always consistent with their capabilities, their knowledge and skills. These circumstances often lead to disappointments during the first year at the university. The main reasons for the inappropriate selection of a Bachelor degree course are:

1) Some parents mislead their children when choosing a profession. Very often, the wishes of the parents do not coincide with the wishes of the children, which leads to negative results;
2) More than 10% of candidate – students from Eastern Europe leave their native country because they do not have complete information about the opportunities to study a given bachelor degree course in their own country [2].

Very often, the methods and technology of training are different from country to country. International students sometimes find it difficult to adapt to these new methods. Besides, another difficulty is that the training is not carried out in the native language. A large part of the students abroad feel nostalgic for Bulgaria and for their relatives and friends. As a result of the circumstances listed, the training ends unsuccessfully, sometimes with health problems and/or with a negative attitude towards the receiving country. The consequences of selecting an inappropriate Bachelor major are significant:

1) Students fail to manage with the study material, fail to pass the exams and this situation demotivates them. As a result of these difficulties, many students drop out after the first year;

2) The Bachelor degree course turns out not to be what the students imagined. Therefore, they look for an alternative to change their major degree during the next academic year. And they are again candidate students.

The challenge of retaining first-year students in the higher education system is becoming one of the most significant issues for the academic lecturers and for the management authorities of the universities in Bulgaria and in Eastern Europe. Several investigation studies have been done concerning the development of models to predict and represent the reasons for students' decline [9], [10], [11], [12], [13] & [14]. The research in the publications listed indicates useful algorithms aiming to help higher education institutions and faculties to provide necessary support for the first-year students in order to increase their academic success and their motivation to study and to learn at the earliest possible stage.

5. Selecting the target group

The target group of the study is determined. The investigation of the preliminary general technical training of future students has been made based upon a statistic research of the number of students and other indicators over the last 7 years. The students have been trained in engineering Bachelor degree course from the professional fields of "Mechanical Engineering" and "General Engineering". When studying a certain subject from the area of Engineering graphics, it is necessary to examine the incoming knowledge of the students in order to be able to improve the methods and means of teaching in accordance with the modern trends in education and in engineering practice. The survey has been conducted for students of engineering bachelor degree courses studying a subject in the area of Engineering Graphics according to the type of upper secondary school completed. The target group of the research consists of first year students from 22 Bachelor degree course indicated in Table 1. Based upon the academic experience of the authors' team, it has been decided to determine what part of the engineering students, studying subjects in the area of "Engineering graphics", have prior technical training in similar topics.

No	Title of the Bachelor degree course
1	Automotive engineering
2	Agricultural engineering
3	Biotechnologies
4	Ecology and Technique for Environmental Protection
5	Electricity and Electrical Equipment
6	Electrical Engineering
7	Electronics
8	Agricultural Technique and Technologies
9	Industrial Management
10	Industrial Engineering
11	Air Conditioning, Hydraulics and Gasification
12	Computer Control and Automation
13	Materials Science and Technology
14	Mechanical Engineering
15	Management and Technical Service
16	Quality Management and Metrology
17	Industrial design
18	Civil Engineering
19	Technology and Transport Management
20	Food technology
21	Transport Equipment and Technologies
22	Chemical Technologies

Table 1. Bachelor degree courses at the University of Ruse delivering Engineering Graphics subjects

The objective of selecting a target group is to find out what percentage of students studying "Engineering Graphics" have learned subjects related to this one in their previous degree of education. This investigation provides extremely important information to the lecturers delivering these subjects concerning the preparation of the introductory teaching materials which depend on the entry level of the first year students. Therefore, this information assists the determination of the scope, contents, structure and topics of the subject.

6. Analysis of results and discussion

An appropriate understanding and assessment of skills and qualifications is essential for achieving a better compliance between students' skills obtained and labour market needs. It is considered possible to support some of the students to acquire skills and update them throughout their lives by moving through different types and levels of education or through education and employment [15], [16] and [17].

The number of students in Engineering Bachelor degree courses, in which subjects from the area of "Engineering graphics" are studied, is presented in Fig. 1. It shows the distribution of engineering students in different years. The investigation has been carried out for the period from 2016 to 2022. Table 2 presents the percentage of first year students according to the type of school of completed prior education: Upper secondary school or Vocational upper school.

Vocational upper schools are divided into non-technically oriented and technical schools. This division in two categories has been made due to the fact that at the technically oriented vocational upper schools the subject "Technical drawing" or a related technical subject has been studied. Therefore, it can be determined what part of first year students have knowledge and prior training in the field of engineering graphics. Based upon the research done, it was found out that 29.8% of the students graduated from Upper secondary school, i.e. graduates have not studied Technical Drawing or a related subject. About 25.9% of the engineering students graduated from Vocational upper schools which are not technically oriented and they have not studied Technical Drawing as well.

On the other side, the remaining 44.3% of the first year engineering students have knowledge of in the area of Technical Drawing. Besides, it becomes clear that the knowledge of each student within these 44.3% (who graduated from Vocational upper school with the study of Technical Drawing) is at a different level. For some of the students, the studies during their upper secondary education have been more thoroughly in this type of subjects due to the contents of the study curricula and syllabi for the relevant profile of the school they graduated from.

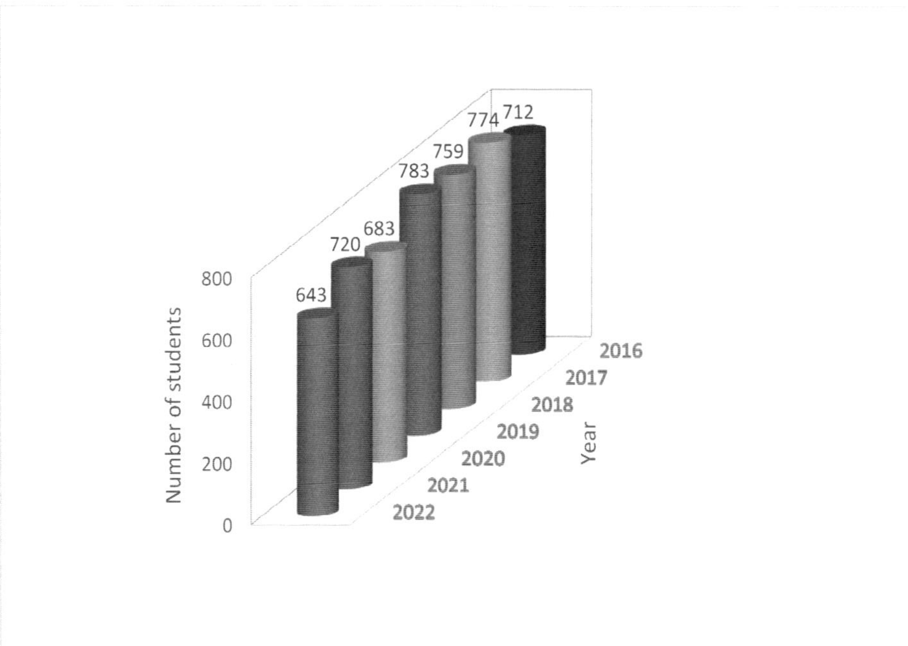

Fig. 1. Number of students in Engineering Bachelor degree courses

Besides, not all upper vocational upper school courses require basic study of Technical Drawing. In some cases, only a section necessary for training in the given course is studied in the Technical Drawing subject. In such specialized classes, only main topics the subject are covered, and in other classes such topics are only mentioned.

Another survey has been carried out concerning engineering students for the period 2016-2022: regarding the time of graduating the upper secondary school. This investigation has been implemented for vocational upper schools which deliver the subject "Technical Drawing".

The students' sample investigated has been divided into three groups in the following way: the first group includes students who graduated from secondary education in the year of admission at the University of Ruse; the second group - students who graduated from upper secondary schools during the last 5 years prior to the year of admission at the University of Ruse; the third group – students who graduated from upper secondary schools 6 or more years before the year of admission at to the University of Ruse. The study is presented graphically in Fig. 2.

From the first group, 41% of the students graduated in the year of enrolment at the University of Ruse. In the second group, 17% of them graduated within the last 5 years of the year of enrolment and in the third group - those who graduated 6 or more years before admission - 42%, Fig. 2.

The first group (graduated from Vocational upper schools studying Technical Drawing or a related subject and graduated during the year of enrolment at the University of Ruse) are 17.9% of the total number of students studying Engineering Graphics.

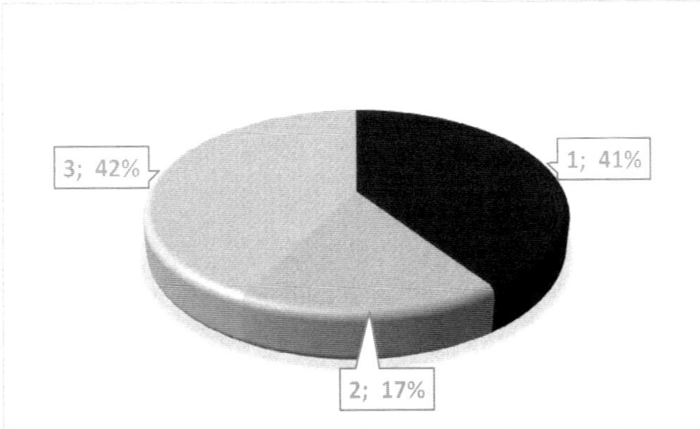

Fig. 2. Distribution (on an average value) of engineering students graduated at Vocational upper schools with "Technical Drawing" depending on the year of graduation (1 – First group; 2 – Second group; 3 – Third group)

The investigation shows that as a matter of fact about 55-60% of first year engineering students do not have basic knowledge of Engineering Graphics. About 20-25% have knowledge in the field, but a certain period of time has passed from the completion of secondary education to their enrolment in the next educational level.

Enrolment Year	Upper secondary school	Non-technical vocational upper school	Technical vocational upper school
2016	28,1%	25,8%	46,1%
2017	25,6%	26,8%	47,6%
2018	26,6%	31,1%	42,3%
2019	32,6%	13,2%	54,2%
2020	31,6%	27%	41,4%
2021	34,6%	30%	35,4%
2022	30,2%	28,7%	41,1%
Average value	29,8%	25,9%	44,3%

Table 2. Distribution of first year students according to the type of their prior education

It can be summarised that the training in the area of engineering graphics should be complete and thorough, covering all the basic concepts and elements in order to enable the first year students to continue to a higher level of acquiring knowledge and skills. This makes it necessary to include additional study materials related to basic concepts in the areas of knowledge of Geometry and Design of drawing documentation when developing and up-grating university curricula and syllabi.

Depending on the specific features of the professional field, the degree of specialization in Engineering Graphics is different for each Bachelor degree course. The subjects in Engineering Graphics are delivered for one or two semesters, which is to be determined through the study curriculum of the relevant Bachelor major. The level of knowledge and the duration of different course units in Engineering Graphics depend on the relevant qualification requirements for the future engineering professionals. The authors' team has been implemented several studies for the past few years. A comparative analysis of number of enrolled students in engineering bachelor degree courses has been made. Based upon these studies, it was found the number of students enrolled in the first year of engineering majors is decreasing compared to the total number of accepted students, [18] and [19]. It has been established that the research question is answered completely and the investigation made provides strategy guidelines for orientation of applicants towards engineering bachelor degree courses.

7. System of additional methods and tools aiming to improve the quality of university education

Based upon the research results, additional methods, models and tools have been determined in order to improve the quality of education in the higher education system. It is possible to provide new and upgraded knowledge using interdisciplinary methods in upper secondary schools. Priority is to be given to these schools in which already STEM teaching laboratories exist. In order to improve the study motivation of first year students and to reduce their retention as university students, it is necessary: to ensure a systematic and sustainable control and distribution of different additional activities within the semester by implementing an appropriate knowledge management system using good practices and opportunities for adapting the teaching methods to the relevant level of knowledge for each student; to structure appropriately the necessary knowledge volume and study contents according to the established entry level of the first-year students; to create opportunities for enterprise visits, through which a realistic idea of the future students' profession will be created; to strengthen the implementing of short term internships and long practice periods in the native country and abroad, which will facilitate the transition between academic environment and industry for future engineers.

The authors' team does not consider that students with good study motivation will receive less consultations and fewer topics for extracurricular work. Different types of additional training and intensive internship programs have been described in details in previous works of the authors. These publications are based mainly on the current achievements of scientific research in the field of STEM, [20], [21], [22], [23] & [24].

The specific characteristics of the intensive internship programs facilitate the achievement of the following results: highly qualified scientific and research human resources with skills to carry out theoretical and experimental activities; providing opportunities to create new scientific products and achieve significant scientific results; building up-to-date theoretical and practical knowledge and skills for carrying out independent research activities; providing opportunities for successful teamwork. The practice of participation of Bachelor, Master and PhD students in Scientific Conferences continues as a result of scientific collaboration between academic community and enterprises.

8. Future work

The following components of future work on the research problem are envisaged: establishment of closer cooperation between the University of Ruse and upper secondary schools; creating programs and practices with these schools initiating discussion about the needs of the surrounding environment and explaining that the world needs hard-working, smart and talented specialists; implementing even more intensive activities related to familiarizing the students with the university buildings, laboratories and with the opportunities that the university education would give them; elaboration of joint projects between upper secondary schools, companies and universities aiming to achieve greater commitment to the young people, for their future professional training and development.

In order to achieve these objectives, it is necessary to develop comprehensive approaches, combining established and new teaching and learning models in interactive way, which will support the final result - improving the quality of study process in the Higher education system.

9. Conclusions

The current challenges in higher education are not limited to training technologies and social issues. They are also problems caused by rapid technological progress and the need to develop basic competencies in pupils and students, encouraging them to develop themselves technically, technologically, analytically and with the ability to think critically. In order to achieve this objective new programs and products are required taking into account the necessary components in higher education. The authors suggest innovations and improvements in two of the four main components that determine the concept and the vision of Education 4.0: advanced educational models – learning methods and tools for increasing the competencies of future engineers.

10. Acknowledgements

The authors would like to express appreciation for the support of the Scientific Research Funds of the University of Ruse, Bulgaria. The research work done is realized in framework of several scientific seminars at the Department of "Machine Science, Machine elements, Engineering Graphics and Physics" at the Transport Faculty at the University of Ruse. The support is gratefully acknowledged.

11. References

[1] Miranda, J., Navarrete, C., Noguez, J., Molina-Espinosa, J., Ramírez-Montoya, M., Navarro-Tuch, S., Bustamante-Bello, M., Rosas-Fernandez, J. & Molina, A. (2021). The Core Components of Education 4.0 in Higher Education: Three Case Studies in Engineering Education. Computers and Electrical Engineering 93 (2021) 107278, pp. 1-13.

[2] Shevtshenko, E.; Karaulova, T.; Igavens, M.; Strods, G; Tandzegolskiene, I., Tutlys V., Tavahodi S. & Kuts, V. (2017). Dissemination of Engineering Education at Schools and its Adjustment to Needs of Enterprises, Proceedings of the 28th DAAAM International Symposium, pp. 0044-0053, B. Katalinic (Ed.), Published by DAAAM International, ISBN 978-3-902734-11-2, ISSN 1726-9679, Vienna, Austria, DOI: 10.2507/28th.daaam.proceedings.006

[3] Graham, R. (2018). The Global State of the Art in Engineering Education. Cambridge, MA: Massachusetts Institute of Technology, pp. 1-170.

[4] Adams, F. G. & Graham, K. W. (2017). Integration, knowledge creation and B2B governance: The role of resource hierarchies in financial performance. Industrial Marketing Management, 63, pp. 179–191.

[5] Abubakar, A. M., Elrehail, H., Alatailat, M. A., & Elci, A. (2019). Knowledge management, decision-making style and organizational performance. Journal of Innovation & Knowledge, 4(2), pp.104-114.

[6] Brix, J. (2017). Exploring knowledge creation processes as a source of organizational learning: A longitudinal case study of a public innovation project. Scandinavian Journal of Management, 33(2), pp. 113–127.

[7] Fitzgerald, D. R., Mohammed, S., & Kremer, G. O. (2017). Differences in the way we decide: The effect of decision style diversity on process conflict in design teams. Personality and Individual Differences, 104, pp. 339–344.
[8] Katalinic, B. (2010). Engineers for Knowledge Based Society, Proceedings of the 21st DAAAM International Symposium, B. Katalinic (Ed.), Published by DAAAM International, ISBN 978-3-901509-73-5, ISSN 1726-9679, Vienna, Austria
[9] Simion, I., Dobre, D., Pascu, N., Adir, V. & Adir, G. (2009). Learning System for Engineering Graphics and Design. Annals of DAAAM for 2009 & Proceedings of the 20th International DAAAM Symposium, Vol 20 (1), ISSN 1726-9679, ISBN 978-3-901509-70-4, Editor B. Katalinic, Published by DAAAM International, Vienna, Austria.
[10] Đambić, G., Krajcar, M. & Bele, D. (2016). Machine learning model for early detection of higher education students that need additional attention in introductory programming courses. International Journal of Digital Technology & Economy, 1 (1), pp. 1-11.
[11] Delen, D. (2010). A comparative analysis of machine learning techniques for student retention management. Decision Support Systems, 49(4), pp. 498-506.
[12] Ram, S., Wang, Y., Currim, F., & Currim, S. (2015). Using big data for predicting freshmen retention.
[13] Chai, K. E., & Gibson, D. (2015). Predicting the Risk of Attrition for Undergraduate Students with Time Based Modelling. International Association for Development of the Information Society.
[14] Jia, J. W., & Mareboyana, M. (2014). Predictive models for undergraduate student retention using machine learning algorithms. In Transactions on Engineering Technologies, Springer, Dordrecht, pp. 315-329.
[15] Thiessen, M., (2019). European Qualifications Framework: supporting learning, work and cross-border mobility, 10, p.39, ISBN 978-92-76-02769-0.
[16] Grigirov, V., Angelov, J. & Savov, S. (2014). Fragmentation and scope of expenditure elements for determining the prime cost of student training. Proceeding of the University of Ruse, Vol 53 (9), pp. 72-75, ISBN 1311-3321.
[17] Sarsar, F., Kale, Ö. A., Andiç-Çakır, O., Gueorguiev, T., Evstatiev, B., Georgieva, T., Kadirova, S., Mihailov, N., Różewski, P., Kieruzel, M., Lipczyński, T., Prys, M. & Leeuwen, M. (2021). Multicultural investigation of the students' acceptance of using digital learning materials in laboratory classes Computer Applications in Engineering Education, 29 (4), 883-896, doi: 10.1002/cae.22322
[18] Nikolaeva, D., Ronkova, V., (2022). Students in engineering bachelor courses and their preliminary general technical background, Proceedings of the University of Ruse, Vol 61 (4.1), pp 34-38, ISSN: 1311-3321
[19] Ronkova, V., (2020). Development of training in Engineering Graphics in higher schools, Publishing Center at the University of Rousse "Angel Kanchev", pp. 1- 127, ISBN: 978-954-712-798-2
[20] Dobreva, A. (2013). Theoretical Investigation of the Energy Efficiency of Planetary Gear Trains. Mechanisms and Machine Science, No 13, pp 289-298.
[21] Dobreva, A. (2013). Methods for Improving the Geometry Parameters and the Energy Efficiency of Gear Trains with Internal Meshing. VDI – Berichte, No 2199.2, 1291 – 1302,
[22] Dobreva, A. & Pavlov. P. (2021). Energy Efficiency of Worm Gear Drives, Annals of DAAAM & Proceedings. 2021, Vol. 10 Issue 2, ISBN 978-3-902734-33-4, ISSN 1726-9679, DOI 10.2507/32nd.daaam.proceedings.016, pp 110-116.
[23] Kamenov, K., Dobreva, A. & Ronkova, V. (2017). Advanced Engineering Methods in Design and Education, IOP Conference Series: Materials Science and Engineering, 252 (1), 012033, DOI 10.1088/1757-899X/252/1/012033, pp. 1-6.
[24] Stoyanov, S. & Dobreva, A. (2021). Systems Analysis and Design of Gear Drives through Innovative Software Approach. In: ISMSIT 2021 - 5th International Symposium on Multidisciplinary Studies and Innovative Technologies, Proceedings, pp 366–370.

DOI: 10.2507/34th.daaam.proceedings.018

DETECTION OF PARKINSON'S DISEASE USING KEYSTROKE DATA AND MACHINE LEARNING ALGORITHMS

Vyacheslav V. Potekhin & Ogul Unal

This Publication has to be referred as: Potekhin, V[yacheslav] & Unal, O[gul] (2023). Detection of Parkinson's Disease Using Keystroke Data and Machine Learning Algorithms, Proceedings of the 34th DAAAM International Symposium, pp.0128-0136, B. Katalinic (Ed.), Published by DAAAM International, ISBN 978-3-902734-41-9, ISSN 1726-9679, Vienna, Austria
DOI: 10.2507/34th.daaam.proceedings.018

Abstract

The human brain is still the most complex organ and still, not fully understood. The brain has ability to adapt, develop, learn, memorize, and perform daily tasks during our survival has inspired us to build artificial intelligent machines and apply machine learning in many fields. Not only does it control our thoughts, emotions, movements, or actions, but the human brain is also responsible for controlling our organs and any unintentional actions. Strong as it is, disorders related to the human brain may be severe and complex to heal or perhaps not curable. Even though we are living at the peak of our technology. However, we could not fully understand how our brain works as well as the reasons for many diseases. One of them is Parkinson's disease. A specific disorder related to dying neurons in our brain due to the unknown existence of proteins in a specific area. Every single corner of the brain is responsible for a specific task therefore any issue happening in the related area can cause a complex health disorder. When the area responsible for creating dopamine which is used for organizing our emotions as well as intentional movement is affected, a person might have Parkinson's disease. The purpose of this work is to analyze the disease and make predictions based on finger movement data that is related to the symptoms of the disease. In this research medical knowledge, medical data, and machine learning algorithms are used to build to model. The models are related one of the the main symptoms of the disease. In this research, different preprocessing techniques are used to increase the accuracy. The suggested models can give in a range of 76-95% accuracy based on medical data for Parkinson's disease.

Keywords: Machine Learning; Data Analysis; Parkinson's Disease; Feature Extraction; Ensemble Methods

1. Introduction

Parkinson's disease is a degenerative neurological disorder that is characterized by the onset of tremor, muscle rigidity, slowness in movement (bradykinesia), and stooped posture (postural instability). Parkinson's is the second-most common neurodegenerative disease after Alzheimer's disease. More than 10 million people worldwide are living with PD. The brain is responsible for many tasks and a combination of different brain regions can perform complex tasks such as movement from the Basal ganglia area. The basal ganglia have direct/indirect pathways to control our movement. When neurodegeneration which means loss of neuron cells occurs, the affected area causes disorders. In the system of basal ganglia, the substantia nigra is a small area responsible for creating dopamine. Dopamine is a neurotransmitter that enables neurotransmission (electrochemical communication between neurons) and affects our behavior with reward feelings as well as during movement.

This means Parkinson's disease is a disease of basal ganglia related to dying neurons in the substantia nigra and depletion of dopamine. There are different symptoms of PD. Those are: Static tremor: or rhythmic shaking, usually begins in a limb, often hand or fingers. The shaking may decrease while performing tasks. Impaired posture: The posture of the patient may become stooped. Rigid muscles: Muscle stiffness may occur in any part of the body. The stiff muscles can be painful and limit the range of motion. Slowed movement (bradykinesia): Over time Parkinson's disease may slow the movement, making simple tasks more difficult to perform. The steps of the patient may become shorter while walking. Speech changes: The patient may speak quickly, softly, or be hesitant before talking. The speech may be more of a monotone rather than having the usual speech behavior. The suggested model analyses finger movements using a keyboard. In this research, after getting raw data, different methods for preprocessing were used, such as IQR for outliers, mean/ std /Variance, and time difference derivations. The ensemble learning methods performed well in this model. The purpose of dealing with different data is; that many models are using one or some of these methods to detect PD. However, in this research, a combination of models was used. Combined models' results will be evaluated by medical experts to make final decisions, hence the system will be able to detect the disease or possibility of the disease.

2. Literature Review

2.1 Human Brain

The brain is a collection of tissue in the head of the body. The brain integrates sensory information takes input generates output, and directs motor responses; it is also the center of learning. The human brain weighs approximately 1.4 kg (3 lb) and is made up of billions of cells called neurons [1]. The connections between neurons, known as synapses, allow the transmission of electrical and chemical messages from one neuron to another in the brain, a process that underlies basic sensory functions and is critical for learning, memory formation and thinking, and other cognitive activities. The transmission is done using electrochemical activities with the help of neurotransmitters.

Neurons communicate through the body and they communicate with one another to transmit signals. However, neurons are not simply connected physically. Each neuron's end is a small gap that is a synapse and to communicate with the next cell, the signal has to be able to cross the space. This process is known as neurotransmission. One of the most important neurotransmitters is dopamine as it is mostly responsible for feelings of reward pleasure and movement. Dopamine is critical for controlling the start and stop of voluntary and involuntary movements [2]. There are many regions in the human brain responsible for specific tasks such as the limbic system [3] which is responsible for emotions, also Basal ganglia, a group of nuclei (clusters of neurons) in the brain that are located deep beneath the cerebral cortex. The basal ganglia specialize in processing information on movement and in fine-tuning the activity of brain circuits that determine the best possible response in a given situation [4]. The basal ganglia also mitigate and control functions ranging from cognitive planning, emotions, and reward functions, and even cognition and learning. The contributions of the Basal Ganglia are still not understood. One hypothesis (direct/indirect pathway model) suggests that the basal ganglia act to facilitate desired movements and inhibit unwanted and/or competing movements [5]. According to the direct/indirect pathway, when there is an intentional movement, a signal to initiate the movement is sent from the cortex to the basal ganglia, arriving at the caudate or putamen. Afterward, the signal follows a circuit in the basal ganglia known as the direct pathway, which leads to the silencing of neurons in the globus pallidus and substantia nigra [5]. This releases the thalamus from the inhibitory effects of the basal ganglia and allows movement to occur. There is also another circuit within the basal ganglia that is the indirect pathway, which involves the subthalamic nucleus and leads to the increased suppression of unwanted movements. It is believed that a balance between activity in these two pathways may facilitate smooth movement.

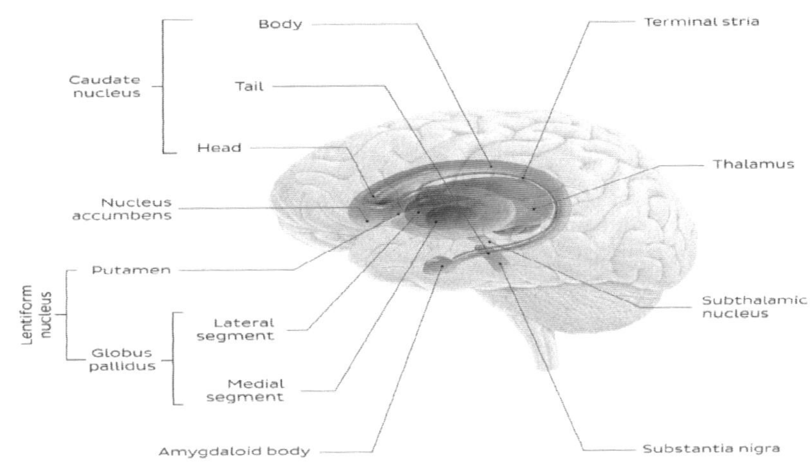

Fig. 1. group of nuclei (Basal Ganglia) and it's position in human brain

The substantia nigra (SN) is a midbrain dopaminergic nucleus that has a critical role in modulating motor movement and reward functions as part of the basal ganglia circuitry [6]. The substantia nigra is classically considered to be the primary input into the basal ganglia circuitry and a critical element to these functions [6]. When these subcortical nuclei are damaged such as during neurodegeneration or a stroke, neurological conditions may result, such as Parkinson's disease, schizophrenia, and Huntington's disease.

Dopamine is produced by neurons in the region at the base of the brain, and tyrosine (amino acid) is converted into another amino acid, named L-dopa. Afterward, L-dopa undergoes another change, as enzymes turn it into dopamine. It is known that low dopamine level causes stiff movements that are the hallmark of Parkinson's disease. Basal ganglia organize movements to occur or stop with the motor cortex therefore, decreased levels of dopamine due to neurodegeneration in the Substantia nigra area causes disorganization of the thalamus and basal ganglia causing Parkinson's disease.

2.2 Neuro Degeneration

Neurodegeneration is the progressive decline of neural tissue, it is a slow and progressive loss of neuronal cells in specified regions of the brain and is the main pathologic feature of Alzheimer's disease, Parkinson's disease, Amyotrophic lateral sclerosis, Huntington's Disease and Lewy Body Dementia. Each neurodegenerative disease of the brain is caused by the progressive accumulation of a specific protein inclusion (proteinopathy) [7]. Over time this accumulation becomes toxic to the brain leading to irreversible degeneration (death) of neurons and atrophy.

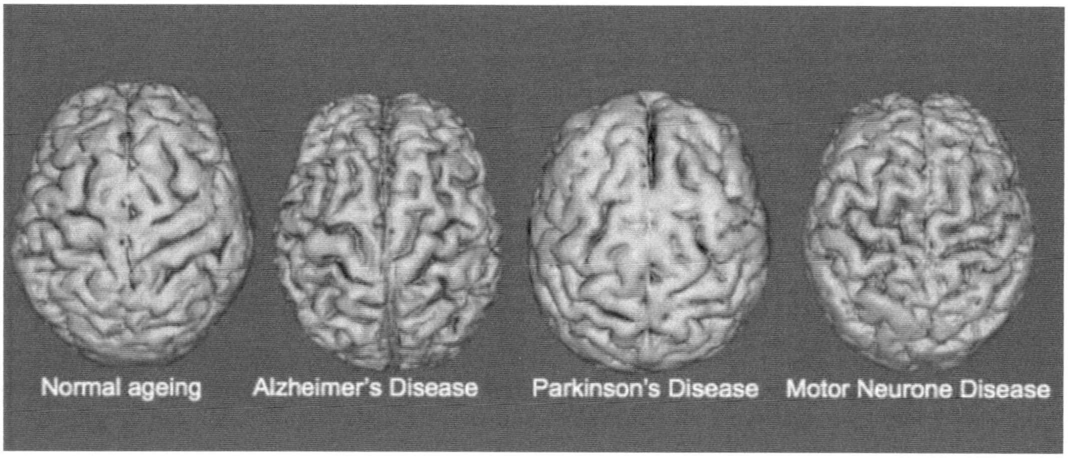

Fig. 2. Example diagram of normal and neurodegenerated brain

2.3 Parkinson's Disease

Parkinson's disease is a progressive disorder which affects the nervous system and the parts of the body controlled by the nerves. Symptoms of PD start slowly. The disease was first described in 1817 by British physician James Parkinson in his Essay on the Shaking Palsy [8]. The first symptom might be a tremor in the hand. Tremors are the common signature of the disease, however, it might also cause slowing of movement or stiffness. In the early stages of PD, the face of the patient may show little or no expression. arms may not swing while walking, the speech may become softer. PD symptoms worsen as the condition of the patient progresses over time. Parkinson's disease can't be cured and the reasons for the disease are still unknown.

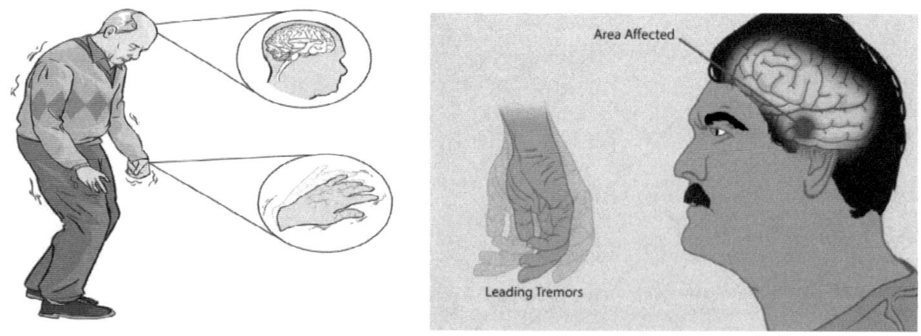

Fig. 3. Parkinson's disease disorder demonstration

Parkinson's is a disease of the basal ganglia. Classic clinical symptoms include bradykinesia (hands or fingers), resting tremors, postural instability, and shuffling gait (during walking). This disease is a result of neurodegeneration of the SNpc (substantia nigra pars compacta) dopaminergic neurons [8]. It is a degenerative neurologic disease marked by a constellation of movement and non-movement symptoms and associated with the presence of Lewy bodies in the brain. As dopamine starts to decrease, signs and symptoms of Parkinson's disease begin to appear [9].

2.4 Parkinson's and Finger movement

In Parkinson's disease, resting tremors most often occur in the hands by rotational movements of the fingers, especially the thumb and index finger. Even though we could observe a resting tremor, sometimes they could assume a pose suppressing the tremor. The finger-tapping test evaluates bradykinesia by focusing on decreasing frequency, amplitude, or both with repetitive actions. The vertical position of the hands during this task may also have clinical significance. Slowing down the speed, loss of amplitude, and pauses when tapping with fingers are consistent with "bradykinesia" — slowness of movement, which is one of the most important symptoms of Parkinson's disease [11]. Research has shown that subjects with Parkinson's disease tapped their fingers "slower and less rhythmically" compared to "healthy" subjects [11]. Based on this knowledge, the keyboard usage datasets were used. There are various kinds of research aiming at early detection of PD. In [10], focuses on vocal detection of Parkinson's disease while in [9] this research focuses on Gait data, and walking patterns during a walk. In [12], hand movements were analyzed, in our research, we analyzed the problem using 2 different datasets and made feature extraction based on outliers in data. Outliers in datasets might represent the moments when bradykinesia occurs.

3. Methods & Materials

3.1 Analysing with Finger movements using Tappy Keystoke Data

In this section, the Tappy keystroke dataset is a raw dataset for the detection of early Parkinson's Disease using multiple characteristics of finger movement while typing [13]. The dataset is created using keystroke logs collected from subjects with and without Parkinson's Disease. The purpose of this implementation is to analyze the routine interaction with computer keyboards can be used to detect changes in the characteristics of finger movement in the early stages of PD [14]. To apply feature extraction, we created additional features extraction as, Flight time and hold time can indicate meaningful information as Parkinson's Disease affects finger movements in early stages [15]. In this section subjects with tremors (mild, medium, or severe) were used for classification. The section aims to classify healthy controls and subjects that have the possibility or already have Parkinson's Disease. Each collected data includes information about the typing time when the participants used their various computer applications (for example email, web browser, and so on). The keystroke logging software ("Tappy") was used to record the activities. The features of this section are below:

- Parkinson's: Whether they have Parkinson's Disease [True/False]
- Whether there is sidedness of movement [None /Left/Right]
- Impact: The Parkinson's disease severity or impact on their daily life [Mild/Medium/Severe] For each subject, a file that contains keystroke details was recorded.
- Hold time: Time between press and release for current key milliseconds
- Direction: Previous to current RL, RR, LL, LR (and S for a space key)
- Flight time: Time between the release of the previous key and the press of the current key in milliseconds

Below, mean (1), standard deviation (2), and variance (3) formulas are shown as they were used to extract new features.

$$\mu = \frac{\sum_{i=1}^{n} xi}{n} \tag{1}$$

$$\sigma = \sqrt{\frac{\sum_{i=0}^{n} (xi - \mu)^2}{n}} \tag{2}$$

$$\sigma^2 = \frac{\sum_{i=1}^{n} (xi - \overline{x})^2}{N} \tag{3}$$

Using the features related to hold time and flight time, new features were created to improve the work. Using hold time, hold time-mean and hold time-std were created. Flight time, mean of flight time Left to Right hand, and Right to left hand were also used. For each, several actions bigger than average are detected using the principle of; if the selected value is bigger than the mean value of the corresponding column. Outliers in this research were detected using the IQR method and manually (explained in the next section). Finally, the percentage of each value was calculated (4) using the formula below. To create the model ensemble learning methods were used.

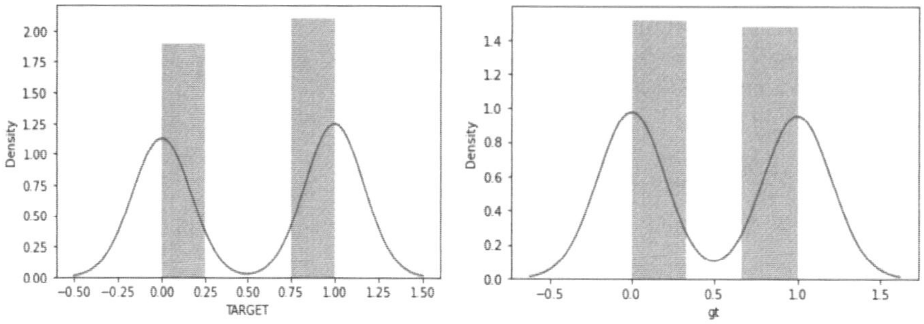

Fig. 4. Class distribution of Tappy keystroke and neuroQWERTY data respectively

$$\text{Percentage} = \frac{\text{Outlier}}{\text{length(flighttimeAction)}} * 100 \tag{4}$$

3.2 Analysing with Finger movements using NeuroQWERTY data

The MIT-CSXPD neuroQWERTY database contains keystroke logs collected from 85 subjects who are healthy or having Parkinson's disease (PD). Each collected data file includes time information collected during typing sessions using a standard Lenovo G50-70 i3-4005U word processor with 4 MB of memory and a 15-inch screen running Manjaro Linux [15]. The subjects were instructed to type as they would normally do at home [15]. Below are the default data columns without further feature extraction.

- pID – Subject ID
- gt – If they had or not they had PD
- updrs108 - Unified Parkinson's Disease Rating Scale part
- afTap - Alternating finger-tapping result
- sTap - Single key-tapping result
- nqScore - neuroQWERTY index
- Typing speed
- file_n - The csv file(s) containing the subject's typing data

Each keystroke data csv file has four columns which give:

1. The key pressed.
2. The hold duration in seconds.
3. The key release time in seconds from time 0.
4. The key press time in seconds from time 0.

In this work, to increase the model accuracy, new features were extracted by detecting outliers that occur during the records. The hold duration in seconds, was used as a feature. Mean, standard deviation, and variance of hold time were added. Subtraction of key release time and Keypress time was added (5). The extracted features belong to each corresponding subject.

$$Subtracted = keyReleaseTime - keyPressTime \tag{5}$$

Parkinson's Disease affects finger movements as delays in hand movements occur [16]. The purpose of this section is to analyze the dataset to extract new columns that represent delays during the finger taps. To detect the outliers the IQR method was used. The quartiles (Q1, Q2, Q3) divide the dataset into four groups, each containing about 25% (or a quarter) of the data points. There are three quartiles: Q1, Q2, and Q3. Q1 (it is the first quartile) is the 25th percentile of the data. Q2 (second quartile) is the 50th percentile (medium part) of the data. Q3 (third or last quartile) is the 75th percentile of the data [17]. The Interquartile Range (IQR) is the distance between the first and third quartiles. Subtract the first quartile from the third quartile to find the interquartile range (6).

$$\text{IQR} = \text{Q3} - \text{Q1} \tag{6}$$

Ensemble learning algorithms (The dataset with default features and the new dataset with extracted features were both used and compared) were applied. Updrs feature has a high correlation with the target column as Updrs is: the most widely applied rating instrument for Parkinson's disease [18]. Some sections of the UPDRS require multiple scores for each limb, with a possible maximum of 199 points.

A score of 199 on the UPDRS scale corresponds to the worst level (complete disability), and a score of zero corresponds to no disability). In the dataset, the UPDRS column was removed to test the model behavior with the actual features. As UPDRS can be used to indicate PD directly.

3.3 Model Evaluation

In this research, Random Forest and Catboost supervised learning algorithms were used to build the model. To test and evaluate; Accuracy: It's the ratio of the correctly labeled subjects to the whole pool of subjects. Accuracy is the most intuitive one. Accuracy (7) answers the following question: How many subjects did we correctly label out of all the patients? (True Positive, True Negative, False Positive, True Negative)

$$\text{Accuracy} = \frac{\text{TP} + \text{TN}}{\text{TP} + \text{FP} + \text{FN} + \text{TN}} \tag{7}$$

If it starts with True then the prediction was correct whether Parkinsonian or not, so a true positive is a PD person correctly predicted & a true negative is a healthy person correctly predicted. Oppositely, if it starts with False then the prediction was incorrect, so a false positive is a healthy person incorrectly predicted as PD & a false negative is a person with PD, incorrectly predicted as healthy.

4. Results

4.1 Analysing with Finger movements using Tappy Keystroke Data

The test results were obtained using feature extraction and machine learning algorithms. In the dataset, records were gathered by pressing, releasing, or pressing and holding a button afterward, measuring the timings between each interaction. Hold time in this case is an important feature. As it may represent stability or instability of the subject finger movement (rigidity, slowness, bradykinesia). PD affects motor functions thus, finger movements are also affected. Patients with Parkinson's disease have their motor activity deteriorate during a task to a greater extent with individual finger movements than with nonindividual finger movements. Thus, Parkinson's disease impairs individual finger movements to a greater extent than general hand movements. This difference reflects the more subtle cortical control required to stimulate and maintain this highly fractionated type of motor activity [20]. Just as during a cold day or when having arthritis can affect movement or cause stiffness, slowness of movement can make it difficult to move the hands and fingers [19]. Parkinson's disease can also cause unstable actions, in which the hand extends into an awkward position, making it difficult and sometimes painful to use. The hold time and flight time are the main features of this work. To increase the features, mean, standard deviation, and variance of hold time were used. Also during the flight time (time between releasing a button and pressing the next button), maximum and minimum values were taken as well as their mean, standard deviation, and variance values. In this section and the next one, outliers are important to have because outliers may represent something unusual as slowness or any unstable action during finger movements. There are two approaches in this work, The first one is because of the possible slowness during movement.

The formulas above represent the from Left to Right and From Right to Left finger movement. The first formula takes values only higher than the mean value of all flight time columns, latter takes 100 values closer to the Maximum value. The final feature except for the target column (0 healthy, 1 subject with PD) is severity as mild, medium, and severe. Ensemble methods offer several advantages over single models, such as improved accuracy and performance, especially for complex and noisy data. They can also reduce the risk of overfitting and underfitting by balancing the trade-off between bias and variance, and by using different subsets and features of the data. Combining multiple models provides overall best accuracies in this and other sections.

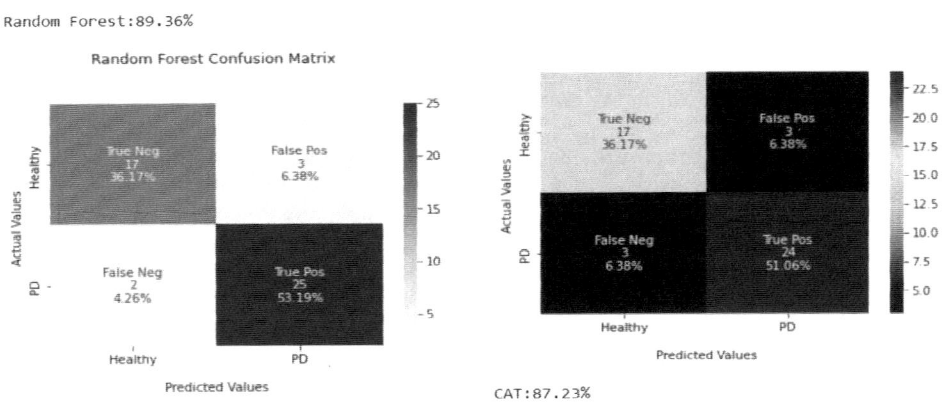

Fig. 5. Results of algorithms (test accuracies of ensemble learning methods)

4.2 Analysing with Finger movements using keyboard Part 2

In this section, the first implementation of NeuroQWERTY without feature extraction gave 76.47% accuracy whereas the final implementation dataset gave 94.12% accuracy. Both results are received using test data. The initial dataset contains similar features as afTap - alternating finger tapping result, nqScore, sTap - single key tapping result, and typing speed. The effects of feature extraction in medical data were experimented with and approved using different datasets and research [9] [21]. For each keystroke, there are .csv files where subject hold times and key press durations were recorded.

In the second implementation that gives over 90 accuracies, the features are below:

1. Gt: Target column to predict
2. afTap (initial feature): alternating finger tapping result
3. sTap (initial feature): single key tapping result
4. nqScore (initial feature): motor index derived from computer-key-hold-time data using an ensemble regression algorithm [68].
5. typingSpeed (initial feature)
6. holdMean (new feature): mean value of hold time
7. holdStd (new feature): standard deviation of hold time
8. holdVar (new feature): variance of hold time
9. subtractionMean (new feature): mean of the difference between release and press time
10. subtractionStd (new feature): std of difference between release and press time
11. subtractionVar (new feature): variance of the difference between release and press time
12. outlierSubtraction (new feature): detected outliers of subtraction of release-press time

In PD, bradykinesia means slowness of movement and speed as movements are continued. Bradykinesia is considered one of the main symptoms of PD. Thus, lack of dopamine affects intentional movement making the subject hesitant and slow during the interaction and these movements are detectable using finger movements data. In this section, outliers were used to create new features as release time and press time of the button were subtracted for each subject, afterwards outliers of the calculation were detected using IQR statistical method. The latter model for PD analysis gave the highest results amongst others with 94.12% test accuracy.

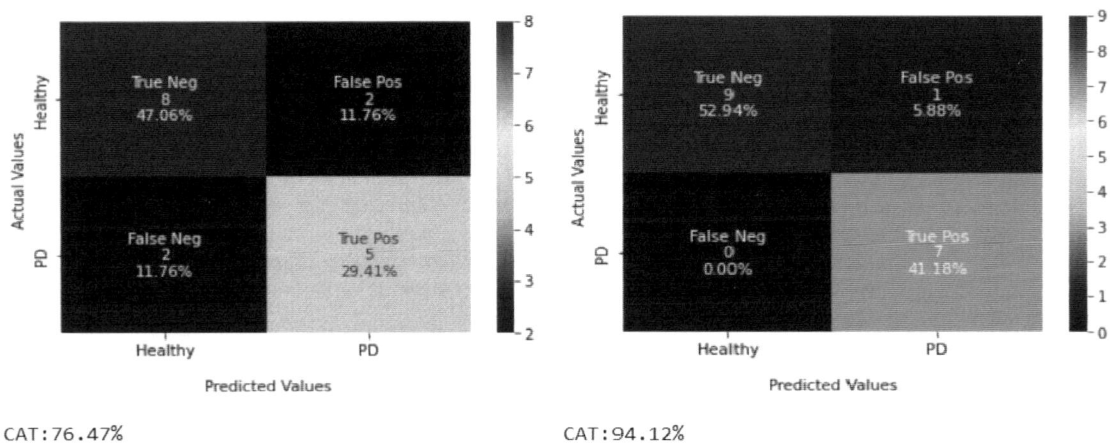

Fig. 6. Results of algorithms (test accuracies of ensemble learning methods) of second implementation

5. Discussion

Below is the chart, a full comparison between Random Forest and Catboost algorithms based on 2 different implementations and also after feature extraction for the second dataset (neuroqwerty2). Any effect coming from Basal Ganglia which is responsible for creating dopamine, can be detected in muscle movements in our body including finger movements. However, depletion of dopamine levels is a long and slow process therefore purpose is to detect the situation in earlier stages. Still, the reasons for Parkinson's Disease are unknown therefore another purpose of this research is to help understand why there is neurodegeneration occurs in the basal ganglia. PD disease is progressive which means in time it gets worse. Therefore it is crucial to detect this as early as possible. Dementia means that there is behavioral and/or cognitive impairment that significantly affects one's activities of daily living. However, it is an umbrella term that encompasses many different etiologies, including neurodegenerative, metabolic, vascular, and infectious diseases. A Lewy body is an abnormal aggregation or clumping of the protein alpha-synuclein.

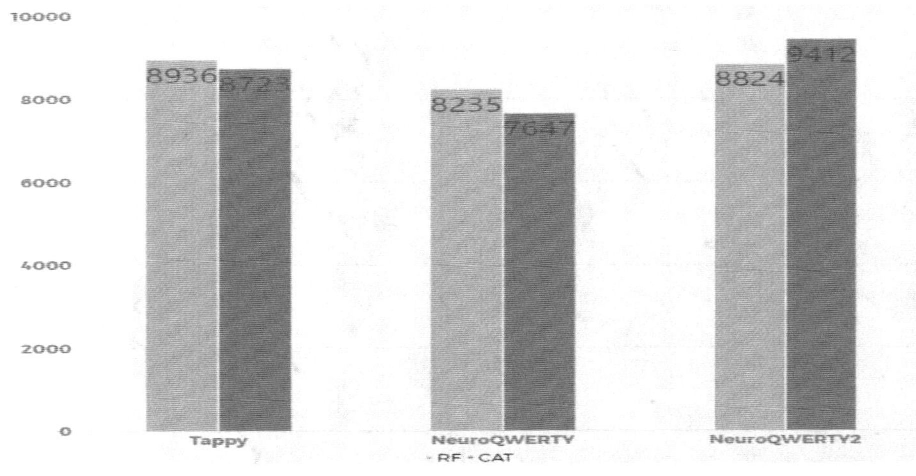

Fig. 7. Comparison chart of the results

There are two main reasons for using ensemble methods, rather than a single model, and these are: A set of learners can make better predictions and achieve better performance from a single contributing model. An ensemble model reduces the spread or scattering of predictions and model performance. Ensemble learning is used to achieve better prediction performance than a single predictive model in an ML implementation. The way is, to add bias to reduce the variance component of the prediction error.

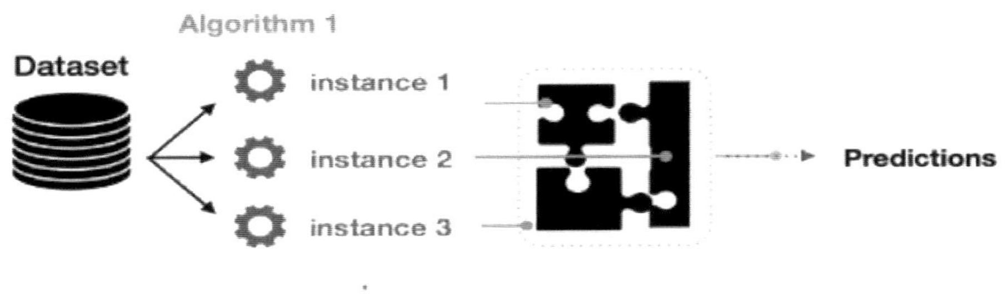

Fig. 8. Ensemble methods were applied in this research

The model can be used as an interpreter between a patient/subject and a medical expert. To test, the subject has to type using a keyboard With the help of collected data, the model can predict the possibility of having PD disease in the subject. Finally, medical experts can analyze the results. It is clear that neurodegeneration in substantia nigra occurs slowly thus it can start at earlier ages of the subject as well. Many people might not be aware of early signs that can, unfortunately, transform into PD disease in the later stages of their life. Providing an awareness and solutions model would help to increase the life quality of the subject as well as predict the disease.

6. Conclusion

The brain regions are responsible for different actions in the human body. Neurons are connected and they communicate with each other all the time during our life. Throughout our life, neurons die and the amount of the neurons decrease. However, when this process is severe and faster than usual, this causes neurodegeneration. The affected area of neurodegeneration causes specific disorders. Ensemble learning algorithms gave the highest accuracies in this research, It is understood that instead of using one model, using multiple learners provides higher accuracies and more reliability (randomness, more realistic approach). One disadvantage of the model is, that the ensembling is less interpretable, and the output of the ensembled model is hard to explain similar to a black box.

The finger movement using the keyboard typing approach was successful with 94.12% accuracy. The initial data gave 76.47%, after using an outlier strategy to add more features, the model was able to perform much better. With the help of this approach, it is understood that PD has many symptoms, one of them being bradykinesia (slowness, hesitation) using outlier detection and creating features based on subtraction of outliers vs. mean/ std values was a successful implementation. The finger movement is also a movement recording data which means, the model is looking for hesitation or slowness during typing. In future work, EMG signals and sleep EEG signals will be added to the main model to make it stronger and more reliable. Even though models gave acceptable results, there is a possibility for False positives or False negatives.

7. Acknowledgments

Firstly, I would like to thank my father Dr. Turgay Ünal for his help in Parkinson's Disease theoretical studies, also would like to thank Associate professor Vyacheslav Potekhin for all his help during my career. Finally, I thank my mom Dr. Seher Gülçin Ünal for her assistance in giving all the support in my life. This research is being conducted at the Peter the Great St. Petersburg Polytechnic University and partially funded by the Ministry of Science and Higher Education of the Russian Federation as part of World-class Research Center programme: Advanced Digital Technologies (contract No. 075-15-2020-934 dated 17.11.2020).

8. References

[1] Britannica, The Editors of Encyclopaedia. "brain". Encyclopedia Britannica, Available from: https://www.britannica.com/science/brain Accessed: 2023-04-19

[2] Marianne, K., Daniella, S., Ariel, R., David, N., Jackson, C. & Ricardo, G. (2019). Dopamine: Functions, Signaling, and Association with Neurological Diseases", Cellular and Molecular Neurobiology, pp.1-30., doi: 10.1007/s10571-x x018-0632-3

[3] Vyacheslav, V., Potekhin, Ogul, U.(2021). Analysis of Emotions using EEG Data and Machine Learning", Proceedings of the 32nd DAAAM International Symposium, pp.0158-0167., doi: 10.2507/32nd.daaam.proceedings.025

[4] Parr-Brownlie, Louise, C, & John, N. "basal ganglia". Encyclopedia Britannica, Available from: https://www.britannica.com/science/basal-ganglion Accessed: 2022-11-03

[5] Lanciego, JL., Luquin, N., Obeso, JA. (2012). Functional neuroanatomy of the basal ganglia", Cold Spring Harb Perspect, doi: 10.1101/cshperspect.a009621, PMID: 23071379

[6] James, S., Vamsi, R., Morris, R., (2022). Neuroanatomy, Substantia Nigra, National Library of Medicine"

[7] Li, D., Liu, C. (2022) Conformational strains of pathogenic amyloid proteins in neurodegenerative diseases", Nat Rev Neuroscience, doi: doi.org/10.1038/s41583-022-00603-7

[8] Kim, S., Sung, JY., Um, JW., Hattori, N., Mizuno, Y., Tanaka, K., Paik, SR., Kim, J & Chung, KC. (2003). Parkin Cleaves Intracellular -Synuclein Inclusions via the Activation of Calpain".

[9] Ogul, U., Potekhin V. (2023). Implementation of machine learning algorithms for Parkinsonian gait data", Computing, Telecommunication and Control, Vol. 16, No. 1, pp. 69–78., doi: 10.18721/JCSTCS.16106

[10] Parisi, L., Ravi, N., Manaog, M. (2018). Feature-driven machine learning to improve early diagnosis of Parkinson's disease", Expert Systems with Applications, vol. 110, pp. 182–190., doi: 10.1016/j. eswa.2018.06.003

[11] Picillo., Marina. (2015). Learning More from Finger Tapping in Parkinson's Disease: Up and Down from Dyskinesia to Bradykinesia", Movement disorders clinical practice vol. 3, pp. 184-187., doi:10.1002/mdc3.12246

[12] Moshkova, A., Samorodov, A., Voinova, N., Volkov, A., Ivanova, E. & Fedotova, E. (2020). Parkinson's Disease Detection by Using Machine Learning Algorithms and Hand Movement Signal from LeapMotion Sensor", 26th Conference of Open Innovations Association, pp. 321-327., doi: 10.23919/FRUCT48808.2020.9087433

[13] Goldberger, A., Amaral, L., Glass, L., Hausdorff, J., Ivanov, P. C., Mark, R. & Stanley, H. (2000). PhysioBank, PhysioToolkit, and PhysioNet: Components of a new research resource for complex physiologic signals", pp. 215–220

[14] Adams, R.(2017). High-accuracy detection of early Parkinson's Disease using multiple characteristics of finger movement while typing". Plos One, doi: doi.org/10.1371/journal.pone.0188226

[15] Giancardo, L., Sánchez-Ferro, A., Arroyo-Gallego, Butterworth, T., Montero, P., Matarazzo, M., Obeso, J., Gray,M. (2016). Computer keyboard interaction as an indicator of early Parkinson's disease", Scientific Reports 6, doi: 10.1038/srep34468

[16] Stegemöller, Elizabeth, L. (2009). Effect of movement frequency on repetitive finger movements in patients with Parkinson's disease" Movement disorders: official journal of the Movement Disorder Society, vol. 24, doi:10.1002/mds.22535

[17] Frederik, D., Cornelis, K., Hendrik, Lopuhaä., Ludolf, Meester., (2005). A Modern Introduction to Probability and-Statistics"

[18] Ebersbach, G., Baas, H., Csoti, I., Müngersdorf, M., Deuschl, G. (2006). Scales in Parkinson's disease", doi: http://dx.doi.org/10.1007/s00415-006-4008-0

[19] Dounskaia, N., Van, Gemmert, AWA., Leis, BC., Stelmach, GE. (2009). Biased wrist and finger coordination in Parkinsonian patients during performance of graphical tasks", Neuropsychologia.

[20] Agostino, R., Currà, A., Giovannelli, M., Modugno, N., Manfredi, M., Berardelli, A. (2003). Impairment of.......individual finger movements in Parkinson's disease", Mov Disord, doi: 10.1002/mds.10313. PMID: 12722170

[21] Potekhin, V., & Ogul, U. (2022). Analysis of Acute Myocardial Infarction using ECG Signals and Machine .Learning Algorithms", Proceedings of the 33rd DAAAM International Symposium, pp.0205-0214., doi:.10.2507/33rd.daaam.proceedings.029

DOI: 10.2507/34th.daaam.proceedings.019

APPLICATION OF MOBILE TECHNOLOGIES TO INCREASE THE EFFICIENCY OF LANGUAGE COMMUNICATIVE SKILLS OF ENGINEERING STUDENTS

Oleg Ipatov, Darina Barinova, Maria Odinokaya, Elena Krylova & Sofia Saramakha

Peter the Great St.Petersburg Polytechnic University

This Publication has to be referred as: Ipatov, O[leg]; Barinova, D[arina]; Odinokaya, M[aria]; Krylova, E[lena] & Saramakha, S[ofia] (2023). Application of Mobile Technologies to Increase the Efficiency of Language Communicative Skills of Engineering Students, Proceedings of the 34th DAAAM International Symposium, pp.0137-0145, B. Katalinic (Ed.), Published by DAAAM International, ISBN 978-3-902734-41-9, ISSN 1726-9679, Vienna, Austria
DOI: 10.2507/34th.daaam.proceedings.019

Abstract

The paper presents the didactic possibilities of mobile technology application in foreign language teaching of engineers. A methodical analysis of the application of mobile learning tools was carried out and the tool that is the most effective for the development of foreign language communicative skills was singled out. Algorithms for using of mobile technology are given in the paper, the results of the experimental research have been analysed and conclusions have been drawn. The article will serve the interest for those who research digital trends in education as well as those who are interested in introducing new teaching approaches into the educational process of engineers.

Keywords: mobile technologies; mobile learning; engineers; foreign language.

1. Introduction

With the advent of digitized education, which provides vast amounts of knowledge and information at any time and place, the need for special competencies in education and learning has arisen. Today, no one questions the fact that it is not just possible, but necessary to introduce digital technologies into the pedagogical process. By the term "mobile technology" handheld devices with mobile applications and Internet connectivity are meant. They allow to learn anywhere and anytime, as well as provide access to up-to-date information and individualized learning. The use of mobile technologies provides flexibility, relevance and efficiency of future engineers training. They contribute to the development of communicative skills adapted to the professional needs of engineers, as well as to the improvement of the quality and efficiency of the educational process. When teaching a foreign language at a university, the most important, in our opinion, is the development of foreign language professionally oriented communicative competence [1], [2]. The aspects of engineering students' communicative skills development are diverse and correlate with different components of the educational process [3], [4].

That is why the priority direction for teachers is modelling of the most relevant and demanded communicative situations with the help of mobile technologies, which will allow to adapt future engineers to the peculiarities of professional communication to the maximum extent possible. The relevance of the problem stated in the study is due to the need to improve the level of students' oral communicative skills using mobile technologies and the lack of research in this area in teaching engineering students. The aim of the research: to develop a set of exercises using mobile technologies in order to develop the oral foreign language communicative skills of engineering students. Practical significance lies in the development of a set of exercises with the help of mobile learning tools that develop oral foreign language communicative skills of engineering students. Scientific novelty of the research is in the scientific and methodological substantiation of mobile technology application, aimed at the development of oral foreign language communicative skills of engineering students, which expands the didactic possibilities of the educational process in learning a foreign language. To achieve the goal, the following tasks were set:

- to study didactic properties and possibilities of mobile technologies application in foreign language teaching;
- to determine the methodological principles of teaching future engineers communicative skills using mobile technologies;
- to analyze the methodological potential of mobile technologies application in teaching foreign language to engineering students;
- to present a system of exercises oriented to the development of oral foreign language communicative skills of engineering students;
- to experimentally test the use of this system and analyze the results obtained.

The didactic essence of the application of these learning tools is to create learning systems that are aimed at the dialogical nature of learning, taking into account the personal characteristics of learners. Thus, mobile learning using mobile devices makes learning truly personalized. This type of learning allows students to choose not only the content of training taking into account their interests, but also the time that students can allocate to training, which makes mobile learning fully student-cantered.

The benefits of using mobile technology for learning:

- Access to learning anywhere, anytime.
- Portability and accessibility of mobile devices.
- Possibility of individualization of learning through personalization of learning content.
- Support for learners with disabilities.
- Constant access to all content, to control and self-control materials, to personal files of their cloud storage.
- Transition from informative and communicative learning to interactive learning, participants of the process independently create content and share knowledge in collective activities.
- Faster dissemination of all types of learning material among students.
- Ease of providing educational communications within the group, with the teacher.

Mobile learning is more flexible, non-formal, interactive and individualized. Thus, the introduction of mobile tools in education allows to expand the scope of the curriculum beyond the walls of the educational institution, as it allows you to travel and learn; increases motivation and interest in the learning process, promotes effective learning, allows people with disabilities to learn; promotes better learning and memorization; supports the research process [5]. Using the advantages of mobile learning expands the possibilities of interaction between all participants of the educational process, allowing teachers to apply various forms of learning activities in both in-class and extracurricular activities, more effectively organize independent work of students [6], which contributes to the development of professional competencies of today's professionals.

The following popular mobile technologies stand out at the moment:

- Calendar-planner for controlling the schedule of classes, events and checkpoints (Google Calendar).
- Messengers that can be used to send notifications and reminders about scheduled training events (WhatsApps, Viber, Telegram), as well as individual and group counselling of students.
- Distribution of electronic materials necessary for the class via e-mail and messengers, as well as access to electronic textbooks, dictionaries and reference books (multimedia learning web-resources).
- Multimedia tools for presentation of lecture material (Miro).
- Educational audio files, video files and educational podcasts that can be used for teaching a foreign language (YouTube).
- Webinars for online lectures and training on platforms such as teachbase.ru.
- Platforms for open e-learning (Stepik, OpenEdu, Coursera).
- Cloud services and applications for doing practical work, creating group projects and organizing teamwork (Google Docs, Yandex.Disk).

Telegram can be a more convenient app for learning a foreign language than some other apps for several reasons. For example, compared to WhatsApp or Viber, Telegram has a built-in translator, which allows you to translate messages directly within the app without having to copy and paste text into a translator. In addition, Telegram provides a more flexible structure for organizing foreign-language communication, where you can communicate in a foreign language with other students and teachers, as well as for accessing materials and assignments.

Compared to Viber, Telegram has a more intuitive and simpler interface, which makes it easier to use as a language learning platform. In addition, Telegram allows you to use bots and integrate other applications that help you learn a foreign language, for example, to improve your vocabulary or to practice pronunciation, to learn technical vocabulary, and to access foreign language materials and video tutorials related to engineering topics.

Compared to Zoom and Skype, Telegram may be more convenient for learning a language on the go, as it does not require installation on a computer and can be used on mobile devices. It provides more opportunities to share materials and homework, as well as to communicate in a foreign language with other students and teachers outside of formal lessons.

Compared to other language learning apps such as Duolingo or Rosetta Stone, Telegram offers a more flexible and adaptive approach to language learning, allowing users to customize the learning process according to their needs. In addition, using Telegram for language learning can be more effective if users are already communicating in a foreign language with native speakers or other learners. Overall, Telegram can be one of the most convenient and effective applications for language learning.

A feature of Telegram is also the ability to create separate topics within a chat room, which makes communication more organized and structured. Each topic can have its own discussion thread, and users can easily find necessary messages and information, even if it was sent long ago, as there is a function to pin the message. These features are especially useful when teaching, as they allow you to create different topics for different aspects of the course, such as grammar, vocabulary, reading, etc. This helps students navigate through the material more easily and facilitates the language learning process.

The effectiveness of using Telegram is also confirmed in studies. First, Telegram has a wide audience of active users. According to statistical data, its Russian audience amounts to 48.8 million people per month. Secondly, researchers note the convenience of Telegram as a platform for creating training courses. For example, the study by T.A. Glebovich and A.D. Novikova emphasizes that Telegram users prefer informative, content-rich text without attention-holding elements, convenient and simple interface that allows creating individual selections of thematic content [7].

The study showed that the Telegram app can be an effective and simple tool for learning English and developing vocabulary. It can improve speaking and writing as well as reading and comprehension of texts. The application can also be used to teach vocabulary words and increase students' interest in learning a foreign language [8].

Telegram has also been confirmed to have a positive effect on improving reading skills in preparation for the IELTS exam. Thus, in the study of Shirinbakhsh&Saeidi, the first group gained knowledge in the traditional way and the second group gained knowledge through Telegram. Statistical analysis of the data showed that the experimental group performed better in the final test. During the study, which is also important, all students agreed that they liked learning through Telegram [9].

Research has shown that Telegram application can facilitate the learning of writing skills. Thus, the result of Alodwan's study, was that there were significant differences between the pre and post test scores of the experimental group, which confirms that using Telegram is an effective method of teaching writing skills. The researcher also noted that teaching using Telegram had a positive impact on the participants because students were required to save and record their writing [10].

A.J. Martin studied the effect of Telegram on foreign language learning motivation, language anxiety, and attitudes toward English language learning. It was found that Telegram instruction can increase the desire and inclination for language learning among English language learners. In addition, Telegram has been shown to have a positive effect on reducing anxiety in international language learners [11].

So, Telegram, a free online application, has it all: a cloud-based file storage system, the ability to organize collaborative online classes, chatbots and the ability to create your own customizable bot for any purpose. Working on different skills in Telegram is very easy, and the app itself is familiar to students and eagerly used by them.

Based on the abovementioned facts the main advantages of Telegram are:

- Messenger is not as boring as, for example, email or the abovementioned social networks: there are stickers, emoji. The ability to add interactive elements makes the messenger more modern and used among students.
- The possibility of microlearning: in their free time, a student can go through a course or answer questions from an instructor or bot, starting or stopping at any moment and at any place without losing progress.
- wide audience: by using Telergam, the student has the opportunity to communicate with a large number of people, bringing together different spheres of life on one platform and gaining knowledge and skills through this communication. The nature of the audience is also important - most of the audience, according to the analyzed sources, seeks self-development and knowledge acquisition. This is due to the fact that there are a huge number of educational channels, educational material and scientific communities in this messenger.
- Research shows that this messenger has a positive impact on students' motivation and reduces anxiety in the process of learning a foreign language.
- The availability of various services and chatbots that enhance the capabilities and functionality of this messenger and provide convenience to users. These tools are able to automate many tasks, such as sending and collecting homework, organizing tests, providing lectures and study materials.
- The ability to create a group chat. Teachers can create a chat group for foreign language students and encourage them to practice speaking with each other. Moreover, students can use messenger to communicate with native speakers of the target language and practice their language skills in a real context.

There is no doubt that for a more effective process of developing students' foreign language communicative skills it is necessary to organize the learning process in such a way that students can easily access information, share knowledge in online format, which is especially relevant after the pandemic period, it is necessary to use the most effective and modern tools for quality training of future engineers. That is why, the use of modern and up-to-date learning tools can increase student engagement, facilitate and qualitatively improve the learning process. Thus, it is mobile learning that is one of the most relevant types of learning today. This is due to the daily use of mobile devices, which consequently affects their active implementation in the field of education.

2. Methods

Theoretical foundations for the study are the methodological principles of teaching a foreign language in the format of mixed and online learning, principles of professionally-oriented teaching of a foreign language, as well as the basic principles of the communicative approach to teaching a foreign approach to foreign language teaching. The material of the study served as general scientific methods of theoretical research: literature analysis of domestic and foreign authors on the topic of research. As the main methods of research were questionnaire survey, analysis of scientific literature.

3. Experiment

When selecting a learning application, we were also interested in the opinion of the study participants. In view of this, a questionnaire was developed to identify the most used online resources by students in order to find the most convenient platform for teaching engineers. The questionnaire contained multiple choice questions on the topics of the most appropriate resources, messengers and social networks in foreign language learning.

The questionnaire was conducted by means of Google.Forms. 128 first-year master's students (56 female, 72 male students) of St. Petersburg Peter the Great Polytechnic University of engineering profile took part in the opinion poll as well as in the experiment. Students' age range: 20-23 years. The direction of training - engineering systems of buildings and structures. The following results were obtained. In response to the question, "What resources do you find convenient to use in your teaching?" the following data was obtained:

Resources handy to use in educational process

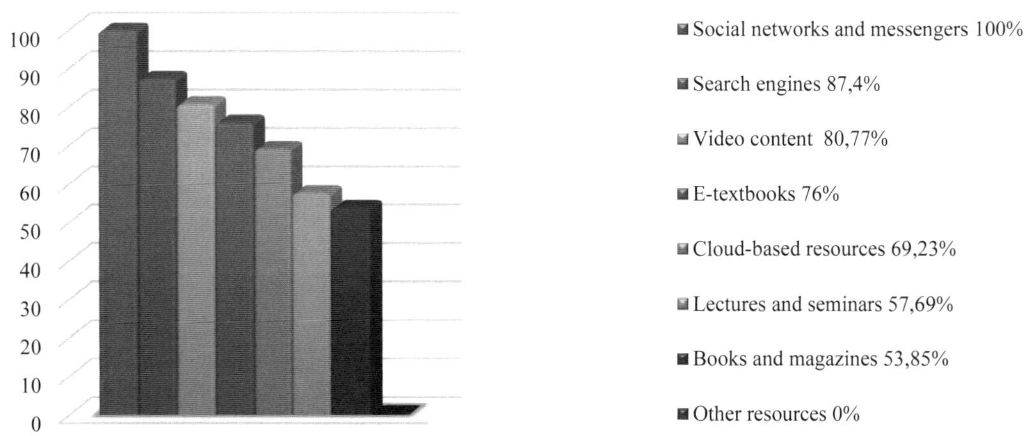

Fig. 1. Results of Questionnaire 1

The data show that the most popular are social networks and messengers, which were chosen by all respondents (100%). Search engines are also widely used (used by 87.4% of students), indicating that students prefer an independent approach to learning. Video content is also popular (80.77% of respondents), which may be due to the fact that visual materials can be more appealing and memorable than textual materials. E-textbooks and cloud resources are also chosen by the majority of students (76% and 69.23% respectively). Books and magazines were chosen by about half of the respondents (53.85%), indicating that traditional methods of learning also remain relevant to some students. Lectures and seminars were chosen by less than 60% of the respondents, which may indicate that students prefer more interactive forms of learning. Analysis of the answer to the question "What messengers and social networks do you prefer to use in educational process?" showed the following:

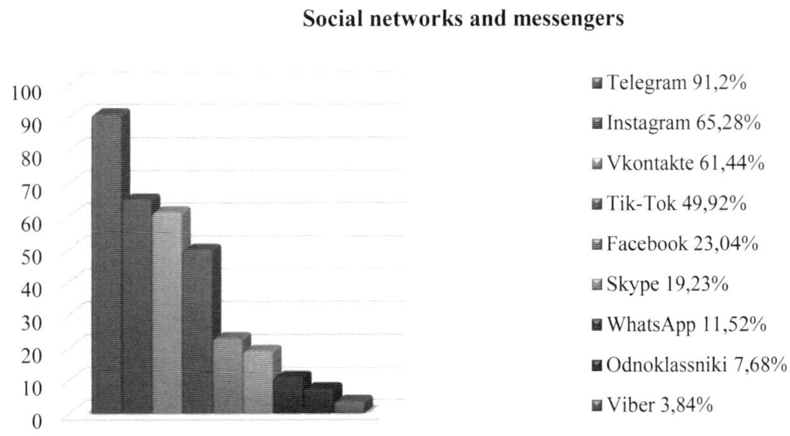

Fig. 2. Results of Questionnaire 2

Telegram is the most popular messenger/social network for use in teaching, among the respondents. It was chosen by 91.2% of respondents, indicating its high popularity in this area. Vkontakte and Instagram are also in high demand among the respondents, with 61.44% and 65.28% choice respectively. This may be due to the fact that they provide the opportunity to communicate with professors and other students, as well as access to various materials and content that can be useful for learning. Tik-Tok also proved to be a relatively popular social network, chosen by 49.92% of respondents. This may be due to the fact that a lot of content related to learning and language learning has recently appeared on the platform. WhatsApp, Viber and Skype proved to be the least popular choices for use in learning, with 11.52%, 3.84% and 19.23% choosing them respectively. This may be because they are less convenient for learning than other messengers/social networks and do not provide as many opportunities to share information and content. Analysis of the answer to the question "Where would it be more convenient for you to communicate with teachers and classmates?" has shown the following:

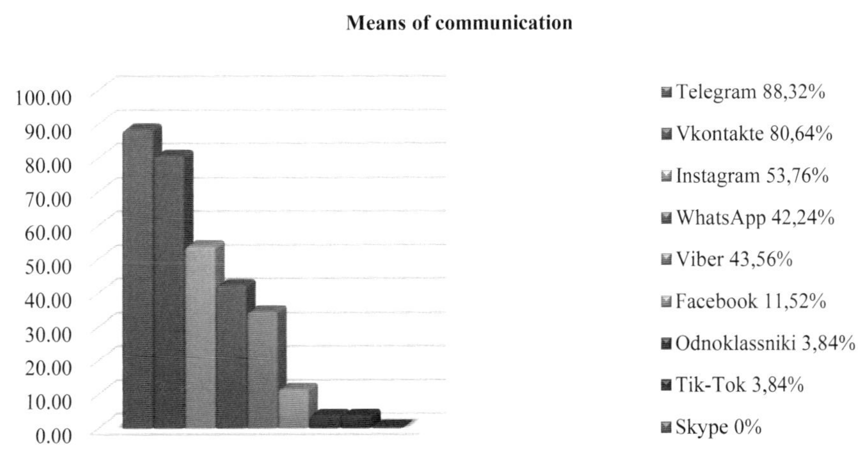

Fig. 3. Results of Questionnaire 3

The data show that the majority of respondents prefer to use messengers and social networks to communicate with the teacher and classmates. In particular, the most popular among respondents was Telegram - 88.32%, which may be due to the fact that this messenger has many features that facilitate the process of communication and organization of group work, including the creation of channels and groups, the ability to send files and exchange messages. The next is Vkontakte, which received the support of 80.64% of respondents. Vkontakte is a popular social network in Russia, and many students may already have an account on this social network, making it a convenient platform for communicating with faculty and classmates. Instagram is next, receiving support from 53.76% of respondents. This may be due to the popularity of this social network among young people and the possibility of using photos and videos for educational purposes. WhatsApp ranks fourth with 42.24% support, while Viber is in fifth place with 34.56% support. Both messengers are known for their ease of use and widespread popularity in Russia, but they probably do not have as extensive opportunities for structuring information as Telegram.

Facebook received 11.52% support, while Tik-Tok and Odnoklassniki received 3.84% support. Facebook and Odnoklassniki do not have the same widespread popularity among young people in Russia as Vkontakte and may be less convenient platforms for communication for educational purposes. Tik-Tok, on the other hand, is not convenient to use for texting purposes. Skype did not receive support among respondents, which may be due to the fact that it is less popular among young people in Russia and does not offer the same convenience as Telegram or Vkontakte. Analysis of the answer to the question "Which resource is the most useful for accessing different educational materials?" showed the following:

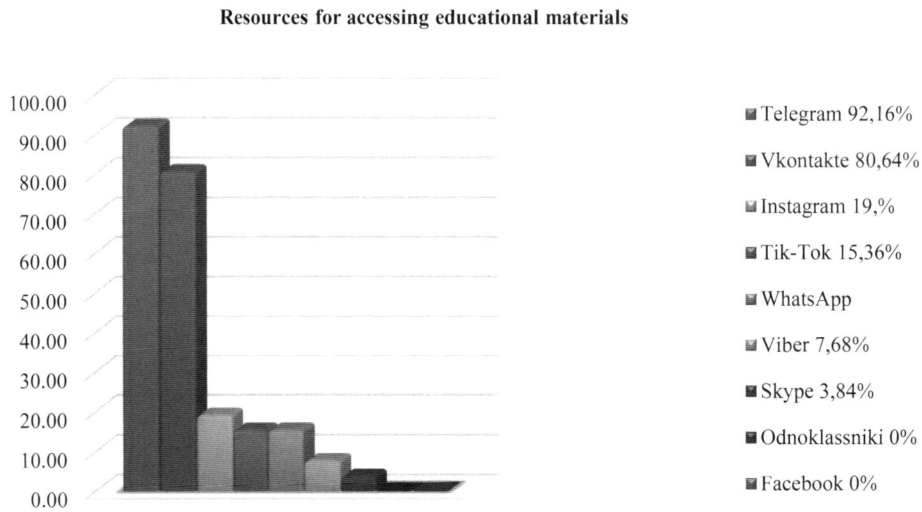

Fig. 4. Results of Questionnaire 4

According to the data, the most useful resource for accessing various educational materials is Telegram with the result of 92.16%. The second place is occupied by Vkontakte with the result of 80.64%, and the third place is occupied by Instagram with 19.2%. WhatsApp and Tik-Tok received approximately the same number of votes, with 15.36% each. Viber and Skype received the least number of votes, respectively, 7.68% and 3.84%. Thus, we can conclude that students prefer to use messengers and social networks to access educational materials, with Telegram and Vkontakte being the most popular among them. WhatsApp and Tik-Tok are used less frequently, and Viber and Skype are practically not used to access learning materials. As the results of the questionnaire have shown, students consider access to such resources as: search engines, electronic textbooks, social networks, messengers convenient for use in learning.

Based on the questionnaire data, we can conclude that Telegram is one of the most popular messengers for teaching a foreign language. More than 88% of respondents believe that Telegram is the most convenient pay-form for teaching foreign language to engineers. This is due to the fact that Telegram has a number of advantages, such as the ability to create separate topics within a chat room, a wide functionality for working with files, the ability to use bots and plug-ins, and a high level of security.

So, Telegram as a platform for developing oral foreign language communicative skills of future engineers was chosen. There is no doubt that for a more effective process of developing students' foreign language communicative skills it is necessary to organize the learning process in such a way that students can easily access information, share knowledge in online format, which is especially relevant after the pandemic period, it is necessary to use the most effective and modern tools for quality training of future engineers.

As a component composition of communicative skills the following key skills necessary for engineering students have been identified:

- Ability to carry out foreign language communication in the course of extracurricular activities.
- Ability to listen to the interlocutor.
- Ability to comment on the received information taking into account the communicative situation.
- Ability to describe and discuss industrial and professional issues.
- Ability to discuss topical issues and express one's point of view.

The tasks engineering students' training are considered, as they reflect the main content of professional training. The complex of exercises included 7 tasks, which were placed in Telegram and all the tasks had professional and communicative orientation.

Tasks of the first stage of training:

Exercise 1. "Speak on your experience in modernizing projects. Have you had to improve or change any projects?" Development rationale:

1. Foreign language communication skills in extracurricular activities: this exercise requires participants to express their thoughts and experiences in English, which helps to develop foreign language communication skills in the professional field.

2. Listening to the interlocutor. Participants of the exercise listen attentively to the stories of other participants posted in the comments.

3. Ability to discuss topical issues and express their point of view. The exercise allows participants to discuss topical issues related to project modernization. Each participant can express his or her point of view based on his or her experience and propose ideas for improving or changing projects.

Exercise 2. "Look at a slide from an engineering training course. Suggest examples of technical improvements that illustrate each one. Are there other items that could be added to the list?" Development rationale:

1. Foreign language communication skills in the course of extracurricular activities. This exercise requires participants to offer examples of technical improvements in English, which helps to develop foreign language communication skills in the professional field.

2. The ability to comment on the information received in relation to the communicative situation. Participants are required to offer examples of technical improvements that illustrate each point on the slide. This helps to develop the skill of analyzing information and commenting on it according to the communicative situation.

3. Describing and discussing production and professional issues. The exercise requires participants to describe specific technical improvements that can be applied in the work environment of engineering students. Participants can discuss the benefits and possible problems associated with the proposed improvements.

Exercise 3. "In the picture you see several professional questions. Your task is to choose one question and answer it, speaking for at least one minute". Development rationale.

1. Ability to carry out foreign language communication in the course of extracurricular activities. Participants are asked to choose one question and answer it in English for at least one minute. This requires them to use a foreign language to express their thoughts and develops their ability to carry out foreign language communication.

2. The ability to comment on the information received, taking into account the communicative situation. When participants answer a selected question, they should comment and develop their point of view, taking into account the communicative situation. They can give explanations, examples or arguments for their answers, which develops the ability to discuss information according to the communicative situation.

3. The ability to describe and discuss production and professional issues. The exercise requires participants to choose a question related to the professional field and give a detailed answer to it. This develops their ability to describe and discuss production and professional issues related to their field of work.

Participants are also asked to listen to a voice message describing our response to the first question proposed.

The addition of a voice message describing the answer to the first suggested question is an addition to Exercise 3. This allows participants to hear an example answer and visualize how their answer to the question can be structured and developed.

1. Listening to the interlocutor. Listening to a voice message allows participants to practice active listening skills and improve their ability to understand foreign language speech. 2. The ability to comment on the information received, taking into account the communicative situation. After listening to the voice message, participants can comment on and analyze the proposed answer, as well as review the structure, phrases and expressions used to develop their ability to understand the information.

Exercise 4. "Look at the pictures. Have you had a similar experience?" (in the sphere of engineering design making process). Development rationale.

1. Ability to carry out foreign-language communication in the course of extracurricular activities. Participants are asked to look at pictures and express their opinions and experiences in English. This promotes the development of foreign-language communication skill and the use of English to describe their experiences and express their thoughts.

2. The ability to comment on the information obtained in relation to the communicative situation. Participants should comment on the pictures and share their experiences, taking into account the context and communicative situation. This develops the ability to analyze and discuss information.

3. Ability to describe and discuss production and professional issues. The exercise allows participants to describe their experiences and discuss pro-industrial and professional issues related to the situations presented in the photographs. Participants can express their opinions, describe their actions and discuss what actions or solutions can be taken in these situations.

Exercise 5. "Choose one quote and tell whether you agree or disagree and why". Development rationale.

1. The ability to comment on the received information taking into account the communicative situation. Participants should comment on the selected quotation, expressing their position and arguing their views in accordance with the communicative situation. They can use different styles and expressions to express their agreement or disagreement.

2. Ability to discuss topical issues and express their point of view. Participants should express their own point of view on a selected quotation and explain the reasons for their agreement or disagreement. This is conducive to development of the ability to discuss current problems and express their own opinions in the professional field.

Exercise 6. "Watch the video "Top 4 Civil Engineering projects" and answer the questions: 1. Which project caught your attention the most from the point of view of engineering construction? 2. How can the process of building tall skyscrapers be complicated? What problems can engineers face?". Development rationale.

1. Ability to listen to the interlocutor. Participants will listen to a story about different engineering projects presented in the video. This will help develop the skill of active listening and understanding the information provided in the video.

2. The ability to comment on the information received, taking into account the communicative situation. Participants should answer questions by expressing their opinions and commenting on the engineering projects mentioned in the video. This develops the skill of analyzing information and commenting on it according to the communicative situation.

3. The ability to describe and discuss industrial and professional issues. The video presents various engineering construction projects and the participants should discuss them, express their opinions about the projects that caught their attention and consider the difficulties engineers face in building skyscrapers.

4. Ability to discuss current issues and express their point of view. Participants will discuss various engineering projects, expressing their views on which project has attracted the most attention and explaining the challenges engineers face in construction. This promotes the ability to discuss current issues and express their point of view on professional issues.

Exercise 7. "When creative thinking is required to solve professional problems, what are the pros and cons of working individually, in small groups, or in large groups? Choose one and explain." Development rationale.

1. Ability to comment on the received information taking into account the communicative situation. Participants should complete a survey expressing their opinions about the work in a group format or individually.

2. Ability to describe and discuss work and professional issues. Participants should explain why working in an individual/group format is better/worse, in case a creative decision needs to be made. They should also give their own arguments to support their choice.

We should also note that all the exercises are communicative in nature. Exercise 1 realizes the principle of communicative orientation, which consists in creating a situation of communication, where students speak on their experience and comment on the speech of other students; in Exercise 2 we observe the principle of connection with life, which is based mainly on professional orientation, so students activate their professional knowledge and the acquired skills are associated by students with the future application of these skills. Two principles can be attributed to Exercise 3 - interactivity and communicative orientation. Interactivity is realised through the fact that the instructor shares his own answer to the question and encourages students to respond to this message. Also, the proposed exercises at the second stage of training also correlate with the principles of communicative orientation (Exercise 4), visibility (Exercise 5) and use of multimedia (Exercises 6, 7). Visibility is provided by the material with various images, and the principle of multimedia is realized through a video about engineering inventions, as well as a survey.

Tasks requiring the use of English for communication contribute to the development of communication skills on professional topics and vocabulary expansion, which develops the ability to carry out foreign language communication in the course of extracurricular activities. Tasks involving listening to audio and video materials help participants to improve their ability to actively listen to and understand information conveyed in English, which develops the ability to listen to the interlocutor. Tasks that require commenting and analyzing information help to develop the ability to argue, express opinions and adapt to different communicative situations, which develops the ability to comment on the received information taking into account the communicative situation. Tasks related to the description and discussion of professional issues help to develop the ability to clearly and accurately convey information using specialized vocabulary and terms, this develops the ability to describe and discuss industrial and professional issues. Tasks that encourage discussion and expression of one's own point of view on topical issues contribute to the development of critical thinking, the ability to argue one's point of view and participate in professional discussions, which has a positive impact on the development of the ability to discuss topical issues and express one's point of view.

4. Conclusion

The theoretical significance of the study is due to the possibility of further use of the results obtained in pedagogy, scientific research in this area, as well as for making recommendations to teachers in preparation for classes with the use of mobile technologies. The development of tasks for the development of communicative skills of engineering students using mobile technologies is an effective approach. A variety of assignments such as describing work experience, discussing technical improvements and answering professional questions contribute to the development of communicative foreign language skills. The use of mobile technologies such as voice messaging and mobile applications enables engineering students to effectively develop communicative foreign language communication, listening to the interlocutor, commenting on the information received and discussing professional issues.

The study has shown that the application of our proposed system of exercises for the development of communication skills using mobile technologies based on Telegram leads to positive changes in academic performance and improvement of communication skills of participants. This research presents the part of a complex experiment and the results of the follow up investigation will be published as soon as the whole experiment is finished. However the are some limitations: the study only included students from a single university, which may limit the generalizability of the findings to other institutions.

5. References

[1] Balyhina, T. & Fedorenkov, A. (2015) Mobile learning as a condition and mechanism for lifelong learning. Mobil'noe obuchenie kak uslovie i mekhanizm nepreryvnogo obrazovaniya // Vestnik Rossijskogo universiteta druzhby narodov, No. 4, pp. 247–253, ISSN 2313-2264

[2] Afzalova, A. (2021) Use of mobile technologies to organize independent work of students. Available from: https://cyberleninka.ru/article/n/ispolzovanie-mobilnyh-tehnologiy-dlya-organizatsii-samostoyatelnoy-raboty-studentov Accessed: 2022-08-25

[3] Ipatov, O.; Saramaha, S.; Barinova, D.; Rubtsova, A. & Odinokaya, M. (2021). English as Medium of Instruction: Language and Content Combination in Digitized Higher Engineering Education, Proceedings of the 32nd DAAAM International Symposium, pp.0168-0176, B. Katalinic (Ed.), Published by DAAAM International, ISBN 978-3-902734-33-4, ISSN 1726-9679, Vienna, Austria DOI: 10.2507/32nd.daaam.proceedings.026

[4] Ipatov, O.; Barinova, D.; Ni, O.; Saramaha, S. & Odinokaya, M. (2022). New Teaching Approaches in Conditions of Globalisation in Digitized Higher Engineering Education, Proceedings of the 33rd DAAAM International Symposium, pp.0215-0220, B. Katalinic (Ed.), Published by DAAAM International, ISBN 978-3-902734-36-5, ISSN 1726-9679, Vienna, Austria DOI: 10.2507/33rd.daaam.proceedings.030

[5] Duhnich, Y. Mobile education. Mobil'noe obuchenie. Available from: http:www.smart-edu.com/mobile-learning.html Accessed: 2022-08-22

[6] Miheeva, S. (2021) Digital technologies in teaching foreign languages at a non-language university Cifrovye tekhnologii v prepodavanii inostrannyh yazykov v neyazykovom vuze. Sovremennye problemy nauki i obrazovaniya, No. 2 (2021) pp. 11-19, ISSN 2070-7428

[7] Glebovich, T. & Novikova D. (2020) Educational blogging in Telegram channels: concepts and forms of content representation. Obrazovatel'nyj bloging v Telegram-kanalah: koncepcii i formy reprezentacii kontenta Izvestiya Ural'skogo federal'nogo universiteta, Vol. 6, No.3 (2020), pp. 57-70, ISSN 2227-2283

[8] Ramya, G., & Madhumathi, P. (2016). Teaching language through mobile technology for kindergarten children in the classroom, Proceedings of International Conference on Innovations in the Teaching of English Language and Literature, pp. 535-541. ISBN: 9789386176189, Tamil Nadu, India.

[9] Sun, J.; Chang, C., Chen, M., & Lin, Y. (2016). Effects of enhanced element managed instruction integrated with tablet PC-based collaborative polling on fifth graders' sight-word reading performance. International Journal of Mobile Learning and Organisation, Vol. 10, No. 1, pp. 102-116. DOI: 10.1504/IJMLO.2016.076192

[10] Alodwan, T. (2021) The Effect of Using Telegram Application on Improving Writing Skill for the Students of English as a Foreign Language. Elementary Education, Vol. 20, No. 6, pp. 847-856., ISSN 1305-3515, DOI: 10.17051/ilkonline.2021.06.092

[11] Martin, A. (2013) Academic buoyancy and academic resilience: exploring "everyday" and "classic" resilience in the face of academic adversity. School Psychology International, Vol. 34(5) (2013), pp. 488-500, DOI: doi.org/10.1177/0143034312472759

DOI: 10.2507/34th.daaam.proceedings.020

INVESTIGATING THE INFLUENCE OF PRINTING ORIENTATION AND FILAMENT DRYING ON TENSILE AND FLEXURAL STRENGTH OF FDM-PRINTED CARBON FIBER-REINFORCED POLYAMIDE COMPOSITES

Adi Pandžić, Edin Kadric & Svetozar Kolesar

This Publication has to be referred as: Pandzic, A[di]; Kadric, E[din] & Kolesar, S[vetozar] (2023). Investigating the Influence of Printing Orientation and Filament Drying on Tensile and Flexural Strength of FDM-Printed Carbon Fiber-Reinforced Polyamide Composites, Proceedings of the 34th DAAAM International Symposium, pp.0146-0153, B. Katalinic (Ed.), Published by DAAAM International, ISBN 978-3-902734-41-9, ISSN 1726-9679, Vienna, Austria
DOI: 10.2507/34th.daaam.proceedings.020

Abstract

The rapid advancement of additive manufacturing (AM), including Fused Deposition Modeling (FDM) technology, hinges significantly upon the continuous evolution of AM materials. To enhance the mechanical characteristics of these materials, filling of carbon or glass fibers has added to them. Consequently, this research delved into the analysis of a carbon fiber-filled polyamide composite. The primary objective was to investigate the impact of printing orientation and pre-printing material drying on the tensile and flexural mechanical properties. Tensile properties were analysed in accordance with ISO 527 standards, while ISO 178 standards were employed for the evaluation of flexural properties. The findings from this study showing the influence of printing orientation and pre-printing drying on both tensile and flexural properties, thereby contributing valuable insights to the field of AM materials development.

Keywords: PA-CF; Composite; 3D Print; FDM; Printed Material.

1. Introduction

Additive manufacturing (AM), also known as 3D printing or spatial printing, is today one of the fastest growing technologies for the production of prototypes as well as fully functional products. In today's competitive industrial landscape, the adoption of additive manufacturing technologies has gained significant traction, offering a pathway to enhance competitiveness and expedite market penetration for novel products. These cutting-edge technologies find applications across diverse functional domains in various industrial sectors. Over recent years, additive manufacturing has emerged as a dynamic force driving innovation in industrial serial production processes. Notably, this transformative trend has been most pronounced within sectors such as mechanical engineering, aviation and space industry, architecture, bioengineering, automobile industry, medical and prosthetic parts, fashion and more [1], [2].

Today's market exhibits significant interest in AM technologies, with AM materials development being their key driver. Among the various AM technologies available, Fused Deposition Modeling (FDM), also known as FDM technology, Fused Filament Fabrication (FFF), or material extrusion, is a prominent and widely used method. This popular, but also complex AM technology offers advantages over traditional methods, enabling the production of simple and complex geometry products. [3], [4].

FDM 3D printing uses a thermoplastic filament from spool fed through a feeder to an extrusion head, where it's melted in the heating chamber and deposited through printing nozzle in lines on the print bed to create the final product geometry layer by layer (Figure 1). Filament materials vary from thermoplastics to composites, ceramics, and metals. Commonly used FDM materials include PLA (Polylactic Acid), PC (Polycarbonate), PET (Polyethylene Terephthalate), PP (Polypropylene), ABS (Acrylonitrile Butadiene Styrene), TPU (Thermoplastic Polyurethane), and Polyamide (Nylon) [5], [6].

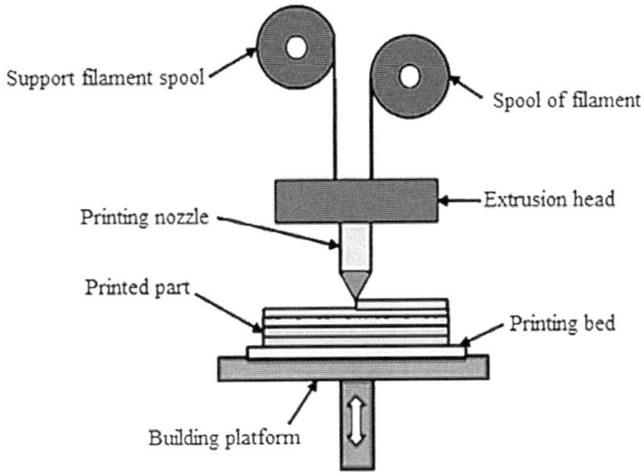

Fig. 1. FDM process and machine setup [5]

FDM excels in material versatility, offering diverse options with distinct mechanical properties. It enables cost-effective concurrent use of various materials in different forms. While polymer production dominates FDM, it can also handle metal printing. The mechanical properties of FDM-produced polymer parts are strongly influenced by printing parameters and the choice of base polymer [7], [8]. In paper [2] mentioned that extensive research in 3D printing materials has resulted to the development of polymers with desired mechanical properties and biocompatibility, where mechanical properties can be „tuned" by adjusting the FDM printing parameters. When FDM printing parameters are mentioned, they are most often: layer height, printing orientation, printing temperature, printing bed temperature, wall thickness, printing speed, raster angle, infill density and pattern [9], [10]. In addition to the printing parameters, the mechanical properties of FDM printed materials are also affected by other factors such as the filament quality (as a base material), color of the material, drying of the material before printing (this is very important for Polyamide - Nylon), heat treatment after 3D printing, and others [11], [12].

This study investigated the effect of printing orientation and pre drying on tensile and flexural strength of FDM printed carbon fiber-filled polyamide composite (with chopped carbon fibers). In general, FDM polyamide offers good combination of strength and toughness compared to other FDM printed materials, and with good impact strength. In research paper [13] stated that polyamide (nylon) is also characterized by properties such as high impact strength, good stress cracking resistance, resistance to oils, fuels, hydraulic fluids and high tensile and flexural strength. In efforts to enhance FDM printed material properties, common reinforcement materials in polymer matrices include carbon, glass, aramid, and natural fibers. Fiber-filled polymer filaments represent an exciting composite material, offering advantages such as better dimensional stability, enhanced mechanical strength, and cost-effectiveness, thereby expanding the potential applications in FDM printing [14].

Previous research has reported a strong correlation between printing orientation and mechanical properties of FDM-printed materials [15], [16]. According to Anubhav et al. [17], part orientation significantly impacts strength and behaviour under various loads in 3D printing. They found that the orientation directly influences the cohesive bonding of deposited materials due to material deposition and layering. Notably, horizontally printed parts exhibit better tensile and flexural properties compared to the vertically printed counterparts. In a study [2] analysing the impact of five printing parameters – printing temperature, infill type and density, layer height, and printing orientation – it was found that printing orientation has the most significant influence on the tensile strength of PET-G and PLA FDM printed materials. In study [13] meintoined that different researchers have established a correlation between build orientation and the mechanical properties of FDM printed materials. Variations in the melt quality between adjacent filaments can lead to a reduction in mechanical properties, such as tensile strength, compressive strength, flexural strength, hardness, and elastic modulus. This effect is particularly pronounced in parts tested perpendicular to the direction of layer construction.

Literature review has also shown that moisture affects the properties of FDM printed PA6 material, unreinforced and reinforced with carbon fibers. In the study by Brown [18], the presence of residual water post-drying could be a factor contributing to the absence of strain softening. In the case of a 50% glass fiber-reinforced PA6, it exhibits behaviour similar to unreinforced PA6. Notably, tensile properties showed improvement after drying, while impact properties experienced a slight reduction. Also, Gong et al. [19] concluded that moisture affects 3D printing with nylon filament, and this applies to other materials as well. Excessive moisture in the filament leads to material degradation, resulting in an uneven surface finish and reduced tensile strength, especially in terms of ductility. Long-term storage without monitoring requires proper drying before reusing the filament.

2. Materials and methods

This research is centered on a systematic experimental exploration aimed to profound impact of two key variables, namely, printing orientation and pre-printing material drying, on the mechanical properties of Carbon Fiber-Filled Polyamide Composite. Experimental methodology is presented on flow diagram below (Figure 2).

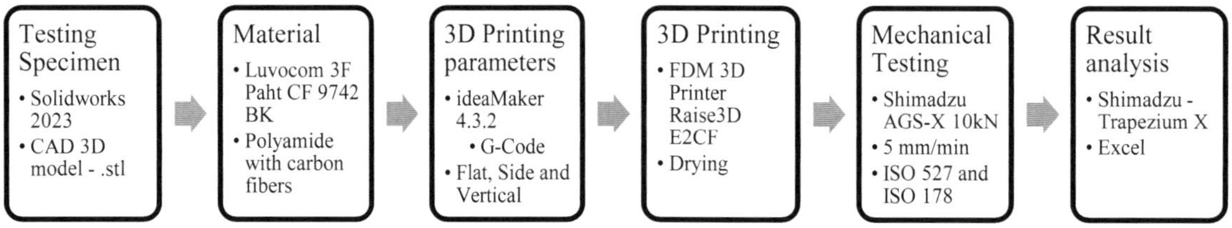

Fig. 2. Experimental methodology

Tensile and flexural testing specimens CAD 3D models were designed using Solidworks 2023. The design for the tensile testing specimen was according to ISO 527-2 standard, while the flexural testing specimen design followed ISO 178 guidelines (see Figure 3). Subsequently, these CAD 3D models were converted into .stl format, facilitating their utilization within the slicer software for the configuration of 3D printing parameters.

Fig. 3. Testing specimen for tensile and flexural testing

In this research "Luvocom 3F Paht CF 9742 BK" material (CF-PA6) was analysed. This material is distinguished as a high-temperature polyamide-based composite, offering the robustness of PA6 while maintaining exceptional printability characteristics. Notably, it demonstrates good resistance to water absorption, with only half the water uptake and a fourfold slower absorption rate when compared to standard PA6. Also, this material does not necessitate a heated chamber during the printing process, and it exhibits minimal warping tendencies. Furthermore, it boasts compatibility with both HIPS and PVOH support materials. With its composition consisting of 15% chopped carbon fiber reinforcement in PA6, "Luvocom 3F PAHT CF 9742 BK" is engineered for industrial applications. This material is tailored for the production of exceptionally robust and rigid components, characterized by a low coefficient of thermal expansion. Consequently, it finds prominent use in diverse sectors such as the automotive industry, textile manufacturing, office machinery, as well as apparatus and precision engineering [20]. Detailed technical specifications are presented in Table 1.

Properties	Test Method	Units	Typical Value
Tensile Strength	ISO 527	MPa	170
Tensile Modulus	ISO 527	GPa	15
Tensile Elongation	ISO 527	%	2
Flexural Strength	ISO 178	MPa	-
Flexural Modulus	ISO 178	GPa	-
Flexural Elongation	ISO 178	%	-
Charpy Impact Strength	ISO 179 1eU	kJ/m^2	47

Table 1. Technical specification of CF-PA6 material by manufacturer - Luvocom 3F Paht CF 9742 BK [20]

Specimens were prepared for 3D printing using the "ideaMaker 4.3.2" slicer software, which generated the necessary G-Code. The specific printing parameters utilized are detailed in Table 2. Specimens were positioned in three different printing orientations: flat, side and vertical.

Printing parameter	Units	Typical Value
Layer Height	mm	0.15
Shells	-	4
Extrusion Width	mm	0.4
Infill Density	%	100
Infill Pattern	-	Grid
Heated Bed Temperature	°C	80
Extruder Temperature	°C	290
Default Printing Speed	mm/s	60

Table 2. Printing parameters

All testing specimens were 3D printed with Raise3D E2CF FDM 3D printer. It is FDM 3D printer with independent dual extruders (IDEX), optimized for 3D printing carbon fiber reinforced materials. This desktop 3D printer is ready for manufacturing, prototyping, and more. Specimens were printed in 3 replicas for every printing orientation: flat, side and vertical, as presented in Figure 4. Also, specimens in flat and side orientations, as ones with better strength compared to vertical orientation, were printed with pre drying filament 80 °C for 12h before 3D printing.

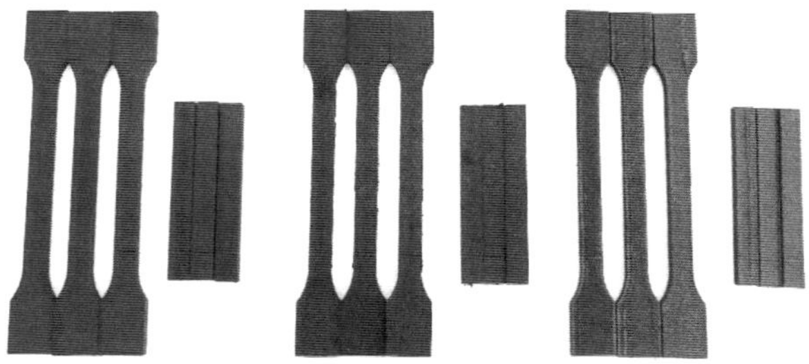

Fig. 4. FDM printed testing specimens (tensile and flexural) in flat, side and vertical orientation

Tensile and flexural mechanical testing were conducted using the Shimadzu AGS-X universal testing machine equipped with a 10 kN load cell. These tests were carried out in accordance with ISO 527 for tensile properties and ISO 178 for flexural properties, maintaining a testing speed of 5 mm/min, as illustrated in Figure 5.

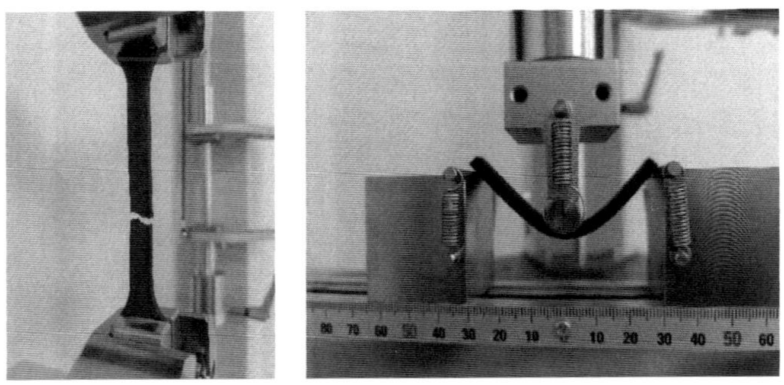

Fig. 5. Tensile (left) and flexural (right) material mechanical properties testing

All data were acquired utilizing Trapezium-X software developed by Shimadzu, subsequently subjected to statistical analysis in Microsoft Excel, stress-strain diagrams were plotted and presented in further text.

3. Results and discussion

After completing the mechanical properties tests, all the results were compiled and presented in Table 3. Tensile testing was performed in accordance with ISO 527 standards, while flexural testing followed ISO 178 standards, with a total of 30 specimens undergoing evaluation. Tensile and flexural strength properties were expressed through maximum stress before break (Rm) and elastic modulus (E), presented in Table 3.

Specimen	Tensile Strength [MPa]	Tensile Modulus [GPa]	Flexural Strength [MPa]	Tensile Modulus [GPa]
Flat	45.2	3.5	43.5	2.6
Side	30.3	2.8	21.9	1.1
Vertical	9.0	0.9	11.1	0.3
Flat with drying	50.5	3.5	54.5	2.8
Side with drying	30.3	2.8	32.6	1.1

Table 3. Results of examined tensile and flexural strength properties

As demonstrated by the stress-strain curves (Figure 6), noticeable differences exist in the strength of FDM-printed materials depending on different printing orientations. Specimens printed in flat and side orientations showing highest tensile and flexural strength, which is expected.

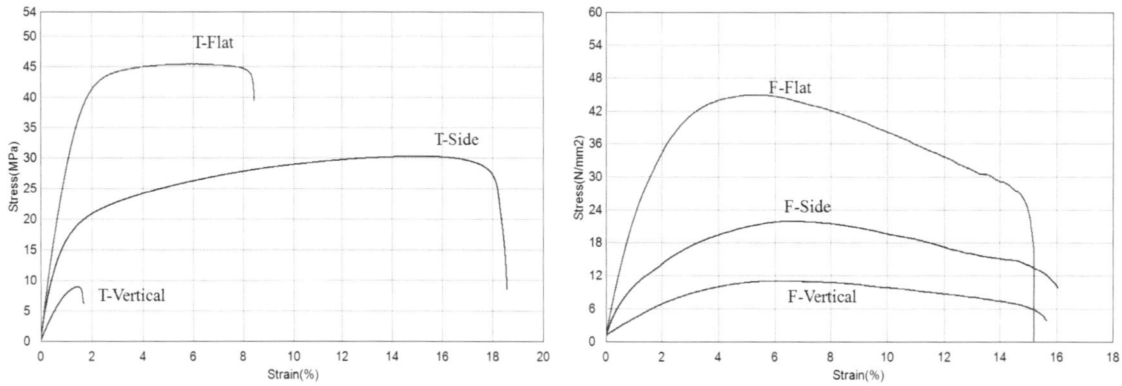

a) b)

Fig. 6. Stress-Strain curves for tensile (a) and flexural (b) testing of FDM-printed CF-PA6 material: Effects of printing orientations (flat, side, and vertical)

Tensile strength and modulus of elasticity differences based on printing orientation are evident in the histograms (Figure 7). Tensile strength is highest in the flat printing orientation (45 MPa), followed by the side (30.3 MPa), and lowest in the vertical (9 MPa) orientation. Similarly, the modulus of elasticity is highest in the flat printing orientation (3.5 GPa), followed by the side (2.8 GPa), and lowest in the vertical (0.9 GPa) orientation.

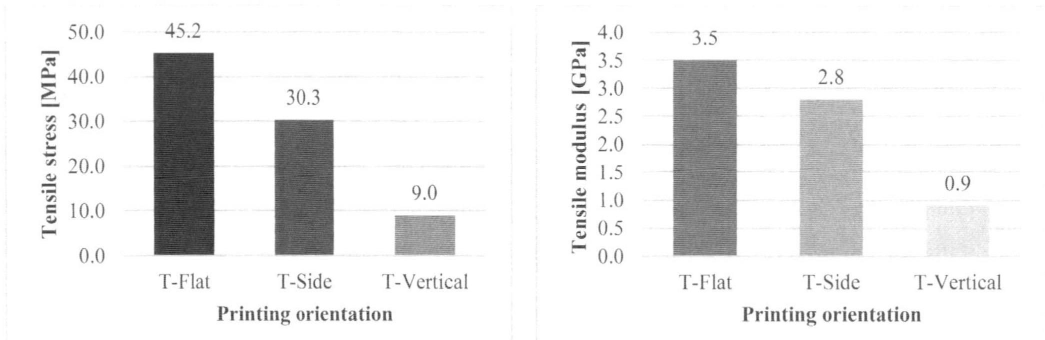

a) b)

Fig. 7. Comparation of tensile strength (a) and modulus (b) for specimens printed in different printing orientation

When analysing the effect of printing orientation on flexural properties, a similar trend to that observed in tensile tests emerges. Flexural strength is highest in the flat orientation (43.5 MPa), followed by the side (21.9 MPa), and lowest in the vertical orientation (11.1 MPa). The flexural modulus exhibits the same pattern, with the highest value in the flat orientation (2.6 GPa), followed by the side (1.1 GPa), and the lowest values in the vertical printing orientation (0.3 GPa).

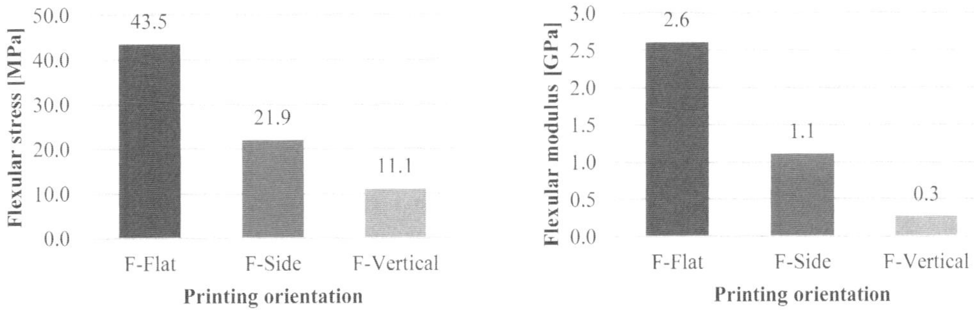

a) b)

Fig. 8. Comparation of flexural strength (a) and modulus (b) for specimens printed in different printing orientations

Previous research and literature recommend pre-drying polyamide (nylon) material before printing. Samples printed in flat and side orientations, which exhibited the highest tensile strength values, were compared with samples printed after filament drying. Stress-Strain diagrams illustrate how pre-drying the FDM printed CF-PA6 material significantly enhanced both tensile and flexural strength (Figure 9).

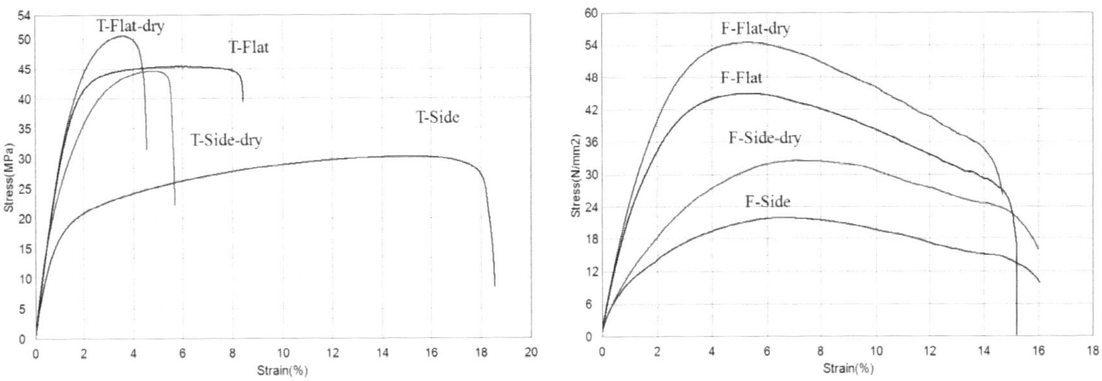

a) b)

Fig. 9. Stress-Strain curves for tensile (a) and flexural (b) testing of FDM-printed CF-PA6 material: Effects of filament drying before printing (orientations flat & side)

Comparing these results using histograms (Figure 10) reveals a noteworthy boost in both tensile and flexural strength when the CF-PA6 filament is dried prior to printing.

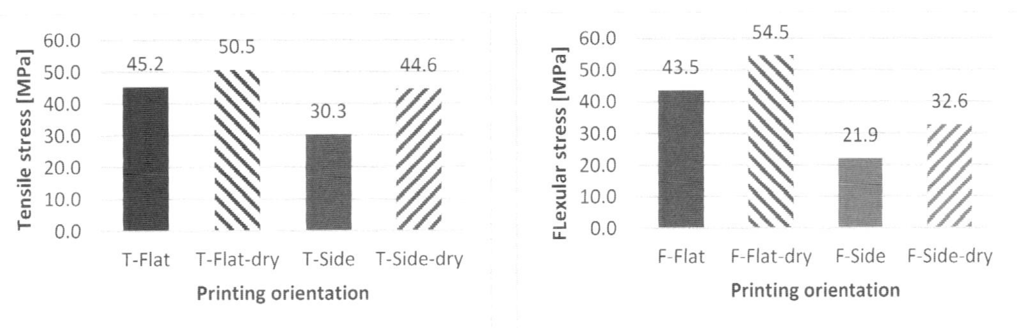

a) b)

Fig. 10. Comparation of tensile and flexural strength for specimens printed with and without drying material before FDM 3D printing

Drying the CF-PA6 filament at 80°C for 12 hours before 3D printing resulted in a 12% increase in tensile strength for flat-oriented specimens and a 47% increase for side-oriented specimens. Similarly, flexural strength saw a 25% increase for flat-oriented specimens and a 49% increase for side-oriented specimens.

4. Conclusion

Aim of this research was to analyse the influence of printing orientation and filament drying before printing on the tensile and flexural mechanical properties of FDM-printed polyamide composite material with chopped carbon fibers. After completing the tests and analysing the results, the following conclusions are:

- The technical specification of the material "Luvocom 3F PAHT CF 9742 BK" from the manufacturer includes only the values of the material's tensile mechanical properties. A comparison of the tested tensile strength and modulus of elasticity with the manufacturer's specifications revealed that the obtained results were significantly lower. The tested tensile strength varied by up to 108%, while the elastic modulus varied by up to 124%.
- Printing orientation had a substantial impact on the mechanical properties of the FDM-printed CF-PA6 material. The highest tensile and bending strength were achieved when the test sample was printed in a flat orientation, while the lowest strength was observed in the case of printing in a vertical orientation.
- Pre-drying the CF-PA6 filament before printing had a positive influence on the increase in tensile and flexural strength. Tensile strength increased by 12% for flat orientation and by 47% for curved orientation. The flexural strength of the samples printed in the flat orientation increased by up to 25%, and in the side orientation, it increased by up to 49%.
- Additionally, it was noteworthy that the samples did not break during the bending test, even in cases where they were printed in a vertical orientation, where breakage was expected.

In future research, it is recommended to focus on the optimization of drying parameters. This optimization is crucial for achieving the highest quality in printed products from this material, characterized by the desired mechanical, physical, and other properties. Additionally, future research should center on the analysis of how printing parameters influence the mechanical and other properties of CF-PA6 materials. This analysis should encompass the examination of additional mechanical properties, such as hardness, impact toughness, compressive strength, and more.

5. References

[1] Hodzic, D. & Pandzic, A. (2019). Influence of Carbon Fibers on Mechanical Properties of Materials in FDM Technology, Proceedings of the 30th DAAAM International Symposium, pp.0334-0342, B. Katalinic (Ed.), Published by DAAAM International, ISBN 978-3-902734-22-8, ISSN 1726-9679, Vienna, Austria, DOI:10.2507/30th.daaam.proceedings.044

[2] Bembenek, M.; Kowalski, L. & Koson-Schan, A. (2022). Research on the Influence of Processing Parameters on the Specific Tensile Strength of FDM Additive Manufactured PET-G and PLA Materials, Polymers, 14, 2446, DOI:10.3390/polym14122446.

[3] Pandzic, A. & Hodzic, D. (2022). Tensile Mechanical Properties Comparation of PETG, ASA and PLA-Strongman FDM Printed Materials With and Without Infill Structure, Proceedings of the 33rd DAAAM International Symposium, pp.0221-0230, B. Katalinic (Ed.), Published by DAAAM International, ISBN 978-3-902734-36-5, ISSN 1726-9679, Vienna, Austria, DOI:10.2507/33rd.daaam.proceedings.031

[4] Hodzic, D. & Pandzic, A. (2019). Influence of Carbon Fibers on Mechanical Properties of Materials in FDM Technology, Proceedings of the 30th DAAAM

[5] Mogan, J.; Harun, W.S.W.; Kadirgama, K.; Ramasamy, D.; Foudzi, F.M.; Sulong, A.B.; Tarlochan, F. & Ahmad, F. (2023) Fused Deposition Modelling of Polymer Composite: A Progress, Polymers 2023, 15, 28. DOI:10.3390/polym15010028

[6] Pandzic, A.; Hodzic, D. & Kadric, E. (2021). Experimental Investigation on Influence of Infill Density on Tensile Mechanical Properties of Different FDM 3D Printed Materials, TEM Journal, Vol. 10, No. 3, pp. 1195-1201., ISSN 2217-8390, DOI: 10.18421/TEM103-25

[7] Alarifi, I. M. (2022). A Performance Evaluation Study of 3D Printed Nylon/Glass Fiber and Nylon/Carbon Fiber Composite Materials, Journal of Materials Research and Technology, Vol. 21, pp. 884-892, DOI:10.1016/j.jmrt.2022.09.085 2

[8] Tutar, M (2023). A Comparative Evaluation of the Effects of Manufacturing Parameters on Mechanical Properties of Additively Manufactured PA and CF-Reinforced PA Materials, Polymers, 15, 38, DOI:doi.org/10.3390/polym15010038

[9] Wang, P.; Zou, B.; Ding, S.; Li, L. & Huang, C. (2020). Effects of FDM-3D Printing Parameters on Mechanical Properties and Microstructure of CF/PEEK and GF/PEEK, Chinese Journal of Aeronautics, Vol. 34, No. 9, pp. 236-246, DOI:10.1016/j.cja.2020.05.040

[10] Farashi, S. & Vafaee, F. (2022). Effect of Printing Parameters on the Tensile Srength of FDM 3D Samples: A Meta-Analysis Focusing on Layer Thickness and Sample Orientation. Progress in Additive Manufacturing, 7, pp. 565–582, DOI:10.1007/s40964-021-00247-6

[11] Jayanth, N.; Jaswanthraj, K.; Sandeep, S.; Mallaya, N.H. & Siddharth, S.R. (2021). Effect of Heat Treatment on Mechanical Properties of 3D Printed PLA, Journal of the Mechanical Behavior of Biomedical Materials, Vol. 123, DOI: 10.1016/j.jmbbm.2021.104764

[12] Hadi, A.; Kadauw, A.; & Zeidler, H. (2023). The Effect of Printing Temperature and Moisture on Tensile Properties of 3D Printed Glass Fiber Reinforced Nylon 6, Materials Today: Proceedings, DOI: 10.1016/j.matpr.2023.04.641

[13] Calignano, F.; Lorusso, M.; Roppolo, I. & Minetola, P. (2020). Investigation of the Mechanical Properties of a Carbon Fibre-Reinforced Nylon Filament for 3D Printing, Machines, 8, 52, DOI:10.3390/machines8030052

[14] Balaji, N. S.; Velmurugan, C.; Kumar, M. S.; Sivakumar, M. & Asokan, P. (2023). Experimental Investigation on Mechanical Properties of FDM-Based Nylon Carbon Parts Using ANN Approach. Surface Review and Letters, 30, DOI:10.1142/S0218625X23500282

[15] Zaldivar, R.J.; Witkin, D.B.; McLouth, T.; Patel, D.N.; Schmitt, K. & Nokes, J.P. (2017). Influence of Processing and Orientation Print Effects on the Mechanical and Thermal Behavior of 3D-Printed ULTEM®9085 Material, Additive Manufacturing, Vol. 13, pp. 71–80, DOI:10.1016/j.addma.2016.11.007

[16] Farashi, S. & Vafaee, F. (2022). Efect Of Printing Parameters on the Tensile Strength of FDM 3D Samples: A Meta-Analysis Focusing on Layer Thickness and Sample Orientation, Progress in Additive Manufacturing, 7, pp-565-582, DOI:10.1007/s40964-021-00247-6

[17] Anubhav; Kumar, R.; Nandi, S. K. & Agrawal, A. (2023). Influence of Build Orientation on Tensile and Flexural Strength of FDM Fabricated ABS Component, Advances in Additive Manufacturing and Metal Joining, Lecture Notes in Mechanical Engineering, pp. 177–187, DOI:10.1007/978-981-19-7612-4_15

[18] Brown, D. M. (2019). Moisture Absorption and Desorption Effects on Mechanical Behavior in Specialty Polyamide Products, Theses and Dissertations, 5637, Lehigh University

[19] Gong, H.; Runzi, M.; Wang, Z. & Wu, L. (2022). Impact of Moisture Absorption on 3D Printing Nylon Filament, Proceedings of the 33rd Annual International Solid Freeform Fabrication Symposium – An Additive Manufacturing Conference, DOI:10.26153/tsw/44131

[20] https://www.luvocom.de/en/products/3d-printing-materials/luvocom-3f/, (2023). Lehmann&Voss&Co – LUVOCOM 3F – Materials for Extrusion Based Processes, Material Technical Data Sheet, Accessed on: 2023-09-05

DOI: 10.2507/34th.daaam.proceedings.021

EFFECT OF FEED AND DEPTH OF CUT ON CHIP SHAPE

Lubos Kroft, Michal Povolny & Jan Mezulianik

This Publication has to be referred as: Kroft, L[ubos]; Povolny, M[ichal] & Mezulianik, J[an] (2023). Effect of Feed and Depth of Cut on Chip Shape, Proceedings of the 34th DAAAM International Symposium, pp.0154-0160, B. Katalinic (Ed.), Published by DAAAM International, ISBN 978-3-902734-41-9, ISSN 1726-9679, Vienna, Austria
DOI: 10.2507/34th.daaam.proceedings.021

Abstract

Machining processes, fundamental to modern manufacturing, require a precise understanding of various parameters, with cutting conditions playing a central role. Optimal feed and depth of cut combinations are critical in achieving efficiency. Tool testing under diverse conditions is imperative to discern effective strategies for productivity enhancement and tool longevity. The chip formation process is a crucial element of production, profoundly influencing machining efficiency. Choice of basic cutting conditions and tool attributes significantly affects chip formation. Feed and depth of cut parameters hold immense importance in machining, impacting chip thickness and morphology. Machining processes encompass turning, milling, drilling, and grinding, each presenting unique challenges related to feed, depth of cut, and chip formation. This review focuses on turning technology, delving into existing research on chip formation, feed rate, and depth of cut. Insights gained from the literature emphasize the need for further research to bridge existing knowledge gaps regarding the intricate relationships between these parameters, particularly under diverse machining conditions. The study highlights the significance of tool testing in achieving optimal machining outcomes and presents a methodology utilizing long-term durability tests to understand the effect of feed rate and depth of cut on chip formation, providing practical insights into tool behaviour and chip morphology.

Keywords: machining; turning tools; chip shape; chip formation

1. Introduction

Machining processes are the foundation of modern manufacturing, with efficiency and accuracy being the most important goals. Achieving these goals requires a thorough understanding of various parameters, among which the choice of cutting conditions plays a key role. The optimum combination of feed and depth of cut are critical factors in achieving efficient machining. Testing tools under different conditions is essential to elucidate the most effective strategies to increase productivity and tool life. The chip formation process during machining is a critical aspect of production and plays a major role in determining the efficiency of the entire machining process. Chip formation is a complex process in which material separation occurs in the cutting plane, leading to chip formation. At the same time, however, primary, secondary, and tertiary plastic deformation occurs. The choice of basic cutting conditions, together with the choice of tool, its geometry and microgeometry, has a major influence on this process [1], [2].

Understanding and optimizing chip formation is essential to increase machining efficiency, tool life and achieve the required surface quality and dimensional accuracy. Feed and depth of cut parameters are of critical importance in machining. Feed is the relative movement between the cutting tool and the workpiece per unit time, which, together with the depth of cut, affects the thickness of the chip to be cut. The cutting speed, on the other hand, represents the speed at which the cutting tool moves over the workpiece material and thus influences the chip formation dynamics. The last of the basic parameters is the depth of cut. While, for example, the cutting speed has a dominant effect on the cutting temperature and thus tool life, the depth of cut is a crucial parameter for chip morphology. The interaction of these parameters significantly influences the morphology of the chip, including its shape, size and thickness [3], [4].

Machining processes are diverse and include various methods such as turning, milling, drilling, and grinding. Each process involves unique challenges and requirements related to feed, depth of cut and chip formation. For example, in turning, the rotating workpiece and stationary cutting tool require careful consideration of feed and depth of cut to achieve optimum chip formation. In milling, where both workpiece and tool are moving, achieving the desired chip morphology requires a different approach [5], [6].

2. Literature search:

This chapter provides an overview of existing research on chip formation, feed rate and depth of cut in turning technology and highlights the need for further investigation to fill the existing gaps in understanding their relationship. In our search for background information for this paper, we looked specifically at sources that addressed chip formation for turning technology. Understanding the relationship between feed rate, depth of cut, and chip formation is essential for optimizing machining processes and improving tool performance. Several studies have addressed these relationships and shed light on the complex dynamics [1], [2], [3], [4], [7].

2.1 Chip formation mechanisms:

Chip formation is a dynamic process influenced by various parameters including feed rate, depth of cut, cutting speed but also other parameters. The basic mechanisms of chip formation are generally divided into continuous and discontinuous chip formation [8]. Continuous chip formation occurs at higher cutting speeds and is characterized by continuous, melting chips. Discontinuous chip formation occurs at lower cutting speeds and results in segmented chips. In turning technology, chip formation is a critical aspect that directly affects machining efficiency and product quality. The basic mechanisms of chip formation in turning include continuous chip formation, discontinuous chip formation, and built-up edge (BUE) formation [9], [10]. Continuous chip formation is pronounced at higher depths of cut, while discontinuous chip formation prevails at lower depths of cut.

2.2 Effect of feed rate on chip formation:

Researche and Rezayi Khoshdarregi [11] highlighted that increasing feed rate in turning leads to greater chip thickness, which affects the chip shape and the forces applied during machining. In addition, higher feed rates can promote the transition from continuous to discontinuous chip formation. Feed rate plays a critical role in chip formation. This research is already very old as it was found in Merchant (1945) that an increase in feed rate leads to thicker chip formation, but the information found is still generally valid. This is attributed to the fact that more material is removed per unit time, which affects the morphology of the chip.

2.3 Effect of depth of cut on chip formation:

The depth of cut significantly affects the chip morphology during turning. At lower depths of cut, the chip is thinner and more continuous, whereas at higher depths of cut, thicker, segmented chips tend to form [12]. Depth of cut also affects the temperature distribution at the tool-chip interface, which has an effect on the mechanisms of chip formation [13].

2.4 Combined effect of feed and depth of cut:

The combination of feed rate and depth of cut significantly affects the chip shape during turning. Studies by Boothroyd and Knight [12] have shown that changes in these parameters lead to changes in chip morphology. Higher feed rates and depth of cut tend to produce thinner, more continuous chips, while lower values lead to thicker, segmented chips. Optimizing the interaction of feed rate and depth of cut is critical to achieving the desired chip morphology and machining efficiency. Research by Tlusty [14] and Barry and Byrne [15] highlighted that the optimum combination of these parameters can lead to improved chip formation, reduced tool wear and improved surface quality.

2.5 Gaps in existing knowledge:

Although extensive research has been conducted on chip formation, feed, and depth of cut in turning, there are significant gaps in the understanding of their complex interplay. Further research is needed to elucidate the exact relationships between feed rate, depth of cut and the resulting chip shape under different machining conditions, considering factors such as tool geometry, workpiece materials and lubrication techniques. Especially recently, these experiments are not sufficiently published, and for modern tool types, actual chip formers with multi-layer coatings provide high machining performances today.

3. Methodology

The process of testing cutting tools involves examining the behavior of the tools under different operating conditions to determine the ideal combination of cutting parameters. The aim of this investigation is to strike a balance between achieving high material removal, reducing tool wear, and improving surface quality. To this end, tool testing is generally divided into two basic approaches: long-term testing and short-term testing. In long-term testing, tools are subjected to long-term machining operations, which allows their durability and performance to be assessed over a longer period under normal machining conditions, as is normally the case in practice. This approach provides insights into the tool's ability to resist wear and maintain cutting performance under continuous use. Short-term testing, on the other hand, involves evaluating tool performance over a shorter period, usually focusing on specific cutting conditions to quickly evaluate tool behavior under controlled conditions. These tests are usually conducted under increasing conditions from very mild to very high. Each of these approaches has advantages and disadvantages. Long-term testing provides a comprehensive understanding of the wear and life characteristics of the tool under realistic production conditions. It offers insights into how tools behave over longer lifetimes, and thus helps in tool selection for longer machining operations. However, long-term testing requires a significant investment of time and resources.

Short-term testing, on the other hand, offers a quick evaluation of tools that enables the rapid identification of suitable cutting conditions for specific applications. This approach is particularly useful for initial tool selection or when exploring new cutting strategies. However, its drawback is the limited insight it provides, as it may not capture long-term wear patterns and sustainability of tool performance. In this paper, we consider the long-term test as it is significantly more revealing than the short-term test in terms of practical application. In the context of finding the optimal combination of feed and depth of cut

It is the aim of this paper to show the practical application of the long-term durability test with the significant effect of the combination of depth of cut and feed on the shape of the resulting chip. The influence of the shape of the formers and the geometry of the cutting edge itself can also be seen in the individual tests. The experimental methodology was designed to investigate the effect of feed rate and depth of cut on chip formation. The setup involved a comprehensive approach including machine specifications, cutting tools, workpiece material, measurement tools and a systematic experimental procedure.

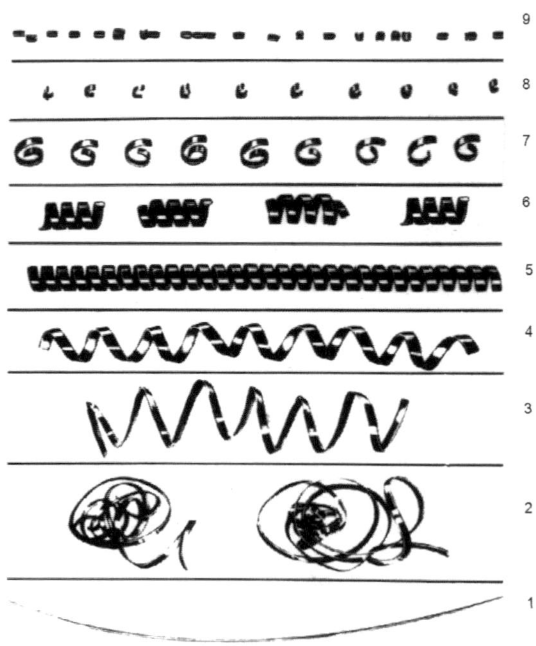

Fig. 1. Chip shape evaluated according to the Seco grading system

4. Preparation of the experiment

4.1. The machine

The experiments were carried out on a DMG CTX BETA 1250 TC A4 machine equipped with controls for precise control of depth of cut and feed during machining. It is a multi-axis machining center that has two spindles as well as a tool spindle and a turret head. The machine provided a stable platform for consistent and controlled experimentation. The specific parameters are shown in Table 1.

Max. workpiece diameter	340 mm
Max. workpiece length	1 185 mm
Max. bar capacity diameter	102 mm
Max. X-axis stroke	490 mm
Max. Y-axis stroke	200 mm
Max. Z-axis stroke	1 200 mm

Table. 1 Parameters of the machine DMG CTX BETA 1250 TC A4

4.2. Cutting tools:

The cutting tools used for the experiments were CNMG 120408-PF interchangeable insert cutting tools from Kenametal, Coromant, Sumitomo, Korloy, Mitsubishi, Cerazizit, and Pramet, which were selected based on their suitability for the workpiece material and their ability to maintain performance throughout the testing period. Such a diverse group of tools was chosen because of the differences, particularly in chip formers, so that differences could be seen.

4.3. Workpiece material:

The workpiece material selected for this study was C45+QT carbon steel due to its importance in industrial applications and its machinability characteristics, which provides a suitable substrate for investigating chip formation. The advantage of this material is its abundant use for testing as well as its well-known mechanical properties.

4.4. Measuring tools:

A suite of precision measurement tools, including a KEYECNE microscope and micrometer, were used to accurately capture and record critical parameters such as chip thickness, chip shape and dimensions.

4.5. Experimental parameters:

4.5.1. Depth of cut:

The depth of cut was varied in the range of 0.3 -2.5 mm to cover the spectrum of machining conditions for both roughing and finishing operations. The chosen depths of cut allowed to observe both the increase in cutting forces and the associated differential effects on chip formation.

4.5.2. Feed rate:

The feed rate was set in the range of 0.1 - 0.4 mm/rot to analyze its effect on chip morphology. Multiple feed rates were chosen to allow a comprehensive investigation of the correlation between feed rate and chip shape.

5. Experiment

The main objective of this experiment is to analyze the effect of variation of feed rate and depth of cut on chip formation during turning. The aim of this study is to provide insights into the relationship between these cutting parameters and chip morphology for a selected group of tools and to compare them with each other. The course of the experiment was as follows setting the initial feed and depth of cut values, a certain distance (L = 50 mm) was always machined. Based on this, chips were taken, and these were then further analyzed. The machine was always cleaned after each test to avoid cross-contamination. The feed rate was systematically varied within a predetermined range while keeping the depth of cut constant. This procedure was repeated, this time with a different depth of cut, for each insert.

Fig. 2. The course of the experiment on the machine DMG CTX BETA 1250 TC A4

The insert was continuously inspected on an optical microscope to prevent wear and thus affecting the experiment.

6. Results and analysis

The result of each experiment is the table shown in Figure 3. In this table, it is quite clear that the effect of depth of cut is clearly visible, especially on the chip area. Here it can be seen that a small depth of cut results in the formation of very thin chips which cannot be broken even by increasing the feed. These chips are not very suitable, particularly in view of their negative effect on the quality of the finished surface. As the depth of cut increases, the primary plastic deformation increases, which increases the tamping, and the chip starts to split more and more. This phenomenon, combined with the increase in feed rate, is clearly visible and it is possible to observe how the combination of feed rate and depth of cut clearly changes both the cross-section of the chip and its shape.

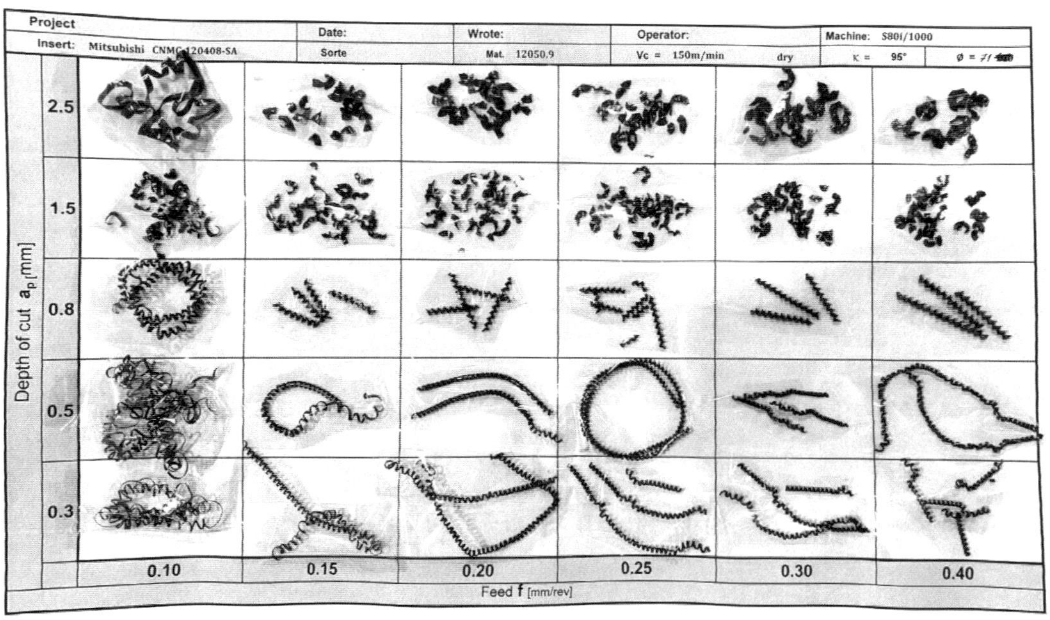

Fig. 3. Snapshot of the effect of feed and depth of cut on chip shape

The separate effect of the displacement is also apparent from the figure, most notably at higher depths of cut. Here it is possible to observe how the chip is gradually packed and thus again the primary plastic deformation ratio increases. As a result, a gradual increase in chip area can be observed. The result of these tests is the detection of problematic chip formation, mainly due to low feed and depth of cut values. It can be seen here that the depth of cut is problematic below ap = 0.8 mm, inclusive. Above this value a more suitable chip formation already occurs, which is however very dependent on the use of the insert. The evaluation of this part is dealt with in Figure 4. where the evaluation of the individual chip shape for the used values and feed rate can be observed. The evaluation was carried out based on the normative of figure 1. Here it can be seen that with increasing depth of cut almost all the insert types are achieving suitable VBD shape. The different geometries of the chip formers are then significant especially at lower depths and with decreasing feed values.

Cutting depth ap [mm]	Insert	0,040 (4%)	0,100 (20%)	0,150 (30%)	0,200 (40%)	0,250 (50%)	0,300 (60%)	0,400 (80%)
2,5	Ceratizit F50		7	8	8	8	8	8
	Korloy LP		2	8	8	8	8	8
	Mitsubishi SA		3	8	8	8	7	7
	Sumitomo LU		7	8	8	8	8	8
	Kennametal FN		2	7	8	8	8	8
	Coromant PF		2	8	8	8	8	8
1,5	Ceratizit F50		4	8	8	8	9	9
	Korloy LP		7	8	8	9	9	9
	Mitsubishi SA		7	8	8	8	8	8
	Sumitomo LU		7	8	8	8	8	7
	Kennametal FN		2	6	8	8	8	9
	Coromant PF		2	8	8	8	8	8
0,8	Ceratizit F50		5	5	5	5	5	5
	Korloy LP		5	6	6	8	8	8
	Mitsubishi SA		4	6	6	6	6	6
	Sumitomo LU		2	6	6	6	6	6
	Kennametal FN		3	5	5	6	6	5
	Coromant PF		5	6	6	6	6	6
0,5	Ceratizit F50		4	4	5	5	6	6
	Korloy LP		4	4	5	5	5	6
	Mitsubishi SA		4	5	5	5	6	5
	Sumitomo LU		4	5	5	5	5	4
	Kennametal FN		2	3	4	4	5	5
	Coromant PF		4	5	5	5	4	4
0,3	Ceratizit F50		4	4	5	5	5	5
	Korloy LP		3	3	3	4	4	5
	Mitsubishi SA		4	5	5	5	5	6
	Sumitomo LU		2	4	4	4	4	4
	Kennametal FN		2	2	4	4	5	5
	Coromant PF		3	5	5	5	4	4
HB		0,040	0,100	0,150	0,200	0,250	0,300	0,400
		4%	20%	30%	40%	50%	60%	80%
		Feed **f** [mm/rev]						

Fig. 4. Table for the evaluation of the dependence of the geometry of the formers on the shape of the chip

The result of these experiments is a comprehensive table that gives clear recommendations for the choice of the selected types of variables per material type used. The overall contribution of this experiment is to determine the effect of the geometry of the chip formers on the shape of the chip. This result is transferable to other material types.

7. Conclusion:

In the pursuit of enhancing machining efficiency, achieving optimal chip formation in turning technology is a critical endeavor. This study has sought to investigate the intricate relationship between feed rate, depth of cut, and chip morphology, with the overarching aim of shedding light on the factors that influence this essential aspect of the machining process. Our research has led to several key findings and insights: Impact of Feed Rate: Through systematic experimentation, it is evident that variations in feed rate play a substantial role in influencing chip thickness, shape, and overall chip morphology. Higher feed rates tend to produce thinner, more continuous chips, while lower feed rates result in thicker, segmented chips. This observation aligns with the existing body of knowledge, emphasizing the importance of feed rate in chip formation. Influence of Depth of Cut: The depth of cut, a parameter often overshadowed by cutting speed and feed rate, has been shown to be a crucial determinant of chip morphology. Our experiments have demonstrated that varying the depth of cut significantly affects chip thickness and shape.

Lower depths of cut lead to thinner and more continuous chips, while higher depths of cut result in thicker, segmented chips. This underlines the importance of considering depth of cut as a fundamental parameter in optimizing chip formation. Combined Effects: The interaction between feed rate and depth of cut is a key aspect of our research. It has been revealed that the combined effect of these parameters plays a pivotal role in shaping the chip. Identifying the optimal combination of feed rate and depth of cut is essential for achieving the desired chip morphology, machining efficiency, and tool longevity. Practical Application: The study has practical implications for the field of turning technology. It provides valuable insights into tool selection and machining strategy by optimizing feed rate and depth of cut. The understanding of these parameters can contribute to improved tool performance, reduced tool wear, enhanced surface quality, and greater dimensional accuracy in machining operations. Research Gaps: While this research has contributed to our understanding of chip formation in turning technology, there remain significant gaps in the current knowledge. Further exploration is warranted to encompass diverse tool geometries, workpiece materials, and lubrication techniques, particularly in the context of modern cutting tools and coatings. In conclusion, this study underscores the importance of feed rate and depth of cut in shaping chip formation during turning technology. By conducting systematic experiments, we have revealed the intricate relationships between these parameters and chip morphology. The results of this research can serve as a foundation for optimizing machining processes and tool selection, ultimately leading to increased productivity and improved tool performance. As the field of turning technology continues to evolve, ongoing research and exploration are essential to address the remaining gaps and refine our understanding of chip formation for enhanced efficiency and precision in machining operations.

8. Acknowledgments

The article contribution has been prepared under project SGS-2022-007 - Research and Development for Innovation in Engineering Technology – Machining Technology IV.

9. References

[1] Zamri, M. F., & Yusoff, A. R. (2019). Effect of Rake Angle and Feed Rate on Chip Segmentation in Machining Carbon Steel 1050. IOP Conference Series: Materials Science and Engineering, 530(1), 012011. https://doi.org/10.1088/1757-899X/530/1/012011

[2] Kuruc, M., Vopát, T., Peterka, J., Necpal, M., Šimna, V., Milde, J., & Jurina, F. (2022). The Influence of Cutting Parameters on Plastic Deformation and Chip Compression during the Turning of C45 Medium Carbon Steel and 62SiMnCr4 Tool Steel. Materials, 15(2), 585. https://doi.org/10.3390/ma15020585

[3] Limin, S. (2019). Correlation between serrated chip formation and turning chatter in the dry turning of superalloy. Ferroelectrics, 549(1), 50–60. https://doi.org/10.1080/00150193.2019.1592543

[4] Kim, D.-G., & Yang, S.-H. (2023). Efficient Analysis of CFRP Cutting Force and Chip Formation Based on Cutting Force Models Under Various Cutting Conditions. International Journal of Precision Engineering and Manufacturing, 24(7), 1235–1251. https://doi.org/10.1007/s12541-023-00822-7

[5] Zatloukal, T., Marsalek, O., & Syrovatka, S. (2022). Machinability of Thermal Spray Stellite 6 with Chemical and Mechanical Analysis. In B. Katalinic (Ed.), DAAAM Proceedings (1. vyd., Roč. 1, s. 0307–0312). DAAAM International Vienna. https://doi.org/10.2507/33rd.daaam.proceedings.042

[6] Peterka, J., Martinovic, J., Wallyson, T. D. S., & Kolesnyk, V. (2022). Analysis of Physical Causes of Machined Surfaces in Selected Machining Methods. In B. Katalinic (Ed.), DAAAM Proceedings (1. vyd., Roč. 1, s. 0034–0041). DAAAM International Vienna. https://doi.org/10.2507/33rd.daaam.proceedings.006

[7] Hnatik, J., Kroft, L., & Kutlwaser, J. (2016). Constant Chip Volume Machining. In B. Katalinic (Ed.), DAAAM Proceedings (1. vyd., Roč. 1, s. 0960–0966). DAAAM International Vienna. https://doi.org/10.2507/26th.daaam.proceedings.135

[8] Shaw, M. C. (2005). Metal cutting principles (2nd ed). Oxford University Press.

[9] Childs, T. (Ed.). (2000). Metal machining: Theory and applications. Arnold [u.a.].

[10] Childs, T. H. C. (2013). Ductile shear failure damage modelling and predicting built-up edge in steel machining. Journal of Materials Processing Technology, 213(11), 1954–1969. https://doi.org/10.1016/j.jmatprotec.2013.05.017

[11] Rezayi Khoshdarregi, M., & Altintas, Y. (2015). Generalized modeling of chip geometry and cutting forces in multi-point thread turning. International Journal of Machine Tools and Manufacture, 98, 21–32. https://doi.org/10.1016/j.ijmachtools.2015.08.005

[12] Knight, W. A., & Boothroyd, G. (2019). Fundamentals of Metal Machining and Machine Tools (0 vyd.). CRC Press. https://doi.org/10.1201/9780429114243

[13] Obikawa, T., Anzai, M., Egawa, T., Narutaki, N., Shintani, K., & Takeoka, E. (2011). High Speed Machining: A Review from a Viewpoint of Chip Formation. Advanced Materials Research, 188, 578–583. https://doi.org/10.4028/www.scientific.net/AMR.188.578

[14] Tlusty, J. (2000). Manufacturing processes and equipment. Prentice-Hall.

[15] Barry, J., & Byrne, G. (2002). The Mechanisms of Chip Formation in Machining Hardened Steels. Journal of Manufacturing Science and Engineering, 124(3), 528–535. https://doi.org/10.1115/1.1455643

DOI: 10.2507/34th.daaam.proceedings.022

METHODOLOGICAL PROCEDURE TO ENSURE THE SAFETY OF TECHNICAL INSTALLATIONS

Dana Prochazkova, Miroslav Rusko, Georg Rockel & Jan Ilko

This Publication has to be referred as: Prochazkova, D[ana]; Rusko, M[iroslav]; Rockel, G[eorg] & Ilko, J[an] (2023). Methodological Procedure to Ensure the Safety of Technical Installations, Proceedings of the 34th DAAAM International Symposium, pp.0161-0171, B. Katalinic (Ed.), Published by DAAAM International, ISBN 978-3-902734-41-9, ISSN 1726-9679, Vienna, Austria
DOI: 10.2507/34th.daaam.proceedings.022

Abstract

The paper deals with meaning the term "safety" and concentrates to its sense for technical installations. It shows relations of terms reliability, resiliency, criticality, risk and safety. It characterizes process safety management and safety management system at technical installations. It describes methodology for ensuring the safety of technical installations. The highest quality of each object (i.e. the State, human community, technical installation etc.) is safety, so the highest priority for the object is the overall (integral) safety. This means that it is not only about partial safety such as internal, external, environmental, fire, economic, etc., but about the safety of the whole, which includes partial safety so that the condition of the whole, is optimal. Due to the systemic nature of today's world, integral safety cannot be built without regard to others. Very important is the safety of basic processes, which ensure products and services for human society.

Keywords: Safety; risk management; process safety management; safety management system; methodology.

1. Introduction

The safe community is now at time of globalisation very dependent on a safety level of complex facilities ensuring the territory by basic service necessary for human live, e.g. the electric energy on which there are dependent supplies of good quality drinking water, utility water, information distribution etc. [54]. Based on current knowledge and experience, it is a fact that the quality of life, health and safety of each person depends on the quality of the human community to which they belong. From the reasons of fulfilment of targets of humans (human security and development) that may be only realised if human communities are in safe territory [28], [29]. Ensuring the safe human system is not easy [30], because the human system is a system of systems [31], i.e. system of several mutually interconnected systems of a different nature. Consequences of interconnections (interfaces) are mutual dependences, the character of which is physical, cyber, territorial and organisational [32]. As a consequence of growing globalisation the new sources of disasters take on force, they also cause complex facility failures [33], [34]. Since, according to the EU [1] and the UN [2], the highest quality of each object (i.e. the State, human community, technical installation etc.) is safety, so the highest priority for the object is the overall (integral) safety. This means that it is not only about partial safety such as internal, external, environmental, fire, economic, etc., but about the safety of the whole, which includes partial safety so that the condition of the whole, is optimal.

Due to the systemic nature of today's world, integral safety cannot be built without regard to others. Very important is the safety of basic processes, which ensure products and services for human society. Climate change, biodiversity loss and pollution - the triple planetary crisis [35] - increasingly threaten the Earth system, necessitating tools such as life-cycle assessment (LCA) that can evaluate the effectiveness of different prevention and mitigation strategies. LCA systematically quantifies the environmental impacts over the whole life cycle of products, processes or policy scenarios [36]. Due to advancements in product design and technology, products in the market today are capable of an extended life cycle by undergoing end-of-life treatments [53]. To enable this, products have to be designed for multiple life cycle (MLC) purposes. Besides that, the presence of technology innovation has led to the generation of multiple products. Therefore, it is crucial to evaluate the life cycle of products in a wider scope. [37]. Design innovation is a key driver to success in the current competitive market by substantially improving product design and features to delight customers' expectations [38]. In view of design innovation, MLC products aim to enhance the value of their life cycle by not only providing satisfactory design functionality, but also proper EOL treatments and a prolongation of their life cycle. This is accomplished by combining relevant strategies of Design for X (DFX), which is a design concept that was recently mentioned as a Design for Multiple Life Cycles (DFMLC). The DFMLC is a sustainable design approach that affords decisive environmental impacts, reduces the amount of waste in landfills, diminishes pollution and the use of natural resources, and protects the quality of natural and built environments [39], [40], [41]. LCA is frequently applied to uncover environmental hotspots and prioritize actions and is increasingly used to assess the environmental impacts of strategy implementation scenarios [42]. Social Life Cycle Assessment (SLCA) is the third assessment dimension and seeks to measure the impact of a product or process on society along the full life cycle. SLCA is similar to LCA in that it follows the LCA four-step procedure [43] providing complementary impact information (social) to the environmental information of LCA and both assessments use a functional unit to define the product system [44]. The differences between them are mainly with respect to interpretations of social impact on people [45]. LCA considers impacts to people indirectly through impacts to the environment, while SLCA more broadly considers many types of impacts directly to people, and the impacts can be positive and negative [44], [46]. Lastly, local and site-specific impacts matter more. The same indicator may show different social impacts to stakeholders based on location [47].

The basic function of the State since its inception has been to ensure the safety of protected assets (interests) of the state and the sustainable development of the State. Protected assets (interests) of the State are the assets of the State that are protected as a priority, i.e. lives, health and security of people, property, environment, public welfare, knowledge, critical technologies and infrastructures. To ensure the safety and development of assets and the State, the State needs to have good management of resources, assets and people under normal, emergency and critical conditions [3].

State safety in its essence is a set of human measures and activities that ensure the security and sustainable development of the State and its assets; i.e. they limit the conditions for realization of danger. To do this, the administration of the State should not only be concerned with survival, power, social compliance and damage prevention, but should solve methodological-conceptual problems and ensure that:

- safety would be considered in a systemic context, i.e. not only in conjunction with a priori defined risks,
- concepts of safety could not be burdened by ideological and political clichés,
- problem solving was based on the results of decision theory,
- the relationship between risk and safety would be correctly understood. The essence of the problem lies in the answer to the question: How are risks and their impacts identified? It is used the answer: They are determined by plausible scenarios. But is there no procedure for how such a scenario should be formed? Usually, the scenario is based on past events and phenomena, and does not consider human violations and the existence of possible surprises caused by the dynamic development of the world and human society,
- the dynamic evolution of the world, which causes changes in the nature of existing risks and brings new risks, would be considered.

The State plays a role that can be described in terms of risk management [3], [4] because it redistributes certain types of risk through decisions about the welfare system/public welfare and health care. The need to consider risks and manage them correctly at all levels of government will prevent failures in the provision of public services.
Risks enter the public domain when they meet any of the following attributes:

- These are externalities that market mechanisms cannot address.
- In connection with legislation, the harmful effects of technology and bad decisions of public administration are imposed on citizens.
- A significant part of the public that disagrees with the concept of managing certain assets is at risk.
- Political decisions, without regard for the safety and development of citizens, give rise to phenomena in which risks are realised.
- Disasters (unacceptable phenomena), which cause unacceptable risks, are distributed in such a way that they disregard political fairness.

Public administrations should therefore analyse the risks both in terms of impacts on society and in terms of impacts on the public administration management system, as a number of examples in the past have shown that public administration decision-making has exacerbated the effects of the emergency situation [5-18].

The steps of the risk management process carried out by public administrations differ from the normal risk management procedure [3] only in that considerable attention should be paid to the formulation of context and risk monitoring from a strategic and procedural point of view, i.e.:

- The capacity of public administrations and other stakeholders to achieve strategic objectives in the areas of safety, security, mobility and environmental conditions (health and environmental risks) should be assessed in a strategic context.
- The problem-solving capacity of public administrations needs to be assessed in an organisational context.
- In the context of risk management, risk thresholds, maximum impact levels and the right priorities for decision-making need to be assessed.

Given the dynamic development of the world, it is necessary for public administration, administrators of all legal organizations of the State (i.e. public institutions and private entities) and individuals to adapt the set of human measures and activities ensuring the security and sustainable development of the State and its assets to the current conditions. This means that permanent risk management is carried out at all levels in favour of integral safety. The present article examines the problems of safety of the public assets of the State, which belong among technical installations. Their safety depends not only on themselves, but also on the quality of their relationship with public administration. Public administration needs them because they provide products and services to citizens and the State, which thus fulfils many of its basic functions. On the contrary, technical installations need public administration, because it sets and controls the conditions of their activity.

2. Safety and Criticality of Technical Installations

Technology is the application of knowledge to achieve practical goals in a specified and reproducible way. It is a systemic use of knowledge for practical purposes. While technology contributes to economic development and human prosperity, it can also have unacceptable impacts, such as pollution or resource depletion, or cause social damage, such as technological unemployment caused by automation. Engineering is the process by which technology is developed and operated. This often requires solving problems under strict constraints because the safety of people and other public assets is at stake. Therefore, the management of the safety of technical installations is essential.

The safety management of technical works, which are complex systems of the "systems of - SoS" types, can be defined as the integrated planning, design, optimization, operation and management of products, processes and services for the benefit of people [3], [4-20]. In terms of competitiveness and economy, their performance must also be ensured. This depends on their resilience, which determines how the technical installation responds to harmful phenomena of all kinds. The theory of system management according to [20] implies that the resiliency of the system is related to robustness, redundancy, ingenuity and speed of starting the correct response, Figure 1.

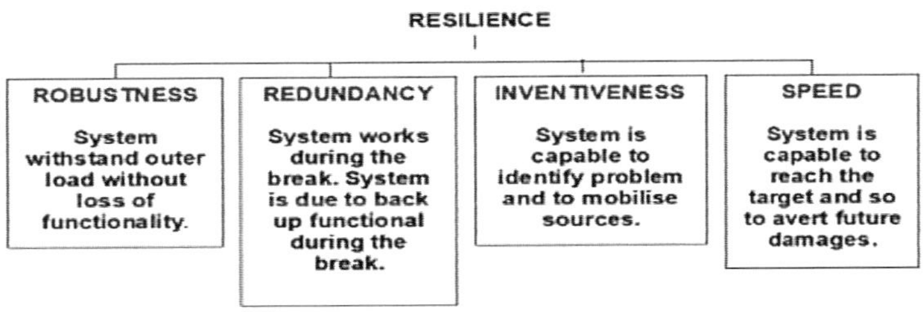

Fig. 1. Context of resiliency of the system with robustness, redundancy, ingenuity and speed [24]

For increasing the safety and reliability of technical installations, it is necessary in terms of references of critical elements and whole to use the disaster sizes with a return period of more than 100 years (current standards consider design disaster for 100 years [20], [24]).

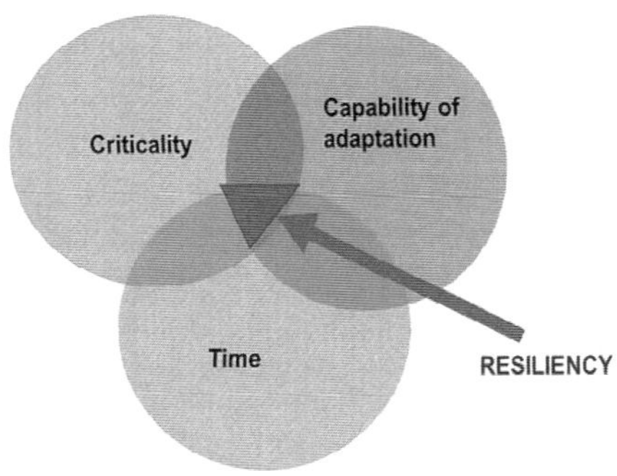

Fig. 2. Critical parameters that are important for resiliency [24]

Mistake is that during the design, it is only carried out the protection of elements and objects only for individual listed disasters, i.e. other some possible and unnamed disasters are not considered - it is necessary to use the application called "All Hazards". As technical facility provides vital functions for humanity, care needs to be taken to ensure the continuity of operations, for which resilience is important. The idea of resilience is portrayed as the intersection of three circles, Figure 2. Resilience means continuously increasing both, the safety and the security while addressing potential conflicts that arise in practice.

Resiliency is the potential of a system, which lies in the specific arrangement of the system that maintains the functions and feedbacks of the system in condition, which include the ability of the system to reorganize itself based on changes induced by failures. It follows that sustainability management must be based on resilience management, which has two objectives:

- To prevent so the system could get into unaccepted conditions due to external failures and external stress.
- To preserve the elements activating systemic reorganization and recovery due to massive changes.

Another important characteristic of technical installations is criticality [20], which is a function of resilience and necessity (importance) for the safety of the system. Criticality can be viewed from two points of view, namely from the technical point of view (failure rate of the elements of the system u, i.e. the technical installation) and social (impacts of the non-functioning of the provision of technical installation services to the population). Both aspects are important from the point of view of national safety. From the point of view of the safety of the operation of the technical installation, the first aspect is important; its value is determined by the design.

3. Types of Safety at Technical Installations

With regard to present knowledge and experience, it is currently monitored process safety and the overall (integral) safety of technical installations.

Process Safety is a set of measures and activities that ensure safe operation, i.e. safe operation of processes, e.g. in the case of chemical processes focus on the prevention of fires, explosions and leakage of hazardous substances into the environment [21]. The specific discipline of Process Safety Management (PSM) has been developing over the last 40 years and its aim is to ensure safe processes that take place in technologies, it is the management of principles and systems for the identification of possible threats, understanding and mastering the processes leading to the implementation of threats. It is a complex procedure and requires a multidimensional approach that combines technology and its management [20]. Process safety management is associated with a safety culture and a checklist is often used for safety assessment [22]. Process control is widely used in factories and other automated environments to ensure production efficiency. In general, process control technology is designed to monitor sensors and set important variables according to measured values. This technology allows a relatively small group of people to manage complex operations and helps ensure that the desired outcome is consistently achieved.

System Safety is a set of measures and activities that ensures safe technical installation and its safe surroundings. The discipline in question originated by the application of a systemic approach in engineering fields. Integral (object) safety has its roots in industry safety engineering, which dates back to the 19th century, and which after World War 2 it applied disciplines: systems engineering and systems analysis to solve new and complex engineering problems. The safety of the system in the monitored concept lies in the application of technical and managerial skills in the identification, analysis, evaluation and management of harmful phenomena and related risks using a systemic approach [4-24].For practical reasons, the approaches used in the area of interest must be effective and affordable. Safety orientation must be part of the company's management system, while respecting the constraints that flow from the outside world. The discipline in question defines:

- technical installation as a system, which is a combination of people, procedures and equipment that are integrated to perform a specific operational task or function in a specific environment,
- the concept of system safety as an application of special technical and organizational skills with the aim of systematically preventing damage and losses to the assets of the system associated with them by identifying threats and managing risks during the entire life of each device or object created and implemented by man.

Over time, the OECD has developed separate concepts for the nuclear and chemical industries [19]. This concept is in work [23] added by cyber security due to increasing role of automatization at present. System safety in technical installations uses systems system theory and system engineering to prevent predictable accidents and minimize the consequences of unpredictable accidents. In the modern concept, it is generally interested in all losses and damages, and not only in fatal accidents or injuries and damage to property, but also in failure to fulfil a mission (mission, purpose), or environmental damage. The key point of the discipline is to consider losses serious enough that enough time, effort and resources are devoted to their prevention. The size of the investment devoted to accident prevention and/or its impact is substantially dependent on social, political and economic factors. Therefore, for technologies that are likely to have serious consequences, the precautionary requirement is imposed by legislation to ensure the protection of public assets [24].

The primary concern of the safety of technical installations is qualified risk management, i.e.: identification of possible hazards; identification and evaluation of risks; and elimination and/or management through design analysis and/or organizational procedures. The programme for the safety of technical installations must provide for a well-defined procedure for methodological control of safety-related aspects and evaluate the design of the equipment in the sense of identifying possible sources of risk and prescribing time- and costly remedial interventions. The objectives of the system safety programme for technical installations is to ensure:

- the safety of technical installation corresponds to its mission, which is built inherently,
- identifying, assessing, eliminating and/or managing risks to an acceptable level of risk for all risks associated with the system, subsystem and individual parts,
- risk management from threats that cannot be eliminated; i.e. risks must be secured in such a way as to protect personnel, facilities and property,
- the use of new materials and/or products and testing techniques involves only minimal risk,
- the need for corrective measures required to improve safety by temporarily incorporating safety factors is minimised during the inception of the system,
- historical safety data generated by similar safety programs shall be considered and used wherever appropriate.

Industries have either adapted systems safety programs from the military or NASA, or have independently developed their own programs based on the experience gained from the construction of nuclear power plants, from the production of complex, dangerous and expensive equipment. Waiting for accidents to occur and then eliminating the causes has become an uneconomical and sometimes even unacceptable way of modifying and improving systems. Building many of today's complex systems requires the integration of parts (subsystems and components) made by various independent suppliers and organizations. Even if each supplier maintains the required quality of its parts, combining subsystems introduces new errors and new hazards that are not visible when viewed as separate parts.

Considering the risks associated with a combination of systems and subsystems is referred to in practice as creating inherent security. It has been confirmed in many industrial sectors that incorporating inherent safety into equipment or products can reduce their overall life-cycle costs and that achieving an acceptable level of safety requires advanced recent systems security approaches [24]. System safety-related activities begin at the earliest stages of system concept development and continue through all design, production, testing, operation and shutdown activities. An essential aspect that distinguishes the system safety approach from other approaches used in the field of safety is the primary emphasis on early identification and classification of hazards so that remedial action can be taken to eliminate or minimize them before the final project decision.

Models trained on large data sets of seismic events can estimate the number of aftershocks better than conventional models do. AI predicts how many earthquake aftershocks will strike — and their strength [48]. Seismologists are finally making traction on one of their most important but challenging goals: using machine learning to improve earthquake forecasts [49]. New machine learning models hold potential for predicting the number of quake aftershocks [50], [51], [52]. Despite the fact that system safety is a relatively new and still evolving discipline, it has its basic ideas, which are preserved in all its manifestations and distinguish it from other approaches to safety and risk management. Principles of system safety management in the recent concept are:

- safety is step by step creating from start of system design and it is not added to the created system,
- safety deals with the system as a whole and not just with subsystems and components,
- safety takes dangers and associated hazards more than just personnel errors,
- safety creating emphasizes analysis rather than experience gained later and standards developed later,
- qualitative approaches are favoured over quantitative ones;
- differentiation of importance of changes and conflicts of goals in a system design is more than system engineering.

The most important aspect of system safety in accident prevention terms is safety management procedures. Effective safety management consists of setting policies and defining safety objectives, i.e. in:

- task and procedure scheduling,
- defining responsibilities and determining competencies,
- documentation and continuous monitoring of threats and resulting hazards, including controls,
- maintenance of the information system for safety management, including feedback and forms of fault/accident reporting, etc.

The safety of the system is responsible for ensuring the security of the system as a whole, including the analysis of the interface between the components. Component safety activities, such as rocket launch pad safety, may be part of general responsibility for system safety, or may be part of component engineering in large and complex projects. For defined types of hazards, such as fires, nuclear safety or explosive atmospheres, a further classification of safety responsibilities may be required. In any gradation of the breakdown of system safety efforts, system safety engineers are responsible for integrating individual security activities and information. The safety of a system is usually linked to corresponding engineering and/or scientific disciplines such as reliability engineering, quality assurance, human factor, etc.

What processes and tasks of system safety will be performed in a particular project depends on its size and the risk level of the designed system [24]. System safety and reliability are closely related [20], [25]. In practice, a safe device or a safe system is reliable, but a reliable device or reliable system may not be safe. Reliability engineering preferentially deals with errors and reducing their frequency. Reliability is defined as a characteristic of a given object expressed in terms of the probability that the object will perform in a specified way the functions that are required of it during a specified time interval and under specified or predicted conditions. Representative reliability engineering techniques aimed at minimizing component (component) errors, and thus complex system failures caused by component errors, are: parallel redundancy; back- up device; safety margin; reducing congestion; and usage time limitation. These techniques have been shown to be effective in increasing reliability, but they do not necessarily increase safety, and may even reduce it under certain circumstances (e.g., incorporating multiple backups into systems can create an environment for unwanted couplings and affinities between components that cause failure under certain conditions). Therefore, risk analyses connected with system safety look at interactions and do not just focus on engineering errors or uncertainties.

Reliability engineers often consider reliability and safety synonymous. This is true only in some special cases. In general, safety has a slightly broader meaning. Normally, reliability and safety have many characteristics in common. However, many crashes occur without a component failing. On the contrary, many times all components in accidents worked as expected and flawlessly [24]. It can also happen that components can fail (fail) without a crash. It is a fact that breakdowns and accidents may be caused by the operation of equipment outside the permitted ranges of parameter values or time limits on which the reliability analyses were based. This means that the system can have high reliability and yet crash. Moreover, generalized probabilities and reliability analyses cannot be directly applied to specific or local conditions. Most importantly, accidents and accidents are often not the result of simple combinations of errors/component failures [24].

Safety is a feature that stands out at the system level when components are operated together. Events leading to an accident can be a complex combination of equipment error, improper maintenance, information and control system problems, human intervention, and design errors. Reliability analyses only look at the probabilities of accidents and error-related accidents; they do not investigate potential damage that may be caused by the correct operation (operation) of individual components. Therefore, it is not possible for reliability engineering to replace system safety engineering, but it can complement it. However, this must be done with the clear knowledge that the ultimate goal is to increase the system's resilience to the risk of accidental errors. It's always better when the device/system is designed in such a way that individually random errors cannot cause an accident.

Great caution should be exercised when applying reliability estimation techniques for safety assessment. Where accidents are not inevitably caused by phenomena which can be expressed in terms of probabilities of occurrence, the degree of probability of the occurrence of a risk cannot be generally applied to them. Probability of occurrence estimates measure the probability of accidental errors occurring, not the probability of the occurrence of the realization of risks, accidents and/or accidents. Practice shows that when a project error is found during the system safety analyses, it is much more effective to fix it than to convince someone using calculated probabilities that the error will never cause a crash. High probability values of reliable behaviour do not guarantee safety, and it is known from practice that safety often does not require ultra-high reliability [24].

Most often, the main drawback of probabilistic models is not what they include, but what they don't. Low values of the probability of unreliable behaviour simply indicate that the system will not fail in the intended way, but says nothing about the fact that the system can fail with a much higher probability in a way that was not considered [24]. Distinguishing accident risk from error is essential to understand the difference between safety and reliability. For practical reasons, system security approaches and programs must be effective and affordable. The cost recovery of the system safety program is achieved when crashes are avoided.

The effectiveness of a system safety program is very difficult to prove, because measuring something that did not happen is difficult. One indirect way to measure the effectiveness of a system safety program, although not entirely satisfactory due to the lack of factors compared, is to compare the operation of systems that had a system safety program with those that did not and crashed. Another way of determining the effectiveness of the system safety program is to report hazards that were corrected by system safety personnel before the accident occurred or was otherwise detected. The third way of estimating the effectiveness of system security programs is to investigate cases in which the system safety recommendations were not respected and accidents occurred.

4. Procedures for Ensuring the Safety of Technical Installations

Based on the current knowledge summarized in the works [5-20,23,24] the safety management system of the technical installations includes the organizational structure, responsibilities, practices, regulations, procedures and resources for determining and applying disaster prevention or at least mitigating their unacceptable impacts in the territory. As a rule, it covers a number of issues, including but not limited to organisation, personnel, identification and assessment of threats and resulting risks, organisation operations, organisational change management, emergency and crisis planning, security monitoring, audits and reviews [23]. Based on the last cited work, the safety management system (SMS) of technical installation is integrated management of7 processes:

- The process for the design and implementation of the concept and management, which is further divided into sub-processes to ensure: overall concepts; safety objectives; safety management/direction; safety management system;

personnel, which are further divided into the following sections: human resources management, training and education, internal communication/awareness, and the working environment; and reviewing and evaluating the achievement of safety objectives.

- The process for implementation of administrative procedures which is further subdivided into: disaster hazard identification and risk assessment; documentation keeping; procedures (including work permit systems); change management; safety in conjunction with contractors; and product safety oversight.
- The process of technical matters, which is further divided into sub-processes for: research and development; design and assembly; an inherently safer process; industry standards; storage of dangerous substances; and maintenance for integrity and maintenance for equipment and objects.
- The process for external collaboration which is further subdivided into: cooperation with administrations; cooperation with the public and other stakeholders (including academic departments); and cooperation with other businesses.
- The process for emergency preparedness and response, which is further divided into sub-processes for: on-site preparedness planning; facilitating the off-site preparedness planning (under the responsibility of public administration); and coordinating as well as departmental organisations' activities in providing emergency preparedness and response.
- The process for processing the reports and investigation of accidents/ near misses, which is further divided into sub-projects for: processing accident reports, near misses and other learning experiences; investigating undesirable phenomena; and responding to and following up on accidents (including application of lessons learned and information sharing).
- The process for physical and cyber security of a technical installation, which is further divided into support for: physical security; and cyber security against hackers and terrorists.

The safety management system (SMS) of a technical installation is based on the concept of disaster prevention or at least their serious impacts [19], [23, [24], which includes the obligation to implement and maintain a management system in which the following problems are considered:

- the roles and responsibilities of persons involved in the management of major disaster threats at all organisational levels of technical work and training measures aligned with identified training needs,
- plans for the systematic identification of major threats from disasters and the resulting risks associated with normal and abnormal conditions and for assessing their likelihood and severity (magnitude),
- plans and procedures to ensure the safety of all components and functions in and around the technical installation, including the maintenance of objects, equipment,
- plans for the implementation of changes in the technical installation, territory, objects and equipment,
- plans to identify foreseeable emergencies by systematic analysis, including the preparation, testing and assessment of contingency response plans for such emergencies,
- plans for the ongoing evaluation of compliance with the objectives clarified in the safety concept, SMS and mechanisms for investigating and implementing corrective actions in case of failures to achieve the set objectives,
- plans for periodic systematic evaluation of the safety concept, effectiveness and suitability of the SMS and criteria for assessing the level of safety by the top team of technical installation personnel.

Safety is a matter for all involved, i.e. managers, employees and randomly present. In this context, we talk about the ***so-called golden rules of all involved*** [24], which are:

- to prevent disasters or at least their unacceptable impacts by means of preventive measures, to ensure preparedness to deal with unacceptable impacts on protected assets (interests) of the technical installation and the effective response of the technical installation,
- to communicate and cooperate with others involved in all aspects of the prevention, preparedness and response of technical installation,
- to know the hazard from disasters and possible risks in the technical installation and its surroundings,
- to implement and respect a "safety culture" that is respected and promoted by all stakeholders at all times,
- to establish safety management systems, monitor and, if necessary, correct their activities,
- to apply the principles of inherent safety in the design, design and operation of technical installation and its equipment,
- to carefully manage changes in the technical installation,
- to be prepared to cope with any harmful phenomena that may occur,
- to assist other stakeholders in carrying out their roles and responsibilities,
- continuously to improve safety,
- to operate in accordance with safety culture, safe practices and training,
- to strive to keep up to date with all information and information and provide feedback to managers,
- to strive to develop, strengthen and continuously improve safety concepts, regulations and directives,
- to guide and motivate all other stakeholders to fulfil their roles and responsibilities,
- to know the risks within the sphere of their own responsibility, appropriately plan measures for their proper management,

- to apply an appropriate and coherent planning and follow-up policy,
- to be aware of the risks in the technical installation and know what to do if they are realized,
- to participate in emergency planning and response to emergencies.

Safety culture means that a person in all his/her roles (manager, employee, citizen or victim of a disaster) adheres to the principles of safety, i.e. behaves in such a way as not to cause the realization of possible risks and when he becomes a participant in the implementation of risks, to contribute to the effective response, stabilization of protected interests and their restoration and to start their further development. An effective safety culture is an essential element of safety. It reflects the concept of safety and is based on the values, opinions and actions of the organization's top managers and their communication with all stakeholders. It is a clear commitment to actively participate in addressing security issues and promotes that all stakeholders act safely and comply with relevant laws, standards and norms. The rules of safety culture must be incorporated into all activities in the technical installation. They are not based on concentrating on punishing the culprits/perpetrators of mistakes, but learning from mistakes and introducing corrective measures so that mistakes cannot be repeated or at least significantly reduce their occurrence.

The tool for ensuring a safe technical installation, i.e. a technical installation in which there is an effective safety culture, is the program to increase the safety of the technical installation [24].The procedure for creating a program to increase the safety of technical installation according to [19] consists of the following steps:

- To define the tasks (partial objectives) and strategic objectives of the technical installation with regard to safety.
- For each technical installation sector (connected with processes and subprocesses given above), to select appropriate target and intermediate indicators for assessing safety levels or to develop specific checklists.
- To create a vocabulary for the needs of managing the integral safety of a technical installation.
- To align standards, good practice practices and local practices.
- To modify the list of target indicators or limit values for checklists according to the conditions in the technical installation in question.
- To modify the list of intermediate indicators or limit values for checklists according to the conditions in the technical installation in question.
- To determine how to evaluate target indicators (i.e. value system) or checklists according to the conditions in the technical installation in question.
- To determine the method of evaluation of interim indicators (i.e. value system) or checklists according to the conditions in the technical installation in question.
- To establish a method/scale for measuring a set of indicators (i.e. a set of values) or a set of checklists and limit limits according to the conditions in the technical installation in question.

Figures 3 and 4 show an idea of entity safety management based on the above risk management insights [20]. The first is simpler and the second shows the decision-making method built into the safety management system of a technical installation.

Fig. 3. Milestones determining the safe or dangerous behaviour of a technical installation in the event of a dangerous situation

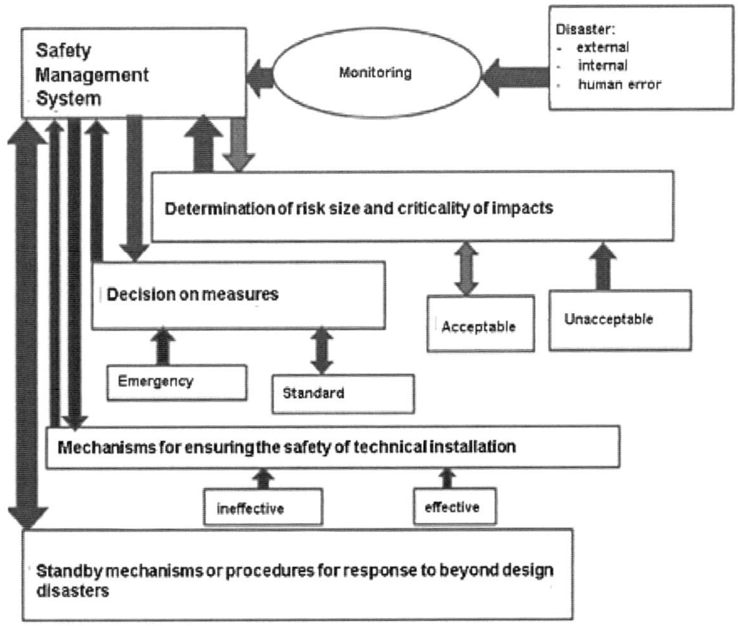

Fig. 4. An idea of the way of managing the safety of a technical installation built into the safety management system

5. Future research plans

Further research is planned to focus mainly on the implementation of the analyzed areas of research in the field of reliability, resistance, criticality, risks and safety in process safety management and especially safety management systems on technical equipment. It will be important to use feedback from practice when applying the described methodology for ensuring the safety of technical equipment and implementing the knowledge gained so far.

6. Conclusion

Safety-oriented engineering also performs safety management tasks, i.e. risk management tasks for the benefit of safety and the development of the human system. In technical slang, we talk about creating inherent safety of a technical installation against design disasters by managing safety. By implementing the precautionary principle, we ensure an increase in resiliency to the unacceptable impacts of over-project disasters that are so unlikely to occur that they are unpredictable. Principles such as "fail safely, perform only specified functions, i.e. if you cannot meet a goal, not do anything" are put into practice in technology[20]. The engineering in question is based on safety management, which is based on specific risk management [20], which is characterized in particular by the following features:

- placement - design - construction - design with minimization of risks, so call: risk-based design, risk-based operation, risk-based maintenance etc.
- operation with the integration of an early warning system and procedures to manage an acceptable level of risk,
- managing the abnormal, emergency and critical conditions during the operation and shutdown.

The specificity of the monitored risk management is that it is the management of risks[4]:

- from all possible disasters at once, with the current list of disasters being determined by the All Hazard Approach [26], [27] (i.e. considering all possible disasters regardless of whether their sources lie inside or outside the system)
- and it is sought optimal solution for relevant possible disasters is sought, while applying the precautionary principle, which includes sustainable development.

Safety-oriented engineering[20]in risks determining the risk uses the principles:

- the risk is determined during the entire life cycle of technical installation, i.e. during the siting, design, construction, operation and decommissioning, and possibly also when the territory is restored to its original state,
- risk determination focuses on user requirements and the level of services provided,
- risks are determined according to the critical impact on processes, services provided and assets determined by the public interest,
- unacceptable risks are mitigated through risk management tools, i.e. technical and organisational measures, standardisation of operational procedures or by automated control.

From the professional point of view, subject engineering is a process that seeks all potential conditions that would endanger the successful functioning of the monitored technical installation at all stages of its lifetime, and identifies possibilities for their management by prevention, preparedness, response and renewal. Safety-oriented engineering requires the use of advanced safety practices. The implementation of the described methodology for ensuring the safety of technical equipment in practice represents an important element for eliminating risks. It can contribute not only to partial safety such as internal, external, environmental, fire, economic, etc., but also to safety as a whole. Such a comprehensive approach based on current knowledge in the context of the safety management process has the potential to contribute to overall integral safety.

7. References

[1] EU. Maastricht Treaty. C 191, 29.7.pp.1–112. Maastricht: EU 1992

[2] UN. Human Development Report. New York 1994, www.un.org

[3] Prochazkova, D. (2011). Strategic Management of Territory and Organization. ISBN 978-80-01-04844-3. Praha: CTU, 483 p.

[4] Prochazkova, D. (2020). Analysis, Management and Settlement of Risks Connected with Technical Facilities. Prague: CTU, 222 p.

[5] Briš, R., Guedes Soares, C. & Martorell, S. [eds.] (2009). Reliability, Risk and Safety. Theory and Applications. ISBN 978-0-415-55509-8. London: CRC Press, 2362 p.

[6] Ale, B., Papazoglou, I., Zio, E. [eds.] (2010). Reliability, Risk and Safety. ISBN 978-0-415-60427-7. London: Taylor & Francis Group, 2448 p.

[7] Bérenguer, C., Grall, A., Guedes Soares, C. [eds.] (2011]. Advances in Safety, Reliability and Risk Management. ISBN 978-0-415-68379-1. London: Taylor & Francis Group, 3035 p.

[8] IAPSAM (2012). Probabilistic Safety Assessment and Management Conference. International. 11th 2012. (Annual European Safety and Reliability Conference). ISBN: 978-1-62276-436-5. Helsinki: IPSAMESRA, 6889 p.

[9] Steenbergen, R., Van Gelder, P., Miraglia, S., Ton Vrouwenvelder, A. [eds.] (2013). Safety Reliability and Risk Analysis: Beyond the Horizon. ISBN 978-1-138-00123-7. London: Taylor & Francis Group, 3387 p.

[10] Nowakowski, T., Mlyňczak, M., Jodejko-Pietruczuk, A., Werbiňska-Wojciechowska, S. [eds.] (2014). Safety and Reliability: Methodology and Application. ISBN 978-1-138-02681-0. London: Taylor & Francis Group, 2453 p.

[11] Podofillini, L., Sudret, B., Stojadinovic, B., Zio, E., Kr̋oger, W. [eds.] (2015). Safety and Reliability of Complex Engineered Systems. ISBN 978-1-138-02879-1. London: CRC Press, 4560 p.

[12] Walls, L., Revie, M., Bedford, T. [eds.] (2016). Risk, Reliability and Safety: Innovating Theory and Practice: Proceedings of ESREL ISBN 978-1-315-37498-7. London: CRC Press, 2942 p.

[13] Cepin, M., Bris, R. [eds.] (2017). Safety and Reliability – Theory and Applications. ISBN: 978-1-138-62937-0. London: Taylor & Francis Group, 3627 p.

[14] Haugen, S., Vinnem, J., E., Barros, A., Kongsvik, T., Van Gulijk, C. [eds.] (2018). Safe Societies in a Changing World. ISBN: 978-0-8153-8682-7. London: Taylor & Francis Group, 3234 p.

[15] Beer, M., Zio, E. [eds.] (2019). Proceedings of the 29th European Safety and Reliability Conference (ESREL). ISBN 978-981-11-2724-3. Singapore: ESRA, Research Publishing, 4315 p.

[16] Baraldi, P., Di Maio, F., Zio, E. [eds.] (2020). Proceedings of the 30th European Safety and Reliability Conference and 15th Probabilistic Safety Assessment and Management Conference (ESREL2020 PSAM15). ISBN 978-981-14-8593-0. Singapore: ESRA, Research Publishing, 5067 p.

[17] Castanier, B., Cepin, M., Bigaud, D., Berenguer, C. [eds.] (2021). Proceedings of the 31st European Safety and Reliability Conference. ISBN 978-981-18-2016-8. Singapore: ESRA, Research Publishing, 3473 p.

[18] Leva, M.C., Patelli, E., Podofillini, L., Wilson, S. [eds.] (2022). Proceedings of the 32nd European Safety and Reliability Conference (ESREL 2022). ISBN 978-981-18-5183-4. Singapore: ESRA, Research Publishing, 3413 p.

[19] OECD. (2002). Guidance on Safety Performance Indicators. Guidance for Industry, Public Authorities and Communities for Developing SPI Programmes Related to Chemical Accident Prevention, Preparedness and Response. Paris: OECD, 191 p.

[20] Prochazkova, D. (2017). Principles of Risk Management of Complex Technological Facility. ISBN 978-80-01-06180-0, e-ISBN:78-80-01-06182-4. Prague: CTU, 364 p. doi:10.14311%2FBK.9788001061824

[21] EU. Seveso III Directive (2012/18/EU). Brussels: EU 2012.

[22] US DOE. DOE -HDBK-110196. Handbook. No 20585- Tennessse: US DOE 1996, 180 p.

[23] Prochazka, J., Prochazkova, D. (2022). Risk Management of Traffic Management Systems. Prague: CTU, 129 p. doi:10.14311/BK.978 80010 69950

[24] Prochazkova, D., Prochazka, J., Lukavsky, J., Dostal, V., Prochazka, Z., Ouhrabka, L. (2019). Risk Management of Processes Associated with the Operation of a Technical Facility during Its Lifetime. ISBN 978-80-01-06675-1. Prague: CTU, 465 p. doi:10.14311%2FBK.978800 1066751

[25] Rausand, M. (2014). Reliability of Safety-Critical Systems: Theory and Applications. - John Wiley & Sons, 421 p.

[26] FEMA. (1996). Guide for All-Hazard Emergency Operations Planning. State and Local Guide (SLG) 101. Washington: FEMA.

[27] EU. FOCUS Project. Brussels: EU 2012 – [on-line] Available on - URL: http://www.focusproject.eu/documents /14976/-5d7 63378-1198-4dc9-86ff-c46959712f8a

[28] Procházková, D., Wessely, E., Rusko, M. & Kralikova, R. (2011). Human System Safety Management and Environmental Management Relation. - In DAAAM International Scientific Book 2011. pp. 103-118. ISSN 1726-9687; DOI: 10.2507/daaam.scibook.2011.09

[29] Necesal, L., Lukas, L. & Jasek, R. (2011). Measures for critical infrastructure protection in the Czech republic. - Annals of DAAAM for 2011 & Proceedings of the 22nd International DAAAM Symposium, Volume 22, No. 1, ISSN 1726-9679, Katalinic, B. [ed.], Vienna, pp. 0843-0844

[30] Lojan, R. (2015). Ľudská dôstojnosť a hodnota ľudského života. - In: Nové horizonty, Bratislava, ISSN 1337-6535, Vol. 9, Issue. 1, pp. 10-14

[31] Prochazkova, D., Prochazka, J., Rusko, M., Mikulova, M. & Ilko, J. (2017). Model for critical infrastructure safety management. - In Annals of DAAAM for 2017, Volume 28, No.1. The 28th DAAAM International Symposium. Zadar, Croatia, 08-11th November 2017. Vienna, ISSN 2304-1382. ISBN 978-3-902734-14-3. DOI: 10.2507/28th.daaam.proceedings.085.

[32] Procházková, D. & Rusko, M. (2017). Relation between human system safety management and environmental management. - In Journal of Environmental Protection, Safety, Education and Management. Vol. 5, No. 9, ISSN 1339-5270

[33] Kralikova, R., Sokolova, H., Wessely, E. & Polak, J. (2013). Approaches to assessment of hot environment. – In DAAAM International Scientific Book 2013. chapter 14. Vienna, pp. 317-328. ISBN 978-3-901509-94-0

[34] Lukas, L. & Hromada, M. (2011). Risk analysis in context of kritical infrastructure protection. - Annals of DAAAM for 2011 & Proceedings of the 22nd International DAAAM Symposium, Volume 22, No. 1, ISSN 1726-9679, Katalinic, B. [ed.] Published by DAAAM International, Vienna

[35] Steffen, W. et al. (2015). Planetary boundaries: Guiding human development on a changing planet. - SCIENCE 15 Jan 2015 Vol 347, Issue 6223, DOI: 10.1126/science.1259855

[36] Kikuchi, Y. (2016). Chapter 24 - Life Cycle Assessment. - Plant Factory An Indoor Vertical Farming System for Efficient Quality Food Production, pp. 321-329, https://doi.org/10.1016/B978-0-12-801775-3.00024-X

[37] Suhariyanto, TT, Wahab, D.A. & M.N. Ab. Rahman (2017). Multi-Life Cycle Assessment for sustainable products: A systematic review. - Journal of Cleaner Production, Volume 165, 1 November 2017, Pages 677-696, https://doi.org/10.1016/j.jclepro.2017.07.123

[38] Moon, H., Miller, D. R. & Kim, S.H. (2013). Product Design Innovation and Customer Value: Cross-Cultural Research in the United States and Korea. - Journal of Product Innovation Management, Volume 30, Issue1, January 2013, pp. 31-43, https://doi.org/10.1111/j.1540-5885.2012.00984.x

[39] Dunmade, I. (2006). Design for Multi-lifecycle: A sustainable design concept applied to an agroindustrial development project. - Conference: 2006 ASAE Annual Meeting, Portland, Oregon, July 9-12, 2006, 068059, DOI: 10.13031/2013.21074

[40] Dunmade, I. (2013). Design for Multi-Lifecycle : A Sustainability Design Concept. - International Journal of Engineering Research and Applications (IJERA), ISSN: 2248-9622, Vol. 3, Issue 2, March -April 2013, pp.1413-1418

[41] Go, T.F., Wahab, D.A. & Hishamuddin, H. (2015). Multiple generation life-cycles for product sustainability: the way forward. - Journal of Cleaner Production, Volume 95, 15 May 2015, pp. 16-29, https://doi.org/10.1016/j.jclepro.2015.02.065

[42] Hellweg, S., Benetto, E., Huijbregts, M.A.J. et al. (2023). Life-cycle assessment to guide solutions for the triple planetary crisis. - Nat Rev Earth Environ 4, 471–486 (2023). https://doi.org/10.1038/s43017-023-00449-2

[43] ISO 14040: 2006 Environmental management — Life cycle assessment — Principles and framework.

[44] UNEP/SETAC (2009) Guidelines for social life cycle assessment of products. Life-Cycle Initiative. -United Nations Environment Programme and Society for Environmental Toxicology and Chemistry, Paris, France, http://www.lifecycleinitiative.org/wp-content/uploads/2012/12/2009 - Guidelines for sLCA - EN.pdf

[45] Hoogmartens, R., Passel, S.V., Acker, K.V. & Dubois, M. (2014). Bridging the gap between LCA, LCC and CBA as sustainability assessment tools. - Environmental Impact Assessment Review, Volume 48, September 2014, pp. 27-33, https://doi.org/10.1016/j.eiar.2014.05.001

[46] Finkbeiner, M., Schau, E.M., Lehmann, A.. & Traverso, M. (2010). Towards Life Cycle Sustainability Assessment. – Sustainability. 2010, 2(10), pp. 3309-3322; https://doi.org/10.3390/su2103309

[47] Dyson, B. (2023). Integration of Life Cycle and Life Cycle Sustainability Assessments Into Decision Analytic Approaches for Sustainable Technologies. - Reference Module in Earth Systems and Environmental Sciences, Elsevier, https://doi.org/10.1016/B978-0-323-90386-8.00057-7

[48] Witze, A. (2023). AI predicts how many earthquake aftershocks will strike and their strength. – Nature, 28 sept. 2023, doi: https://doi.org/10.1038/d41586-023-02934-6

[49] Hall, S. (2023). What Turkey's earthquake tells us about the science of seismic forecasting. – Nature 615, 388-389 [2023], doi: https://doi.org/10.1038/d41586-023-00685-y

[50] Dascher-Cousineau, K., Shchur, O., Brodsky, E. E. & Günnemann, S. (2023). Using Deep Learning for Flexible and Scalable Earthquake Forecasting. - Geophys. Res. Lett., Vol. 50, Issue 17, https://doi.org/10.1029/2023GL103909

[51] Stockman, S., Lawson, D. J. & Werner, M. J. (2023). Forecasting the 2016–2017 Central Apennines Earthquake Sequence With a Neural Point Process. - Earths Future, Vol.11, Issue 9, https://doi.org/10.1029/2023EF003777

[52] Zlydenko, O. et al. (2023). A neural encoder for earthquake rate forecasting. - Scientific Reports, vol. 13, 12350 [2023], https://doi.org/10.1038/s41598-023-38033-9

[53] Rusko, M., Kralikova, R., Mikulova, M. & Ilko, J. (2016). Labeling of Products From the Context of Environment, Quality and Safety. - In DAAAM International Scientific Book 2016, pp.419-434, B. Katalinic (Ed.), ISBN 978-3-902734-09-9, ISSN 1726-9687, Vienna, Austria. Chapter 37, pp. 419 - 434. DOI: 10.2507/daaam.scibook.2016.37

[54] Prochazkova, D., Prochazka, J., Rusko, M., Kollar, V., Majernik, M. & Ilko, J. (2020). Safety Management of Complex Facilities.- In Proceedings of the 31st DAAAM International Symposium, pp.0366-0375, B. Katalinic (Ed.), ISBN 978-3-902734-29-7, ISSN 1726-9679, Vienna, Austria, DOI: 10.2507/31st.daaam.proceedings.051

DOI: 10.2507/34th.daaam.proceedings.023

METHODOLOGY FOR SELECTING A SOURCE OF ELECTRICAL ENERGY SUITABLE FOR THE SITE

Jan Prochazka, Dana Prochazkova, Miroslav Rusko, Georg Rockel & Jan Ilko

This Publication has to be referred as: Prochazka, J[an]; Prochazkova, D[ana]; Rusko, M[iroslav]; Rockel, G[eorg] & Ilko, J[an] (2023). Methodology for Selecting a Source of Electrical Energy Suitable for the Site, Proceedings of the 34th DAAAM International Symposium, pp.0172-0182, B. Katalinic (Ed.), Published by DAAAM International, ISBN 978-3-902734-41-9, ISSN 1726-9679, Vienna, Austria
DOI: 10.2507/34th.daaam.proceedings.023

Abstract

To ensure a reliable supply of electricity in a given territory, it is necessary to choose the source that is most suitable in all aspects, i.e. the availability of resources, technology, costs and personnel requirements. Since the requirements are incommensurable, a multi-criteria approach is chosen. Used procedure is described in detail. To ensure the development and safety of the State, each State needs raw materials, energy, transport, industry, agriculture and services to ensure the basic functions of the State. In the article, we focus on energy production, which is the basis for the extraction of raw materials, industry, transport, the operation of critical infrastructures and services that ensure the protection of the lives and health of the population, the protection of property and the protection of the environment.

Keywords: Power sources; availability; risks; safety; multi-criteria approach.

1. Introduction

The first UNDP *Human Development Report* stated that, 'The basic objective of development is to create an enabling environment for people to enjoy long, healthy, and creative lives', and defined human development as 'a process of enlarging people's choices' [63]. Human development focuses on improving the lives people lead rather than assuming that economic growth will lead, automatically, to greater wellbeing for all. Income growth is seen as a means to development, rather than an end in itself [71]. To ensure the development and safety of the State, each State needs raw materials, energy, transport, industry, agriculture and services to ensure the basic functions of the State.

Safety and security represent many things, including a stable income, consistent housing, clothing, and food supplies as part of the predictability of daily life, protection from crime, and psychological security [58]. Based on current knowledge and experience, it is a fact that the quality of life, health and safety of each person depends on the quality of the human community to which they belong. A number of studies, notably the World Bank's Voices of the Poor report [33], show that safety, security and justice are major concerns for citizens. From the reasons of fulfilment of targets of humans (human security and development) that may be only realised if human communities are in safe territory [34], [52]. Safety and security represent many things, including a stable income, consistent housing, clothing, and food supplies as part of the predictability of daily life, protection from crime, and psychological security [66]. Ensuring the safe human system is not easy [26], because the human system is a system of systems [48], i.e. system of several mutually interconnected systems of a different nature.

Consequences of interconnections (interfaces) are mutual dependences, the character of which is physical, cyber, territorial and organisational [51]. As a consequence of growing globalisation the new sources of disasters take on force, they also cause complex facility failures. [22], [27]. A Million Voices report, on findings from the post-2015 development agenda consultation process, highlights security and justice as key elements missing from the Millennium Development Goals (MDGs) [70]. There is a growing body of evidence demonstrating that shortfalls in safety, security and justice contribute to both poverty and underdevelopment [19]. Conversely, the presence of safety, security and justice can contribute to development outcomes including virtuous cycles of security and development, 'with high levels of security leading to development and development further promoting security' [63]. There is a strong association between justice and development, in the sense that prosperous countries generally have more complex legal systems and deliver a higher quality of security and justice services to citizens [5]. However, while evidence suggests that safety, security and justice are associated with development, establishing direct causality is complex [55], [56]. Safety, security and justice are seen as moral rights and intrinsic to development [3].

Mitigating climate change and the clean energy transition

The safe community is now at time of globalisation very dependent on a safety level of complex facilities ensuring the territory by basic service necessary for human live, e.g. the electric energy on which there are dependent supplies of good quality drinking water, utility water, information distribution etc. [47]. Energy production for each State must be sufficient in terms of need and must be cost-effective and it must not damage public assets, so as not to harm the public good. This can be achieved through an appropriate combination of cost, quality, time, flexibility and innovation [23]. Daas et al. [6] developed a Decision Support System (DSS) for designing business models. A Decision Support System (DSS) consists of a design process that follows different design methods. Its application ensures compliance with legislative requirements.

The Paris Agreement duly reflects the latest scientific understanding of systemic global warming risks. Limiting the anthropogenic temperature anomaly to 1.5–2 °C is possible, yet requires transformational change across the board of modernity [1]. After some 20 years of negotiations under the auspices of the United Nations Framework Convention on Climate Change (UNFCCC), a historic, binding climate agreement was reached in Paris. At the twenty-first Conference of the Parties (COP21), 195 nations committed1 to "holding the increase in the global average temperature to well below 2 °C above pre-industrial levels and to pursue efforts to limit the temperature increase to 1.5 °C above pre-industrial levels, recognizing that this would significantly reduce the risks and impacts of climate change" [59]. The release of carbon dioxide (CO_2) and other greenhouse gases (GHGs) due to human activity is increasing global average surface air temperatures, disrupting weather patterns, and acidifying the ocean. Left unchecked, the continued growth of GHG emissions could cause global average temperatures to increase by another 4 °C or more by 2100 and by 1.5 to 2 times as much in many midcontinent and far northern locations [64]. Although our understanding of the impacts of climate change is increasingly and disturbingly clear, there is still debate about the proper course for policy [36].

Mitigating climate change is unavoidably linked to developing affordable low-carbon energy technologies that can be adopted around the world [67]. In the face of the Paris climate agreement, a combined transition to clean energy and acceleration of decarbonization goals will require the refocusing international research and deployment schemes to promote energy R&D (research and development) [30]. Dramatic cost declines in solar and wind technologies, and now energy storage, open the door to a reconceptualization of the roles of research and deployment of electricity production, transmission, and consumption that enable a clean energy transition [31]. The solar photovoltaic industry has undergone a dramatic evolution over the past decade. [76]. While basic research remains a vital element to address a clean energy transition, increasingly an interdisciplinary approach is needed. Deeper integration with policies that build market growth [36] and cutting-edge business models will enable far faster uptake of critical research programme outputs. Accelerating the development and deployment of energy technologies is a pressing challenge. Doing so will require policy reform that improves the efficacy of public research organizations and strengthens the links between public and private innovators [2]. The clean energy transition requires a co-evolution of innovation, investment, and deployment strategies for emerging energy storage technologies [21].

2. Energy sources and their requirements

In general, energy in a particular locality can currently be provided by the construction of a technical facility, i.e.: hydroelectric power plants; large nuclear power plants; gas-fired power plants; thermal power plants; geothermal power plants; diesel engine systems; wind power plant systems; photovoltaic power plant systems; and small modular reactor (SMR) systems [44]. A basic analysis of the conditions of individual technologies shows that for the implementation of the mentioned technical facilities it is necessary to have at your disposal for:

- hydroelectric power plant - river with a sufficient amount of water throughout the year,
- large nuclear power plant - territory that meets the requirements for location of large nuclear power plant, water source for cooling and available and safe nuclear technology,
- gas-fired power plant - stable and affordable gas supply,
- thermal power plant - stable and affordable supply of coal, wood or others biomass,

- geothermal power plant - a stable and affordable local source of geothermal energy,
- diesel engine system - stable and affordable supply of diesel,
- wind power plant systems - stable and sufficient wind,
- photovoltaic power plant system - stable and sufficient source of solar energy,
- small modular reactor (SMR) system -territory that meets the location requirements for SMR.

Since only SMRs are energy sources that are not yet so well known, we will discuss them in more details in the next chapter.

3. Small modular reactors (SMRs)

Small modular reactors (SMR) have been in development for decades [12], [13], [14], [15], [17], [37], [38], [42], [57], [74]. The International Atomic Energy Agency [17] defines small, medium and large reactors according to output electrical performance; reactors up to 300 MWe are classified as small reactors. They are increasingly used in practice, as they are cheaper and their area of emergency planning is smaller compared to large nuclear power plants [42], [57], [62]. In the Czech Republic, we work on the Energy Well reactor [68], [69] which we want to use as energy sources in technical installations producing the energy for: train and ship drive; operation of processes as reverse osmosis; hydrogen production and hydrogen storage; and mining the minerals in remote regions [8], [32].

The advantages of SMRs are mainly in the low installed performance. This advantage is conditioned by a high degree of inherent safety of the system. Due to the low power, passive cooling and the possibility of connecting to low-voltage networks are often talked about in connection with SMRs. If it were not a passive cooling system, batteries could be used for emergency power instead of diesel generators. Due to the low installed capacity, it is possible to reduce the emergency planning zone and generally reduce the licensing period due to the greater simplicity of the system in comparison with large nuclear power sources [43]. With outputs of less than 60 MW, it would be possible to cool the generator with air [18].

Another advantage is the size of the device. Due to this fact, it is possible to build nuclear facilities that are, compared to conventional nuclear sources, more resistant to earthquakes. At the same time, it is possible to renew older ideas about the operation of a nuclear reactor under the Earth's surface, which would be economically unacceptable in the case of conventional nuclear power plants [8]. Burying the SMRs would increase the safety of the system against windstorms, tornadoes and hurricanes, for example, as well as the possibility of a terrorist attack. At the same time, however, access to the equipment during the shutdown work or accident is noticeably reduced. Burying it under the Earth's surface would also be lucrative for military use, drawing on an extremely small built-up surface. However, for military purposes, the installation process would need to be extremely accelerated. At the same time, underground systems could be less sophisticated, reducing the price of SMRs in direct proportion [43].

Furthermore, the relatively small dimensions of the entire system would not cause much difficulty in transport or in the selection of suitable locations. Possibility of connection where it is not possible to build large power plants for many reasons (logistics, subsoil, cooling, etc.) [43]. According to the IAEA, the advantage is, of course, lower investment risk, improved cash flow or shorter construction time [11]. With smaller component dimensions, the price also decreases due to greater competitiveness among suppliers potentially bidding for contracts [18].

The largest economic indicator for nuclear power plants is LCOE (Levelized Cost of Electricity), which is the ratio of the final price of a nuclear power plant to the installed electrical capacity. From the point of view of SMR optimization, it is, therefore, essential to reduce LCOE [8]. Opinions on the method of reduction vary, one of them is, for example, a decrease in installed capacity, which should also have an impact on greater simplicity of the system, and therefore, lower investment costs [17]. The estimated costs according to [73] are as follows: SMART - construction costs 5000 $/kWe, operation and maintenance 6.1 cents/kWh (which is lower than hydropower), NuScale - construction costs 4000 $/kWe (40 months of construction), fuel costs 5.5 $/MWh.

According to [18], it is reality that in SMRs around 15 systems and components necessary for LOCA disaster resolution are eliminated and, for example, in NuScale units, they are replaced by only one system due to the low electrical power of 50 MWe. As a small facility, SMRs have also benefit in terms of supply chains. The production of smaller pressure vessels gives the opportunity to more companies, including the domestic ones. At the same time, there are no complications in dealing with large manufacturers of heavy components, respectively with their workload. The same approach applies to the production of turbines and generators. A larger group of possible manufacturers encourages a more vigorous competition in a competitive market, and therefore, a possible reduction in acquisition costs. In addition, as already mentioned, smaller equipment is easier to transport, maintain and possibly dispose of. In the case of turbogenerators with lower outputs than 50 MWe, it would be possible to transport the entire equipment on a "separate pallet" and, in addition, it would be possible to cool it with air [18]. The impacts on investment decisions can be summarized in 11 classes, which include, for example, the stability of the electricity grid, public acceptance, technical constraints on location, project risks, national industrial system, time of market launch, competences required for operation, effects on employment, design robustness or political relations [25]. In a 2016 document [39], the OECD argues that the problem still remains that capital costs, operating and maintenance (O&M) costs and fuel costs are not yet known.

On the other hand, it confirms the advantages of SMRs, among which it states: increased nuclear safety and implementation of unique passive elements; reducing the number of systems and simplifying the energy conversion; easier financing; better network flexibility, e.g. load monitoring modes; lower transmission system requirements due to lower power outputs; identical SMRs are advantageous in terms of human resource management; easier decommissioning; and the ability to avoid downtime by help of suitable SMR configurations. At the same time, the report [39] states that the following will be key for customers: GDP and its possible growth; electricity consumption per person; credit rating; self-sufficiency of electricity supply; environmental protection; national membership of the IAEA; electricity price; and specifics of the electrical network (size, voltage, quality, interconnection, load).

For SMR manufacturers, according to [39], it is a matter of ensuring: technology readiness and demonstration possibilities; financial background; and supply chain and public procurement.
At the same time, the report [39] mentions a comparison of the aviation industry with nuclear facilities, which is a frequently used parallel, but it is not as simple as it may seem to others. In particular, the efficient full assembly of SMRs at the factory may not allow the inspectors of supervision office to check all the steps (an on-site inspector could be preferred to optimize the assembly process at the factory) [7]. With regard to the deployment of SMRs, the innovative licensing system should separate the general approval or licensing of production equipment for SMRs, the SMR itself and the site for SMRs [39].

It can be argued that SMRs have reached a certain degree of maturity and are competitive with other energy sources [7]. In addition, large nuclear power plants and SMRs are expected to have the same LCOE of USD 70/MWh (at a reasonable weighted average cost of capital) with a load factor of 85 %, but a slightly different distribution between fixed and variable costs [57]. This leads to the idea that SMRs may have lower investment costs due to factory production, shorter construction times, simpler financing and the like, but higher variable costs due to higher O&M costs per MWh due to the fixed component of O&M costs, lower fuel efficiency, etc. In this example, SMRs are most competitive with load factors of 60-85 %, replacing the coal and large nuclear power plants to this extent [39]. The DoE [7]. states that the safety of SMRs might not be as perfect as it is generally claimed. The main argument concerns passive safety systems and elements that are not infallible. A smaller containment is disadvantageous in terms of the ppm value of hydrogen, which would be enough for an explosive concentration in the containment area. If the reactor were to be buried, earthquake resistance could be increased, but flood safety decreased. At the same time, safety is reduced due to fewer operating staff or in the event that the manufacturer decides to reduce costs [43].

The advantages of SMRs according to DoE [7]. are modularity, the ability to build units in a factory and transport them to the site. It is suitable for small electric markets, places with low logistical support and places with smaller industrial plants. At the same time, it is possible to replace old coal-fired power plants with green sources, which are SMRs. There is also the possibility of connecting SMRs with other sources of electricity, which could increase the stability of the network or the safety of the transmission system. A large part of the SMRs is planned in such a way that there would be no situ betting, but it would be done directly in the factory [43].

In the case of multiple SMRs, this technology could be less safe than in the case of a large nuclear power plant, when the SMR manufacturer fails in one project [43]. However, this presumption is applicable not only to SMRs or large nuclear units, but to series production in general. The advantage of SMRs in terms of safety is also the so-called integral arrangement, where the reactor and the components of the primary circuit are placed in one pressure vessel [18]. This eliminates the amount of cooling pipe and its dimensionality. Due to this, the sleeves leading out of the TNR are also drastically narrowed, which has a positive impact on the course of the LOCA accident. Heat exchangers are usually seated higher than the core, thereby contributing to the natural circulation of the refrigerant. However, some systems, due to the natural circulation of the refrigerant, completely eliminate the main circulation pumps and the associated possibility of a LOCA accident [18].

Sources of risk were monitored in work [46]. Other factors that are sources of high risk, which is common to a capital-intensive industry, are according to [4], [39], large initial expenses; the uncertainty of the return on initial investment; the danger that the work will not be completed, or changes will be necessary, which will require enormous additional costs for completion; long building and construction time; sensitivity to demand; reliability, availability and load factor of the power plant; electricity price; unstable public support; low acceptance of nuclear power installations with the public (although nuclear technologies used in medicine and food, which are not as secure as energy [46], do not create public disfavor); decisive influence of the regulator; and decommissioning - decommissioning and subsequent decontamination of equipment and surface treatment, which does not yet have a clear procedure, although there is experience [16] and civil procedures [49].

According to the work [4], the advantages of the reduced power plant size, complexity and simplified design offered by SMRs would allow: better control over construction; less risk associated with suppliers; and better control over the cost of equipment design. According to works [8], [18], [32], [35] SMR can be used for: long-range heating; desalination and water purification; advanced oil extraction and oil refining processes; production of hydrogen for the enrichment of liquid fuels and, where appropriate, the use of fuel cells; advanced energy conversion processes such as coal liquefaction and petrochemical production; general process heat for chemical or manufacturing processes; the standby power of a nuclear power plant; data centers; military bases; mining sector; remote island operations; industrial complexes; production and liquefaction of hydrogen; steelworks; oil and gas terminals; large chemical plants; desalination of seawater; and propulsion of ships.

4. Methodology for selection of a suitable source of electrical energy in real site

In any case, it is a question of choosing a technology for which there are conditions in the territory and which is safe, which means reliable and functional and procure quickly, while its demands on safety, operation, human resources and other service systems are such that it will be cost-effective. This means that the benefits for the territory while ensuring a level of safety will be optimal. The optimal solution is obtained by assessing possible variants [45], [49].

According to [49], any suitable solution for meeting the specified goal is considered a variant solution, i.e.:

- various localization of the construction site and traffic route management,
- various technological processes,
- variant type of activity, e.g. choice of import instead of domestic production,
- different implementation timetables,
- substitution of raw materials,
- various solutions for the disposal of waste, emissions, etc.

Variant generation is a creative model of thinking that depends on the criteria of the goal. In a State governed by the rule of law, the objective must be consistent with legislation that promotes the security and development of the State. The goal in terms of development and cost in any case is to select a technology that will be: safe; tried-and-true; feasible within an acceptable time; easy to maintain, and have reasonable demands on fuel, self-consumption, personnel and finances necessary for safe operation. In the world, a procedure called "technology assessment", which is codified in each country [50], is used for this purpose. The form of technology evaluation is determined by legislation in each country; the goal is the same, but the form of application varies from country to country [9], [10], [28], [29], [41], [50], [53], [54], [60], [61], [65], [72], [77].

According to [9], [10], [28], [29], [41], [50], [53], [54], [60], [61], [65], [72], [77] the evaluation in question is a comprehensive interdisciplinary expert evaluation of planned technical facilities, which considers both, the possibilities of the investor and the impacts of current and future on the areas of technology, the environment, social, social and economic; in Europe, it began to be used in the early 90s. The evaluation in question is not directed against technologies; its aim is to detect problems and prevent damage caused by uncritical application and commercialization of new technologies. The results of the evaluation are intended for investors (in the case of public projects for politicians), who ultimately decide to enable the implementation of a technical facility. When making a decision, based on the evaluation of technology, there is a dilemma:

- correctly appreciate the impacts of the planned technical facility, which cannot be easily predicted until the technical facility is extensively developed and used,
- it is difficult to manage or modify a technical facility once it is widely used.

To do a decision is difficult, because in a particular case:

- it is difficult to estimate the cost of externalities and internalities,
- it is not easy to select indicators to assess the benefits and impacts of the planned technology,
- it is not easy to convert damages and injuries into money,
- there are also ethical barriers.

Based on the above quotes, "technology assessment" is mainly used in the following areas: information technology; hydrogen technologies; nuclear technology; molecular nanotechnology; pharmacology; organ transplantation; genetic technology; artificial intelligence; internet, etc. On the basis of the OTA (Office for Technology Assessment) documents collected in the database [41], the specific assessment is carried out in two steps. First, a screening of possible variants is carried out, i.e. an evaluation of essential factors to exclude major unsatisfactory variants, and a detailed evaluation is carried out for the remaining variants, which forms the basis for the decision of the investor (or politicians, in the case of public projects). Based on the documents cited above, an evaluation based on the following criteria shall be used for screening:

- Is the operation of the technology tested?
- Is the power supply safe, i.e. reliable and functional for life?
- Is the technology available to the investor?
- Is the performance of the technology stable in the long term?
- Can the technology be implemented quickly?
- Is the cost of applying the technology acceptable?
- Can a source with this technology quickly (less than 1 year) meet the needs of the investor for the entire duration of operation of the building for which it is being built?

The comparison of suitable variants of technologies determined by screening for an investor or supplier of an electricity source in the monitored case for understandable reasons (it is to ensure the needs of a company) does not include or only to a limited extent includes the needs for the territory in which the energy source is located and put into operation (e.g. public welfare, employment, services in the territory, etc.).

Table 1 shall be used for the detailed evaluation of variants. The optimal solution is a variant that:

- has the most benefits,
- does not have too large specific requirements for operation,
- according to [20] losses caused by accidents or failures will not reduce the expected annual return below value 0.7 x total useful income / lifetime.

Since nothing is absolute, no technology meets the requirements absolutely. Since this is a complex decision-making problem in which it is necessary to consider criteria from many areas that are often incommensurable, it is necessary to apply a multi-criteria evaluation [45], [20] based on the philosophy put into practice in the work [40]. For the selection of source of electricity production, in accordance with the knowledge [6], [20], [24], [45], [50], [75], we have constructed a decision support system for evaluating the contributions of individual criteria to the integral safety of the technical facility (energy source) and its surroundings; Table 1. We have created the criteria on the basis of the knowledge gained from the critical analysis of [20], [24], [41], [45], [50], [61], [72], [75], [77] so that "the higher the value of the valuation, the higher the contribution to integral safety" is paid. The evaluation of the criteria is carried out by assigning points as follows:

0 point - the criterion is met at less than 5 %, i.e. it does not contribute to ensuring the integral safety,
1 point - the criterion is met at 5- 25 %, i.e. it contributes little to ensuring the integral safety,
2 points - the criterion is met at 25 – 45 %, i.e. only moderately contributes to ensuring the integral safety,
3 points - the criterion is met at 45 – 70 %, i.e. it contributes highly to ensuring the integral safety,
4 points - the criterion is met at 70 – 95 %, i.e. it contributes very highly to ensuring the integral safety,
5 points - the criterion is met at more than 95%, i.e. it contributes extremely highly to ensuring the integral safety.

Since we use incommensurable items in the evaluation, we use the auxiliary scale, shown in Table 2, to ensure commensurability [45].

Criterion	**Evaluation**	**Note**
Safety of technology		
The rate in whichtechnologyhasinherentsafety.		
The rate in which technology is a clean source of energy.		
The rate in which the technology is able to ensure a stable supply of electricity. This means that it is still operational and does not depend on frequent specific deliveries or specific external conditions.		
The rate in which the technology has the ability to operate without trouble.		
The rate in which the technology does not require frequent repairs.		
The rate in which the demands of the technology operation do not require qualified operator intervention.		
The rate in which the impacts of the operation of the technology on employees are acceptable.		
The rate in which the impacts of the operation of the technology on the surrounding environment are acceptable.		
Material demands on the feasibility of technology in a certain place		
The rate in which technology includes measures to manage emergency situations.		
The rate in which the operation of the technology can be done without specific knowledge.		
The rate in which the installation of the technology requires a local object.		
The rate in which the technology can do without frequent supplies of raw materials for operation.		
The rate of demands of technology on a specific and expensive location.		
The rate of demands of the technology on technical tasks during the commissioning.		
The rate in which the operation of the technology requires the skill of the operator.		
The rate in which the technology needs energy for its own consumption to operate.		
The rate in which technology requires a large information provision to operate.		
The rate in which the technology has a specific I&C for operations.		
The rate of uninterrupted performance of the technology due to maintenance and repairs.		
The rate of the quality of operating regulations for normal, abnormal and critical conditions.		
The rate of level of protection of the technology against local natural disasters.		
The rate in which technology protects the lives and health of operators.		
Míra, v jaké technologie neznečišťuje životní prostředí.		
The rate in which the technology does not require on-site waste management.		
The rate in which technology is protected against insiders and terrorists.		
Accessibility and competitiveness		
The rate of acceptability of the installation time of the technology.		

The rate of the technology's capability to ensure power generation for the operation of the monitored entity for 10 years or more.		
The rate of acceptability of the price of the technology.		
The rate in which the technology is such that it does not reduce the expected annual return below the acceptable value according to [20].		
The rate in which the transport of technology and the supply of spare parts are feasible.		
The rate in which the financial demands on the technology and its long-term operation are acceptable.		
The rate of difficulty of granting an operating permit.		
The rate in which specific objects are not needed for waste disposal at a given location.		
The extent to which the technology does not require site-specific equipment to reduce contamination of environmental components.		
The rate of acceptability of the scope of the emergency planning zone.		
LEVEL OF INTEGRALSAFETY		

Table 1. A tool for evaluating the integral safety of variants in order to obtain the optimal variant

Domain	**Risk rate**	**Classification criterion**
Social	*By accident or failure of technical facility, it is affected:*	
	0	less than 50 humans
	1	50 - 500 humans
	2	500 - 5000 humans
	3	5 000 – 50 000 humans
	4	50 000 – 500 000 humans
	5	more than 500 000 humans
Technical and Economic	*Accident or failure of technical facility causes damages:*	
	0	less than 0.05 p
	1	equal to p
	2	between p and 0.05 ABT
	3	between 0.05 ABT and 0.075 ABT
	4	between 0.75 ABT and 0.1 ABT.
	5	higher than 0.1 ABT.
Environment	*Accident or failure of technical facility causes:*	
	0	very low damages of environment
	1	damages of environment with which the nature cope during the acceptable time
	2	moderate damages of unrenewable resources of nature and natural reservations.
	3	medium damages of unrenewable resources of nature and natural reservations
	4	unreturnable damages of unrenewable resources of nature and natural reservations
	5	devastation of landscape, unrenewable resources of nature and natural reservations

Table 2. Scale for determination of rate of risk that planned technical facility means for its surroundings (rate of coexistence disruption); by analogy to scales in [50]; p - annual insurance, ABT-the annual budget of territory governance (safety rate = 1 - risk rate)

The evaluation of Table 1should be performed by specialists from different domains (technology, region safety, public administration, investor, emergency service) [50]. The resulting value is the median for each criterion, and in cases of great variance of the values in one criterion it is necessary, so that the worker of public administration responsible for territory safety may ensure further investigation, on which each assessor shall communicate the grounds for his / her review in the present case, and on the basis of panel discussions or brainstorming session, the final safety rate value is determined. Judgement of acceptability of integral safety is performed according to Table 3 [50].

The level of safety	Values in % n/N
Extremely high - 5	More than 95 %
Very high - 4	70 - 95 %
High - 3	45 - 70 %
Medium - 2	25 - 45 %
Low - 1	5 - 25 %
Negligible - 0	Low than 5 %

Table 3. Value scale for determining the safety rate; n = the number of criteria with YES; N = five times the number of criteria in Table 1.

At selection of variant, it is possible in simple case to use the level integral safety according to result in Table 1; the higher number, the better variant for a given territory. The variant with maximal integral safety rate is the best one. From safety reasons it is better to use a two-step procedure:

- based on Table 1 and Table 3 to exclude all variants where safety is less than high,
- from the rest variants to select this which is better from economic or time viewpoint.

More precisely sophisticated procedure [50] is the following:

- for each variant which have according to result in Table 1 minimally high integral safety according to Table 3 to calculate benefitsof variantfor the given territory,
- for each from these variantsto determine the differences Dsbetween annual benefits ofa technical facility for the given territory and maximum annual losses of technical facilities caused by trade-off with risks. In this case optimal variant is the variant with the maximum D.

5. Future research plans

Further research is planned to focus mainly on the application of the proposed methodology aimed at reliable electricity supplies in the given territory so that the source that is most suitable in all respects is chosen during the selection process, i.e. availability of resources, technologies, costs and personnel requirements. It will be important to use feedback from practice when applying the described methodology, which aims to ensure the protection of the lives and health of the population, the protection of property and the protection of the environment.

6. Conclusion

Since no objective decision-making on real matters is usually black and white, the benefits and impacts caused by the risks taken must always be weighed. The paper shows a method that is in line with professional knowledge and procedures in the EU and developed countries. The methodology considers all important aspects that are important from the point of view of long-term sustainability of the solution. It is based on the assessment of incommensurable factors by a team of experts who have an auxiliary scale for categorizing risks. Therefore, evaluations also need to be carried out by a team of experts who are not under political, financial or other pressure. In case of disagreement among experts, a way to proceed is proposed. Three ways of selection of optimal variant are shown; the most precise selection is based on comparison of annual benefits of a technical facility for the given territory and maximum annual losses of technical facilities caused by trade-off with risks. The implementation of the described method of ensuring energy production, which is the basis of the extraction of raw materials, industry, transport, operation of critical infrastructure and services, aims to ensure the protection of the lives and health of the population, the protection of property and the protection of the environment. Such a comprehensive approach based on current knowledge in the context of the security management process has the potential to contribute to overall integrated security and the elimination of potential risks in the given territory.

7. References

[1] Adoption of the Paris Agreement FCCC/CP/2015/L.9/Rev.1 (UNFCCC, 2015).

[2] Anadon, L., D.,Chan, G., Bin-Nun, A. Y. & Narayanamurti, V. (2016). The pressing energy innovation challenge of the US National Laboratories. - Nat. Energy 2016, 1, 16117. - [on-line] Available on - URL: https://doi.org/10.1038/nenergy.2016.117

[3] AusAID. (2012). Building on Local Strengths: Evaluation of Australian Law and Justice Assistance. - Canberra: Australian Agency for International Development. - [on-line] Available on - URL: https://www.dfat.gov.au/development/performance-assessment/aid-evaluation/program-evaluations/building-on-local-strengths-evaluation-of-australian-law-and-justice-assistance

[4] Carelli, M. D. & Ingersoll, D. T. (2015). Handbook of Small Modular Nuclear Reactors. - ISBN 978-012823-91-62. Amsterdam: Woodhead Publ.

[5] Cox, M. (2008). Security and Justice: Measuring the development returns. - London: Agulhas. – [on-line] Available on - URL: https://agulhas.co.uk/app/uploads/2020/10/DFID_2008_Security_and_justice_measuring-_he_development_returns-.pdf

[6] Daas, D., Hurkmans, T., Overbeek, S., Bouwman, H. (2013). Developing a Decision Support System for Business Model Design. - Electronic Markets, 23(3), 251–265. – [on-line] Available on - URL: https://doi.org/10.1007/s12525-012-0115-1.

[7] DOE. Advanced Small Modular Reactors. – [on-line] Available on - URL: https://www.energy.gov/advanced ; https://www.energy.gov/science/advanced-and-sustainable-energy

[8] Dostál, V., Štěpánek, J. & Syblík, J. (2020). Business plán reaktoru EW. - Zpráva pro ÚJV. Praha: ČVUT 2020.

[9] EPTA. – [on-line] Available on - URL: https://eptanetwork.org/index.php

[10] Genus, A. & Coles, A. (2005). On Constructive Technology Assessment and Limitations on Public Participation in Technology Assessment. - *Technology Analysis and Strategic Management,* 17(2005), 4, pp. 433-443
[11] IAEA. - [on-line] Available on - URL: https://www.iaea.org/
[12] IAEA. (2011). Status of Small and Medium Sized Reactor Designs: a Supplement to the IAEA Advanced Reactors Information System (ARIS). - Vienna: IAEA.
[13] IAEA. (2013) Approaches for Assessing the Economic Competitiveness of Small And Medium Sized Reactors. - Vienna: IAEA.
[14] IAEA. (2013) Status of Innovative Fast Reactor Designs and Concepts. - Vienna: IAEA.
[15] IAEA. (2013). Advances in Small Modular Reactor Technology Developments: A Supplement to the IAEA Advanced Reactors Information System (ARIS). - Vienna: IAEA.
[16] IAEA. (2015). The Fukushima Daiichi Accident. - ISBN 978-92-0-107015-9. Vienna: IAEA, 218 p.
[17] IAEA. (2020). Considerations for Environmental Impacts Assessment for Small Modular Reactors. - IAEA-TECDOC-2915. Vienna: IAEA, 48 p.
[18] Ingersoll, D. T. (2016). Small Modular Reactors: Nuclear Power Fad or Future? - NY: Woodhead Publishing, ISBN 978-0081002520.
[19] Ismail, O. & Hendrickson, D. (2009). What is the case for a security and justice focus in development programming? An assessment of the existing literature and evidence. - Birmingham: Global Facilitation Network for Security Sector Reform, University of Birmingham.
[20] Keeney, R. L & Raiffa, H. (1993). Decision with Multiple Objectives. - Cambridge: Cambridge University Press, 569 p.
[21] Kittner, N., Lill, F. & Kammen, D. (2017). Energy storage deployment and innovation for the clean energy transition. - Nat Energy 2, 17125 [2017]. https://doi.org/10.1038/nenergy.2017.125
[22] Kralikova, R., Sokolova, H., Wessely, E. & Polak, J. (2013) Approaches to assessment of hot environment. - DAAAM International Scientific Book 2013. chapter 14. Vienna. DAAAM International. pp. 317-328. ISBN 978-3-901509-94-0
[23] Lim, M. (2010). Environment-Strategy-Structure-Operations (ESSO) Business Model. - In: Knowledge Management Module at Bangor University.- Wales. - [on-line] Available on - URL: http://www.communitiesofinnovation.com/L/Dr_Michael_Lim_%20Business_Model_Knowledge_Creation_and_Innovation_Lecture%209.ppt
[24] Liu, Z. & Fan, J. (2014). Technology Readiness Assessment of Small Modular Reactor (SMR) Designs. - Progres in Nuclear Energy. 70 (2014), pp. 20-28.
[25] Locatelli, G., Mancini, M. (2011). The Role of the Reactor Size for an Investment in the Nuclear Sector: An Evaluation of Not-Financial Parameters. - Progress in Nuclear Energy. ISSN 0149-1970. 53 (2011), 2, pp. 212-222. Doi:10.1016/j.pnucene.2010.10.002
[26] Lojan, R. (2015): Ľudská dôstojnosť a hodnota ľudského života. - In: Nové horizonty, Bratislava, ISSN 1337-6535, Vol. 9, Issue. 1, pp. 10-14
[27] Lukas, L. & Hromada, M. (2011). Risk analysis in context of kritical infrastructure protection. - Annals of DAAAM for 2011 & Proceedings of the 22nd International DAAAM Symposium, Volume 22, No. 1, ISSN 1726-9679, Katalinic, B. [ed.], Vienna, pp. 1469 - 1470
[28] Machleidt, P. (1994). Nové demokratické formy hodnocení techniky za účasti veřejnosti. - In: Teorie vědy 3 (1994), 1-2.
[29] Machleidt, P. (2005). Hodnocení techniky (Technology Assessment) aneb jak zvládat technická rizika. - In: Člověk a rozhodování - rizika a nejistoty. Praha: ČVUT.
[30] Margolis, R. M. & Kammen, D. M. (1999). Underinvestment: the energy technology and R&D policy challenge. - Science 285, no. 5428, pp. 690–692 (1999). http://www.jstor.org/stable/2898477.
[31] Mileva, A., Nelson, J. H., Johnston, J. & Kammen, D. M. (2013). SunShot solar power reduces costs and uncertainty in future low-carbon electricity systems. - Environ. Sci. Technol. 2013, 47, pp. 9053–9060. https://doi.org/10.1021/es401898f
[32] Mysak, L., Prochazkova, D., Dostal, V. & Prochazka, J. (2022). Proposal of Reduction of Emissions of Marine Shipment. - In: Understanding and Manageing Risk and Reliability for a Sustainable Future. ISBN 978-981-18-5183-4. Singapore: Research Publishing. DOI:10.3850/978-981-18-5183-4_R15-02-063-cd
[33] Narayan, D., Chambers, R., Shah, M.K. & Petesch, P. (2000). Voices of the Poor: Crying Out for Change. - Washington D.C.: The World Bank.
[34] Necesal, L., Lukas, L. & Jasek, R. (2011) Measures for critical infrastructure protection in the Czech republic. - Annals of DAAAM for 2011 & Proceedings of the 22nd International DAAAM Symposium, Volume 22, No. 1, ISSN 1726-9679, Katalinic, B. [ed.], Vienna, pp. 0843-0844
[35] NUVIA. (2016). Market And Technical Assessment of Micro Nuclear Reactors. - Washington: Nuvia Limited, 56p.
[36] Obama, B. (2017). The irreversible momentum of clean energy. – Science 2017, Vol. 355, Issue 6321, pp. 126–129. DOI: 10.1126/science.aam6284

[37] OECD. (2015). Nuclear New Build Insights into Financing and Project Management. - Paris: OECD.
[38] OECD. (2016). Small Modular Reactors: Nuclear Energy Market Potential for Near-term Dep-loyment. - Paris: OECD.
[39] OECD. (2021). Small Modular Reactors: Challenges and Opportunities. - Paris: OECD, 81 p.
[40] OECD: OECD Statistics.- [on-line] Available on - URL: https://stats.oecd.org
[41] Office of Technology Assessment. – [on-line] Available on - URL: http://ota.fas.org/otareports/
[42] Pannier, C. P. & Skoda, R. (2014). Comparison of Small Modular Reactor and Large Nuclear Reactor Fuel Costs. - Energy and Power engineering. 6 (2014), 4, pp. 82-94. DOI: 10.4236/epe.2014.65009
[43] Pedraza, J. M. (2017). Small Modular Reactors for Electricity Generation. - New York, NY, Berlin Heidelberg: Springer. ISBN 978-3-319-52215-9.
[44] Prochazkova, D. & Prochazka, J. (2020). Analysis, Management and Trade-off with Risks of Technical Facilities. - Praha: ČVUT, ISBN 978-80-01-06714-7. 172p. Doi:10.14311%2FBK.9788001067147
[45] Prochazkova, D. (2011). Strategické řízení bezpečnosti území a organizace. - Praha: ČVUT, ISBN 978-80-01-04844-3. 483 p.
[46] Prochazkova, D., Prochazka, J. & Dostal, V. (2021). Risks of Power Plants with Small Modular Reactors. - In: Proceedings the 31st ESREL Conference. Singapore: Research Publishing Services, ISBN 978-981-18-2016-8, pp. 124-131, Doi:10.3850/978-981-18-2016-8_125-cd
[47] Prochazkova, D., Prochazka, J., Rusko, M., Kollar, V., Majernik, M. & Ilko, J. (2020). Safety Management of Complex Facilities. - Proceedings of the 31st DAAAM International Symposium, pp.0366-0375, B. Katalinic (Ed.), ISBN 978-3-902734-29-7, ISSN 1726-9679, Vienna, Austria DOI: 10.2507/31st.daaam.proceedings.051
[48] Prochazkova, D., Prochazka, J., Rusko, M., Mikulová, M. & Il'ko, J. (2017). Model for critical infrastructure safety management. - In Annals of DAAAM for 2017, Volume 28, No.1. The 28th DAAAM International Symposium. Zadar, Croatia, 08-11th November 2017. pp. 0602 – 0610. Vienna : DAAAM International, ISSN 2304-1382. ISBN 978-3-902734-14-3. DOI: 10.2507/28th.daaam.proceedings.085.
[49] Prochazkova, D., Prochazka, J., Říha, J., Beran, V., Procházka, Z. (2018). Řízení rizik spojených s ukončením provozu technického díla a s předáním území do dalšího užívání. - ISBN 978-80-01-06527-3. Praha: ČVUT, 114p. Doi:10.14311%2FBK.9788001065273
[50] Prochazkova, D., Prochazka, J., Říha, J., Beran, V., Procházka, Z. (2018). Řízení rizik procesů spojených se specifikací a umístěním technického díla do území. - Praha: ČVUT, ISBN 978-80-01-06467-2. 134p. DOI:10.14311%2FBK.97880010 64672
[51] Prochazkova,, D. & Rusko, M. (2017). Relation between human system safety management and environmental management. - In Journal of Environmental Protection, Safety, Education and Management. Vol. 5, No. 9, ISSN 1339-5270
[52] Prochazkova,, D., Wessely, E., Rusko, M. & Králiková, R. (2011). Human System Safety Management and Environmental Management Relation. - In DAAAM International Scientific Book 2011. pp. 103-118. ISSN 1726-9687; DOI: 10.2507/daaam.scibook.2011.09
[53] Projects – EPTA Network. – [on-line] Available on - URL: https://eptanetwork.org/database/projects
[54] Rip, A., Misa, T. & Schot, J. (eds). (1995). Managing Technology in Society: the Approach of Constructive Technology Assessment. - London: Pinter 1995.
[55] Roseveare, C. (2013). The rule of law and international development. Department for International Development (DFID), London, UK (2013) 87 pp.
[56] Roseveare, C. (2013). The rule of law and international development. – Literature review – not policy. – [on-line] Available on - URL: https://assets.publishing.service.gov.uk/media/57a08a0be5274a27b20003c7/Literature_Review_RoL_DFID-GSDCH-PEAKS_FINAL.pdf
[57] Rosner, R. & Goldberg, S. (2011). Small Modular Reactors – Key to Future Nuclear Power Generation in the U.S. - Chicago: University 2011, 81 p.
[58] Rusko, M., Kralikova, R. & Mikulova, M. (2016). The Management Tools and Methods Application of the Organization to Environmental Protection. - Proceedings of the 27th DAAAM International Symposium, pp.0283-0288, B. Katalinic (Ed.), ISBN 978-3-902734-08-2, ISSN 1726-9679, Vienna, Austria, DOI: 10.2507/27th.daaam.proceedings.042
[59] Schellnhuber, H. J., Rahmstof, R. & Winkelmann, R. (2016). Why the right climate target was agreed in Paris. - Nature Climate Change, 2016, vol. 6, issue 7, pp. 649-653, https://doi.org/10.1038/nclimate3013
[60] Schot, J. (2001). Towards New Forms of Participatory Technology Development. - *Technology Analysis and Strategic Management* 13(2001), 1, pp. 39–52.
[61] Schot, M. & Rip, A. (1997). The Past and Future of Constructive Technology Assessment. - Technological Forecasting & Social Change 54 (1997), pp. 251–268.
[62] Stenberg, C. (2020). Energy Transitions and the Future of Nuclear Energy: A Case for Small Modular Reactors. - Washington Journal of Environmental Law & Policy. Vol.11, Iss.1(2020)

[63] Stewart, F. (2004). Development and security. - Conflict, Security & Development, 4:3, 261-288, DOI: 10.1080/1467880042000319863
[64] Stocker T.F. et al. (2013). Climate Change (2013): The Physical Science Basis. - Contribution of Working Group I to the Fifth Assessment Report of the Intergovernmental Panel on Climate Change. Cambridge Univ. Press, New York, 1585 p. (pp. 33-115).
[65] Tarr, J. A. (ed.). (1977). *Retrospective Technology Assessment – 1976.* - San Francisco: The San Francisco Press.
[66] The importance of safety, security and justice. - [on-line] Available on - URL: https://gsdrc.org/topic-guides/safety-security-and-justice/concepts/the-importance-of-safety-security-and-justice/
[67] Trancik, J. E. et al. (2015). Technology Improvement and Emissions Reductions as Mutually Reinforcing Efforts: Observations from the Global Development of Solar and Wind Energy. - MIT, 78 p.
[68] UJV. (2017). Design of the Active Zone and Feasibility Study of a Small Modular Reactor Cooled by Molten Salt.- Řež: Centrum výzkumu 2017, 48 p.
[69] UJV. (2022). Plan of Use of Constructed Energy Well Reactor in Practice. - Řež: Centrum výzkumu 2022, 205p.
[70] UNDG. (2013). A Million Voices: The World We Want. - New York: United Nations Development Group.
[71] UNDP: What is human development?. – [on-line] Available on - URL: https://hdr.undp.org/about/human-development
[72] Van Eijndhoven. (1997). Technology Assessment: Product or Process? Technological Forecasting and Social Change 54 (1997), pp. 269–286.
[73] Vujić, J., Bergmann, R. M., Škoda, R. & Miletić , M. (2012). Small Modular Reactors: Simpler, Safer, Cheaper?Energy. - Proceedings of the ICE - Energy 45(1), pp. 288–29, 5ISSN 0360-5442. DOI: 10.1016/j.energy.2012.01. 078.
[74] World Nuclear Association: Small Nuclear Power Reactors. (Updated July 2023). - [on-line] Available on - URL: https://www.world-nuclear.org/information-library/nuclear-fuel-cycle/nuclear-power-reactors/small-nuclear-power-reactors.aspx
[75] Zhang, Z. & Sun, Y. (2007). Economic Potential of Modular Reactor Nuclear Power Plantas based on the Chinnese HTR-PM Project. - Nuclear Engineering and Design. 237 (2007), pp. 2265-2274.
[76] Zheng, C. & Kammen, D. M. (2014). An innovation-focused roadmap for a sustainable global photovoltaic industry. - Energy Policy 2014, 67, pp. 159–169. https://doi.org/10.1016/j.enpol.2013.12.006
[77] Rip, A. (1999). STS in Europe. - *Science, Technology and Society* 4(1999), 1, pp. 73-80.

DOI: 10.2507/34th.daaam.proceedings.024

CONSTRUCTION OF THE MATHEMATICAL MODEL OF THE STRUCTURE OF THE ACTUATOR AND DESIGN OF CONTROL SYSTEM

Yuriy Konoplev, Alexey Krivushov, Fedor Mitin & Alexey Yuev

This Publication has to be referred as: Konoplev, Y[uriy]; Krivushov, A[lexey]; Miltin, F[edor] & Yuev, A[lexey] (2023). Construction of the Mathematical Model of the Structure of the Actuator and Design of Control System, Proceedings of the 34th DAAAM International Symposium, pp.0183-0192, B. Katalinic (Ed.), Published by DAAAM International, ISBN 978-3-902734-41-9, ISSN 1726-9679, Vienna, Austria
DOI: 10.2507/34th.daaam.proceedings.024

Abstract

This article discusses the problem of constructing a mathematical model of an actuator for specifying the shape of the radio-reflective surface of a large-sized space-based reflector using a cable-stayed structure. For large-sized structures deployment in outer space, an important performance indicator is the ability to minimize the resulting vibrations. So, to configure a radio-reflective net, it is necessary to select an actuator and ensure its reliable control that satisfies the conditions of the task. A DC motor is selected as the actuator. Its mathematical model is presented. A solution to the problem of controlling a DC motor is presented based on an algorithm for optimal correction of the parameters of the control structure. The problem of filtering with optimal adjustment of the shape of the radio-reflecting surface of the frontal network of a transformable reflector has been solved. The results of numerical simulation are presented, showing the possibility of using the proposed optimal control algorithm in real time. Recommendations are given for applying the developed algorithm in real devices.

Keywords: Large-sized transformable reflector; Radio-reflective net; DC motor; Optimal control; Simulation; Modeling.

1. Introduction

One of the actively developing approaches to creating space-based reflectors is the creation of large-sized systems [1], [2]. One of the criteria for their operation is the presence of a simple and reliable structure capable of receiving and transmitting signals in various frequency ranges. This requires [3], [4], [5] the presence of a radio-reflective net, maintaining geometric accuracy during operation, minimizing vibrations of the entire structure as a whole and the radio-reflective net in particular [6]. This construction must have a small volume when stowed in the launch vehicle, withstand the overloads that occur when launching a mission into a given orbit, and at the same time have a large working surface during operation [7], [8], [9]. One of the possible solutions to this problem is the use of a cable-stayed structure [10], [11], [12], [13], [14].

Figure 1 shows the spoke 1, the two main radial cords of the back 3 and front 2 nets and the cable 4 between them. From the right end, the spoke 1 is divided into two parts to set the desired shape of the radio-reflecting surface. In each cable 4 there is an actuator 5, for which a DC motor is accepted in this task. DC electric motors are widely used in various fields, including the space industries [15] due to the simplicity of design, low manufacturing cost, ease of control and maintenance.

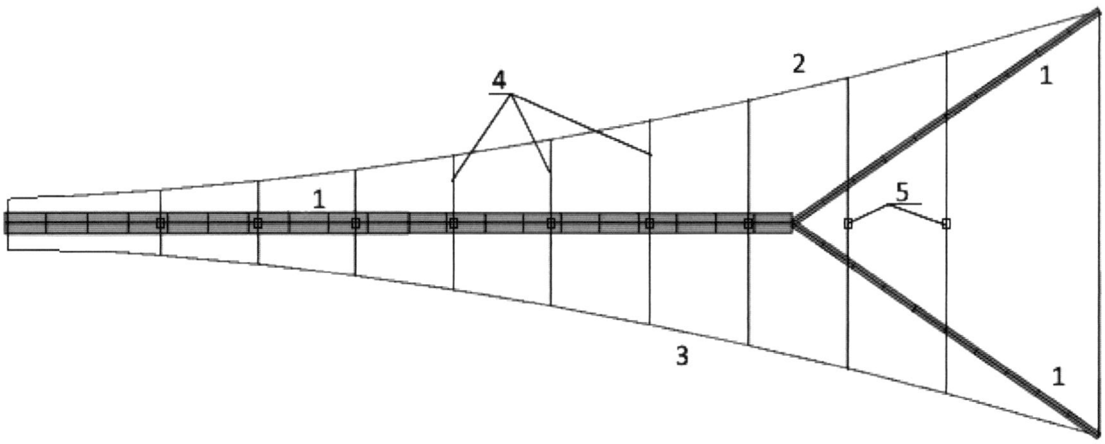

Fig. 1. The scheme of the spokes of the space-based reflector

The following limitations and difficulties should be taken into account when solving the problem of form configuration:

- The energy reserve on a space-based reflector is limited, the question of minimizing energy costs to perform the assigned tasks arises separately
- It is necessary to minimize the vibrations of the entire structure as a whole and of the radio-reflecting net in particular.

The problem solved in this study is to set the shape of the radio-reflecting net, determined by the frontal net 2, by tuning the length of the cables 4 using control of the actuators 5 taking into account minimizing energy costs and minimizing vibration of the structure. At present, control is carried out according to the proportional-integral-derivative (PID) control structure, which makes it possible to achieve the required parameters in the terminal part. The use of a PID control structure is not optimal for minimizing energy consumption and minimizing design fluctuations. At the same time, the use of optimal control algorithms will minimize energy costs and minimize fluctuations in the net. This can be achieved by smoothly stopping the engine when using the developed algorithm proposed in this paper.

2. Construction of a mathematical model of the structure of the actuator and modeling of its operation

In order to represent and predict the operation of the actuator used in the development, a mathematical model of the brushless motor included in the MRB-32 [16] assembly was built. The simulation was carried out in the Simulink application package using elements of the SimPowerSystems library. Automatic based on time function was chosen as the control method. A diagram of a model of a DC drive with automatic control by a function of time, in a simple approximation, is presented in Fig 2.

Ideal switches simulate the operation of contactors, and control occurs using "Stair Generator" blocks, which contain time functions that simulate real control signals. So, for key 1, the mode is as follows: 0...1.5 s – the key is open; 1.5...3.5 s – the key is closed. The "Step" block is needed to create reactance. Scope virtual instruments are installed in the system as tools for visual interpretation of signals and construction of oscillograms. The circuit also contains a power element that simulates the energy source for the system under consideration. The power supply of the field winding and the armature winding of the motor were chosen to be controllable in order to implement reverse functions. Data on the output parameters of the engine is received via a combined bus, which is decomposed into components for further use through the connector module.

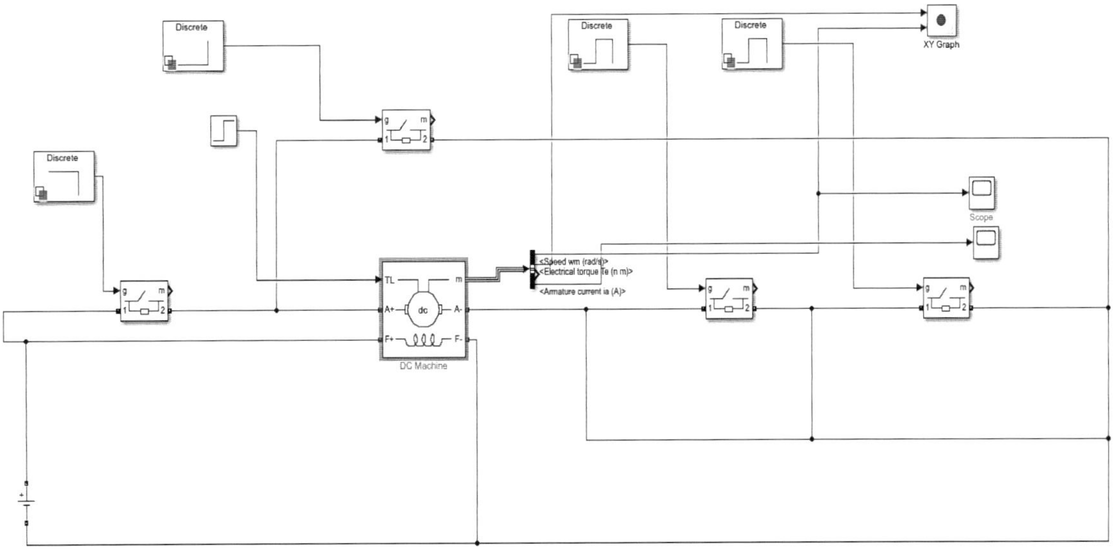

Fig. 2. Diagram of a DC drive model with automatic time control

Next, Fig. 3-5 shows part of the simulation results using the example of the processes of starting, braking and reversing the used motor, taking into account the setting of the following parameters: supply voltage 24 V; rotor winding resistance 3.45 Ohm; rotor winding inductance 0.0380 H; rotor inertia $17*10^{-6}$ kg*m^2. Fig. 3 shows the dependence of the rise and fall of shaft rotation on time.

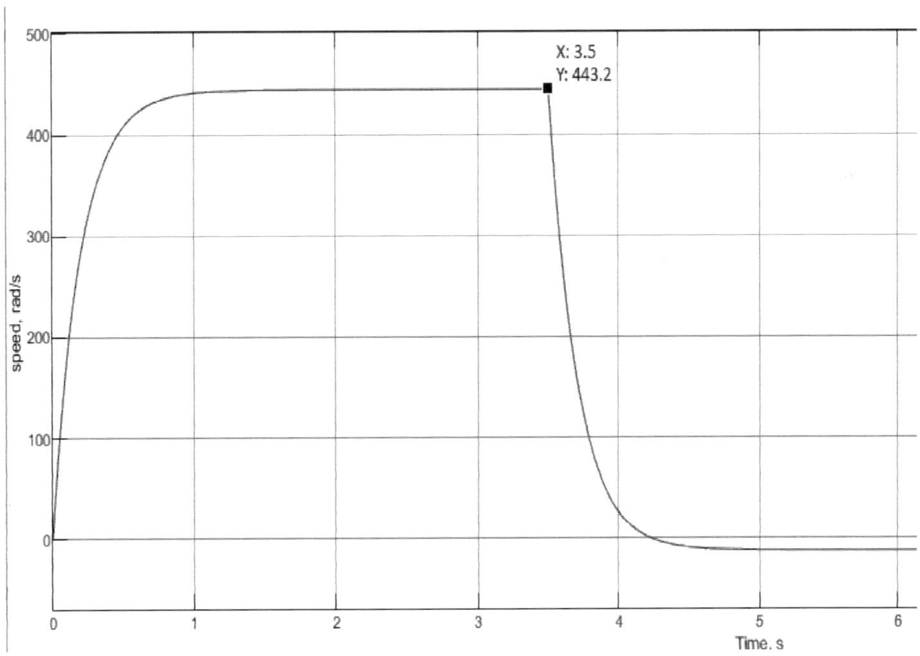

Fig. 3. Dependence of the rise and fall of shaft rotation over time

As can be seen from Fig. 3, the shaft from a state of rest in 0.5 seconds picks up an angular velocity close to 400 rad/s, which in terms of angular frequency is about 3820 rpm and, in general, corresponds to the engine data stated in the technical specification. We can also conclude that the shaft returns to a resting state approximately 1 second after the power is removed. Fig. 4 shows the dependence of the motor armature current on time.

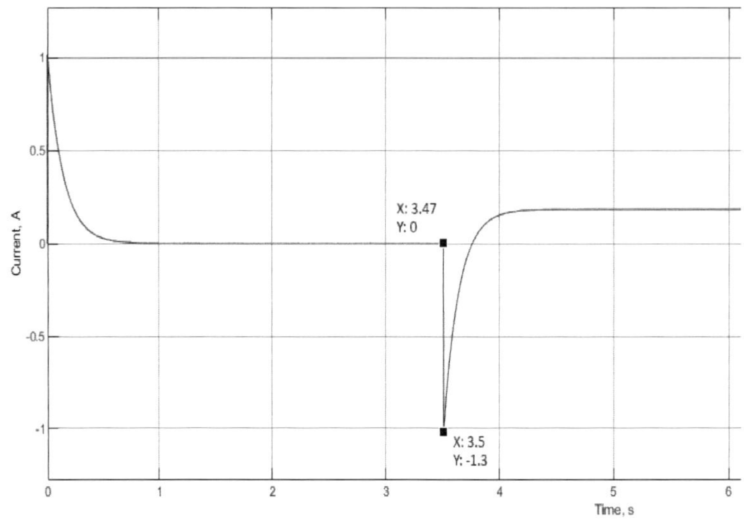

Fig. 4. Dependence of motor armature current on time

The starting current reaches a value of about 1 A and decreases as the engine accelerates to a value close to zero (it should be borne in mind that theoretically the starting process has not yet ended), with a steady angular velocity of 443.2 rad/s. Also shown in Fig. 5 is the dependence of changes in engine torque over time.

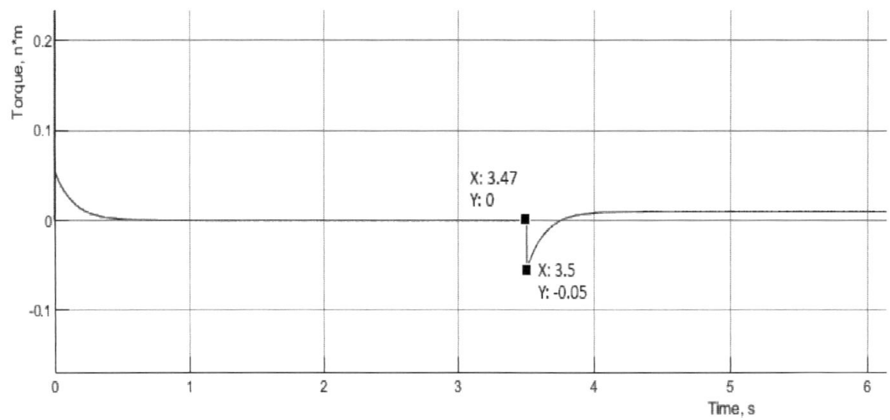

Fig. 5. Dependence of changes in engine torque over time

The resulting model is simplified, however, even in the presented configuration, through proper selection and input of input parameters, it becomes possible to get an idea of the aspects of the operation of the actuator. In order to obtain a much more accurate and high-quality picture of the operation of the drive used, it is planned to significantly supplement this model with gearbox units, additional sensors and other tools that allow for a comprehensive add-on to the actuator system.

3. Equations of DC motor

When developing the control algorithm for the MRB-32 geared motor, a mathematical model of the electric motor was used, described in [17], [18], [19] in the form

$$\frac{dI(t)}{dt} = \frac{U_a - R \cdot I(t) - k_E \cdot \omega(t)}{L} + \xi_I, \tag{1}$$

$$\frac{d\omega(t)}{dt} = \frac{k_M \cdot I(t) - M_{ex}}{J} + \xi_\omega, \tag{2}$$

$$\frac{d\varphi(t)}{dt} = \omega(t) + \xi_\varphi, \tag{3}$$

where I is motor current, U_a is armature (rotor) voltage, R is armature (rotor) active resistance, L is armature (rotor) winding inductance, k_E is coefficient of proportionality, called back EMF constant, ω is angular velocity of the motor shaft., k_M is motor torque constant, M_{ex} is electromagnetic torque of the motor, J is rotor moment of inertia, φ is the rotation angle, ξ_I, ξ_ω, ξ_φ is noise values corresponding to current, angular velocity and shaft rotation angle of the gearmotor, respectively.

As a first approximation, in the MRB-32 gearmotor, an instrument gearbox with a gear ratio is selected $k_g = 1/i_g$, where i_g = 150 is gear ratio of instrument gearbox.

Due to the need to solve the problem of controlling and filtering noise that arises during the operation of the system for tuning and maintaining the shape of the radio-reflective net of a large reflector, described in Section 1, we will use the generally accepted practice of posing and solving the problem of joint synthesis of optimal control. The formulation of the problem of combined synthesis of optimal control of systems is based on the separation theorem. According to this theorem, an optimal control system consists of a filter that forms estimates of the system's state vector, and an optimal controller that determines control in a deterministic setting under the assumption that the state vector is known exactly. The result of this theorem, proven for linear systems, is quite reasonably transferred to nonlinear systems with small estimation errors [20], [21]. Otherwise, an already complex and time-consuming task becomes dramatically more complicated. In this case, it is possible, for example, due to the duality of optimal control and estimation problems, to apply an approach to combined control synthesis with consideration of the hierarchy of corresponding functionals [22].

The angle of rotation of the shaft φ and the current I of the gearmotor MRB-32 are available for measurement. For this purpose it is supposed to use the 4GNSS sensor OS-213 [23]. For the problem of controlling a geared motor, we consider the observation equations in the form

$$z = h(x,t) + \xi_z. \tag{4}$$

In Equation (4) $z = [z_1 \quad 0 \quad z_3]^T$, $h = [I \quad 0 \quad \varphi]^T$, $z = [\xi_I \quad \xi_\omega \quad \xi_\varphi]^T$ are random processes such as white noise with intensity $B_z = diag(B_{z1}, B_{z2}, B_{z3})$.

In this article, a linearized Kalman filter and an algorithm for optimal correction of control structure parameters are used to construct control [24], [25].

The choice of an algorithm for optimal correction of control structure parameters to solve the control problem is explained by the fact that the use of a classical PID control structure does not take into account the minimization of energy costs, which is extremely important in conditions of energy limitation. The works [16], [17] show that the algorithm for optimal correction of control structure parameters is more stable in terms of convergence compared to classical optimal control methods, such as the Newton and Krylov-Chernousko method.

The control task is to drive the DC shaft of the MRB-32 gearmotor (1) – (3) from the initial position $\mathbf{x}(t_0) = [0,0,0]^T$ to a given end state $\mathbf{x}(t_f) = [0,0,250\pi]^T$ under the control constraint u, which is taken as the external supply voltage $U_{a\,\max}$=±24 V for a given time t_f. This rotation angle φ will provide a change in the length of the cable.

4. Development of a control algorithm of DC motor

In accordance with the separation principle, the control task is preceded by the task of estimating the state vector from incomplete data specified by equation (1) – (3). The optimal estimate can be obtained using the Kalman filter, the equations of which for this problem will take the form of the equations

$$\dot{\hat{I}} = \frac{U_a - R\cdot\hat{I} - k_E\cdot\hat{\omega}}{L} + R_{11}\cdot B_{z_1}^{-1}(z_1 - \hat{I}_z) + R_{13}\cdot B_{z_3}^{-1}(z_3 - \hat{\varphi}), \tag{5}$$

$$\dot{\hat{\omega}} = \frac{k_M\cdot\hat{I} - M_{ex}}{J} + R_{21}\cdot B_{z_1}^{-1}(z_1 - \hat{I}_z) + R_{23}\cdot B_{z_3}^{-1}(z_3 - \hat{\varphi}), \tag{6}$$

$$\dot{\hat{\varphi}} = \hat{\omega} + R_{31}\cdot B_{z_1}^{-1}(z_1 - \hat{I}_z) + R_{33}\cdot B_{z_3}^{-1}(z_3 - \hat{\varphi}), \tag{7}$$

$$\dot{R} = f_x\cdot R + R\cdot f_x^T - R\cdot h_x^T\cdot B_z^{-1}\cdot h_x\cdot R + B_x,\ R(t_0) = R_0. \tag{8}$$

In (5) – (8) $f_x = \frac{\partial f}{\partial x}$, $h_x = \frac{\partial h}{\partial x}$, $h_x = \begin{bmatrix}1 & 0 & 0\\ 0 & 0 & 0\\ 0 & 0 & 1\end{bmatrix}$, $f_x = \begin{bmatrix}-\frac{R}{L} & -\frac{k_E}{L} & 0\\ \frac{k_M}{J} & 0 & 0\\ 0 & 1 & 0\end{bmatrix}$, $R_{ii}(t_0) = R_0$, $i = \overline{1,n}$, other elements initial covariance matrices of estimation errors were taken equal to zero.

From an engineering point of view, the main task that arises during filter synthesis is to define mathematical models of noise. According to [15], system noise can be specified in the form of white noise with a power spectral density equal to unity. We use the maximum principle to search for optimal control, taking into account the formulation of the problem of minimizing energy problems over the entire control time. As a target, we consider the Boltz functional:

$$J = V_f(x, t_f) + \int_{t_0}^{t_f} f_0(x, u, t)dt, \quad (9)$$

where $V_f = 0{,}5\left[x(t_f) - x_f\right]^{\mathrm{T}} \rho\left[x(t_f) - x_f\right]$; $f_0 = 0{,}5k_u U_a{}^2/R^2$; $\rho = \mathrm{diag}(\rho_1, \rho_2, \rho_3)$; k_u, ρ_1, ρ_2, ρ_3 are given coefficients; $x_f = (I_f\ \omega_f\ \varphi_f)^{\mathrm{T}}$ is given final values of the corresponding variables in (1) – (3). Since the task was set to minimize energy costs, the expression for the consumed supply voltage $U_a$over the entire modeling interval was taken as the function f_0.

Solving this problem using the maximum principle using classical methods allows us to identify the structure of optimal control. Using the moment of time $\tau_i$of switching the control structure as parameters, we represent the control structure in the form

$$u(t) = u_1(t) + \Delta\mathbf{u}_i^{\mathrm{T}}(t)\boldsymbol{l}(t, \tau_i), \quad (10)$$

where $i = \overline{1, r}$, i is control structure section number, r is number of control switchings in the structure $\boldsymbol{l}^{\mathrm{T}}(t, \tau_i) = [l(t, \tau_1)\ l(t, \tau_2) \dots l(t, \tau_r)]$, $\Delta\mathbf{u}_i^{\mathrm{T}} = [-u_i + u_{i+1}, \dots, -u_{i-1} + u_r]$, $u_i(t) = u_1(t)$ for $t<\tau_1$. Here u_1, u_{i+1} are control in the previous and subsequent sections of this structure relative to τ_1; $l(t, \tau_1)$ are Heaviside functions.

To use the algorithm for correcting the parameters of the control structure, the equations are added to the original system (1) – (3) $\dot{\boldsymbol{\tau}} = \mathbf{w}$, where $\boldsymbol{\tau} = (\tau_1\ \tau_2 \dots \tau_r)^{\mathrm{T}}$; $\mathbf{w} = (w_1\ w_2 \dots w_r)^{\mathrm{T}}$ are vectors of new variables taken as controls. Then the equations for the generalized state vector will be written as

$$\dot{\mathbf{x}} = \boldsymbol{f}(\mathbf{x}, u, t); \ \dot{\boldsymbol{\tau}} = \mathbf{w}. \quad (11)$$

This problem can be solved by minimizing the new criterion

$$J_1 = J + \frac{1}{2}\int_{t_0}^{t_f} \mathbf{w}^{\mathrm{T}}\mathbf{k}^{-2}\mathbf{w}dt + \frac{1}{2}\int_{t_0}^{t_f} \mathbf{w}_0^{\mathrm{T}}\mathbf{k}^{-2}\mathbf{w}_0 dt\,, \quad (12)$$

where $\mathbf{k}$ is diagonal matrix of given coefficients, $\mathbf{w}_0$ is optimal value $\mathbf{w}$.

By introducing a new control vector $\mathbf{w}$, the original optimization problem by defining u is solved indirectly through finding this vector $\mathbf{w}$.

5. Results of testing the proposed control algorithm of DC Motor

To assess the influence of disturbances on the system, it is necessary to determine the magnitude of external disturbances and measurement noise. In addition to external disturbances, a change in the initial and final states can be observed, caused by impacts on the structure when delivering the reflector into orbit.

Let us assume that the filtered measurement noise does not exceed the errors of the measuring sensors. External disturbances (the influence of load, ambient temperature, solar pressure, radiation, etc.) affect the entire vector of state variables. As a rule, external disturbances are random, uncorrelated and uniformly distributed over a given range. In outer space, the external influence exerted on the reflector is quite long-lasting and slowly growing, so over the considered time interval (no more than 10 s) it can be considered quasi-stationary. Let us take the disturbances to be equal to ±1% of the maximum values of the corresponding variables. In general, the magnitude of disturbances from emergency situations is difficult to predict.

Having analyzed the control structure obtained in [17], [18], we can describe function (10) taking into account restrictions in the form

$$U_a(t) = U_{a\,\max}\mathrm{sign}(P_I) - 2U_{a\,\max} l(t, \tau_1) + U_{a\,\max} l(t, \tau_2), \quad (13)$$

where P_I is conjugate variable in I arising from the solution of the optimal control problem.

Based on (12), two structure changes were chosen at time τ_1 and τ_2, respectively. Numerical modeling of the transition of the system from the initial state $\mathbf{x}(0) = (0\ 0\ 0)^{\mathrm{T}}$ to the final state $\mathbf{x}(t_f) = (0\ 0\ 250\pi)^{\mathrm{T}}$ while minimizing the energy $P_e = U_a \cdot I$ in time $t_f = 3$ seconds was carried out by the Euler method with a step $\Delta t = 0.001$ seconds.

In Fig. 6 presents the results of the simulation: graphs of the dependences $\varphi(t)$ and $\omega(t)$. It can be seen that it was possible to solve the task, rotate the shaft of the MRB-32 gearmotor to a given angle $\varphi = 250\pi$ rad. Moreover, for all modeling cases, the error in the shaft rotation angle φ at the final moment of time did not exceed 0.1 rad.

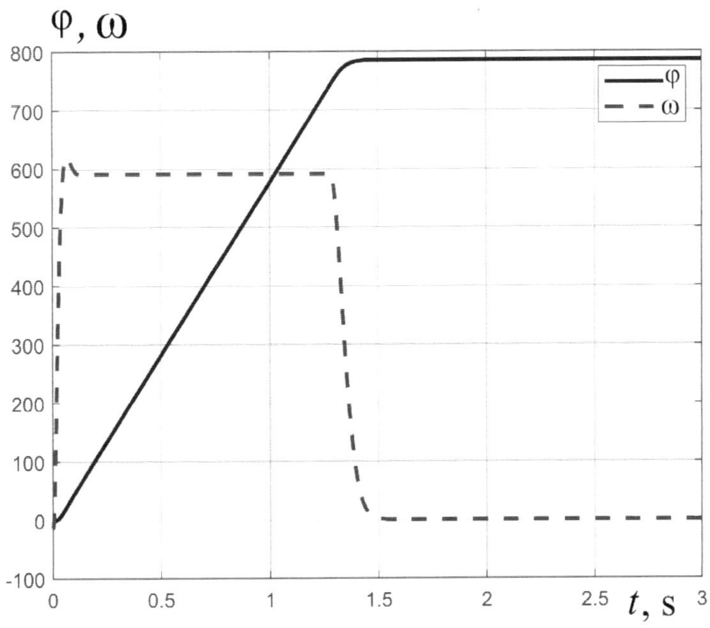

Fig. 6. Dependence $\varphi(t)$ and $\omega(t)$

For selected disturbances and measurement noise, optimal filtering and control algorithms successfully solve the problem. A feature of this algorithm is the good convergence of the structure change time τ_1 and τ_2 to the optimal value when choosing these variables in the modeling interval t=[0, 3] and subject to the conditions $\tau_1(0)$< and $\tau_2(0)$. In Fig. 7 shows graphs of the control of the switching moments τ_1 and τ_2 versus time, clearly demonstrating that the moment τ_1 converges to the optimal value, and τ_2 tends to t_f.

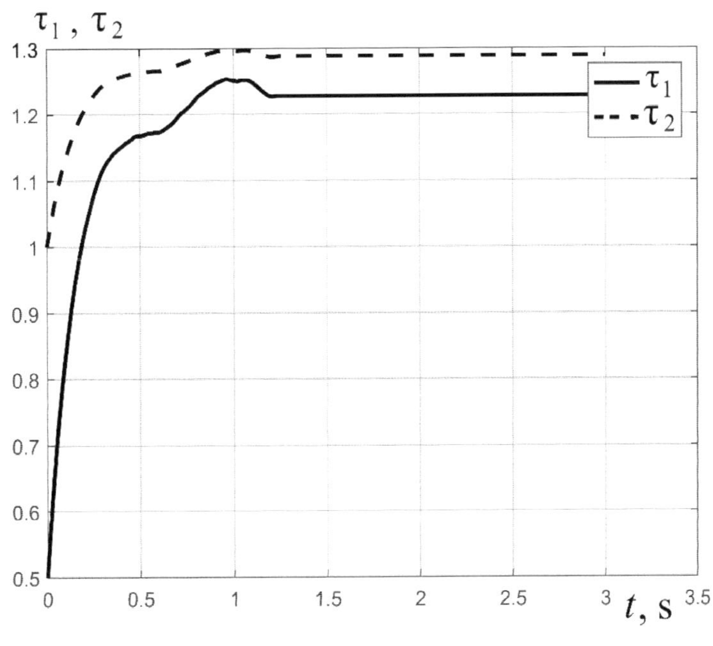

Fig. 7. Structure change time graphs $\tau_1(t)$ and $\tau_2(t)$

As can be seen from the presented graphs, τ_2 is located quite close to τ_1. In some cases, due to the presence of disturbances, their intersection occurs. Based on this, it is advisable to determine the condition $|\tau_2-\tau_1|<\varepsilon$, where ε is a given small value, at which only one switching remains. Time τ_1 allows one to successfully work out the disturbances exerted on the system before the moment of structure change occurs. After reaching time τ_2, the algorithm does not have a control effect on system (1)-(3), which leads to failure of the terminal conditions.

To solve this problem, it is necessary to control the final simulation time t_f, that is, determine $\tau_2 = t_f$. Next, it is necessary to apply a stabilization algorithm, for example, in the form of a PID (proportional-integral-differentiating) structure. Due to the proximity of the state vector **x** to the specified final conditions, one adjustment of the PID controller will be required without changing its coefficients, regardless of the operating mode, which greatly simplifies its use.

In Fig. 8 shows the diagonal elements of the covariance matrix at $R_{11}(0) = 0.00055$; $R_{22}(0) = 1.9$; $R_{33}(0) = 0.0004$. Quite quickly they come to steady-state values. Simulation was carried out with different initial values of the diagonal elements of the covariance matrix and different intensity of measurement noise, the Kalman filter successfully worked out all options.

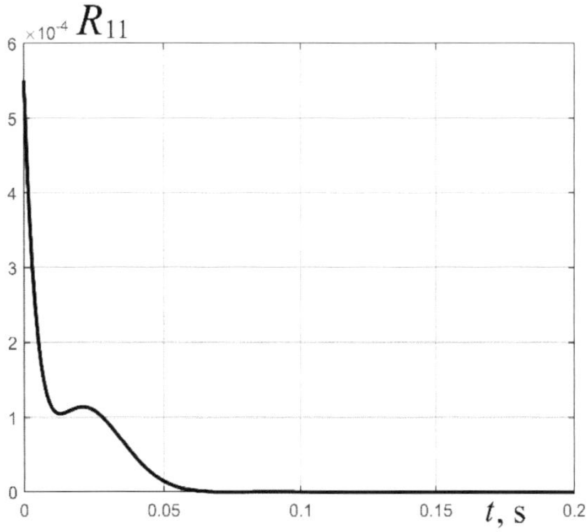

a)

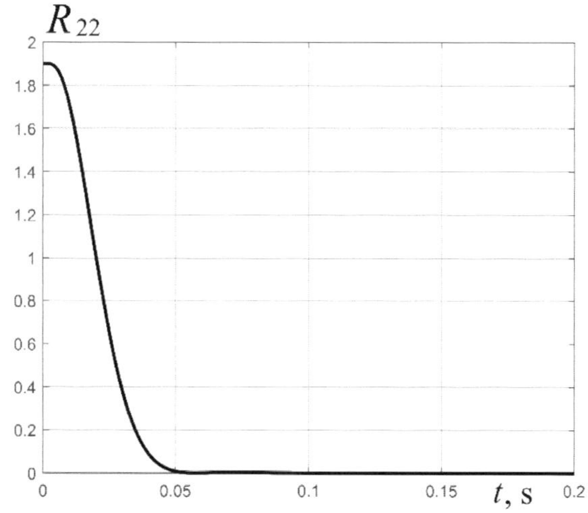

b)

Fig. 8. Graphs a) R_{11} (t); b) R_{22} (t)

When the intensity of disturbances increases to values $\mathbf{B}_x = diag(0.005, 0.1, 0.5)$, vibrations of the structure and the operation of the control algorithm become noticeable. In Fig. 9 and fig. 10 shows the graphs $\varphi(t)$ and $U_a(t)$, respectively.

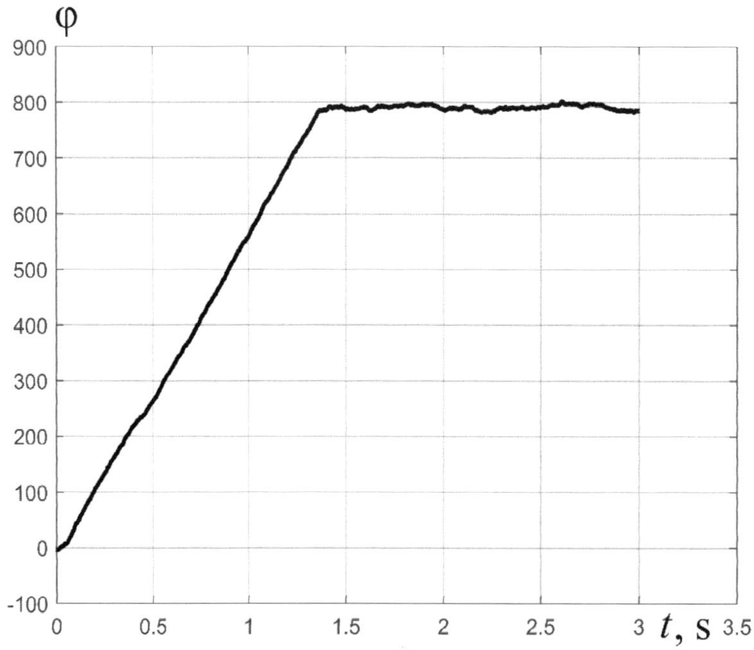

Fig. 9. Graph (t) under the influence of disturbances

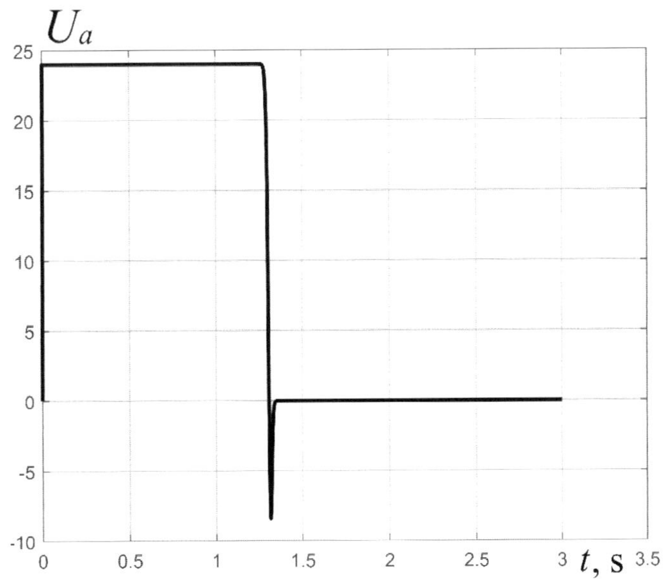

Fig. 10. Graph U_a (t)

In this case, the control U_a does not have time to completely change the structure, but reaches a level of ~ -9 V and returns to the final value of 0 V. This is due to the choice of the continuous function $l(t, \tau_i)$. The transition between control levels in (12) does not occur by relay switching, but rather smoothly. This allows you to customize the control algorithm for a real system with a given delay.

6. Conclusion

As a result of the work done, the mathematical model of the selected DC motor MRB-32 was confirmed. A solution to the problem of optimal control of a stochastic DC motor model using incomplete data using the separation principle is presented. The use of an algorithm for optimal correction of control structure parameters with the obtained data evaluated using a Kalman filter made it possible to solve the problem posed for various intensities of disturbances and measurement noise.

To use the algorithm for optimal correction of control structure parameters in real time, it is recommended to switch to the PID control structure when approaching the final simulation time. This will allow us to solve the problem of optimal control of the actuator, as well as solve the problem of stabilization. In this case, due to small deviations of the state vector at the final moment of time from the specified terminal ones, only one adjustment of the PID controller parameters will be required.

Taking into account the disturbances and noises of the proposed sensors acting on the system under study leads to complex problems of processing the measurement results and their planning. It is expedient to use the algorithm of correcting the control structure parameters studied in the article in more complex solutions using measurement processing and optimization of observation intervals.

The results of these studies can be used to calculate energy costs, select the actuator and power plant of the space-based reflector. The application of the developed optimal control algorithm requires significant computational costs, which must be taken into account when designing on-board equipment. With further research, it is planned to conduct an additional analysis of possible electronic components (microcontrollers, FPGAs, etc.), as well as conducting prototype tests of the actuator and confirm the effectiveness of the developed algorithm on the layout.

7. Acknowledgments

The research was supported by grant No. 22-79-10112 from the Russian Science Foundation, https://rscf.ru/en/project/22-79-10112/.

8. References

[1] Zhengrong Chu et al. (2014). Modeling and Analysis of a Large Deployable Antenna Structure, Acta Astronautica, Vol. 95, No. 1, pp. 51–60, ISSN 0094-5765, DOI 10.1016/j.actaastro.2013.10.015

[2] Guarriello, A., Loison, R., Bresciani, D., Legay H. & Goussetis, G. (2023). Structural and Radio Frequency Co-Design and Optimization of Large Deployable Reflectarrays for Space Missions, IEEE Transactions on Antennas and Propagation, Vol. 71, No. 5, pp. 3916-3927, doi: 10.1109/TAP.2023.3247940

[3] Imbriale, W. A., Gao, S., Boccia, L. (2012). Space Antenna Handbook. John Wiley & Sons. 744 p

[4] Galimov, G.K. (2017). General theory of mirror antennas. M.: Advance Solutions, Vol. 6. 704 p

[5] Kapoor, A., Nagaraj, B. P., Kumar, A., Kumar, H. S., Dinesh, B. V. S., Viswanatha, N., ... & Muthu, B. K. (2023). A novel antenna steering mechanism with self hold-down feature for communication spacecrafts. CEAS Space Journal, pp. 1-22

[6] Tishkov A. I., Konoplev Y. V. , Yuev A. A., Shevtsov I. V. (2019). Development of a Combined Electric Power Storage Device, Russian Aeronautics, Vol. 62, No. 2, 2019, pp. 321-325, doi: 10.3103/S106879981902020X

[7] Reznik, S. V., Chubanov, D. E. (2018). Large-sized transformable space antenna reflector made of composite materials dynamic modeling process, RUDN Journal of Engineering Researches, Vol. 19, No. 4, pp. 411-425. ISNN 2312-8151, DOI 10.22363/2312-8143-2018-19-4-411-425

[8] Bakulin ,V.N., Borzyh, S.V. (2020). Modeling the dynamics of the process of deployment large-sized transformable space structures, Izvestiya vysshih uchebnyh zavedenij, Aviacionnaya tekhnika, No. 4, pp. 50-56. ISSN 2782-5760, DOI 10.31772/2712-8970-2022-23-4-602-614

[9] S. Matveev, A. Gorbunov, A. Shirshov [et al.] (2020) Control of the system of piezoelectric actuator devices for precision drive systems, International Journal of Mathematical, Engineering and Management Sciences, Vol. 5, No. 2. 2020, pp. 319-327, DOI 10.33889/IJMEMS.2020.5.2.026

[10] Yiqun Zhang, Wenrui Ru, Guigeng Yang, Na Li. (2016). Deployment analysis considering the cable-net tension effect for deployable antennas, Vol. 48, pp. 193-202, ISSN 1270-9638, DOI 10.1016/j.ast.2015.11.012

[11] Yiqun, Z.; Na, L.; Guigeng, Y. & Wenrui, R. (2017). Dynamic analysis of the deployment for mesh reflector deployable antennas with the cable–net structure, Acta Astronautica, Vol. 131, pp 182–189, ISSN 0094-5765, DOI 10.1016/j.actaastro.2016.11.038

[12] Evstigneeva, E.I., Lekanov, A.V., Matveev, S.A. et al. (2019). Development of the Wireless Control System for Shape of Large Transformed Structures Using Precision Drives. Russ. Aeronaut. Vol. 62, pp. 502–507, ISSN 1934-7901, DOI 10.3103/S1068799819030206

[13] Kabanov, S.A., Zimin, B.A., Mitin, F.V. (2020). Development and Research of Mathematical Models of Deployment of Mobole Parts of Transformable Space Construction. Part II. Mekhatronika, Avtomatizatsiya, Upravlenie, Vol. 21, No. 2, pp. 117-128, ISSN 1684-6427, DOI 10.17587/mau.21.117-128

[14] Kabanov, S.A. & Mitin, F.V. (2021). Optimization of the Processes of Deployment and Shape Generation for a Transformable Space-Based Reflector, Journal of Computer and Systems Sciences International, Vol. 60, No. 2, pp. 283–302, Print ISSN 1064-2307, Online ISSN 1555-6530, DOI: 10.1134/S1064230720060052

[15] Kabanov, S.A. & Mitin, F.V. (2020). Optimization of the stages of deploying a large-sized space-based reflector, Acta Astronautica, Special Issue on 6th SFS 2019, Vol. 176, pp. 717-724, ISSN 0094-5765, DOI: 10.1016/j.actaastro.2020.04.066

[16] https://electroprivod.ru/brushless-motor-mrb-32.htm, (2023). Brushless motors with planetary gearbox MRB-32, Accessed on: 2023-09-14

[17] Mitin, F. & Krivushov, A. (2018). Application of Optimal Control Algorithm for DC Motor, Proceedings of the 29th DAAAM International Symposium, pp. 0762-0766, B. Katalinic (Ed.), Published by DAAAM International, ISBN 978-3-902734-20-2, ISSN 1726-9679, Vienna, Austria, DOI: 10.2507/29th.daaam.proceedings110

[18] Mitin, F. & Krivushov, A. (2022). Optimal Control of the DC Motor in the Tuning of the Space-Based Reflector of the Radar-Reflecting Netting, Proceedings of the 33rd DAAAM International Symposium, pp.0247-0253, B. Katalinic (Ed.), Published by DAAAM International, ISBN 978-3-902734-36-5, ISSN 1726-9679, Vienna, Austria DOI: 10.2507/33rd.daaam.proceedings.034

[19] Mitin, F.V., Krivushov, A.I. (2023) Improving the efficiency of DC motor control to adjust the reflective surface of the space-based reflector, Russian Aeronautics (Iz. VUZ), Vol. 66, No 1, pp. 162-169, ISSN 1934-7901, DOI 10.3103/S1068799823010221

[20] Krasovsky, A. A. (1987). Handbook on the theory of automatic control, Nauka, Moscow.

[21] Kabanov, S. A. (2008). Optimization of the dynamics of systems under the action of disturbances, Moskva: Fizmatlit, 200 p.

[22] Kabanov, S. A. (2000). Combined synthesis of optimal control as a hierarchical differential game, Dokl. Akad. Nauk, Vol. 372, pp. 317–318.

[23] https://orsyst.ru/receivers/oc-213, (2023). 4GNSS OC-213, Accessed on: 2023-09-14

[24] Kabanov, S. A., (2014). Dynamic system optimization using correction of control structure parameters, Vest. SPbGU, Ser. 1: Mat. Mekh. Astron, Vol. 1, No. 2, pp. 254–260.

[25] Kabanov, S.A.; Krivushov, A.I. & Mitin, F.V. (2017). Modeling of joint deployment of units of the large-sized transformable reflector of space basing, SPIIRAS Proceedings, Vol.54, No 5, 2017, pp. 130-151, Print ISSN 2078-9181, Online ISSN 2078-9599, DOI: 10.15622/sp.54.6

DOI: 10.2507/34th.daaam.proceedings.025

ENHANCING REALISM IN ROBOTIC FLIGHT SIMULATORS: A CASE STUDY ON THE IMPLEMENTATION ANALYSIS OF A WASHOUT FILTER

Alina-Ioana Chira, Marian-Ciprian Bîlu, Răzvan-Ionuț Bălașa & Cătălin Iordache

This Publication has to be referred as: Chira, A[lina-Ioana]; Bilu, M[arian] C[iprian]; Balasa, R[azvan] I[onut] & Ioradche, C[atalin] (2023). Enhancing Realism in Robotic Flight Simulators: A Case Study on the Implementation Analysis of a Washout Filter, Proceedings of the 34th DAAAM International Symposium, pp.0193-0202, B. Katalinic (Ed.), Published by DAAAM International, ISBN 978-3-902734-41-9, ISSN 1726-9679, Vienna, Austria
DOI: 10.2507/34th.daaam.proceedings.025

Abstract

The present work provides a comprehensive analysis of the implementation of a washout filter in an anthropomorphic-robotic flight simulator. It discusses the significance of adding a 7th degree of freedom (DOF) through a linear unit to enhance the pilot's perception within the flight simulator. The primary objective of this research was to develop a flight simulator on an anthropomorphic robot, serving as both a dynamic simulation system for pilot-in-the-loop applications and a pilot training platform. This simulator was designed to faithfully reproduce the sensations a pilot typically experiences during more demanding maneuvers, such as changes in G-forces in aircraft flight tests. A critical requirement for such a simulator is the accurate representation of its motion system, achieved through the proper implementation of the washout filter. This paper presents experimental results that validate the proposed solutions.

Keywords: robotics; flight simulation; washout filter; motion cueing; stability; 7th DoF.

1. Introduction

Nowadays, the aerospace industry places a high focus on safety procedures and equipment, leading to the development of increasingly durable and stable systems. While avionics systems themselves are becoming safer, the human factor is another element that can significantly impact the overall safety of a flight and should be taken into account. In recent years, aeronautics designs have become increasingly complex as the certification process has demanded higher levels of demonstration. Due to the increasing demands of the certification process, aeronautics designs have become more intricate in recent years. This is because certain systems in an aircraft are required to demonstrate failure rates of less than 10^{-9} flight hours over the operational life of the aircraft. To reduce the risks and costs of aircraft development ([1], [2]), flight simulation environments have been created to improve flying quality while maintaining a consistent level of operational performance. A summary of flight simulation history, as well as a list of applications and benefits of flight simulators, is presented [1]. As the aircraft design stages progress, more sophisticated and representative simulation environments are created [3]. Building on the above, the present research introduces a real-time flight simulator that enhances the fidelity of the effects of various aircraft dynamics on potential pilots and focuses training sessions on improving human performance. RoFSim (Dynamic Simulation System for pilot-in-the-loop applications – Robotic Flight Simulator) aims to design and implement a complex control system for simulating the overall dynamics of aerospace vehicles by utilizing a robotic manipulator arm [4] as a motion platform for use in motion perception test studies and studies to optimize training procedures for civilian and military pilots. The six degrees of freedom flight simulator comprises an industrial robotic arm on which the cockpit is placed, a central processing unit running the software, command instruments, and a Virtual Reality (VR) headset that serves as a display for the user (pilot). The architecture of RoFSim is designed to simulate, with high fidelity, changes in attitude and velocity when performing certain manoeuvres, helping the pilot understand the effects of their actions in-flight on a real aircraft. Furthermore, because the simulator's display is a VR headset, various aircraft models and weather condition settings for the training session can be chosen. Compared to static platform flight simulators, RoFSim is intended to be a versatile, low-cost tool designed to improve human performance. RoFSim is a flight simulator mounted on an ABB 7600 robot arm [5] with six degrees of freedom and an additional 7th degree of freedom provided by a linear track, offering a distinct advantage over the Stewart platform. While it can move within a broader dynamic range, certain limitations still exist for safety reasons. These limitations stem from external factors, internal considerations, and safety requirements, all of which strictly restrict the joint angle ranges of the robot arm. To ensure the entire system can move in a specified direction without encountering nearby objects and to maintain the safety of the pilot and the entire setup, it becomes imperative to redefine the hardware and software limits for each joint of the robotic arm. This work outlines an approach to optimize the workspace of the robotic motion platform through three key aspects: hardware joint angle limitations, collision avoidance measures, and software-based joint angle restrictions. By applying these constraints, the robot's working space is carefully limited to create a secure and safe simulation environment.

Fig. 1. RoFSim Flight Simulator

The paper's first section provides a description of the procedure for scaling the workspace of the six degrees of freedom simulator. It emphasizes three critical boundary conditions necessary for ensuring proper and safe simulator operation: hardware joint angle limitations, collision avoidance, and software-imposed joint angle restrictions. The primary focus of this work is the implementation of a washout filter (WF) on the RoFSim simulator, which translates aircraft dynamics into robot movements. One of the key components for achieving realistic motion is the software responsible for transforming aircraft accelerations into simulator motion cues. The objective is to enhance the simulator's realism and fidelity.

Experiments conducted with a simplified prototype of the flight simulator have revealed the limitations of the classical washout filter (CWF) architecture. Chapter 3 introduces the implemented motion algorithm for the RoFSim simulator, providing detailed explanations for each motion channel and specific system conditions. Chapter 4 delves into the testing and validation scenarios of the system, encompassing flight plans, simulator configurations, and parameters extracted from the simulated aircraft dynamic model, all integrated with the defined CWF motion algorithm. Finally, Chapter 5 presents the experimental data and results obtained from the testing phase of the RoFSim simulator.

2. Related work

As stated in the introduction, this study aims to develop, implement, and validate a washout filter algorithm that faithfully replicates the acceleration sensations a pilot experiences in a real aircraft, utilizing a motion platform with limited workspace. The Washout filter, an essential motion algorithm, serves the critical function of translating real acceleration into the motion experienced within a flight simulator ([6], [7], [8]). Its core purpose is to provide motion cues that align with human perception [9] while adhering to the constraints and limitations inherent in the proposed flight simulator [10]. The significance of this filter is most pronounced in flight simulators employing the Stewart parallel robot mechanism [11]. While CWFs are renowned for their simplicity and adaptability [12], they have faced challenges in maintaining the fidelity of human perception. In recent years, new theories for implementing washout filters have gained prominence, with optimal ([14], [17]), adaptive ([7], [15], [16]), and robust ([13], [18]) theories emerging as pivotal areas of focus in this field.

An adaptive washout filter (AWF) was introduced by Nahin & Reid [19], building upon the CWF using a self-tuning algorithm. An optimal washout filter (OWF) is rooted in optimal theories [20], solving the Algebraic Riccati Equation (ARE) to meet specified requirements.

In our study, we also employ a classical washout filter, introducing fundamental changes to the required transfer functions. We leverage evolutionary algorithms to identify the optimal configuration within this framework. The primary objective is to minimize perception errors and align the simulator's actions more closely with those experienced in the actual workspace of a flight simulator.

Notably, one of the key innovations in our approach is the realistic representation of transient G-force losses/gains, achieved through the effective implementation of the translational channel of the classical washout filter. The adjustability of the robotic platform's movement constraints is detailed in the subsequent section.

3. RoFSim Workspace Scaling Procedure

To establish a suitable workspace for the RoFSim six-degree-of-freedom simulator, we defined hardware and software limitations. For safety reasons, we strictly constrained the robot's movement range to ensure the pilot's safety when using the simulator, preventing collisions with nearby objects. These adjustments resulted in additional constraints that reduced the robot's effective workspace. A comparison between the workspace of the IRB 7600 500/2.55 robot and the workspace when coupled with the RoFSim cabin reveals that the simulator's workspace is more restricted due to its larger dimensions.

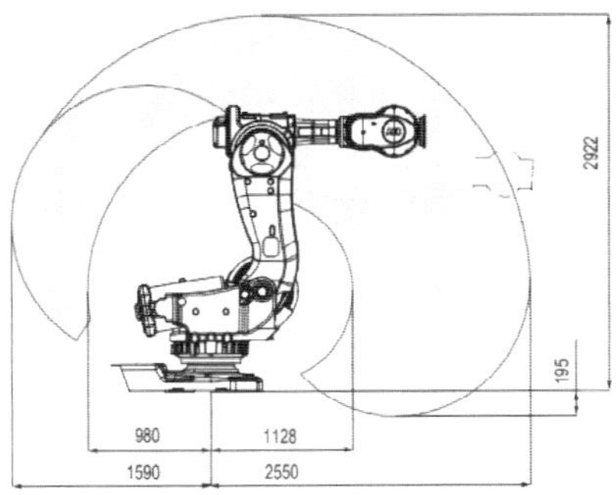

Fig. 2. Working Range IRB 7600-500/2.55 [5]

Depending on the specific application of the RoFSim simulator, different workspace constraints are established. It's important to note that there is a significant interdependency among joint angles, meaning that constraining one joint will affect the others. As a result, it is not possible to independently maximize the range of each joint angle. Consequently, we formulate the following optimization problem:

$$max\ \vec{\omega} \cdot \vec{f}_{kin}(\vec{q})\ ,\ \vec{q}\ \in\ R^6 \tag{1}$$

$$\vec{f}_{kin}(\vec{q}) = \{\Delta x, \Delta y, \Delta z, \Delta\phi, \Delta\theta, \Delta\psi\}^T \tag{2}$$

where x, y, z are the cartesian positions, ϕ, θ, ψ the rotation angles (roll, pitch, gyration), and $\vec{\omega} \in R^6$ the weighting factor vector. In the process of formulating the workspace scaling procedure for the RoFSim system, the following essential conditions are considered: hardware limitations of joint angles, collision avoidance, and software-imposed limitations of joint angles.

A. Hardware Limitations of Joint Angles

The manufacturer provides specifications for the robot joint angle intervals in the data sheet. These hardware limitations are further complemented by software restrictions set by the manufacturer, which come into effect before reaching the hardware limits.

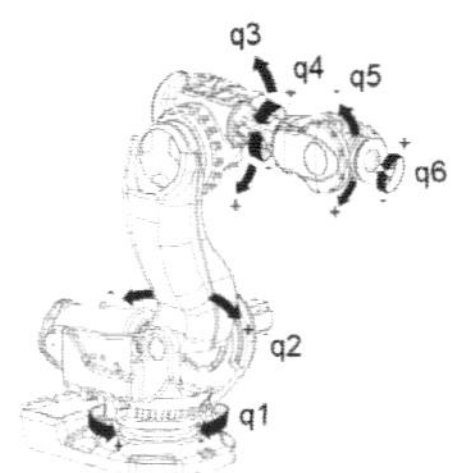

Fig. 3. Robot Motion Axes IRB 7600 - manufacturer configuration [5]

$$\vec{q} \in [\vec{q}_{min}, \vec{q}_{max}] \tag{3}$$

IRB 7600 Robot	Axes	$\vec{q}_{min}$	$\vec{q}_{max}$	Speed limit	Acceleration limit
A	q_1	- 180 [°]	+ 180 [°]	75 [°/s]	-
B	q_2	-60 [°]	+85 [°]	50 [°/s]	-
C	q_3	-180 [°]	+60 [°]	55 [°/s]	-
D	q_4	- 300 [°]	+ 300 [°]	100 [°/s]	-
E	q_5	- 100 [°]	+100 [°]	100 [°/s]	-
F	q_6	- 360 [°]	+ 360 [°]	160 [°/s]	-
Track Gudel TMF- 4	q_7	-10 [mm]	20010 [mm]	90 m/min	86 [m/s^2]

Table 1. Defined IRB 7600 robot limits

B. Collision avoidance

The design of the RoFSim cabin significantly impacts the definition of joint motion ranges. When establishing collision avoidance criteria, we considered the following parameters: RoFSim cabin size, flange position, flange angle, and the dimensions of the subject (pilot). Safety must be ensured throughout the robot's entire range of motion to prevent collisions. The following objects were taken into account: the floor, the Gudel TMF-4 track, the IRB 4600 robot, the IRC5 controller of the IRB 4600 robot, the IRC5 controller of the IRB 7600 robot, and the nearby wall.

$$V(obiect_i, \vec{q}) \cap V(obiect_k, \vec{q}) = \emptyset \tag{4}$$

where $obiect_i = \{cab\ RoFSim, robot\ IRB\ 7600, pilot, \dots\}$, and $obiect_k = \{robot\ IRB\ 4600, track\ Gudel\ TMF - 4, controller\ IRC5, wall, \dots\}$.

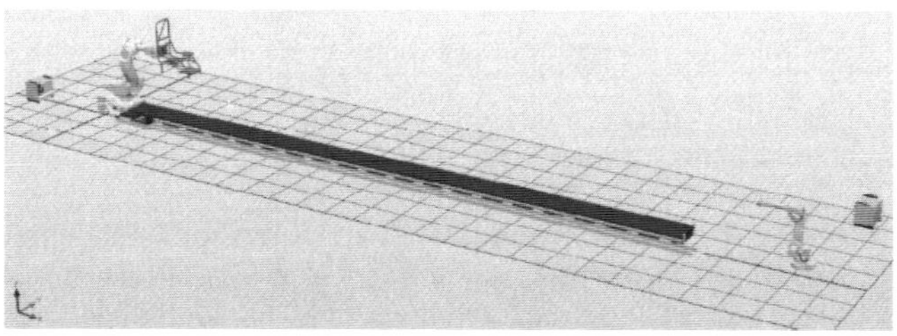

Fig. 4. RobotStudio Simulation – RoFSim and Objects in the Vicinity of the System

C. Software-Imposed Joint Angle Restrictions

Depending on the simulation type, various constraints and workspace limitations are applied. The following decisions were made following the analysis:

- Installation of SafeMove Hardware/Software Module: this module serves as an additional safety measure. If the IRB 7600 robot's path collides with surrounding objects (TMF-4 track, lab floor, IRC5 controller, or other areas/objects near the RoFSim workspace), it activates and blocks robot movement.
- Software Implementations - RoFSim Controller Module: to enhance safety during robot movements, the RoFSim Controller module has been programmed not to exceed +/- 90° (1.57 radians) on all three axes: X (Roll), Y (Pitch), and Z (Yaw). To prevent collisions with the Gudel track, the linear motion on these axes was constrained as follows: X (North-South) to +/- 3 meters, Y (East-West) to 0.4 meters, and Z (Vertical) to 0.4 meters. It's important to note that the simulator does not accept any motion commands while it is on the ground.
- Software Implementations in IRC5 Controller: additional modifications have been implemented to ensure that the q_5 axis does not exceed +/- 90° (1.57 radians).
- This constraint is necessary to avoid singularity errors during angular motions, and the q_5 axis must remain different from 0°.
- Singularities of the robotic arm occurs in any configuration when the wrist centre reaches the q_1 axis.

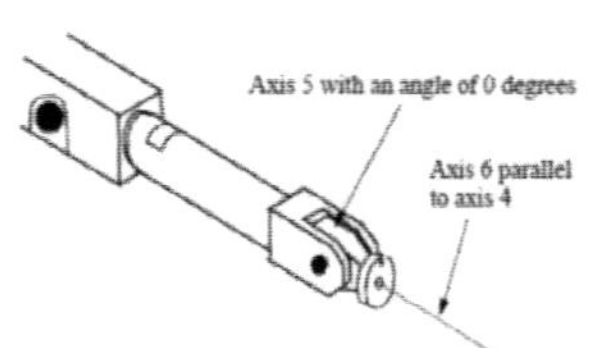

Fig. 5. Wrist Singularity [5]

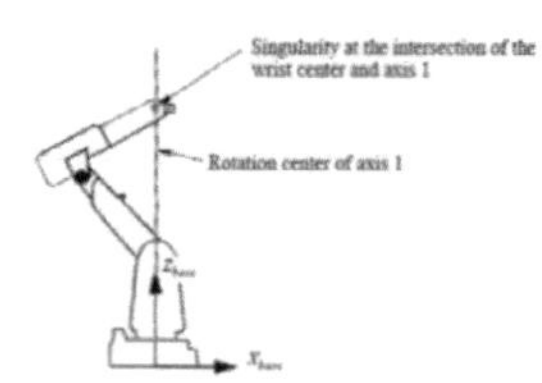

Fig. 6. Arm Singularity [5]

Figure 7 illustrates the differences between the two working envelopes (2D format - white marking, and 3D format - green marking) with and without the RoFSim system coupled.

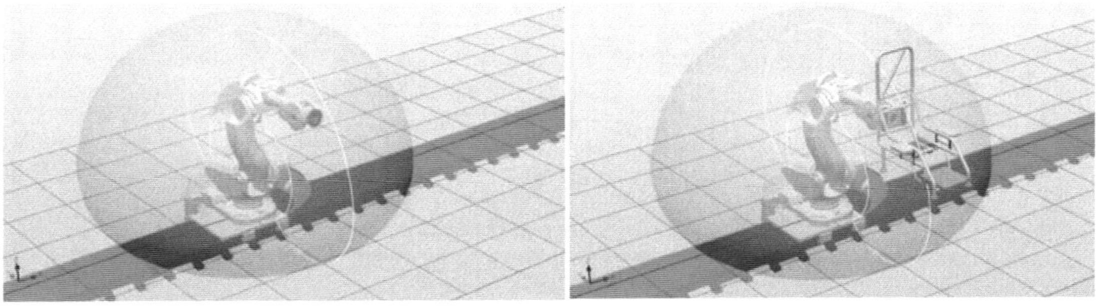

Fig. 7. IRB 7600 Robot Workspace Envelope: uncoupled (left) / coupled (right) – details from RobotStudio

Optimizing joint angle limits is a complex procedure as changes to one joint impact others. To simplify the process, we've divided the multidimensional optimization problem into several sub-problems. For each sub-problem, we've identified constraints and implemented measures to ensure simulator safety and integrity. This involved software modifications at various levels (RobotStudio/Rapid, Visual Studio/C# programming language) and hardware adjustments for each axis.

4. Implementation CWF

Flight simulators serve various purposes, including understanding human-machine interaction, studying pilot behaviour, and providing training. The effectiveness of a flight simulator largely depends on its ability to generate motion cues. By employing motion-cueing algorithms, we create a simulation environment. The RoFSim's effectiveness is measured by its capability to generate realistic motion cues. Pilots in the cockpit should perceive motion cues similar to those experienced in a real aircraft during manoeuvres. However, due to the system's limited workspace, the RoFSim simulator cannot directly mimic aircraft motion. The platform would quickly reach its physical limits, rendering it unable to provide motion cues to the pilot.

To address this limitation, motion algorithms have been developed ([6] – [20]), including one tailored for the RoFSim simulator. This algorithm governs the generation of motion cues while keeping the robotic platform within its physical boundaries. The primary objectives of these motion algorithms are twofold: to provide lifelike motion cues to the pilot and to ensure the motion platform stays within its physical limits. This chapter presents the motion algorithm designed specifically for RoFSim. It integrates mathematical models of vehicle dynamics and pilot behaviour into the overall system modeling framework.

The use of the robotic manipulator arm as a motion platform for closed-loop simulation raises two engineering challenges. Firstly, it requires kinematic schemes capable of replicating real-time, operator-commanded random system motions. Secondly, the entire system responsible for faithful motion replication must be adapted to fit within the specific motion envelope of the IRB 7600 robotic arm.

Central to the flexible and modular design of the proposed simulator is the control interface. This interface facilitates the connection between a realistic simulation environment, a simulation platform, and a virtual environment. Simultaneously, the control system supervises safety devices and provides manual and automatic control of the simulator. The scaling procedure developed in the previous chapter, aimed at assessing the user's perception of induced motion, is a critical aspect of the simulation system's development.

The motion reproduction algorithm's functionality involves filtering/scaling the ideal trajectory of the simulated vehicle. This trajectory is generated numerically based on the operator's commands. The algorithm's goal is to adapt the trajectory to fit within the motion platform's working space while providing a perception of motion as close as possible to that of a real vehicle.

The algorithm relies on the use of a smoothing filter or a combination of several smoothing filters. These filters are responsible for reproducing high-frequency motions and tilt coordination algorithms to replicate low-frequency motions. Essentially, vehicle motion is divided into low and high-frequency components. High-frequency components are replicated by physically moving the platform since they generate small, manageable displacements. In contrast, low-frequency components (linear accelerations) are not achieved by physically moving the platform but rather by utilizing the gravitational acceleration vector as a source of sustained acceleration.

The figure below provides a representation of the experimental motion perception system along with its specific subcomponents. The motion controller is responsible for transferring the pilot's commands to both the robotic motion platform and the visual system. Advanced control devices designed for the aircraft are utilized to transmit commands to the RoFSim simulation software module.

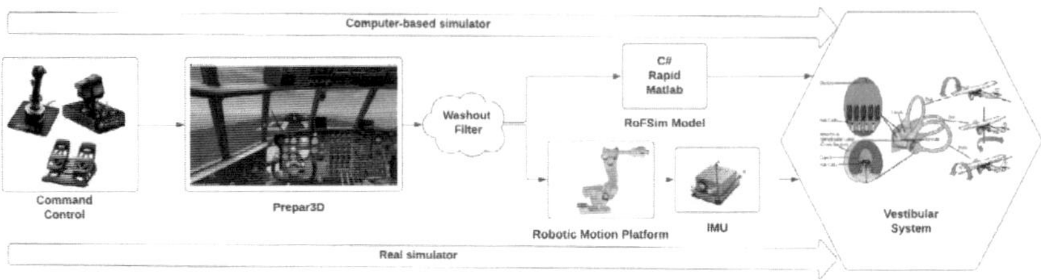

Fig. 8. Representation of the Experimental Motion Perception System Based on Subcomponents

Before delving into the detailed explanation of the implemented motion algorithm, it's crucial to establish a clear understanding of the relevant reference systems used in this study:

- Aircraft Reference System (S_A);
- RoFSim Simulator Reference System (S_S);
- Pilot - Aircraft Reference System (S_{AP});
- Simulator - Pilot Reference System (S_{AS}).

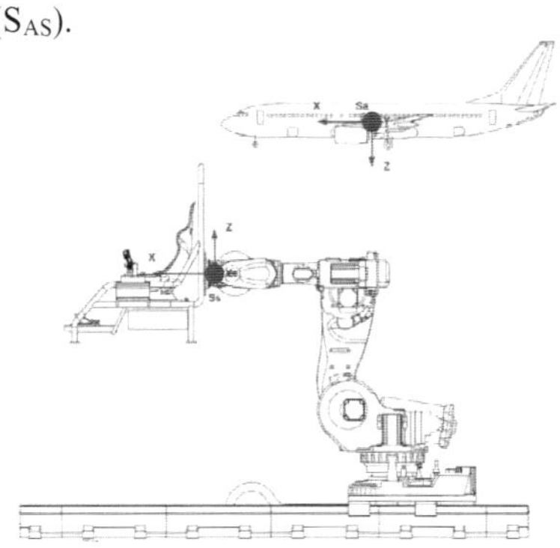

Fig. 9. RoFSim Reference Systems

When applying the motion algorithm, the transition from the aircraft reference system to the simulator reference system is accomplished using a matrix of directive cosines, employing a Z-Y-X rotation method.

$$\overrightarrow{S_A} = DCM_{ZYX} \cdot \overrightarrow{S_S} \tag{5}$$

where

$$DCM = \begin{bmatrix} \cos(\theta) \cdot \cos(\psi) & \cos(\theta) \cdot \sin(\psi) & -\sin(\theta) \\ sin(\phi) \cdot sin(\theta) \cdot cos(\psi) - \cos(\phi) \cdot \sin(\psi) & sin(\phi) \cdot sin(\theta) \cdot sin(\psi) + \cos(\phi) \cdot \cos(\psi) & sin(\phi) \cdot cos(\theta) \\ cos(\phi) \cdot sin(\theta) \cdot cos(\psi) + \sin(\phi) \cdot \sin(\psi) & cos(\phi) \cdot sin(\theta) \cdot sin(\psi) - \sin(\phi) \cdot \cos(\psi) & cos(\phi) \cdot cos(\theta) \end{bmatrix} \tag{6}$$

Within the operation of the motion algorithm, we applied both constraints and scaling to the input signals for the translational and rotational channels. This constraint and scaling uniformly change the input signal's amplitude across all frequencies. Limiting, a non-linear process, modifies the signal to restrict it to a specific magnitude, ultimately reducing the motion response of the simulator. In our approach, we used a third-order polynomial for scaling, which was integrated into the general motion algorithm scheme. Let x represent the input, y the output, x_{max} the desired maximum input, y_{max} the maximum output, and s_0 and s_1 the slopes at $x = 0$ and $x = x_{max}$ respectively. The third-order polynomial used to achieve specific characteristics is in the following form:

$$y = c_3x^3 + c_2x^2 + c_1x^1 + c_0 \tag{7}$$

Parameter scaling for the translational channel:

$$max\ input\ (x_{max}) = 6\ m/s^2 \tag{8}$$

$$max\ output\ (y_{max}) = 6\ m/s^2, \text{ for } X \tag{9}$$

$$max\ output\ (y_{max}) = 0.8\ m/s^2, \text{ for } Y \text{ and } Z \tag{10}$$

and the coefficients are:

$$s_0 = 1.5, \text{ for X}$$
$$s_0 = 0.2, \text{ for Y and Z} \tag{11}$$
$$s_1 = 0, \text{ for X, Y and Z}$$

Parameter scaling for the rotational channel:

$$max\ input\ (x_{max}) = 3.14 \text{ rad/s} \tag{12}$$
$$\text{max output}\ (y_{\text{max}}) = 1.57 \text{ rad/s} \tag{13}$$

and the coefficients are:

$$s_0 = 0.785, \text{ for X, Y and Z}$$
$$s_1 = 0, \text{ for X, Y and Z} \tag{14}$$

Linear accelerations are limited to 6 m/s² on all axes for the translational channel, and the limitation for the rotational channel is 3.14 rad/s. The structure of the aircraft simulation system, based on a robotic platform motion system, is depicted in the figure below. Operator (pilot) inputs are first directed to the aircraft dynamic model, which generates the aircraft state vector. The aircraft state vector is then processed through the motion algorithm, producing the desired motion cues and the states of the robotic platform that provide the motion to the simulator. Subsequently, the desired states of the robot platform are transformed from the degree of freedom space to the axis workspace, generating commands for the six robot axes. These axis motion commands are used as input for the robotic platform's dynamics, resulting in the actual motion of the simulator.

Fig. 10. RoFSim simulator structure

The primary concept of the CWF is to replicate specific forces and rotations, simulating sensations similar to those experienced in a real cockpit. The filter's quality is assessed by minimizing the „feel" error, which reflects the difference between the specific forces and rotations perceived from the simulator and those of the aircraft. The CWF algorithm includes three channels: translational, rotational, and tilt coordination, as shown in Figure 11.

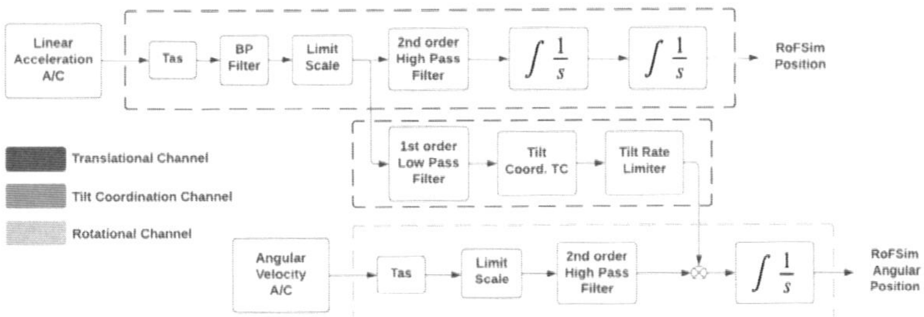

Fig. 11. Motion Perception Structure for RoFSim simulator

A. Translational Channel

The translational channel manages linear acceleration by transitioning from the aircraft reference system (S_A) to the RoFSim simulator reference system (S_S). A band-pass filter is applied, where values less than a user-defined threshold (0.06) are set to zero. This is followed by the limiting and scaling phase, explained in the previous chapter, ensuring RoFSim simulator movement accuracy to within 1 mm.

To address the robotic platform's inability to replicate low-frequency linear accelerations, a high-pass filter is used. The filter takes the form of:

$$y = \alpha \cdot (y_o + x - x_0) \tag{15}$$

where: y is current output, $\alpha = 0.95$, y_o is previous output, x is current input, x_0 is previous input.

When the acceleration input remains constant for a period, the robotic platform gradually returns to the zero position, providing a stable starting point for the next acceleration. The subsequent step involves double integrating the filtered accelerations to calculate the position in three directions: *X*, *Y*, and *Z* of the RoFSim simulator:

$$s = s_o + v_o \cdot t + a \cdot \frac{t^2}{2} \tag{16}$$

where: s is final position, s_o is initial position, v_o is initial velocity, a is acceleration and t – time.

B. Tilt Coordination Channel

The Tilt Coordination channel converts accelerations into Euler angles, allowing the pilot to perceive gravitational force as translational acceleration. This transformation is achieved through a transfer function that switches from accelerations to angles.

$$\phi = arcsin\left(\frac{X}{g}\right), \text{ where } g = 9.803\ \frac{m}{s^2} \tag{17}$$

$$\theta = arcsin\left(\frac{-Y}{g \cdot cos\theta}\right)$$

$$\psi = 0$$

where ϕ, θ, ψ represent roll, pitch and yaw.
Input received from the translational channel is passed through a low pass filter:

$$y = y_o + \alpha \cdot (x - x_0) \quad (18)$$

where: y is current output, $\alpha = 0.05$, y_o is previous output, x is current input and x_0 represents previous output.
A rate limit of 0.2 rad/s is applied to restrict rotational speeds. This ensures that the pilot does not perceive rotation but only experiences the gravitational force decomposed into the two axes (roll and pitch), allowing the pilot to feel linear accelerations.

$$\begin{aligned} &if\ |x_1 - x_0| > limit \\ &\qquad if\ x_0 < x_1 \rightarrow x_0 = x_0 + limit \\ &\qquad else\ x_0 = x_0 - limit \\ &else\ x_0 = x_1 \end{aligned} \quad (19)$$

where: $limit = 0.2$, x is current input and x_0 represents previous input.

C. Rotational Channel

The rotational channel is directly responsible for the rotational movements of the robotic platform. This channel takes the angular velocities of the aircraft as its primary input and the Euler angles from the tilt coordination channel as its secondary input.
In the first phase, the input is limited and scaled as described above. Then, a high-pass filter, identical to the one used for the translational channel, is applied. The resulting output after filtering is summed with the output received from the tilt coordination channel.
The final step involves integrating the angular velocities to obtain the angular position of the RoFSim simulator.
The CWF algorithm described above was implemented in the C# programming language. In the graphical user interface of the RoFSim Controller module, both inputs (linear accelerations, angular velocities) and outputs (*X, Y, Z* position, and angular position of the RoFSim simulator) are displayed in real-time in graphical format.

5. Materials and methods

The flight plan for the final testing of the CWF algorithm includes taking off from Henri Coandă Otopeni International Airport - LROP, runway 08R, following the VFR flight procedure, with a climb to 2500 ft, and then making a 90° turn with a compass heading of 170.
During the testing and validation phase of the CWF algorithm, the MTi-G-710-2A8G4 inertial measurement system (IMU) was positioned in the subject's (pilot's) head area. The collected data include linear accelerations and Euler angles. Various parameters were recorded and are presented in Tables 2 to 4.

Parameters IMU	**Unit**
Time	UTC
Acceleration X	m/s²
Acceleration Y	m/s²
Acceleration Z	m/s²
Roll	m/s²
Pitch	deg
Yaw	deg

Table 2. Parameters recorded by the IMU system

Parameters A/C Model	**Unit**
Time	UTC
Pitch	deg
Roll	deg
Acceleration world X	m/s²
Acceleration world Y	m/s²
Acceleration world Z	m/s²
Velocity Body X	feet/s
Velocity Body Y	feet/s
Velocity Body Z	feet/s
Acceleration Body X	m/s²
Acceleration Body Y	m/s²
Acceleration Body Z	m/s²
Plane Latitude	deg
Plane Longitude	deg
Plane Altitude	deg
Plane Heading Degrees True	deg
Incidence Alpha	deg
Incidence Beta	deg
Airspeed Mach	Mach
Airspeed True	Knots
Airspeed True Calibrate	Knots
Max G Force	G Force
G Force	G Force
Aileron Average Deflection	rad
Elevator Deflection	rad
Rudder Deflection	rad

Table 3. Parameters recorded from the dynamic model of the simulated aircraft

Parameters CWF	Unit
Time	UTC
Input Acceleration X	m/s^2
Input Acceleration Y	m/s^2
Input Acceleration Z	m/s^2
Robot Input Acceleration X	m/s^2
Robot Input Acceleration Y	m/s^2
Robot Input Acceleration Z	m/s^2
Robot Output Position X	m
Robot Output Position Y	m
Robot Output Position Z	m
Input Rotational Velocity X	rad/s
Input Rotational Velocity Y	rad/s
Input Rotational Velocity Z	rad/s
Robot Input Rotational Velocity X	rad/s
Robot Input Rotational Velocity Y	rad/s
Robot Input Rotational Velocity Z	rad/s
Robot Output Angle X	rad
Robot Output Angle Y	rad
Robot Output Angle Z	rad

Table 4. Parameters recorded from the CWF algorithm

6. Results

The test scenario was conducted over 20 flight sessions to validate the functionality of the CWF motion algorithm governing the RoFSim simulator's motion. Data from test number 8 were used to graphically display the simulation results.

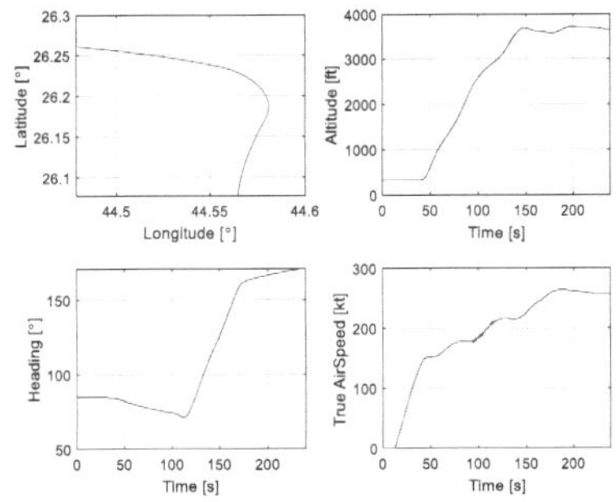

Fig. 12. Lat-Long; Heading, Altitude, TAS vs time

In order to validate the implemented motion algorithm, the linear accelerations of the robot and those recorded by the IMU system were compared, as well as the Euler angles (roll, pitch and yaw).

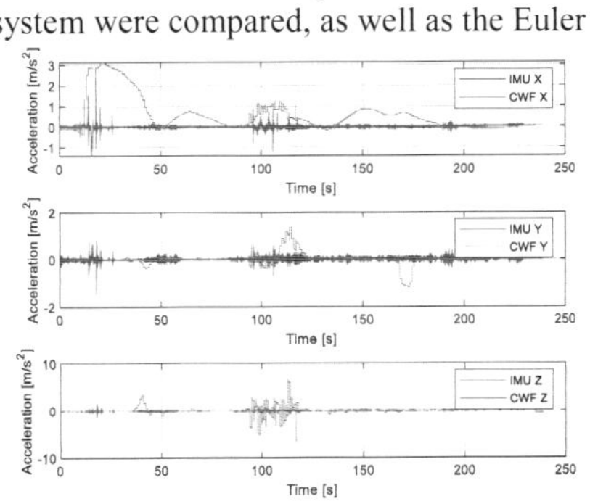

Fig. 13. Comparison of Linear Accelerations Recorded by IMU and CWF algorithm

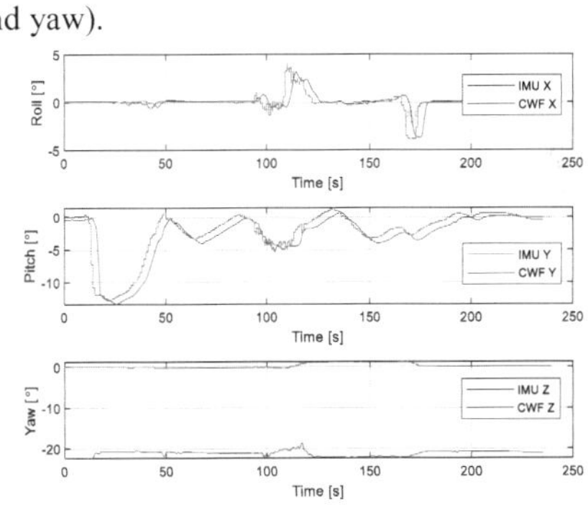

Fig. 14. Comparation of Euler Angles Recorded by IMU and CWF algorithm

7. Conclusion

In this paper, we implemented a washout filter in the RoFSim simulator to replicate sensations similar to those experienced by a pilot during real flight. Firstly, we devised a workspace scaling procedure for the RoFSim simulator to ensure it possesses an adequate workspace. We approached this as an optimization problem with three primary conditions: hardware joint angle constraints, collision avoidance, and software-imposed joint angle restrictions. For each of these issues, we identified constraints and implemented appropriate measures to ensure the simulator's safety and integrity. We made modifications at multiple levels, including software adjustments using RobotStudio and programming in C#/Rapid languages. These modifications underwent rigorous testing and validation through both offline and real-time simulations. Additionally, we developed and implemented a motion algorithm aimed at providing realistic motion feedback to the pilot within the cockpit while adhering to the specified limits of the robotic motion platform. This motion algorithm, referred to as the Classical Washout Filter, underwent testing in a dedicated campaign.

Its functionality was confirmed as it successfully replicated the motion of the RoFSim simulator within the confines of its workspace, thereby offering the pilot a faithful perception of motion similar to what is experienced in a real cockpit. To validate the motion algorithm, a total of 20 flights were conducted in the test campaign. This validation process involved comparing the data recorded by the IMU system with the parameters recorded by the motion algorithm. Graphical representations of the results clearly demonstrate that the simulator behaves accurately, and the motion algorithm effectively conveys motion cues to the pilot, closely resembling those experienced in a real cockpit. The extensive series of tests performed underscores the stability of the RoFSim simulator and confirms the effectiveness of the developed configuration. As part of our future work, we aim to enhance the motion algorithm model to achieve a higher fidelity in motion indices by implementing an optimal algorithm.

8. Acknowledgements

The work has been funded by the Romanian Government through the Ministry of Research, Innovation and Digitisation (MCID), under the "AEROEXPERT 2019-2022" program, financial agreement no. 8N/2019.

9. References

[1] Brain, C.J.; Clayton, T.D.; Ward, N.J. (1996). Organizing safety-related aircraft simulation. Presented at: 2nd Test and Evaluation International Aerospace Forum; London, UK, DOI: 10.2514/6.1996-3331

[2] Allerton, D.J. (2010). The impact of flight simulation in aerospace. The Aeron Journal 114(1162):747-756

[3] Mendonça, C.B.; Silva ET, Curvo M, Trabasso LG (2013). Model-based flight testing. Journal of Aircraft 50(1):176-186, DOI: 10.2514/1.C031778

[4] Afloare, A.I.; Apostolescu, N.; Chira, A.I.; Munteanu, C.E. (2019). Inverse Kinematic Solution of a 6 DoF Serial Manipulator, Proceedings of the 30th DAAAM International Symposium, pp.0628-0631, B. Katalinic (Ed.), Published by DAAAM International, ISSN 1726-9679, Vienna, Austria

[5] ABB Robotics - Product specification IRB 7600, Available from: https://new.abb.com/products/robotics/robots/articulated-robots/irb-7600, Accessed: 2023-10-02

[6] Grant, P. R.; Reid, L. D. (1997). Motion washout filter tuning: Rules and requirements, Journal of Aircraft, 34(2), 145–151, DOI: 10.2514/2.2158

[7] Parrish, R. V.; Dieudonne, J. E.; Bowles, R. L.; Martin Jr, D. J. (1975). Coordinated adaptive washout for motion simulators, Journal of Aircraft, 12(1), 44–50, DOI: 10.2514/3.59800

[8] Sivan, R.; Ish-Shalom, J.; Huang, J. (1982). An optimal control approach to the design of moving flight simulators. IEEE Transactions on Systems, Man, and Cybernetics, 12(6), 818– 827, DOI: 10.1109/TSMC.1982.4308915

[9] Song, C.C.; Liaw, D.C.; Chung, W.C. (2002). Washout-filter based bifurcation control of longitudinal flight dynamics. IEEE Region 10 Conference on Computers, Communications, Control and Power (pp.1646–1649), Beijing, China

[10] Wang, S. C.; Fu, L. C. (2004). Predictive washout filter design for VR-based motion simulator, IEEE International Conference on Systems, Man and Cybernetics, SMC 2004, 6291–6295, The Hague, Netherlands

[11] Affan, M.; Ahmed, S. U.; Manek, A. I.; Uddin, R. (2019). Design and implementation of the washout filter for the Stewart-Gough Motion Platform, 2019 International Conference on Computational Intelligence and Knowledge Economy, (pp. 415–419), Dubai, UAE

[12] Liao, C.S.; Huang, C.F.; Chieng, W.H. (2004). A novel washout filter design for a six degree-of-freedom motion simulator, JSME International Journal Series C Mechanical Systems, Machine Elements and Manufacturing, 47(2), 626–636

[13] Asadi, H.; Mohamed, S.; Lim, C. P.; Nahavandi, S. (2016). Robust optimal motion cueing algorithm based on the linear quadratic regulator method and a genetic algorithm, IEEE Transactions on Systems, Man, and Cybernetics: Systems, PP (99), 1–17

[14] Gharib, M. R. (2020). Comparison of robust optimal QFT controller with TFC and MFC controller in a multi-input multioutput system, Reports in Mechanical Engineering, 1(1), 151– 161

[15] Huang, C.; Fu, L. (2006). Human Vestibular Based (HVB) Senseless maneuver optimal washout filter design for VR-based motion simulator, IEEE International Conference on Systems, Man and Cybernetics, pp. 4451–4458, Taipei, Taiwan

[18] Moavenian, M.; Gharib, M. R.; Daneshvar, A.; Alimardani, S. (2011). Control of human hand considering uncertainties, International Conference on Advanced Mechatronic Systems, pp. 17–22, IEEE

[19] Nahon, M. A., Reid, L. D., & Kirdeikis, J. (1992). Adaptive simulator motion software with supervisory control. Journal of Guidance, Control, and Dynamics, 15(2), 376–383

[20] Telban, R., Cardullo, F., & Houck, J. A. (2002). Nonlinear, human-centered approach to motion cueing with a neurocomputing solver. In AIAA Modeling and Simulation Technologies Conference and Exhibit, pp.1–10, Monterey, CA, USA

DOI: 10.2507/34th.daaam.proceedings.026

DESIGN OF AN EXPERIMENTAL MODEL OF A SYSTEM FOR PRODUCT SORTING THROUGH IMAGE PROCESSING

Vlad Gheorghiță & Florin Adrian Geantă

This Publication has to be referred as: Gheorghita, V[lad] & Geanta, F[lorin] A[drian] (2023). Design of an Experimental Model of a System for Product Sorting Through Image Processing, Proceedings of the 34th DAAAM International Symposium, pp.0203-0208, B. Katalinic (Ed.), Published by DAAAM International, ISBN 978-3-902734-41-9, ISSN 1726-9679, Vienna, Austria
DOI: 10.2507/34th.daaam.proceedings.026

Abstract

During the production process, a lot of undesirable or defective things are also produced. Since they cannot be used or distributed on the market, these things need to be weeded out. In modern industrial applications, checking the integrity and quality of the product is one of the most important procedures, being also the result of the optimization and efficiency of the production process. The automatic product verification and sorting system is the fastest and most efficient system that achieves compliance with the quality standards of one or more products. These systems are classified as critical in some applications because replacing them with human personnel requires a large number of employees and related costs. This paper presents an experimental model with the help of electronic and 3D printed components that through image processing will manage to sort different kind of products.

Keywords: Product design; image processing; sorting; quality; LabVIEW

1. Introduction

Object sorting is one of the most important phases in industrial production lines. During the production process, a lot of undesirable or defective things are also produced. Since they cannot be used or distributed on the market, these things need to be weeded out. It takes a lot of time, money, accuracy, and effort to manually classify those inappropriate commodities. With the use of machine vision, this task can be done with greater accuracy in less time and at a lower cost. However, technological problems could appear infrequently.

Both cameras and software for machine vision are freely accessible. Due to the effectiveness of the programme, automated inspection has already found applications. The user merely needs to provide inputs to the software. However, users are free to create their own programmes. A machine vision application is analogous in theory and application [7]. Applications can also be created to become more and more crucial as a user's experience grows. There are optical devices to detect and measure various characteristics of objects, such as size, shape, color, imperfections: X-rays that uses electromagnetic radiation to check defects that are difficult to detect by the human eye; ultrasound uses sound waves that are sent through the material and returned in case of detection of an imperfection or gaps encountered; video camera, the simplest, use the optical sensor to capture the image that can be analyzed later for various defects.

The paper [1] presents the application of a 4DOF fruit sorting robot based on color and size in a packaging system. The sorting is made possible by image processing where color is recognized by HSV analysis, and the diameter is known in the grayscale image and setting the thresholding. An algorithm for indirect egg defect recognition using image processing and also a developed graphical user interface for recognizing defective eggs in the MATLAB environment, based on computer vision are proposed in [2]. The values of the egg samples are collected by image processing and HSV and YIQ color spaces are used for egg defects recognition.

Hafiz [3] proposed a thorough analysis of recent work on the automated grading and sorting of agricultural products. On the basis of image processing, a complete end-to-end system for automatic and effective fruit sorting and grading is provided. The preliminary experimental results demonstrate the proposed structure's efficacy. Two CCD cameras, two capture cards, a suitable lighting system, and other mechanical components make up the system designed for fruit sorting that is presented [4]. The fruit is initially removed from the backdrop by the algorithm. In front of the cameras are samples of various lemon grades that have been calibrated offline. The retrieved data is then saved in a database along with estimates of the fruit quantities and HSI color values. The final grade of the passing fruits is calculated by comparing the data collected during the sorting step with the data that is present inside the database.

To contribute to a better understanding what and how is changing in manufacturing in the era of AI, the paper [9] describes five innovation challenges that will, in the view of the authors, shape the future of manufacturing. Focusing on changes in innovation focus, importance of data, need for increased collaboration, impact of customer experience and shortage of talent, valuable insights are offered on how manufacturing companies should approach innovation challenges in the next decade. With the help of the images the camera takes, the software created [6] will run algorithms to sort the objects based on their attributes and save the locations of the sorted objects in a database. The data from the sorted objects may then be used for pick-and-place operations that work in tandem with the actuator controller. The steps in the sorting process include detecting the object, figuring out color, size, and shape, finding the sorted object, and calculating the actual gripping position.

In this paper the purpose was sorting a group of 3D printed toothed gears through image analysis according to their characteristics (number of teeth, outer diameter, bore diameter, shape and degree of tooth damage). The system used for sorting is created with the help of 3D printing at a low cost. The technical and dimensional characteristics are taken from the execution drawing of a toothed wheel that is part of an assembly with a pinion and a worm coupling.

2. Materials and Methods

The control of the characteristics of a piece in the field of machine technology has a critical importance. With the purpose of designing a simple and efficient system at a low cost, the paper presents a solution for product sorting. The system can be used for various components in the manufacturing industry, by only changing the image processing algorithm and functions for color analysis or particle analysis can be implemented. For prototyping the sorting system, a versatile additive manufacturing technology was chosen, capable of using biodegradable plastics and allowing parts obtainment which are feasible to be used in a prototype, FDM technology (Fused Deposition Modeling). For the 3D printing Ultimaker S5 3D printer was operated using the above-mentioned technology with a filament of 2,85 mm made out of PLA (polylactic acid) and a precision of ± 0,1 mm. The technological parameters used were the printing temperature of 220°C, the printing speed of 40 mm/s, the filling density of 100% and the deposition angle of 0°.

The components of the proposed system are: 1. The DC motor, 2. The DC motor driver, 3. 12V power supply, 4. The servo motor, 5. Stroke limiter, 6. Structural framework, 7. Forward arm, 8. Redirected Product, 9. Video camera, 10. Belt stretching system, 11. The ultrasonic sensor, 12. The Arduino microprocessor, 13. Conveyor belt. The belt stretching subsystem has the following main components: the sliding screw, the nut on the right side, the bearing inside it, the shaft from the cylinder that sets the system in motion, the support on which the screw slides. Thus, when the nut is tightened, the screw moves forward with the shaft, respectively the belt (fig. 1). Some limitations of the system are given by the dimensions of the conveyor belt: length 800 mm and width 400 mm.

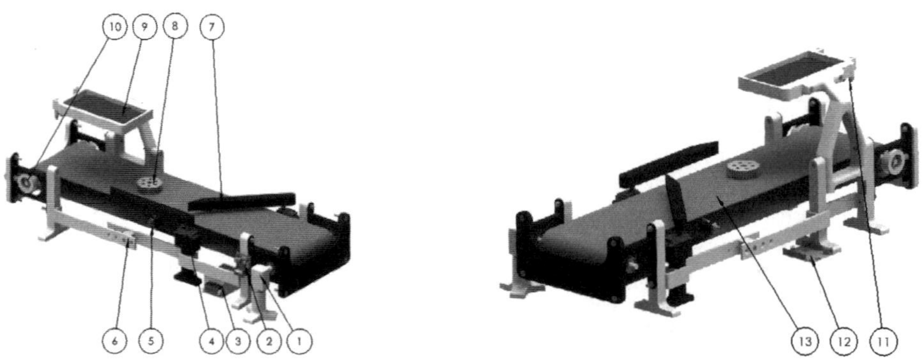

Fig. 1. 3D model of the experimental system

The camera used is ESP32, a system-on-chip microcontrollers with integrated Wi-Fi. The sorting designed works thus: the product is placed on the conveyor belt, when the sensors detect the presence of a part, the camera takes a picture and then the image is processed through NI Vision Assistant and LabVIEW programs The product is redirected from the conveyor according to the signal given by the applications as can be seen in fig. 2.

Fig. 2. Separation subsystem

The separation subsystem redirects the products from the belt in two distinct directions through a redirection arm that is actuated by a servo motor. When the product is removed from the belt, due to the weight of the product and the flexibility of the belt, it will actuate a limiter that will signal the servomotor to bring the arm to its initial position. The conveyor motion subsystem is formed by: motor, coupling between the motor shaft and the cylinder axis and the cylinder that sets the system in motion. Vision Assistant is a tool for prototyping and testing image processing applications. The Vision Assistant scripting functionality was used to create an algorithm for image processing. Every stage of the processing algorithm is documented through the scripting feature. Through image processing, a product in this case a gear wheel is checked from the point of view of the diameter, the number of teeth, the number and size of the bores. A digital image can be obtained in many ways, such as a sequence of images or an image at a certain time [4], in the graphical programming environment LabVIEW with different installed modules (Vision Acquisition, Vision Assistant). The algorithm for analyzing the images for the products starts from the product image (toothed wheel) and follows a series of functions to process the image and the export the data to LabVIEW software.

In order to distinguish the product from the background more easily, a suitable plane is chosen based on how well it is highlighted in the image. After a plane has been extracted, the image is used for subsequent tasks like particle classification or fundamental morphology [8]. The most crucial phase in this kind of non-contact analysis is image acquisition. The creation of a space with the right lighting has the biggest impact on the purchase process. The photos are acquired using a smart vision camera, which digitises and processes the images in real time. Application data monitoring is done using the LabVIEW programme. Before picking the best algorithm for measuring or inspecting components, the advantages of each image processing technique must be weighted.

The application for image acquisition and processing uses the image acquisition function from the IMAQdx library from LabVIEW and writes this image into a new PNG file using the IMAQ Write File function. The image reaches the image processing program through the Express Vision Assistant and has as output the data retrieved after the image processing. In NI Vision Assistant, the script was made using functions for Color Plane Extraction, Image Calibration, Threshold, Set Coordinate System and Find Edge [5]. The data are exported and received as in fig. 3, but there are also data that need processing to reach the information relevant to the application.

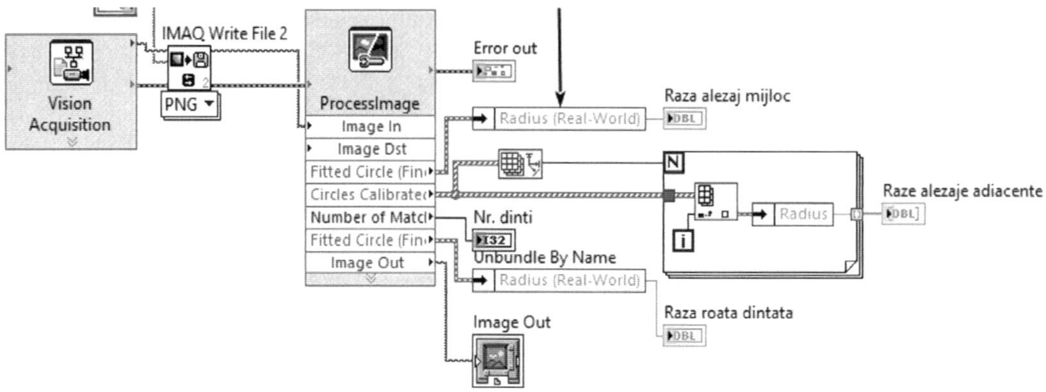

Fig. 3. LabVIEW application

The gear wheel radius is checked through a comparison with two manually entered values that are taken from an array and it is checked if the radius falls between the two values. Thus the radius may have a dimensional tolerance entered manually. The radius of the middle bore is checked similarly. The number of teeth does not need a margin of error and thus a simple check is made between what is received from image processing and what is entered manually. The process of checking the adjacent bores is a little more complex, the data taken from the image being some arrays that must be checked individually (fig. 4).

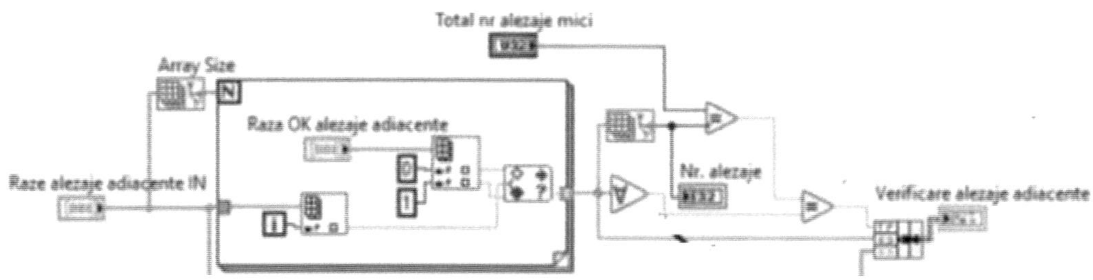

Fig. 4. Checking the characteristics of the analyzed product

The programming of the electronic components and the control of the motors were also done with the help of the LabVIEW program. For the DC motor control, the entries are used by the Digital Write function and set to N channel, the motor is considered being a switch that builds an array with its value and the negation of this value. Another array shows the pins that the driver uses to control the sense. The PWM Set Duty Cycle function controls the motor speed and is used by pin 5 on the Arduino board. At the top, a button deactivates the switch for motor direction if the speed is greater than 0, thus the direction cannot be changed during operation (fig. 5).

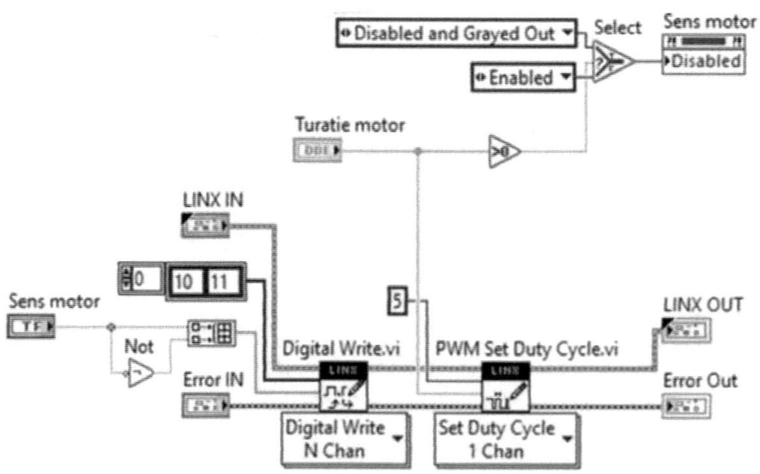

Fig. 5. VI for the control of DC motor

The part detection program presented in fig. 6 uses LINX's Read function from LabVIEW to read the distance from the ultrasonic sensor. It has as inputs the session, error and the pins used by the sensor. The output shows the session, the error and the distance in centimeters that is converted into millimeters, and if this distance is less than 80 millimeters, the program detects the part after it waits 8 seconds to take the photo.

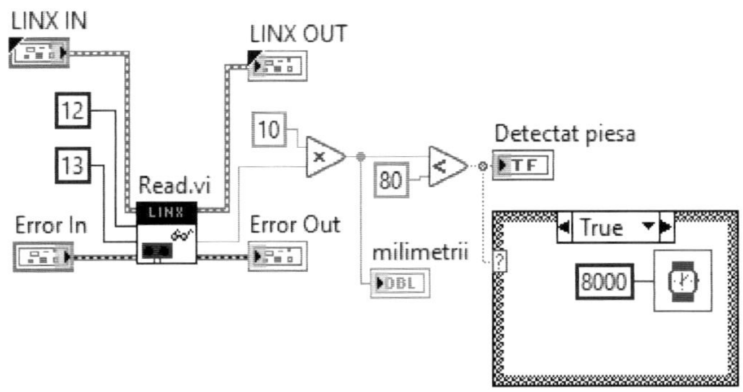

Fig. 6. Product detection algorithm

The results of the sorting process are then saved and converted into signals that are used to communicate with the actuator control system to carry out the sorting tasks. The control of the servomotors is done with the help of a sequential structure that performs in 3 steps: the arm is set at 45 degrees, the product is expected to leave the belt and the program will realize that it has done so when the system activates the stroke limiter. The While structure closes and it can move to the last step: the arm is reset and repositioned to the initial state, 0 degrees (fig. 7).

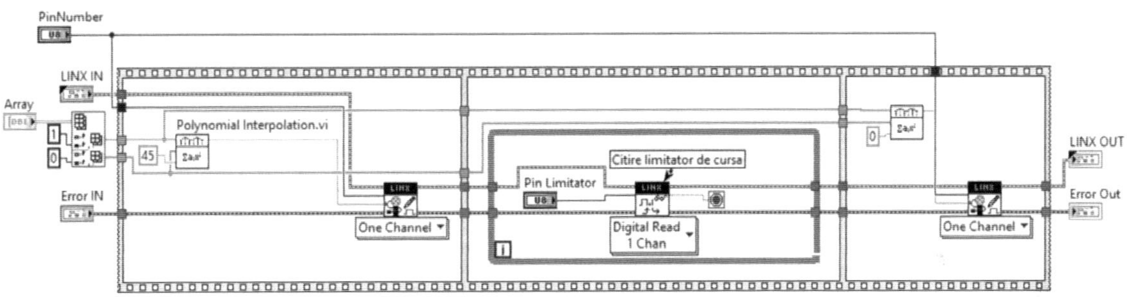

Fig. 7. Servomotor control

All the programs previously presented as subVIs with modified icons are integrated in the main VI in LabVIEW that is used for sorting the products, as suggested in fig. 8.

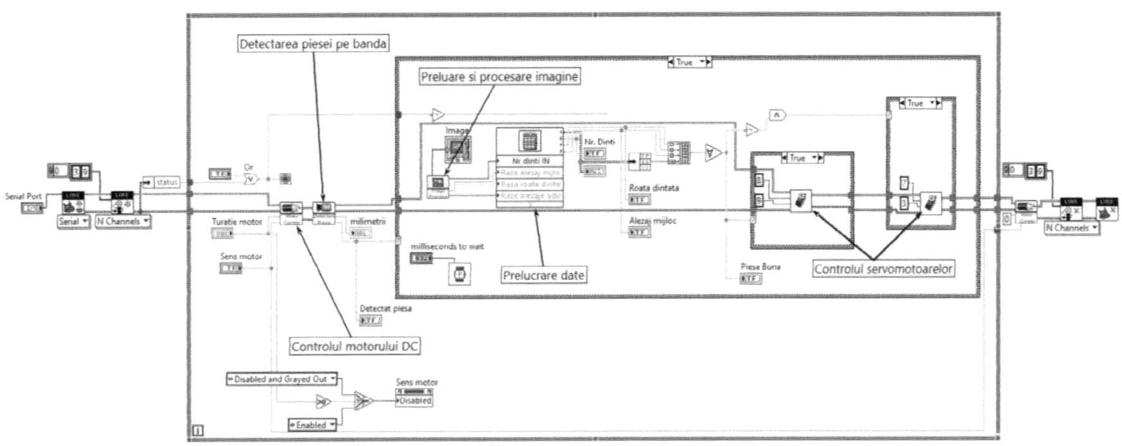

Fig. 8. Main VI in LabVIEW

3. Conclusion

The disadvantages of current commercial sorting systems are their high cost, complexity, dimensions and maintenance because they require specific equipment to function properly. This paper presents an experimental model of a system for product sorting using image analysis and developed graphical user interface for sorting in the LabVIEW environment, based on computer vision. The user interface is modular, allowing to upgrade the processes and algorithms used.

In order to ensure the highest possible functionality and reduce errors during system operation, the following key points of the system were analyzed and tested: constant engine speed to ensure that everything will run optimally; establishing an optimal speed for running the system; development of the algorithm for image processing, which will result in more accurate identification; training the recognition algorithm with as many qualitative images as possible; commanding the servo motor based on the information generated by the algorithm; detection of the shape and position of the product; determining the optimal position of the camera in front of the conveyor belt; the possibility that the algorithm does not recognize the object; finding a solution for the case of non-identification. This type of system can be programmed to perform various tasks, increasing versatility and it can gather useful information while operations are running, giving important insights to improve operations and make data-driven decisions. The future research plans are represented by testing the system and algorithms on various products to reduce the response time and by using the system to identify and separate waste with distinct textures.

4. References

[1] Dewi, T.; Risma, P.; Oktarina, Y. (2020). Fruit Sorting Robot based on Color and Size for an Agricultural Product Packaging System, Bulletin of Electrical Engineering and Informatics Vol. 9, No.4, pp. 1438-1445, ISSN: 2302-9285, DOI: 10.11591/eei.v9i4.2353

[2] Georgieva, T; Stefanov, E; Alikhanov, J; Shynybay, Z; Kulmakhambetova, A.; Daskalov, P. (2019). Approach for Egg Defects Assessment Using Image Analysis, Proceedings of the 30th DAAAM International Symposium, pp.1102-1106, B. Katalinic (Ed.), Published by DAAAM International, ISBN 978-3-902734-22-8, ISSN 1726-9679, Vienna, Austria, DOI: 10.2507/30th.daaam.proceedings.154

[3] Hafiz, M.T.; Shakoor, U.; Khan, M.J. (2019). Automated Sorting and Grading of Agricultural Products based on Image Processing, 8th International Conference on Information and Communication Technologies (ICICT), Date of Conference: 16-17 November 2019, DOI: 10.1109/ICICT47744.2019.9001971

[4] Khojastehnazhand, M.; Omid, M.; Tabatabaeefar, A. (2010). Development of a Lemon Sorting System based on Color and Size, African Journal of Plant Science Vol. 4 (4), pp. 122-127, ISSN 1996-0824

[5] Klinger, T. (2003). Image Processing with LabVIEW and IMAQ Vision, Prentice Hall, ISBN 9780130474155, United States of America

[6] Phuoc, T.; Truong, T.N.; Huy B.N. (2016), Design and Development of the Vision Sorting System, 2016 3rd International Conference on Green Technology and Sustainable Development (GTSD), DOI: 10.1109/GTSD.2016.

[7] Ready, S.; Kwon, K.S. (2015). Practical Guide to Machine Vision Software, Wiley VCH, ISBN: 978-3-527-33756-9, Germany

[8] Relf, C. (2003). Image Acquisition and Processing with LabVIEW (Image Processing Series), CRC Press, ISBN 9780849314803, United States of America

[9] Tekic, Z.; Cosic, I.; Katalinic, B. (2019). Manufacturing and the Rise of Artificial Intelligence: Innovation Challenges, Proceedings of the 30th DAAAM International Symposium, pp.0192-0196, B. Katalinic (Ed.), Published by DAAAM International, ISBN 978-3-902734-22-8, ISSN 1726-9679, Vienna, Austria, DOI: 10.2507/30th.daaam.proceedings.025

DOI: 10.2507/34th.daaam.proceedings.027

ANALYSIS OF EXISTING DESIGN TEMPERATURE FOR HEATING IN SARAJEVO IN LIGHT OF CLIMATE CHANGES

Rejhana Blažević, Armin Teskeredžić & Melisa Zečević

This Publication has to be referred as: Blazevic, R[ejhana]; Teskeredzic, A[rmin] & Zecevic, M[elisa] (2023). Analysis of Existing Design Temperature for Heating in Sarajevo in Light of Climate Changes, Proceedings of the 34th DAAAM International Symposium, pp.0209-0218, B. Katalinic (Ed.), Published by DAAAM International, ISBN 978-3-902734-41-9, ISSN 1726-9679, Vienna, Austria
DOI: 10.2507/34th.daaam.proceedings.027

Abstract

Designers of heating systems and planners of the district heating consider large number of variables that influence the final proposal of the system which will meet the heating needs. Size of the system and investment costs depends on different parameters but are mostly influenced by the outside design temperature. Engineers are aware of the climate change influence but during design process they need either legal or scientific proof that they can use new/updated values of the design temperatures at the specific location. In this paper, an analysis of hourly air temperature values for the city of Sarajevo, Bosnia and Herzegovina (B&H) was conducted in the latest available 20 years dataset. Data on the outside air temperature were obtained by the Federal Hydrometeorological Institute of Bosnia and Herzegovina. Five different methods were briefly explained and used for determination of the new design temperature for Sarajevo based on the period 2001–2020. Results of the analysis demonstrate that the currently valid and official outside design temperature was low and needs to be revised. As a small-scale example, calculation of the heating needs of one residential unit was made in order to demonstrate the influence of different design outside temperature on the heating needs. These numbers were then extrapolated on the district heating system and the benefits of the proposed approach were underlined. Performed analysis suggest an urgent change in the design temperature for all cities in Bosnia and Herzegovina as it was demonstrated on the example of Sarajevo in this paper.

Keywords: outside design temperature; heating system; heat load; climate change;

1. Introduction

The negative impact of man on the environment is increasing, and global warming has become an everyday topic. In the past 100 years, and especially the last twenty, the average global temperature of the Earth's surface has increased significantly. The period from 2011 to 2020 is the warmest decade since outside temperature measurements have been made [1]. Global warming, which is defined as the warming of the earth due to human activities is one of the biggest problems today [2], [3]. The increasingly frequent weather disasters are precisely the result of climate change (extremely hot summers with more and more hot days and heat strokes, fires due to dry soil, hurricanes and tornadoes in the Western Hemisphere, floods due to heavy rains) [3], [4].

Considering that the human factor has played a major role in the deterioration of the climate, countries together with world organizations are working to reduce the human impact in the future. An increasing number of countries are turning to renewable energy sources and shutting down fossil fuel power plants with the aim of reducing greenhouse gas emissions into the atmosphere. Recent research has shown that if the global temperature were to rise by 2°C by 2050, it would have a terrible impact on agriculture and people's lives in general, and all the previous deteriorations would become even more pronounced and would make life on Earth more difficult. Therefore, the current trend is towards an increase of 1,5°C in order to reduce the impact of climate change [5], [6]. Data for the European Union (EU) show that around 40% of the total final energy consumption goes to the building sector [7]. In terms of CO_2 emissions, residential and commercial buildings account for 36% of total emissions. Energy is used in households for various purposes (hot water, cooking, household appliances), however 70% of the total energy consumption in buildings is used for heating [8].

The climate varies from place to place and directly affects the investment and exploitation costs of the plant, so it is important for the designer to know the outside design parameters, in order to be able to take the influence of the climate into account in an appropriate way. Many studies have investigated the various possible effects of climate change, from rising outside temperatures to changes in precipitation. Spinoni at al. [2] in their study investigated whether energy demand for cooling and heating buildings can be expected to increase or decrease under climate change. For this purpose, they considered two indicators of weather-related energy consumption for heating and cooling buildings: heating degree-days and cooling degree-days. The results they obtained show a projected general decrease in heating degree-days over Europe and increase in cooling degree-days. Vucijak et al. [9] presented practical usability of the one of the statistical process control methods for the evaluation of the climate change effects to the precipitation patterns.

They compared the precipitation patterns from the 1961-1990 reference period with the precipitations from the 2000-2010 period in the Bihać area in Bosnia and Herzegovina. As a conclusion they stated that the monthly precipitation patterns have changed, but the observed changes in the total monthly precipitation are statistically expected. Different studies, for different locations, have shown a constant decrease in the number of degree-days for heating. For example, Cvitan et al. [10] analysed climate change for the area of Croatia from the aspect of changing needs for heating and cooling indoor living spaces. As a result of the research, they found a decrease in the need for heating and an increase in the need for cooling in Croatia in the period 1901-2008. Zaninovic et al. [11] analysed in their work relevant meteorological parameters in the design of thermal protection of buildings for 15 locations in Croatia for the 1961 - 1990 period in order to harmonize Croatian norms with norms in European countries. Popovic et al. [12] investigated climate change in Serbia as a result of global warming and assessed the future climate in accordance with the results of climate modelling according to the most used Intergovernmental Panel on Climate Change (IPCC) scenarios. Their assessments indicate that the territory of Serbia will be exposed to influences that may have consequences for the entire society. Muhamedagic et al. in their earlier work [13] announced the need to revise the outside design temperature. They analysed temperature values for the period from 2001 to 2010 and calculated outside design temperature by using only one method.

In this paper, the influence of climate change on the outside design temperature in Sarajevo, and because of that on the required annual energy for heating the interior space, is analysed. For determining new outside design temperature, data on hourly temperature values in Sarajevo for the period from 2001 to 2020, which were obtained by the Federal Hydrometeorological Institute for the Bjelave measuring station, were used. Different approaches for determining the outside design temperature for the winter period were considered, and it was shown that the values of the outside design temperature are highly dependent on the methodology by which they are calculated. The investment value of the building and the power of the central heating system depend on the outside design temperature, because the amount of heat is directly proportional to the difference between the indoor and outside design temperature. The calculated amount of heat required to heat the building directly affects the selection of heating plant equipment (boilers, pipelines, pumps), and reliable determination of the outside design temperature is of great importance when it comes to properly sizing the heating system.

2. Indicators of climate change in Bosnia and Herzegovina

At the beginning of this millennium, the average increase in the mean annual temperature in the territory of Bosnia and Herzegovina was 0,6°C for the period of the last hundred years, and now that increase is more than 1°C. An example of a hundred-year time series shows that the impact of climate change on the climate of our country is evident. In the Strategy for adaptation to climate change and low-emission development of Bosnia and Herzegovina for the period 2020 - 2030 [4], it is stated that according to the analyses of meteorological data for the period 1961 - 2014, the average annual temperature maintains a continuous increase in the entire territory. A positive linear trend was observed in the average annual temperature, which was particularly pronounced in the last 30 years [14]. The increase in air temperature on an annual basis, ranges from 0,4 to 1°C, and during the growing season from April to September, up to 1,2°C. Annual temperature trends at all analysed stations are statistically significant, and the changes are more pronounced in the continental part. The most significant change in this period is observed in the number of cold days and the number of warm days. At all meteorological stations, the number of cold days has a negative trend [4]. In Figure 1 it is shown increase in average annual temperature in the last decade (1990-2000) compared to the reference period (1961-1990) in B&H expressed in °C.

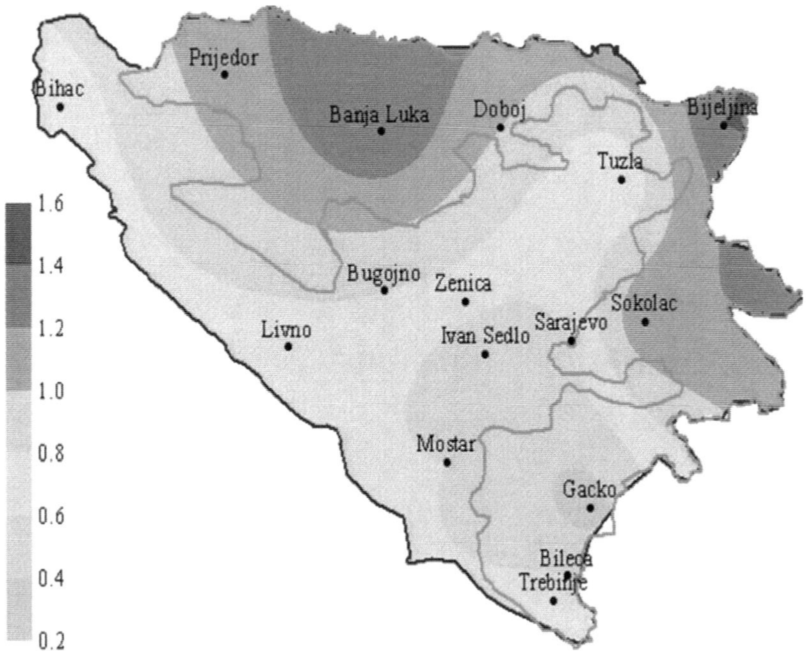

Fig. 1. Increase in average annual temperature in the last decade (1990-2000) compared to the reference period (1961-1990) in B&H expressed in °C [15]

Figure 2 shows the change in average annual outside air temperature for the twenty-year time period 2001-2020 for Sarajevo. An indicator of the severity of the impact of global warming is a linear increase in temperature.

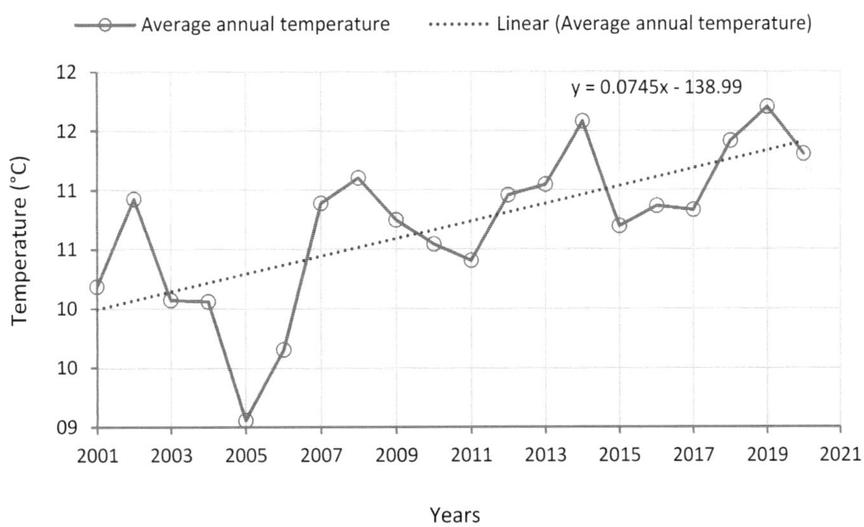

Fig. 2. Average annual temperature trend for Sarajevo in the period 2001-2020.

In addition to the increase in temperature, changes are also noted when it comes to the snow cover. Specifically, for Sarajevo, it is possible to monitor two parameters:

- number of days with snow (sleet, snow) and
- number of days with registered snow cover.

Figure 3 shows the data related to the number of days with snow and the number of days with a registered snow cover in Sarajevo for the period from 1992 to 2020. Also, a linear decrease in the value of the number of days with snow is shown.

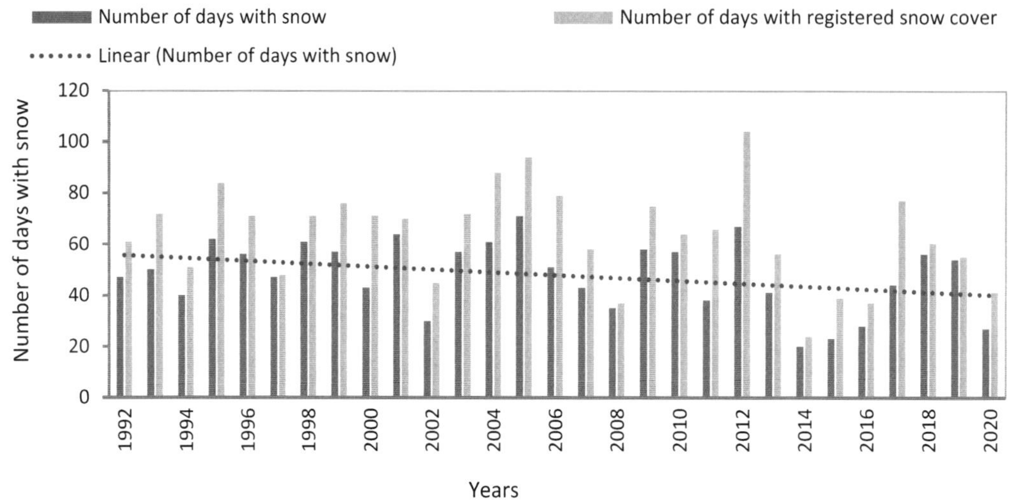

Fig. 3. Number of days with snow and number of days with registered snow cover for Sarajevo (1992-2020)

3. Methodology for determining the outside design temperature

There is no single way to define and determine the outside design temperature for now. It could conditionally be defined as "the lowest temperature of the outside air, at which the HVAC system ensures the pre-required parameters of thermal comfort inside the treated space, regardless of the duration of such condition of the outside air" [16], [17]. However, the lowest temperature that appeared in that place in an observed period is not taken as the outside design temperature of a place, because it occurs very rarely and lasts for a short time. The heating system, which would be designed based on such an absolute minimum, would be over-dimensioned, investment expensive and exploitation uneconomical, because it would very rarely work at full capacity. For the heating system not to be oversized, short-term lowering of the temperature in the building is allowed at extremely low outside temperatures. There are several methods for determining the outside design temperature. They are mainly based on the statistical processing of outside temperatures over a longer period, 10, 20 years or longer.

For the purpose of writing the paper and making calculations, data on hourly values of outside air temperature obtained from the Federal Hydrometeorological Institute for the period 2001-2020 were used. To calculate the outside design temperature, three different periods were analysed, namely two periods lasting 10 years and one period lasting 20 years in order to get a better general picture of the advantages and disadvantages of certain methods. The current value of the outside design temperature used in the design of the heating system for Sarajevo is -18°C and it was obtained according to B. M. Chaplin for a series of 30 years (1956-1985). However, due to global warming and climate change, it is to be expected that this value has changed. The methods used to calculate the new outside design temperature are described below.

3.1. B. M. Chaplin method

The most widespread method for determining the outside design temperature in the areas of the former Yugoslavia is the method named after the Russian scientist B. M. Chaplin. This method was used for determination of the existing and still valid outside design temperatures for all larger places in Bosnia and Herzegovina. According to the Chaplin methodology, the outside design temperature is determined depending on the average temperature of the coldest month in a period of 10 or 20 years (t_m) and the absolute minimum (t_{min}) from the same period, which should be the most current. The ratio of the participation of these two temperatures in the value of the outside design temperature, t_{ODT}, is given by the following expression [16], [17]:

$$t_{ODT} = 0{,}4t_m + 0{,}6t_{min} \quad (1)$$

where $t_m[°C]$ is the mean monthly temperature of the coldest month in the observed period (20 or 10 years) and $t_{min}[°C]$ is the absolute minimum temperature in the observed period (at least 10 years), that is, the lowest temperature that occurred in the observed time period. Table 1 shows the average monthly air temperatures for the analysed period. From Table 1 it can be seen that for the analysed time period the coldest month is January and that the value of the mean monthly temperature of the coldest month in the observed period is $t_m = 0{,}39$ °C.

	2001-2020											
Month	J	F	M	A	M	J	J	A	S	O	N	D
t_m (°C)	0,39	2,33	6,26	10,75	14,76	19,41	21,47	21,47	15,88	11,17	6,51	1,20

Table 1. Average monthly air temperature (2001-2020)

The second parameter required by the method is the absolute minimum, which specifically in this case means that it is necessary to find the lowest temperature that occurred in the given period (for a series of 20 years, the lowest temperature occurred on January 8, 2017 at 6:00 AM with a value -22.2°C, and it lasted for only an hour). By including in expression 1, the outside design temperature for Sarajevo for the period 2001-2020 is obtained, which is -13,16°C.

3.2. DIN 4701, 1955

This approach to determining the outside design temperature, as well as the ASHRAE approach, starts from the duration of the outside temperature or the frequency of its occurrence. According to DIN 4701, the outside design temperature is the lowest two-day mean value of the outside air temperature that has been reached or failed ten times in a period of twenty years, or in average once every two years [18].

3.3. NOVAK triad

In 1987, the scientist Novak together with his collaborators has developed a method for the calculation of the outside design temperature for the needs of the former Yugoslavia, which is why this method is also called JUS U.J5.600. This method defines the outside design temperature as the mean value of the coldest triad, that is, the coldest three consecutive days of the year for the time series of the considered years [17], [19].

3.4. ASHRAE

The ASHRAE (American Society of Heating, Refrigerating and Air-Conditioning Engineers) approach for determining the outside design temperature is based on the length or duration of the outside temperature. Therefore, according to American regulations, the outside design temperature is defined as the lowest temperature that occurred for a certain number of hours or days at the observed location. This approach considers the percentage of occurrence of hourly air temperature values above some reference value for a long period of time. The first temperature value that reaches or exceeds the defined threshold (reference value) is considered the outside design temperature. This method is widely accepted in America. According to the American regulations (ASHRAE), there are two different approaches, namely, the old approach that observes the three coldest months of the year (December, January and February, i.e., 2160 h/year) and the new approach that takes into account the whole year during the calculation (8760 h/year) [20], [21].

According to the old ASHRAE approach, there are two different criteria: 99% and 97,5% of hourly temperature values during the 3 coldest months (December, January and February, which is a total of 2160 h/year) when the outside air temperature is equal to or higher from the reference. This practically means that according to the 99% criterion, only in 1% of cases is it allowed for the outside air temperature to be lower than the outside design temperature, i.e., can be overcome in an average of about 22 h/year. The 97,5% criterion is somewhat milder compared to the 99% criterion, and its application results in the outside design temperature being 1 to 2°C higher than the value of the temperature obtained according to the 99% criterion. According to the 97,5% criterion, the outside air temperature will be exceeded on average about 54 h/year, i.e., it is allowed that in 2,5% of cases the outside air temperature is lower than the outside design temperature [20].

A new ASHRAE approach reported in ASHRAE Fundamentals from 2009 [21] looks at the distribution of temperature throughout the year. This approach has two criteria, stricter and milder:

- The stricter criterion requires the representation of the outside temperature in 99,6%, but calculated according to hourly values for the whole year and not for the three coldest months, and this means that the outside temperature must be lower than the outside design temperature 35 h/year. For a period of 20 years, this would mean that the outside air temperature should be the same or lower for 700 hours during the 20 years, while a minimum of 350 hours should occur in a period of 10 years.
- The milder criterion is with hourly representation of outside temperature 99%, and it means that 88 h/year the outside temperature is lower than the outside design temperature [20].

3.5. Method based on minimum annual temperature

This method can be characterized as the simplest that was used in this work and is not related to the duration of the phenomenon. The method based on the minimum annual temperature is based on the determination of the mean values of the absolute minimum annual temperatures for a multi-year series. The expression that defines this method is:

$$t_{ODT} = \frac{\sum_{i=1}^{n} t_{GOD,min,i}}{n} \tag{2}$$

where $t_{GOD,min,i}[°C]$ is absolute minimum annual temperature and n represent number of years.

The calculated outside design temperatures according to different methods were used in the calculation of total heat losses according to standard EN 12831. For the purposes of the calculation, a family house located in Sarajevo, with a total heating area of 62,72m^2, was analysed. Figure 4 shows the geometry of observed residential unit with all internal and external dimensions.

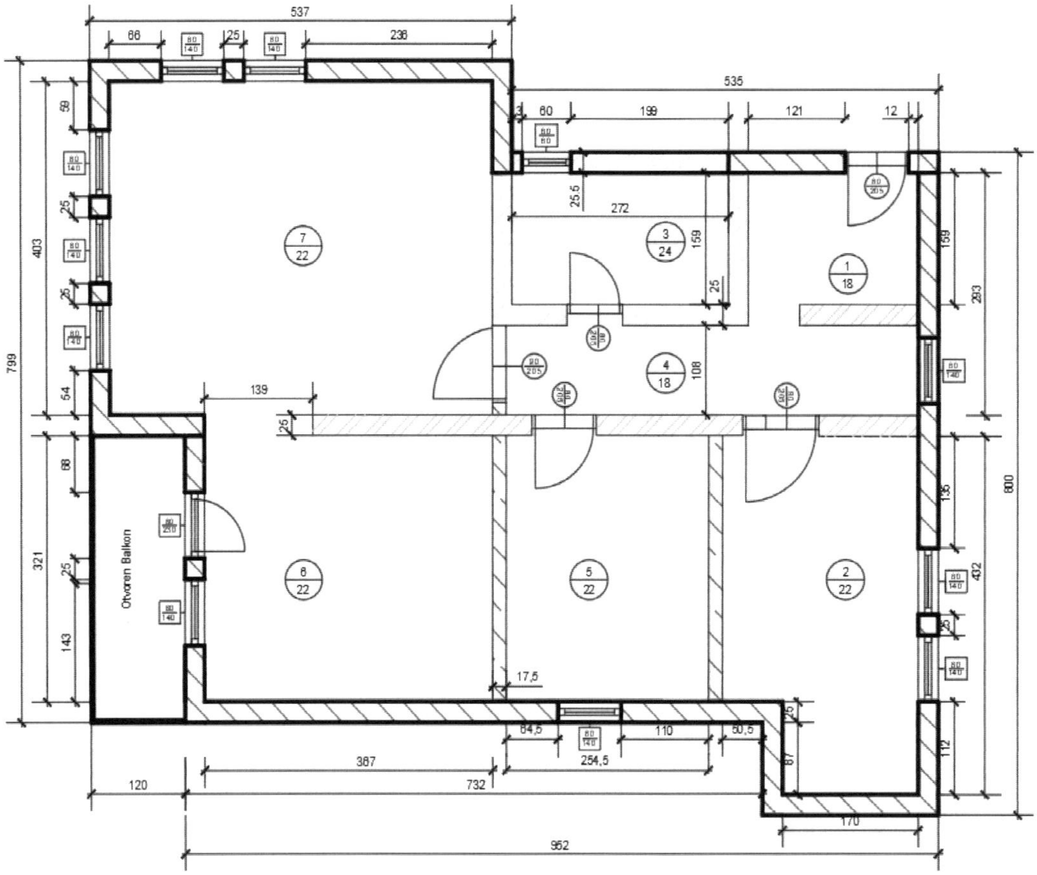

Fig. 4. The geometry of observed residential unit

Heat transfer coefficient for construction elements is given in Table 2.

Heat transfer coefficient [W/m^{2}K]	Outer wall	Inner wall 1	Inner wall 2	Inner wall 3	Floor	Ceiling	Windows	Outer door	Inner door
	0,36	1,71	1,13	2,13	2,09	3,0	1,9	3,5	2,0

Table 2. Heat transfer coefficient for construction elements

4. Results and discussion

The annual temperature change determines the length of the heating period, that is, the number of working days of the heating system. The official value of the average annual temperature for Sarajevo according to measuring data from station Bjelave for the period 1931-1960 is 9,7°C, and that value is still used for calculation [22]. Figure 5 shows the annual temperature trend in Sarajevo for the period from 2001 to 2020, where the value of the mean annual temperature according to the data of the Bjelave measuring station is 10,7°C. Based on the annual temperature variations in Sarajevo, a trend of increasing temperature can be observed, especially in the last 20 years.

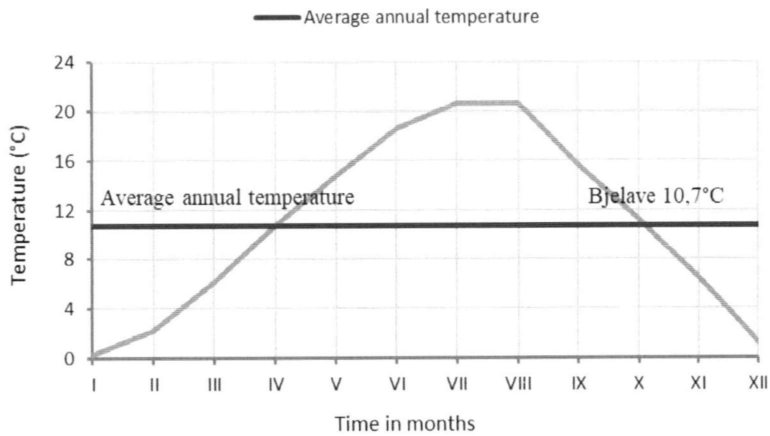

Fig. 5. The annual air temperature trend for Sarajevo for the period 2001-2020.

Since the absolute minimum temperatures in the observed period are used to determine the outside design temperature according to B. M. Chaplin, Figure 6 shows three different temperature values by month for the time period from 2001 to 2020: absolute maximum, mean outside temperature and the absolute minimum.

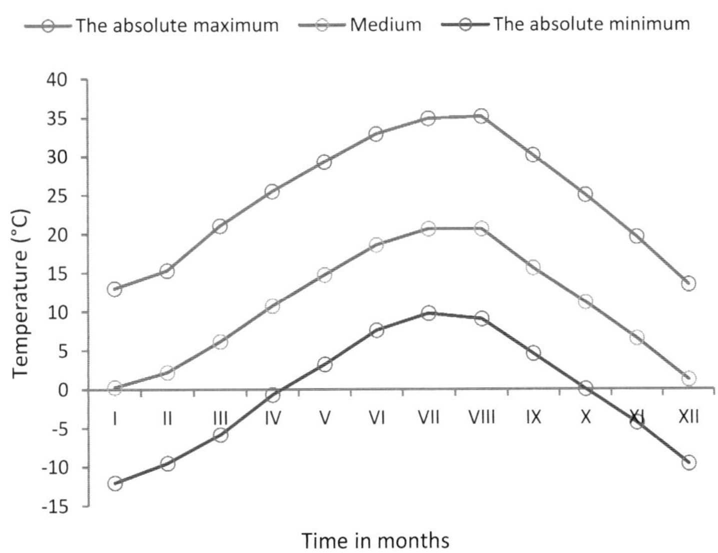

Fig. 6. Absolute maximum, mean outside temperature and absolute minimum for a 20-year time series for Sarajevo

Table 3 provides an overview of the calculated values of the outside design temperature according to several methods. Based on the obtained results, the lowest value of the outside design temperature for a period of twenty years was obtained using the Novak triad and the method based on minimum annual temperatures, while the highest was calculated using a new criterion according to the ASHRAE approach with 99% representation of hourly temperature values during the whole year. Based on the obtained results, it can be seen that the stricter criteria (99,6%) according to the new ASHRAE approach approximately corresponds to the earlier criterion of 99%, while the milder criteria (99%) according to the new ASHRAE approach approximately corresponds to the earlier criterion of 97,5%.

When it comes to the outside design temperature calculated according to B. M. Chaplin, the temperature values are lower if the time period observed is longer. However, an increase in the outside design temperature is evident, from -16°C as calculated for the 20-year time series from 1966 to 1985 according to [17] to -13,16°C for the 20-year time series from 2011 to 2020. It is important to note that when designing the heating system, the official outside design temperature for Sarajevo is still -18°C, which was calculated according to B. M. Chaplin for a series of 30 years (1956-1985) [19]. It can be observed, also, that there is a good agreement between the outside design temperature obtained according to B. M. Chaplin, DIN 4701 and the old ASHRAE 99% approach.

B. M. Chaplin	DIN 4701	Novak-triade	ASHRAE old approach		ASHRAE new approach		Absol. min. temp.
			99%	97,5%	99,6%	99%	
Time period: 2001-2020							
-13,16	-12	-15,13	-12	-9,6	-11	-8,4	-14,04
Time period: 2001-2010							
-10,85	-11,2	-12,14	-11,8	-9,7	-11,2	-8,7	-14,66
Time period: 2011-2020							
-13,07	-12,5	-15,13	-12,1	-9,4	-10,7	-7,7	-13,41

Table 3. Comparison of outside air temperature design parameters for heating determined by different approaches for Sarajevo (2001-2020)

By analysing the data, it was determined that the temperature lower than or equal to -12°C occurs 443 hours during 20 years, and that would be an average of 23 hours a year, which represents only one day a year. The coldest year in the considered period of 20 years was 2017, and in that year the temperature lower than or equal to -12°C occurred 90 hours during the year, which represents an incomplete 4 days. Therefore, a heating system dimensioned according to a temperature lower than the outside design temperature obtained according to the DIN 4701 method (-12°C), ASHRAE 99% (-12°C) or B. M. Chaplin (-13°C), especially in residential buildings, would be oversized.

Figure 7 shows a frequency curve based on which it is possible to see how many days or hours a year the mean two-day outside temperature is below or above a certain value. The two-day mean value of the outside temperature, which was reached or undershot 10 times in a period of 20 years, which means on average once every other year, is -12°C. It can be seen from Figure 7 that the periods of outside temperature duration are longer if the outside temperature values are higher in cold periods.

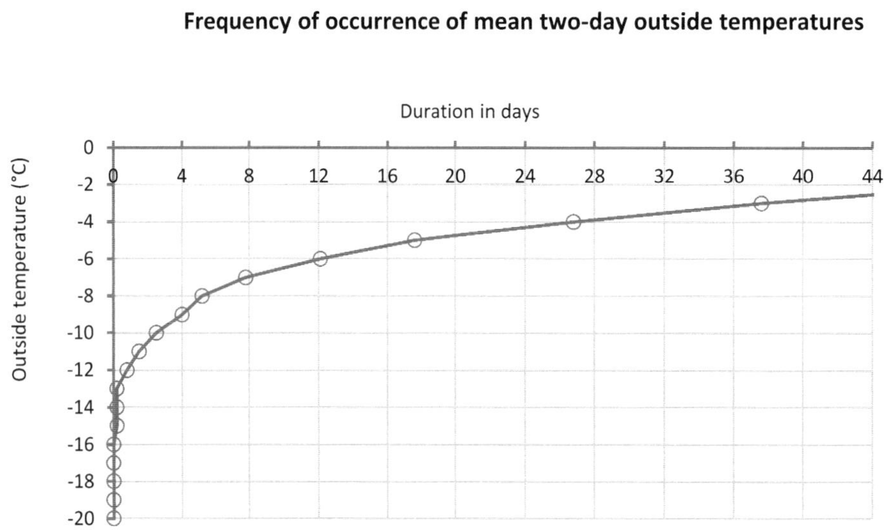

Fig. 7. Outside temperature as a function of duration and frequency

Table 4 shows the heat load for the analysed residential unit when the currently valid outside design temperature for Sarajevo (-18°C) is used in the calculation, as well as the heat load obtained using new outside design temperatures calculated according to different methods. It can be noted that the total heat load at the currently valid outside design temperature of -18°C is 10,95 kW. The total heat load of the building using the new outside design temperature calculated according to the Chaplin methodology is 9,6 KW, which represents a reduction of 12,3% compared to the value of the total heat load at the currently valid outside design temperature.

If the new outside design temperature calculated according to the Novak-triad (-15,13°C) is used for the calculation of the total heat load of the object, it is also obtained that the heat load is less compared to the heat load obtained with the valid outside design temperature about 7,3%. Using the ASHRAE standard, five different values of the outside design temperature were obtained, however, one value of the outside design temperature obtained by the old and one by the new approach was used for the calculation of the heat load.

Heat load	Official for Sarajevo - 18°C	B. M. Chaplin - 13,16°C	DIN 4701 - 12°C	Novak-triade - 15,13°C	ASHRAE old (99%) - 12°C	ASHRAE new (99%) - 8,4°C	Absol. min. temp. - 14,04°C
Transmission heat load $Q_{T,i}$ W	9766,26	8570,47	8283,87	9057,19	8283,87	7394,43	8787,88
Ventilation heat load $Q_{V,i}$ W	1185,10	1041,19	1006,70	1099,77	1006,70	899,65	1067,35
Total heat load $Q_{i,}$ W	**10951,37**	**9611,65**	**9290,57**	**10156,95**	**9290,57**	**8294,08**	**9855,24**

Table 4. Total heat load at different values of the outside design temperature calculated for a 20-year time period

By comparing the heat losses calculated according to the ASHRAE (old and new approach) and the heat losses obtained with the currently valid outside design temperature (-18°C), it is possible to see that the heat losses are lower by 15,2% with the new outside design temperature according to the old ASHRAE approach, or 24,3% with the outside design temperature according to the new ASHRAE approach. The same percentage reduction of the heat load of 15,2% is obtained with the outside design temperature calculated according to DIN 4701 because the same value of the outside design temperature is obtained as according to the old ASHRAE 99% approach. The last method used to calculate the outside design temperature is the method based on the minimum annual temperature and the value of the outside design temperature was obtained -14,04°C. The total heat load calculated with the outside design temperature obtained according to this method is 10% less than the heat load obtained with the valid outside design temperature. Finally, it can be concluded that the calculated heat load is less in the range of 10% to 24.3% obtained according to the new value of the outside design temperature according to any of the methods compared to the currently valid outside design temperature for Sarajevo.

Chosen example is a small-scale project and differences in the heat demand are not clearly visible and convincing. But if we upscale the level of abstraction and apply the same principle at the district heating of the small municipality we will end up with more convincing numbers. If we assume that instead of originally calculated heat demand of one municipality with 60 MW by using -18°C design temperature, we now apply the value according to DIN 4701 (-12°C, the new heat demand would be 50,5 MW. Now we are talking about large numbers. Consequently, we have less investment in the heat source facility (solar collectors and backup heat pump) but also in the pipeline and substations equipment. The potential savings will be clearly visible and the tightly sized system will operate smooth in all conditions. In our analysis we neglected the thermal accumulation of the buildings where they can play the role of the heat sources and sinks and, we did not mention weather predictor methods in control of the heating systems. These combined approaches can help new and modern systems to maintain the thermal comfort even if the temperature drops below designed value.

5. Conclusion and further work

Different methods for calculating new values of outside design temperature were presented in this paper. It has been shown that different methods result in different values of the outside design temperature. However, according to each of the analysed methods, higher values of the outside design temperature compared to the actual value were obtained for Sarajevo. This can be attributed to the climate change process in Bosnia and Herzegovina. By using the currently valid outside design temperature initially determined according to B. M. Chaplin for a period of 30 years (1956-1985) and new value of the outside design temperature calculated according to B. M. Chaplin for a more recent time period (2001-2020) on example of one residential unit presented in the paper, the heat load was lower by 12,3%. While comparing the heat load calculated with the outside design temperature obtained according to the DIN 4701, a reduction of 15,2% was visible. Therefore, the outside design temperature calculated according to any of the analysed methods, compared to the currently valid, resulted in a reduction of the heating needs. From the obtained results, it can be concluded that there is an urgent need for revision of existing and calculation of new outside design parameters. Since energy consumption in buildings in B&H is more than 50% of the overall final energy consumption, this topic becomes very important.

Based on conducted analysis, it can be concluded that the heating systems designed according to the valid outside design parameters are over-sized and there is a need to correct the outside design parameters. Oversized systems are not only more expensive but are also more energy consuming due to their control difficulties. Further research will be complemented by the analysis of the other outside parameters that were not included in this work, such as heating degree days and heat load duration curves. These values complemented with the concept of thermal inertia of buildings where building can be treated as heat source and/or sink, together with an advanced weather predictor method, should result in optimal size of the heating systems for individual and collective housing or for district heating projects as well.

6. References

[1] Zhang, Y., Zou, M., Yang, S., Lin, Y. & Ye, C. (2017). Colder-winter monthly-temperature forecasting in general trend of global warming via multiple sine functions decomposition. In 36th Chinese Control Conference (CCC) Dalian, China, pp. 4551-4555, DOI: 10.23919/ChiCC.2017.8028073.

[2] Spinoni, J., Vogt, J. V., Barbosa, P., Dosio, A., McCormick, N., Bigano, A., & Füssel, H. M. (2018). Changes of heating and cooling degree-days in Europe from 1981 to 2100. International Journal of Climatology, Vol. 38, No. S1, e191-e208, https://doi.org/10.1002/joc.5362

[3] Reddy, M. P., Aneesh, A., Praneetha, K., & Vijay, S. (2021). Global Warming Analysis and Prediction Using Data Science. In 2021 Fifth International Conference on I-SMAC (IoT in Social, Mobile, Analytics and Cloud) (I-SMAC) (pp. 1055-1059). IEEE. DOI: 10.1109/I-SMAC52330.2021.9640944

[4] Radusin, S., Jevtić, I., Cero, M., Abdurahmanović, I., & Cupać, R. (2020). Strategy for adaptation to climate changes and low-emission development of Bosnia and Herzegovina for the period 2020–2030. UNDP.

[5] Hoegh-Guldberg, O., Jacob, D., Bindi, M., Brown, S., Camilloni, I., Diedhiou, A., ... & Zougmoré, R. B. (2018). Impacts of 1.5 C global warming on natural and human systems. Global warming of 1.5° C.

[6] Steffen, W., Rockström, J., Richardson, K., Lenton, T. M., Folke, C., Liverman, D., ... & Schellnhuber, H. J. (2018). Trajectories of the Earth System in the Anthropocene. Proceedings of the National Academy of Sciences, Vol. 115, No. 33, pp. 8252-8259, https://doi.org/10.1073/pnas.1810141115

[7] Directive 2010/31/EU of the European Parliament and of the Council of 19 May 2010 on the energy performance of buildings (EPBD). Off. J. Eur. Union, (2010). 153, pp. 13–35.

[8] De Rosa, M., Bianco, V., Scarpa, F., & Tagliafico, L. A. (2014). Heating and cooling building energy demand evaluation; a simplified model and a modified degree days approach. Applied energy, Vol. 128, pp. 217-229. https://doi.org/10.1016/j.apenergy.2014.04.067

[9] Vucijak, B., Midzic Kurtagic, S., Ceric, A., Kupusovic, T., & Spago, S. (2012). Assessment of climate change effects to precipitation patterns using statistical process control methods. Annals of DAAAM for 2012 & Proceedings of the 23rd International DAAAM Symposium, Volume 23, No.1, ISSN 2304-1382, ISBN 978-3-901509-91-9, Ed. B. Katalinic, Published by DAAAM International, Vienna, Austria, EU, pp. 277-280

[10] Cvitan, L., & Sokol-Jurković, R. (2011). Changes of heating and cooling demands in Croatia during the period 1901-2008. Croatian Meteorological Journal, Vol. 46, No. 46, pp. 27-33.

[11] Zaninović, K., Gajić-Čapka, M., Žibrat, Z., & Šimetin, V. (2001). Meteorological parameters in the design of thermal protection of buildings. Građevinar, Vol. 53, No. 10, pp. 619-629.

[12] Popović T, Đurđević V, Živković M, Jović B & Jovanović M. (2009). Climate change in Serbia and expected impacts. Environment towards Europe, Fifth Regional Conference – Belgrade, pp. 6 – 11.

[13] Muhamedagic, R., Martinovic, S., & Lulic, H. (2013). Global warming and outdoor design temperature, 17th International Research/Expert Conference "Trends in the Development of Machinery and Associated Technology" TMT 2013, pp. 289 – 292, Istanbul, Turkey.

[14] Radusin, S., Medić, V., Cero, M., Abdurahmanovic, I., & Avdić, S. (2016). Third National Report and Second Biennial Report on Greenhouse Gas Emissions of Bosnia and Herzegovina under the United Nations Framework Convention, UNDP, Sarajevo.

[15] Vukmir, G., Stanišljević, LJ., Cero, M., Cacan, M., Marković, M., Rudež, M., Laganin, O., Kostić, R., Oprašić, S., Ćatović, S., & Lukić, T. (2009). Initial National Communication of Bosnia and Herzegovina under the United Nations Framework Convention on Climate Change, UNDP, Banja Luka.

[16] Todorović, B. (2009). Designing of central heating plants. Projektovanje postrojenja za centralno grejanje, Univerzitet u Beogradu, Mašinski fakultet, 11. izdanje, ISBN 978-86-7083-678-5, Beograd.

[17] Kulić, E. (2002). Building energy use – situation in B&H and worldwide - part II specific energy consumption characteristics development of domestic standards on heat loss calculations. Mašinstvo Vol. 2, No. 6, pp. 111 – 124.

[18] DIN 4701-1 (1997). Rules for calculating the heating load of buildings, Part 1: Basics of the calculation. Regeln für die Berechnung der Heizlast von Gebäuden, Teil 1: Grundlagen der Berechnung, DIN Taschenbuch 84, Beuth, Berlin.

[19] Kulić, E. (2019). Energy use in building construction. Zbornik Međunarodnog Kongresa o KGH, Vol. 32, No. 1, pp. 135-147.

[20] American Society of Heating, Refrigerating and Air-Conditioning Engineers. (2001). ASHRAE Handbook of Fundamentals, Chapter 27. Climatic Design Information, ASHRAE, Atlanta.

[21] American Society of Heating, Refrigerating and Air-Conditioning Engineers. (2009). ASHRAE Handbook of Fundamentals, Chapter 14. Climatic Design Information, ASHRAE, Atlanta.

[22] Kulić, E. (1989). Principles of designing heating systems. Principi projektovanja sistema grijanja, Mašinski fakultet Univerziteta u Sarajevu, Sarajevo.

DOI: 10.2507/34th.daaam.proceedings.028

APPLYING THE DESIGN OF EXPERIMENTS TO MODEL THE ENERGY CONSUMPTION OF TYPICAL SINGLE-FAMILY HOUSE IN BOSNIA AND HERZEGOVINA

Džana Kadrić, Rejhana Blažević, Emina Zukić & Edin Kadrić

This Publication has to be referred as: Kadric, D[zana]; Blazevic, R[ejhana]; Zukic, E[mina] & Kadric, E[din] (2023). Applying the Design of Experiments to Model the Energy Consumption of Typical Single-Family House in Bosnia and Herzegovina, Proceedings of the 34th DAAAM International Symposium, pp.0219-0228, B. Katalinic (Ed.), Published by DAAAM International, ISBN 978-3-902734-41-9, ISSN 1726-9679, Vienna, Austria
DOI: 10.2507/34th.daaam.proceedings.028

Abstract

The building sector in Bosnia and Herzegovina is a major energy consumer due to the age, inadequate maintenance, and poor energy characteristics of the majority of buildings. More than 90% of the residential buildings are categorized as individual, single-family houses, accounting for more than 80% of energy consumption. It is of great importance to model the impact of architectural, construction elements, and other relevant parameters on the building's energy consumption. Using the statistical method of design of experiments, a mathematical model is developed a model for the annual energy consumption of a typical residential building based on the building's relevant parameters. The model considers climatic conditions, architectural and construction features, and the characteristics of the installed heating system. This model enables identification of key factors influencing energy consumption and a rapid evaluation of potential energy savings through the implementation of energy efficiency measures. The model is developed for a typical building from the Typology of Residential Buildings of Bosnia and Herzegovina, specifically in the category of single-family houses constructed between 1981 and 1990, representing more than 200000 buildings. The validation of the model demonstrated its effectiveness in predicting the energy consumption of a representative building, therefore the analysis results can be extrapolated to all buildings within the same category.

Keywords: residential sector; design of experiments (DOE); full factorial design (FFD); energy consumption; energy efficiency

1. Introduction

The building sector has the largest share in energy consumption and global greenhouse gas emissions. In the European Union, energy production and consumption account for 75% of greenhouse gas emissions, with approximately 40% attributed to the building sector. In 2021, Bosnia and Herzegovina (B&H) consumed 48,73 GWh of final energy, with the building sector being the largest consumer, followed by transport and industry [1], as shown in Figure 1. Within the residential sector of B&H, 72% of total energy consumed is used for heating.

The building stock and thermo-technical systems in B&H exhibit energy characteristics that offer significant energy-saving potential, through the implementation of energy efficiency measures. By examining the Typology of residential buildings in the country, where buildings are categorized by construction period and architectural-constructional characteristics, it is possible to gain insights into the prevalence of specific building types and their energy performance [2]. According to the Typology of Residential Buildings in B&H, 93,11% of residential buildings in B&H are single-family houses, while high-rise buildings are the least represented with 0,3%. Numerous studies focused on reducing energy consumption in B&H building stock [3], [4], and the benefits of improving thermo-technical systems [5], [6], due to the poor energy characteristics of buildings, low efficiency of typical systems, and the considerable potential for reduction of energy consumption.

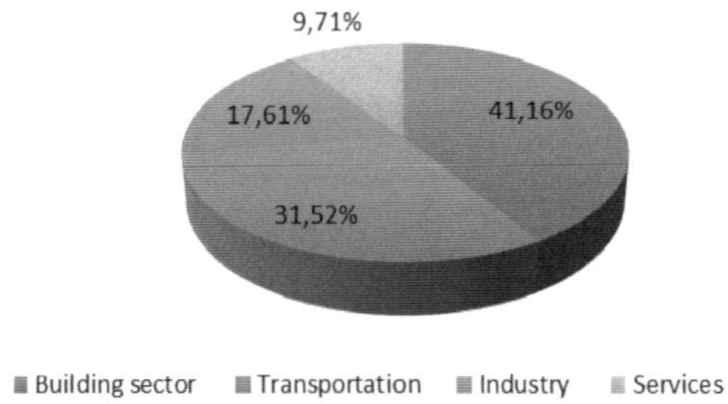

Fig. 1. Distribution of final energy consumption across various sectors in B&H, 2021[1]

Energy performance of buildings is influenced by various factors, including climatic conditions, architectural and construction characteristics, devices utilized, their usage time, and occupancy patterns. Zahraee et al. (2014.) investigated methods for enhancing energy savings in residential buildings by examining three climate factors: temperature, humidity, and airflow. The research employs building simulation and Design of Experiment (DOE) techniques to evaluate the impact of these factors on energy savings and cooling load. Results have shown that temperature and humidity have the most significant influence on energy savings, where optimal values of factors are determined to achieve the highest energy savings within the model's range [7]. Calculation methods and software adhering to relevant standards are frequently employed to examine the impact of energy efficiency measures on residential buildings. Shabunko et al. (2018) presented the results of an analysis of electricity consumption for over 400 residential buildings in Brunei Darussalam using EnergyPlus [8]. The models showed that the specific annual energy consumption of this type of buildings is around 64,2 to 47,8 kWh/m^{2}annual, where HVAC systems have the largest share in electricity consumption. This research highlights the effectiveness of engineering models in establishing benchmarks and simulating conditions for enhancing energy performance in buildings. Garcia-Cuadrado et al. (2022) utilized the Response Surface Method (RSM) to develop a model for calculating energy consumption for heating and cooling in single-family homes across three different climate regions [9]. Their research involved varying set temperature for heating and cooling and altering heat transfer coefficients of external walls by changing thermal insulation thickness. Energy simulation results from DesignBuilder are compared with estimates from the simplified model derived from the RSM analysis. The results showed a deviation of less than 3% between the two methods, demonstrating that the mathematical prediction models are suitable for studying the energy performance of buildings while saving computational time, costs, and associated human resources. In a study presented in Pekdogan et al. (2020), thermal performance of external walls is examined employing a three-level full factorial statistical experimental design. The research focused on an opaque wall in low-energy buildings to analyse the impact of various factors, such as location (A), orientation (B), insulation location at the wall cross section (C), and month of the year (D), on heat loss or gain. The study encompassed different climatic regions of Turkey (Erzurum, Ankara, and Izmir) with conducted 81 experiments. The resulting Analysis of Variance (ANOVA) table indicated that the month of the year, location, and orientation are significant factors for heat transfer through brick walls. Additionally, the interaction effects of factors AB, AD, BD and CD is shown to be significant [10].

The main objective of this study is to develop a mathematical model for analysing the energy characteristics of a representative residential building for quick assessment of its energy consumption and energy-saving potential. A full factorial design is employed to evaluate the impact of selected factors on energy consumption for heating. The analysis is based on dynamic simulation of energy characteristics for a representative single-family house, constructed in the period between 1981 and 1990, whereby this category includes approximately 28% of the total number of buildings in the building sector of B&H.

2. Methodology

2.1. Representative building from the Typology of residential buildings in Bosnia and Herzegovina

In the "Typology of Residential Buildings in Bosnia and Herzegovina," residential buildings are categorized into six distinct construction periods and four different building types, as shown in Figure 2 [2].

Representative buildings are statistically selected for each category, serving as typical examples of their respective categories based on their architectural characteristics. More than 90% of the residential buildings are categorized as individual, single-family houses (SFH), accounting for more than 80% of energy consumption. The focus of the presented analysis is on SFH category, constructed during the period from 1981 to 1990. This category comprises 236075 buildings, which is approximately 28% of the total number of buildings in the B&H residential sector.

Fig. 2. Classification of residential buildings according to the "Typology of residential buildings in Bosnia and Herzegovina" [11]

The energy characteristics of the representative building, marked with red in Figure 2, are examined in this study. The representative building has two floors. There is a garage in the basement and an impassable attic above first floor ceiling, both of which are treated as unheated areas. The building's envelope consists of hollow brick blocks with 5 cm of thermal insulation, wooden windows with double glazing, and wooden exterior doors. The roof is a two-sided wooden structure with clay roof tiles. The interior is divided by inner walls into separate rooms, each representing a distinct thermal zone. The corridors on both floors are connected by an internal staircase and treated as a single thermal zone. To analyse the energy performance of a building, the study used an designed internal temperature of 20°C for the living room, kitchen, and bedrooms, and 24°C for the bathrooms, as specified by the BAS EN 12831 standard. Table 1 provides basic information about the building's geometry, such as gross area of the base, the net heated surface, and the building compactness ratio. The heat transfer coefficients of characteristic construction elements are shown in Table 2.

Number of floors	2
The gross area of the base of the building	69,66 m^2
Net area of the heated space	101,44 m^2
The volume of the heated space	250,15 m^3
Building compactness ratio	0,83

Table 1. Basic data on building geometry

Heat transfer coefficient	Outer wall 1	Outer wall 2	Ceiling 1	Ceiling 2	Ceiling 3	Roof	Floor	Window	Outer door
[W/m^{2}K]	0,499	0,643	0,433	1,564	0,344	0,615	0,531	2,90	2,9

Table 2. Heat transfer coefficient for particular construction elements

2.2. Modelling the energy performance of a representative building

Building energy performance is modelled using DesignBuilder software, which incorporates EnergyPlus simulation tool. The study analyse how energy consumption for heating varies based on values of selected factors: the wall heat transfer coefficient (U_{wall}), heat transfer coefficient of the ceiling to the unheated attic (U_{ceiling}), the solar heat gain coefficient (*SHGC*), and the heating system efficiency (η_{H}). Simulations are conducted according to a design matrix, and the mathematical relationship is established using the Minitab software version 19. Figures 3 and 4 display a visual representation of the selected single-family home (SFH) constructed between 1981 and 1990, as well as the graphical interface of the model in DesignBuilder.

Fig. 3. The visual representation of the selected representative building object from TABULA registry SFH 1981–1990 [2]

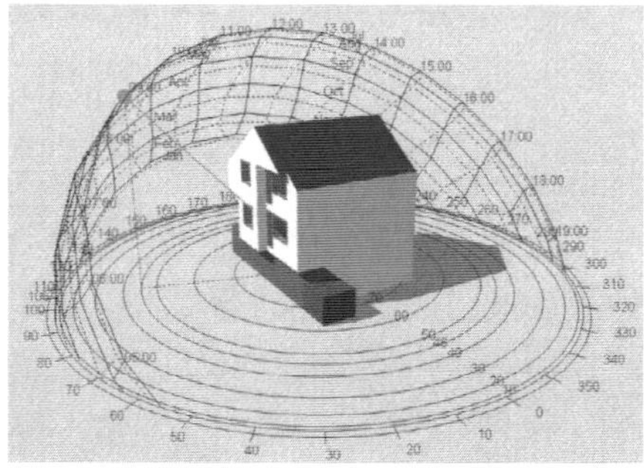

Fig. 4. Graphical interface of the representative object SFH 1981–1990 as designed in DesignBuilder.

EnergyPlus employs hourly data to define external conditions when simulating the energy performance of buildings. National Meteorological Service-monitored data, such as air temperature, atmospheric conditions, solar radiation, wind speed, and direction, are incorporated into the analysis [12]. The SFH examined in this study is located in Banja Luka, B&H, north climate region known for its warm summers, cold and snowy winters, and partial cloudiness throughout the year. Figures 5 and 6 display the average monthly values of the highest and lowest daily temperatures and the average daily short-wave solar energy reaching the ground. Air temperatures fluctuate between -3°C and 28°C annually, seldom dropping below -10°C or exceeding 33°C. July is the hottest month, with an average highest daily temperature of 27°C, while January is the coldest month, with an average lowest daily temperature of -3°C [13].

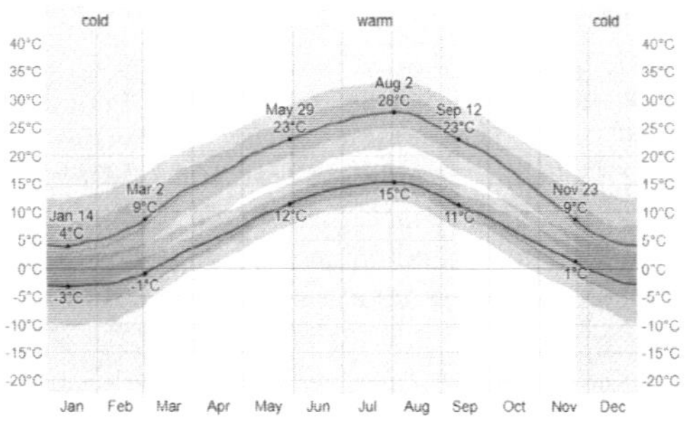

Fig. 5. Average daily external air temperatures

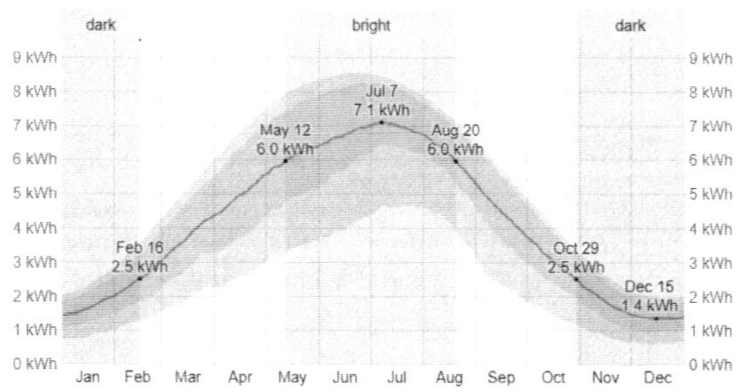

Fig. 6. Average daily shortwave solar energy reaching the ground

For the purpose of the analysis, it is assumed that a family of four lives in the house, with two working adults and two children. This assumption helped establish the occupancy schedule for a week, taking into account that weekdays and weekends have different occupancy rate. As shown in Figure 7, two separate schedules are created for weekdays and weekends. For a typical family of four, most members are absent from the house on weekdays between 07:00 and 17:00, and it is assumed that all members are present from 22:00 to 07:00 during the night period.

The occupancy of the residential unit, or the time spent by people in the building, influences the usage of lighting, electrical appliances, and other household devices, thereby affecting the intensity of internal heat gains. In analysing the internal heat gains of this SFH, factors such as the operation time of the electrical water heater, household appliances, various electrical devices, and lighting are considered. A specific installed lighting power of 5 W/m^2 is used for calculating the internal heat gains. The typical building is equipped with individual solid fuel stoves (wood and coal), with the efficiency of 50% [2], indicating substantial energy losses in such systems. In the DesignBuilder software, a heating system schedule is developed, as shown in Figure 8.

The schedule is set according the assumption that the heating system operates at full capacity from 06:00 to 22:00, seven days a week. The heating system's operation is governed by the internal temperature, meaning it will continue to function until the room temperature reaches the preset internal temperature value.

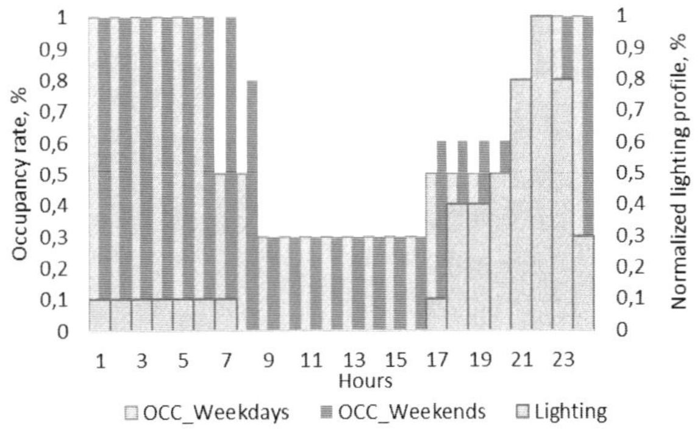

Fig. 7. Occupancy rate and lighting hourly profile

For: Weekdays SummerDesignDay,
Until: 06:00, 0,
Until: 22:00, 1,
Until: 24:00, 0,
For: WinterDesignDay,
Until: 24:00, 1,
For: Weekends,
Until: 06:00, 0,
Until: 22:00, 1,
Until: 24:00, 0,
For: Holidays,
Until: 06:00, 0,
Until: 22:00, 1,
Until: 24:00, 0,
For: AllOtherDays,
Until: 24:00, 0;

Fig. 8. Heating system operation schedule modelled in DesignBuilder

2.3. Research method

Design of experiments (DOE) is a statistical method employed to plan and analyse experiments results, as well as to interpret the results obtained either experimentally or through simulations. To understand the effect of multiple factors and their interactions on a system or process's response, experiments must be systematically conducted using factorial designs, where several factors are altered during each experimental run. Factorial designs can be executed in two ways: full factorial design (FFD), which analyses all possible factor combinations, and fractional factorial design, which examines only a portion of the possible combinations [14].

In this study, the FFD is employed for statistical analysis to examine the impact of specific factors on the response, which in this case is the energy consumption for heating. The analysis considered four factors, each on two levels. Following factors are considered: the wall heat transfer coefficient (U_{wall}), heat transfer coefficient of the ceiling to the unheated attic ($U_{ceiling}$), the solar heat gain coefficient (SHGC), and the heating system efficiency (η_H). The low and high levels of each factor are determined based on data from the Typology of Residential Buildings in B&H. The low level (-1) represents the poorest energy characteristics found in this category of buildings, while the high level (+1) represents the best energy performance. The values for the (+1) level are complied with the Rulebook on Minimum Requirements for Energy Characteristics of Buildings [15]. The wall heat transfer coefficient is calculated by varying the thickness of the external walls' thermal insulation, with the (-1) level referring to a wall without insulation and the (+1) level referring to a wall with 20 cm of insulation. A similar procedure is applied to determine the ceiling heat transfer coefficient factor level. The building currently has double-glazed windows with wooden frames, but according to the Typology of Residential Buildings in B&H, 26,2% of buildings in this category have single-glazed windows. Therefore, the SHGC factor value at the (-1) level is characteristic of wooden windows with single glazing [2], while the value at the (+1) level is characteristic of PVC windows with triple glazing. The heating system efficiency factor value at the (-1) level represents the efficiency of solid fuel stoves typically installed in the building, while the value at the (+1) level represents the efficiency of a newly installed central heating system with pellet boiler. Factors considered in this study and their levels are shown in Table 3.

Factor	**Level**	
	-1	1
U_{wall}	1,454	0,18
SHGC	0,858	0,242
$U_{ceiling}$	2,046	0,178
η_H	0,5	0,9

Table 3. Considered factors and their levels

In this study, a 2^4 factorial design is employed. The experimental matrix contains 16 simulations, where each simulation run represents specific combination of factor levels and corresponding energy consumption for heating.

3. Results and discussion

Upon establishing the required input parameters, the building's energy characteristics are calculated using DesignBuilder software. Following, the FFD, is employed to analyse the impact of selected factors on the building energy consumption for heating.

3.1. Energy performance of a representative building

Using DesignBuilder, energy characteristics are calculated by considering various factors such as the local climate data, building orientation, occupancy, energy consumption of electrical devices and lighting, and their typical usage periods. Table 4 provides a breakdown of heat losses through individual building elements, and shows that the total heat losses are 9,22 kW. The largest heat losses occur due to infiltration through joints and gaps, followed by losses through glazing and external walls.

Heat losses		
Heat losses through glazed surfaces	2,33 kW	24,01 W/m^2
Heat losses through external walls	2,27 kW	23,41 W/ m^2
Heat losses through the ceiling	0,48 kW	4,96 W/ m^2
Heat losses through the roof	0,39 kW	4,01 W/ m^2
Heat losses through the floor towards the unheated basement	0,14 kW	1,43 W/ m^2
Heat losses to the ground	0,43 kW	4,47 W/ m^2
Heat losses due to infiltration	3,18 kW	32,73 W/ m^2
Total heat losses	9,22 kW	95,02 W/ m^2

Table 4. Building heat load

A simulation of energy consumption in the analysed residential building is conducted, considering factors such as the building's heat losses, the heating system operating time, and the effects of occupancy, lighting, and various electrical appliances. The daily energy consumption patterns, temperatures, and heat balance for a heating season are shown in Figure 9.

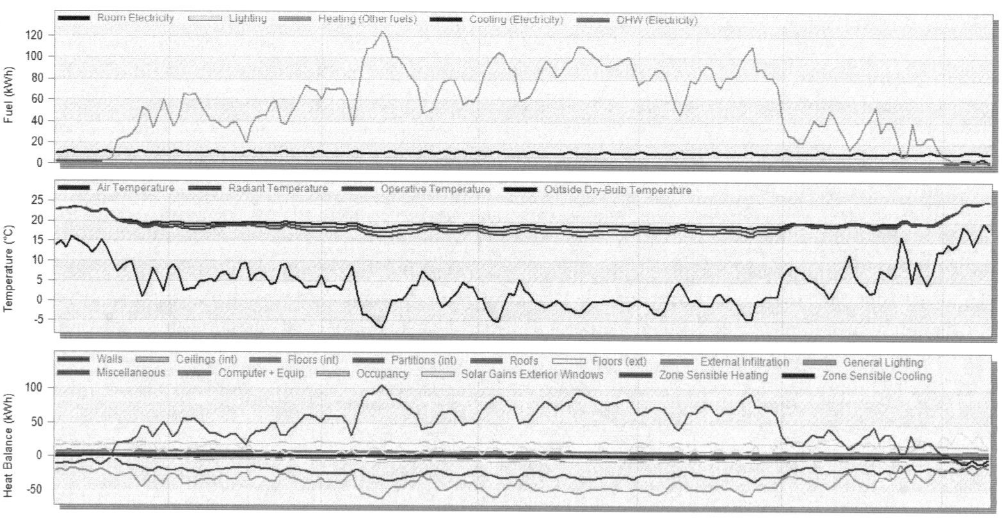

Fig. 9. Daily diagrams of energy consumption, temperature and heat balance

From Figure 9, it is visible that the dominant building energy consumption is for heating. The peak energy consumption occurs on December 13, reaching 125,25 kWh. The energy consumption pattern corresponds with external temperature fluctuations. The heat balance diagram demonstrates that heat losses through the building's envelope are counterbalanced by the heat gains from lighting, electrical devices, occupants, and solar gains and heating system, as well. For a more comprehensive analysis, Figure 10 presents a half-hourly energy consumption and temperature chart for the coldest month in Banja Luka region.

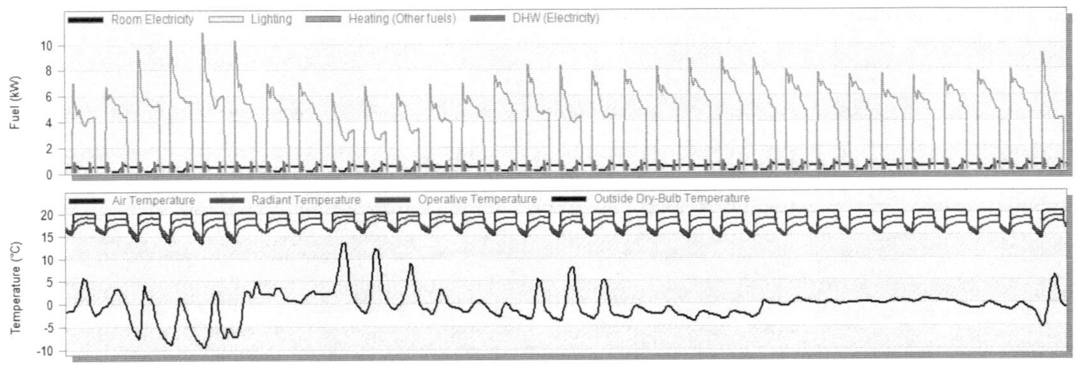

Fig. 10. Half-hourly diagram of energy consumption and temperatures for the month of January

In Figure 10, the operating parameters of the heating system are shown. The energy consumption curve for building heating (orange line) exhibits abrupt increases, indicating the activation of the heating system at 06:00. Following a night time pause, the system resumes operation at full capacity to counterbalance the heat loss that occurred overnight and achieve the designed room temperature. Subsequently, the heating system's performance stabilizes, operating at reduced capacity to maintain the set thermal zone temperature. The diagram of temperature profiles corresponds with the energy consumption pattern. A decline in room temperature is observed during the night when the heating system is not in function.

3.2. Statistical analysis using design of experiments

Statistical analysis is performed using Minitab software version 19, with the aim of generating regression model for prediction of building energy consumption for heating. The experimental matrix, with coded factors and the computed building energy consumption, for each simulation, is presented in Table 5. Building energy consumption for heating, is calculated using DesignBuilder for each simulation run.

Exp	U_{wall}	SHGC	$U_{ceiling}$	η_H	$E_{H,del}$, (kWh/m²ann)
1	-1	-1	-1	-1	394,93
2	1	-1	-1	-1	226,92
3	-1	1	-1	-1	285,55
4	1	1	-1	-1	122,02
5	-1	-1	1	-1	333,79
6	1	-1	1	-1	159,19
7	-1	1	1	-1	223,32
8	1	1	1	-1	55,66
9	-1	-1	-1	1	219,40
10	1	-1	-1	1	126,06
11	-1	1	-1	1	158,65
12	1	1	-1	1	67,79
13	-1	-1	1	1	185,44
14	1	-1	1	1	88,44
15	-1	1	1	1	124,06
16	1	1	1	1	30,92

Table 5. Experimental matrix

In the DOE methodology, the statistical analysis relies on the Analysis of Variance (ANOVA), which helps to determine the significance of each factor in terms of its impact on the response variable [16]. ANOVA in this study is performed at 0,05 significance level. The preliminary analysis revealed that three and four-factor interactions are not statistically significant, therefore only main effects and effects of two-factor interactions are considered. Table 6 displays the results of ANOVA, with only statistically significant terms and their corresponding F and p-values.

Based on ANOVA results, it can be concluded that all model terms are statistically significant ($p < 0{,}05$), indicating significant impact of all model terms (main effects and two-factor interaction effects) on energy consumption for heating.

Source	DF	Seq SS	Contribution	Adj SS	Adj MS	*F*-Value	*p*-Value
Model	7	155176	99.97%	155176	22168.0	4314.12	0.000
Linear	4	146488	94.38%	146488	36622.1	7127.02	0.000
U_{wall}	1	68662	44.24%	68662	68662.3	13362.38	0.000
SHGC	1	27739	17.87%	27739	27738.9	5398.27	0.000
$U_{ceiling}$	1	10025	6.46%	10025	10025.0	1950.97	0.000
η_H	1	40062	25.81%	40062	40062.0	7796.47	0.000
2-Way Interactions	3	8688	5.60%	8688	2895.9	563.58	0.000
$U_{wall} \cdot \eta_H$	1	5605	3.61%	5605	5604.8	1090.74	0.000
$SHGC \cdot \eta_H$	1	2265	1.46%	2265	2264.8	440.75	0.000
$U_{ceiling} \cdot \eta_H$	1	818	0.53%	818	818.2	159.24	0.000
Error	8	41	0.03%	41	5.1		
Total	15	155217	100.00%				

Table 6. ANOVA of the model for prediction of building energy consumption for heating $E_{H,del}$

Figure 11 presents a Pareto chart of standardized effects, where the bar length represents the standardized effect of each model term (factors and their interactions). Statistical significance of all standardized effects is estimated at 0,05 significance level. According to standardized effects shown in Figure 11, it can be observed that wall heat transfer coefficient (U_{wall}) has the largest, while the least impact on energy consumption for heating has two-factor interaction of heat transfer coefficient of the ceiling to the unheated attic ($U_{ceiling}$) and heating system efficiency (η_H).

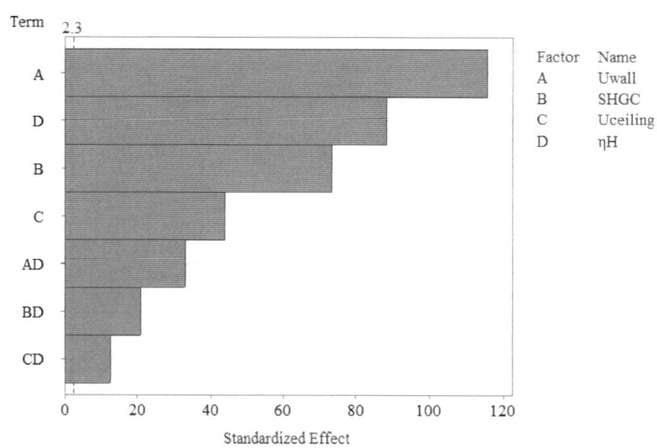

Fig. 11. Pareto chart of standardized effects

Scatter plots in Figure 12 show relationship between building energy consumption for heating, as response, and wall heat transfer coefficient (U_{wall}), heat transfer coefficient of the ceiling to the unheated attic ($U_{ceiling}$), the solar heat gain coefficient (SHGC), and the heating system efficiency (η_H), as factors, respectively. From Figure 12 it can be observed that the response linearly decreases as each factor is increased from low (-1) to high level (+1). These plots clearly show that improving the thermal characteristics of building elements (external walls, windows, ceiling towards the unheated attic and heating system) would result in a reduction of energy consumption for heating.

Using regression analysis, model for prediction of energy consumption for heating, for building located in Banja Luka, is developed and given by:

$$E_{H,del} = 175.134 - 65.509\,U_{wall} - 41.638\,SHGC - 25.031\,U_{ceiling} - 50.039\,\eta_H + 18.716\,U_{wall} \cdot \eta_H + 11.898\,SHGC \cdot \eta_H + 7.151\,U_{ceiling} \cdot \eta_H \quad (1)$$

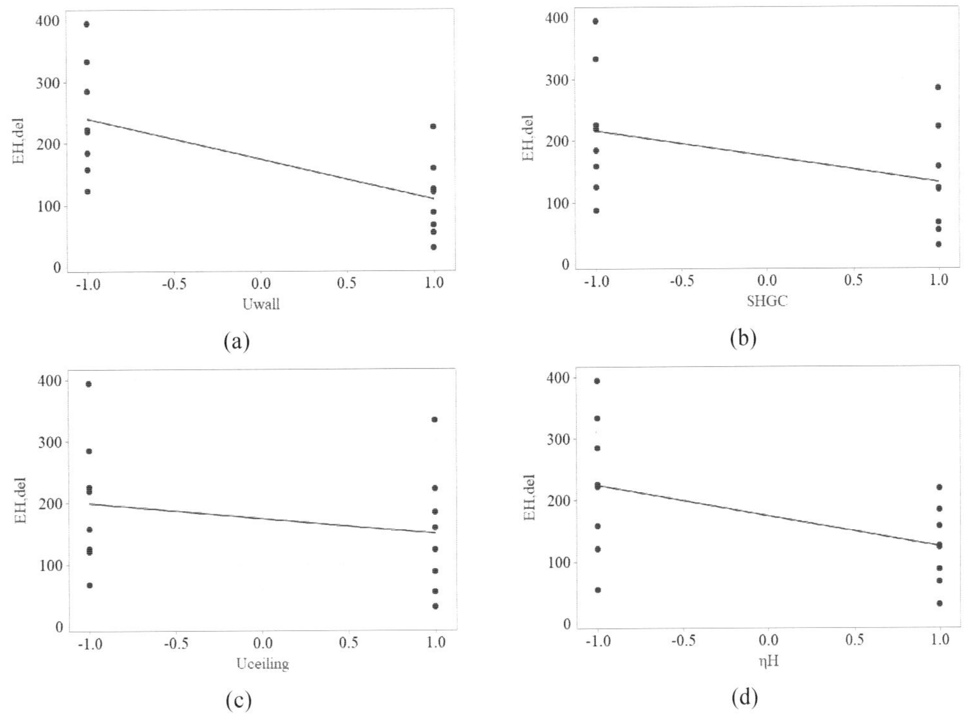

Fig. 12. Scatter plots of $E_{H,del}$ versus (a) U_{wall}, (b) SHGC, (c) $U_{ceiling}$, and (d) η_H

After performing ANOVA, it is necessary to validate the developed regression model using residual analysis. For the model to be considered adequate, it is necessary that residuals are normally distributed, and additionally, identically and independently distributed. To determine if the residuals follow a normal distribution, the Anderson-Darling (AD) test is performed at 0,05 significance level. If the AD test p-value is less than 0,05, than residuals do not follow normal distribution. Conversely, if the AD test p-value is greater than 0,05, than residuals follow a normal distribution. For obtained AD test statistic of 0,309 and its corresponding p-value of 0,522, it can be concluded that residuals are normally distributed. The normal probability plot of the residuals is displayed in Figure 13, where all data points (residuals) lie within the 95% confidence limits (outer red lines).

The second condition requires that residuals are identically and independently distributed. By examining the graphical representation of the residual's distribution around the reference line (mean of residuals), it is possible to determine whether residuals are identically and independently distributed. Figure 14 shows that residuals are randomly distributed around the reference line, indicating identical and independent distribution of residuals.

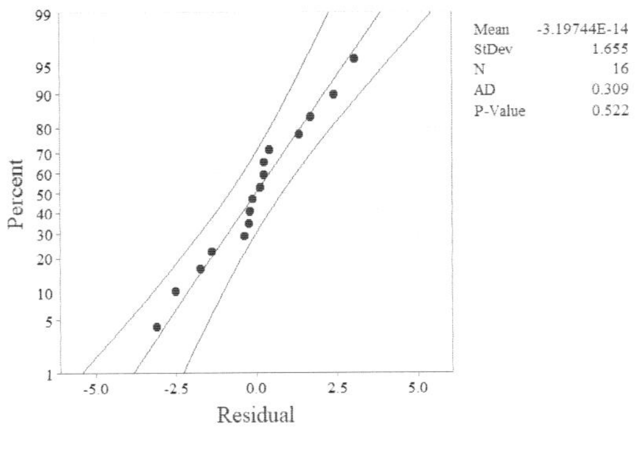

Fig. 13. Normal probability plot of residuals

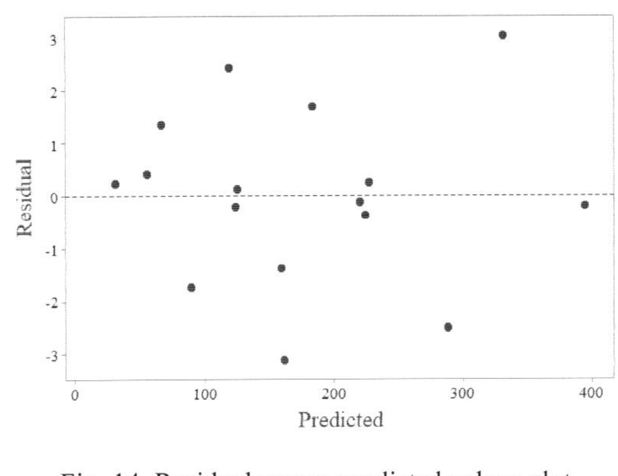

Fig. 14. Residual versus predicted values plot

Based on the model validation, it is evident that the developed model is applicable in real-world situations. This implies that the model can effectively estimate the annual energy consumption for heating of buildings in a specific category, such as SFH, constructed between 1981 and 1990.

4. Conclusion and further work

The impact of considered factors on the building energy consumption for heating is examined using statistical method, design of experiments. The research focused on single-family houses constructed between 1981 and 1990, represented via typical building from Building Typology. Selected building is located in Banja Luka in the north climate region. Dynamic simulations with very detailed input data are used to calculate building energy consumption. The analysis resulted in a regression model that establishes the relationship between the energy consumption for heating and the considered factors. This model reveals how variations in the thermal characteristics of building elements, such as heat transfer coefficient of external walls, windows, ceilings, and heating system efficiency, affect the building's energy consumption. The most significant factor influencing the energy consumption for heating is heat transfer coefficient of the external wall, followed by the heating system efficiency.

The model is validated, confirming its applicability for predicting the energy consumption for heating of single-family houses built between 1981 and 1990 in Bosnia and Herzegovina. This research can serve as a foundation for determining potential for energy savings for both heating and cooling of residential buildings, which can be explored in future studies. Additionally, further research will focus on developing regression model for predicting annual energy consumption for cooling, particularly relevant for the southern climate region, characterized by high average daily temperatures throughout the year.

5. References

[1] Ministarstvo vanjske trgovine i ekonomskih odnosa Bosne i Hercegovine (2021). 5th Annual Report under the Energy Efficiency Directive Prepared by: Ministry of foreign trade and economic relations of Bosnia and Herzegovina

[2] Arnautović, A. D., Burazor, M., Delalić, N., Gajić, D., Gvero, P., Kadrić, D., & Salihović, E. (2016). Typology of residential buildings in Bosnia and Herzegovina.

[3] Kadrić, Dž., Aganovic, A., Martinović, S., Delalić, N. & Delalić-Gurda, B. (2022). Cost-related analysis of implementing energy-efficient retrofit measures in the residential building sector of a middle-income country – A case study of Bosnia and Herzegovina, Energy and Buildings, Vol. 257, 111765, https://doi.org/10.1016/j.enbuild.2021.111765

[4] Kadrić, Dž., Delalić, N., Delalić, B., Kadrić, E. & Bajrić, H. (2018). Benchmarking energy efficiency of school buildings, Annals of DAAAM & Proceedings of 29th International DAAAM Symposium, pp. 0299–0305, DOI: 10.2507/29th.daaam.proceedings.043

[5] Blazevic, R., Teskeredzic, A. & Jugo, E. (2019). Optimal heat flux in intermittently heated buildings, Annals of DAAAM & Proceedings of 30th International DAAAM Symposium, 0701–0708, DOI: 10.2507/30th.daaam.proceedings.096

[6] Teskeredzic, A., Blazevic, R. & Dzino, S. (2020). Application of air-water heat pump in refurbished building with existing radiator system, Annals of DAAAM & Proceedings of 30th International DAAAM Symposium, 0510–0519, DOI: 10.2507/31st.daaam.proceedings.071

[7] Zahraee, S. M., Khademi, A., Khademi, S., Abdullah, A. & Ganjbakhsh, H. (2014). Application of design experiments to evaluate the effectiveness of climate factors on energy saving in green residential buildings, Journal Teknoli, (Sciences and Engineering),Vol. 69, No.5, pp. 107–111, DOI: 10.11113/jt.v69.3215

[8] Shabunko, V., Lim, C. M. & Mathew, S. (2018). EnergyPlus models for the benchmarking of residential buildings in Brunei Darussalam, Energy and Buildings, Vol. 169, pp. 507–516, https://doi.org/10.1016/j.enbuild.2016.03.039

[9] García-Cuadrado, J., Conserva, A., Aranda, J., Zambrana-Vasquez, D., García-Armingol, T. & Millán, G. (2022). Response Surface Method to Calculate Energy Savings Associated with Thermal Comfort Improvement in Buildings Sustainability, Vol. 14, No.5, 2933, https://doi.org/10.3390/su14052933

[10] Pekdogan, T., Akkurt, S. & Basaran, T. (2020). A full 34 factorial experimental design for the low energy building's external wall, Thermal Science, Vol. 24, No. 2, pp. 1261–1273, https://doi.org/10.2298/TSCI180726345P

[11] Institut Wohnen und Umwelt GmbH 2016 TABULA WebTool Natl. Build. Typologies

[12] DesignBuilder Software Ltd 2019 DesignBuilder 6.1 - User's manual 90

[13] The Weather Year-Round Anywhere on Earth - Weather Spark, Available from: https://weatherspark.com/y/83139/Average-Weather-in-Sarajevo-Bosnia-&-Herzegovina-Year-Round

[14] Jankovic, A., Chaudhary, G. & Goia, F. (2021). Designing the design of experiments (DOE) – An investigation on the influence of different factorial designs on the characterization of complex systems, Energy and Buildings, Vol. 250, 111298, https://doi.org/10.1016/j.enbuild.2021.111298

[15] Official Gazette of Federation of Bosnia and Herzegovina N 81/19. (2019). Rulebook on minimum requirements for energy performance of buildings

[16] Shina, S. (2022). Industrial design of experiments: A case study approach for design and process optimization, Springer Nature Switzerland, ISBN 978-3-030-86266-4, ISBN 978-3-030-86267-1 (e-book).

DOI: 10.2507/34th.daaam.proceedings.029

STATISTICAL ANALYSIS OF HYDROLOGICAL PARAMETERS IN THE BOSNA RIVER BASIN

Ismira Ahmović, Halima Hadžiahmetović, Rejhana Blažević, & Mirela Alispahić

This Publication has to be referred as: Ahmovic, I[smira]; Hadziahmetovic, H[alima]; Blazevic, R[ejhana] & Alispahic, M[irela] (2023). Statistical Analysis of Hydrological Parameters in the Bosna River Basin, Proceedings of the 34th DAAAM International Symposium, pp.0229-0236, B. Katalinic (Ed.), Published by DAAAM International, ISBN 978-3-902734-41-9, ISSN 1726-9679, Vienna, Austria
DOI: 10.2507/34th.daaam.proceedings.029

Abstract

The occurrence of extreme precipitation, floods, and dry periods, according to all forecasts, will probably become more frequent both in Europe and in Bosnia and Herzegovina. For this reason, it is necessary to consider climate change when designing infrastructure facilities. Following this, this paper aims to assess the flow rate data based on the daily measured values obtained at the Reljevo hydrological station on the Bosna River. The analyses are conducted for two periods (1961-1990 and 2000-2010) to obtain the temporal distribution of relevant flow rate data. For observed periods, analysed data are compared, and the coincidence of hydrological parameters along the perceived changes is observed. Significant monthly and yearly variations are observed based on the analysis of variance (ANOVA) applied to the flow rate data for observed periods. As a result, maximum flow rates are found in the summer period (March and April mainly), which should be considered in flood management.

Keywords: climate change; analysis of variance (ANOVA); hydrological parameters; two-factorial experimental design method

1. Introduction

The hydrological cycle is part of the climate system and is sensitive to climate change. Sometimes, insignificant climate changes lead to significant changes in hydrological processes [1]. Global warming refers to the increase in the average planet's surface temperature caused primarily by the emission of gases from fossil fuel combustion since the Industrial Revolution. However, climate change refers to long-term (a period of several decades or longer) changes that concern changes in temperature, precipitation and wind [2]. According to the Intergovernmental Panel on Climate Change (IPCC), climate change is observed as statistically significant changes expressed by changes in mean values and variability of observed properties, which can last for a period of time (a decade or more) [3]. The observed climate changes, published in the sixth report of the IPCC, are as follows: from 2011 to 2020, global surface temperature increased by 1,09°C (0,95°C to 1,20°C). In that period, more noticeable increase in temperature was over land (by 1,59°C) than over the ocean (by 0,88°C). Since the 1950s the frequency and intensity of heavy precipitation events have increased over most land area for which observational data are sufficient for trend analysis, with even faster rate of increase since the 1980s.

In the period from 2010 to 2019 compared to the period 1979-1988, there was an increase in the average amount of precipitation on land and a decrease in the Arctic sea ice area of about 40% in September and about 10% in March. From 1901 to 2018, as a consequence of the increase in temperature, there was a rise in the mean sea level of 0,2 m. Rise in the mean sea level during the early 1900s was due to natural factors, however since 1970s, dominant cause of global average sea level rise were human activities [4]. Hydrological cycle parameters such as evapotranspiration, soil moisture and runoff are sensitive to small changes in temperature and precipitation [5], [6], [7]. River flow forecasts are essential for extreme water management, policy-making, and sustainable water resource management practices [8], [9].

Generally, climatic changes have been observed in Bosnia and Herzegovina during all seasons since 1981. Based on previous research, from 1961 to 2010, the temperature has increased in all regions of Bosnia and Herzegovina [10]. In the last hundred years, the average annual temperature has increased by about 0.6 °C [11]. As for the precipitation quantity, from 1981 to 2010, a slight increase in annual values has been observed compared to 1961-1990 [11]. Trends of five to twenty-day periods of extreme hot/cold weather have been observed; the same trends are noticed for intense drought/rainfall periods. Intensive droughts have been recorded in 2000, 2003, 2007, 2011 and 2012. In addition to observed temperature and precipitation changes, future climate scenarios predict significant changes in the magnitude and frequency of climatic extremes, such as storms accompanied by hail, devastating effects of wind, floods, long-term droughts, heat waves, and extremely high and low temperatures. The Bosna River forms at the foot of the Igman Mountain at an altitude of about 500 meters. Figure 1 shows the basin of the Bosna River, which is 273 km long and 35-170 m wide. Its average drop is 1,48 m/km, and the average flow rate is 100 m^3/s. The main tributaries on the right side of the river Bosna are Miljacka (35,9 km), Krivaja (101 km), Željeznica (26.9 km), Spreča (137,5 km) and Stavnja (30,4 km), while on the left side are the Usora (82 km), Lašva (49,4km) and Fojnička river (46km).

Fig. 1. River Bosna basin [12]

In this paper, the impact of climate change on the flow rate on the Bosna River is analysed. For this purpose, a statistical analysis of the flow parameters was carried out based on the measured values of daily flows at the Reljevo hydrological station on the Bosna River. The analysis was conducted for two different time periods (1961-1990 and 2000-2010) to obtain the temporal distribution of relevant flow rate data.

2. Research method

Different techniques are used to analyse the data sets collected during long periods (10 or 30 years). For the study, the two-factor factorial experimental design techniques was used where one factor was years (a-level), and the other was months (b-level). Statistical analysis provides a temporal distribution of relevant flow rate in two periods, 1961-1990 and 2000-2010, and warns of possible scenarios in the future [13], [14]. The experiment's purpose is to check whether a certain variable affects a clearly defined phenomenon as the dependent variable.

An independent variable (experimental variable) is a factor that is intentionally varied in the experiment to see if it affects the dependent variable. The analysis includes flows at the Reljevo hydrological station. The statistical analysis of the two-factor factorial involves only two factors, say A and B, one factor being years (a-level) and the second factor being months (b-level). Analysis of variance (ANOVA) is shown in Table 1. It is assumed that A and B are fixed factors, and a is the level of factor A and b is the level of factor B, specially selected by the researcher and the conclusions are limited to these levels only.

Source of variation	**Sum of Squares**	**Degrees of Freedom**	**Mean value**	F_0
A Treatment	SS_A	$(a-1)$	$MS_A = \frac{SS_A}{a-1}$	$\frac{MS_A}{MS_E}$
B Treatment	SS_B	$(b-1)$	$MS_B = \frac{SS_B}{b-1}$	$\frac{MS_B}{MS_E}$
Interaction	SS_{AB}	$(a-1)(b-1)$	$MS_{AB} = \frac{SS_{AB}}{(a-1)(b-1)}$	$\frac{MS_{AB}}{MS_E}$
Error	SS_E	$ab(n-1)$	$MS_E = \frac{SS_E}{ab(n-1)}$	
Total	SS_T	$(abn-1)$		

Table 1. ANOVA table two-factor factorial experimental design [12]

The hypothesis that were tested are [15]:

1. H_0: $\tau_1 = \tau_2 = ... = \tau_a = 0$ (no effect of factor A)
   H_1: at least one $\tau_i \neq 0$
2. H_0: $\beta_1 = \beta_2 = ... = \beta_b = 0$ (no effect of factor B)
   H_1: at least one $\beta_i \neq 0$
3. H_0: $(\tau\beta)_{11} = (\tau\beta)_{12} = ... = (\tau\beta)_{ab} = 0$ (no interaction)
   H_1: at least one $(\tau\beta)_{ij} \neq 0$

Analysis of Variance (ANOVA) is a hypothesis testing procedure used to decompose the total variability of the data into its components and then compare the different decomposed elements. The total variability is calculated according to the expression (1) [15]:

$$SS_T = SS_A + SS_B + SS_{AB} + SS_E \tag{1}$$

From the previous equation, it can be seen that the overall sum of squares (SS_T) consists of the sum of squares for factor A (SS_A), the sum of squares for factor B (SS_B), the sum of squares for the interaction between factors A and B (SS_{AB}) and residual sum of squares (i. e. the sum of squares error, SS_E). The total number of degrees of freedom is $(abn-1)$. The main effects of factors A and B have $(a-1)$ and $(b-1)$ degrees of freedom, while the interaction effect of factor AB has $(a-1)(b-1)$ degrees of freedom. In each of the ab cells, in Table 1, there are $(n-1)$ degrees of freedom between n repetitions. Observations in the same cell may differ only due to random error, and we have $ab(n-1)$ degrees of freedom for error. Therefore, the degrees of freedom are distributed according to equation (2) [15]:

$$abn - 1 = (a-1) + (b-1) + (a-1)(b-1) + ab(n-1) \tag{2}$$

If each sum of squares on the right side of the equation is divided by the corresponding number of degrees of freedom, the mean square value for AB, interaction and error is obtained [15].

$$MS_A = \frac{SS_A}{a-1} \qquad MS_B = \frac{SS_B}{b-1}$$
$$MS_{AB} = \frac{SS_{AB}}{(a-1)(b-1)} \qquad MS_E = \frac{SS_E}{ab(n-1)} \tag{3}$$

In order to test that all effects of factor A are equal to zero ($H_0\colon \tau_i = 0$), the following relationship is used:

$$F_0 = \frac{MS_A}{MS_E} \tag{4}$$

which has F – distribution with $(a-1)$ and $ab(n-1)$ degrees of freedom if $H_0\colon \tau_i = 0$ is true.

This null hypothesis is rejected at α significant level if $f_0 > f_{\alpha,a-1,ab(n-1)}$. Similarly, to test the hypothesis that all effects of factor B are equal to zero (H_0: $\beta_i = 0$), the relationship is used:

$$F_0 = \frac{MS_B}{MS_E} \tag{5}$$

which has F – distribution with $(b-1)$ and $ab(n-1)$ degrees of freedom if H_0: $\beta_i = 0$ is true.

This null hypothesis is rejected at α significant level if $f_0 > f_{\alpha,b-1,ab(n-1)}$. Finally, to test the hypothesis H_0: $(\tau\beta)_{ij} = 0$, i.e. hypothesis that all interaction effects are equal to zero, the relationship is used:

$$F_0 = \frac{MS_{AB}}{MS_E} \tag{6}$$

which has F – distribution with $(a-1)(b-1)$ and $ab(n-1)$ degrees of freedom if H_0: $(\tau\beta)_{ij} = 0$ is true. This null hypothesis is rejected at α significant level if $f_0 > f_{\alpha,(a-1)(b-1),ab(n-1)}$.

It is usually better to test for interaction first and then estimate main effects. If the interaction is not significant, interpretation of tests for main effects is required. However, when the interaction is significant, the effects of the main factors involved in the interaction may not have significant practical interpretive value [16].

2.1. Application of t-test

The t-test is used to determine if there is a statistically significant difference between the means of two samples and is valid only when both samples are obtained from a normally distributed population. It is true in many practical cases where there has been a change in working conditions with a need to determine if there is a significant change in those conditions. The first step is to determine the means of $\overline{x}_1$ and $\overline{x}_2$ for the two samples. The standard deviation is obtained based on the expression:

$$s^2 = \frac{\sum(x_1 - \overline{x_1})^2 - \sum(x_2 - \overline{x_2})^2}{n_1 + n_2 - 2} \tag{7}$$

Where x_i refers to any experimental value in the first sample and x_2 to any in the second sample, and the sample sizes are n_1 and n_2, respectively.

Usually, it is easier to calculate with a modified expression:

$$s^2 = \frac{\sum x_1^2 + \sum x_2^2 - \frac{1}{n_1}(\sum x_1)^2 - \frac{1}{n_2}(\sum x_2)^2}{n_1 + n_2 - 2} \tag{8}$$

After determining standard deviation, ***t*** is obtained as follows:

$$t = \frac{\overline{\mathrm{x}_1} - \overline{\mathrm{x}_2}}{\mathrm{s}\sqrt{\frac{1}{\mathrm{n}_1} + \frac{1}{\mathrm{n}_2}}} \tag{9}$$

The obtained value for t is compared with the tabled value for t_α, at the required confidence level with $(n_1 + n_2 - 2)$ degrees of freedom. If $t > t_a$ there is a significant difference between the mean values.

2.2. Hydrological and meteorological analysis

The analysed temperature and precipitation data in [12] didn't indicate extreme climatic events, like heat waves or heavy precipitation. The analysis of climatic conditions for the previously discussed periods shows moderate high temperatures in summer and moderate low temperatures in winter, with seasonal variations. Also, this affects the hydrological cycle (occurrence of droughts and floods). Water and climate variability already pose a risk due to flooding. The Bosna River basin is exposed to occasional floods, which cause significant damage to the economy.

The Reljevo hydrological station is located on the downstream side of the bridge over the Bosna River in the Reljevo settlement. It was founded in 1904. It stopped working in 1991 and was rebuilt at the end of 1998 with the same profile and the same ground-level zero. The area of the basin is 1094 km^2 [12]. There are anthropogenic effects on the Bosna River upstream of the Reljevo hydrological station, but there are insignificant on the river flow at the Reljevo site. Željeznica River, the right tributary of the Bosnia River (ca. 8,5 km upstream from Reljevo hydrological station), is affected by the Bogatići hydroelectric power plant.

3. Results and discussion

The cumulative monthly and annual flow rate data at the hydrological Reljevo station are investigated. By comparing data for two periods, possible changes in distribution can be observed, as well as significant changes that occurred in the past in the observed area. The mean flow rate values during the two analyzed periods, 1961-1990 and 2000-2010, are shown in Tables 2 and 3.

Num.	WB / vatercourse	I	II	III	IV	V	VI	VII	VIII	IX	X	XI	XII	Qm.a.
1	Reljevo / Bosna	29,2	33,4	40,3	51,5	43,4	24,2	15,8	11,7	13,9	20,5	28,8	34,8	28,9

Table 2. Average monthly and annual flows (m^3/s) from 1961-1990

Table 2 shows the values of average monthly and annual flows for the period from 1961 to 1990. Based on a thirty-year analysis, it is evident that the average monthly flow rate values are high in the spring season (March, April and May), followed by a gradual decrease in the summer season. The maximum mean monthly flow rate value is in April, 51,5 m^3/s. After a gradual drop in flow rates in June, July and August, the increase is observed again in September. The increase in flow rates continues in October, November and December. A slight drop in flow rates is noticed in January, after which flow rate values increase in February. The average annual flow rate based on a thirty-year analysis is 28,9 m^3/s.

Num.	WB / vatercourse	I	II	III	IV	V	VI	VII	VIII	IX	X	XI	XII	Qm.a.
1	Reljevo / Bosna	32,5	28,5	49,9	54,5	31,1	20,4	13,0	10,8	16,0	24,3	29,0	40,6	29,2

Table 3. Average monthly and annual flows (m^3/s) from 2000-2010

For the period from 2000-2010, the average monthly and annual flow rates are shown in Table 3. Based on the ten-year analysis, it is evident that the average monthly flow rate values follow the same trend as previously discussed: higher values in March, April and May, with a maximum mean monthly flow rate in April (54,5 m^3/s), followed by a gradual decrease during the summer season. The growth trend continues during September, October, November and December, followed by a slight drop in flow in January. The average monthly flow based on a ten-year analysis is 29,2 (m^3/s).

According to the current Water Law [17], the ecologically acceptable flow rate is determined as the minimum average monthly flow is 95 % of the probability of occurrence based on the hydrological characteristics of the water body for observed seasons. The ecologically acceptable flow rate for the Reljevo hydrological station is 4,96 m^3/s.

By comparing the monthly mean and annual mean flow rate values for the observed periods at the Reljevo hydrological station, a decreasing trend in mean values can be observed from May (V) to August (VIII) in the 2000 to 2010 period and the opposite trend from March (III) to April (IV) and then from September (IX) to December (XII). However, no significant differences are observed in average annual flow rates for 1961-1990 compared to the 2000-2010 period.

Statistical analysis factorial experiment includes only two factors A and B. For factor A it is always 36 (number of months, $b = 12$, times number of repetitions, $n = 3$). For factor B, it is 90 or 30 (number of years $a = 30$ or 10 and $n = 3$). The variance analysis (ANOVA) of the flow rate data measured at the Reljevo hydrological station for the two analysed periods from 1961 to 1990 and from 2000 to 2010 is shown in Tables 4 and 5.

In Tables 4 and 5, in addition to the obtained F_o values for individual factors, values for F_t were also calculated (two-tailed t-test). The obtained F_o values are higher than F_t values, and there is a significant difference at the new confidence level of 95 % (α= 0.05). The F_o values for treatment B are higher than F_t as expected because changes in precipitation quantity are significant during the observed period.

Source of variation	**Sum of Squares**	**Degrees of Freedom**	**Mean value**	**F_o**	**F_t**
A Treatment	39939,952	29	1377,240	6,732	1,47
B Treatment	152299,701	11	13845,427	67,672	1,80
Interaction	184252,684	319	577,595	2,823	1,18
Error	147308,847	720	204,596		
Total	523801,184	1079			

Table 4. ANOVA two-factor factorial experimental design (flow rate for the period 1961-1990)

Source of variation	Sum of Squares	Degrees of Freedom	Mean value	F_o	F_t
A Treatment	7186,780	9	798,531	2,764	1,92
B Treatment	62152,786	11	5650,253	19,561	1,81
Interaction	46928,783	99	474,028	1,641	1,31
Error	69324,880	240	288,854		
Total	185593,229	359			

Table 5. ANOVA two-factor factorial experimental design (flow rate for the period 2000-2010)

All cumulative values presented in the paper were obtained as part of the output in the programming language C++ for the two-factorial experimental design. Figure 2 shows the cumulative yearly flow rate values at the Reljevo hydrological station from 1961 to 1990 period. The minimum cumulative flow rate recorded in 1990 is 644, while the maximum cumulative flow value recorded in 1978 is 1599,3.

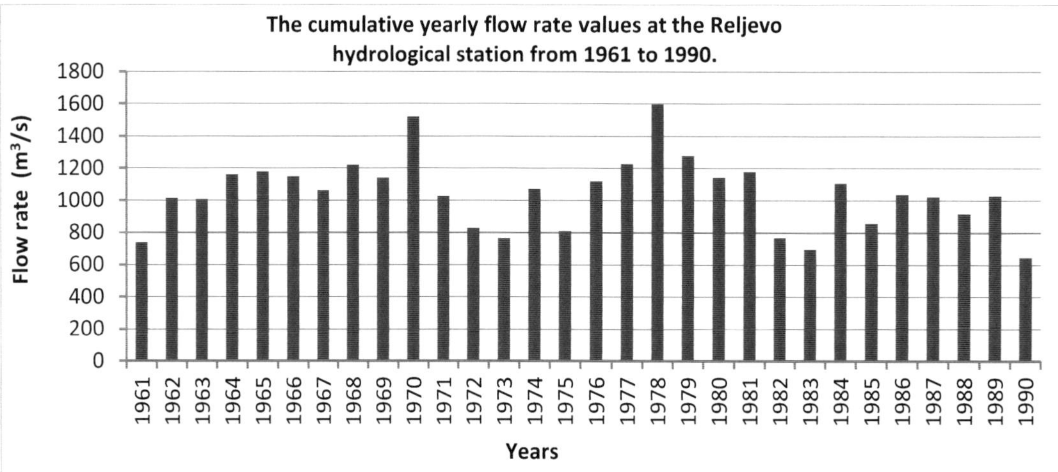

Fig. 2. The cumulative yearly flow rate values at the Reljevo hydrological station from 1961 to 1990

Figure 3 shows the cumulative flow rate values at the Reljevo hydrological station from 2000 to 2010. The maximum cumulative value flow rate recorded in 2006 is 1369,5 m^3/s, while the minimum value recorded in 2004 is 788 m^3/s.

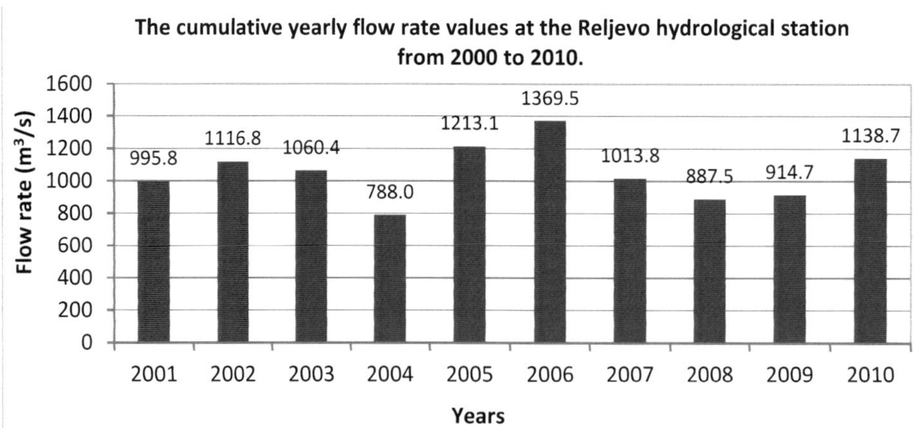

Fig. 3. The cumulative yearly flow rate values at the Reljevo hydrological station from 2000 to 2010

Based on the analysis of both periods, the change in absolute maximum flows from 1961 to 1990 hasn't been observed. Figure 4 shows the cumulative monthly value at the Reljevo hydrological station from 1961 to 1990. In January, the measured value of the flow was 2638,9 m^3/s, followed by a slight increase in the flow rate in February to 3011,5 m^3/s.

In March and April, the trend of increasing flow rates continues. The cumulative flow value from 1961-1990 reached its maximum in April as 4630,9 m^3/s. The possible reason for the increase in the flow rate is the rise in temperatures and the snow melting. From May to September, the flow rate gradually decreases. There is a slight increase in the flow rate due to rainfall in September. The trend of increasing flow rates continues in October 1847,6 m^3/s, November 2596,3 m^3/s and December 3126,2 m^3/s.

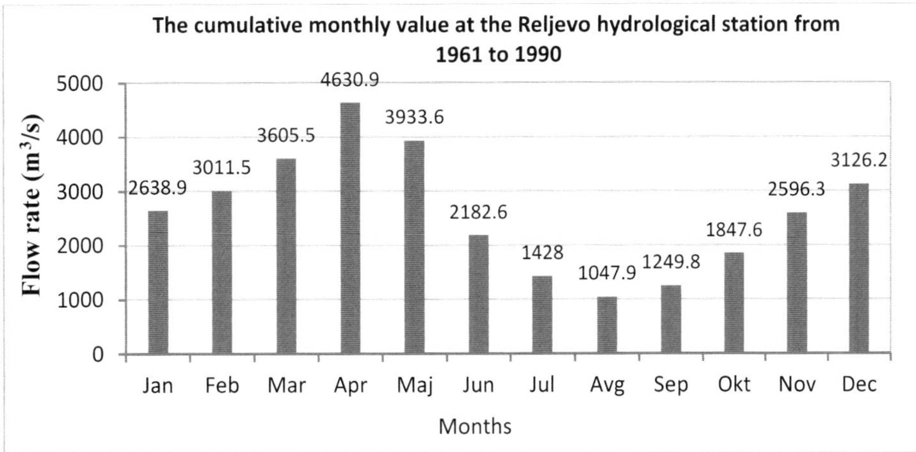

Fig. 4. The cumulative monthly value at the Reljevo hydrological station (1961-1990)

Figure 5 shows the cumulative monthly value at the Reljevo hydrological station from 2000 to 2010. In January, the measured value of the flow rate was 971,9 m^3/s, followed by a slight decrease in the flow rate in February to 856,5 m^3/s. In the following months (March and April), the flow rate values were increased to 1484,7 m^3/s and 1634,9 m^3/s, respectively. The flow rate values reached their maximum in March and April due to the sudden rise in temperatures and melting snow. In May, the measured value of the flow rate was 938,9. In June (611 m^3/s), July (391,3 m^3/s) and August (323 m^3/s), the trend of decreasing flow rates continues. There was a slight increase in flow rate in September (479 m^3/s). This increasing trend continues from October (724,1 m^3/s) until the end of the year.

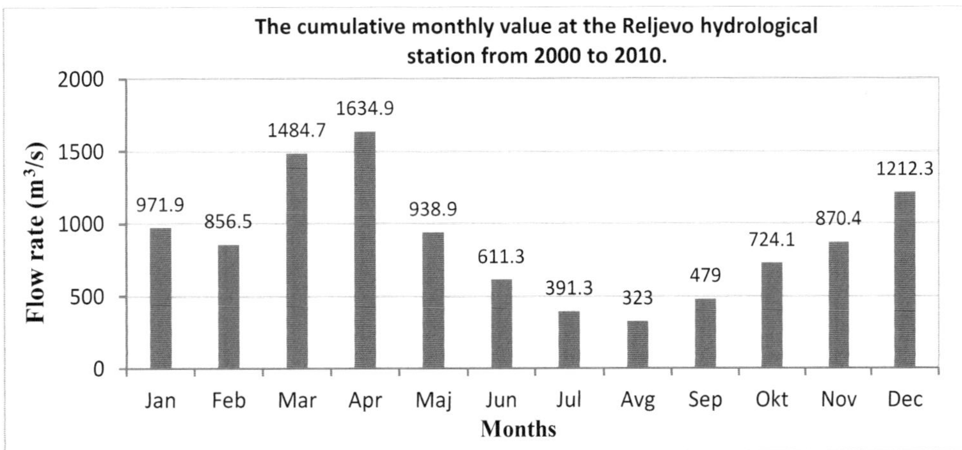

Fig. 5. The cumulative monthly value at the Reljevo hydrological station (2000-2010)

4. Conclusion and further work

Climate change represents one of the main challenges of today. Many relevant institutions (IPCC) have recognized the problem of climate change with its complexity. Bosnia and Herzegovina is involved in several studies, but a regional climate assessment model still does not exist. Since climate models have advantages and disadvantages, we cannot expect them to "predict" the future. However, the global models of the future climate for Bosnia and Herzegovina indicate significant changes in climate conditions. During the 20th century, a trend of decreasing precipitation was observed, and it is difficult to understand if it is the result of natural climatic fluctuations or is due to human influence. Water and climate variability already represent a risk due to floods, and Bosnia and Herzegovina have been exposed to occasional floods that cause significant damage to the economy.

In this paper, based on the ANOVA tables for flow rates for both observed periods in Reljevo station, there have been variations by year and more significant by month. The approximation of the flow rates by a parabola is not satisfactory due to the existence of maxima and minima, and for this purpose, a third-order polynomial should be used. The analysis showed that the maximum flows are mainly in March and April. The increase in flow rates in spring months is due to increased precipitation and snow melting due to a rise in air temperature. As a result of statistical analysis, the increasing flow rate is observed from 2000 to 2010 compared to the period from 1961 to 1990.

The cause of the increase in precipitation in most cases, and therefore water rise and floods, is climate change due to the increase in the concentration of greenhouse gases. The construction of large hydrotechnical systems causes changes in water regimes and also influences the occurrence of floods. Authorities that prepare water resource management plans should include the possible effects of climate change in their planning. This problem may require further development of specific information – such as regional climate models development in flood protection planning, groundwater recharge and river flows. There is very little data on the effects of climate variability, either direct or indirect, so in the future, it would be useful to analyse other climate variables (temperature and precipitation) to get a clearer picture of climate yearly variability.

5. References

[1] Hattermann, F. F.; Vetter, T., Breuer, L.; Su, B., Daggupati, P.; Donnelly, C.; Fekete, B.; Florke, F.; Gosling, S. N.; Hoff-mann, P.; Liersch, S.; Masaki, Y.; Motovilov, Y.; Muller, C.; Samaniego, L.,; Stacke, T.; Wada, Y.; Yang, T. & Krysnaova, V. (2018). Sources of uncertainty in hydrological climate impact as-sessment: A cross-scale study. Environmental Research Letters, Vol. 13, No. 1, 015006. https://doi.org/10.1088/1748-9326/aa9938.

[2] Leiserowitz A.; Maibach E.; Rosenthal S.; Kotcher J.; Carman J.; Neyens L.; Myers T.; Goldberg M.; Campbell E.; Lacroix K. & Marlon J. (2022). Politics & Global Warming, April 2022. Yale University and George Mason University. New Haven, CT: Yale Program on Climate Change Communication.

[3] IPCC, 2007: Climate Change 2007: Synthesis Report. Contribution of Working Groups I, II and III to the Fourth Assessment Report of the Intergovernmental Panel on Climate Change [Core Writing Team; Pachauri, R.K. & Reisinger, A. (eds.)]. IPCC, Geneva, Switzerland, 104 pp.

[4] IPCC, 2021: Climate Change 2021: The Physical Science Basis. Contribution of Working Group I to the Sixth Assessment Report of the Intergovernmental Panel on Climate Change [Masson-Delmotte, V., P.; Zhai, A.; Pirani, S. L.; Connors, C.; Péan, S.; Berger, N.; Caud, Y.; Chen, L.; Goldfarb, M.I.; Gomis, M.; Huang, K.; Leitzell, E.; Lonnoy, J. B. R.; Matthews, T. K.; Maycock, T.; Waterfield, O.; Yelekçi, R. Yu & B. Zhou (eds.)]. Cambridge University Press, Cambridge, United Kingdom and New York, NY, USA, 2391 pp. doi: 10.1017/9781009157896.001

[5] Chattopadhyay, N. & Hulme, M. (1997). Evaporation and Potential Evapotranspiration in India under Conditions of Recent and Future Climate Change. Agricultural and Forest Meteorology, Vol. 87, No. 1, pp. 55-73. http://dx.doi.org/10.1016/S0168-1923(97)00006-3.

[6] Milly, P.C.D.; Dunne, K.A. & Vecchia, A.V. (2005). Global Pattern of Trends in Streamflow and Water Availability in a Changing Climate. Nature, Vol. 438, pp. 347-350. http://dx.doi.org/10.1038/nature04312

[7] Seneviratne, S. I.; Corti, T.; Davin, E. L.; Hirschi, M.; Jaeger, E. B.; Lehner, I.; Orlowsky, B. & Teuling, A. J. (2010). Investigating soil moisture-climate interactions in a changing climate: A review, Earth Science Reviews, Vol. 99, No. 3-4, pp. 125–161, doi:10.1016/j.earscirev.2010.02.004.

[8] Meresa, H. K. & Gatachew, M. T. (2019). Climate change impact on river flow extremes in the Upper Blue Nile River basin. Journal of Water and Climate Change, Vol. 10, No. 4, pp. 759-781. https://doi.org/10.2166/wcc.2018.154.

[9] Meresa, H. K. & Romanowicz, R. J. (2017). The critical role of uncertainty in projections of hydrological Extremes. Hydrology and Earth System Sciences, Vol. 21, No. 8, pp. 4245–4258. https://doi.org/10.5194/hess-21-4245-2017

[10] Ahmovic, I.; Dzaferovic, E. & Hadziahmetovic, H. (2018). Analysis of Temperature Data in the Area of the City of Sarajevo (Bjelave) in Period From 2000-2010, Proceedings of the 29th DAAAM International Symposium, pp. 0260-0266, B. Katalinic (Ed.), Published by DAAAM International, ISBN 978-3-902734-20-4, ISSN 1726-9679, Vienna, Austria. doi: 10.2507/29th.daaam.proceedings.037

[11] Radusin, S.; Oprašić, S.; Cero, M.,; Abdurahmanović, I. & Vukmir, G. (2013). Climate change adaptation and low-emission development strategy for Bosnia and Herzegovina. Sarajevo: Ministry of Foreign Trade and Economic Relations, 82, pp. 17-21.

[12] Ahmović, I. (2014). Climate change and their influence on the water regime in Bosnia and Herzegovina, with special focus on the Sarajevo field, Master thesis, Faculty of Mechanical Engineering Sarajevo, Sarajevo, Bosnia and Herzegovina.

[13] Bajramović, R. (2003). Determination of critical values in normal and t - distribution, Mašinstvo, Vol. 2, No. 7, pp. 93-104.

[14] Hayter, A.J. (2012). Probability and Statistics for Engineers and Scientists, Thomson Brooks/Cole, ISBN 9781133112143, Boston, USA.

[15] Montgomery, D.C. & Runger, G.C. (2018). Applied Statistics and Probability for Engineers, 7th Edition, John Wiley & Sons, Inc., ISBN 978-1-119-40036-3, New York, USA.

[16] Montgomery, D.C. (2001). Design and Analysis of Experiments, 5th Edition, John Wiley & Sons Ltd., ISBN 978-0471316497, New York, USA.

[17] Zakon o vodama FBiH (2006). Službene novine FBiH", br. 70/06.

DOI: 10.2507/34th.daaam.proceedings.030

WIND FARM ENERGY PRODUCTION IN BOSNIA AND HERZEGOVINA: LEVELIZED COST OF ENERGY

Rejhana Blažević, Halima Hadžiahmetović, Džana Kadrić & Mirela Alispahić

This Publication has to be referred as: Blazevic, R[ejhana]; Hadziahmetovic, H[alima]; Kadric, D[zana] & Alispahic, M[irela] (2023). Wind Farm Energy Production in Bosnia and Herzegovina: Levelized Cost of Energy, Proceedings of the 34th DAAAM International Symposium, pp.0237-0246, B. Katalinic (Ed.), Published by DAAAM International, ISBN 978-3-902734-41-9, ISSN 1726-9679, Vienna, Austria
DOI: 10.2507/34th.daaam.proceedings.030

Abstract

This paper analyses the levelized cost of energy (LCOE) for wind farms currently operating in Bosnia and Herzegovina (B&H). An overview of existing wind farms and planned projects in B&H is provided. The methodology for calculating LCOE is presented, where LCOE represents the electricity selling price needed for wind farm investors to recover capital costs. Using this methodology, LCOE values were calculated for three operational wind farms in B&H - Mesihovina, Jelovača, and Podveležje. A sensitivity analysis was also conducted by varying key parameters like discount rate, plant lifetime, operation and maintenance costs (O&M), and energy output. The results showed LCOE is most sensitive to discount rate variations, while changes in operations and maintenance costs have the least impact. This analysis provides valuable economic assessment of wind power investments in B&H. The calculated LCOE values can inform policy-making and incentives aimed at further growth of cost-competitive wind energy to support B&H's renewable energy and decarbonization goals aligned with the European Union (EU).

Keywords: renewable energy sources; wind farm; levelized cost of energy (LCOE); decarbonization; electricity production; wind atlas;

1. Introduction

The energy sector is responsible for over 70% of global CO_2 emissions, representing the greatest potential for reducing greenhouse gases [1]. The power sector, as the single largest source of CO_2, must undergo significant transition to reduce emissions. Currently, Bosnia and Herzegovina (B&H) energy production relies heavily on coal (≈66%) and hydropower (≈33%). The share of wind farms and solar photovoltaic power plants is less than 3% [2]. To align with the EU, B&H must increase system energy efficiency, energy security, transition from coal, and accelerate transition to renewable energy sources. The EU aims to reduce greenhouse gases for 80-95% below 1990 levels by 2050, requiring complete decarbonization. The strategy for achieving these goals is clear, namely the transition to sustainable, renewable energy sources and a significant improvement in energy efficiency [3]. B&H path to decarbonization have made significant strides, particularly with the signing of the Sofia Declaration on the Green Agenda for the Western Balkans in November 2020.

This commitment aligns the country with the EU goals for energy transition and carbon neutrality by 2050, as well as establishing an energy transition plan until 2030 [2], [3], [4]. To achieve the necessary reduction in greenhouse gas emissions by 2030, B&H will primarily need to shut down outdated and inefficient thermal power plants and replace them with renewable energy sources. Additionally, the country will need to introduce an emissions trading system and implement numerous energy efficiency measures to reduce energy consumption.

Bosnia and Herzegovina strategic commitment to include renewable energy sources in the energy production system, particularly wind farms and solar power plants, is evident in various studies analysing the country wind potential and the development of wind farms. The country has set ambitious targets for increasing energy efficiency and the use of renewable energy sources, with a goal of achieving a 43.6% share of renewables in gross final consumption by 2030. The forthcoming National Climate and Energy Plan will help Bosnia and Herzegovina to ensure energy security while improving its long-term resilience to climate change.

Numerous studies have analysed the wind energy potential in various countries and the particularities of legislation, challenges, and perspectives for wind farm development [5]. Several studies presented assessment of wind potential at specific locations in B&H. For instance, the wider area of Sarajevo area was examined, concluding insufficient wind potential for wind-powered electricity generation in Sarajevo, while the available wind energy potential near Sarajevo on Bjelašnica mountain is very high [6], [7]. When evaluating wind farm profitability, the electricity production cost for a given site is analysed. Tizgui et al. assessed wind potential and compared electricity production costs from nine wind farms in Morocco, finding lower production costs than the guaranteed electricity purchase price [8]. Similarly, Bahrami et al. utilized performance indicators to evaluate technical and economic feasibility of systems in Uzbekistan, determining the critical specific investment cost to equalize energy production cost and the electricity tariff [9].

Onshore wind farms account for 90% of the total wind energy capacity in EU, and most of the new wind energy capacity under construction is also onshore. In B&H, new onshore wind farms are being constructed to expand renewable energy capacity. Across EU, onshore wind energy represents the most cost-competitive form of new electricity generation, as demonstrated by levelized cost of energy (LCOE) analyses. Figure 1 illustrates the wide range of LCOE for different technologies in Germany in 2021. Onshore wind is on the lower end of the cost spectrum, demonstrating its competitiveness [10].

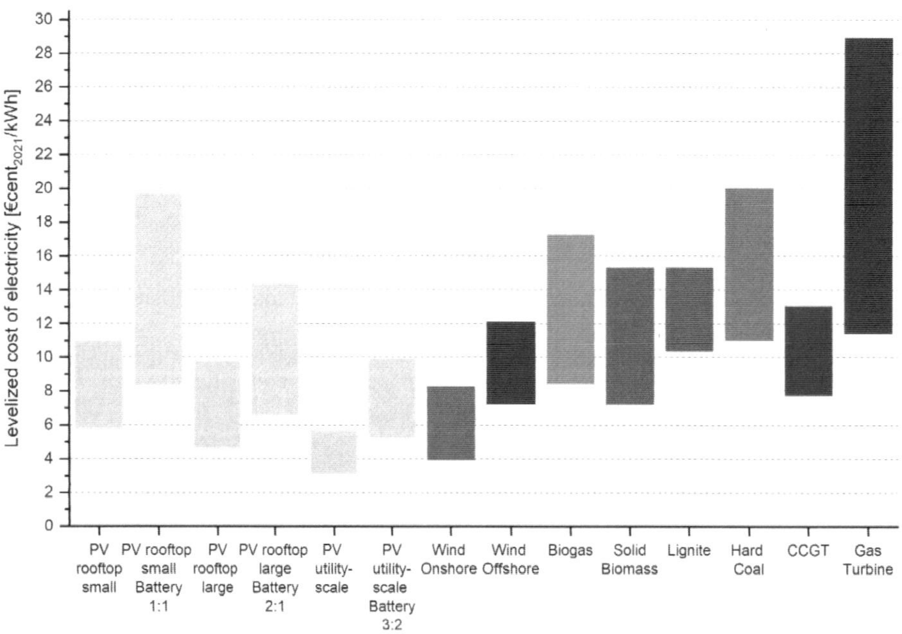

Fig. 1. Levelized cost of energy of renewable energy technologies and conventional power plants in Germany in 2021 [10]

All renewable energy technologies continue to record decreasing LCOE. From Figure 1 it can be seen that the LCOE of onshore wind power plants in 2021, with specific plant costs ranging from 1.400 to 2.000 €/kW, were between 3,94 and 8,29 €cent/kWh, or 39,4 and 82,9 €/MWh [10].

The main goal of this research is to calculate and present the levelized cost of electricity (LCOE) values for three wind farms in B&H that are currently in operation. The obtained LCOE results are analysed and compared. An appropriate LCOE calculation methodology was developed and is presented in this paper. As part of the analysis, the sensitivity of the LCOE to variations in key parameters like the discount rate, plant lifetime, operation and maintenance costs, and energy output variability was also investigated.

2. Wind potential in Bosnia and Herzegovina, an overview of existing systems

In the electricity development strategy of B&H until 2030, wind energy occupies a significant place. Wind energy has significant potential in B&H, especially in the southern and western regions. Measurements conducted at around 30 potential wind farm locations found average annual wind velocities of 6-8 m/s at a height of 50 meters. The total potential of the observed locations, from the point of view of the available space, is estimated at about 900 MW, Figure 2. If the availability of suitable spaces (connection to the network, environmental protection, etc.) is not taken into account, the total technical wind potential is significantly higher and is estimated at about 2000 MW [11]. However, a detailed wind atlas mapping wind velocities and power densities at different heights is lacking, which creates uncertainty in potential assessments. The selection of the optimal location for constructing a wind power plant depends on technical, economic, and environmental constraints. The initial technical analysis involves assessing the wind potential at a given site. By conducting long-term wind measurements and using topographical maps of the terrain, wind potential maps can be created, including maps of mean annual wind velocity (m/s) and mean annual wind power density (W/m^2) at different heights.

A major challenge for B&H is the lack of a wind atlas, which serves as the foundation for detailed research and, together with other spatial data like topographic maps and satellite images, enables credible evaluation of the country wind energy potential. Precise determination of wind potential necessitates the most accurate indicators, i.e. a detailed wind atlas, which requires continuous multi-year measurements per standards and regulations. Creating a wind atlas involves long-term assessment of the wind regime and terrain through field measurements and numerical modelling [12]. Wind atlases provide wind velocity and direction data in the form of maps, time series, and distributions at different heights. They are used to pre-select wind farm sites by providing high-resolution data on wind resources. The Global Wind Atlas offers freely accessible global wind resource data to facilitate site selection and planning. Overall, a detailed wind atlas formed by sustained standardized measurements over several years is essential for accurately evaluating wind potential and siting wind power plants.

Since wind atlas data and wind maps are not available for B&H, the assessment of wind energy potential in the country is primarily based on the spatial distribution of average annual wind velocity and power. These data are the result of applying the global atmospheric model formulated in the World Wind Atlas [11], [13]. The World Wind Atlas utilizes global meteorological information, and the results obtained from this model have not been verified with ground measurements. While on-site measurements would provide more accurate results, this wind atlas can be considered sufficiently representative for an initial, indicative evaluation of wind potential in B&H [13].

Figure 3, taken from [13], shows the general spatial distribution of average annual wind velocity across B&H. It indicates that the highest wind potential is in the southwestern part of the country. According to world wind atlas, almost 50% of the land area of B&H has an average wind velocity above 6 m/s. Figure 4 depicts the surface roughness length in B&H, while Figure 5 shows the orography. Surface roughness greatly impacts wind velocity – as roughness increases, so does drag, slowing down the wind. With increasing height above ground level, the influence of roughness decreases and wind velocity increases. Figure 4 shows that roughness in southwestern Bosnia and Herzegovina ranges from 0,03 to 0,1, corresponding to roughness classes 1 and 2 per the European Wind Atlas. Figure 5 indicates that B&H has complex terrain. Elevation in the southwest, where wind potential is highest, varies from 300 to 1700 meters above sea level.

Fig. 2. The area with the highest wind potential [11]

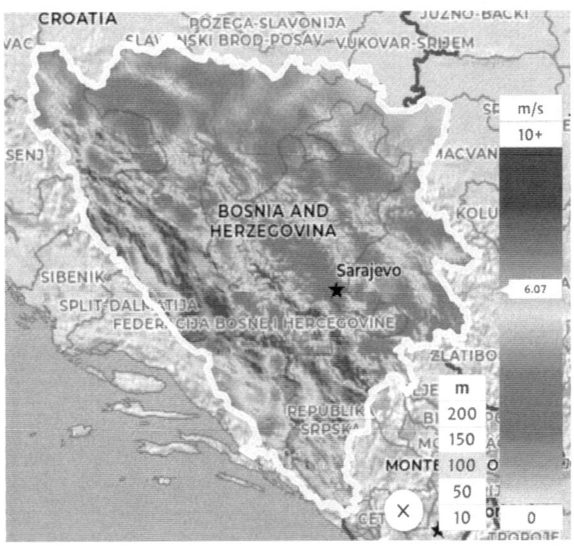

Fig. 3. Wind potential in Bosnia and Herzegovina [13]

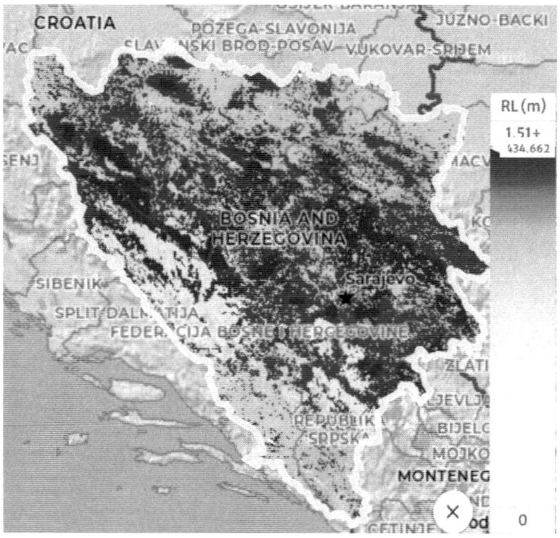

Fig. 4. Roughness length in Bosnia and Herzegovina [13]

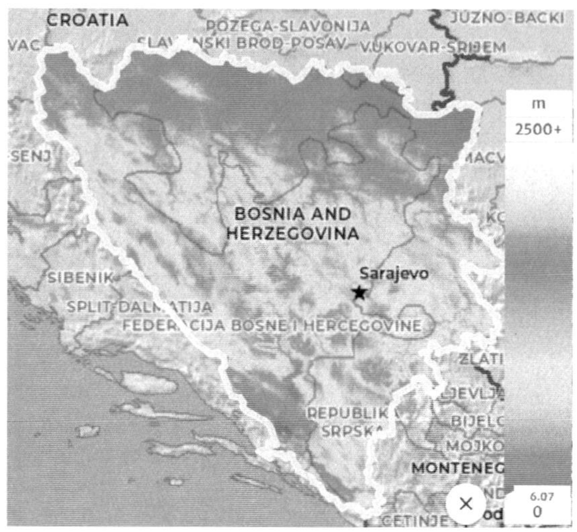

Fig. 5. Orography in Bosnia and Herzegovina [13]

Bosnia and Herzegovina has recently begun constructing wind farms at several locations, with some already completed and operational. The country currently has three wind farms producing electricity, representing a total investment of around €200 million. Two wind farms are located in the Tomislavgrad area (the Mesihovina wind farm operational since 2018 and the Jelovača wind farm since 2019), while a third in the Podveležje plateau above Mostar started producing electricity in early 2021. Two wind farms are owned by the state and one is privately owned. Funding was provided primarily through loans from the German Development Bank. Together, these three wind farms generate approximately 400 GWh of electricity per year on average.

The Mesihovina wind farm, located near Tomislavgrad, has a total of 22 wind turbines of type SWT-2.3-108 manufactured by Siemens Gamesa, with a total installed capacity of 50,6 MW. The Jelovača wind farm, also located near Tomislavgrad, consists of 18 wind turbines from Siemens Gamesa's 2.0 series, with a total installed capacity of 36 MW. The Podveležje wind power plant near Mostar has 15 wind turbines and a total installed capacity of 48 MW. Considering technological advancements and global trends in wind power utilization, it is realistic to expect increased construction of wind farms and their integration into B&H energy system. The extent of wind power development in B&H will largely depend on the policies and actions of state and entity administrations regarding wind energy utilization. Table 1 provides an overview of existing wind farms in Bosnia and Herzegovina [2].

Wind farm	**Electric Utility of B&H**	**Total Power**	**Power of aggregate**	**Location of wind farm**	**Annual production of electricity**	**Commissioning**
-		MW	MW	-	GWh	-
Mesihovina	PE " Electric Utility HZ HB"	50,6	22 x 2,3	Municipality Tomislavgrad	165	2018
Jelovača	FL WIND d.o.o.	36	18 x 2	Municipality Tomislavgrad	110	2019
Podveležje	PE " Electric Utility B&H"	48	15 x 3,2	Mostar city	130	2021

Table 1. Wind farms in Bosnia and Herzegovina [2]

Public Enterprise Electric Utility of B&H is working on the Bitovnja wind power project near Konjic, which plans to install 25 wind turbines with a total capacity up to 60 MW and estimated annual production of 110 GWh. Additionally, there are plans for the 50 MW Vlašić wind power project with 18 turbines and expected annual production of 115 GWh. The 2019-2029 Development Plan of Public Enterprise Electric Utility of Republic Srpska defines activities for implementing renewable energy projects including three wind farms - Hrgud, Donja Trusina, and Radimlja. New laws were passed in the Republic Srpska entity in 2022 and the Federation entity in 2023 to promote renewable energy growth through measures like auctions for feed-in tariffs, renewable energy communities, and simplifying administrative procedures, therefore it is expected that number of wind farms will grow in the future.

3. Methodology for estimating levelized cost of energy production from wind farms

Levelized cost of energy (LCOE) is a widely utilized economic assessment metric used to calculate the average cost per unit of electricity generation (kWh) expressed in net present value (NPV) throughout the lifetime of the plant. In other words, LCOE is an economic estimate of the total cost of construction and operation of an electricity production plant during its lifetime divided by the total amount of electricity produced [14]. It represents the break-even price for an asset over its lifetime and is widely used for investment decision-making and policy analysis in the energy sector. The LCOE enables consistent comparison of different electricity generation technologies across key economic and technical parameters such as capital costs, capacity factors, fuel costs, operating life, and financing costs. However, the LCOE methodology has limitations, as it does not fully capture integration costs, risk, and other externalities. The LCOE calculation methodology and key assumptions can significantly impact results, so transparency is critical. Various methodologies for calculating LCOE can be found in different literatures [15].

LCOE can be expressed as [15], [16]:

$$LCOE = \frac{C_{annual,total}}{E_{system,output}} \tag{1}$$

where $C_{annual,total}$ is total annual investment costs of the system and $E_{system,output}$ is total annual output power produced by the wind farm during lifetime.

The annual energy produced by the system can be estimated by following expression:

$$E_{system,output} = 8760 \cdot P_{inst} \cdot CF \tag{2}$$

where P_{inst} is total plant capacity in kW and CF is capacity factor.

The LCOE method is based on the net present value (NPV), i.e. every future cost must be adjusted to the present value due to inflation and other factors. Therefore, the basic expression for LCOE is [17], [18], [19]:

$$LCOE = \frac{\sum_{t=1}^{n} \frac{I_t + M_t + F_t}{(1+r)^t}}{\sum_{t=1}^{n} \frac{E_t}{(1+r)^t}} \tag{3}$$

where $LCOE$ is levelized cost of energy which is expressed in monetary units per kWh or MWh, I_t is investment costs in the year t in euros, M_t is operation and maintenance cost in the year t in euros, F_t is fuel cost in the year t in euros, E_t is electricity generation in the year t in MWh, r is discount rate in percentage, n is lifetime of the system (project evaluation period) in years and t is year of lifetime (1, 2, ..., n).

Investment costs are the main expenditures in the process of constructing a power plant in relation to the installed power of the plant. They relate to the planning, design and construction of the plant, and vary depending on the type of plant. Power plant construction costs are usually expressed as a specific investment per unit of installed power according to the expression:

$$I = I_{spec} \cdot P_{inst} \tag{4}$$

where I_{spec} is specific value of the investment in €/kW P_{inst} is total plant capacity in kW.

The total investment value for the three analysed wind farms is known, so this value was utilized in the levelized cost of energy (LCOE) calculation.

Fuel costs do not exist for plants producing energy from renewable sources. Instead, there are only operating and maintenance costs as periodic expenses. Although cost analysis can be very detailed, this paper simplifies it as a sum of fixed and variable costs regardless of energy produced. These costs are already expressed relative to plant power, following the methodology of the Rulebook on determining guaranteed electricity purchase prices from renewable energy and efficient cogeneration facilities. This was adopted by the Regulatory Commission for Energy in the Federation of Bosnia and Herzegovina (FERK) per Table 2 [20].

The wind turbine classification in Table 2 by installed power aligns with the Regulation on encouraging renewable energy and efficient cogeneration production and incentive fees ("Official Gazette of the Federation of B&H " number 48/14).

It also follows the Rulebook on acquiring qualified electricity producer status ("Official Gazette of the Federation of B&H" number 53/14). Table 2 shows investment costs, O&M costs, as well as the guaranteed purchase price of electricity depending of the size of the wind power plant, valid from September 1, 2020. The category to which the wind farms analysed in this paper belong, whose power ranges from 36 to 50,6 MW, is marked in gray in Table 2.

Wind farms	Power	Specific value of the investment	O&M costs	Fuel costs	Reference price of electricity	Guaranteed price
	kW	€/kW	€/kW	€/kWh	€/kWh	€/kWh
Micro	23	1.417	115	0	0,0645	0,1177
Mini	150	1.417	61	0	0,0645	0,0963
Small	1.000	1.331	34	0	0,0645	0,0809
Medium	10.000	1.171	22	0	0,0645	0,0682
Large	> 10.0000	1.112	20	0	0,0645	0,0645

Table 2. Guaranteed electricity purchase prices valid from September 1, 2020. [20]

Studies [21], [22] and [23], were utilized to determine the unit cost of the investment as well as the operation and maintenance expenses. The studies found that the unit price of wind power plants, depending on the size of the plant, ranges from 1110 €/kW to 1.415 €/kW, while the operation and maintenance costs range from 20 €/kW for large plants to 115 €/kW for micro plants.

All wind farms built so far in B&H generate revenue by selling electricity at contracted and guaranteed purchase prices, known as feed-in-tariffs. To determine the guaranteed purchase prices, the form of primary energy, the contracted purchase period of 12 years, the technology used, the commissioning date of the plant, and the installed capacity of the plant are taken into account according to FERK. Producers of renewable energy electricity from plants like small hydro power plants, wind farms and solar power plants in the Federation of Bosnia and Herzegovina (FB&H) sign 12-year contracts with utility companies, while contracts in the Republic of Srpska (RS) last for 15 years. Other plants have 5-year contracts. In addition to these benefits, producers also receive incentives for producing so-called clean energy by collecting fees through electricity bills paid by citizens of B&H.

To calculate the net present value (NPV) of energy investment projects, future cash flows must be discounted by the rate of return of comparable alternative investments (the opportunity cost), such as simply depositing funds in banks at a certain interest rate, investing in certain securities or some other more profitable projects. Thus, for investors, the discount rate equals the interest rate generated by investing capital in a certain market segment, including market risk. Per the recommendation of World Bank experts, a 7% discount rate was used for the analyses [24]. Sensitivity analysis can also be done for other discount rate values. It should be noted that this analysis did not account for various incentives and subsidies when calculating LCOE, but taking these into account would further reduce the LCOE for all three wind farms.

4. Results and discussion

The levelized cost of energy (LCOE) represents the electricity price at which revenues equal costs, including return on investment equal to the discount rate. Higher electricity selling prices yield better returns while lower prices incur losses. Using known data on installed capacity, investment, and estimated annual electricity production for three wind farms in B&H, the LCOE calculation was made. Considering that plants using renewable energy sources for electricity production are capital-intensive technologies, the discount rate will be the factor with the greatest impact on levelized electricity costs. This explains the LCOE differences for similar technologies across the EU. Operation and maintenance costs are estimated as 20 €/kW (20.000 €/MW) for these >10 MW wind farms per in accordance with Table 2. An annual 0,5% escalation rate is applied, although different rates could be used if costs are expected to increase over time. The lifetime of the plant is also one of the input data when estimating the LCOE. The lifetime of the plant is an estimate of how many years a certain plant will be able to work effectively, i.e. bring a revenue. A 25-year lifetime is assumed.

A calculation tool in MS Excel was developed for LCOE assessment purposes. Using the assumed value of the discount rate, the growth rate of operation and maintenance costs as well as the lifetime of the turbines, and based on the available data on the investment, the estimated annual electricity production, the estimated LCOE values for the three wind farms in B&H that are in operation are obtained. The input parameters used for LCOE calculation are presented in Table 3 for all three wind farms as well as the overview of costs during wind farm life cycle which can be extrapolated to other wind farms in Figure 6. The capital costs of onshore wind power project can be divided into the following categories: the turbine costs including blades, tower and transformer in range from 65% to 84 % of the total investment, construction works including construction costs for site preparation and the foundations for the towers in range from 4% to 16% of the total investment, grid connection costs with share from 9% to 14% and other capital costs with share from 4% to 9% in the total investment which can include the construction of buildings, control systems, project consultancy costs, etc [25]. In this paper, the growth rate of operating and maintenance costs of 0,5% per year is assumed.

Assumptions	Wind farms in B&H		
	Podveležje	Mesihovina	Jelovača
Total Plant Capacity (MW)	48	50,6	36
Initial Investment Cost (€)	77.000.000	82.000.000	40.000.000
Operations and Maintenance Cost (€/MW)	20.000	20.000	20.000
Operations and Maintenance Cost (€)	960.000	1.012.000	720.000
O&M Growth rate (%)	0,5	0,5	0,5
Annual Fuel costs	0	0	0
Annual Electricity Output (MWh)	130.000	165.000	110.000
Project Lifetime (years)	25	25	25
Discount Rate (%)	7	7	7

Table 3. Input data for three wind farms in Bosnia and Herzegovina

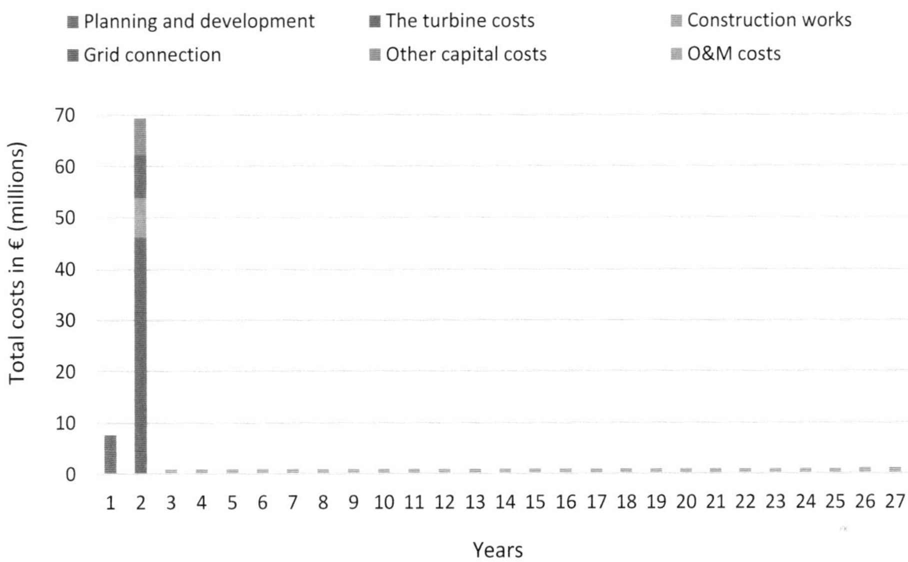

Fig. 6. The overview of costs over the wind farm life cycle

The levelized cost of energy (LCOE) analysis shows that for the Podveležje, Mesihovina and Jelovača wind farms to reach breakeven, where revenue equals expenditures, the electricity price needs to be 62,09 €/MWh, 52,04 €/MWh and 40,23 €/MWh respectively, as shown in Figure 7. These prices align with cost trends in Germany, presented in Figure 1. At these prices, the NPV of each project would be zero.

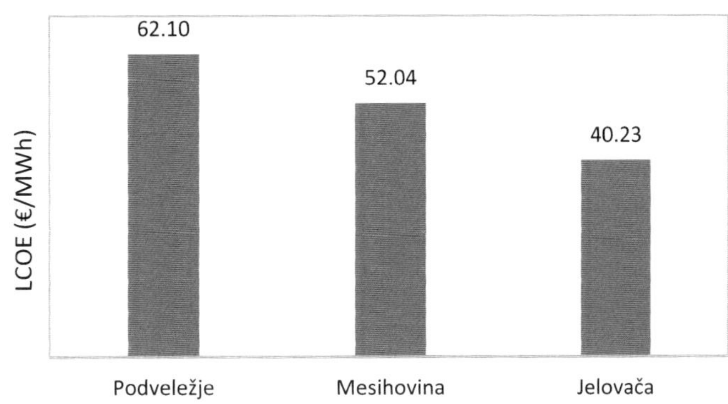

Fig. 7. Estimated LCOE for analysed wind park in €/MWh

The LCOE estimates illustrate that Podveležje has the highest MWh price of 62,10 €/MWh, while Jelovača has the lowest of 40,23 €/MWh. Podveležje's higher LCOE of 62,10 €/MWh compared to Mesihovina's 52,04 €/MWh reflects the differences in their investment costs and energy production. Podveležje has a unit investment of 1.604 €/kW and estimated annual production of 130 GWh. In contrast, Mesihovina has a lower unit investment of 1.620 €/kW but higher estimated annual production of 165 GWh. Jelovača has the lowest LCOE at 40,23 €/MWh because of its substantially lower unit investment cost of 1.111 €/kW, despite its lower estimated annual production of 110 GWh.

Reductions in the LCOE improve the economic viability of wind power plants, while increases in LCOE reduce profitability. Specifically, factors that decrease capital or operating costs, or increase electricity production, reduce LCOE and improve profitability. Conversely, factors that increase capital or operating costs, or reduce electricity production, increase LCOE and reduce profitability. The analysed wind power plants are economically attractive investments because their LCOE is below the government's guaranteed purchase price of 0,0645 €/kWh (64,5 €/MWh). When the LCOE is lower than the guaranteed electricity price, wind power projects can potentially generate profits.

In order to perform a sensitivity analysis of the estimated LCOE for the analysed wind farms, the discount rate, O&M costs, output energy and project lifetime will be individually varied below. Figure 8 shows the variation in LCOE by varying the discount rate in the interval from 1% to 10%. Based on the results shown in Figure 8, it can be concluded that LCOE strongly depends on the variation of the discount rate, i.e. an increase in the discount rate leads to an increase in LCOE, while a decrease in the discount rate results in a decrease in LCOE.

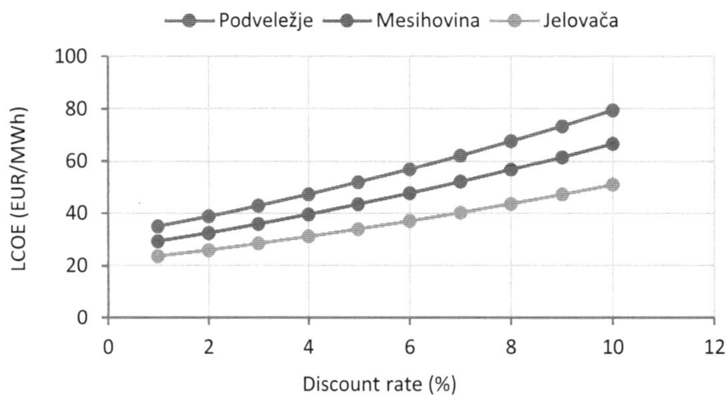

Fig. 8. Variation of the LCOE depending on discount rate

Figure 9 shows the influence of varying plant lifetime on LCOE. The variation was made in the range of 15 to 30 years. Extending the life of the project leads to higher energy production, which affects the reduction of LCOE. In contrast, reducing the lifetime of projects causes a large increase in LCOE.

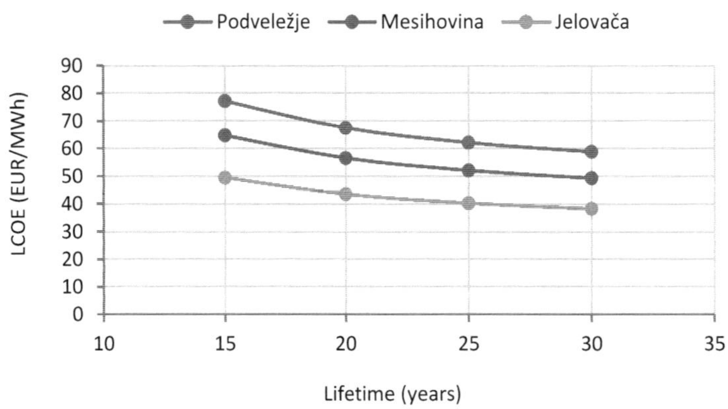

Fig. 9. Variation of the LCOE over the project lifetime

Figure 10 presents the impact of the variation of operation and maintenance costs on LCOE. From Figure 10, it can be seen that for all wind farms, operation and maintenance costs have a very small impact on LCOE. By varying the O&M costs in the range of −60% to +60%, no significant change in LCOE is observed. Such results can be explained by the fact that the share of annual operating and maintenance costs in comparison with the total investment costs for the first years of operation of wind farms is very small and ranges from 1,23% to 1,8%.

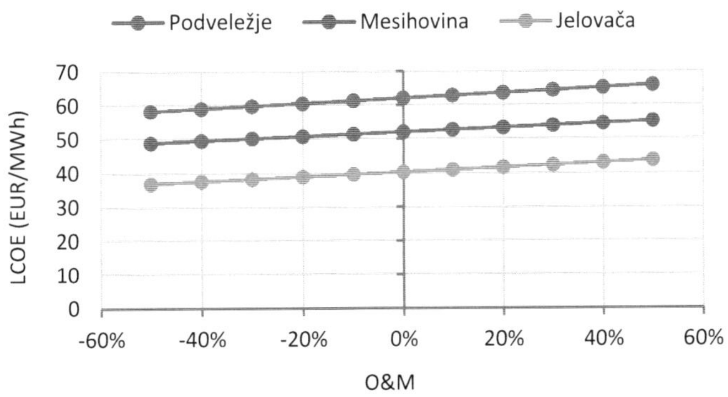

Fig. 10. Variation of the LCOE depending on O&M

Figure 11 shows the dependence of LCOE on output energy. For the analysed wind farms, the produced energy was varied in the range from -20% to +20%. Based on the obtained results shown in Figure 11, it is clear that LCOE decreases with increasing electricity production.

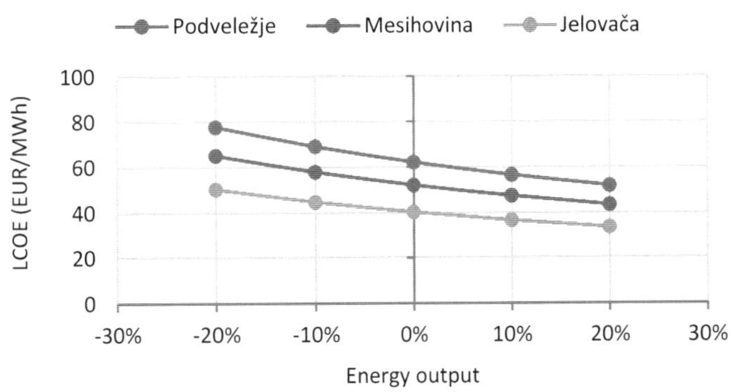

Fig. 11. Variation of the LCOE depending on the electricity production

Two key variables when calculating the LCOE are total plant costs and energy output. Essentially, anything that increases the energy output over the lifetime of the plant and reduces costs results in a lower LCOE. This is why it is very important at the beginning of a wind farm project to thoroughly assess the wind resources in the planned location and optimally position the turbines to maximize annual electricity production.

5. Conclusion and further work

The paper provides an overview of the development of wind farms in Bosnia and Herzegovina from the first constructed wind farm to the present day. It presents data on the total power, estimated electricity production, and total investment for the constructed wind farms. The paper introduces a methodology for calculating the levelized cost of energy (LCOE), which can be applied to plants with different electricity production technologies. Using this methodology, the LCOE was calculated for individual wind farms in Bosnia and Herzegovina, showing the approximate price of electricity produced and its competitiveness with the country reference electricity price. A sensitivity analysis of LCOE to the discount rate, operation and maintenance costs, lifetime, and energy output was conducted.

The results showed electricity production costs for all analysed wind farms are lower than Bosnia and Herzegovina guaranteed purchase price of 0,0645 €/kWh (64,5 €/MWh) for large wind farms. This suggests the government should encourage more wind farm investments as current farms have limited lifetimes, helping energy independence and reducing fossil fuels usage and greenhouse gas emissions. The sensitivity analysis revealed that LCOE is most affected by changes to the discount rate and least by changes to operation and maintenance costs.

As incentives, subsidies, and actual rather than planned annual production were not considered in the LCOE calculations, future research could incorporate this data and compare electricity production profitability across different systems. Also, considering that the LCOE method is often used to compare the costs of energy production from different sources and different electricity generation technologies, it would be of great usage to compare the profitability of electricity production from different production systems in the further research.

6. References

[1] Radusin, S., Senad, O., Cero, M., Abdurahmanovic, I., & Vukmir, G. (2012). The second national report on Bosnia and Herzegovina in accordance with the framework convention of the United Nations. Sarajevo: Council of Ministers of Bosnia and Herzegovina, Sarajevo

[2] NOSBIH (2021). Indicative production development plan 2022-2031. Indikativni plan razvoja proizvodnje 2022-2031 (in Bosnian)

[3] http://ec.europa.eu/energy/energy2020/roadmap/doc/roadmap2050_ia_20120430_en.pdf, (2011). Energy Roadmap 2050 Impact assessment and scenario analysis, EUROPEAN COMMISSION, Accessed on: 2023-08-10

[4] Council, R. C. (2020). Sofia Declaration on the Green Agenda for the Western Balkans. Sofia: Regional Cooperation Council

[5] Talarek, K., Knitter-Piątkowska, A., & Garbowski, T. (2022). Wind Parks in Poland—New Challenges and Perspectives, Energies, Vol. 15, No. 19, 7004, https://doi.org/10.3390/en15197004

[6] Blažević, R., Hadžiahmetović, H., & Ahmović, I. (2021). Statistical analysis and assessment of wind energy potential in Sarajevo, Bosnia and Herzegovina, Technical gazzete, Vol. 28, No. 5, pp. 1511-1518, https://doi.org/10.17559/TV-20200720163605

[7] Blazevic, R., Hadziahmetovic, H., Kurtagic, S. M., & Alispahic, M. (2022). Assessment and analysis of wind energy potential in the mountain area around Sarajevo. Annals of DAAAM for 2022 & Proceedings of the 33rd International DAAAM Symposium, Volume 33, No.1, ISSN 1726-9679, ISBN 978-3-902734-36-5, Ed. B. Katalinic, Published by DAAAM International, Vienna, Austria, EU, pp. 0058-0067, DOI: 10.2507/33rd.daaam.proceedings.009

[8] Tizgui, I., El Guezar, F., Bouzahir, H., & Vargas, A. N. (2018). Estimation and analysis of wind electricity production cost in Morocco, International Journal of Energy Economics and Policy, Vol. 8, No. 3, pp 58-66, ISSN: 2146-4553

[9] Bahrami, A., Teimourian, A., Okoye, C. O., & Shiri, H. (2019). Technical and economic analysis of wind energy potential in Uzbekistan, Journal of cleaner production, Vol. 223, pp. 801-814, https://doi.org/10.1016/j.jclepro.2019.03.140

[10] Kost, C., Shammugam, S., Fluri, V., Peper, D., Memar, A. D., & Schlegel, T. (2021). Levelized Cost of Electricity-Renewable Energy Technologies, pp. 1–45, Fraunhofer institute for solar energy systems ISE

[11] ESSBIH (2008). Study of the energy sector in Bosnia and Herzegovina; Module 12 - Consumption management, energy saving and renewable energy sources. Studija energetskog sektora u BiH; Modul 12 – Upravljanje potrošnjom, štednja energije i obnovljivi izvori energije (in Bosnian), Energetski institut Hrvoje Požar, Hrvatska.

[12] Simić, V., Pavlović, S., & Cvijić, S. Renewable energy sources, energy efficiency and environmental protection in Bosnia and Herzegovina. Obnovljivi izvori energije, energetska efikasnost i zaštite životne sredine u Bosni i Hercegovini (in Bosnian), SVAROG, No.5, pp. 221-236, UDK 620.9:502.131.1(497.6), DOI br. 10.7251/SVR1205221S

[13] https://globalwindatlas.info/en, (2007). World Wind Atlas, Sander&Partner GmbH, Accessed on: 2023-09-05

[14] Ragheb, M. (2017). Economics of Wind Power Generation. In: Letcher, Trevor M. (Ed.), Wind Energy Engineering. Academic Press, pp. 537–555

[15] Arinaldo, D. & Pujantoro, M. (2019). Levelized Cost of Electricity in Indonesia, Institute for Essential Services Reform (IESR), Jakarta

[16] Campbell, M., Aschenbrenner, P., Blunden, J., Smeloff, E., & Wright, S. (2008). The drivers of the levelized cost of electricity for utility-scale photovoltaics, SunPower corporation, pp. 1-21

[17] IRENA (2019). Renewable Power Generation Costs in 2018, Abu Dhabi: International Renewable Energy Agency

[18] Kost, C., Mayer, J. N., Thomsen, J., Hartmann, N., Senkpiel, C., Philipps, S., ... & Schlegl, T. (2013). Levelized Cost of Electricity–Renewable Energy Technologies, pp. 1-47, Fraunhofer institute for solar energy systems ISE

[19] Williams, A. J., & Rubert, T. (2019). Levelised cost of energy–A theoretical justification and critical assessment. Energy policy, Vol. 124, pp. 169-179, https://doi.org/10.1016/j.enpol.2018.10.004

[20] https://www.ferk.ba/_ba/index.php/tarife/naknade-za-prikljucenje/2-uncategorised/19320-odluke-o-garantovanim-i-referentnim-cijenama, (2021). Decision on the guaranteed purchase price of electricity (2021). Odluka o garantovanoj otkupnoj cijene električne energije (2021) (in Bosnian). Regulatorna komisija za energiju u Federaciji Bosne i Hercegovine – FERK, Accessed on: 2023-09-20

[21] IRENA (2021). Renewable Power Generation Costs in 2020, International Renewable Energy Agency, Abu Dhabi.

[22] REN21 (2021). Renewables 2021 Global Status Report, ISBN 978-3-948393-03-8

[23] Green-X (2004). Deriving optimal promotion strategies for increasing the share of RES-E in a dynamic European electricity market, Final report of the project Green-X, Research project within the 5th Framework Programme of the European Commission, DG Research

[24] ESSBIH (2008). Study of the energy sector in Bosnia and Herzegovina; Module 3 – Production of electricity. Studija energetskog sektora u BiH; Modul 3 – Proizvodnja električne energije (in Bosnian), Energetski institut Hrvoje Požar, Hrvatska

[25] IRENA (2012). Renewable energy technologies: Cost analysis series, wind power, International Renewable Energy Agency

DOI: 10.2507/34th.daaam.proceedings.031

NUMERICAL SIMULATION OF FLUID FLOW THROUGH THE VALVE

Halima Hadžiahmetović, Rejhana Blažević & Sanda Midžić Kurtagić

This Publication has to be referred as: Hadziahmetovic, H[alima]; Blazevic, R[ejhana]; & Midzic Kurtagic, S[anda] (2023). Numerical Simulation of Fluid Flow Through the Valve, Proceedings of the 34th DAAAM International Symposium, pp.0247-0252, B. Katalinic (Ed.), Published by DAAAM International, ISBN 978-3-902734-41-9, ISSN 1726-9679, Vienna, Austria
DOI: 10.2507/34th.daaam.proceedings.031

Abstract

The main goal of this paper is to present an analysis of the complex problem of fluid flow through the valve with a seat type A 10 FS1.0364 of a variable cross-section and the obtained results. For this purpose, a geometry was created in Ansys Fluent for an "open" valve configuration with different types of meshes, tetrahedral and hexahedral. The paper also presents a comparison of different configurations and meshes, and a comparison of the results obtained in this paper using numerical simulations and ANSYS Fluent, with the previously published results of the group of authors obtained from Solid Works Flow Simulation. Numerical simulations carried out in this paper using Fluent showed a good agreement of the results with previously obtained results of the group of authors in Solid Works Flow Simulation. Also, an interpretation of the results with a discussion is given and the main findings are highlighted in the conclusion.

Keywords: valve; numerical mesh; volume flow; velocity magnitude; numerical simulation;

1. Introduction

In modern hydraulic and pneumatic components for direct and control, the valves are defined as devices whose main task is to direct and control hydraulic energy in the process of transfer from the source (hydraulic pump) to the hydraulic motor. At variable flow, there is an additional mechanism that adjusts the size of the valve opening. An essential element of the hydraulic system is the working fluid. The main function of fluid in the hydraulic system is the transfer of energy. Fluid currents transfer energy from the source (hydraulic pump) to the consumers (motors). The working fluid lubricates the moving parts in the hydraulic system so they do not have special equipment for lubrication.

In addition, the working fluid also serves as a means of preserving the hydraulic system, but also washes outflow channels in the hydraulic system. Different fluids can be used as working fluids in the hydraulic system. They were selected so that their physical and chemical properties comply with the conditions of exploitation [1]. Today, as a working fluid in hydraulic systems, 90% of mineral oil used is pure either with water or a mixture of synthetic fluids. Therefore, the common name for the working fluid of hydraulic systems is hydraulic oil.

Limited capabilities of analytical methods, in practice very often lead to the use of experimental solutions to fluid flow problems, i.e. an approach to measuring velocity, pressure, etc. real progress, or in their model, which uses so-called similarity theory. But in some cases, it is very difficult to conduct an experiment and sometimes even impossible. Metering equipment can disrupt the flow, or it is such a fluid flow that it is impossible to set the measuring sensors. Progress in technology and increased competition require optimized design and high technology such as a mechanism which will predict the fluid flow. Experimental design and its realization require considerable financial support and time to get results. An alternative experiment, or at least a complementary method, is the advent of fast, inexpensive digital computers, which also leads to a significant (although most of the key ideas were founded more than a century) application of numerical solutions of partial differential equations. When the possibility of using computers in fluid flow analysis was recognized, interest in numerical methods increased significantly. Solving the equations of fluid mechanics using computer programs has become an important field of research known as Computational Fluid Dynamics (CFD) analysis [2], [3]. Numerical simulations are used in many industrial areas, such as power plants, automotive, chemical engineering, aviation, petroleum refining, etc [4], [5], [6]. Therefore, many studies have used numerical simulations and various computer programs to solve the problem of fluid flow through a valve.

Vedova et al. [5] conducted research on the 3D model of the full geometry of the valve (OpenFOAM) that has been developed to predict the distribution of pressures (hence forces) inside it to make an optimization process possible. Simic and Herakovic [7] investigated the reduction of the flow forces in a small hydraulic seat valve in detail by using a commercial simulation tool, Ansys CFX. Banaś et. al. [8] compared the results of numerical calculations obtained by using ANSYS® R17.2 with experimental studies for a prototype of a hydraulic throttle valve made from transparent plastic. Szpica [9] presented the numerical research methodology to designate the throttle valve flow characteristics using SolidWorks Flow Simulation as the alternative to long and expensive experimental research. Polášek et. al. [10] investigated fluid flow through the throttle valve for the laminar model, RANS models, and LES model using numerical simulation in ANSYS Fluent. Hodzic et al. [11] investigated the possibility of applying numerical simulations to solve the problem of calculation and determination of the fluid flow through the hydraulic valve. Results of numerical simulation by using Solid Works Flow Simulation of hydraulic oil flow through a valve with seat surface type A 10 FS1.0364 were taken from paper [11], and the own results obtained in this paper are compared with them.

So, this paper aims to present the results obtained by numerical simulations in Fluent and compare them with the previously published results by Hodzic et al. obtained in Solid Works Flow Simulation [11] for the same valve type with the same input data for the analysis, and to provide interpretation of the results with discussion.

2. Methodology

The finite volume method (FVM) is a numerical method used to solve the problem of mineral oil flow. According to this method, the calculation domain is divided into a finite number of control volumes that do not overlap (numerical mesh), each of which is represented by the value of the physical size of the node that is within the control volume (usually in the middle).

There are two different approaches to generating numerical mesh. The first approach generates the appropriate mesh domain, and then the nodes (calculation points) are placed in the center of the control volume. This approach is used more often. The second approach is based on the creation of computational points and a mesh of control volumes created so that the surface of the control volume lies in the middle between two calculation points [12].

The numerical algorithm consists of the following stages:

- proper integration of the main equations of fluid flow overall (finite) control volumes domain solutions,
- discretization, which involves solving some integral in discrete form (by use of integral approximation), which converts the integral equation into a system of algebraic equations for the nonlinear modelling of fluid flow,
- solving algebraic equations using final iterative methods.

Numerical simulation was carried using the Ansys Fluent CFD software. Fluent is a state-of-the-art computer program for modelling fluid flow and heat transfer in complex geometries. Fluent provides complete mesh flexibility, including the ability to solve flow problems using unstructured meshes that can be generated in complex geometries with relative ease. Supported mesh types include 2D triangular/quadrilateral, 3D tetrahedral/hexahedral/pyramid/wedge/polyhedral, and mixed (hybrid) meshes. Fluent also allows the use of refined or coarse grids based on the flow solution [12].

3. Results and discussion

In the paper, the problem of the flow of mineral oil through the valve was analysed, therefore, certain assumptions were given that were used in solving the problem. The mineral oil HL-ISO VG 32 is released into the atmosphere at pressure $p = 1$ bar. Density and viscosity coefficient were assumed as $\rho = 887$ kg / m3, $\mu = 0{,}057655$ Pa·s, respectively. Volume flow has fixed values of 5, 9, 13, 16 and 20 l/min. These are necessary input data for numerical simulations in Fluent. The first step is to calculate velocity for volume flow values of 5, 9, 13, 16 and 20 (l/min). The following expression can be used for calculation:

$$\dot{V} = u \cdot A \quad \Longrightarrow \quad u = \frac{\dot{V}}{A} \tag{1}$$

where u is velocity in m/s and A is area m^2.

The second step is to calculate the Reynolds number (Re) for volume flow values 5, 9, 13, 16 and 20 l/min with known velocities. The equation for calculating Reynolds number is:

$$\mathrm{Re} = \frac{\rho \cdot u \cdot d}{\mu} \tag{2}$$

where ρ is density in kg/m^3, d is diameter with value d= 0,007 m and μ is dynamic viscosity in kg/m·s.

Table 1 shows velocities values and Reynolds numbers for different values of volume flow.

Volume flow [l/min]	**Velocity [m/s]**	**Re-number**	**Flow**
5	2,19	235,84	laminar
9	3,96	426,461	laminar
13	5,716	615,56	laminar
16	7,037	757,830	laminar
20	8,797	947,36	laminar

Table 1. Velocities and Reynolds numbers for different values of volume flow

Based on the obtained values of velocities and Reynolds numbers, it can be concluded that laminar flow occurs. Laminar flow occurs when the calculated Reynolds number is less than 2300.

The analysis of fluid flow through the valve was performed for the "open" valve configuration and for two different meshes, tetrahedral and hexahedral. Figure 1 and Figure 2 show the geometry and tetrahedral mesh "open" configuration.

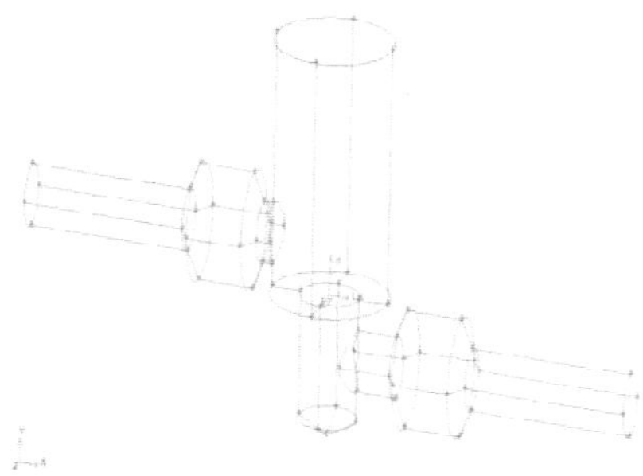

Fig. 1. Geometry

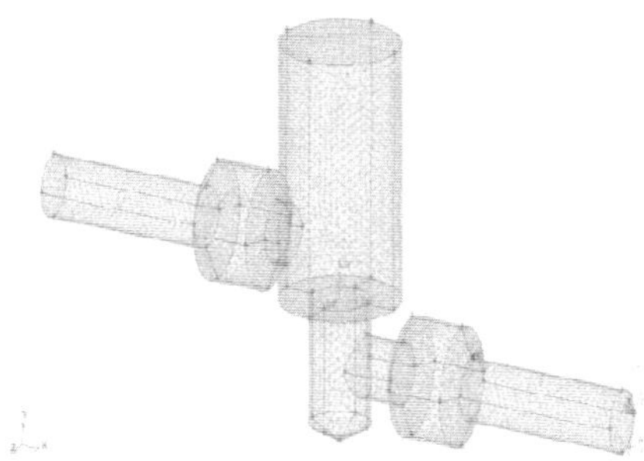

Fig. 2. Tetrahedral mesh "open" configuration

Figure 3 shows the quality of the mesh with a total number of elements of 91.000 tetrahedral cells. Figure 4 shows the position of the worst element and a quality value of 0,8037.

Fig. 3. Quality of tetrahedral mesh

Fig. 4. Position the worst element in geometry

Figure 5 and Figure 6 show the geometry and hexahedral mesh "open" configuration. Geometry with hexahedral mesh is possible and complex.

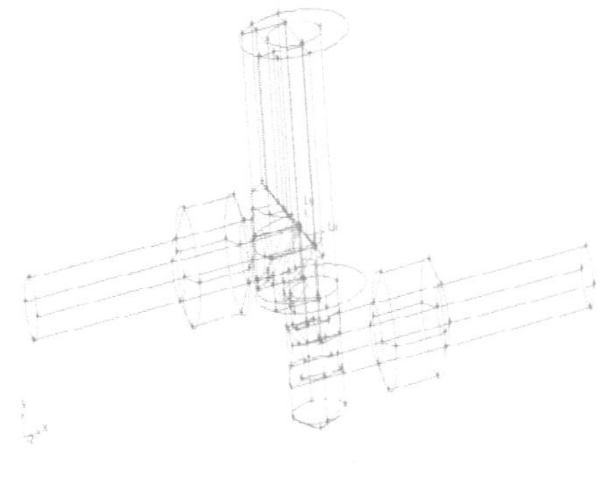

Fig. 5. Complex geometry

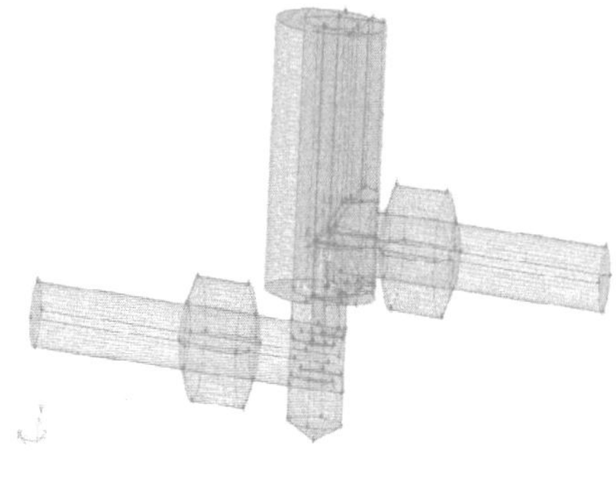

Fig. 6. Hexahedral mesh "open" configuration

Figure 7 shows the quality of the mesh with a total number of elements of 151.894 hexahedral cells. Figure 8 shows the position of the worst element and the quality value of 0,7858.

Fig. 7. Quality of hexahedral mesh

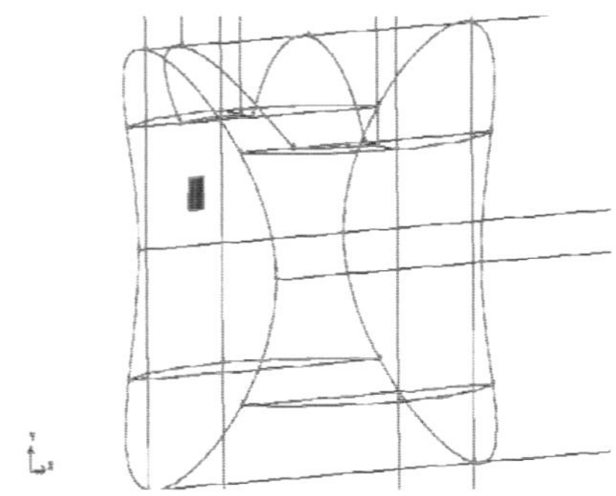

Fig. 8. Position the worst element in geometry

Figure 9 shows the contours of the velocity magnitude results in tetrahedral mesh. The minimum velocity value is 0 m/s, while the maximum is 3,58 m/s. From Figure 10, it can be seen the contours of velocity magnitude results in hexahedral mesh. The minimum value of velocity magnitude is 0 m/s, while the maximum is 3,97 m/s.

When comparing velocities magnitude, in both cases, the minimum velocity is 0 m/s, and the difference between the maximum velocities is very small, velocities are almost the same (3,58 m/s for tetrahedral mesh and 3,97 m/s for hexahedral).

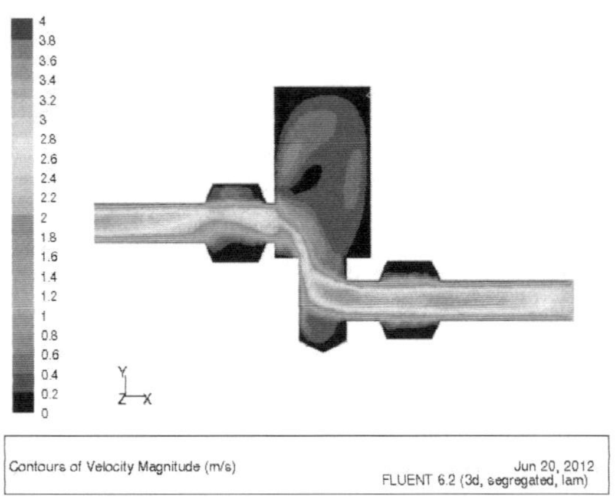

Fig. 9. Contours of velocity magnitude in tetrahedral mesh

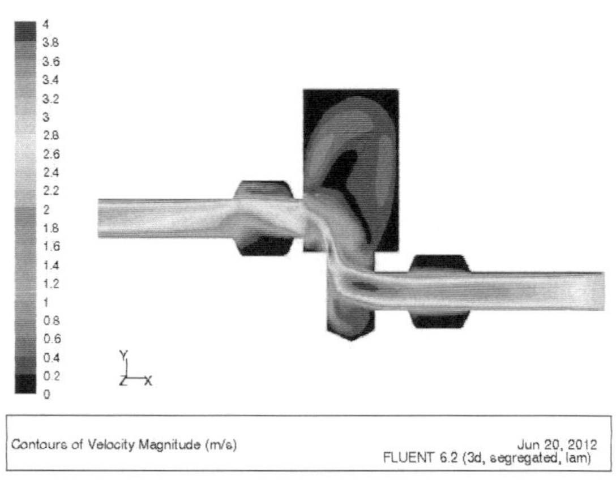

Fig. 10. Contours of velocity magnitude in hexahedral mesh

Table 2 shows the results of the mass flow rate for the inlet and outlet, the area-weighted average for static pressure in Pascal, and the area-weighted average for velocity magnitude in m/s, for two different meshes, and as it can be seen the difference between them is not significant.

		Tetrahedral (worst geometry)	Hexahedral (complex geometry)
Mass flow rate (kg/s)	Outlet	-0,07352	- 0,07450
	Inlet	0,07353	0,07450
	Net mass flow	1,9803643e-05	-6,7055225e-08
Area-weighted average static pressure (Pa)	Outlet	0	0
	Inlet	16529	15791
	Cross-section	8892	8584
Area-weighted average velocity magnitude (m/s)	Outlet	1,84	2,11
	Inlet	2,19	2,19
	Cross-section	1,34	1,36
Velocity magnitude (m/s)	Min.	0	0
	Max.	3,58	3,97
Static pressure (Pa)	Min.	0	0
	Max.	17099	16611

Table 2. Calculation results for two different meshes and volume flow 5 l/min

Figure 11 shows the diagram of pressure drop and volume flow dependencies obtained by numerical simulation in Fluent for hexahedral and tetrahedral mesh and the results of numerical simulation Solid Works Flow Simulation [11]. The green line shows the results of the hexahedral mesh, while the red line shows the results of the tetrahedral mesh. When comparing these lines, the difference is not significant. Also, the results obtained in Fluent for both meshes have a good match with the existing results from Solid Works Flow Simulation. However, the results obtained in Fluent with the hexahedral mesh have a better match with the results from Solid Works Flow Simulation.

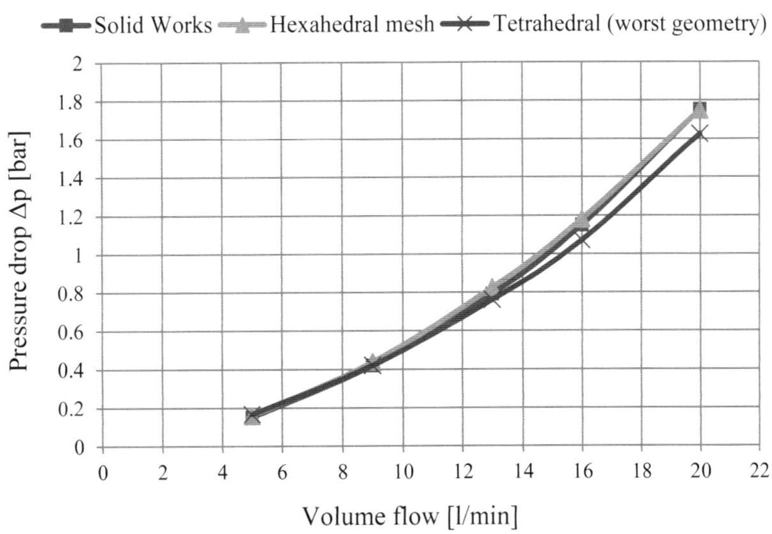

Fig. 11. Diagram of pressure drop and volume flow dependencies, comparison numerical simulation results from Fluent with existing results from Solid Works

4. Conclusion and further work

This paper provides the numerical calculation results of pressure drop obtained in Fluent for two different meshes, hexahedral and tetrahedral, for fixed volume flow values 5, 9, 13, 16 and 20 l/min through the valve. The main characteristics of the results of numerical simulations through the valve can be highlighted as:

- By comparing the results obtained for two different meshes, less deviations between them can be noticed. Minimum velocity values for both meshes are the same (0 m/s), while difference between the maximum velocity values is very small, about 9,8% (for tetrahedral mesh maximum velocity is 3,58 m/s and for hexahedral 3,97 m/s).

- When comparing results for configuration "open" valve with hexahedral mesh and tetrahedral mesh, the results of the mass flow rate for inlet and outlet, the area-weighted average for static pressure, and the area-weighted average for velocity magnitude, it can be observed that the difference between them is not significant.
- The results obtained in Fluent for both meshes have a good match with the existing results from Solid Works Flow Simulation, but the results obtained with the hexahedral mesh have a better match.

In some of the following research, as further work, other models will be developed and experimental studies will be included.

5. References

[1] Hadziahmetovic, H; Dzaferovic, E. & Cohodar, M. (2010). Description Of Technological Process Of Hydraulic Transport From Thermal Power Plant, Annals of DAAAM & Proceedings, 20-23rd October 2010, Zadar, ISSN 1726-9679, Katalinic, B. (Ed.), pp 685-686, DAAAM International Vienna Audience, Zadar, Croatia.

[2] Ferziger, J.H. & Peric, M. (1999). Computational Methods for Fluid Dynamic, Third Edition, Springer-Verlag Berlin Heidelberg, ISBN 3-540-42074-6, Germany

[3] Krieger, M.; Mehrle, A. & Thumfart, M. (2012). Practical Training in Computational Fluid Dynamics, Institute of Fluid Mechanics and Heat Transfer, Johannes Kepler University Linz, Linz, Austria

[4] Gevorkov, L.; Smidl, V. & Sirovy, M. (2018). Stepper Motor Based Model of Electric Drive for Throttle Valve, Proceedings of the 29th DAAAM International Symposium, 24-27th October 2018, Zadar, ISBN 978-3-902734-20-4, ISSN 1726-9679, Katalinic, B. (Ed.), pp.1102-1107, Published by DAAAM International, Vienna, Austria, DOI: 10.2507/29th.daaam.proceedings.157

[5] Vedova, D.; Maggiore, M.D.L. & Riva, P.G. (2017). A new CFD-Simulink based systems engineering approach applied to the modelling of a hydraulic safety relief. International Journal Mechanics 11, pp. 43–50

[6] Hadziahmetovic, H.; Hodzic, N.; Kahrimanovic, D. & Dzaferovic, E. (2014). Computational fluid dynamics (CFD) based erosion prediction model in elbows. Tehnički Vjesnik – *Technical Gazette*, 21 (2), 275–282.

[7] Simic, M. & Herakovic, N. (2015). Reduction of the flow forces in a small hydraulic seat valve as alternative approach to improve the valve characteristics, Energy Conversion Management, 89, 708–718, https://doi.org/10.1016/j.enconman.2014.10.037.

[8] Banaś, M.; Antoniak, P.; Marciniak, L. & Stryczek, J. (2018). *Visualization of flow phenomena in hydraulic throttle valves of plastics*, MATEC Web Conferences, 211, 19001, DOI: https://doi.org/10.1051/matecconf/201821119001.

[9] Szpica D. (2015). Simplified numerical simulation as the base for throttle flow characteristics designation, Mechanika, 21(2), pp. 129–133.

[10] Polášek, T.; Ledvoň, M.; Hružík, L. & Bureček. A. (2023). Numerical simulation and experimental verification of flow through throttle valve. AIP Conference Proceedings, 2672 (1): 020017, https://doi.org/10.1063/5.0119995

[11] Hodžić, N.; Spahić, D. & Skopljaković, M. (2010). Numerical simulation of hydraulic oil flow through a valve with seat area type A 10 FS1.0364, 14th International Research / Expert Conference "Trends in the Development of Machinery and Associated Technology" TMT 2010, Mediterranean Cruise, 11-18 September 2010., ISSN 1840-4944, pp. 533 - 536, University of Zenica, Faculty of Mechanical Engineering in Zenica, Bosnia and Herzegovina.

[12] Fluent Inc. FLUENT 6.3 User's Guide. Fluent Inc., 2006.

DOI: 10.2507/34th.daaam.proceedings.032

AUTOMATED ROBOTIC COMPLEXES IN ENGINEERING, ADJUSTMENT, PRODUCTION AND EDUCATION PROCESSES

Sergey Stazhkov, Boris Feshin, Lubov Shpakova,
Kuralay Tokhmetova & Yelena Kalashnikova

This Publication has to be referred as: Stazhkov, S[ergey]; Feshin, B[oris]; Shpakova L[ubov]; Tokhmetova, K[uralay] & Kalashnikova, Y[elena] (2023). Automated Robotic Complexes in Engineering, Adjustment, Production and Education Processes, Proceedings of the 34th DAAAM International Symposium, pp.0253-0261, B. Katalinic (Ed.), Published by DAAAM International, ISBN 978-3-902734-41-9, ISSN 1726-9679, Vienna, Austria
DOI: 10.2507/34th.daaam.proceedings.032

Abstract

Distributed monitoring and control systems are developed for automating the control of large-scale and complex industrial systems. They consist of a combination of digital and mechanical parts operating in complex processes, both deterministic and stochastic, inside and outside of the systems. A feature of monitoring and control systems in large-scale and complex industrial systems is ability to collect, analyze and act based on a large volume of information in limited time periods. At the same time, large-scale and complex industrial systems have the properties of a flexible distributed production line. The article analyses the modernization tasks of control and monitoring systems of large-scale objects by focusing on "Industry 4.0" technology. The prerequisites for creating and designing digital twins for components of such systems are considered. New technologies of automated electric drives setting up are proposed. The theoretical principles to assess an efficiency and probability to reduce an energy consumption in large-scale systems affected by significant external disturbances and interference is shown. It demonstrates the results of students and undergraduate education in the field of management of large-scale industrial systems within the framework of the international educational project "Synergy". The long-term objectives of artificial intelligence application for further research and improvement of large-scale systems are identified.

Keywords: large objects of control and monitoring; digital twins; robotic systems; design; commissioning; production; education; "Industry 4.0" technology; artificial intelligence.

1. Introduction

Features of the "Industry 4.0" technology, announced the same as "The Fourth Industrial Revolution", are the transition to fully automated digital production, controlled by intelligent systems in real time in constant interaction with the external environment, going beyond the boundaries of a single enterprise [1, 2, 3, 4]. The characteristics of "Industry 4.0" technology has practically unlimited important options in terms of performance measures: volume, speed, and methods of information flows transmission.

Within the framework of distance international learning associations (for example, the Synergy type [5]), it makes sense to consider, analyze and examine elements of control and monitoring of large-scale and complex industrial systems. It is worth mentioning that not the number of technologically connected subobjects, but unlimited volumes of a priori, a posteriori and virtually obtained information is important. This sets and solves optimization problems that were previously limited for any reason. Further as monitoring and control (M&C) system of large-scale industrial objects the combination of the following automation objects is considered:

- underground coal-mining complex with supervisory multilink control system for electric drives of machines and mechanisms [7], as part of the complex: longwall panel, consisting of hydraulic sections, with jacks and special constructions, fixing the excavated space in the mine face up to 200m long, that can contain up to 200 sections of support; coal-mining machine with controlled electric drives for cutting and feeding; oil pumping station; scraper conveyors along the mine face; belt conveyors on air-feeding slices up to 1000 and more meters long; special point for mined coal reloading; ventilation system; local control and monitoring systems; special software and hardware multilink supervisory control system; created elements of the digital twin of underground coal mining complex. In coal mining modes the complex is considered as an automated robotic complex.
- FESTO laboratory robotic installation with the unique remote-control system [8, 9] - robot manipulator, system of movement and transportation of structural elements of the equipment. Local control subsystem and hierarchical system with an open software and hardware subsystem that allows developing unique digital control systems. The FESTO laboratory unit is an automated robotic complex, and a digital twin of this unit is studies within the framework of the current work.
- FESTO laboratory bench with a tracking electric drive [10] is an electromechanical tracking drive of the working mechanism. Software and hardware component compatible with LabVIEW and MatLAB application program package. Local control subsystem and hierarchical system with an open hardware and software subsystem that allows developing unique digital control systems. The FESTO laboratory bench is an automated robotic system, and a digital twin of this unit is studies within the framework of the current work.
- laboratory bench of frequency-controlled DC and AC electric drives with specialized software and hardware system MexBIOS Development Studio [8, 11] is a laboratory bench of frequency-controlled DC and AC electric drives with specialized software and hardware system MexBIOS Development Studio [8, 11, 12]. The bench allows to investigate, and study automated electric drive systems.

2. Design of monitoring and control (M&C) system of large-scale industrial objects in the context of "INDUSTRY 4.0" technology

It is proposed to consider the process of M&C of complex industrial systems designing in the context of "Industry 4.0" technology using the example of underground coal mining complex. Despite the different approaches of the solution, the main problem of underground mining has previously been and remains for now is providing of safe extraction of rock. As part of solving the problem, the purpose, designation, features and characteristics of technological process objects are determined, in particular, a priori information concerning the characteristics of the mining face and the structural elements of technological equipment is carefully collected and studied [13]. A priori information is the starting material for the develop process of any technological object.

The analysis of technological process of coal mining is based on scientific papers of the Deputy Director of Ugleservis company, the Coal Department of the Arcelor Mittal Temirtau, Joint Stock Company, Associate Professor, D.E.Sc., Avdeyev L.A. [14, 15]. According to the technological process, the coal mining machine cuts the slice of coal face and moves it along the scraper conveyor to the upper part of the production face. In the face, along its entire length, sections of longwall panel with many structural elements to support the mined-out space, jacks for moving the conveyor and support sections, and routing of hydraulic drive and control hoses for the support sections are sequentially installed.

The Germany made Eickhoff SL coal mining machine with controlled cutting and feeding electric drives has been applied on flat and inclined coal seams of medium and above medium thickness in the mines of the Coal Department of Arcelor Mittal Temirtau, JSC (hazardous due to gas and dust). The coal-mining machine SL 300 model is used for mining coal from seams with a thickness of 2 to 6 m. It is a complex engineering device weighing more than 150 tons, dimensions between the cutting drums axes - 12112 mm, diameter of two cutting drums - 2200 mm each, weight is 41 tons each, total installed power of electric drives is 688 kW, special systems for cooling electric motors and moving drums using hydraulic jacks, two feed mechanisms using gears and racks that are installed on a scraper conveyor stand.

The theoretical and practical problems for creating a supervisory multilink control system for electric drives of a coal mining machine were solved [16]. Based on the mentioned works [17], a hierarchical structure of an automated electromechanical underground coal-mining complex was proposed. The digital twin of underground coal-mining complex is designed within the framework of a functional hierarchical block diagram [17] and consists of many mathematical models and software analogues of machines, installations, electric drives and local control systems of the coal-mining complex. Some mathematical models and software analogues (for coal mining machines) were developed in [16]. The issue of replenishing the set requires efforts of qualified specialists.

The peculiarity of operating modes of technological elements of the coal-mining complex, defined in [16], is that in the process of mining and transporting coal the machines are subject to stochastic disturbances, i.e. normal operating modes are stochastic, and the startup and stop modes of electric drives of machines in this case should be considered as close to the modes of step, pulse and harmonic actions classified in the theory of automatic control as typical actions on dynamic objects. In relation to the coal-mining complex, we will call such start-up and stop processes standard machine modes. Specialized software "Digital Twin" allows you to virtually implement algorithms for the synthesis of optimal controls and multilink supervisory control, both in divided and real time modes.

The digital twin is a fundamentally new design since individual models of elements must be combined into a single whole and scenario compositions created that simulate processes in a coal mining face. The number of compositions will be the set M_{sd} = {NRE, TRM, ARE}, reflecting all possible situations under normal (NRM), standard modes (TRM) and emergency modes (ARE) of the operation of a mining machine. The tools for developing, adjustment, configuring and operating multiple operating modes are the MatLAB subsystems: MatLAB-Simulink, LabVIEW, Statictica, as well as the software for situational modeling SimEvents and Stateflow [18, 19].

"Industry 4.0" technology largely removes technical, software-algorithmic, computational and methodological restrictions for the design process. Here it is necessary to seek a compromise between the possible, truly necessary and sufficient level of decisions made. The path to achieving sustainable results lies in setting technology-based priorities. In relation to the underground coal-mining complex, this is primarily: production safety; control of dynamic load, service life, reliability and energy saving of machines; quality of coal and the cost of mined coal subject to the fulfillment of production volumes within the framework of the necessary requests.

As part of the study of FESTO benches and installations and for the bench by the MexBIOS system with a rigid structure, but with software-algorithmic components allowing for variation in the design process, there can be solved the problems of creating and adjustment new intelligent software-algorithmic tools of control and monitoring individual components and objects as a whole.

3. Setting up and adjustment of M&C system of large-scale industrial objects in the context of "INDUSTRY 4.0" technology

In the review article [20], methods for setting up and adjustment regulators of automated electric drives are divided into two groups. The first group, called traditional, is focused on modular and symmetrical optimums, and is performed as a function of the time constants of the drive motor. This technique, developed in [17, 18, 19] works, was and remains a necessary stage in preparing an automated electric drive for operation.

The second group, associated with progress in the fields of power electronics and microprocessor technology, as well as with the advent of powerful specialized application software packages for instance MatLAB-Simulink [18, 19], MexBIOS [11, 12], contributed to the mass introducing of various variants of frequency-controlled electric drives, which significantly complicate the drive circuit technique and require recording when setting up and adjustment not only the dynamic properties of electric motors, but also frequency converters, analog-to-digital converters (ADC), digital-to-analog converters (DAC), data input-output, microprocessor devices of various types, programmable logic controllers (PLC) and PC devices.

Demand determines supply; accordingly, new power technology for automated electric drives required new setting up and adjustment technologies. Within the second group of techniques, we highlight two approaches:

1. The use of specialized bench-analogs for conducting physical experiments with subsequent large-scale transfer of the results obtained to an industrial automated electric drive [11, 12];
2. Setting up and adjustment of industrial electric drives using hardware and software simulators of a real-life electric drive [24, 25, 26].

We describe below in more detail both approaches for technology setting up and adjustment the power equipment of an automated electric drive.

3.1 Hardware and software simulators of real-life electric drives

Professor from the Ural Federative University, A.M. Zyuzev, developed a technology for setting up and adjustment an automated electric drive using hardware and software tools called "Controlled Electric Drive Simulators" [24, 25, 26].

Replacing the power part of an electric drive during setting up and commissioning processes using software and hardware simulators is effective in terms of maintaining the functionality of the new electrical equipment, reducing the cost of the setup process and the time for preparing the drive for operation. There are a large number of units and installations for which presence of the noted capabilities is the primary factor in operation processes (for example, oil production installations operating in harsh climatic conditions [24, 25, 26]).

Using the materials of work [25], which describes the creating process of a simulator of the power part of a modernized locomotive, which includes real-life mathematical models of the electric drive power parts. In [25], the electric drive simulator is a set of tools that ensure interaction of a real-life electric drive control system with hardware and software tools used to simulate the electric drive system in real time at a frequency of 1 MHz. Depending on the assigned tasks, electrical equipment simulators are developed to simulate in real time the operation of an electric motor, converter, converter-motor or converter-motor-mechanism systems.

For example, to set up a converter control system, a simulator of the power part of an electric drive can be used, which receives control signals from the converter keys, models operation of the keys and the electric motor, and produces feedback signals necessary for the control system operation.

A hardware-software simulator may be based on a computer running a real-time operating system. Such simulators are fast enough to simulate systems containing a thyristor converter. Simulating the operation of a transistor converter requires higher performance, which can be provided by systems based on a digital signal processor (DSP) or programmable logic device (PLD). The latest version of simulators implementation has become widespread due to the high speed of PLD and the availability of convenient programming tools.

3.2 Setting up and adjustment technology of industrial automated electric drive using simulator benches

We propose an algorithm for setting up and adjustment a specific electric drive, for example, a pumping station of a heat supply system in a megalopolis or an electric drive of an oil-pumping station of an underground coal-mining complex equipped with asynchronous, frequency-controlled electric drives.

1. The nominal parameters of the original electric motor and electric drive (rated power, supply voltage, current, active and inductive resistance of the stator winding, energy conversion efficiency, rotor torque of inertia, actuator, gearbox) are evaluated.
2. The time constants of the electric drive, the reduced moment of inertia is calculated.
3. The design scheme of the electric drive is drawn up.
4. The electric drive operating modes are determined and the characteristic of braking torque on the motor shaft is evaluated.
5. Based on the similarity theory, the conversion factors of the original motor parameters into the motor simulator bench parameters are determined from the assumption of existence, in the first approximation of a linear dependence of parameters of the original and installed on the simulator bench asynchronous electric motors.
6. Subsequent actions of a commissioning engineer come to the development in the MexBIOS™ Development Studio environment of simulation models of an automated frequency-controlled electric drive with a load characteristic similar to that of the original motor for all possible options for constructing automatic control systems for an asynchronous motor of the simulator bench for subsequent transformation (using calculated similar coefficients) to the automated electric drive of a working mechanism.
7. According to the simulator bench operating technology, the results of simulation studies are verified through the physical implementation of each structural diagram of the simulation model on the real equipment of the simulator bench, followed by physical experiments in modes similar to the operating modes of the original electric drive of a working mechanism.

4. Developing of M&C system of large-scale industrial objects in the context of "INDUSTRY 4.0" technology

4.1 Prerequisites for the need to assess the M&C Systems durability.

In industrial production, a very significant factor (sometimes the most important) is the stochasticity of processes associated with the external environment of large-scale industrial systems. As a result, it is necessary to operate with probabilistic characteristics. Some theoretical results are shown in [27, 28, 29].

4.2 Assessment of durability ranges of technological machines and installations

The analysis object will be the elements of production plants and machines with rotating shafts driven by an electric drive and subject to stochastic disturbances [16]. These can be machines (for example, technical systems containing means of energy supply and movement of substances), stationary pumping and compressor stations, cutting drums of coal-mining machines, flexible automated production, which involves robot manipulator and conveyors are involved in interconnected work operations on structural elements of production (parts). It is necessary to forecast the actual service life of machines.

Rotating machines operate in dynamic modes. In the works [16, 30], it is proved that in the function of deviations of rotational speeds and torque of motor and machine shafts it is possible to realize estimation of the spent resource of motos and machines as a whole. In [16, 30], the machine reliability functionality is defined as the dependence of the probability of failure-free operation for the period of time between two overhauls from a variety of varying factors - machine operating conditions. The durability features are defined as the dependence of the time frame (years, hours), in the continuation of which with a probability of not less than the minimum possible level of machine reliability – σ_{Nmin} (on verge of operation risk), the machine will be in operating condition and will not require major repairs. We were talking about mining machines operating in significantly stochastic conditions. It is economically and technically sufficient to ensure the established design levels of minimum reliability and durability in the established period of machines operation [31].

In the works [16, 30, 31], it is established that machine's reliability and durability depends on intensity of dynamic processes occurring in gearboxes, electric drives and structural shells (cases) of machines, but there are no analytical dependencies of the reliability functionality of machines and the durability features from reliability level, durability period and other variable factors.

If in some way it is possible to determine the actual values of the reliability level (σ_N) and the service life period (T_D) of the machine for specific operating conditions, then these values can be used to predict operating modes of the machine that ensure that σ_N and T_D are maintained at the level $\sigma_N \geq \sigma_{Nmin}$, $T_D \geq T_{Der}$, where T_{Der} is the period of cost-effective durability of the machine.

Subsequently, the individual values of σ_N, T_D, as a function of the operating conditions existing for them, can be approximated graphically or analytically and used in the control system of machine operating modes. The machines operating mode in technical systems is generally conditioned by the random character of changing the main disturbance (machine load), and then the angular velocities (ω) and torques (M) of parts are random (stochastic) processes. These processes' characteristics (mathematical expectations, dispersions, standard deviations) depend on control quality of the local automated electric drive or complex optimal control system. Here are additional reserves for increasing durability of controlled machines.

By putting the information about the duration of machine switching operating (t_{jt}) into the database of the hierarchical control system, M&C, it is possible to control the machine resource. The data arrays $Y_j(t)$ contain information about the states of moments and angular velocities of elements and the duration of machines switching t_{jt}. The array data consists of two groups of elements: the first, corresponding to deterministic operating modes (on, off, hard work...); the second is stochastic, corresponding to normal operating conditions, is often characterized as stochastic.

In deterministic operating modes of technical systems of electrical complexes, the prevention of emergency situations must be ensured by hardware protection. If, in this case, the shaft rotation frequencies $\omega_j(t)$ and torques $M_j(t)$ do not exceed the limit values, then these situations should not lead to a decrease in service life in machines of electrical complexes, therefore, in the future, it makes sense to evaluate only the stochastic group $Y_j(t)$, t_{jt}. for reliability and durability of machines in technical systems of electrical engineering complexes.

4.2.1. Prerequisites for analyzing the energy efficiency of the M&C system of large-scale industrial facilities

The issues of reducing energy consumption are a priority for both individual producers and society as a whole. The costs for consumed energy affect not only the cost of a unit of finished products, but also carry a certain environmental burden [32, 33].

4.2.2. Initial provisions for the object of analysis

In the flexible automated process under study, robotic manipulators perform "point" operations to apply markings to structural elements of production. Tripod robots carry out transfer of volumetric structural elements onto a conveyor belt and their subsequent installation in specialized boxes. Several lines of interconnected robotic arms and robotic tripods operate in multiple. Local robot control systems solve problems of positioning, automatic control and logical control. Events occurring in flexible automated manufacture are, to a first approximation, Markov processes with discrete states and continuous time.

Taking into account the properties and characteristics of robots, conveyors, local and global networks, flows of structural elements, it is proposed to carry out optimization of electric consumption by parameterization of automatic control systems of electric drives. Processes for searching optimal parameters of control systems are virtually solved using methods of mathematical, simulation modeling and planning of factor experiments on the hierarchical control system server using flexible automated production.

The hierarchical control system includes telecommunication networks, servers and specialized hardware and software and performs complex control of technological structural elements and solves system tasks of optimizing energy consumption of production process. Mathematical models of large-scale industrial objects are built based on multi-mass calculation schemes [16, 28]. Using ratios and dependencies from [14, 16], a priori, as a function of the nominal parameters of electric drive objects, indicators of the properties and characteristics of large-scale industrial system elements can be calculated, for example: oscillation frequencies, resonant frequencies, damping coefficients, logarithmic decrements, frequencies of free undamped oscillations, resonant amplification coefficients, dynamic coefficients, electromagnetic and electromechanical time constants. For manipulator robots, the electric drive objects are 6 (six) DC motors. For tripod robots, the electric drive objects are an electromechanical rotary actuator and three servomotors.

For the conveyor plant, the electric drive objects are asynchronous electric drives with three-mass design schemes, taking into account the rigidity, dissipative losses and reduced clearances of gearbox gears, with a significant lag depending on the length of the carrier and idle conveyor belts. In accordance with the technological process, flows of input structural elements (semi-finished products) are supplied for processing by elements of flexible automated production in stochastic sequences, the patterns of which are known but can change. Information on element flows is transferred over global and local telecommunication channels.

The quality criteria of the M&C large-scale industrial system depend on the size of structural elements, and the concept of "flexible automated production" assumes the existence of different batches of semi-finished products, for example, according to geometric characteristics or required technological processing operations. The latter conditions require storing information about batches of semi-finished products in the database of flexible automated production, and this information is the basis for individual control subsystems of the technological complex as a whole. Processing operations by robotic manipulators and robotic tripods of each structural element and their groups, as well as movement of elements by conveyor installations are characterized as deterministic processes. This suggests that ranges of quality criteria for controlling robots and conveyors can be estimated, both by direct and integral estimates, as well as the stability conditions of a large-scale industrial system with their automatic control systems.

4.2.3. Algorithms and sources of power consumption minimization in large-scale industrial systems

In deterministic automatic control systems, it makes no sense to search for some unique methods and means for saving power consumption [34]. This is a technical problem of parametric optimization under different variants of electric drive and structure of automatic control systems. Some originality of this problem can be given in the robust formulation and, possibly, in the analysis of system of large-scale industrial objects as dynamic objects with incomplete information according to the theory of Professor N.V. Skibitskiy. [35].

The stochastic nature of semi-finished goods arrival creates the problem of estimating the ranges of possible energy consumption values in flexible manufacturing. Assuming that the characteristics of the flow of semi-finished products to various lines of robotic manipulators are known and that these flows are characterized by known distribution laws (for example: normal, Poisson, etc.). There is a possibility that a defective structural element – a semi-finished product will enter the production line. This probability is stochastic and must be described by the equations of probability theory. The selected types of electric drives and the structure of automatic control systems for flexible automated production are the means of ensuring minimization of electrical energy consumption, and the probabilities of a possible level of energy consumption are determined by the existing patterns of stochastic processes in flexible automated production and are the object of research in [29].

In [29], there is presented mathematical calculations of energy consumption minimization ranges for four operating modes of flexible automated production, including:

- Mode 1. Computer vision systems record the moments when a semi-finished product arrives for processing by a robotic manipulator, a tripod robot and a conveyor line. Information from sensors of computer vision system is transmitted via global and local telecommunication channels, placed in database and can be processed in the server of a hierarchical monitoring and control system. All processes are considered as deterministic. Then the range of energy consumption is within the boundaries between the maximum energy consumption (of a non-optimized system) and the sum of the minimum (optimized) consumption and energy costs for calculating the optimal operating mode of mechatronic devices;
- Mode 2. Due to the fact that semi-finished products arrive at the production line unevenly, the system is considered as a Markov process with discrete states and continuous time. The patterns corresponding to the flow of semi-finished products characterize it as the simplest (or stationary Poisson flow). In this case, the minimum energy consumption value will take into account the probability of a semi-finished product entering the production line at any time;
- Mode 3. The operation of flexible automated production using computer vision systems (i.e., as in mode 1) is influenced by the possible appearance of defects in elements of semi-finished products, which leads to skipping its processing. The operating time of the production line and the processing cycle of semi-finished products remain unchanged, but the total number of semi-finished products supplied for processing will decrease. Consequently, in the range of possible optimization according to the energy consumption criterion, the probability of the appearance of a defective semi-finished product is taken into account, which leads to a decrease in the limit of maximum (non-optimized) energy consumption;
- Mode 4. Functioning of flexible automated production is considered as a Markov process with an uneven flow of semi-finished products onto the line and the impact on the production line operation of the probable occurrence of defective semi-finished elements is taken into account (as in the mode 3).

4.3. Prerequisites for creating a universal methodology for effective management of power consumption in robotic production

The above theoretical provisions and algorithms for assessing power consumption, as well as structural and algorithmic proposals for implementing the processes of control, monitoring, regulation, automatic and automated control of an electric drive with parameterization of actuators according to the criteria for minimizing power consumption, suggest the possibility of creating a universal methodology invariant to the type of products being created in flexible automated production, and the laws of flows distribution of semi-finished products entering the production line.

It is fundamentally important to represent the production line as a "big" system with "big" data, in which the problem of uncertainty in the state of objects and systems is eliminated by concentrating in the database of a hierarchical system of control, monitoring and management of actually measured physical information by means of control and monitoring subsystems, forecasting physically unobservable information by subsystems of mathematical and simulation modeling [29]. The concept of "State of electrical complexes of aggregated installations of flexible automated production", which is defined as a set of information flows is described below:

- coordinates physically observed in real time are elements of the output signals that make up the vector Z(t);
- physically observable and unobservable state space coordinates are elements of the vector Y(t), which are obtained by simulation of mathematical models of electric drive elements and electric drive control systems;
- set of indicators of standard modes of electric drive elements, electric drive control systems;
- set of design data of electric drive, electric drive control system in flexible automated production;
- estimation of parameters of electric consumption by M&C electric drives for a fixed and declared period of time;
- estimation of resource and reliability of elements of the automated electric drive and large-scale system.

Characteristics of resource and reliability together with the evaluation of electric consumption analyzed above can be considered as an integral complex of technical state and operation of large-scale system elements, electric drive elements, electric drive control systems in flexible automated production. There are components in the information flow of states of electrical engineering complexes that allow us to take advantage of theoretical developments from [29, 34, 36] for estimating resource and reliability of automated electric drive elements.

In [7, 16], a methodology for assessing the durability of machine-building products with rotating shafts as a function of the ratios of deviations of minimum and maximum angular velocities and torques on the shafts for the analysed period was developed. Coordinates Z(t), Y(t) – from the databases of complex of technical state and operation of large-scale objects of control and management allow, according to the method [16], to estimate the resource and reliability of the elements of the automated electric drive and, together with the estimation of electric consumption, to obtain the estimation of the integral complex of technical state and operation of the M&C large-scale system of the electric drive elements.

5. Research of M&C large-scale systems in the context of "INDUSTRY 4.0" technology

Effective research in real- and divided-time modes of the statics and dynamics of large-scale system components is possible using methods for conducting active and passive experiments [37] using specialized software and hardware LabVIEW and programming environment MatLAB-Simulink [18, 19] in the M&C server. The results of research are shown in [10, 28].

6. Training monitoring and control system of large-scale industrial objects in the context of "INDUSTRY 4.0" technology

The training process, more often understood as educational (process 1) for various categories of specialists, as well as students, undergraduates, graduate students and doctoral students, within the framework of the "Industry 4.0" technology is the process 2, aimed at adapting the M&C system of large-scale industrial objects using means and methods of artificial intelligence to the stochasticity of the environmental characteristics of large-scale systems and its elements. In the process 1, significant results were achieved, including within the framework of the International Network Distance Learning Program "Synergy" [38]. Experience for realization of the process 2 is to be accumulated and there is important to create monitoring systems of state data about large-scale system components, storing and processing stochastic information flows, setting optimization intellectual tasks and their technical implementation [39].

7. Conclusion

This work "Automated robotic complexes in engineering, adjustment, production and education processes" solves the current scientific problem of creating a methodology, principles for constructing, research and improving of flexible automated production with robotic complexes driven by electric drives. Additionally, there are presented training projects for creating the M&C large-scale systems in the context of "Industry 4.0" technology within the framework of educational programs and research works of students and undergraduates majoring in "Automation and Control" and "Electric Power Engineering" based on the international educational project "Synergy". The research work results in the field of determining the optimization of an automated electric drive according to the criterion of energy consumption and calculations of resource characteristics and reliability of elements of automated production systems are shown.

8. References

[1] K. Shvab (2020) Chetvertaya industrialnaya revolyuciya (The fourth industrial revolution) https://www.youtube.com/watch?v=dCtuSThn9fo&t=2792s

[2] A.Zuyev (2018) Promyshlennaya revolyuciya. Industriya 4.0. (Industrial Revolution. Industry 4.0) https://www.youtube.com/watch?v=7um5warRsQs&feature= youtu.be

[3] Cifrovizaciya promyshlennosti. (2022) (Industrial digitalization) https://center2m.ru/tsifrovizatsiya-promishlenosti

[4] Tesic, S; Bajic, B; Cosic, I & Rikalovic, A (2021). Frugal innovations in Industry 4.0: The Identification of Key Indicatiors, Proceedings of the 32nd DAAAM International. Symposium, pp.0262-0269, B. Katalinic (Ed.), Published by DAAAM International, ISBN 978-3-902734-33-4, ISSN1726-9679, Vienna, Austria, DOI: 10.2507/32nd.daaam.proceedings.037

[5] Polivka, M. & Dvorakova, L. (2021). The Current State of the Use of Selected Industry 4.0 Technologies in Manufacturing Companies, Proceedings of the 32nd DAAAM International Symposium, pp.0652-0659, B. Katalinic (Ed.), Published by DAAAM International, ISBN 978-3-902734-33-4, ISSN1726-9679, Vienna, Austria DOI: 10.2507/32nd.daaam.proceedings.092

[6] Stazhkov S.M. (2021). International university network project “Synergy”: 15 years of cooperation. Automation. Informatics, No. 1-2, p. 92-95, ISSN 1560-7305.

[7] Feshin B.N. (2011). Supervisory multi-connection control systems for electrical complexes of mining enterprises. Giga Trade, ISBN 978-601-879-107-3, Almaty.

[8] Tokhmetova K.M. (2021). Development of a remote-control system for the RTK “TRIPOD” in the FAM structure. Automation. Informatics, No. 1-2, p. 14-23, ISSN 1560-7305.

[9] Tokhmetova K.M., Kobylandy A.A., Baigul A.S., Suleimenova A.B. (2021). Development of a control system using the FESTO Mechatronic Line bench. Automation. Informatics, No. 1-2, p. 38-45, ISSN 1560-7305.

[10] Feshin B.N., Kalashnikova Ye.V. (2013). Monitoring system for FESTO servo and step electric drives. Automation. Informatics, No. 2 (33), p. 71-75, ISSN 1560-7305.

[11] http://mexatronica-pro.com. Matronika-Pro RPC. MSV-02 - Adjustment kit for a developer of electric motor control systems. Technical description. Tomsk. - 2014.- p.5.

[12] http://mexatronica-pro.com. Matronika-Pro RPC. MexBIOS Development Studio. Getting started - a guide for new users. Description of integrated environment. Tomsk. - 2014.- p.50.

[13] Wang, D. L.; Zeng, X. T.; Wang, G. F. & Li, R. (2021) Stability of a face guard in a large mining height working face, International Journal of Simulation Modelling, 20 (2021), pp. 547-558, Publisher: DAAAM International Vienna, DOI 10.2507/IJSIMM20-3-572, ISSN:1726-4529

[14] Video lecture, Feshin B.N. Joint inter-university course of the network distance learning program Synergy “Intelligent Control Systems” for master’s students majoring in “Automation and Control”, “Electric Power Engineering”. http://y2u.be/uwz_hM3nGnE

[15] Avdeev L.A., Kostin M.I. (2017). Karaganda Scientific Research Coal Institute (structural transformations and achievements in the field of automation of production processes). Automation. Computer science, 2(41), pp.140-148, ISSN: 1560-7305

[16] Feshin B.N. (2018). Multi-connected supervisory control systems for coal-mining machines: Monograph. Karaganda: KSTU. ISBN 978-601-315-470-1.

[17] Feshin, B.N., Shpakova, L.G., Tokhmetova, K.M. (2023). Hierarchical Management System of the Coal-Winning Complex. Lecture Notes in Mechanical Engineering, Publisher: Springer Nature, pp. 220–230, ISSN: 2195-4356, DOI: 10.1007/978-3-031-38126-3_23

[18] Dyakonov V.P. (2004). MATLAB 6/6.1/6.5/ + SIMULINK 4/5®. Application basics [Text]: complete user guide, SOLON-Press, ISBN 5-98003-007-7, Moscow.

[19] Terekhov V.B. (2022). Computer modeling of DC and AC electric drive systems in Simulink: textbook. Yurait Publishing House, ISBN 978-5-5-534-06858-0, Moscow.

[20] Breido I.V., Zyuzev, Feshin B.N. (2016). New technologies for setting up and adjustment automated electric drives. Automation. Informatics, No. 2 (39), p. 72-84, ISSN 1560-7305.

[21] Basharin A.V., Novikov V.A., Sokolovsky G.G. (1982). Control of electric drives: textbook for universities. L.: Energoizdat, - p. 392.

[22] Belov M.P., Zementov O.I., Kozyaruk A.Ye., Kozlova L.P., Novikov V.A., Savva S.V., Chernigov L.M., Gorokhov S.G., Tatarintsev N.I. Engineering of electric drives and automation systems: textbook. M.: Academy Publishing center, ISBN 5-7695-2448-0, Moscow.

[23] Basharin A.V., Postnikov Y.V. (1990). Examples of calculation of the automated electric drive on the computer: [Textbook for universities majoring in “Electric drive and automation of industrial plants and technological complexes”]. L.: Energoatomizdat. Leningrad Branch, ISBN 5-283-04468-8, St. Petersburg.

[24] Zyuzev A.M., Mudrov M.V., Nesterov K.Ye. (2014). FPGA-simulator of real-time AC electric drive. Energy. Innovative directions in the energy sector. Cals technologies in the energy sector, №1, pp. 229-238

[25] Nesterov K.E., Mudrov M.V., Zelentsov V.I. (2016), Simulator of the shunting locomotive's electric drive 2nd international conference on industrial engineering, applications and manufacturing, 6 2016 Chelyabinsk, 19–20 мая 2016 года, Chelyabinsk, doi: 10.1109/icieam.2016.7911523

[26] Zyuzev A.M., Bubnov M.V., Mudrov M.V. Simulator of electric drive of the rod deep-well pumping unit. II International Scientific and Technical Conference "Prom-Engineering". 2016. – p.4.
[27] Feshin B.N., Shpakova L.G., Tokhmetova K.M. (2018). Hierarchical control and monitoring systems of large automated technological complexes. University works, No. 2, p. 114-120, ISSN 1609-1825.
[28] Kalashnikova Y., Feshin B. and Peshko M. (2021). Monitoring of High-Voltage Electric Drives of Pumping Units of Heat Supply Complexes of Megacities. XVIII International Scientific Technical Conference Alternating Current Electric Drives ACED, May 24-27, 2021, Ekaterinburg, Russia, pp. 1-5, ISBN 978-1-6654-1248-3, DOI: 10.1109/ACED50605.2021.9462310.
[29] Feshin B.N., Shpakova L.G., Tokhmetova K.M. (2019). Algorithms of energy saving in flexible automated productions (part 1). University works, No.3, p.145-147, ISSN 1609-1825.
[30] B.N. Feshin, K.M. Tokhmetova, Ye.V. Kalashnikova and L.G. Shpakova. (2023) Optimization of power consumption and equipment operating life in large mechatronic and robotic facilities. EAI Endorsed Transactions on Energy Web, unpublished.
[31] Moldavskiy L.A., Finkelshtein E.L., Verklov B.A. (1981). Types of damage and durability of mining machine transmissions. Moscow: Nedra.
[32] P. Waide, C. Brunner, (2011) Energy-efficiency policy opportunities for electric motor driven systems. Int. Energy Agency, p.132. ISSN: 2079-2581 https://doi.org/10.1787/20792581.
[33] Jollands, N., Waide, P., Ellis, M., Onoda, T., Laustsen, J., Tanaka, K., … Meier, A., (2010) The 25 IEA energy efficiency policy recommendations to the G8 Gleneagles Plan of Action. Energy Policy, Volume 38, Issue 11, pp. 6409–6418. doi.org/10.1016/j.enpol.2009.11.090.
[34] Feshin B.N., Shpakova L.G., Tokhmetova K.M. (2019). Optimal robots control in flexible automated manufacturing. University works, No. 1, p. 145-148, ISSN 1609-1825.
[35] Skibitskiy N.V. (2005) Interval methods and problems of constructing models of objects and control processes: Dissertation of Doctor of Engineering Sciences: 05.13.01. Department of Management and Informatics, National Research University Moscow Energy Institute, Moscow, Russia.
[36] Feshin B.N., Shpakova L.G., Tokhmetova K.M., Kalashnikova E.V. (2021). Optimization of power consumption and service life of machines in large production complexes of mechatronic-robotic type. Abstracts of the XI international scientific and technical Internet conference for young scientists "Automation, mechatronics, information technologies" (AMIT), May 2021, Omsk, ISBN 978-5-8149-3306-5, p. 37-40.
[37] Breido I.V., Feshin B.N. (2003). Simulation modeling and parametric optimization of automated electric drives. Ministry of Education and Science of RK, Karaganda State Technical University, ISBN 9965-606-56-0, Karaganda.
[38] Breido I.V., Yeliseyev A.S., Katalinich B., Kramar V.A., Pashkov E.V., Stazhkov S.M., Feshin B.N., Homchenko V.G., Shtol V. (2015). International university network project "Synergy": modern possibilities of combining and using educational resources of engineering universities in cooperation with the leading European concern FESTO. Automatics. Informatics, No. 1, P. 16-20, ISSN 1560-7305.
[39] Jones M.T. (2011). Artificial intelligence programming in applications. DMK Press, ISBN 978-5-94074-746-8, Moscow.

DOI: 10.2507/34th.daaam.proceedings.033

A NEW METHOD FOR CONTACTLESS MEASUREMENT OF SMALL MODULE GEARS WITH ASYMMETRIC PROFILE

Genadi Tsvetanov †, Miroslav Kokalarov, Tsanko Karadzhov & Ivan Georgiev

This Publication has to be referred as: Tsvetanov, G[ennady]; Kokalarov, M[iroslav]; Karadzhov, T[sanko] & Georgiev, I[van] (2023). A New Method for Contactless Measurement of Small Module Gears with Asymmetric Profile, Proceedings of the 34th DAAAM International Symposium, pp.0262-0267, B. Katalinic (Ed.), Published by DAAAM International, ISBN 978-3-902734-41-9, ISSN 1726-9679, Vienna, Austria
DOI: 10.2507/34th.daaam.proceedings.033

Abstract

In the field of mechanical engineering, achieving precise measurements of small-module gears and transmissions is critical. This paper presents an innovative non-contact method for assessing the accuracy of these components, based on optic-electronic signal processing principles. It offers superior precision in evaluating gear geometry, pitch, and profile. With recent advancements in gear manufacturing, there is a growing need for efficient and precise measurement methods to ensure optimal performance and durability. Traditional tactile methods and double-flank inspections have limitations, including potential subjective errors. This paper addresses these challenges by providing a comprehensive solution that not only enhances measurement accuracy but also sheds light on the relationship between production variables and gear precision. This study is significant for delivering high-precision gear assessments, revealing production-precision links, reducing subjective errors, highlighting automation potential, and contributing to the discourse on mechanical precision in gear-dependent industries. Drawing from foundational research in gear measurement, optic-electronic signal processing, and manufacturing automation, its primary aim was to determine if non-contact methods improve small-module gear measurement. Through rigorous analysis, this study unequivocally affirms this question, advancing gear and transmission measurement and control.

Keywords: Small module gears; centroidal enveloping method; asymmetric cog profile; kinematic error; contactless measurement.

1. Introduction

Gears, integral components in a multitude of mechanical systems, play a pivotal role in ensuring the efficiency and functionality of machinery across various industries. Their manufacturing, however, is riddled with potential pitfalls stemming from the inherent challenges of achieving precise formations. Harmonic oscillations, a predominant concern, accumulate during manufacturing, leading to periodic errors that jeopardize the optimal functioning of the gears [1], [3], [10]. While radial errors are attributed to factors like product basing errors, tool inconsistencies, and the stability of the Machine, Adaptation, Tool, Detail (MATD) system, tangential errors often emerge from discrepancies within the kinematic chain's elements, such as worm pairs and dividing discs.

The intricate interplay of these individual errors in the technological process manifests in various ways on the gear, affecting kinematic accuracy indicators like operational fluency, contact quality, and clearances. These complex errors, in turn, influence the gears' overall operational quality [2]. Moreover, understanding elemental errors is not just a theoretical exercise but has tangible implications for technological control. Against this backdrop, this study seeks to address two fundamental questions:

How can we better understand the sources and implications of these harmonic oscillations and errors in gear manufacturing? Is there a more accurate and efficient method for evaluating and mitigating these errors, particularly in the realm of small-module gears and transmissions? In pursuit of these questions, we introduce a novel non-contact method and device, aiming to shed light on the relationship between manufacturing processes and resultant gear accuracy, while also offering potential solutions for improved production and control processes.

2. Discussion

Analyzing and controlling the precision of gears is an intricate endeavor. The intricate geometric configurations, multifaceted performance metrics, and rigorous precision criteria inherent to gears make their evaluation a formidable task. Typically, two predominant strategies dominate this sphere: the elemental evaluation techniques, which furnish direct insights into the functional proficiency of gears and transmissions, and those that present both technological and functional data regarding gear quality. Both streams of information are imperative to optimize the manufacture and functionality of gears. Hence, a pressing need arises for methodologies and tools that facilitate a seamless and intuitive bridging between these distinct data realms.

Small-module gears, defined by modules under 1 mm, intensify these challenges. Their minuscule dimensions and distinct tolerance ranges, juxtaposed with their medium and large-module counterparts, further amplify the complexity. This milieu catalyzed the inception of a comprehensive measuring instrument designed to amalgamate insights on both functional accuracy metrics and detailed technological discrepancies, along with underlying causatives. This invention took shape at the "Mechanical Engineering and Equipment" department of the Technical University - Gabrovo [11].

The said device harnesses the prowess of photoelectric technology, amalgamated with the computational capability of personal computers. As illustrated in Figure 1, the mechanism hinges on the modulation of a laser's light beam (laser 1) by the teeth of the gear under scrutiny (gear 3), which consistently rotates at a predetermined angular velocity. This modulated light, when channeled through an aperture (aperture 4), interacts with a photoelectric sensor, thereby being transmuted into a discernible signal. This signal, subsequently processed by the Analog-to-Digital Converter (ADC) 8, is relayed to a computer (computer 9). The unwavering angular velocity is upheld by a motor in conjunction with the photo raster converter (PRC) [9]. To mitigate potential positioning discrepancies, the gear under examination is meticulously positioned on a high-precision spindle system. The device's affinity for laser technology augments its resolution, thereby elevating the precision of the measurements. Moreover, the integration of a computerized system facilitates swift, real-time processing of the captured signals, proffering insights into functional viability, discernible inaccuracies, and the underlying technological rationale [12].

This expanded discussion provides a detailed and academic perspective on the complexities and solutions involved in gear measurement and control.

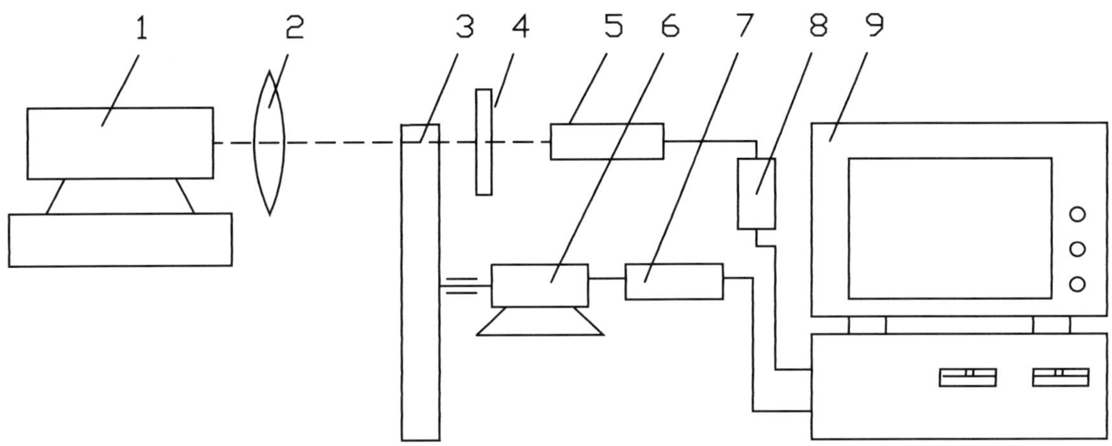

Fig.1. Principle of operation of a complex control device of small modular gears
1. Laser, 2. Optical system, 3. Controlled gear, 4. Aperture, 5. Optic-electronic receiver, 6. Motor, 7. Photo-raster converter, 8. Analog-to-digital converter, 9. Personal computer

The signal obtained on the Optic-electronic receiver as a voltage is proportional to the displacement of the tooth profile points on the line of engagement, so it is proportional to the accumulated error in the fundamental pitch. This signal can be measured either as time for passing points from two homonymous profiles of adjacent teeth or as pulses on the photoelectric converter 7. In the first case, the controlled wheel must rotate at strictly constant angular velocity along the conditional working part of the tooth profile.

While in the latter case, this is not necessary, but a high sensitivity PRC should be used. When the signal is measured as time, it represents a periodic function (voltage - time), or a period determined by the cog frequency. The envelope of this curve expresses the eccentricity of the gear, and the amplitude deviations inform about errors in the base step.

When the signal is received as a number of PRC pulses, the identification depends on the parameter measured. When determining the accumulated error in the base step (**f_{ptz}**), the expression is used for each part of the cog profile when the gear is reversible and requires a minimum dead stroke:

$$F_{ptz} = T_o - T_z = \frac{2\pi}{N} r_b j_z = q j_z; \; F_{ptz} = T_0^* - T_z^* = \frac{2\pi}{N} r_b^* j_z^* = q j_z^* \qquad (1)$$

where:
T_o и T^*_0- are the initial impulses of the PRC when passing the first cog of the gear with an asymmetric profile;
T_z и T^*_z - the impulses of the PRC when passing the controlled homonymous profile;
N - the number of discrete PRC signals per rotation;
r_b и r^*_b - the radii of the fundamental circles of the gear;
j_z и j^*_z - the number of pulses counted by the PRC when measuring the gear with an asymmetric profile;
q - the value of the pulse in μm.

For the non-working part of the tooth profile, a dependency (1) is used, but the values of the non-working involute are placed in it. With a significant difference in the two profile angles (over 20°), it is recommended to use a different pitch of discretization and a different number of pulses for the two profiles, which ensures approximately equal accuracy.

In [4], [5], [6], [7], [8] the determination of the rectangular coordinates of the transitional curve for a symmetrical cog profile is given, solving the following system of partial derivatives for the instrument tip:

$$\begin{vmatrix} \dfrac{\partial f_1}{\partial t} \cdot \dfrac{\partial f_2}{\partial t} \\ \dfrac{\partial f_1}{\partial \gamma} \cdot \dfrac{\partial f_2}{\partial \gamma} \end{vmatrix} = 0 \qquad (2)$$

where the coordinates **x** and **y** of the transition curve are functions of **t** – the wrapping angle and γ - the angle characterizing points from the rounding at the top of the instrument. In the case of a preset classical instrument for making gears by the method of centroidal traversing, in a polar coordinate system, coinciding with the geometric center of the gear 1 or 2 and external engagement, the two evolvents are determined in the front section from:

- *First possibility of formation:*

$$\left| \begin{array}{l} inv\,\gamma_t = tg\,\gamma_t - \gamma_t \\ r_{tb_{1,2}} = \dfrac{m_t . z_{1,2}}{2} . cos\ \alpha_t \\ r_{te_{1,2}} = \dfrac{r_{tb_{1,2}}}{cos\ \gamma_t} \\ \phi_{te_{1,2}} = \left(\dfrac{\pi}{2 . z_{1,2}} + \dfrac{2 . x_{1,2} . tg\ \alpha}{z_{1,2}} \right) + inv\ \alpha_t \end{array} \right. ; \quad \left| \begin{array}{l} inv\,\gamma_t^* = tg\,(-\gamma_t^*) + \gamma_t^* \\ r_{tb_{1,2}}^* = \dfrac{m_t . z_{1,2}}{2} . cos\ \alpha_t^* \\ r_{te_{1,2}}^* = \dfrac{r_{tb_{1,2}}^*}{cos\ \gamma_t^*} \\ \phi_{te_{1,2}}^* = \left(\dfrac{\pi}{2 . z_{1,2}} + \dfrac{2 . x_{1,2} . tg\ \alpha^*}{z_{1,2}} \right) + inv\ \alpha_t^* \end{array} \right. \qquad (3)$$

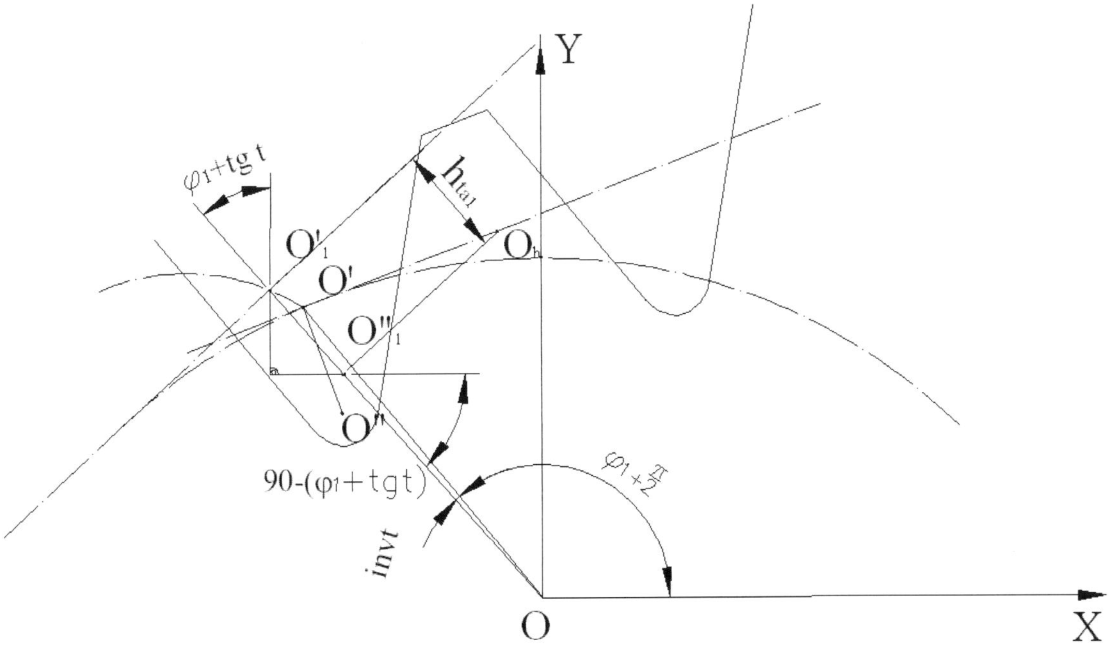

Fig. 2. Computational scheme for determining the parameters of the transitional curve

- *Second possibility of formation:*

$$\begin{cases} inv\,\gamma_t = tg\,\gamma_t - \gamma_t \\ r_{tb_{1,2}} = \frac{m_t.z_{1,2}}{2}.cos\ \alpha_t \\ r_{te_{1,2}} = \frac{r_{tb_{1,2}}}{cos\ \gamma_t} \\ \phi_{te_{1,2}} = \left(\frac{\pi}{2.z_{1,2}} + \frac{2.x_{1,2}.tg\ \alpha}{z_{1,2}}\right) + inv\ \alpha_t \end{cases} ; \quad \begin{cases} inv\,\gamma_t^* = tg\,(-\gamma_t^*) + \gamma_t^* \\ r_{tb_{1,2}}^* = \frac{m^*{}_t.z_{1,2}}{2}.cos\ \alpha_t^* \\ r_{te_{1,2}}^* = \frac{r_{tb_{1,2}}^*}{cos\ \gamma_t^*} \\ \phi_{te_{1,2}}^* = \left(\frac{\pi.m_{t1,2}^*}{2..m_t.z_{1,2}} + \frac{2.x_{1,2}.tg\ \alpha^*}{z_{1,2}}\right) + inv\ \alpha_t^* \end{cases} \tag{4}$$

$$m_{t_1}^* + m_{t_2}^* = 2.m_t$$

where $\mathbf{r_{te1,2}}$ and $\mathbf{r^*_{te1,2}}$ are radius-vectors of the involutes in the polar coordinate systems, on **3K 1** and **2** of the gear transmission;

$\varphi_{te1,2}$ and $\varphi^*_{te1,2}$ – the initial angles of the involutes on the base circles, in the frontal section;

γ_t and γ^*_t – the wrap angles when the lines of engagement roll without slipping on the base circles of the gears of the transmission;

α_t and α_t^* - profile angles of the tool with which the gears are made by the method of centroidal wrapping in the frontal section;

$\mathbf{z_{1,2}}$ – the number of teeth on gear 1 or 2 of the transmission;

$\mathbf{m^*_t}$ - is the frontal module of the output contour with non-uniform thickness, corresponding to the profile angle αt;

The obtaining of Cartesian coordinates xte1,2, x*te1,2, yte1,2 and yte1,2 in the absolute coordinate system of the theoretical involutes is performed by:

$$\begin{cases} x_{te_{1,2}} = r_{te_{1,2}}.sin\left(inv(t) + \frac{\pi}{2} + \phi_{te_{1,2}}\right) \\ y_{te_{1,2}} = r_{te_{1,2}}.cos\left(inv(t) + \frac{\pi}{2} + \phi_{te_{1,2}}\right) \end{cases} ; \quad \begin{cases} x_{te_{1,2}}^* = r_{te_{1,2}}^*.sin\left(inv(t^*) + \frac{\pi}{2} + \phi_{te_{1,2}}^*\right) \\ y_{te_{1,2}}^* = r_{te_{1,2}}^*.cos\left(inv(t^*) + \frac{\pi}{2} + \phi_{te_{1,2}}^*\right) \end{cases} \tag{5}$$

For better accuracy, it is recommended to calculate 100 points in Cartesian coordinates for each of the two dissimilar involutes.

The significance of the profile error ($\mathbf{f_{fri}}$) is determined within the active section of the line of engagement at a given angle of rotation separately for the both parts of the gear profile:

$$\varphi = 2\pi j_i/N; \varphi^* = 2\pi j^*_i/N^* \quad (6)$$

The error is determined as the difference between the nominal and actual displacement for each of the two different profiles:

$$f_{fri} = T - T_{gi} = q(j_i - j_{gi}) \quad ; \qquad f^*_{fri} = T^* - T^*_{gi} = q^*(j^*_i - j^*_{gi}) \quad (7)$$

where $\mathbf{j_i}$ and $\mathbf{j^*_I}$ are the initial values, equal to the number of pulses on the RPM at the start signal;
$\mathbf{j_{gi}}$ and $\mathbf{j^*_{gi}}$ - the number of pulses on the RPM between the initial pulse and the actual passage of the corresponding gear profile.
The discrete value of the kinematic error is determined, consecutively for the two profiles, from the expression:

$$F_{iz} = F_{ptz} + f_{fri} = q[j_z + (j_i - j_{gi})]; F^*_{iz} = F^*_{ptz} + f^*_{fri} = [j^*_z + (j^*_i - j^*_{gi})] \quad (8)$$

Post-processing of the signal can be carried out by computer analysis of frequency characteristics, rapid Fourier Transformation, Cepstral Analysis, in order to seek the influence of the technological process on the accuracy of gears.

3. Conclusion

The advancement and introduction of the proposed non-contact method and device for gauging the accuracy of small-module gears and transmissions mark a pivotal evolution in the domain of precision engineering. The significance of this innovation can be enumerated in the following facets: Enhanced Precision and Dependability: At its core, the method transcends conventional paradigms, offering unparalleled accuracy in gauging the geometrical intricacies of gears. This isn't limited to mere pitch and profile discernment but spans the realms of kinematic precision and operational fluidity. Such meticulousness not only underscores its potential but sets a precedent in measurement standards.

Technological Insights: By bridging gear manufacturing processes with their resultant accuracy metrics, the method lends itself to a profound analytical prowess. This capability doesn't just illuminate the intricacies of gear production but paves the way for iterative enhancements in both production and quality assurance protocols. Mitigating Human-induced Anomalies: One of the perennial challenges in precision measurements has been human-induced errors. The contactless nature of the method, juxtaposed against conventional contact and double-flank methodologies, drastically reduces these anomalies, thereby bolstering the reliability quotient of the results.

Advancing Automation: Automation isn't merely about efficiency but about repeatability and consistency. The new device, by virtue of its design, elevates automation standards in measurement processes, translating to enhanced throughput, consistency, and resource optimization. Versatility in Application: Beyond its intrinsic merits, the method's adaptability across diverse industries, be it automotive, aerospace, or heavy machinery, underscores its universal relevance. This wide applicability augments its value proposition, hinting at its potential ubiquity in the coming times.

Holistic Measurement Capability: The device, in its essence, isn't just about passive measurement. It actively processes signals to discern gear pitch, profile, and kinematic fidelity. Moreover, its ability to draw correlations between technological facets and gear accuracy, all the while negating the need for reference gears and specialized apparatus, is laudable. In retrospection, this groundbreaking method and apparatus don't just represent an incremental addition to the gear and transmission measurement cosmos. They signify a paradigm shift, providing tools that are not only precise and reliable but are poised to redefine manufacturing and control processes for gears and transmissions.

Future Directions: As industries evolve and the demand for precision grows, the implications of the findings in this study present several exciting avenues for further research: Integration with Machine Learning and AI: Leveraging artificial intelligence could further optimize the process, allowing for predictive analyses, anomaly detection, and self-calibration. Scalability: Exploring how this technology can be scaled or adapted to measure larger module gears or other complex mechanical components would be beneficial. Material Interaction Analysis: It would be interesting to see how different gear materials interact with the non-contact method, potentially leading to tailored measurement techniques for diverse materials. Interdisciplinary Applications: The potential of applying this method to realms beyond gears, possibly in sectors like microelectronics or nanotechnology, could be explored. These prospective directions not only stem from the core findings of this research but are a testament to the profound impact the study could have on the broader scientific community.

4. Acknowledgments

The publication of this study is supported by Project BG05M2OP001-1.001-0004 UNITe, funded by the Operational Programme "Science and Education for Smart Growth", co-funded by the European Union trough the European Structural and Investment Funds.

5. References

[1] Sotirov, B.; Ivanov, S. & Tonev, D. (2014). Development and perspectives of cog wheel and gear control and measurement, XXIV National scientific symposium with international participation "Metrology and metrological assurance", Collection of reports, no. 1, pp. 486-493, ISSN-1113-9126, Technical University - Sofia, Bulgaria

[2] Sotirov, B.; Ivanov, S. & Zhelezarov, I. (2015). Determining the values of differential accuracy indicators of spur gears based on the results of the Double Flank Gear Test Information Processing System, issue 2 (127) Measurement uncertainty: scientific, normative, applied and methodical aspects, 61-63, ISSN 1681-7710, Kharkov, Ukraine

[3] Ivanov, S. (2015). Application of complex double-flank inspection for technological control of cylindrical gears, Ph.D. Dissertation, Ruse University, Ruse, Bulgaria

[4] Mladenov, Y. (1999). Transitional curves when cutting cylindrical gears with centroidal tools, Scientific works of Ruse University, vol 37, series 2, Ruse University, Ruse, Bulgaria

[5] Sakharov, G. (1983). Rolling Tools, "Machinery", Moscow, Russia

[6] Ordinartsev, I. (1987). Toolmaker's Handbook - ed., "Machinery", Leningrad, Russia

[7] Sukhorukov, Y. (1992). Modification of involute gears, "Technique", Kiev, Ukraine

[8] Sabchev, P. (1982). Metal cutting tools, "Technics", Sofia, Bulgaria

[9] Zaharieva, S.; Borodzhieva, A. & Mutkov, V. (2019). Metrological Assessment of an Electronic System for Monitoring the Geometrical Parameters of Cylindrical Profiles, IEEE 25th International Symposium for Design and Technology in Electronic Packaging (SIITME), pp. 404-407, Cluj-Napoca, Romania, DOI: 10.1109/SIITME47687.2019.8990845

[10] Paulins, K.; Irbe, A. & Torims, T. *, (2013). Spiral Bevel Gears with Optimized Tooth-End Geometry, 24th DAAAM International Symposium on Intelligent Manufacturing and Automation, Selection and peer-review under responsibility of DAAAM International Vienna, Vienna, Austria, DOI: 10.1016/j.proeng.2014.03.003

[11] Sotirov, B.; Ivanov, S. & Sakakushev, B. (2016). Investigation of a complex method of measurement, based on the classical double-flank control on straight-toothed cylindrical gears. KNOWLEDGE International journal scientific papers, Vol 13.1, 355-356, ISSN 1857-92, Skopje, Macedonia

[12] Zaharieva, S.; Stoev, I.; Borodzhieva, A. & Petrova, T. (2023). Assessment of Gross Errors and Uniformity of Readings of Sensors for Relative Humidity Measurement in Indoor Premises, 2023 22nd International Symposium INFOTEH-JAHORINA (INFOTEH), pp. 1-5, East Sarajevo, Bosnia and Herzegovina, DOI: 10.1109/INFOTEH57020.2023.10094132

[13] Sakakushev, B., (2019). Measuring Systems - Theory and Practice - Monograph, Faculty of Aviation, National Military University "Vasil Levski", p. 118, ISBN 978-954-713-129-3, Veliko Tarnovo, Bulgaria

DOI: 10.2507/34th.daaam.proceedings.034

METROLOGICAL ASPECTS IN MAINTENANCE OF ENERGY-SAVING AUTOMATED SYSTEMS FOR CONTROL OF PUMP UNITS IN PUMPING STATIONS

Svitlana Artiukh, Miroslav Kokalarov, Viktoria Kniazieva,
Ivan Georgiev, Anastasiia Artiukh & Mihail Milchev

This Publication has to be referred as: Artiukh, S[vitlana]; Kokalarov, M[iroslav]; Kniazieva, V[iktoria]; Georgiev, I[van]; Artiukh, A[nastasiia] & Milchev, M[ihail] (2023). Metrological Aspects in Maintenance of Energy-Saving Automated Systems for Control of Pump Units in Pumping Stations, Proceedings of the 34th DAAAM International Symposium, pp.0268-0273, B. Katalinic (Ed.), Published by DAAAM International, ISBN 978-3-902734-41-9, ISSN 1726-9679, Vienna, Austria
DOI: 10.2507/34th.daaam.proceedings.034

Abstract

In the specialized realm of hydrocarbon transport engineering, this article presents a focused exploration into refining the current regulatory paradigms, with a cardinal objective: to enhance the operational efficacy of pumping units within oil pumping stations throughout their standard functional modalities. Historically, there exists a significant lacuna concerning comprehensive solutions capable of attenuating energy dissipation during oil conveyance and achieving benchmarked efficiency metrics intrinsic to oil pumping infrastructure. This pivotal matter, if addressed rigorously, stands to elevate both the technical and economic dimensions of primary oil conduits. The article's central aim, underpinning its entire analytical discourse, is to accentuate the energy efficiency of pumping units operational within oil transport hubs. To achieve this, the research introduces an avant-garde methodology primed for the calibration of pumps' dynamics and their associated network functionalities, optimizing their consumption paradigms in the process. Given the tangible fiscal constraints around the modernization of both primary and auxiliary equipment within oil pumping stations, there emerges an even more pressing impetus to amplify the paradigm of energy conservation. In a technologically progressive era where oil transport processes are rapidly automating, the development of automated energy-conservation-centric control systems is paramount. Such systems, by design, should ensure the reduction of energy losses, underscoring the article's overarching aim of bolstering energy efficiency within the hydrocarbon transportation sector.

Keywords: oil pumping station; efficiency; main pump; automated control system.

1. Introduction

The share of oil and gas in the energy balance of industrialized countries is about 80%. Oil transportation requires significant energy consumption to drive pumping units at oil pumping stations. In this regard, the urgent task is to reduce energy consumption by developing and implementing energy-saving automatic control systems (ACS) for oil pumping units, which are an integral part of integrated ACS for oil transportation and ensure maximum efficiency [1].

That is why improving the regulatory framework to increase the efficiency of pumping units at oil pumping stations remains an important scientific and technical task. Oil and gas continue to dominate the energy landscape of industrialized nations, accounting for approximately 80% of their energy mix. Such a substantial reliance underscores the magnitude of energy necessitated for oil transportation, particularly for driving the pumping units integral to oil pumping stations. Given this context, there is an exigent imperative to minimize energy usage. One salient approach to address this pressing concern is through the design, development, and deployment of energy-conservative ACS tailored for these pumping units. These systems not only form the backbone of integrated automated control platforms for oil transportation but are also pivotal in ensuring their peak operational efficiency, as referenced in [1].

Given this backdrop, refining the regulatory framework to amplify the performance efficiency of pumping units within these oil transport hubs emerges as both a crucial scientific endeavor and a technical imperative. This, in turn, further accentuates the significance of our research, aimed at revolutionizing the standards and practices in the oil transportation sector.

2. Discussion

Trunk pipelines are designed in such a way that their design capacity values are provided by pumping (for oil and oil products) or compressor (for gas) stations. At the same time, it is assumed that the pumping units of the stations operate at rated modes corresponding to the rated speed of their rotors and maximum efficiency. At the same time, the analysis of existing regulatory and technical documentation shows that today the regulatory and technical and regulatory and methodological documentation does not fully formulate the requirements for the creation and operation of control systems for main oil pipelines that would ensure maximum efficiency of pumping units and maintain their operation at an appropriate level.

The need to regulate the operation mode of the oil pumping stations - main oil pipeline system, is due to energy savings for pumping a unit volume (mass) of oil through the main oil pipeline. Methods of regulation of main oil pipelines can be divided into two classes: methods determined by the pressure balance and methods determined by the smoothness of regulation. Methods determined by the pressure balance are divided into methods related to changes in the parameters of the NPS and methods related to changes in the parameters of the main pipeline.

The main element that determines the efficiency of an individual pumping unit or an oil pumping station as a whole is the reference model of the control object and the energy-saving control program. The energy-saving control program is an algorithm for finding the required values of the regulated quantities at which the efficiency of the control object will be maximized while maintaining the required process parameters. For the tasks of energy-saving control of main oil pipelines, the program is actually a procedure for minimizing the function of total energy losses in the oil pipeline-oil pumping station system with the subsequent issuance of relevant information to the regulators or diagnostic information to the control panel (in case of changes in pump or network parameters and the impossibility of ensuring energy-efficient operation without personnel intervention, for example, necessary repairs, etc.). For an oil pumping station consisting of groups of pumps connected in series and parallel, a generalized functional diagram will look like the one shown in Figure 1 [2], [11].

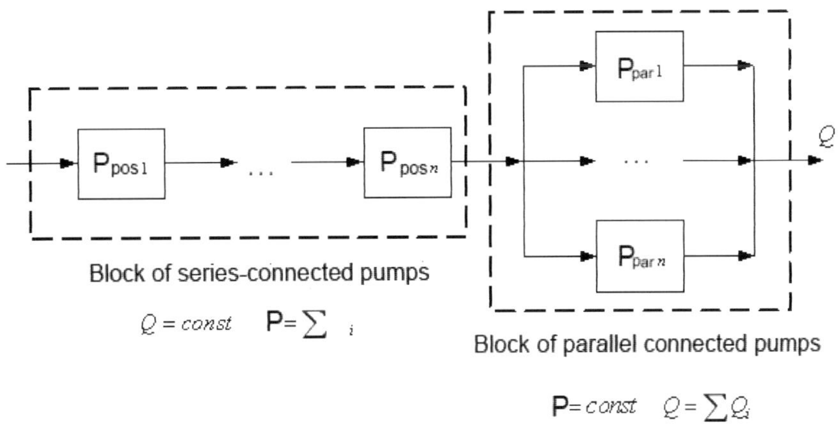

Fig. 1. Generalized functional diagram of an oil pumping station consisting of pumps connected in series and parallel

The use of a reference mathematical model of the oil pumping complex equipment, which can change its parameters in real time depending on specific conditions, allows for a more accurate and complete determination of the energy loss function at any time and identifies ways to reduce it without compromising the main technological indicators of the oil pumping equipment.

The positive effect of implementing the energy management system is as follows: power and energy losses at oil pumping stations are reduced, energy efficiency of oil pumping stations and the relevant section of the main oil pipeline is increased, and the cost of oil transportation through main oil pipelines is reduced [3]. Improving the measurement accuracy in determining the average oil flow rate in main oil pipelines is carried out through an integrated approach that includes the application and improvement of measurement methods and techniques, re-equipment of the technical park of measuring equipment, and the use of modern mathematical tools.

The measured parameter $X_{(t)}$, which passes through the measuring mechanism, has instability even when the adjustment of the regulatory bodies is unchanged. The presence of this instability causes errors in parameter measurements [4], [5], because if the phases of the moments of the beginning and end of the interval do not coincide, averaging over the periods of the parameter's pulsations, its average value X0, measured with any accuracy, will differ from the average over time X_{sr}, i.e., $X_{sr} - X_0 = \Delta_{nf}$. The nature of this phenomenon is illustrated in Fig. 2.

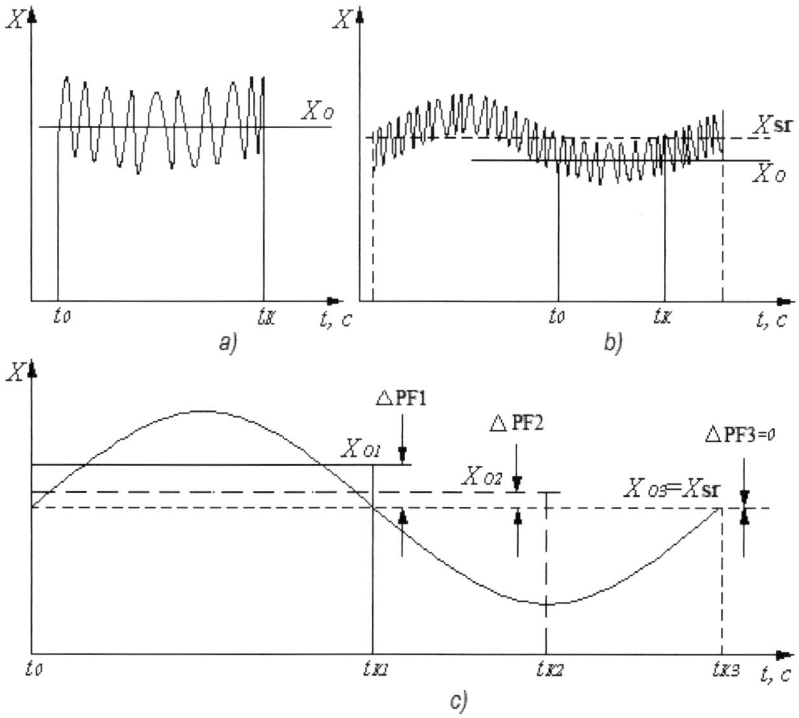

Fig. 2. Illustration of the nature of the out-of-phase error (a, b) and its dependence on the choice of the averaging (measurement) time interval $T = t - t_0$ (c)

The following model of the measured parameter was used to study the nature and estimate the out-of-phase error Δ_{nf}:

$$X(t) = X_0 + \tilde{X}, \tag{1}$$

X_0 and $\tilde{X}$ – is a constant and periodic variable of the constituent parameter, respectively.

A periodic process can be viewed as a superposition of harmonious components, i.e:

$$\tilde{X} = X_m sin(\omega t + \varphi) \tag{2}$$

X_m, ω *and* φ - are the amplitude, circular frequency, and initial phase of the parameter oscillations, respectively. The average value over the integration time T, taking into account (1) and (2), is represented as follows [6], [7], [8]:

$$X_{sr} = \frac{1}{T}\int_0^T X(t)\,dt = X_0 - \frac{X_m}{\omega T}[\cos(\omega T + \varphi)] \tag{3}$$

In this case, the relative error of the out-of-phase δnf of the measurement is determined:

$$\delta_{nf} = \frac{\Delta_{nf}}{\bar{X}_{sr}} = \frac{\bar{X}_{sr} - \bar{X}_0}{\bar{X}_{sr}} = \frac{\bar{X}_m}{\bar{X}_{sr}}\frac{\cos(\varphi) - \cos(\omega T + \varphi)}{\omega T} = AK \tag{4}$$

$A = \frac{\bar{X}_m}{\bar{X}_{sr}}$ – is the relative amplitude of the parameter oscillations;

$K = \frac{\cos\varphi - \cos(\omega T + \varphi)}{\omega T}$ - phase coefficient;

ω– is the circular frequency of the parameter oscillations, rad/s; φ– is the initial phase of the oscillations, rad.

It follows from (4) that the out-of-phase error depends not only on the initial phase of the parameter oscillations but also on the relative amplitude of the periodic component X_m/X_{sr}, the circular frequency ω, and the integration time T. The nature of the change in the phase coefficient K from the oscillation phase ωT at $0 < \varphi < \pi/2$ is shown in Fig. 3 - a). The initial phase of oscillations φ is a random variable, so to simplify calculations, we can set $\varphi=0$, while the family of curves $K = f(\omega T, \varphi)$ degenerates into the graph shown in Fig. 3 - b).

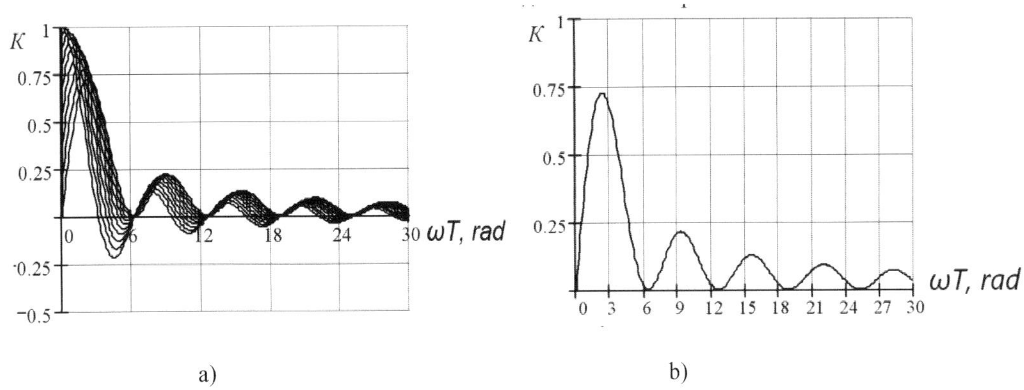

Fig. 3. Dependence of K on ω T at a) φ = random variables, b) φ = 0 rad

Measurements of hydraulic parameters (flow rate, head, pressure) are the basis for determining the characteristics of all pumps, which are the main consumers of auxiliary needs of main oil pipelines. The periodic processes occurring in the equipment of an oil pumping station are characterized by high inertia, which increases the out-of-phase error [9]. The criterion for the optimal operation of the energy-saving system of automated control of pumping units of the oil distribution system is the minimum total energy losses in the main oil pipeline while maintaining the required oil flow parameters at a given level determined by the technological process and reliability indicators. Based on the results of solving the problem of energy-saving control of pumping units of oil pumping stations, a mode map of adjustable parameters is constructed over the entire range of changes in oil flow (Fig. 4) [10].

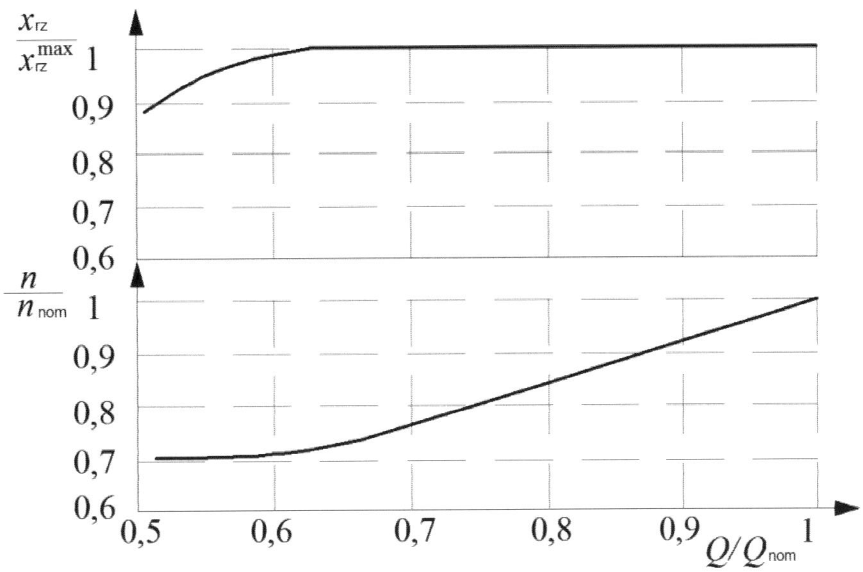

Fig. 4. Mode map of energy - efficient operation of an oil pumping station pump

Over the entire range of oil flow rates, the values of the positions of the control valves xrz and rotational speeds n are determined, at which the total energy losses in the main oil pipeline will be minimal. These dependencies should be used in the automated control devices of the pumping units of the Oil Pumping Station and periodically updated in real time. This will ensure minimal energy losses in the pumping unit and accurate maintenance of oil flow rates regardless of the technical condition of the pump and oil viscosity.

3. Conclusion

In light of the ever-increasing global emphasis on energy efficiency and conservation, this article introduces a meticulously crafted structural blueprint for the energy-saving automated control system for main oil pipeline pumping stations. Anchored in cutting-edge scientific methods tailored for energy-saving control, our research not only puts forth a pioneering methodology to bolster the energy efficiency of pumping unit controls but also foregrounds a strategic approach to minimize energy wastage during the intricate process of oil transportation. The novel analytical models proposed herein - which elucidate the intricate dependencies between energy losses in pumps and their respective input parameters - serve as a foundational cornerstone for the evolution of control systems. Such systems, underpinned by our core principle of minimizing energy losses, are instrumental in pushing the boundaries of contemporary practices.

Moreover, the tangible benefits of these innovations extend well beyond mere technical enhancements. By integrating our findings, oil industry enterprises stand to gain significantly, from ensuring consistent and disruption-free oil transportation to realizing substantial economic efficiencies. Furthermore, our approach champions a paradigm shift towards a more environmentally conscious oil industry. By advocating for the predominant use of oil pipelines, we underscore their pivotal role in mitigating environmental hazards, notably by diminishing the traditional reliance on rail and maritime oil transportation avenues. In essence, our research not only paves the way for operational advancements within the oil sector but also underscores the intertwined narrative of technological innovation, economic prudence, and environmental stewardship.

Future Research Recommendations: Technological Innovations Analysis: While the article introduces advanced methods of automation and control, future research could delve deeper into the latest technological innovations specific to oil transport systems, evaluating their efficiency, and feasibility.

- Environmental Footprint Assessment: Beyond energy efficiency, studies could be conducted on the environmental footprint of the introduced technologies and methods, examining emissions, waste, and other ecological impacts.
- Economic Implications: A detailed economic study focusing on the Return on Investment (ROI) for the new technologies and methods proposed in the article would be invaluable for stakeholders in the industry.
- Cross-industry Comparison: Analyze methods and technologies utilized in parallel industries to discern potential insights or technologies that could be integrated into the oil transportation sector.
- System Sustainability and Reliability: Future studies can focus on the long-term resilience and reliability of the proposed systems, especially under changing climatic conditions or other extreme circumstances.
- Education and Training: Create educational and training programs for personnel handling the new technologies, ensuring their effective and safe operation.
- Artificial intelligence, big data, neural networks and fuzzy logic [12]: Investigate the potential of integrating artificial intelligence and big data analytics for optimizing operations at oil transport stations.
- Internet of Things (IoT) Application: Research opportunities to integrate IoT devices for real-time monitoring and control of various parameters within pumping stations.

These suggestions for future research stem from the content and emphasis of the provided article and are aimed at a more profound understanding and development of practical solutions for the industry. Complex analysis and evaluation of hazardous environment in case of oil man-made disaster [13]. Research and prediction of the need for individual anti-chemical protection in case of technological accidents with oil pipelines [14].

4. Acknowledgments

The publication of this study is supported by Project BG05M2OP001-1.001-0004 UNITe, funded by the Operational Programme "Science and Education for Smart Growth", co-funded by the European Union trough the European Structural and Investment Funds.

5. References:

[1]. Kniazeva, V. N.; Kaniuk, H. Y.; Mezeria, A. Iu. & Andreev, A. V. (2019). Analyz normatyvnikh dokumentov po obespechenyiu effektyvnoi rabotoi nasosnikh ustanovok mahystralnыkh nefteprovodov (Analysis of normative documents providing the effective work of the pumping stations on main oil supply pipelines). Visnyk Natsionalnoho tekhnichnoho universytetu (Newspaper of the National technical university) «KhPI». Seria: Novi rishennia v suchasnykh tekhnolohiiakh (Series: New decisions in modern technologies). Vol.5 (1330). p. 27-34;

[2]. Topliceanu, L.; Gabriel, P. & Furdu, I. (2016). Functional problems and maintenance operations of hydraulic turbines. TEM Journal Vol. 5, No. 1, (February 2016), p. 32-37, DOI: 10.18421/TEM51-05;

[3]. Cherniuk, A. M. & Kniazieva, V. M. (2016). Povishenye urovnia enerhosberezhenyia pry upravlenyy nasosnыmy ustanovkamy nefteperekachyvaiushchykh stantsyi (Increasing the level of energy saving in management of pumping installations of oil pumping stations). Proceedings of «Informatyka, Matematyka, Avtomatyka» naukovo-tekhnichnoi konferentsii («Informatics, Mathematics, Automation» technical science conference). Sumy.Ukraine;

[4]. Stoev, I. I.; Zaharieva, S. L. & Mutkov, V. A. (2019). Evaluation of gross errors in measured temperature with an electronic system for management of residential energy systems. Proceedings of 27th telecommunications forum (TELFOR), pp. 1-4, doi: 10.1109/TELFOR48224.2019.8971309, Belgrade, Serbia;

[5]. Zaharieva, S.; Stoev, I.; Borodzhieva, A. & Petrova, T. (2023). Assessment of gross errors and uniformity of readings of sensors for relative humidity measurement in indoor premises. Proceedings of 22nd international symposium INFOTEH-JAHORINA (INFOTEH), pp. 1-5, doi: 10.1109/INFOTEH57020.2023.10094132, East Sarajevo, Bosnia and Herzegovina;

[6]. Dichev, D.; Diakov, D.; Zhelezarov, Y.; Nikolova, H.; Kupriyanov, O. & Dicheva, R. (2022). Accuracy evaluation of flat surfaces measurements in conditions of external influences. Proceedings of 2022 XXXII international scientific symposium "Metrology and Metrology assurance" (MMA), (September 2022), pp. 1-7. DOI: 10.1109/MMA55579.2022.9992334, Sozopol, Bulgaria;

[7]. Dichev, D.; Zhelezarov, I. & Madzharov, N. (2019). System for measuring the attitude of moving objects, using a Kalman filter and MEMS sensors. Comptes rendus de l'academie bulgare des sciences, Vol.72, No. 11, pp. 1527-1536. DOI: 10.7546/CRABS.2019.11.10. Sofia, Bulgaria;

[8]. Dichev, D.; Kogia, F.; Koev, H. & Diakov, D. (2016). Method of analysis and correction of the error from nonlinearity of the measurement instruments. Journal of engineering science & technology review, Vol.9, No. 6, pp. 116 – 121. DOI: 10.25103/jestr.096.17. Sofia, Bulgaria;

[9]. Andreev, A. V.; Cherniuk, A. M. & Kniazeva, V. N. (2017). Effektivnost upravlenyia nasosnimy ahrehatamy nefteperekachyvaiushchykh stantsyi. Udoskonalennia turboustanovok metodamy matematychnoho i fizychnoho modeliuvannia (Effectiveness of management of pumping aggregates of oil pumping stations. Improvement of pipeline stations by methods of mathematical and physical modeling). Proceedings of VI Mizhnar. nauk.-tekhn. konf. (VI-th international technical science conference), NAN Ukraine, International problem mashynobuduvannia imeni A. N. Pidhirnoho NAN Ukraine, Turboatom, Enerhoprohres, Ukraine;

[10]. Pulov, D.; & Karadzhov, Ts. (2021). Optoelectronic device for measuring the power of laser radiation, Proceedings of the 13th International scientific and practical conference "Environment. Technology. Resources.". Vol. 3, p. 286-290, Online ISSN 2256-070X, DOI: 10.17770/etr2021vol3.6590, Rezekne, Latvia;

[11]. Kaniuk, H. Y.; Mezeria, A. Iu. & Laptynov, Y. P. (2014). Model enerhosberehaiushcheho upravlenyia nahnetatelnimy ustanovkamy teplovikh elektrostantsyi (A model of energy saving management of pressurizing devices in thermal power stations). Visnyk NTU "KhPI": Enerhetychni ta teplotekhnichni protsesy y ustatkuvannia (Newspaper: Power and thermal technical processes and equipment). Vol. 12 No 1055, p. 90-97. National Technical University, Kharkiv, Ukraine.

[12]. Veselov, O.; Veselov, A. & Svetushenko, S. (2019). The use of fuzzy logic in automatic control systems, Proceedings of the 30th DAAAM international symposium, pp.0897- 0903, B. Katalinic (Ed.), Published by DAAAM international, ISBN 978-3-902734-22-8, ISSN 1726-9679, Vienna, Austria DOI: 10.2507/30th.daaam.proceedings.125

[13]. Burdzhiev, S.; Stoychev, K.; Bratanov, D.; Dimitrov, D.; Panevski, V.; Damyanov, G. & Georgiev G. (2014) Comprehensive analysis and assessment of a hazardous environment. IN: fourth national conference with international participation metallurgy, hydro- and aerodynamics, national security '214, pp. 323-330, Bulgarian Academy of Sciences, IMCTCX-BAS, ISBN 1313-8308

[14]. Georgiev, G. & Burdzhiev, S. (2013) Individual anti-chemical and anti-radiation protection. In: Collection of materials from the third national conference with international participation "Metallurgy, hydro- and aerodynamics, national security, pp. 453-457, Sofia, ISBN 1313-8308

DOI: 10.2507/34th.daaam.proceedings.035

METHODOLOGY FOR TESTING: LABORATORY COMPARISON OF CLIMATE CHAMBERS IN THE AUTOMOTIVE INDUSTRY

Nikola Skřivanová, Martin Melichar & Kateřina Kašová

This Publication has to be referred as: Skrivanova, N[ikola]; Melichar, M[artin] & Kasova, K[aterina] (2023). Methodology for Testing: Laboratory Comparison of Climate Chambers in the Automotive Industry, Proceedings of the 34th DAAAM International Symposium, pp.0274-0280, B. Katalinic (Ed.), Published by DAAAM International, ISBN 978-3-902734-41-9, ISSN 1726-9679, Vienna, Austria
DOI: 10.2507/34th.daaam.proceedings.035

Abstract

This paper describes the design of an experimental methodology for a laboratory comparison of two climate chambers in the context of performing a test with the attributes "Test: Humidity heat, cyclic 12+12 hours" according to DIN EN 60068-2-30: 2006-06. The objective was to verify the consistency of operation of these chambers for identical test setups in the environment of two accredited laboratories focusing on testing electronic components in the Automotive industry. The results of this research are a key element in ensuring consistency and reliability of test results for electronic components for the automotive industry. Further, this testing methodology and the comparison itself enhances the credibility of both laboratories. These results provide important insights for further development and optimization of climate chamber operations. This paper offers valuable information for automotive professionals involved in the process of testing and verifying electronic components under extreme climatic conditions.

Keywords: climate chamber; methodology; laboratory comparison; automotive industry.

1. Introduction

Electronic components in the automotive industry face many challenges that require thorough testing and verification of their durability in different climatic conditions. DIN EN 60068-2-30: 2006-06 provides a framework of testing and comparison conditions, but several key issues remain open for practical deployment. While the standard defines the ambient conditions, it does not provide a specific methodology for performing individual comparison tests. [1]; [2]; [3]. It is in this context that this research was undertaken to design and validate a laboratory comparison methodology for the test "Test Db: Damp Heat, Cyclic 12+12 hours" in accordance with the standard. In addition, it was necessary to ensure that this methodology was reproducible, which is a key element for reliable test results. The objective of this study was to examine the testing process and develop a methodology for laboratory comparison of two identical climate chambers in two accredited testing laboratories. This inter-laboratory comparison would then provide evidence and verification that the two laboratories were operating identically and increase the credibility of the results. [1]; [7]. This study not only provides new insights into the testing of automotive electronic components, but also highlights the importance of a properly set-up laboratory benchmarking methodology to achieve results and its execution.

In component testing, this laboratory comparison strengthens the credibility of the result, as the agreement of the results in both laboratories indicates the consistency of the test environment. [4]; [5]. With these key factors in mind, this paper seeks to shed light on the importance of reproducibility of laboratory comparison tests and provide a comprehensive view of the development of a methodology for laboratory comparison of the performance of two climate chambers within two different laboratories in accordance with DIN EN 60068-2-30: 2006-06. [1].

2. Description of the proposed methodology

As mentioned above both accredited laboratories were asked to propose a methodology for laboratory comparison of two identical climate chambers that would reflect the requirements of this standard DIN EN 60068-2-30: 2006-06. The laboratory comparison was designed so that the stresses performed were identical and the procedure was reproducible. This laboratory comparison will be used to verify the reproducibility of the test "Test Db: Moisture heat, cyclic 12+12 hours" according to DIN EN 60068-2-30. The proposed test methodology for the laboratory comparison of two climate chambers will now be presented. [1]; [6]; [8].

	Number of pieces	**Marking**	**Date of receipt**	**Producer**
Test object	5 pcs	Stainless steel balls	10.1.2023	External supplier
	5 pcs	Funnel	10.1.2023	External supplier
	5 pcs	Measuring cylinder	10.1.2023	External supplier
	5 pcs	Stainless steel balls	24.1.2023	External supplier
	5 pcs	Funnel	24.1.2023	External supplier
	5 pcs	Measuring cylinder	24.1.2023	External supplier
The information and test object designation is based on information from the customer. Unless otherwise stated, all tests were conducted under the following ambient conditions: Temperature: 23 °C ± 5 °C Relative humidity: 25 % to 75 %. Results refer only to test objects tested in an accredited test laboratory. The uncertainty of measurement is given when the declaration of conformity is used or can be requested for each test method. The expanded uncertainty of measurement has been determined in accordance with JCGM 100 'Guidance for the quotation of uncertainty of measurement' and is given with a coverage factor of 2. This means that the measurand lies within the assigned interval of values with a probability of 95 %.				
Test specifications	Designation of the standard DIN EN 60068-2-30			Date of issue 2006-06

Table 1. Test specifications

2.2. Providing a test facility

The two accredited test laboratories, laboratories X and Y, had five identical test samples each. Five stainless steel balls weighing 874 g each served as test specimens. Figure 1 shows an example of a test specimen.

Fig. 1. Test sample

2.3. Test assumptions and parameters

Since the two test laboratories involved in the laboratory comparison had identical test samples, homogeneity is guaranteed. In addition, the tests in both laboratories were carried out in two identical and technically identical climatic test chambers. In each of the two chambers there were five test samples with identical test set-ups. Five stainless steel spheres weighing 874 g each were used as test specimens and were placed above the measuring cylinder. A funnel was placed between the spheres and the graduated cylinder. The undersides of the balls were placed at a distance of 50 mm to 100 mm above the entrance to the funnel. These assemblies were positioned in the climatic chamber as shown in Figure 2, and their positions were fixed, with the spheres in positions 1, 2, and 5 symmetrical to position 4. Using these assemblies, the amount of condensate that accumulates in the measuring cylinders during climatic stress was measured. In addition to the graphical evaluation of the stresses, this measured quantity should provide another indication of the reproducibility of the method. The five funnels were weighed with their corresponding graduated cylinders before and after loading and the weight was recorded in grams to determine the amount of condensate removed. The average values of the quantities of condensate removed must not have differed by more than 3 % according to the internal specification. The test was carried out in both test laboratories with the same predefined test parameters.

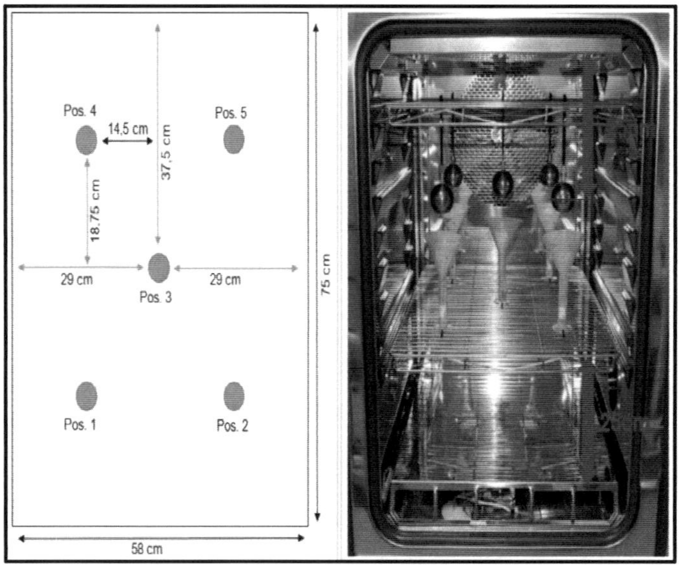

Fig. 2. Location of test assemblies in the climate chamber: top (left) and front (right)

3. Experimental verification

This chapter documents the tests of the two participating Test laboratories X and Y.

3.1. Test Db: Moisture heat, cyclic 12+12 hours - Test laboratory X

The test was carried out according to DIN EN 60068-2-30. The test specimens were placed in the test chamber at standard climatic conditions (Figure 3) and conditioned at 25 °C and 60 % rpm for 75 minutes to achieve thermal equilibrium. The humidity was then increased to 98 % relative humidity within 45 minutes while the temperature in the chamber remained constant. Subsequently, stress testing was performed according to DIN EN 60068-2-30: 2006-06, variant 1. The upper test temperature was 55 °C. A total of two cycles were carried out. After the exposure, a post-treatment was carried out in which the chamber temperature was maintained at 25 °C while the humidity was controlled from 98 % RH to 75 % RH during the first 25 minutes. The test specimens remained in these conditions for a further 95 minutes (Figure 4). [1]; [6]; [8].

Laboratory comparisons were made by:	Technician A
Location of laboratory comparison:	Test laboratory X
Test facility:	ClimeEvent Climate Stress Screening Chamber Analytical balance
Test parameters:	Upper test temperature: 55 °C
Relative humidity:	98 %
Variant:	1
Number of cycles:	2

Test setup:

Fig. 3. Climate chamber – Test laboratory X

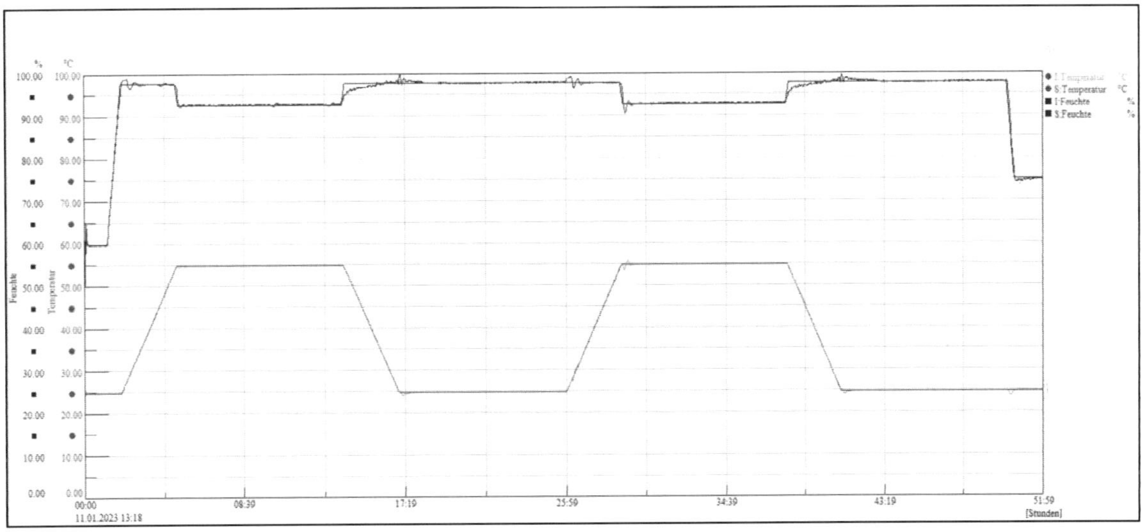

Fig. 4. Climate profile – Test laboratory X

Result: The weights of the graduated cylinder funnels before and after loading are documented in Table 2.

Climate chamber 6 – Test laboratory X			
Position in the chamber	**Weight of graduated cylinder and funnel before test[g]**	**Weight of graduated cylinder and funnel after test [g]**	**Amount of condensate removed [gr]**
Pos. 1	29.20	33.49	4.29
Pos. 2	28.90	32.82	3.92
Pos. 3	29.20	33.33	4.13
Pos. 4	29.40	33.28	3.88
Pos. 5	29.30	32.89	3.59

Table 2. Results – Test laboratory X

3.2. Test Db: Moisture heat, cyclic 12+12 hours - Test laboratory Y

The test was performed as described in section 3.1.

Laboratory comparisons were made by:	Technician B
Location of laboratory comparison:	Test laboratory Y
Test facility:	ClimeEvent Climate Stress Screening Chamber Santorius AX6202 precision balance Set of weights
Test parameters:	Upper test temperature: 55 °C
Relative humidity:	98 %
Variant:	1
Number of cycles:	2
Test setup:	

Fig. 5. Climate chamber – Test laboratory Y

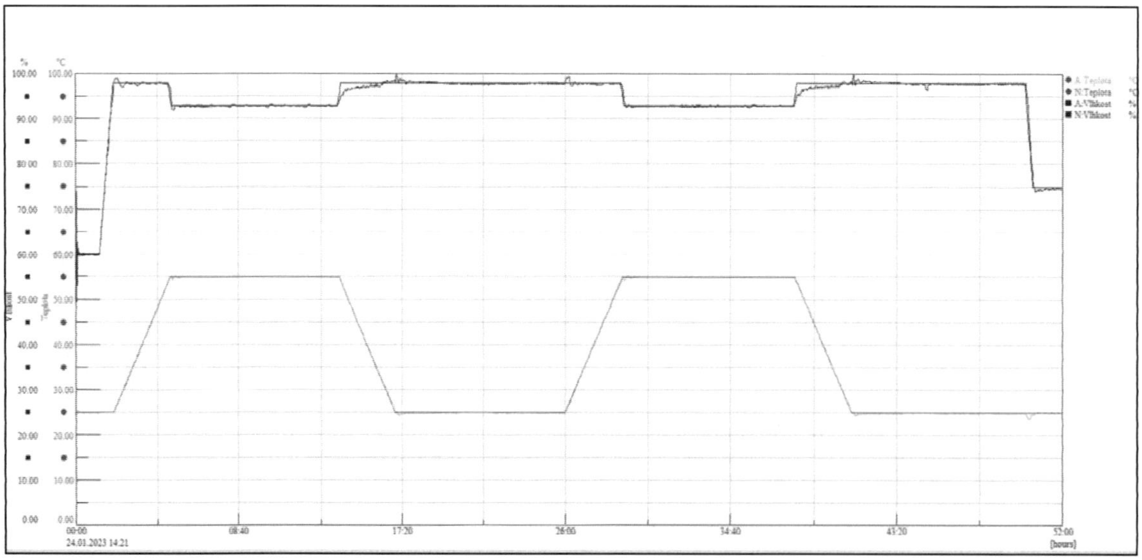

Fig. 6. Climate profile – Test laboratory Y

Result: The weights of the graduated cylinder funnels before and after loading are documented in Table 3.

Climate chamber 1 – Test laboratory Y			
Position in the chamber	**Weight of graduated cylinder and funnel before test[g]**	**Weight of graduated cylinder and funnel after test [g]**	**Amount of condensate removed [gr]**
Pos. 1	29.20	33.42	4.22
Pos. 2	28.90	32.90	4.00
Pos. 3	29.20	33.30	4.10
Pos. 4	29.40	33.24	3.84
Pos. 5	29.30	32.98	3.68

Table 3. Results – Test laboratory Y

4. Conclusion

The results of this study clearly show that both participating test laboratories successfully achieved identical ratings with identical test samples in identical climatic test chambers, while fully complying with the specified parameters. This careful adherence to the specified temperature range and handling of the test samples gives confidence that the measurement results are credible and reliable. The deviation of the average values of the respective sampled condensate quantities, which was within the predetermined range of 3 % (see Table 4), clearly confirms that the procedure used is stable and reproducible.

Comparison of results of the test – Test laboratory X and Test laboratory Y			
Position in the chamber	**Amount of condensate removed [gr] – Test laboratory X**	**Amount of condensate removed [gr] – Test laboratory Y**	**Difference [%]**
Pos. 1	4.29	4.22	-1.66
Pos. 2	3.92	4.00	2.00
Pos. 3	4.13	4.10	-0.73
Pos. 4	3.88	3.84	-1.04
Pos. 5	3.59	3.68	2.45
Average value	3.96	3.97	0.15

Table 4. Comparison of results of the test – Test laboratory X and Test laboratory Y

This research pushes the frontiers of knowledge in the field of testing electronic components for automobiles. An efficient methodology for laboratory comparison has been developed and validated in the context of the test "Db: Damp Heat Test, cyclic 12+12 hours" according to DIN EN 60068-2-30: 2006-06. This methodology represents a key step towards ensuring reliability and consistency in testing electronics under extreme climatic conditions. The next natural step in this research field could be to extend the testing methodology to other types of climate chambers for other ranges and testing parameters. This research provides important insights for industry and the research community and holds promise for future innovation and improved reliability of electronic components in the automotive industry.

5. Acknowledgments

The article contribution has been prepared under project SGS-2022-007 - Research and Development for Innovation in Engineering Technology - Machining Technology IV.

6. References

[1] DIN EN 60068-2-30:2006-06. Environmental Testing - Part 2-30: Tests - Test Db and Guidance: Damp Heat, Cyclic (12 h + 12 h Cycle). German Institute for Standardization, 2006.

[2] Skrivanova, Nikola, Melichar, Martin (2021). Evaluation of Effect of Attributive Measurements in a Clime Compartment on the Value of Dimensional Measurement Uncertainty in an Accreted Test Laboratory, Proceedings of the 32nd DAAAM International Symposium, pp.0289-0293, B. Katalinic (Ed.), Published by DAAAM International, ISBN 978-3-902734-33-4, ISSN 1726-9679, Vienna, Austria, DOI: 10.2507/32nd.daaam.proceedings.041

[3] Bicova Katerina, Bebr Lukas (2018). Analysis and Dependability of Production Processes for the Automotive Industry, Proceedings of the 29th DAAAM International Symposium, pp.0416-0420, B. Katalinic (Ed.), Published

by DAAAM International, ISBN 978-3-902734-20-4, ISSN 1726-9679, Vienna, Austria, DOI: 10.2507/29th.daaam.proceedings.061

[4] Dudackova Simona, Bicova, Katerina (2019). Verification of Process and Machine Capability for Precision Automotive Production, Proceedings of the 30th DAAAM International Symposium, pp.0962-0966, B. Katalinic (Ed.), Published by DAAAM International, ISBN 978-3-902734-22-8, ISSN 1726-9679, Vienna, Austria, DOI: 10.2507/30th.daaam.proceedings.133

[5] Skrivanova, Nikola, Melichar Martin (2019). The Comparison of Destructive and Non-Destructive Forms of Measurement in the Automotive Industry, Proceedings of the 30th DAAAM International Symposium, pp.0995-1002, B. Katalinic (Ed.), Published by DAAAM International, ISBN 978-3-90273422-8, ISSN 1726-9679, Vienna, Austria, DOI: 10.2507/30th.daaam.proceedings.138

[6] https://www.testsysteme.cz/nova-generace-teplotnich-klimatickych-komor-climeevent-2, JD Dvořák - Test technique. (nd). JD Dvořák, sro - Test technique, Accessed on: 2023-10-10

[7] ČSN EN ISO/IEC 17025:2018. Zkušební laboratoře a kalibrační laboratoře - Obecné požadavky na kompetenci (ISO/IEC 17025:2017, IDT). Úřad pro technickou normalizaci, metrologii a státní zkušebnictví, 2018.

[8] https://pdf.aeroexpo.online/pdf/weiss-umwelttechnik-gmbh/clime-event/173964-6414.html, (2023). Weiss Umwelttechnik GmbH. (n.d.). Clime Even, Climate test, chamber, Accessed on: 2023-10-10

DOI: 10.2507/34th.daaam.proceedings.036

HARDWARE AND SOFTWARE PROTECTION OF THE POWER SUPPLY AND CONTROL OF THE SERVICE ROBOTS

Vadim Chernyshev, Valentin Pryanichnikov, Mariia Soloveva, Radomir Tarasov & Aleksandr Telegin

This Publication has to be referred as: Chernyshev, V[adim]; Pryanichnikov, V[alentin]; Soloveva, M[ariia]; Tarasov, R[adomir] & Telegin, A[leksandr] (2023). Hardware and Software Protection of the Power Supply and Control of the Service Robots, Proceedings of the 34th DAAAM International Symposium, pp.0281-0287, B. Katalinic (Ed.), Published by DAAAM International, ISBN 978-3-902734-41-9, ISSN 1726-9679, Vienna, Austria
DOI: 10.2507/34th.daaam.proceedings.036

Abstract

The article is devoted to the development of power supply systems for mobile service robots. An important aspect of the power supply system for autonomous devices is tracking the battery parameters and monitoring the charging and discharging process. For example, an intelligent battery system has been developed for the AMUR-307 robot, which allows you to track all the most important battery parameters and set charging parameters. The system provides a variety of hardware and software protection schemes for current, voltage, temperature, and the state of internal balancing of batteries in assemblies, as well as overload protection of drives. Incorrect actions of operators during supervisory control are analyzed and blocked. The system supports the smart battery specification - SBS v1.1, which provides standardized access to the system from the operating system space. This approach presents significant advantages for the implementation of devices designed for the underwater robotic complex (RC) under development, consisting of a walking base vehicle and autonomous sensor carrier satellites. RC is designed for inspection of underwater objects, delivery, and use of cargo in the conditions of the shelf zone and the unprepared bottom with extreme conditions.

Keywords: Mobile service robots; power system; smart battery system; SBS v1.1; Linux power supply.

1. Introduction

This article discusses the design and subsequent integration of the robot's power supply system. To integrate this system, the AMUR-307 service robot is used, designed to transport objects between rooms with the ability to open doors with standard rotary handles [1]. In the process of manufacturing a prototype of the robot, some shortcomings were revealed, one of which was in the power system. In this model, it was planned to install two lithium-ion batteries of the 4S2P topology, each having a battery monitoring module, which are connected in parallel to provide the required current for the operation of the robot kinematics. A significant disadvantage is the inability to track the following important parameters: current, voltage, current charge, and remaining battery capacity, as well as plot the dependencies of all key battery variables. The power supply system developed and installed earlier is shown in Figure 1.

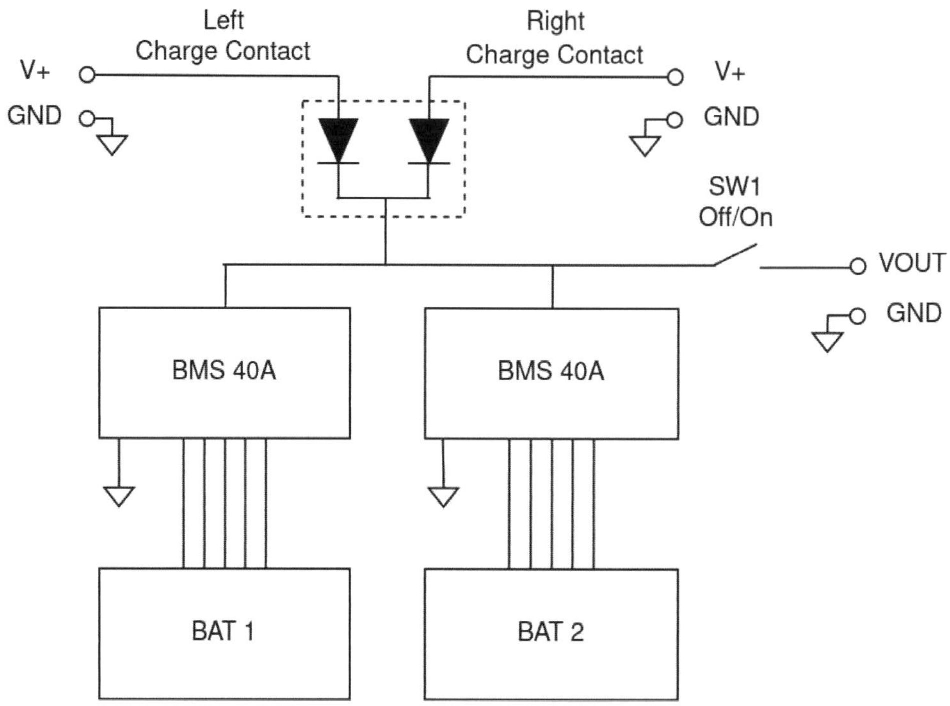

Fig. 1. Structural diagram of the electrical system of the robot

These batteries were supposed to be charged with an external charger, which is connected to the current and voltage regulation module based on a step-down pulse adjustable voltage converter XL4016. As it turned out, this is not a sufficiently optimal and convenient solution. To solve all the identified shortcomings, the task was set to develop an original charging/discharging module and a battery monitoring module with small dimensions. Due to the existing design of the robot, we have limitations in the size of the control electronics and battery charge in 25x30x160 mm. A direct current source is connected to the terminals to charge lithium-ion batteries. GND serves as the total mass. The SW1 toggle switch is used to ensure that the robot is powered on and off by batteries. VOUT is the power output from the battery pack.

2. Smart battery system

To correct the shortcomings of the mobile robot described earlier, a unified system of intelligent chargers was selected based on the SBS v1.1 specification, which includes Smart Battery, Charger, Selector, and System manager. The SBS specification defines the communication between system components via the SMBus interface. SMBUS is a serial data exchange protocol for power supply devices, which is based on the I2C bus, but at the same time uses a low level of signal voltage (3.3V). This protocol requests information from the battery and then uses it in the power management scheme of the system and/or uses it to provide the user with information about the current state and capabilities of the battery. The SMBus node will also receive critical messages from the Smart Battery when a problem is detected [2]. Smart Battery Charger is one of the elements of a comprehensive system solution for rechargeable batteries used in portable electronic equipment, designed for use with batteries conforming to the Smart Battery Data specification. Smart Battery Charger has charging parameters that are controlled by the battery itself and the host [3].

The combination of Smart Battery and Smart Battery Charger provides clear advantages in terms of security, performance, and cost of the system, namely:

1. Charging characteristics and safety limits are an integral part of the battery itself, ensuring proper charging algorithms that match specific cell types. Each smart battery defines a safe charging scheme that best matches its chemical composition and capacity, maximizing useful energy with each charge, reducing charging time, and increasing the number of charging cycles.
2. The cost of the system is reduced because the charger must provide only the charging voltage and current set by the battery itself, without duplicating the measuring and control electronics already available in the Smart battery.

The block diagram of the nodes and connections of the power supply system in compliance with the SBS specification is shown in Figure 2.

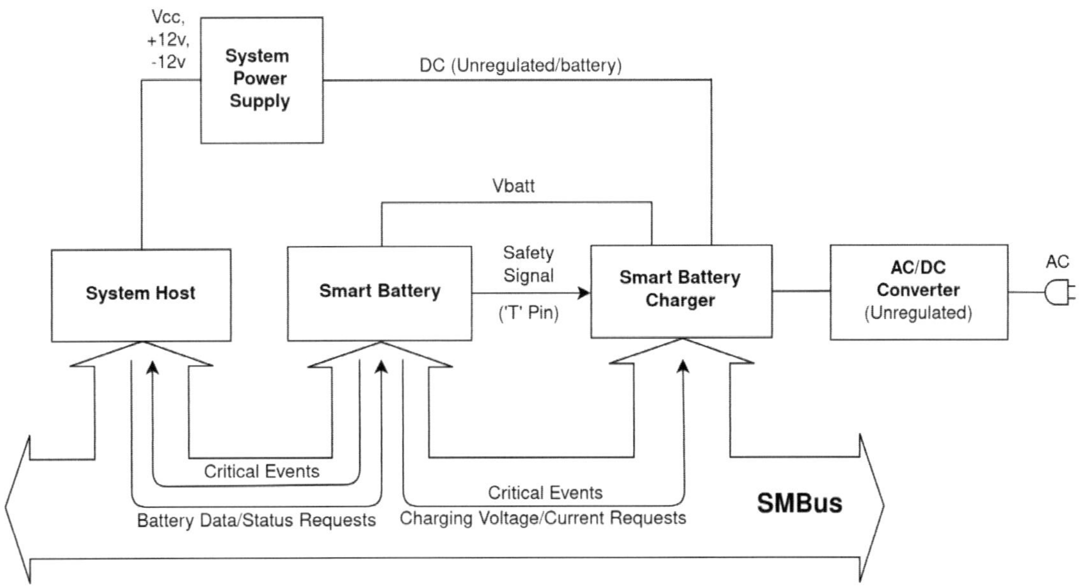

Fig. 2. Block diagram of nodes and connections of the SBS-based system

The choice of this specification has a special advantage over the previous robot power system since it becomes possible to control all processes related to charging and discharging the robot's battery. Often, for electronic systems using two or more batteries, an active battery selection node is used, but this approach is not suitable for powering the robot, since the battery selection unit appears with this approach involves alternating switching between batteries and the system, and for the robot in question, simultaneous operation with both batteries is assumed to remove large currents.

The considered components of the specification are represented in the Linux OS kernel by power system drivers. To work with charging devices in Linux OS, the Charger Manager utility is used, which provides management of all power devices in the kernel space [4]. This software can provide management and status information on all chargers, batteries and other power devices. Charger Manager can support uevent system notifications, suspension polling in RAM, handling premature battery discharge events, and uevent-notify support. Since there are multiple batteries, multiple instances of Charger Manager use the same structure of power device descriptors, namely charge_global_desc, which will manage standby monitoring for all instances of Charger Manager.

Charger Manager provides management of power devices inside the core. This software can provide management and status information on all chargers, batteries, and other power devices. Charger Manager can support uevent notifications, suspend polling in RAM, handle premature battery discharge events, and uevent-notify support. Since there are multiple batteries, multiple instances of Charger Manager use the same structure of power device descriptors, namely charge_global_desc, which will manage standby monitoring for all instances of Charger Manager. For each independently charged battery, its instance of Charger Manager is launched.

3. BMS module

To meet the SBS specification and the required battery parameters, it was necessary to select a battery management controller. When choosing, such parameters were taken into account as control of battery parameters via the SMBus interface via the Linux kernel, completeness of the protection system, permissible charge and discharge currents, battery balancing current, compactness of the resulting device, and the ability to change the address of the controller since it is necessary to use two battery monitoring boards, independently for each battery. In the projected battery monitoring system, the necessary monitored protection parameters were determined: the temperature of the battery cells, the temperature of the charge and discharge keys, and the voltage and current of the battery. The sufficient balancing current for the battery is 100 mA. Due to the limitations of the robot's design, the maximum allowable dimensions of each battery control board were 25x80 mm. After a thorough study of the existing mass-produced battery controllers at the time of development, the BQ40Z50-R2 was chosen as the optimal controller, which is ideally suited for the task of developing a battery control board for the AMUR-307 robot. This controller has many protection mechanisms for the monitored battery parameters: protection against battery discharge below the threshold value, protection against overcharging above the threshold value, protection of the battery from overheating based on four thermistors for each battery assembly, protection against overheating of the charge and discharge keys, protection against overheating of the controller based on the built-in thermistor, the maximum charge time, stopping the charging process by reducing the charge current to a threshold value, protection against exceeding the charging current, short circuit protection.

This controller has a built-in EEPROM memory with a configuration table of all parameters used, it is programmed via the SMBus bus, which makes it possible to update the software of battery controllers and charge controllers, as well as change configuration parameters directly on the robot if necessary [5]. To understand the principle of operation of the BQ40Z50-R2 battery controller, a simplified electrical diagram is shown in Figure 3.

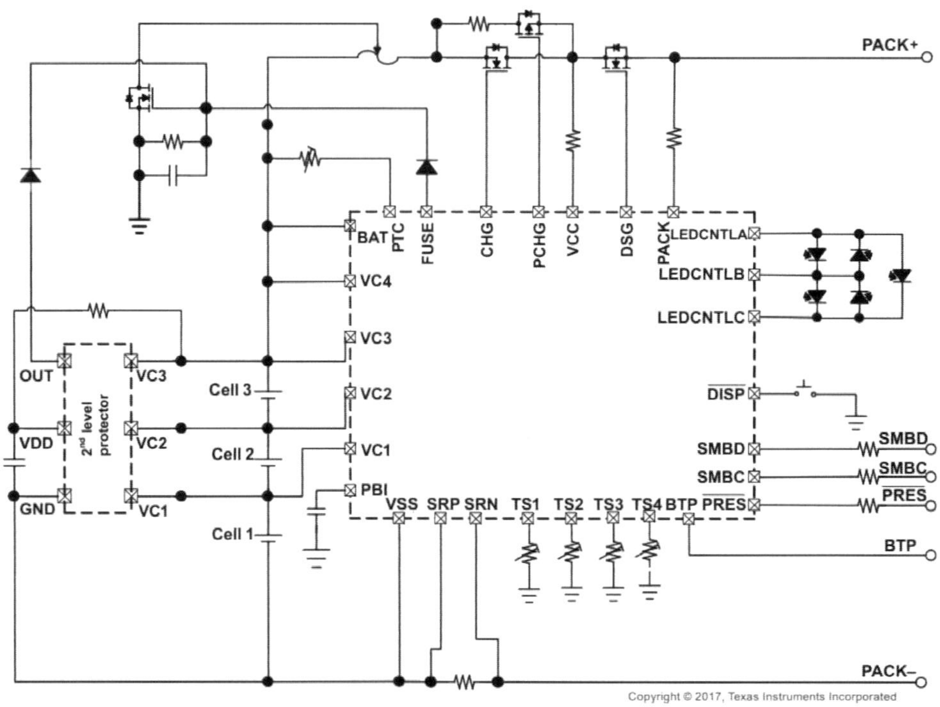

Fig. 3. Simplified electrical diagram of the battery control board

Powerful N-type MOSFETs were used - CSD16570, which has an extremely low pass resistance of 0.49 mOhm, which allows you to remove significant currents from the battery with extremely low heat dissipation. Due to the extremely low heat dissipation of the elements and the simplicity of the circuit, it was possible to develop a battery control board with dimensions of 22x60 mm (Fig. 4).

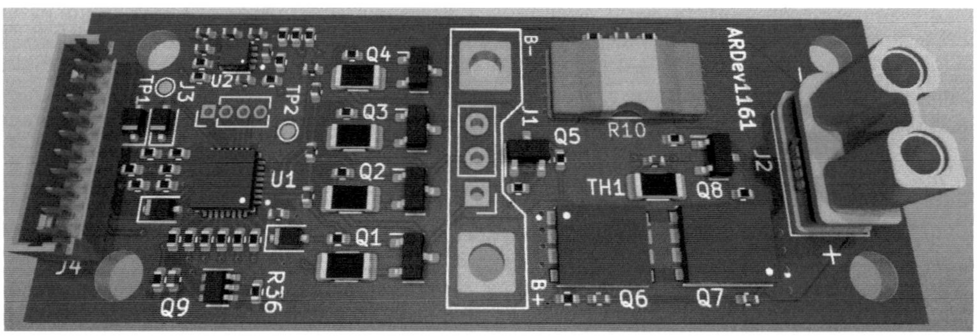

Fig. 4. Appearance of the battery control board

A second-level protection chip, the BQ296100, was also added. This chip protects the battery from overcharging if the BQ40Z50-R2 has not stopped the charging process for some reason, and one or more cells are already charged to maximum values. It is triggered with a programmable delay, usually 6.5s. Also, according to a similar principle, this chip monitors the process of battery discharge to the lower threshold.

The designed board will be mechanically connected to the robot's charge board common to the two batteries, and the board is soldered to the battery and is not mechanically disconnected, since the smart battery controller stores some battery information and monitors its condition. Also, the battery charge and control boards are mechanically attached using inter-board racks to prevent involuntary loss of contact between the boards from vibration.

4. Charger module

After developing the battery control board, it was necessary to develop a robot battery charge board. As a design solution, the development of a cross-board was chosen in which the battery monitoring modules described in the previous section will be installed. After a thorough analysis of the charge controllers, the choice was made on two LTC1760 and BQ24735 chips. The LTC1760-based solution offers the possibility of independent, but simultaneous charging and discharging of two batteries without the problem of energy flows between the batteries, such a solution would allow installing batteries on the robot regardless of their degree of charge, so in the worst case, when one battery is completely discharged and the other is charged, there would be no overflows that would violate the protection of the boards. However, this controller has a significant drawback, this is the maximum charging current of 4A. The BQ24735-based solution initially works with only one battery, so there is a problem with large currents between the batteries, but this controller has a maximum charge current of 8A, which is great for charging both batteries at the estimated time.

The problem with overflows between batteries can be solved in three ways:

1. Recommendation for installers of robot batteries, which says that robot batteries should differ from each other in charge level by no more than 10%. Then the currents between the batteries will not exceed the permissible charging currents for the batteries and spoil their life.
2. Using two separate charge controllers, with parallel connection of power outputs and charging inputs between them.
3. Duplicate keys for two batteries, but also control the circuit of one controller.

We actively use the first method, but human error is permissible during maintenance, as a result, it is necessary to use another method. The method using separate charge controllers is supported by the Linux kernel, however, the scheme is greatly complicated because all suitable controllers have a non-configurable I2C address, therefore, the cost increases, the method of protection based on duplicate keys allows using one controller, however, the charging process will occur simultaneously on both batteries and is limited only by the battery status controller. Due to the described features of each approach, the optimal solution is to use independent lines BQ40Z50-R2 + BQ24735 + CP2112. The developed block diagram of the power supply system is shown in Figure 5.

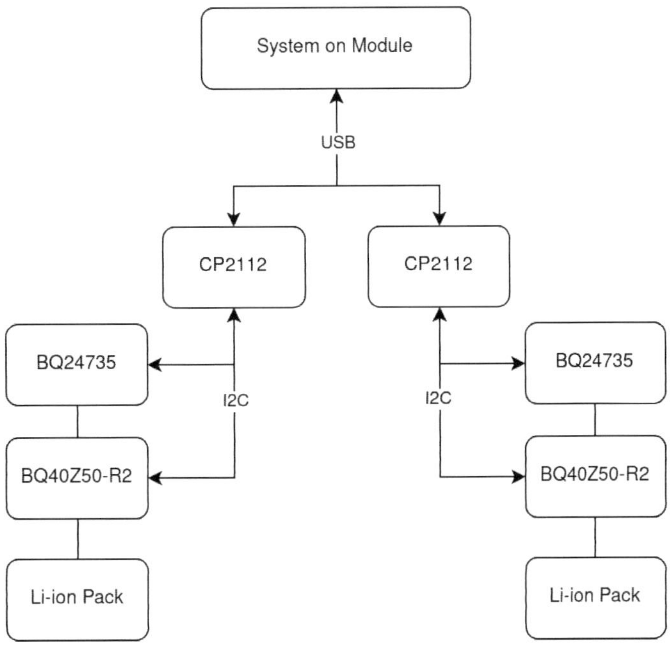

Fig. 5. Block diagram of the battery control board

Also in this version, a broadcast mode can be used, where the BQ40Z50-R2 sends commands and status parameters directly to the BQ24735 charge controller bypassing processor processing, which increases the system's response speed, as well as its security [9]. After forming the structure of the robot's battery monitoring and charging system, as well as selecting components, a schematic electrical diagram and a charge control board for the robot were designed. The developed power system will also be integrated on the underwater walking vehicle platform with the increase of independent power circuits from two to four [12], [13].

5. Docking station

For the process of charging the robot, a robot charging dock was developed and manufactured, a 20V 8A power supply unit was installed on board the dock station, a mechanism for disconnecting the power supply in the absence of a connected robot, and contact pads for charging the robot. Two groups of spring-loaded contacts are installed on the robot from the right and left sides to ensure confident contact with the docking station even in the event of a robot's arrival with an angle to the normal of the contact plane on the front panel of the docking station up to 15 degrees. The appearance of the AMUR-307 series robot docking station is shown in Figure 6.

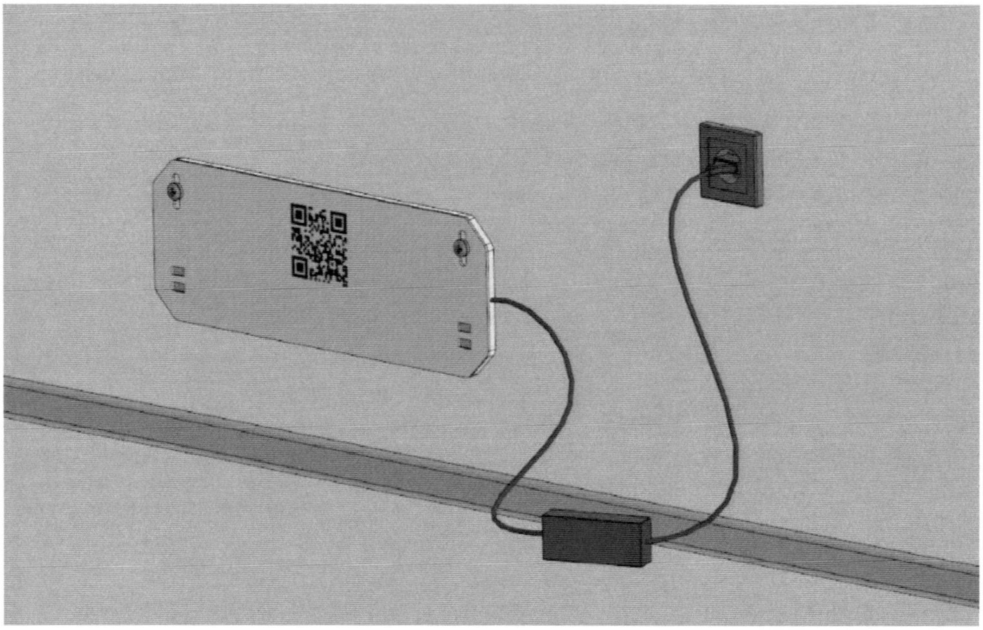

Fig. 6. Appearance of the AMUR-307 series robot docking station

The mechanism for determining the presence of a connected robot to the docking station occurs by measuring the resistance on the contact pads, and the power supply is switched on using a relay controlled by a mechanical switch. The housing of the docking station is manufactured using laser cutting technology, then the parts are connected to the finished housing of the device. The front panel of the docking station has adjustable mounting holes for adjusting the plate in height relative to the floor to more accurately align with the spring-loaded contacts of the robot. To determine the docking station by the robot, a QR or ArUco code with the station identification number is placed on the docking station, computer vision algorithms are used on the robot to search and read data from the code [6], [7], [8]. The spatial position of the station relative to the robot is calculated to determine the optimal trajectory of the robot's arrival at the docking station [10], [11].

6. Conclusion

In the AMUR-307 robot, weaknesses in the power system were identified, namely the lack of battery interfaces and insufficient protection mechanisms. As a result of the research and development of the power supply system and its modules, several tasks were solved: managing the battery charging process, obtaining battery status information, and adding multi-level battery protection for a variety of parameters. This system is based on the SBS v1.1 specification for easy connection to Linux OS based on standard drivers. The battery monitoring module reads all important parameters from the battery and transmits them directly to the charge controller, which reduces the computational load on the processor.

The charge controller module has 2 independent control circuits and each independently charges its battery, which in turn increases the reliability of the system as a whole, since if any component fails, the robot will continue to work on one battery. For the automatic charging process, a docking station was designed and manufactured, which has an optical marker to identify the docking station and calculate its position to calculate the optimal arrival trajectory. The developed system provides a unified modular assembly, which has also been applied for use on an underwater walking vehicle. In the future, it is planned to develop software for the current sensors of the robot's motor drivers and analyze the force of each motor based on the data to obtain additional space parameters and surface characteristics for locomotion.

7. Acknowledgments

The study was partially funded by the grant from the Russian Science Foundation № 23-29-00720 and KIAM support.

8. References

[1] Pryanichnikov, V[alentin]; Ksenzenko, A[lexander] & Plotnikov, A[leksey] (2022). Development of the Intelligent Robotic Systems, Proceedings of the 33rd DAAAM International Symposium, pp.0029-0033, B. Katalinic (Ed.), Published by DAAAM International, ISBN 978-3-902734-36-5, ISSN 1726-9679, Vienna, Austria DOI: 10.2507/33rd.daaam.proceedings.005.

[2] Ricco, Mattia & Meng, Jinhao & Gherman, Tudor & Grandi, Gabriele & Teodorescu, Remus. (2019). Smart Battery Pack for Electric Vehicles Based on Active Balancing with Wireless Communication Feedback. Energies. 12. 3862. 10.3390/en12203862.

[3] Petreus, Dorin & Patarau, Toma & Szilagyi, Eniko & Cirstea, Marcian. (2023). Electrical Vehicle Battery Charger Based on Smart Microgrid. Energies. 16. 3853. 10.3390/en16093853.

[4] Wen, Y., Wolski, R., Krintz, C. (2005). Online Prediction of Battery Lifetime for Embedded and Mobile Devices. In: Falsafi, B., VijayKumar, T.N. (eds) Power-Aware Computer Systems. PACS 2003. Lecture Notes in Computer Science, vol 3164. Springer, Berlin, Heidelberg. https://doi.org/10.1007/978-3-540-28641-7_5

[5] Davydov D.V., Eprikov S.R., Kirsanov K.B., Pryanichnikov V.E. (2017). Service Robots Integrating Software and Remote Reprogramming, Proceedings of the 28th DAAAM International Symposium, pp.1234-1240, B. Katalinic (Ed.), Published by DAAAM International, ISBN 978-3-902734-11-2, ISSN 1726-9679, Vienna, Austria DOI: 10.2507/28th.daaam.proceedings.172.

[6] Pryanichnikov V. E. (2023)Development of remote sensing and control systems using image convolutions / XV Scientific and Practical Conference "Recording and reproduction of volumetric images in cinematography, science, education, media and other fields", Moscow, 3–5 April 2023.: Materials and documents // edited by O. N. Raev. — Moscow: IPP «KUNA», 2023. — 347 s., S.87-101, ISBN 978-5-98547-144-1.

[7] Volden, Ø., Stahl, A. & Fossen, T.I. (2022). Vision-based positioning system for auto-docking of unmanned surface vehicles (USVs). Int J Intell Robot Appl 6, 86–103. https://doi.org/10.1007/s41315-021-00193-0

[8] Garrido-Jurado S. (2014). Automatic generation and detection of highly reliable fiducial markers under occlusion//Pattern Recognition. №6 (47). C. 2280 – 2292

[9] R.B. Tarasov, O.I. Davydov, V.E. Pryanichnikov (2021). Software-control shell for service robots with parallel computing // Robotics and technical cybernetics. – T. 9. – № 2. – S. 99-105. – DOI 10.31776/RTCJ.9203. – EDN EUOQJP.

[10] Nuchter A., Hertzberg J. (2008). Towards Semantic Maps for Mobile Robots. //Robotics and Autonomous Systems, vol. 56, no. 11, Nov. 2008, pp. 915–926.

[11] Huang W. H., Beevers K. R. (2004). Topological Mapping with Sensing-limited Robots. //In Proc. of the 6th International Workshop on the Algorithmic Foundations of Robotics (WAFR), pp. 367-382.

[12] Pryanichnikov V.E., CHernyshev V.V., Arykancev V.V., Aryskin A.A, Ksenzenko A.YA., Petrakov M.S., Travushkin A.S., Eprikov S R. (2019). Supervisory control of a walking underwater base with autonomous robotic satellites. Intelligent adaptive robots // Information-measuring and control systems. T. 16. № 1–2. S. 49–56. ISSN 2070-0814.

[13] Chernyshev V. V., Pryanichnikov V. E., Arykantsev V. V. (2020). Algorithms of Self-control of the Underwater Walking Device According to Information on a Collision of Feet with an Unorganized Support Surface. In: Yuschenko A. (Eds) Modern Problems of Robotics. MPoR2020. Communications in Computer and Information Science, vol 1426. Springer, Cham. DOI https://doi.org/10.1007/978-3-030-88458-1_14, Print ISBN978-3-030-88457-4 Online ISBN978-3-030-88458-1.

DOI: 10.2507/34th.daaam.proceedings.037

METHODS AND ALGORITHMS FOR ANALYSING AND PREDICTING ERRORS IN SERVER SYSTEMS

Plamena Koleva-Stoynova & Velko Iltchev

This Publication has to be referred as: Koleva-Stoynova, P[lamena] & Iltchev, V[elko] (2023). Methods and Algorithms for Analysing and Predicting Errors in Server Systems, Proceedings of the 34th DAAAM International Symposium, pp.0288-0293, B. Katalinic (Ed.), Published by DAAAM International, ISBN 978-3-902734-41-9, ISSN 1726-9679, Vienna, Austria
DOI: 10.2507/34th.daaam.proceedings.037

Abstract

Logging is important in monitoring and predicting server failures. The information is stored in log files, which contains (usually as plain text) information about messages, error reports, file requests and file transfers. Reasons for logging are diagnostic, audit, capturing stack traces or data. The purpose of this work is to develop a software architecture for analysis and prediction of errors in server systems. First, messages are collected from client servers and stored on the main server. Then a message parser recognizes the lexical and syntactic structure by using context-free grammar. Key components are stored in a relational database. A synthesizer is connected to a relational database. It is used to recreate the original messages when a user needs to read them. The most important part of this work is the recognition of interconnections between messages and environment with the help of artificial intelligence, which will thereafter be used to predict failures.

Keywords: logging; detection of errors; parsing; prediction.

1. Introduction

Logging is an essential practice, but it comes with potential issues that can interfere with its effectiveness. One common problem is the lack of storage space, leading to a low performance. Data acquisition generates a significant amount of information, consuming large storage space and impacting server performance. This information can also be sensitive, making proper handling crucial to avoid security issues. An illustrative instance is detailed in publication [1], wherein a swift expansion of the database can ultimately lead to disk capacity reaching full utilization, consequently giving rise to a severe issue. Another issue is the log noise, where excessive or irrelevant logs make it challenging to diagnose actual problems effectively. For example, log entries might include messages that are not directly related to system health, security, or performance. These irrelevant messages obscure critical information. Additionally, excessive logging can impact an application's performance and response time. Furthermore, logs may not always provide sufficient context, making it harder to understand the full context of an issue. Finally, analysing a large volume of logs can be complex and time-consuming, which requires significant resources. An approach to address these challenges involves the utilization of artificial intelligence techniques capable of discerning pertinent data from the unstructured log entries. An integral component of this solution is the anomaly detection. Through the analysis of individual data points, future errors can be readily anticipated and rectified promptly [2].

To perform an accurate analysis of the data log entries, log text parsing is necessary. It's noteworthy that in [3], a novel approach to anomaly detection based on log data known as NeuralLog is employed, which eliminates the necessity for log parsing. The reason for creating this work is to develop an architecture for analysis and prediction of errors in server systems. The proposed paper is organized as follows: section 2 describes related to the research works, section 3 describes the proposed method with topology and the connection with neural network. The conclusion closes the article.

2. Related Works

2.1. The platform LogPAI

Using advanced algorithms for predicting and parsing methods, scientists from the Chinese University of Hong Kong developed a platform for collecting, parsing, and analysing logs from server systems. This project consists of large collection of system log datasets for AI-powered log analytics (loghub), a toolkit for automated log parsing (log parser), a log analysis toolkit for automated anomaly detection (loglizer) and log-based impactful problem identification using machine learning(log3C). All these are components of the same platform.

At first, logs are generated at runtime and aggregated into a centralized place with a data streaming pipeline, such as Flume and Kafka [4]. Then a log parser converts unstructured log messages into a map of structured events, based on which sophisticated machine learning models can be applied. [5] Afterwards the structured logs can be sliced into short log sequences through interval window, sliding window, or session window. Then, feature extraction is performed to vectorize each log sequence, for example, using an event counting vector. [6] Finally, the anomaly detection models are trained to check whether a given feature vector is an anomaly or not. [7]

2.2. Software Logstash

Another software example is Logstash. It is an open-source data collection engine with real-time pipelining capabilities. It can dynamically unify data from disparate sources and normalize the data into destinations. It serves as a crucial component in the ELK (Elasticsearch, Logstash, Kibana) stack, which is widely used for log management and data analysis. As a data collector, Logstash can ingest data from diverse sources such as files, syslog, redis, and beats. Once the data is collected, Logstash processes it through a series of filters in a pipeline. These filters can parse, structure, modify, and enrich the incoming data, making it more organized and suitable for analysis.

At first, inputs are used to get data into Logstash. These inputs are file, syslog, redis, beats. Then, filters process information in pipeline. Used filters are: grok, mutate, drop, clone and geoip. Grok-filter parses and structures the unstructured log. Mutate-filter performs general transformations on event fields, such as rename, remove or modify. Drop-filter drops an event completely. Clone-filter makes a copy of an event. Geoip-filter adds information about geographical location of IP addresses.

Outputs are the final phase of the Logstash pipeline. An event can pass through multiple outputs, but once all output processing is complete, the event has finished its execution. With its flexible and configurable nature, Logstash allows users to handle a wide variety of data formats and sources, making it a powerful tool for managing large-scale log data and other data streams in real-time. [8]

2.3. Software Fluentd

Fluentd is another software example. It is an open-source data collector, which unifies the data collection and consumption for a better use and understanding of data. It collects, processes and aggregates logs. Fluentd was designed to work as a unified layer for logging a central component that aggregates data from multiple sources that vary formatted data in JSON objects and to different output destinations can forward. One of Fluentd's key strengths lies in its pluggable architecture, which allows users to extend its capabilities with custom plugins for data enrichment, filtering, and transformation. This enables seamless integration with different data storage and analytics systems, such as Elasticsearch. [9]

3. Proposed Method

The objective of this development is to define an architecture for the analysis and prediction of errors within server systems. The proposed approach delineates the interrelationship connecting client servers, the primary server, and a relational database. Client Servers establish a connection with the main server, facilitating the transmission of log messages from client servers to the primary server for aggregation. Utilizing a parser, log messages' lexical and syntactic structure is discerned through the application of context-free grammar. Consequently, facilitating the distribution and storage of key components into designated tables within the relational database is important. A synthesizer possesses authorization to access this database. In instances where a user seeks to access a specific message, the synthesizer reconstructs it by applying a template alongside pertinent components sourced from the database.

Subsequently, using an artificial intelligence, the neural network identifies linkages between messages and their contextual surroundings, thereby enhancing comprehension of their origin and implications. In the fourth figure presented below, the depicted approach can be observed.

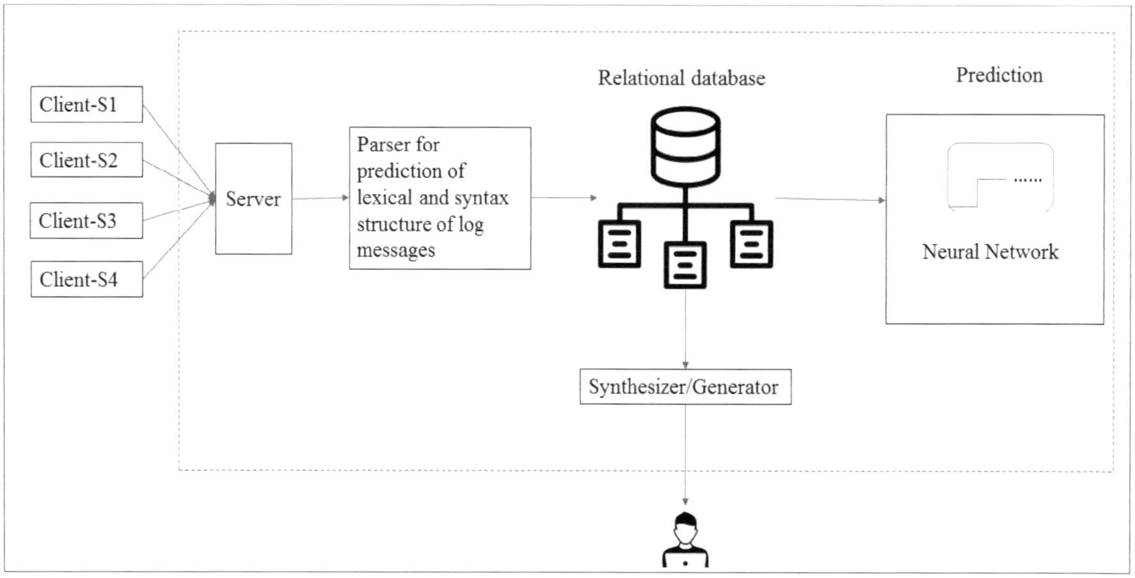

Fig. 1. Proposed Architecture

An essential aspect of this study involves investigating the interrelationships within this architecture and comprehending the behavioural patterns of data. The interplay between neural network analysis and a relational database offers a holistic approach to data-driven decision-making, prediction, and understanding complex systems. This integration capitalizes on the strengths of both methodologies, leading to more accurate, contextually aware analyses and predictions.

3.1. Extraction of facts

Important point in this work is the extraction of facts from messages. The process involves identifying and isolating specific structured information within the log data. This procedure encompasses the recognition of typical patterns within log message formats, achieved through techniques like regular expressions or advanced sequence modelling. Pertinent details such as timestamps, error codes, user identifiers, and IP addresses within log messages are targeted for extraction through algorithmic means. Log messages encompass a combination of structured and unstructured data. While structured data, such as timestamps and numerical values, is relatively straightforward to extract, unstructured data like error descriptions necessitates the application of advanced natural language processing methods. Fact extraction can leverage grammar expressions for recurring patterns, while custom rules can be tailored to capture domain-specific information by identifying distinct keywords or phrases. Extracted facts find utility across various analytical pursuits, including trend analysis, anomaly detection, predictive modelling, and system monitoring.

```
1 [appbf]/var/lxc/cXXXX/var/perp/app/bruteforce/mon/current
2
3 2021-06-08 20:04:06.932583 gotdata: {"ua":"Mozilla/5.0 (Windows NT 10.0; Win64; x64) AppleWebKit/537.36(KHTML,like Gecko) Chrome/83.0.4087.0 Safari/537.36","app":"wordpress","ip":"1.2.3.4","type":"comments","path":"/home/customer/www/autoexample.com/public_html","login":"unknown","host":"www.autoexample.com"}
```

Fig. 2. Example of Log Message

The provided example on figure 2 serves as a representation of a simple textual format messages. Line 1 shows the name of file and its full path destination. Line 3 shows the plain text message with key components. To extract the fundamental elements from the message, using of the context-free grammar is needed. Therefore, the example will be depicted as following graph:

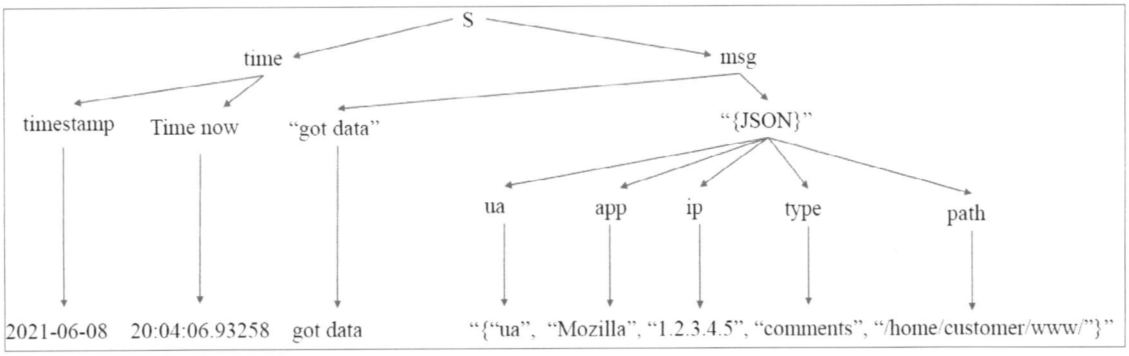

Fig. 3. Syntax tree of log message example

The represented syntax tree illustrates the message dissected into its constituent elements. The entire message is denoted by the abbreviation "S" and subsequently divided into two segments, namely "time" and "msg." The acronym "msg" signifies "message." Subsequently, the "time" segment is further subdivided into three components: "timestamp," "current time," and "data received" indicated by "got data." The data received is structured in JSON (JavaScript Object Notation) format, enhancing accessibility to key elements. For instance, for the "app" key within messages of the same type, the corresponding value imparts insights about the utilized application.

3.2. Allocation of facts

Another significant aspect within this study pertains to the dissemination of factual elements across a database. Numerous tables will be established for the purpose of retaining values derived from the log message. These tables will be linked via a primary identifier. This connectivity is advantageous for facilitating convenient retrieval during message reconstruction or data analysis endeavours. Thereafter, a neural network will be connected to this database with the aim to be trained to analyse and predict future errors.

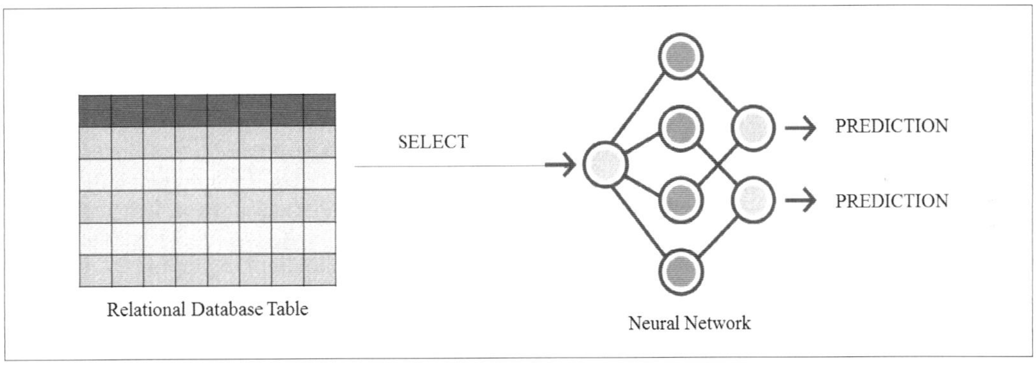

Fig. 4. Communication between table from database and neural network

The depiction in Figure 4 showcases the process. By utilizing the "Select" command, structured data within the database is extracted and enters the initial phase of the neural network. The neural network is pre-trained to assess outcomes generated by such commands. The message is then stored in an intermediate format, but with a distinct word order.

3.3. Prediction of future events

Collection of facts is needed for analytics of the whole picture. The use of time series forecasting with log messages involves the use of statistical and mathematical techniques to predict future patterns and trends within the log data. By treating log messages as sequential data points ordered by time, time series forecasting methods can be applied to anticipate potential future events, anomalies, or changes in the log data. This approach leverages historical log message data to establish patterns and relationships over time. The collected information is then used to develop predictive models that can project how the log messages might evolve in the future. This could include predicting potential system failures, performance issues, or unusual activities based on past patterns and trends observed in the log messages.

Time series forecasting techniques, such as ARIMA (Autoregressive Integrated Moving Average), Exponential Smoothing, or more advanced machine learning algorithms like LSTM (Long Short-Term Memory) neural networks, can be applied to log messages. These methods analyse temporal dependencies, seasonal variations, and other recurring patterns to generate predictions. The incorporation of time series forecasting with log messages is particularly valuable in proactive system maintenance, resource optimization, and anomaly detection. It aids administrators in identifying potential issues before they escalate, thus enhancing overall system reliability and performance. The method described herein finds application in the subsequent scholarly work titled "The importance of time series data filtering for predicting the direction of stock market movement using neural networks" [10]. Searching for errors in such huge files is slow and inefficient. The solution we offer, we can quickly find errors of one type that occurred in a certain time interval from certain servers. A neural network training will allow to analyse these errors and therefore predict future ones.

4. Conclusion and future work

The core issue discussed in this article pertains to server logging, specifically the challenge of excessive disk space use due to logging activities. This problem can be attributed to overly detailed logging, necessitating a focus on capturing essential information. Furthermore, the practice of log rotation, which entails the periodic creation of new log files accompanied by the compression or removal of older log files, is identified as another factor contributing to increased disk space consumption. Additionally, misconfigurations of logging levels, such as the use of INFO or DEBUG, may exacerbate this issue.

While logging remains a vital component of system monitoring, its storage in raw text format can impose a substantial space burden. To mitigate this, it is imperative to selectively extract and retain only the crucial data, subsequently storing it within a database. This structured approach enables the generation of templates, facilitating enhanced comprehension of log messages by clients seeking to discern the underlying issues. This paper introduces a novel approach to analyse and forecast errors within server systems. To achieve this objective, an architecture is built to establish connections between client servers and the central server. A parser is employed to extract and disseminate information into a database, which serves as the foundation for making predictions regarding future errors. The primary emphasis of this research lies in the examination of interconnections within this architectural framework.

Future research will encompass the development of advanced machine learning models for more accurate error anticipation and classification. It will also focus on creating real-time predictive systems to minimize service disruptions. Interdisciplinary collaboration, involving fields like cybersecurity and data analytics, will yield innovative methods for improving system reliability. Research will explore human-machine interaction for enhanced error diagnosis and resolution, particularly in complex server environments. Additionally, scalability in cloud-based systems, privacy and security considerations, economic impact assessments, and open-source collaboration will be central to shaping the future of error analysis and prediction.

The exploration of methods and algorithms for analysing and predicting errors in server systems yields valuable insights into enhancing system reliability and performance. By meticulously investigating various techniques, this study contributes to the advancement of error detection, diagnosis, and prevention strategies. The integration of sophisticated algorithms and predictive models equips server administrators with proactive tools to mitigate disruptions, minimize downtime, and optimize system operations. As technology continues to evolve, the continuous refinement of these methods promises a more resilient and efficient server ecosystem, enabling organizations to uphold robust performance standards and deliver seamless user experiences.

5. Acknowledgments

This paper is dedicated to my grandmother, Ing. MSc. Paraskeva Ilieva, for her constant faith in my abilities. Her influence on my academic and personal growth has been profound, and I am grateful for the wisdom and support she has provided. This work stands as a tribute to her memory, which continues to shape and guide my endeavors.

6. References

[1] Korbar D, "Problem with rapid transaction log growth in SQL server databases," in DAAAM International Vienna, 2010

[2] Min Du, Feifei Li, Guineng Zheng, Vivek Srikumar, "DeepLog: Anomaly Detection and Diagnosis from System Logs," in CCS '17: Proceedings of the 2017 ACM SIGSAC Conference on Computer and Communications Security, New York, 2017

[3] Van-Hoang Le, Hongyu Zhang, "Log-based anomaly detection without log parsing," in Proceedings of the 36th IEEE/ACM International Conference on Automated Software Engineering, Melbourne, 2021

[4] WangJun, Wang Wenhao, Chen Renfei, „Distributed Data Streams Processing Based on Flume/Kafka/Spark" in 3rd International Conference on Mechatronics and Industrial Informatics (ICMII 2015), October 2015, Zhuhai, China

[5] A. Zbiciak and T. Markiewicz, "A new extraordinary means of appeal in the Polish criminal procedure: the basic principles of a fair trial and a complaint against a cassatory judgment," Access to Justice in Eastern Europe, vol. 6, p. 1–18, March 2023

[6] S. He, J. Zhu, P. He and M. R. Lyu, "Experience Report: System Log Analysis for Anomaly Detection," in 27th IEEE International Symposium on Software Reliability Engineering, ISSRE 2016, Ottawa, ON, Canada, October 23-27, 2016, 2016

[7] Shilin He, Qingwei Lin, Jian-Guang Lou, Hongyu Zhang, Michael R. Lyu, Dongmei Zhang, "Identifying impactful service system problems via log analysis," in ESEC/FSE 2018: Proceedings of the 2018 26th ACM Joint Meeting on European Software Engineering Conference and Symposium on the Foundations of Software Engineering, 26 October 2018

[8] "Parsing Logs with Logstash," [Online]. Available: https://www.elastic.co/guide/en/logstash/current/advanced-pipeline.html. Accessed on: 24 August 2023

[9] "What is Fluentd?," Cloud Native Computing Foundation (CNCF), [Online]. Available: https://www.fluentd.org/architecture. Accessed on: 24 August 2023

[10] Ive Botunac, Ante Panjkota & Maja Matetic, "The importance of time series data filtering for predicting the direction of stock market movement using neural networks," in 30th DAAAM international symposium on intelligent manufactoring and automation, Vienna, Austria, 2019

DOI: 10.2507/34th.daaam.proceedings.038

HYPERSPECTRAL IMAGES AS A TOOL FOR SOIL FIELD RECOGNITION

Tsvetelina Georgieva, Stefka Atanasova, Georgi Manchev,
Stanislav Penchev, Eleonora Kirilova & Plamen Daskalov

This Publication has to be referred as: Georgieva, T[svetelina]; Atanasova, S[tefka]; Manchev, G[eorgi]; Penchev, S[tanislav]; Kirilova, E[leonora] & Daskalov, P[lamen] (2023). Hyperspectral Images as a Tool for Soil Field Recognition, Proceedings of the 34th DAAAM International Symposium, pp.0294-0298, B. Katalinic (Ed.), Published by DAAAM International, ISBN 978-3-902734-41-9, ISSN 1726-9679, Vienna, Austria
DOI: 10.2507/34th.daaam.proceedings.038

Abstract

Hyper-spectral imaging provides an opportunity for high-precision and detailed spectral measurements of the object of study with a wide range of applicability - from laboratory imaging of samples through field to remote satellite measurements. The paper presents a study in which the average spectral characteristics for each sample were obtained from the hyperspectral images of soil samples from three different pre-treatment fields. Spectral data were processed in the Pirouette software environment and principal component analysis was applied. Using the first three principal components, the soil fields are clearly separable. Interclass distances were also calculated. From the resulting discriminating power, wavelengths can be derived that are informative about the content of the soil fields. The sections that were determined from the modelling power from 1200 to 1250 nm; 1300 to 1350 and 1400 to 1450 nm, the wavelengths can be defined that would be used as inputs to the classifiers. The future work is developing of models for predicting micro- and macro-elements in soil using hyperspectral images.

Keywords: Hyper-spectral images; Soil field; Principal component analysis; Pirouette software

1. Introduction

Soil type is essential for farmers to determine its fertility status [1],[2]. Soil evaluation encompasses various chemical processes that determine the amount of nutrients available in the soil. Based on the results of the evaluation, recommendations are made for treating the soil fields with appropriate means - machines or using appropriate fertilizers to improve soil fertility for the respective crop. In the recent years computer vision systems, ultrasound and infrared signals and tomographic imaging are used for quality assessment of different parameters in agriculture [3]. The computer vision-based quality inspection comprises of four main steps, namely, acquisition, segmentation, feature extraction and classification. The artificial intelligence techniques such as fuzzy logic are also widely used for the automatic analysis and detection of objects [4]. Our previous research works is related to quality assessment of soil using different sensor for measurement of the main quality soil parameters [5] and research of the influence of external factors on the measurement of a basic soil quality parameter [6].

These methods are time – consuming and are not suitable for express assessment of the whole field. Hyper-spectral analysis is not so popular method in agriculture. The purpose of the research is to assess the possibility for soil field recognition using hyper-spectral imaging technique.

2. Soil samples and obtaining of the hyper-spectral images

Soil samples were obtained from the three typical fields form Ruse region in Bulgaria. The 110 samples are collected and analysed. Hyper-spectral imaging (HSI) provides the possibility of highly accurate and detailed spectral measurements of an object of view with a wide range of applicability - from laboratory imaging of samples through field to remote satellite measurements. In the case of HSI, it represents a three-dimensional hyper cube, which is a carrier of spatial and spectral information, where each pixel of the spatial part is represented by spectral information from each spectral band of the sensor. A SPECIM model N17E sensor with 320 hyper-spectral pixels with 256 spectral bands ranging from 850nm to 1700nm was used in this study. The experiment was conducted in laboratory conditions, and 110 soil samples were taken from the three fields.

The experimental setup shown in Figure 1 consists of:

1. movable base driven in one direction by precision servo motor;
2. four broad spectrum light sources;
3. hyper-spectral sensor SPECIM N17E;
4. soil sample.

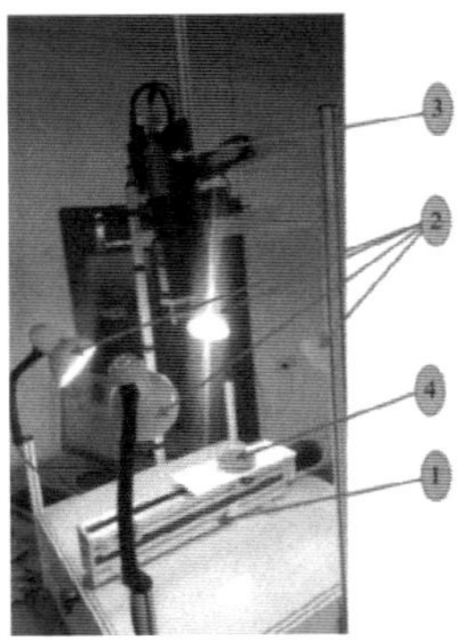

Fig. 1. Experimental setup with soil sample

Equal weight amounts were taken from all soil samples, each placed in a Petri cuvette and one hyper-spectral image is obtained. Since HSIs received from the sensor are in raw form, this requires additional processing. Due to the influence of the environment (temperature, humidity, etc.) and specific features of the sensor, there is minimal static noise in the data, expressed in a non-uniform static offset with respect to the wavelength for each spectral band. To remove static sensor spectral noise from each HSI of the samples, a black HSI captured with the lens fully closed with the same sensor and under the same conditions during the experiment was used.

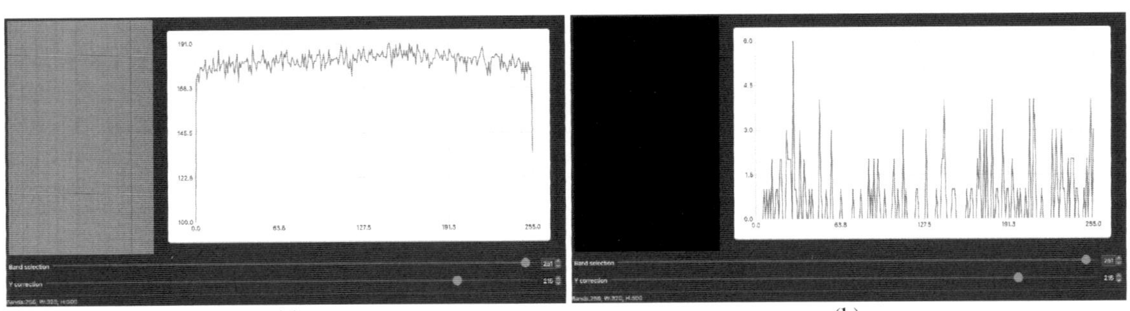

(a) (b)

Fig. 2. Spectral characteristics of one pixel
(a) – with static noise; (b) – with removed noise

3. Spectral data processing

The correct selection of an area of the photographed soil sample is essential for the quality of the research result. It should include only the subject of study without the background or side objects present in the frame of the original image. There are various ROI determination methods based on mathematical analysis and transformation, aiming to form a mask to remove unwanted pixels.

Techniques based on Principal Component Analysis (PCA), Cluster Analysis, K-means algorithm, Fuzzy C-means algorithm, similarity between spectral characteristics (similarity) based on correlation coefficients, etc. are widely used. [7].

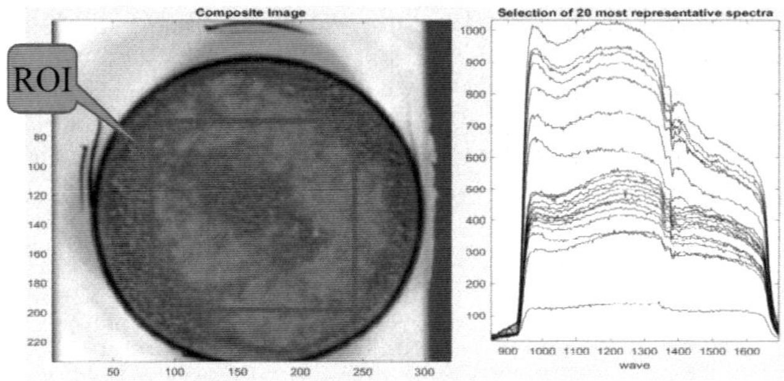

Fig. 3. Selection of ROI from soil sample image

Since the captured soil samples in this study are relatively similar, a manual ROI determination method was chosen, by which regions of the soil samples with approximately equal areas were isolated from the frame. In Matlab, by means of the HyperTools 3 tool [8], a region of interest (ROI) covering only a homogeneous area with a soil sample is separated from each HSI.

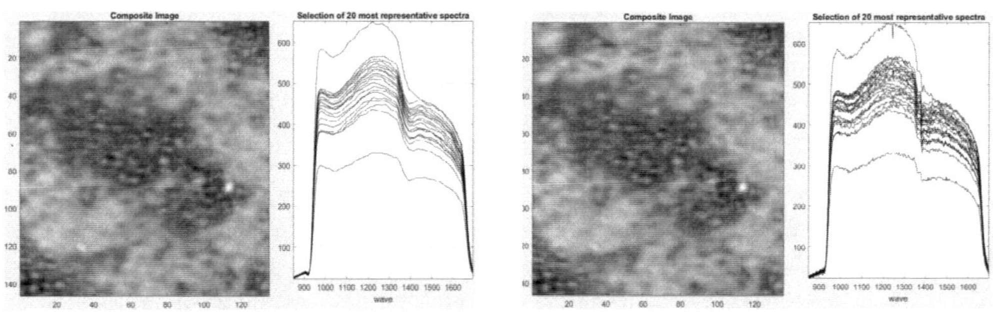

(b) (b)

Fig. 4. Image ROI and spectral characteristics
(a) - no filtration applied, (b) – with applied filtration

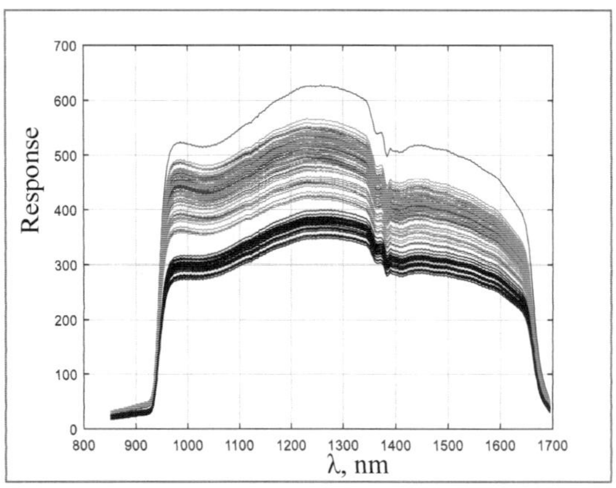

Fig. 5. Averaged spectral characteristics for each sample from the three soil fields

In order to reduce the spectral noise, while preserving the spectral shape to the maximum extent, a minimal additional spectral filtering was applied through a combination of smoothing/derivatives (Savitszki-Golay) and Standard Normal Variance (SVN) algorithms. For the needs of the subsequent analyses for each soil sample an average spectral characteristic of the ROI was calculated. In fig. 5 presents the averaged spectral characteristics for each sample, for the three soil fields.

As a classification procedure we used SIMCA classifier which is integrated in Pirouette software and principal component analysis as a data pre-processing procedure. The result from principal component analysis with the first three components is shown on Fig. 6. It is clearly visualised separability between the classes.

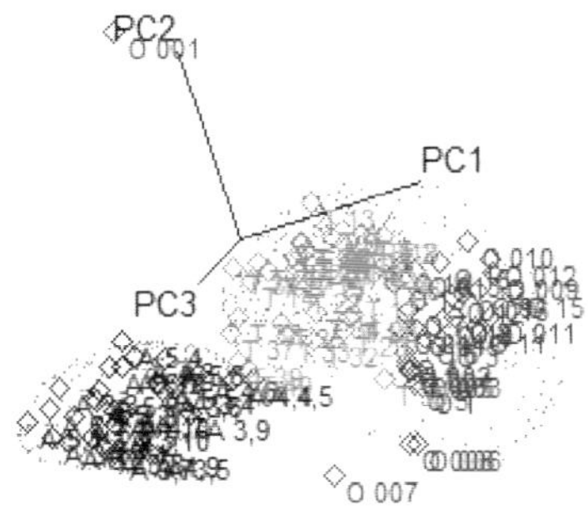

Fig. 6. Result from principal component analysis with the first three components

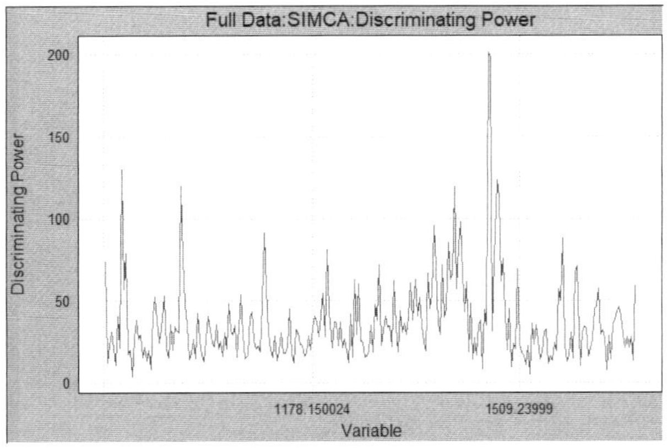

Fig. 7. Discriminating power for the three soil fields

It is also important to know which variables are best at discriminating between classes. For each wavelength, the average residual variance for each of the classes fit to all other classes and the residual variance of all classes fit to themselves are compared. This shows how important a given wavelength is to the difference between "correct" and "incorrect" classification. This is possible using Discrimination Power (DP). A value close to 0 indicates low discrimination ability in a wavelength, while a value much larger than 1 implies high discrimination power. The discriminating power is shown on Fig. 7. In the Fig. 7, the wavelengths between 860 and 1000 nm; 1200 and 1550 nm provide the most discriminating power for the classes. Total Modeling Power determined which wavelenghts have importance for each class in the set. The ranges are from 0 to 1.

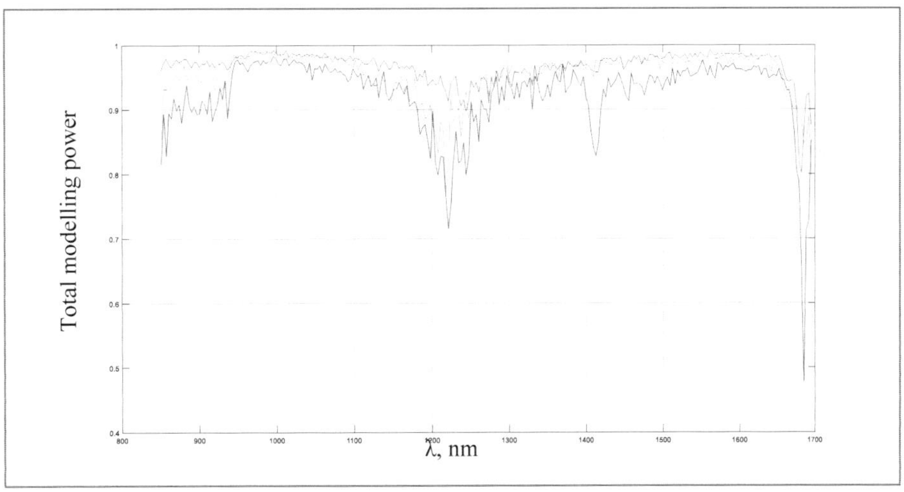

Fig. 8. Total modelling power for the three classes.

The sections of the total modelling power that were determined from 1200 to 1250 nm; 1300 to 1350 and 1400 to 1450 nm, the wavelengths can be defined that are informative and would be used as inputs to the classifiers. The interclass distances presented in the table 1 indicate sufficiently separable classes.

	CS1@8	CS2@9	CS3@8
CS1	0.000000	3.565116	5.395718
CS2	3.565116	0.000000	5.654512
CS3	5.395718	5.654512	0.000000

Table 1. Interclass distances

4. Conclusion

The hyper-spectral images of the soil could be used as a method for express soil field recognition and assessment. The analysis of the obtained results shows that SIMCA classifier and principal component analysis as a pre-processing procedure can be used for successfully separate the soil fields. This is of significant importance for the subsequent treatment of soil fields with a view to obtaining high yields and low production losses. From the experimental results and value of the modelling power are defined informative wavelengths for the three soil fields - 1200 to 1250 nm; 1300 to 1350 and 1400 to 1450 nm. In the future research work wavelengths will be determined for each class of data for which classification and recognition accuracy of soil fields will be evaluated. It would be a precondition for development of farmers information systems.

5. Acknowledgments

This research is supported by the National Scientific Program "Intelligent Plant Breeding" of the Ministry of Education and Science, approved by Decision of the Council of Ministers No. 866 / 26.11.2020.

6. References

[1] Tiwari, T.; Sharma, S.; Kumar, K.; Ravindra, S. (2023). Importance of Soil Testing in Sustainable Agriculture, In: Emerging Trends in Agricultural Sciences, Publisher: P.K. Publishers & Distributors, pp. 174-185

[2] Buzoianu, D-A. & Coloja, M. P. (2012). Diagnostic analysis of soils contaminated with oil and salt water, strands for their remediation, Annals of DAAAM for 2012 & Proceedings of the 23rd International DAAAM Symposium, Volume 23, No.1, ISSN 2304-1382, ISBN 978-3-901509-91-9

[3] Bhargava, A.; Bansal, A. (2021). Fruits and vegetables quality evaluation using computer vision: A review, Journal of King Saud University - Computer and Information Sciences, Vol., 33, Issue 3, 243-257, ISSN 1319-1578

[4] Gheorghe, C.A.; Cicic, D.T. (2015). Industrial Image Processing Using Fuzzy-Logic, 25th DAAAM International Symposium on Intelligent Manufacturing and Automation, DAAAM, Procedia Engineering, Vol. 100, 492 – 498

[5] Remzi, S., Georgieva, Ts., Paskova, N., & Daskalov, P. (2019). Performance of web based microprocessor system for measurement of soil quality parameters, Proceedings of the 30th DAAAM International Symposium, pp.1097-1101, B. Katalinic (Ed.), Published by DAAAM International, ISBN 978-3-902734-22-8, ISSN 1726-9679, Vienna, Austria

[6] Georgieva, Ts., Remzi, S., Paskova, N., Stefanov, E., Sigrimis, N. & Daskalov, P. (2019). Research of the influence of external factors on the measurement of a basic soil quality parameter, Proceedings of the 30th DAAAM International Symposium, pp.1097-1101, B. Katalinic (Ed.), Published by DAAAM International, ISBN 978-3-02734-22-8, ISSN 1726-9679, Vienna, Austria

[7] Amigo, J.M., Cruz, J., Bautista, M., Maspoch, S., Coello, J. & Blanco, M. (2008). Study of pharmaceutical samples by NIR chemical-image and multivariate analysis, TrAC Trends in Analytical Chemistry, Vol. 27, Issue 8, 696-713

[8] Mobaraki, N., Amigo, J. M. (2018). HYPER-Tools. A graphical user-friendly interface for hyperspectral image analysis., Chemometrics and Intelligent Laboratory Systems, Vol. 172, 174-187

DOI: 10.2507/34th.daaam.proceedings.039

MODELING OF THE RELATIONSHIP OF THE COLOR FEATURES AND AGROCHEMICAL SOIL INDICATORS USING DIGITAL IMAGES

Antonina Mihaylova, Tsvetelina Georgieva & Plamen Daskalov

This Publication has to be referred as: Mihaylova, A[ntonina]; Georgieva, T[svetelina] & Daskalov, P[lamen] (2023). Modeling of the Relationship of the Color Features and Agrochemical Soil Indicators Using Digital Images, Proceedings of the 34th DAAAM International Symposium, pp.0299-0307, B. Katalinic (Ed.), Published by DAAAM International, ISBN 978-3-902734-41-9, ISSN 1726-9679, Vienna, Austria
DOI: 10.2507/34th.daaam.proceedings.039

Abstract

In this report, alternative tools for determining agrochemical parameters of soil - content of organic matter (humus), pH, phosphorus, potassium and nitrogen, obtained from colour characteristics of digital images from optical devices by statistical methods are proposed. Regression models were developed using two methods: single factor linear regression analysis LR and multifactor linear regression analysis MLR Stepwise. The models were compared and evaluated using accuracy assessment criteria – coefficient of determination (R2), root mean square error (RMSE) and residual prediction deviation (RPD). With the resulting models from data from the camera device, pH can be determined with an accuracy of 99% and phosphorus P with an accuracy of 90%. With the camera of a mobile phone, the content of organic matter can be determined with the accuracy of 96%. The proposed tools could be used in modern smart agriculture as a proposed method for express monitoring of soil agrochemical indicators, as well as implemented in programming mobile, web-based applications for determining the content of organic matter (humus), phosphorus, pH and potassium. Future work involves using other linkage modelling methods to increase the accuracy rate for nitrogen determination.

Keywords: Agrochemical soil indicators, Prediction, Regression models, Digital Images.

1. Introduction

The express and timely monitoring of the soil implies the determination of a large number of indicators - macro and microelements, pH, humus, humidity at each stage of the phenological development of the cultivated crops. A characteristic feature of the modern farm in the application of sustainable smart agriculture is the expansion of the possibilities for the application of digital and optical measuring devices, based on the extremely rapid development of digital microprocessor and computer technology. The integration of high-speed data processing enables the implementation of algorithms and applications in agricultural production related to obtaining fast and timely information from the field, improving the accuracy of measurements and their transmission from a distance, making them a fast working tool for the modern farmer. The massive use of smart information and communication systems requires the development of mobile, web-based and cloud-based technologies that are accessible and easily applicable [1].

Therefore, it is necessary to develop and implement methods for express monitoring of soil composition and properties, which will contribute to finding the much-needed overall balance between preserving the environment, natural resources and meeting the growing needs of agricultural products. Maintaining the ecological condition of soils is particularly imperative under the conditions of significant climate changes reported in recent years. With modern smart agronomic technologies for growing agricultural crops, accurate and fast information from the field and transmission on mobile devices is increasingly required.

In traditional laboratory measurements, soil samples are examined with analytical methods that include combustion, oxidation, use of chemical substances and harmful products [2], [6], [9]. This type of laboratory measurement is time consuming and requires trained personnel. These are prerequisites to look for innovative, express methods for measuring soil quality directly in the field [10], [12]. Known diagnostic tools for rapid assessment of soil quality are based on spectral analysis in the UV-VIS-NIR region, satellite imagery, unmanned aerial systems UAV and others [3], [4]. Contact-free, non-destructive methods for soil analysis directly in the field have not been proposed yet.

One of the ways to obtain a fast, reliable and inexpensive evaluation of the parameters of the cultivated soil is by modeling information about the color of the soil obtained from optical devices according to algorithms based on mathematical models through statistical methods. The application of static methods for qualitative analysis and computer technology allows the extraction of information from difficult-to-interpret optical data. In the literature, there are a number of examples of finding relationships between soil color and indicators such as the presence of moisture, soil organic matter, iron impurities, etc. [5], [6], but despite research in this field, there are still aspects not covered and opportunities to search for effective and rational solutions for remote determination of soil quality by its color.

For the conditions in Bulgaria, similar methods for express monitoring of agrochemical indicators of the soil - content of organic matter (humus), phosphorus, pH and potassium, based on computer vision and image analysis techniques have not been applied. All this gives reason to undertake a scientific study on the possibility of modeling the relationship of the color characteristics of digital soil images obtained from optical devices and the agrochemical indicators of the soil as an alternative and rapid measurement method. The previous research access of the influence of external factors on the measurement of a basic soil quality parameter [16] but the method is based on lab equipment and standardized methods and is time consuming and expensive.

The aim of the research presented in this paper is to find mathematical models that describe the relationship between the agrochemical parameters of the soil, the content of organic carbon-humus H, pH, nitrogen N, phosphorus P, potassium K and the selected color characteristics obtained from the images of soil samples by selected optical devices – document camera (Dd), camera (Df), mobile phone camera (Dt) and colorimeter (Dk), from those presented in press [7].

2. Material and methods

The most frequently applied methods in modeling when looking for a quantitative relationship between the values of the predicted quantity and the input characteristics are the regression methods. This study seeks to describe the relationship and create mathematical models between the values of the soil parameters organic carbon-humus content H, pH, phosphorus P, potassium K and the selected informative color characteristics in a previous study in press [7] obtained from selected optical devices Dd, Df, Dt and Dk.

The tasks to be solved are related to finding a regression model that describes the relationship between the conditional mean E[Yk] of the soil parameters E[Yk] = E[H, pH, N, P, K], [14] and the selected informative color characteristics listed in Table 1.

Optical devices	Optical drive code	Selected informative color characteristics and indexes
document camera	Dd	R'd, Sd, Sdix, Cdix
camera	Df	R'f, Sf, bf, Sfix
mobile phone camera	Dt	R't, bt, Vt, Btix
colorimeter	Dk	R'k, Skix, Ckix, Ik

Table 1. Selected informative color characteristics and indices obtained from optical devices

The regression models were built using two methods: single-factor linear regression analysis LR and multi-factor linear regression analysis MLR Stepwise [14], [15].

Obtaining linear regression models by the LR method with one regression factor.

The regression factor in this study is the common to all devices color component R' of a model (RGI) [11], which is a normalized red color value. Linear regression models by the LR method were obtained in the MATLAB programming environment with the Curve Fitting Tool function. Fig. 1 shows a screenshot of the selection of variables involved in the regression equation for determining humus H with the **R'**f color component for a camera.

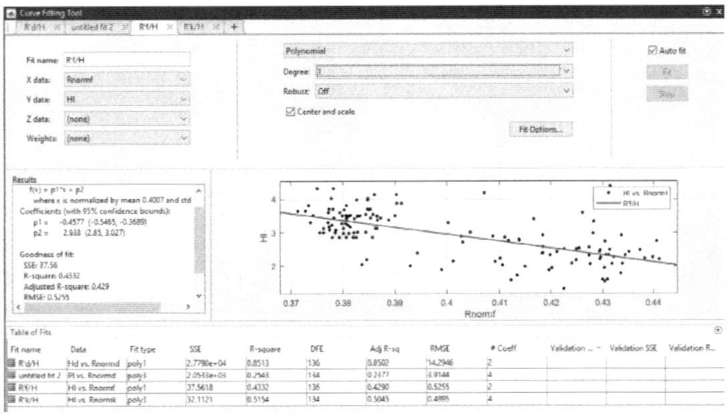

Fig. 1. Screen of obtaining linear regression models with the Curve Fitting Tool function in MATLAB R2021a.

The general form of the regression equation has the form:

$$Y = b_0 + b_1 R'n$$

where: n is the type of optical device; b_0 and b_1 are regression coefficients, [13]. The correlations were obtained between the values of R' from a model (RGI) of the soil samples obtained by the optical devices Df, Dt, Dd, Dk and the values of the soil indicators presented in Table 2.

Soil index	Correlation matrix			
	R'f	**R't**	**R'd**	**R'k**
Content of organic carbon - humus H, measured by the Turin method	-0,77	0,87	-0,84	-0,80
Mobile forms of phosphorus P_2O_5 – P, determined spectrophotometrically, by the Egner-Riem acetate-lactate method	0.62	-0.56	0.60	0.59
Mobile forms of potassium K_2O – K determined by Egner-Riem acetate-lactate method with flame photometer	-	0.44	-0.49	-0.32
Values of Soil reaction pH, determined potentiometrically in a soil suspension, with the soil : water (e. H_2O) ratio being 1 : 2.5	0,43	-0.26	0.34	0.32
The mobile forms of nitrogen N-NH3+N-NH4, determined by the Kjeldahl method	-	-	-	0,21

Table 2. Correlation coefficients for determining the relationship between soil indicators and color component R' from a model (RGI) for individual optical devices

Correlation coefficients with values from -0.77 to 0.87 indicate that there is a strong relationship between soil indicator humus H and color component R' from all optical devices. The relationship with the correlation coefficients between the indicator mobile forms of phosphorus P2O5 – P and color components R' from all optical devices with values in the range from -0.56 to 0.62 is significant. Indicators pH and potassium K2O – K have a moderate dependence with color components R = 0.34 R'f from the camera R = 0.43 and R'd from the document-camera device R = 0.49 respectively. For an indicator of absorbable forms of nitrogen, there is a weak dependence only with the color component R'k. Regression models were built on data from the strong and significant correlations.

Obtaining MLR Linear Regression Models with Multivariate Linear Regression Analysis and Stepwise Technique.

In order to check the possibility of predicting the soil indicators with more color components included, besides the common to all devices color component R', the data of the already determined color components with the greatest informativeness were applied (Table 1). The method of multivariate linear regression analysis and stepwise regression technique (MLR Stepwise) which is applied when multicollinearity must be avoided without losing information about the input color data. The general form of the required regression model is:

$$Y_k f = b_0 + \sum_{1}^{n} b_l \, f(X_n)$$

where Y_k are the values of the required soil parameters; b_o , b_l – regression coefficients, $f(X_n)$ is the color data for n number of color variables for individual f devices. The regression equation has the form:

$$Y_k f = b_0 + f b_1 x_1 + f b_2 x_2 + \cdots + f b_k x_n,$$

A multivariate linear regression analysis was performed for the soil parameters organic carbon-humus H content, pH, nitrogen N, phosphorus P, potassium K and the selected color characteristics obtained from the images of soil samples from the selected optical devices Dd, Df, Dt and Dk (Table 1). Regression models were obtained with the STATISTICA StatSoft package and Stepwise method selection shown in Fig. 2.

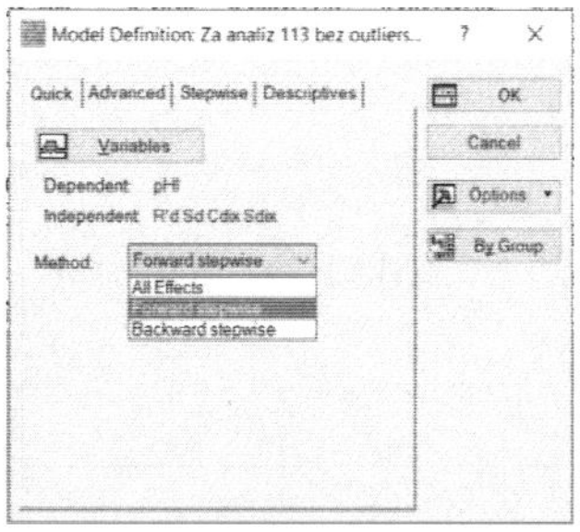

a)

	Regression Summary for Dependent Variable: K R= 0,32 R^2= 0,10 Adjusted R^2= 0,08 F(3,108)=4,0191 p					
	b*	Std.Err. of b*	b	Std.Err. of b	t(108)	p-value
Intercept			196,95	75,29	2,62	0,01
I'k	-0,42	0,21	-0,44	0,21	-2,06	0,04
Ckix	4,45	2,64	160,44	95,34	1,68	0,10
R'k	-4,25	2,67	-320,95	201,92	-1,59	0,12

b)

Fig. 2. a) Stepwise method selection from STATISTICA StatSoft package
b) regression analysis results

Obtaining regression models for determining the content of organic carbon-humus H by the LR and MLR method with the Stepwise technique

The study involved soil samples from three investigated arable fields with different organic carbon content from 1.31% low stocked to 4.40% very high stocked [8] (Table 3).

Soil index	**Organic carbon, %**			
	Mean	**Minimum**	**Maximum**	**Std.Dev.**
Content of organic carbon - humus, measured by the Turin method	2,94	1,31	4,40	0,69

Table 3. Descriptive statistics for organic carbon content

The average content of organic carbon in the samples of the three soil types was 2.94%. The larger range of organic carbon content in the calibration group is an important factor in obtaining a calibration equation, which with proven high accuracy of determining this parameter, could be successfully applied in determining organic carbon of new soil samples, characterized with a wide range of organic carbon content.

A regression model for the determination of humus in soil	**R**	**R^2**	**RMSE**	**F**	**p-value**
Hd =5,70-6,77. R'd	-0.83	0.68	0.34	F(1,136)=155,17	p < 0,05
Hk =3,97-3,04. R'k	-0.70	0.49	0.50	F(1,136)=128,86	p < 0,05
Hf=11,24-20,71. R'f	-0.66	0.43	0.53	F(1,136)=103,92	p < 0,05
Ht = 25,92. R't -6,89	0.85	0.75	0.32	F(1,136)=101,39	p < 0,05

Table 4. Values of the criteria for determining the adequacy of the obtained models

Table 4 shows the regression models with the adequate regression coefficients for predicting the content of organic carbon-humus in the soil by the individual optical devices. The significant coefficients of the models were calculated and analyzed for adequacy depending on the value of the Fisher coefficient and p-value. Nonsignificant coefficients, those for which p values > 0.05 were not included in the models. From the given example in figure 2 b) it is clear that the coefficients in front of the color parameters Ckix and R'k are insignificant and will not be applied in the model.

The significance of the regression coefficients and the adequacy of the entire model were assessed. The correlation coefficient R, coefficient of determination R2, Fisher's criterion, the probability p-value at the significance level $\alpha=0.05$ and the value of the root mean square error of the models RMSE were determined, which are criteria for determining the adequacy of the obtained models. In the obtained models for humus determination, the p-value is below 0.05, which indicates their adequacy. The RMSE values are relatively low, in the normal range of 0.32 to 0.53. The coefficient of determination R^2 indicates that 73% of the humus variation is due to the color parameter R't obtained from a mobile phone camera and described by the linear model. Highest values of coefficient of determination 0.85, correlation coefficient 0.73 and lowest value of RMSE=0.32 shows the model obtained for mobile phone camera. A model was also built for this device using the MLR Stepwise method. The regression models for soil humus index obtained from the color data by the mobile phone camera using linear regression LR and multivariate linear regression analysis MLR Stepwise and their characteristics are summarized in Table 5.

linear regression method LR	**Model evaluation criteria**			**linear regression with method MLR Stepwise**	**Model evaluation criteria**		
Organic carbon - humus	R	R^2	RMSE	Organic carbon - humus	R	R^2	RMSE
Ht = 30,41. R't - 8,63	0.85	0.73	0.32	Ht = 1,99+0,04.bt	0.87	0.75	0.31

Table 5. Regression models and characteristics for humus soil indicator obtained from the color data from mobile phone camera device by two methods: LR and MLR Stepwise

From the higher value of the coefficient of determination R=0.87 and the coefficient of determination R^2 and the lower value of RMSE, it is evident that the model built by the MLR Stepwise method improves the model built by the LR method.

The distributions of the residuals for these models were estimated, thus verifying that the prerequisites for regression analysis were met. The normal probability plot of the residuals of the obtained models for humus H from a mobile phone camera by the two methods are shown in Fig. 3 a) and b).

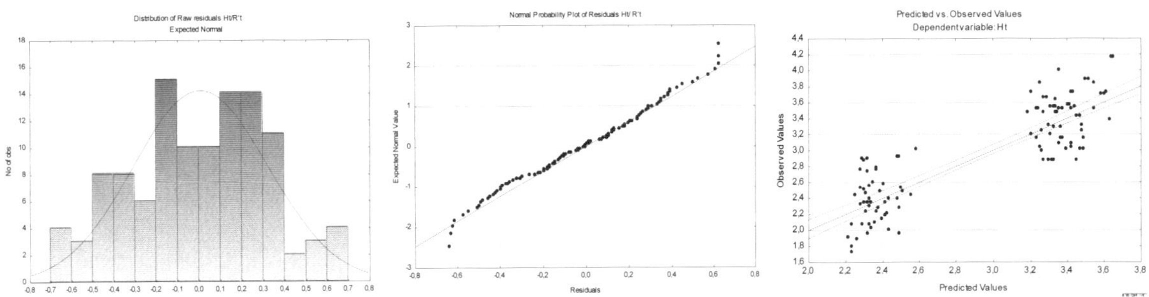

a)

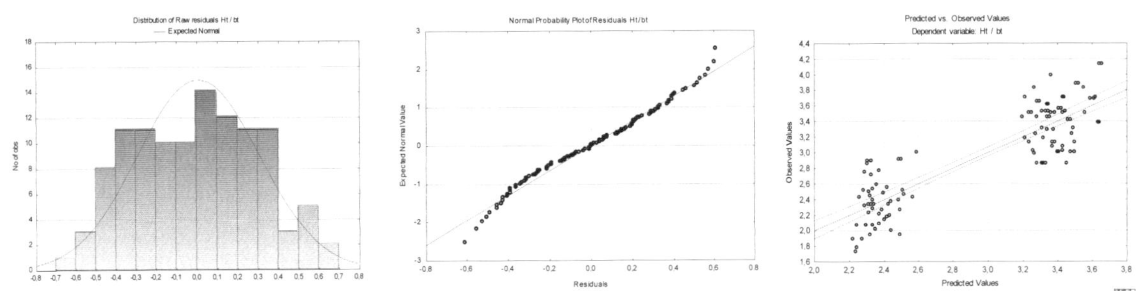

b)

Fig. 3. Distribution of residuals, normal probability plot and graphical representation of the models for determination of humus H a) linear regression method LR; b) linear regression with method MLR Stepwise

From the graphs in Fig. 3 it can be concluded that the distribution of the residuals follows a normal law and their location is close to the regression line. The confidence region for the predicted value of H from data by the MLR Stepwise model is narrower, suggesting a higher accuracy of the model. This is evident from the Cook's method inspection graphically presented in Fig. 4.

In the LR method, the systematic deviations of the measured data by the Cook method are reaching 0.06 in samples 1, 49 and 80 and are 40% greater than those in the MLR Stepwise method, which are within 0.04 (Fig. 4).

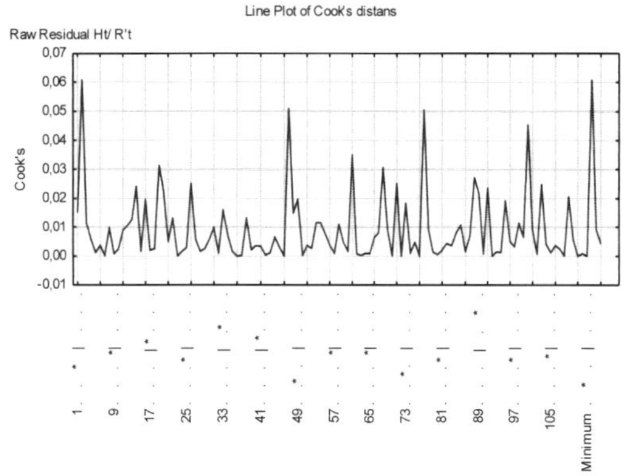

a)

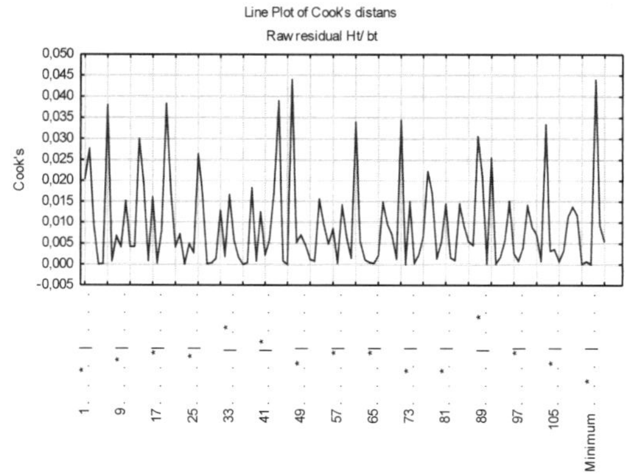

б)

Fig. 4. Systematic deviations of measured data by Cook's method for a) linear regression method LR; б) linear regression method MLR Stepwise.

Obtained regression models for determining the content of pH, nitrogen N, phosphorus P, potassium K by methods LR and MLR Stepwise

Table 6 summarizes the characteristics of the built regression models for soil indicators humus, phosphorus, pH, potassium and nitrogen by methods LR and MLR Stepwise.

linear regression method LR	**Model evaluation criteria**			**linear regression with method MLR Stepwise**	**Model evaluation criteria**		
organic carbon - humus	**R**	**R^2**	**RMSE**	**organic carbon - humus**	**R**	**R^2**	**RMSE**
Ht = 30,41. R't - 8,63	0.85	0.73	0.32	Ht = 1,99+0,04.bt	0.87	0.75	0.31
Phosphorus				**Phosphorus**			
Pf= -29,89+107,59. R'f	0.60	0.36	3.58	Pf=0,63.bf	0.62	0.38	3.58
Pk = 7,80+16,69. R'k	0.58	0.34	3.64	Pk =13,07 + 5,83.Skix	0.59	0.34	3.64
Pd = 29.11. R'd	0.49	0.24	3.93	Pd =11.24 +15.8.Cdix	0.58	0.33	3.67
Pt = 60.06 -123.56. R't	-0.48	0.23	3.94	Pt = 16,39 - 0,26.bt	-0.50	0.25	3.90
pH				**pH**			
pHd = 5,33+2,88R'd	0.34	0.12	0.75	pHd=104,45+11,28Sd-297,22R'd + 66,91Cdix + 65,09Sdix	0.44	0.20	0.72
pHk =6,08+1,26R'k	0.25	0.06	0.80	pHk =6,06+0,02Ik	0.35	0.13	0.74
pHf =3,11+8,57.R'f	0.43	0.18	0.72	pHf =77,29-206,61R'f+41,66Sfix	0.48	0.23	0.71
pHt = 10,09-9,45R't	-0.20	0.04	0.81	pHt =7,07-1,16Vt	-0.26	0.08	0.80
Potassium				**Potassium**			
Kt = 17,43+127,29R't	0.24	0.06	11.50	Kt =-5,4B**tix**-4,73bt+1374,75Rt	0.44	0.20	10.80
Kd = 80,695169-36,83R'd	0.30	0.09	11.28	Kd =69.49-12.95.Sd	0.49	0,24	10.60
Kk = 71,29-16,59R'k	0.22	0.05	12.69	Kf= -585,25-585,25R'f-376,40. Sfix	0,32	0.10	12.46
Nitrogen				**Nitrogen**			
There is no adequate model		-	-	Nk=67,99 -135,87R'k+ 1,72I'k	0.21	0.05	41,83

Table 6. Summary characteristics of the built regression models for soil indicators humus, phosphorus, pH, potassium and nitrogen by two methods LR and MLR Stepwise for the individual optical devices

For the obtained models, the level of significance and adequacy was assessed by determining the correlation coefficient R, determination coefficient R^2 and the value of the root mean square error RMSE. RMSE values were lowest for humus and pH models for both methods, but with MLR Stepwise errors being 5% lower. For phosphorus determination models, the RMSE values were 3.8% lower with the MLR Stepwise method. The coefficient of determination R^2 shows a significant increase in the MLR Stepwise models for pH and K parameters.

A cross-validation procedure was conducted, which aims to test how well the resulting models predict soil parameters by comparing the actual measured laboratory values of humus, phosphorus, potassium, and pH with the model-derived values. If the model results with the test set are close to those with the calibration subset, the output model is assumed to be valid and its accuracy can be determined. When creating the models, the option was chosen to work with the values of the soil indicators for all 138 soil samples, and the test sample was formed with 30% of them using the Monte Carlo method [13], [15]. The method is implemented in the package Statistica StatSoft and in Fig. 5 a screen of selection is shown.

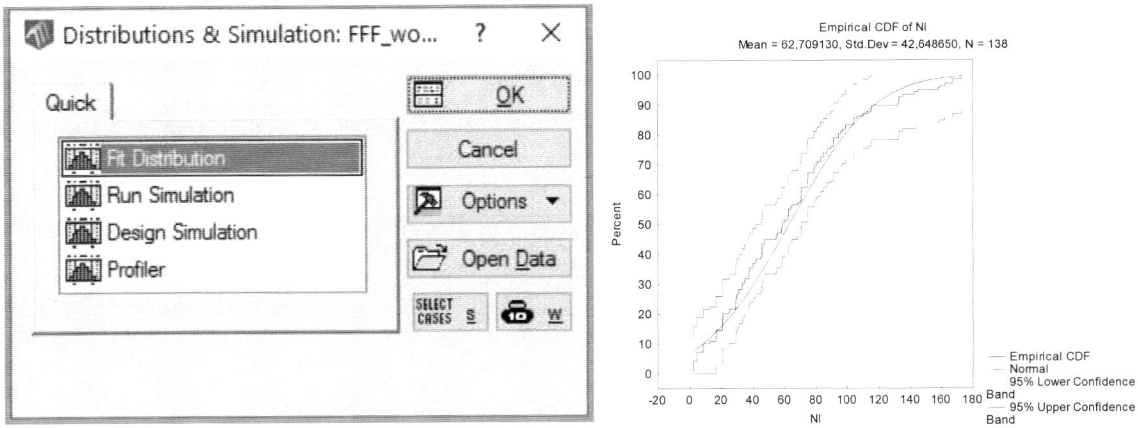

Fig. 5. Monte Carlo method selection settings

For this purpose, all adequate models obtained by both methods were checked for 40 observations from the test set. As criteria for evaluating the calibration predictive models, root mean square error (RMSE) obtained was calculated by the formula (1):

$$RMSE = \sqrt{\frac{1}{n}\sum_{i=1}^{n}(y_{i\ izm} - y_{ipred})^2} \tag{1}$$

and residual prediction deviation ***RPD***, which is obtained by the formula (2):

$$RPD = \frac{SD}{RMSE} \tag{2}$$

where ***SD*** is the standard deviation of the values (3):

$$SD = \sqrt{\frac{\sum |Yi\ izm - \bar{Y}\,|}{N}}^{\,2} \tag{3}$$

The results of the calculated model evaluation criteria are summarized in Table 7. Post-validation statistical parameters describing the calibration equations obtained with MLR analysis indicate the highest accuracy in determining soil response pHf with data from the camera device, with correlation coefficient values $R^2 = 0.99$ and the lowest accuracy in determining nitrogen with $R^2 = 0.65$ (Fig. 6).

The high values of the relation to the measured and obtained by model indicators of soil pH and phosphorus, R2=0.99, 0.90 according to data of the camera device and for indicators, potassium K, R2=0.97, and organic carbon - humus, measured with a camera of mobile phone with strong and moderate dependence on color data shows an excellent possibility of mobile phone camera and camera to be applied as an alternative measuring device for measuring agrochemical parameters of soil directly in the field or farm.

High values of RPD, which are above 3 [10] in the interval 4.56-5.07, classify the obtained models as models with very high - excellent accuracy. From the accuracy results of the models, it was found that the color data from the camera device could be used to determine potassium phosphorus and pH. To determine humus and potassium, device camera on a mobile phone. The colorimeter device shows the ability to measure phosphorus and available forms of nitrogen with high error. The document-camera device shows a high accuracy of 0.99, 0.97 for the determination of pH, potassium and humus with linear regression models, which is most likely due to the local homogeneous illumination.

Criteria calibration model				Post-validation criteria		
soil index	R	R^2	RMSE model	Accuracy model R^2	RMSE model	RPD
pH						
pHf	0.43	0.18	0.72	0.99	0.80	4.84
pHdSw	0.35	0.12	0.75	0.99	0.83	4.65
Potassium						
KtSw	0.44	0.20	10.80	0.97	11.97	4.95
KdSw	0.49	0.24	10.60	0.97	11.69	5.07
organic carbon - humus						
HtSw	0.87	0.75	0.31	0.96	0.56	5.02
HdSw	0.84	0.70	0.33	0.77	1.83	1.64
Phosphorus						
PfSw	0.62	0.38	3.58	0.90	4.69	4.59
PkSw	0.59	0.34	3.64	0.90	4.73	4.56
Nitrogen						
Nk	0.21	0.04	43.11	0.65	41.52	4.83

Table 7. Criteria for evaluating the accuracy of models after validation.

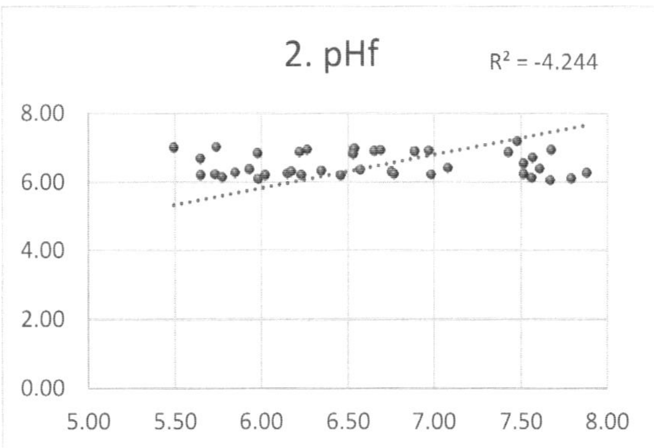

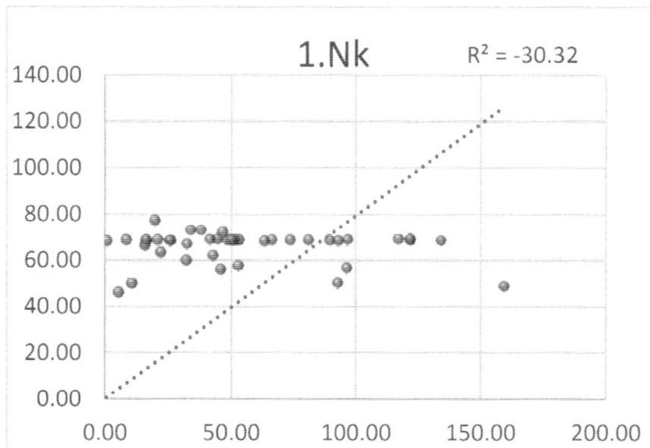

Fig 6. Graphical representation of the correlation coefficient pHf, R2 = 0.99 and nitrogen Nk with R2 = 0.65.

According to the coefficient of determination R2, the Stepwise regression method supported the models by an average of 13.5% when determining K and pH with a device camera and a mobile phone camera, and showed a weak or weak relationship with absorbable forms of nitrogen by color data from a colorimeter device . The (RMSE) values of the humus and pH models are within normal limits in the range 0.56 – 0.80. For the indicators, phosphorus has a value of 4.69 and potassium 11.69, which leads to errors.

3. Conclusion

The experimental results shows that linear regression models using two regression methods (LR, MLR- Stepwise) could describe the relation between soil agrochemical indicators H, pH, P, K, N and color characteristics obtained from images of soil samples from optical devices. The values of error (RMSE) of the models created by the Stepwise method for the parameters potassium and pH are up to 10% lower than those created by the linear regression LR method. Absorbable forms of nitrogen can be modelled with the stepwise regression method only with color characteristics obtained using colormeter device. Analysis of agrochemical parameters of the soil in the field or on the farm using a camera and a mobile phone camera can allow obtaining stand-alone results in real time. The application of mobile devices as a measuring instrument in agriculture would facilitate the work of the farmer as a complementary method to chemical laboratory measurements to determine soil quality for on-site analysis.

4. Acknowledgments

This research is supported by the National Scientific Program "Intelligent Plant Breeding" of the Ministry of Education and Science, approved by Decision of the Council of Ministers No. 866 / 26.11.2020.

5. References

[1] Georgieva, Tsv.; Mihaylova, A.; Stefanov, E.; Daskalov, P.; Sigrimis, N. (2020). Knowledge engineering in smart agriculture to secure food and the environment, Proceedings of University of Ruse, Vol. 59, 69-76.

[2] Bünemann, K.; Bongiorno, G.; Bai, Z.; Creamer, R.; De Deyn, G. (2018). Soil quality – A critical review, ELSEVIER, Soil Biology and Biochemistry, Vol. 120, 105–125.

[3] Mouazen, A.; Maleki, M.; De Baerdemaeker J.; Ramon, H. (2007). On-line measurement of some selected soil properties using a VIS–NIR sensor, ELSEVIER, Soil & Tillage Research, Vol. 93, 13–27.

[4] Rossel V.; Bouma, J. (2022). Soil sensing: A new paradigm for agriculture, ELSEVIER, Agricultural Systems Vol. 148, 71–74.

[5] Heil, J.; Jörges, C.; Stumpe, B. (2022).Evaluation of using digital photography as a cost-effective tool for the rapid assessment of soil organic carbon at a regional scale, ELSEVIER, Soil Security, Vol. 6, 1-10.

[6] Lavanya, G.; Rani, C.; Ganeshkumar, P. (2020). An automated low cost IoT based Fertilizer Intimation System for smart agriculture, ELSEVIER, Sustainable Computing: Informatics and Systems, Vol. 2, 1–12.

[7] Mihaylova, A.; Georgieva, Tsv.; Daskalov, P. (2023). Impact evaluation of the optical devices for obtaining soil color characteristics, International Conference Automatics and informatics`2023, October 05 - 07, 2023, Varna, Bulgaria (ICAI'23), in press.

[8] Giurov, G.; Artinova, N. (2001). Soil science, (in Bulgarian), Plovdiv, Makros.

[9] Wang, Y.; Huang, T.; Liu, J.; Lin, Z.; Li, S.; Wang, R.; Ge, Y. (2015). Soil pH value, organic matter and macronutrients contents prediction using optical diffuse reflectance spectroscopy, ELSEVIER, Computers and Electronics in Agriculture, Vol. 111, 69-77.

[10] Rossel, V.; Minasny, B.; Roudier, P.; McBratney, A. (2007). Colour space models for soil science, ELSEVIER, Geoderma, Vol. 133, 320– 337.

[11] Rasras, R.; Ibrahiem, M.; El Emary, M.; Skopin D. (2006). Developing a New Color Model for Image Analysis and Processing, Computer Science and Information Systems, Research Gate, Vol. 4, Number 1, 43-55.

[12] Levin, N.; Ben-Dor, E.; Singer, A. (2005). A digital camera as a tool to measure colour indices and related properties of sandy soils in semi-arid environments, International Journal of Remote Sensing, Vol. 26, No. 24, 5475–5492.

[13] Mitkov, A. (2011). Theory of experiment (in Bulgarian), Ruse, Danube press.

[14] Bratoev, K.; Beloev, H.; Mitkov A.; Mitev, G. (2020). On the possibility of conducting fast and reliable soil tests (in Bulgarian) ", Mechanization in agriculture & Conserving of the resources, Vol. 66, Issue 2, 71-76.

[15] Mitkov A.; Bratoev, K. (2023). Statistical methods in agriculture and agricultural technology, Ruse University Academic Publishing House.

[16] Georgieva, Ts., Remzi, S., Paskova, N., Stefanov, E., Sigrimis, N. & Daskalov, P. (2019). Research of the influence of external factors on the measurement of a basic soil quality parameter, Proceedings of the 30th DAAAM International Symposium, pp.1097-1101, B. Katalinic (Ed.), Published by DAAAM International, ISBN 978-3-02734-22-8, ISSN 1726-9679, Vienna, Austria

DOI: 10.2507/34th.daaam.proceedings.040

SPECTRAL ANALYSIS FOR PREDICTION OF SOIL AGROCHEMICAL INDICATORS

Tsvetelina Georgieva, Stefka Atanasova, Antonina Mihaylova, Stanislav Penchev & Plamen Daskalov

This Publication has to be referred as: Georgieva, T[svetelina]; Atanasova, S[tefka]; Mihaylova, A[ntonina]; Penchev, S[tanislav] & Daskalov, P[lamen] (2023). Spectral Analysis for Prediction of Soil Agrochemical Indicators, Proceedings of the 34th DAAAM International Symposium, pp.0308-0311, B. Katalinic (Ed.), Published by DAAAM International, ISBN 978-3-902734-41-9, ISSN 1726-9679, Vienna, Austria
DOI: 10.2507/34th.daaam.proceedings.040

Abstract

The composition of the soil is essential for higher yields in agriculture. The assessment of agrochemical soil indicators is carried out in laboratories using standardized chemical methods, which are time-consuming and relatively expensive. In the present study, an approach for predicting agrochemical indicators (nitrogen, phosphorus, potassium, humus and acidity) using spectral characteristics of soil samples was investigated. 110 soil samples from fields in the area of the city of Ruse, Bulgaria were analysed. The classification with the SIMKA method is carried out in the Pirouette software platform, and three procedures for pre-processing the data - first derivative, second derivative and multiplicative scatter correction (MSC) - have been compared. The results obtained for the standard error of validation show that for the nitrogen the minimum SEV is for MSC and 10 factors. For phosphorus - 2-nd derivative and 6 factors. For potassium - MSC and 9 factors. For humus - MSC and 10 factors. For acidity - 2-nd derivative and 10 factors. The future work will be focus on models for prediction of the indicators and assessment of the accuracy.

Keywords: Spectral characteristics; Soil agrochemical indicators; Prediction; SIMCA; Multiscatter correction; Pirouette software

1. Introduction

The soil content is very important factor for farmers and for hire yield production [1]. In the recent years different techniques are used for quality assessment of different parameters in agriculture [2]. The computer vision-based quality inspection comprises of four main steps, namely, acquisition, segmentation, feature extraction and classification. The artificial intelligence techniques such as fuzzy logic are also widely used for the automatic analysis and detection of objects [3]. Our previous research works is related to quality assessment of soil using different sensor for measurement of the main quality soil parameters [4] and research of the influence of external factors on the measurement of a basic soil quality parameter [5]. These methods are time – consuming and are not suitable for express assessment of the whole field. Some of soil properties as organic matter [6], fungal enzyme activities [7] and acidity [8] are objects of the other researches. Spectral analysis is not so popular method for assessment and prediction of the soil content.

The purpose of the research is to assess the possibility for soil agrochemical indicators prediction using spectral analysis.

2. Spectral characteristics acquisition

Spectral characteristics of the soil samples were acquired using NIR spectrophotometer GraiNit. The samples (110) are from the Ruse region, Bulgaria. The spectrophotometer is AlbaNir. It is a portable Nir analyser incorporating the most advanced technology developed and designed for the agro-food and renewable energy sectors. The design of the device meets the requirements for a portable analyser for field measurements in any weather condition. Single InGaAs sensor is used with DLP technology (Texas Instrument) and programmable between 950-1650nm spectral range. The sample is placed in a glass cuvette, covered with a transparent foil, on which the spectrophotometer is placed (Fig. 1).

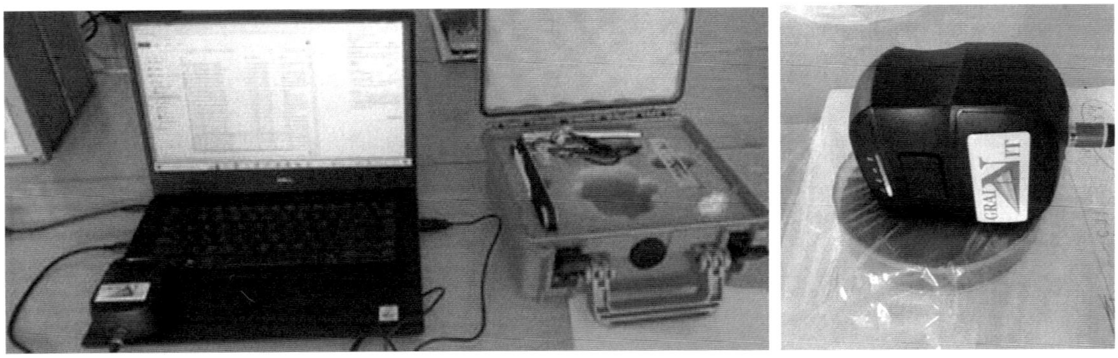

Fig. 1. Spectrophotometer and soil sample

Spectral characteristics of the soil from three different field in Ruse region are shown in Fig. 2. The variable is the wavelength in nm.

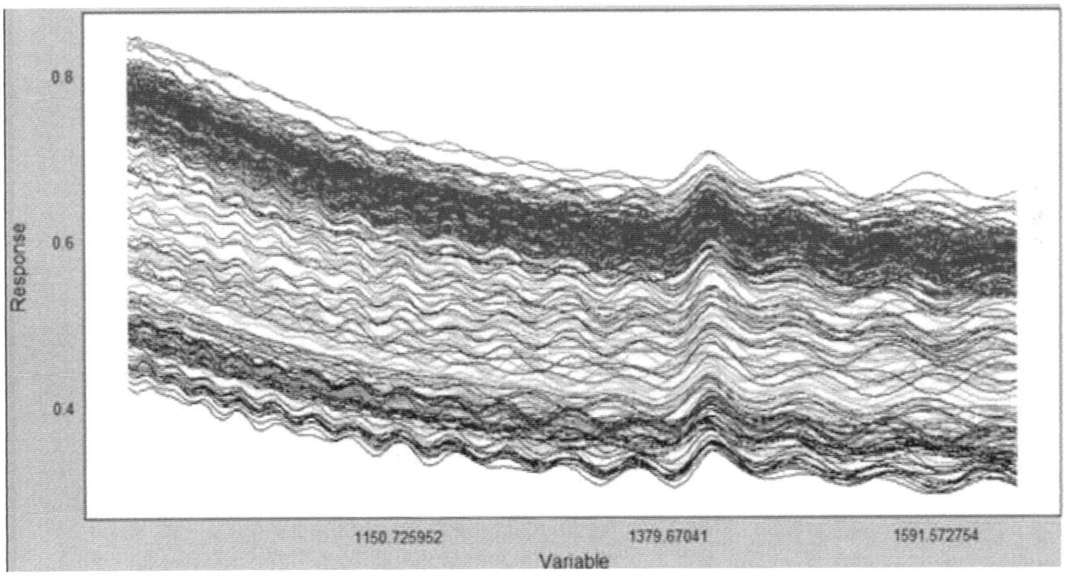

Fig. 2. Spectral characteristics of one pixel

3. Pre-processing of spectral characteristics

Analysis of the spectral characteristics is done in software platform Pirouette. Partial Least Squares (PLS) regression models of the data is done. Three pre-processing procedures are used – first derivative, second derivative and multiplicative scatter correction (MSC). MSC is a standard approach to compensating for scattering by solids in NIR spectrometry. Each sample spectrum is regressed linearly against an ideal spectrum to yield a slope and intercept. The sample spectrum is then "corrected" at each wavelength by first subtracting the intercept, then dividing by the slope. The ideal spectrum is most often simply the mean of included samples.

4. Results from analysis of the data from pre-processing procedure

The results were obtained for each of the agrochemical indicators: nitrogen, phosphorus, potassium, humus and acidity. The function standard error of validation from number of factors is shown on Fig. 3. The results show that for the nitrogen the minimum SEV is for MSC and 10 factors. For phosphorus - 2-nd derivative and 6 factors. For potassium - MSC and 9 factors. For humus - MSC and 10 factors. For acidity - 2-nd derivative and 10 factors. It is shown on Table. 1.

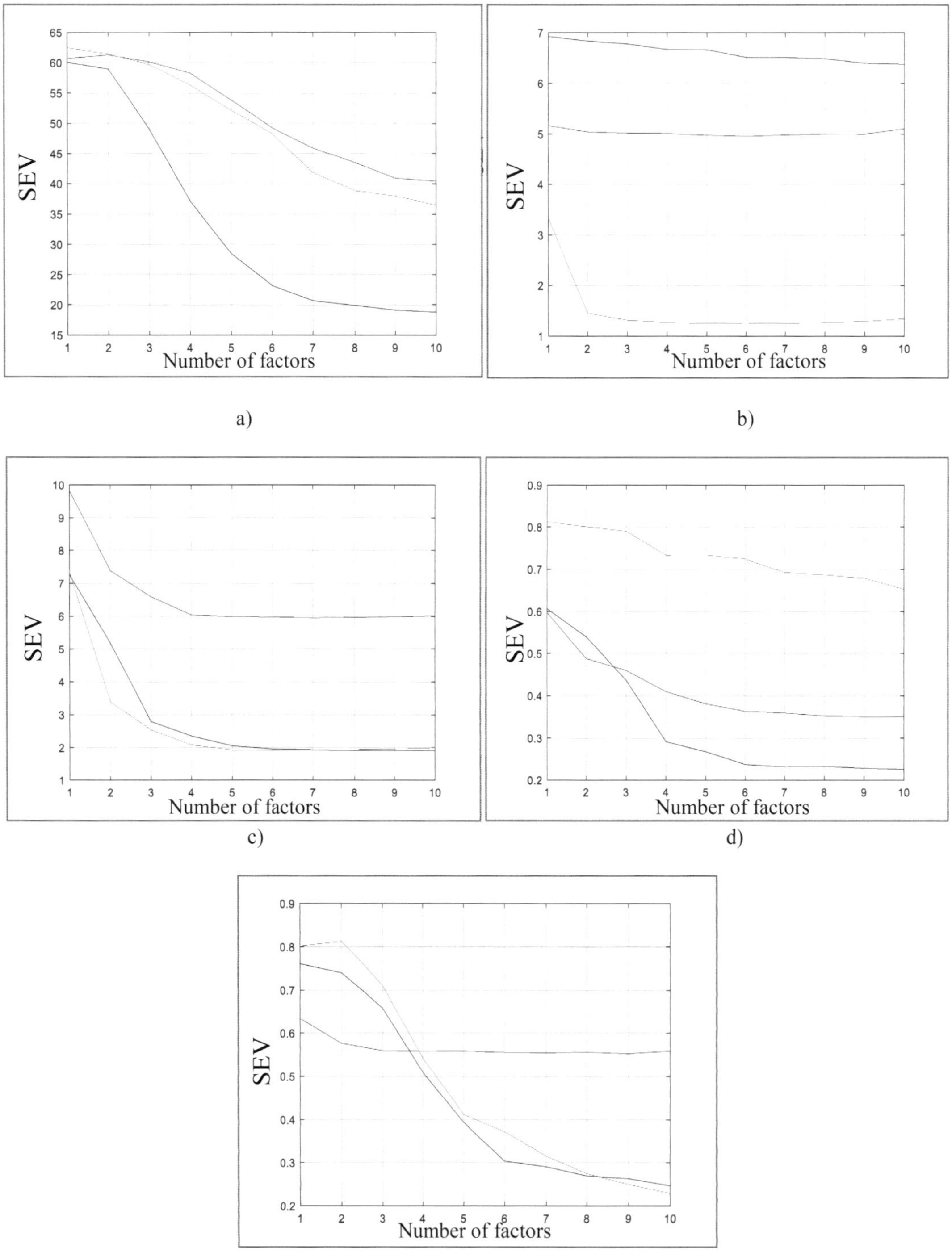

Fig. 3. Standard error of validation as a function of number of factors
a) for nitrogen, b) for phosphorus, c) for potassium, d) for humus, e) for acidity

	1-st derivative	2-nd derivative	MSC
nitrogen	40.4184 (10)	36.4910 (10)	18.7865 (10)
phosphorus	4.9550 (6)	1.2469 (6)	6.3757 (10)
potassium	5.9512 (7)	1.9233 (7)	1.9099 (9)
humus	0.3507 (10)	0.6531 (10)	0.2251 (10)
acidity	0.5526 (9)	0.2291 (10)	0.2465 (10)

Table 1. SEV and factors

5. Conclusion

The spectral characteristics of the soil could be used for soil agrochemical indicators prediction. The experimental results shows that method PLS and appropriate pre-processing procedures as 1-st derivative, 2-nd derivative and MSC could successfully define the soil indicators. For the agrochemical indicator nitrogen, the minimum SEV is achieved for MSC pre-processing procedure and 10 factors; for phosphorus - 2-nd derivative and 6 factors; for potassium - MSC and 9 factors; for humus - MSC and 10 factors; for acidity - 2-nd derivative and 10 factors. These results are of significant importance for the subsequent treatment of soil fields with a view to obtaining high yields and low production losses. The future work will be focus on models' development for prediction of the indicators and assessment of the accuracy and development of smart farmers information system.

6. Acknowledgments

This research is supported by the National Scientific Program "Intelligent Plant Breeding" of the Ministry of Education and Science, approved by Decision of the Council of Ministers No. 866 / 26.11.2020.

7. References

[1] Lehmann, J., Bossio, D.A., Kögel-Knabner, I. et al. (2020). The concept and future prospects of soil health. Nat Rev Earth Environ 1, https://doi.org/10.1038/s43017-020-0080-8, 544–553

[2] Aloui, S., Mazzoni, A., Elomri, A., Aouissi, J., Boufekane, A. & Zghibi A. (2023). A review of Soil and Water Assessment Tool (SWAT) studies of Mediterranean catchments: Applications, feasibility, and future directions, Journal of Environmental Management, Vol. 326, Part B, 116799, ISSN 0301-4797,

[3] Bhargava, A.; Bansal, A. (2021). Fruits and vegetables quality evaluation using computer vision: A review, Journal of King Saud University - Computer and Information Sciences, Vol., 33, Issue 3, 243-257, ISSN 1319-1578

[4] Remzi, S., Georgieva, Ts., Paskova, N., & Daskalov, P. (2019). Performance of web based microprocessor system for measurement of soil quality parameters, Proceedings of the 30th DAAAM International Symposium, pp.1097-1101, B. Katalinic (Ed.), Published by DAAAM International, ISBN 978-3-902734-22-8, ISSN 1726-9679, Vienna, Austria

[5] Georgieva, Ts., Remzi, S., Paskova, N., Stefanov, E., Sigrimis, N. & Daskalov, P. (2019). Research of the influence of external factors on the measurement of a basic soil quality parameter, Proceedings of the 30th DAAAM International Symposium, pp.1097-1101, B. Katalinic (Ed.), Published by DAAAM International, ISBN 978-3-02734-22-8, ISSN 1726-9679, Vienna, Austria

[6] Yilin, B., Meng, X., Ustin, S., Wang, X., Zhang, X., Liu, H. & Tang, H. (2020). Vis-SWIR spectral prediction model for soil organic matter with different grouping strategies, CATENA, Vol. 195, ISSN 0341-8162

[7] Thiele-Bruhn, S., Schloter, M., Wilke, B., Beaudette, L., Martin-Laurent, F., Cheviron, N., Mougin, C. & Römbke, J. Identification of new microbial functional standards for soil quality assessment, Soil, 6, 17–34

[8] Yang, M., Xu, D., Chen, S., Li, H. & Shi, Z. (2019). Evaluation of Machine Learning Approaches to Predict Soil Organic Matter and pH Using vis-NIR Spectra. Sensors, Vol. 19, 263. https://doi.org/10.3390/s19020263

DOI: 10.2507/34th.daaam.proceedings.041

USING THE QUALITY FUNCTION DEPLOYMENT METHOD IN THE DESIGN OF DIE CASTING CELLS

Stiliyan Nikolov, Reneta Dimitrova, Ivo Malakov & Velizar Zaharinov

This Publication has to be referred as: Nikolov, S[tiliyan]; Reneta, D[imitrova]; Malakov, I[vo] & Zaharinov, V[elizar] (2023). Using the Quality Function Deployment Method in the Design of Die Casting Cells, Proceedings of the 34th DAAAM International Symposium, pp.0312-0319, B. Katalinic (Ed.), Published by DAAAM International, ISBN 978-3-902734-41-9, ISSN 1726-9679, Vienna, Austria
DOI: 10.2507/34th.daaam.proceedings.041

Abstract

The present work proposes, by using the Quality Function Deployment (QFD) method, to account for customer requirements in the design of specialized cells for die casting. The proposed method allows for the evaluation of the client's requirements related to the degree of automation, working parameters, and the cost of the designed cell. Considering the specifics of machine casting, two QFD matrices have been developed and the expert evaluations necessary for their use have been determined. Application of the method allows precise definition of the design assignment and reduction of the risk of significant changes in the cell construction at later stages of the design and related costs.

Keywords: Design; Die Casting Cells; Quality Function Deployment Method; Industrial Automation

1. Introduction

In the industry, in the large-scale and mass production of parts of metals and alloys with a low melting temperature, the methods of machine casting under pressure, in metal moulds (Die) are mainly used [2]. For this purpose, special machines are used, which are designed to hold open and close the two halves of the mould, to keep the mould closed while the liquid metal is fed into the cavity and to provide the necessary pressure to fill the mould cavity of the mould with molten metal [1]. Industrial robots allow the operation of these machines to be fully automated by creating specialized cells for machine casting under pressure. The use of such cells improves the quality of manufactured products, safety, working conditions and reduces the cost of manufactured parts.

Currently, a number of companies offer machines for pressure casting [10], industrial robots for automating various stages of the operation of these machines, and specialized cells for the machine casting of high pressure [8], [9], [11]. This great variety poses the question of finding an optimal solution for the conditions of a specific company wishing to automate the production of details from metals and alloys with a low melting temperature. The present work proposes, by using the Quality Function Deployment (QFD) method, at the design stage of specialized cells for machine Die casting, to determine the requirements of the user and to select the main components of the cell, which will significantly reduce the time for design, manufacture, and commissioning of the cell.

2. Design of die casting cells

The main component in the structure of an automated Die casting cell is the Die casting machine, which performs the main process operation. The automation of the operation of this machine can include the following different levels:

- Partial automation of the operation of the machine - only some of the auxiliary operations related to the operation of the machine are automated: supply of molten metal to the machine, lubricating of the form, removal of the casting from the machine;
- Full automation of the machine's operation - all auxiliary operations related to the machine's operation are automated;
- Full automation of the operation of the machine with additional processing of the obtained casting - all auxiliary operations related to the operation of the machine are automated and inclusion in the cell of additional automated operations for processing the obtained casting such as: cooling, trimming (removal of the back scrap), control, etc.

Depending on the goals pursued with automating the operation of the Die casting machine, a cell for the Die casting may include a different combination of the components shown in Fig. 1.

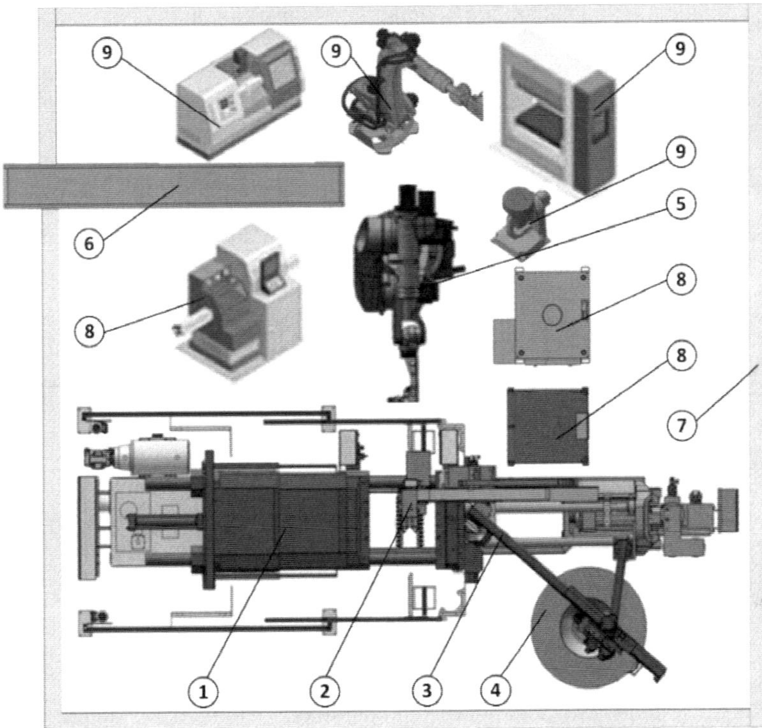

Fig. 1. Die casting cell

These components are:

1. Die casting machine
2. Sprayer robot for lubricating of Die - Sprayer
3. Robot for supplying molten metal - Feeder
4. Melting furnace
5. Robot for extracting the casting from the Die - Extractor
6. Conveyor for removing the casting from the cell
7. Protective fences
8. Components for primary processing of the resulting casting - cooling, trimming, control
9. Components for additional processing of the resulting casting - industrial robots, processing machines, reorienting devices, etc.

Many publications are devoted to methodologies and algorithms for the design of various products [3], [5] and cells for Die casting [4], [6], [7]. They examine the main stages of design, the relationship between these stages and the possibilities for optimizing the designed products. The main stage in the design of the cell is the selection of the necessary components, determining their location in space and the functional connections between them.

This requires a clear formulation of the tasks that will be solved with the help of the designed cell since changing the already selected components at a later stage of the design is associated with significant losses of time and money.

3. Using the quality function deployment method in the design of die casting cells

To solve the tasks related to the selection of the necessary components in the Die casting cell structure, the QFD method is proposed, which uses a series of matrices to define the customer's requirements to be embodied in the final product. The first proposed matrix Fig. 2 aims to determine what are the necessary components in the structure of a Die casting cell, depending on the level of automation that is desired by the customer.

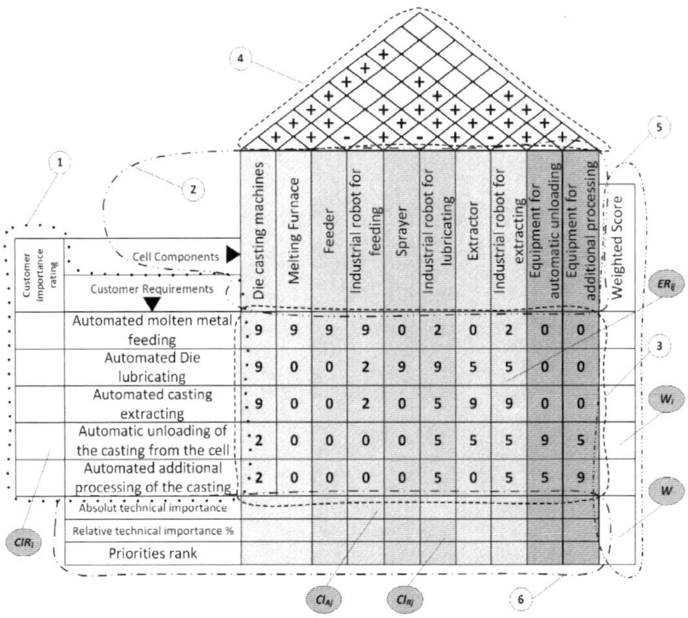

Fig. 2. The first proposed matrix

The individual sections in this matrix are:

- **Customer Requirements (1)** - all the options for automating the operation of the Die casting machine discussed above are recorded here. The customer gives an assessment of these possibilities, according to the desired degree of cell automation. Customer ratings range from ***5*** (mandatory automation of the relevant process) to ***0*** (no need to automate the relevant process).
- **Cell Components (2)** - here are written the groups of components that make up an automated Die casting cell, according to the scheme shown in Fig. 1.
- **Relationships between customer requirements and cell components (3)** - here an expert assessment of the importance of individual components to satisfy customer requirements is given (***ER***). A scale from ***9*** (the component is mandatory for the implementation of the corresponding function) to ***0*** (the component has no connection with the implementation of the corresponding function) was used.
- **Interaction between the components of the cell (4)** - here the defined relationships between the components necessary to realize the customer's requirements are shown. Possible connections are:
  (+) - positive - possible integration of the components and harmonization of the functions performed by them;
  (-) - negative - mutually exclusive components performing the same function;
  (**empty**) - lack of interaction between components.
- **Comparing components (5)** - here, based on the client's desire and the given expert assessments, the weight of realizing the degree of automation desired by the client with the necessary components is determined. Equation (1) is used for this purpose:

$$W_i = CIR_i * \sum_{j=1}^{10} ER_{ij} \qquad (1)$$

where:

W_i - Weight of realization of the ith degree of automation desired by the client;
CIR_i - Customer Importance Rang of the ith degree of automation desired by the customer;

ER_{ij} - Expert Rating of the importance of achieving the i^{-th} degree of automation desired by the customer with the j^{-th} component.

The total weight for achieving the degree of automation desired by the customer is determined by (2):

$$W = \sum_{i=1}^{5} W_i \tag{2}$$

where:

W - Total weight of realizing the degree of automation desired by the client;

W_i - Weight of realization of the i^{-th} degree of automation desired by the client.

- **Importance (6)** - here, based on the degree of automation desired by the client and the given expert assessments, the absolute and relative importance of individual components to achieve the client's wishes is determined. Equations (3) and (4) are used for this purpose:

$$CI_{Aj} = \sum_{i=1}^{5} CIR_i * ER_{ij} \tag{3}$$

where:

CI_{Ai} - Absolut technical importance of the j^{-th} component.

$$CI_{Rj} = \frac{CI_{Aj}}{W} * 100\% \tag{4}$$

where:

CI_{Ri} - Relative technical importance of the j^{-th} component.

Based on the relative importance value obtained (CI_{Ri}), the components of the cell are ranked, which allows to determine their importance for realizing the user's desire.

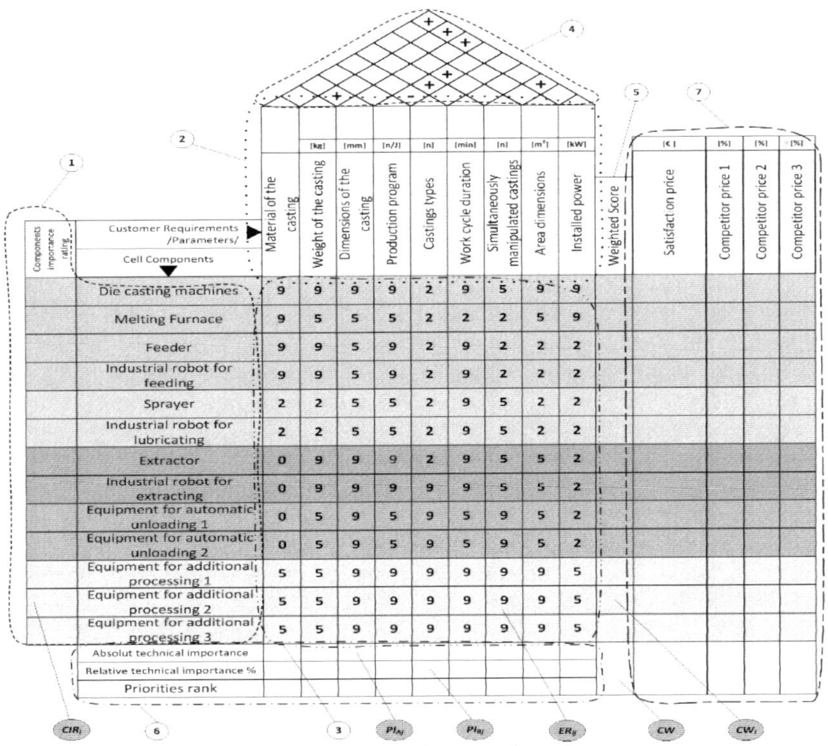

Fig. 3. The second proposed matrix

The second proposed matrix, Fig. 3, aims to determine the influence of the working parameters of the cell desired by the customer on the components of the cell determined with the help of the first matrix.

The individual sections in this matrix are:

- **Cell components (1)** - here are recorded the cell components determined using the first matrix. The assessment of the importance of these components (Component's importance rating)is according to their ranking in the first matrix (Priorities rank Fig. 1). Ratings range from ***5*** (component ranked first) to ***1*** (component ranked last).
- **Customer requirements (2)** - here are recorded the operating parameters of the cell desired by the customer, and their specific values are indicated.
- **Relationships between customer requirements and cell components (3)** - here an expert assessment of the relationship between the cell components and its operating parameters required by the customer is given (***ER***). A scale from ***9*** (the parameter determines the selection of the corresponding component) to ***0*** (the parameter does not influence the selection of the corresponding component) is used.
- **Connection between cell parameters (4)** - here are shown the defined connections between the working parameters of the cell required by the customer. The possible connections are:
  (+) - positive - changing of one of the parameters does not complicate the construction of the cell;
  (-) - negative - changing of one of the parameters leads to the complication of the cell construction;
  (**empty**) - lack of interaction between parameters.
- **Comparing components (5)** - here, based on the assessment of importance for the respective component and the given expert evaluations, the weight of the individual components is determined. Equation (5) is used for this purpose:

$$CW_i = CIR_i * \sum_{j=1}^{9} ER_{ij} \tag{5}$$

where:
CW_i - Weight of the i^{-th} component;
CIR_i - Customer Importance Rang of the i^{-th} component;
ER_{ij} - Expert Rating of the relationship of the i^{-th} component with the j^{-th} cell parameter
The total weight of the components to achieve the customer's wishes is determined by (6):

$$CW = \sum_{i=1}^{13} CW_i \tag{6}$$

where:
CW - Total weight of the components to realize the wishes of the client;
CW_i - Weight of the i^{-th} component.

- **Importance (6)** - here, based on the cell parameters desired by the client and the given expert assessments, the absolute and relative importance of the individual parameters on the cell construction is determined. Equations (7) and (8) are used for this purpose:

$$PI_{Aj} = \sum_{i=1}^{13} CIR_i * ER_{ij} \tag{7}$$

where:
PI_{Ai} - Absolut technical importance of the j^{-th} parameter.

$$PI_{Rj} = \frac{PI_{Aj}}{CW} * 100\% \tag{8}$$

where:
PI_{Ri} - Relative technical importance relative of the j^{-th} parameter.

- **Price (7)** - here, based on market research, an estimate is given of the price of marketed components necessary for the construction of the cell and the funds that the customer has foreseen for these components.

Based on the values obtained for the relative importance (PI_{Ri}), the parameters of the cell are ranked, which allows to determine their influence on the construction of the cell. Analyzing the data on the cost of the components currently offered on the market, necessary to build the cell and the determined influence of the working parameters of the cell desired by the customer, on its construction, the assignment for the design of the cell is prepared.

The proposed method allows, before the start of the actual design of die casting cells, to clarify the wishes of the client related to the degree of automation, working parameters and expected cost of the cell. The correct definition of the customer's wishes significantly reduces the risk of changes in the basic structure of the cell, at a later stage of the design process, and the associated costs.

4. Examples of using the quality function deployment method in the design of die casting cells

Exemplary applications of the proposed method for two cases of Die casting cells design with different degrees of automation are given below. The price data of the cell components used in the compilation of the matrices are illustrative and do not correspond to the actual ones.

4.1 Partial automation of the operation of Die casting machine

The first of the proposed matrices is shown in Fig. 4.a. Taking into account the customer's wishes for the automatic loading of molten metal into the machine and automatic lubricating of the Die, the components necessary for building the cell were determined and ranked.

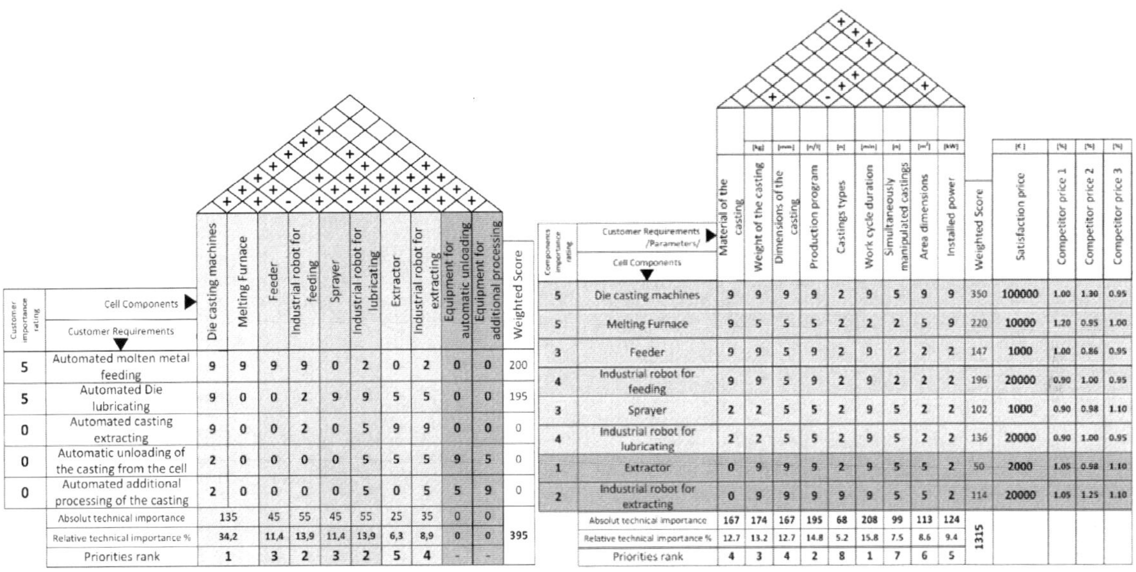

a)

Customer importance rating	Cell Components ▶ / Customer Requirements ▼	Die casting machines	Melting Furnace	Feeder	Industrial robot for feeding	Sprayer	Industrial robot for lubricating	Extractor	Industrial robot for extracting	Equipment for automatic unloading	Equipment for additional processing	Weighted Score
5	Automated molten metal feeding	9	9	9	9	0	2	0	2	0	0	200
5	Automated Die lubricating	9	0	0	2	9	9	5	5	0	0	195
0	Automated casting extracting	9	0	0	2	0	5	9	9	0	0	0
0	Automatic unloading of the casting from the cell	2	0	0	0	0	5	5	5	9	5	0
0	Automated additional processing of the casting	2	0	0	0	0	5	0	5	5	9	0
	Absolut technical importance	135		45	55	45	55	25	35	0	0	
	Relative technical importance %	34,2		11,4	13,9	11,4	13,9	6,3	8,9	0	0	395
	Priorities rank	1		3	2	3	2	5	4	-	-	

b)

Components importance rating	Customer Requirements /Parameters/ ▶ / Cell Components ▼	Material of the casting	Weight of the casting [kg]	Dimensions of the casting [mm]	Production program	Castings types [n]	Work cycle duration [min]	Simultaneously manipulated castings [n]	Area dimensions [m²]	Installed power [kW]	Weighted Score	Satisfaction price [€]	Competitor price 1 [%]	Competitor price 2 [%]	Competitor price 3 [%]
5	Die casting machines	9	9	9	9	2	9	5	9	9	350	100000	1.00	1.30	0.95
5	Melting Furnace	9	5	5	5	2	2	2	5	9	220	10000	1.20	0.95	1.00
3	Feeder	9	9	5	9	2	9	2	2	2	147	1000	1.00	0.86	0.95
4	Industrial robot for feeding	9	9	5	9	2	9	2	2	2	196	20000	0.90	1.00	0.95
3	Sprayer	2	2	5	5	2	9	5	2	2	102	1000	0.90	0.98	1.10
4	Industrial robot for lubricating	2	2	5	5	2	9	5	2	2	136	20000	0.90	1.00	0.95
1	Extractor	0	9	9	9	2	9	5	5	2	50	2000	1.05	0.98	1.10
2	Industrial robot for extracting	0	9	9	9	9	9	5	5	2	114	20000	1.05	1.25	1.10
	Absolut technical importance	167	174	167	195	68	208	99	113	124	1315				
	Relative technical importance %	12.7	13.2	12.7	14.8	5.2	15.8	7.5	8.6	9.4					
	Priorities rank	4	3	4	2	8	1	7	6	5					

Fig. 4. Example matrices in partial automation

The second of the proposed matrices is shown in Fig. 4.b. There the operating parameters of the cell desired by the client are shown. Taking into account the ranking of the components necessary for the construction of the cell, determined in the first matrix, the assessment of their importance is given (from ***5*** to ***1***). An estimate of the expected price of the cell is made. The obtained data show that, in the specific case, the defining parameters for the structure of the cell, desired by the user, are ***1*** - duration of the work cycle; ***2*** - production program and ***3*** - casting weight.

4.2 Full automation of the operation of Die casting machine with additional processing of the resulting cast

The first of the proposed matrices is shown in Fig.5.a. Taking into account the wishes of the client for full automation of the operation of the Die casting machine and the possibility of additional processing of the received castings, the components necessary for the construction of the cell have been determined and ranked. The second of the proposed matrices is shown in Fig.5.b. There the cell operating parameters desired by the client are shown. The additional components of the cell are defined, which will use to:

- automated removal of the casting from the cell - a conveyor for moving the castings to other cells for further processing and an industrial robot for moving the castings between the positions of the cell;
- -automated additional processing of the castings - cooling, trimming and control of the castings.

Taking into account the ranking of the components necessary for the construction of the cell, determined in the first matrix, the assessment of their importance is given (from ***5*** to ***1***). An estimate of the expected price of the cell is made. The obtained data show that, in the specific case, the determining parameters for the structure of the cell, desired by the user, are ***1*** - the size of the casting; ***2*** - the duration of the work cycle and ***3*** - the production program. Analyzing the obtained results, the client defines the design task by determining which of the mutually exclusive components performing the same function (matrix 1) should be included in the cell structure.

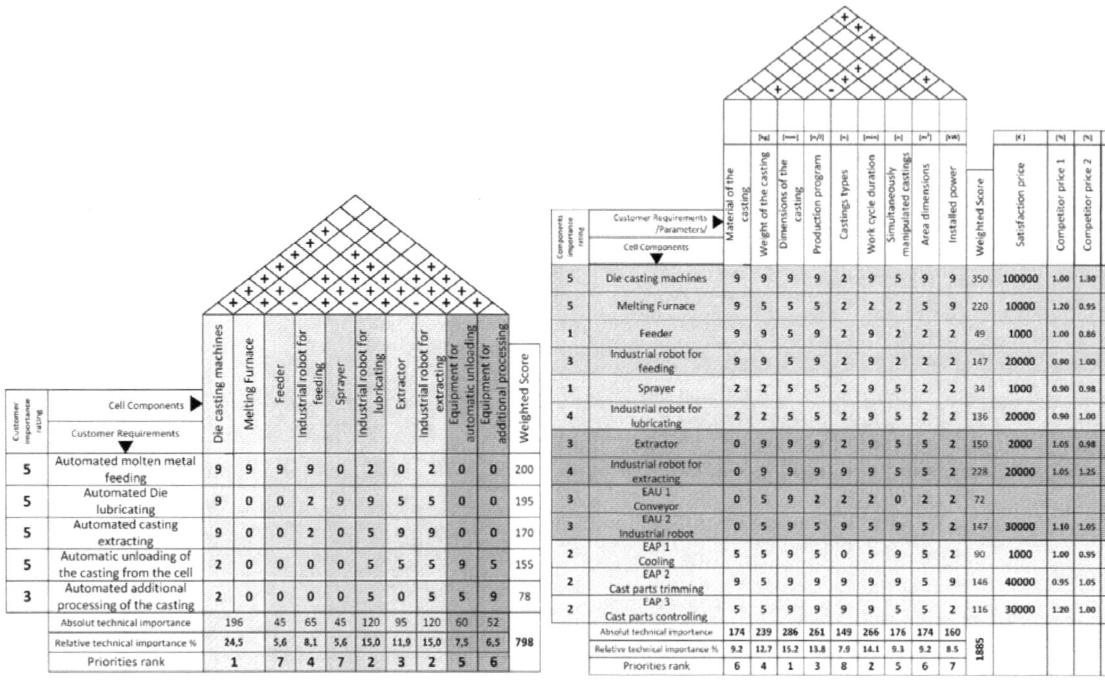

Fig. 5. Example matrices in full automation

In case of large discrepancies in the expected cost of the cell (components of the cell described in matrix 2 determine more than 70% of the cost of the cell) and the funds planned for investment by the customer, he can change the degree of automation or some of the parameters, to fulfil the financial restrictions.

5. Conclusions

The wide variety of currently available Die casting machines, industrial robots for automating various stages of the operation of these machines, and specialized cells for Die casting creates problems when choosing a solution for automating the production of metal and alloy parts with a low melting point, in the conditions of a specific company. The method proposed here, of using two QFD matrices, at the task definition stage for the design of specialized Die casting cells, allows the creation of a task that most precisely corresponds to the degree of automation desired by the customer and his financial opportunities, taking into account the variety of solutions offered on the market.

For the application of the proposed method, expert ratings of the relationships between the degree of automation, the operating parameters and the components that make up the cell are given. The interactions between the cell components and the relationships between the customer's desired cell parameters are defined. The expert ratings and interactions between the cell components are based on a study of publications related to the design and operation of Die casting cells.

The cell's components price data used in the work is illustrative, as the actual prices depend on the current state of the market and are tied to a specific time period of delivery of the desired component. The use of the proposed method significantly reduces the risk of changes in the basic structure of the designed Die casting cells and the associated costs, simultaneously reducing the time for designing, manufacturing and commissioning of the cell. It is possible to develop the method through the development of type matrices (the second matrix), in which various options for additional processing of the obtained cast can be offered, such as drilling holes, milling, marking, etc.

6. Acknowledgments

This work was supported by the European Regional Development Fund within the Operational Programme "Science and Education for Smart Growth 2014 - 2020" under the Project CoE "National center of mechatronics and clean technologies" BG05M2OP001-1.001-0008

7. References

[1] Groover. M.; (2013). Fundamentals of modern manufacturing materials, Processes, and systems, Fifth edition, John Wiley & Sons, Inc., IBSN 978-1-118-231463, USA

[2] Andresen. B.; (2004). Die casting engineering, A hydraulic, thermal and mechanical process, CRC Press, IBSN 978-0-203-99739-0, USA

[3] Kamberov, K.; Todorov, G.; Sofronov, Y.; Nikolov, N. (2021). Methodology for designing, manufacturing and integration of personalized spinal implants for surgical treatment of the cervical spine, AIP Conference Proceedings, ISSN:0094-243X, E-ISSN:1551-7616

[4] Zhang. J.; Fang. X.; (2017). Challenges and key technologies in robotic cell layout design and optimization, Journal of Mechanical Engineering Science, ISSN: 0022-2542, 2017, Vol. 231(15), DOI: 10.1177/0954406216642473, pp. 2912–2924

[5] Zaharinov. V.; Malakov. I.; Nikolov. S.; Dimtrova. R.; Stambolov. G.; (2017). Classification of parts used in mechatronic products and produced by permanent-mold casting methods, 9th International Scientific Conference "TechSys 2020" – Engineering, Technologies and Systems 14-16 May 2020, Plovdiv, Bulgaria

[6] Zaharinov, V.; Malakov, I.; Nikolov, S.; Dimitrova, R.; Stambolov, G.; (2020). Choosing an optimal size range of pneumatically actuated linear modules for sprayer robots, Annals of DAAAM and Proceedings of the International DAAAM Symposium, 2020, 31(1), pp. 159–167

[7] Lim Z.; Ponnambalam G.; Izui K.; (2016). Nature inspired algorithms to optimize robot work cell layouts, Applied Soft Computing 49 (2016), ISSN: 1568-4946, pp. 570–589

[8] http://www.spesima.eu, Automation of Horizontal Die Casting Machines, Accessed on: 2023-05-10

[9] https://www.yizumi.com/en, DM Series Cold Chamber Die Casting Machine, Accessed on: 2023-05-10

[10] https://www.frech.com, Hot Chamber Die Casting Technology, Accessed on: 2023-05-10

[11] https://www.kuka.com, KUKA Casting Solutions Die Casting, Accessed on: 2023-05-10

DOI: 10.2507/34th.daaam.proceedings.042

ADDITIVE MANUFACTURING OF THIN-WALLED INCONEL 718 BRACKETS FOR EXTREME CONDITIONS

Jindrich Sykora & Lubos Kroft

This Publication has to be referred as: Jindrich, S[ykora] & Lubos, K[roft] (2023). Additive Manufacturing of Thin-Walled Inconel 718 Brackets for Extreme Conditions, Proceedings of the 34th DAAAM International Symposium, pp.0320-0323, B. Katalinic (Ed.), Published by DAAAM International, ISBN 978-3-902734-41-9, ISSN 1726-9679, Vienna, Austria
DOI: 10.2507/34th.daaam.proceedings.042

Abstract

The demand for components capable of withstanding extreme conditions, including high and low temperatures and corrosive environments, continues to grow. In the aerospace industry, where weight is a critical factor, there is a pressing need for lightweight components made from materials like Inconel 718. Furthermore, there is high economic as well as ecological cost associated with subtractive manufacturing of light components from advanced materials. The high material removal rates are taxing on the tools and a significant amount of raw material is unutilized. This paper presents an approach to manufacturing thin-walled components from Inconel 718 using additive manufacturing techniques, which greatly reduces material waste. Additionally, with this approach, no complex fixturing is needed. Variants of the additive manufacturing approach are proposed and compared to subtractive manufacturing. The effectiveness of 3D printing is among other things dependent on the part orientation, therefore, several print orientations are compared. Finally, a sample component is manufactured using the DMLS method. The load case is simulated, and the simulation is verified in a use case. In conclusion, we propose an innovative approach for manufacturing thin-walled aerospace components from Inconel 718 with reduced reliance on skilled labour. This approach offers significant advantages in terms of material efficiency and cost-effectiveness.

Keywords: Inconel 718; additive manufacturing; 3D printing; aerospace components

1. Introduction

The highly demanding applications in the aerospace industry require the production of components that can withstand extreme temperature variations. These aerospace components must be lightweight for fuel efficiency while possessing exceptional mechanical properties to ensure safety and performance. Currently, materials from the ISO S group are commonly used in the construction of such components. In scenarios that demand high mechanical stress resistance across fluctuating temperatures and superior creep resistance, nickel superalloys like Inconel 718 are the preferred choice. [1], [2]. However, machining Inconel 718 presents a significant challenge. It places considerable stress on both the machine tools and cutting tools due to the immense forces required for material removal. Tool wear rates are much higher compared to materials from other ISO groups, necessitating lower cutting speeds and resulting in significantly reduced material removal rates.

This combination leads to substantial cost escalation in the manufacturing of components from ISO S materials. Moreover, these costs are not just financial; they also have environmental implications, as the subtractive manufacturing of such parts consumes a significant amount of CO^2. Additionally, aerospace components often require high material removal rates to achieve the desired weight savings. [3], [4]

This is where additive manufacturing comes in as a transformative solution for aerospace components. Additive manufacturing has the advantage of using materials almost entirely in the final part, thereby eliminating wastage. Furthermore, it completely eliminates tooling costs, reducing the overall expense of production. Various methods can be employed to manufacture specimens from Inconel 718, with fused deposition modelling (FDM) and direct metal laser sintering (DMLS) being among the most prevalent. FDM requires debinding and sintering post-3D printing but provides slightly lower resolution compared to DMLS. DMLS, on the other hand, offers more precision, with layer heights as low as 0.02 mm, and delivers detailed parts. Although the DMLS process generally results in a final part, a heat treatment step is usually carried out after the build process. [5], [6]

This study focuses on the process of manufacturing a thin-walled bracket component. The primary objectives include achieving maximum weight reduction while maintaining performance and ensuring operational safety. Additionally, the bracket must endure extreme temperature variations and exhibit resistance to corrosion. In pursuit of these goals, Inconel 718 was selected as the material, and through an additive manufacturing process, we explored the production of this aerospace component.

2. Materials and methods

The process of additive manufacturing for the specimen utilized the EOS M290 DMLS metal 3D printing machine. Inconel 718 powder, provided by EOS GmbH, was the chosen build material. A combination of Materialise Magics and EOS Print software facilitated the print date preparation. Post 3D printing and heat treatment, the material's yield strength stands at a minimum of 1145 MPa. Initially, the specimen underwent solution annealing, which involved a one-hour exposure to a temperature of 954°C (1750°F), followed by an argon-cooled environment. Subsequently, the specimen experienced ageing treatment (Step 2), entailing an 8-hour hold at 718°C (1325°F). The next phase included furnace cooling to 621°C (1150°F) and a subsequent 18-hour maintenance at this temperature for total precipitation. The final phase incorporated cooling the specimen to room temperature in an argon atmosphere. [7]

Properties	Values
Yield strength – Rp0.2 [MPa]	1145
Tensile strength – Rm [MPa]	1375
Hardness, HRC	47

Table 1. Mechanical properties of 3D printed and heat-treated Inconel 718 sample in the vertical direction. [7]

The component's design and load case simulation were conducted using Fusion 360 software. Since the part was designed for additive manufacturing, it includes many features that would be difficult to produce using other methods. The comparison between additive manufacturing and machining was based on the speeds and feeds recommended for machining this material, as per the tooling catalogue from Iscar. As the final step, the component underwent testing in one of its intended use cases. It was subjected to a 6 Kg payload at room temperature and maintained under this load for 180 hours. Two basic manufacturing methods, additive manufacturing and subtractive manufacturing (milling) are considered in this research.

3. Results and discussion

The subtractive manufacturing method was considered first, utilizing a raw material block of Inconel 718 with a 1 mm allowance on three sides and 4 mm on the last side. This setup allows the part to be held in a standard machining vice. The machining time was estimated at 28 minutes, based on the cutting conditions from the tool manufacturer. A significant disadvantage of this process is the high material removal rate. Using the raw material mentioned above results in the removal of 61,591 mm^3 of material, equating to a 97.17% material removal rate.

It is important to note that the component design heavily relies on additive manufacturing. Consequently, some features are not ideal for milling and would require rework. Additionally, a second operation would likely involve complex fixturing, and the extensive material removal could complicate the machining process due to the cutting forces generated by milling. The wall thickness is designed to be 1 mm. [8]

For the DMLS additive manufacturing process, the specimen must be connected to the steel build platform, either directly or using support structures. The latter is chosen here to minimize raw material use. Two orientations were considered (See Fig. 1): orientation a) requires 758 mm^3 of support material, while orientation b) needs only 595 mm^3. Despite using the least amount of support material, functional surfaces must also be considered (critical surfaces highlighted in Fig. 1). Support structure removal is labour-intensive, and achieving high-quality surfaces can be challenging. For these reasons, orientation a) was deemed more suitable and used.

The additive manufacturing process results in 758 mm3 of material wasted in support structures, while the subtractive process removes 61,591 mm3 of material as chips. From the standpoint of material waste, the additive approach is superior.

Energy consumption is closely related to the carbon footprint, a crucial consideration in modern manufacturing. A significant portion of energy consumption in Inconel 718 manufacturing is associated with raw material production. The raw powder for additive manufacturing requires additional atomization of the bulk material, using more energy. The embodied energy of bulk Inconel 718 is approximately 321 MJ/kg. The additive manufacturing process adds approximately 55.6 MJ/kg for atomization. Therefore, raw Inconel 718 powder for 3D printing requires approximately 376.6 MJ/kg of energy. [9] The energy needed for the manufacturing process itself was not considered here, as there would be significant differences based on the device used. The energy consumption of machining centres can vary by orders of magnitude. The total energy consumption for the raw material for the subtractively manufactured specimen would be 167.76 MJ, while the additively manufactured part would require 7.79 MJ of energy.

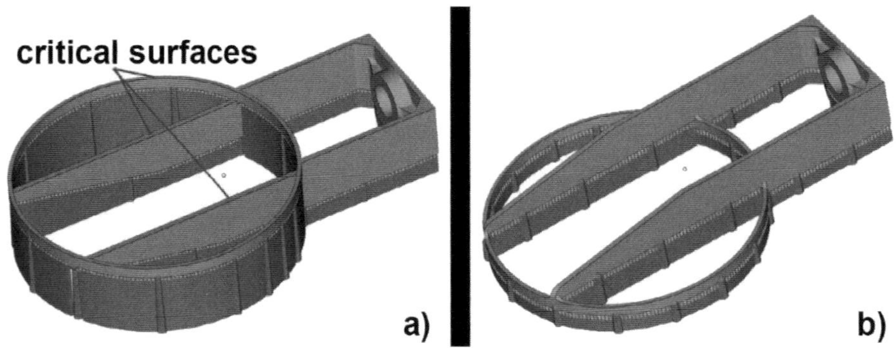

Fig. 1. Possible sample orientations with blue support structures

The integrity of the thin-walled design was rigorously evaluated using a static stress simulation. This step was crucial to ensure that the design could withstand the operational conditions it was intended for. The load case for the simulation was crafted based on the specific application requirements, where the component was expected to handle a 6 Kg cylindrical payload. To accurately mimic the real-world conditions, the bracket's connection to the rest of the assembly was also factored into the simulation. This connection was achieved through a single M6 bolt. The forces that this bolt would experience and transfer to the bracket during the operation were crucial parameters in the simulation. To maintain consistency and reliability in the results, a force of 1000 N was applied in the simulation. This value was derived from the ISO standard for a 10.9 class M6 bolt, ensuring that the simulation adhered to widely accepted engineering norms.

The results of the simulation indicate that the design was capable of supporting the specified 6 Kg payload. Moreover, the design demonstrated a minimum safety factor of 13.78. This safety factor, a measure of the structural capacity of the design beyond the expected loads, provided assurance that the component could reliably perform its function even under unforeseen dynamic events or slight overloads. In summary, the expanded analysis through static stress simulation not only validated the design's capability to handle the specified payload but also provided confidence in its structural integrity and safety under various conditions.

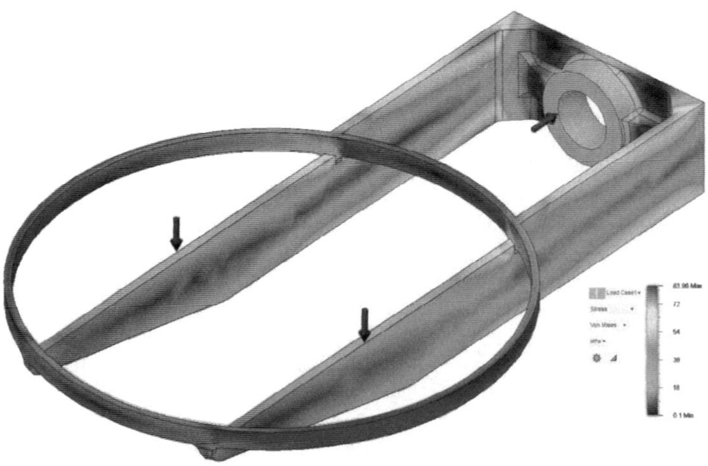

Fig. 2. Simulation of the sample load case

4. Conclusion

To conclude, a comparative analysis between additive and subtractive manufacturing processes was conducted for the fabrication of an aerospace bracket made from Inconel 718. The additive manufacturing approach was found to be superior in terms of energy consumption, requiring only 4.64% of the energy needed for milling. However, it's important to note that raw powder for additive manufacturing has about 14.76% higher embodied energy than the material for subtractive manufacturing. The energy savings for a general sample would be highly dependent on the material removal rate required. For future research, thorough mechanical testing of the specimen must be conducted, with a particular focus on investigating the creep behaviour of the part. When considering the carbon footprint of a manufacturing process, various factors such as the geographical location of the manufactured raw materials, energy sources, and more need to be considered. A detailed future study focused on this topic is planned to gain a comprehensive understanding of the carbon footprint.

5. Acknowledgments

This article has been prepared under the project SGS-2022-007—Research and Development for Innovation in Engineering Technology—Machining Technology IV.

6. References

[1] H. Qi, (2012), 'Review of INCONEL 718 alloy: Its history, properties, processing and developing substitutes', Cailiao GongchengJournal Mater. Eng., pp. 92–100.

[2] R. Schafrik, D. D. Ward, and J. Groh, (2001), 'Application of Alloy 718 in GE Aircraft Engines: Past, Present and Next Five Years', doi: 10.7449/2001/SUPERALLOYS_2001_1_11.

[3] I. A. Choudhury and M. A. El-Baradie, (1998), 'Machinability of nickel-base super alloys: a general review', J. Mater. Process. Technol., vol. 77, no. 1, pp. 278–284, doi: 10.1016/S0924-0136(97)00429-9.

[4] A. Celaya et al., (2019), 'Influence of cutting edge radius on tool life in milling Inconel 718', presented at the AIP Conference Proceedings, p. 080019. doi: 10.1063/1.5112627.

[5] O. Bodur, V. Stepanek, E. Walcher, and N. Durakbasa, (2020), 'Precision in Additive Manufacturing, Optimization and Evaluation of the Accuracy of 3D Printer based on GPS System', Annals of DAAAM for 2020 & Proceedings of the 31st International DAAAM Symposium , pp. 0963–0972. doi: 10.2507/31st.daaam.proceedings.134.

[6] I. Zetková, L. Kučerová, M. Zetek, J. Cesanek, P. Hanzl, and M. Dana, (2017), 'Evaluation of Metal Powder for Additive Manufacturing of Margaring Steel', Annals of DAAAM for 2017 & Proceedings of the 28th International DAAAM Symposium, pp. 0410–0416. doi: 10.2507/28th.daaam.proceedings.057.

[7] 'Nickel Alloys for Industrial 3D Printing | EOS', (2023), [Online], Available: https://www.eos.info/en/3d-printing-materials/metals/nickel-alloys

[8] G. Sridhar and P. Ramesh Babu, (2018), 'Effect of material removal rate on machining distortion of thin-wall thin-floor components', Mater. Today Proc., vol. 5, no. 13, Part 3, pp. 27018–27027, 2018, doi: 10.1016/j.matpr.2018.09.005.

[9] C. Fredriksson, (2019), 'Sustainability of metal powder additive manufacturing', Procedia Manuf., vol. 33, pp. 139–144, doi: 10.1016/j.promfg.2019.04.018.

DOI: 10.2507/34th.daaam.proceedings.043

ANALYSIS AND MODERNIZATION OF THE DUAL MANIPULATOR OF SERVICE ROBOTS

Valentin Pryanichnikov, Mariia Soloveva,
Vadim Chernyshev & Yaroslav Kalinin

This Publication has to be referred as: Pryanichnikov, V[alentin]; Mariia, S[oloveva]; Chernyshev, V[adim] & Kalinin, Y[aroslav] (2023). Analysis and Modernization of the Dual Manipulator of Service Robots, Proceedings of the 34th DAAAM International Symposium, pp.0324-0329, B. Katalinic (Ed.), Published by DAAAM International, ISBN 978-3-902734-41-9, ISSN 1726-9679, Vienna, Austria
DOI: 10.2507/34th.daaam.proceedings.043

Abstract

The purpose of the study is to investigate the method of modernization of the manipulator of the AMUR-307 service robot in order to increase its load capacity. The uArm robot was chosen as a prototype for structural solutions. A new scheme of a two-link parallel-plane manipulator was proposed, which orients the end effector of the manipulator in a horizontal position. A 3D design of the upgraded manipulator for the AMUR-307 service robot was created. Based on the analysis of the resulting 3D model, the workspace of the manipulator was determined and a kinematic diagram of the manipulator's link movements was constructed. The use of a cyclo gear box in the rotation drive of the manipulator's links instead of a two-stage spur gear has allowed for an increase in torque and a reduction in the drive's dimensions. The manipulator is also supposed to be used on crawler and walking robots.

Keywords: Service mobile robots; parallel-plane kinematics of manipulators; cyclo gear box & parallelogram mechanisms and it's 3D workspace.

1. Introduction

Keldysh Institute of Applied Mathematics with partners is engaged in the development of service robots to assist medical personnel in moving objects between rooms within a single building. The developed mobile service crawler robot, AMUR-307, can capture and hold objects weighing 5-7 kg. The manipulator's design consists of two rigidly connected manipulators that have parallel-plane kinematics (see Figure 1).

The purpose of the study is to investigate the method of upgrading the manipulator of the service robot "AMUR-307" in order to increase its load capacity to the specified parameters in the technical requirements. Identify the workspace of the manipulator.

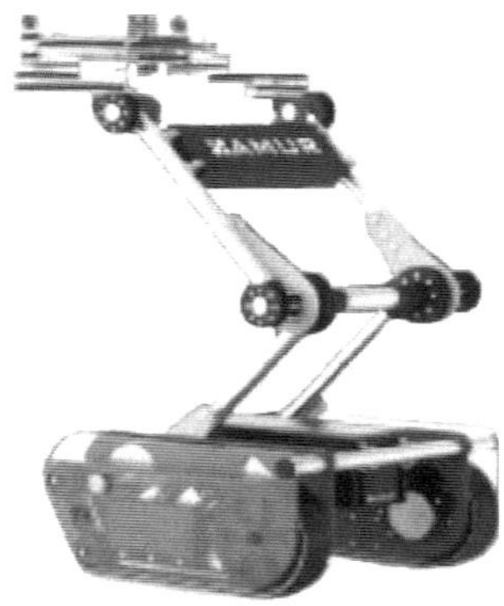

Fig. 1. 3D model of the AMUR-307 service robot

2. Description of kinematics

The manipulator has 2 links, which are swivels with angular movements [3]. We use an angular coordinate system. This design is compact due to the ability to fold almost completely without protruding beyond the dimensions of the robot base. The lower link performs planar angular displacement using a drive installed in the robot's base. This drive consists of a worm gear motor connected to a two-stage spur gear. The upper link performs angular displacement using a cable transmission connected to a similar drive installed in the robot's base. Another drive is required for the angular displacement of the manipulator's gripping device.

Since the cable transmission in the considered manipulator is open and the distance from one pulley to another is about half a meter, it poses a danger to the user. Based on the identified shortcomings, a search for existing more advanced analogues was carried out. As a result of the search, a two-link robot manipulator uArm, produced by the Chinese company Ufactory, was identified (see Figure 2).

Fig. 2. uArm Swift Pro robot manipulator as a prototype.

The feature of this manipulator's design is that the drives of the first and second links are located in the robot's base. The driving motion is transmitted through a linkage, that connects the drive to the link. This mechanism is similar to a parallelogram. This allows the use of a less powerful drive compared to the option, when the drive is installed in the link's hinge. The working tool of the manipulator is connected to the hinge of the first and second links by a linkage fixed to the base. This allows the working tool always maintain a horizontal position and eliminates the need for an additional drive with angular displacement.

3. Modernization of the manipulator

Based on this manipulator, an upgraded design was developed (see Figure 3). This manipulator consists of two linked manipulators. All drives of these manipulator links are installed in the robot's base to reduce its weight and increase its load capacity. The lower link is connected directly to the drive. The upper link is connected to the drive through a linkage. The described mechanism represents a set of parallelogram mechanisms (see Figure 4). The working tool of the manipulator is connected by two linkages, one end of which is connected to the triangle through hinges. The other end of the linkages is connected to the base and the working tool through hinges. In this case, the working tool always maintains a position close to horizontal.

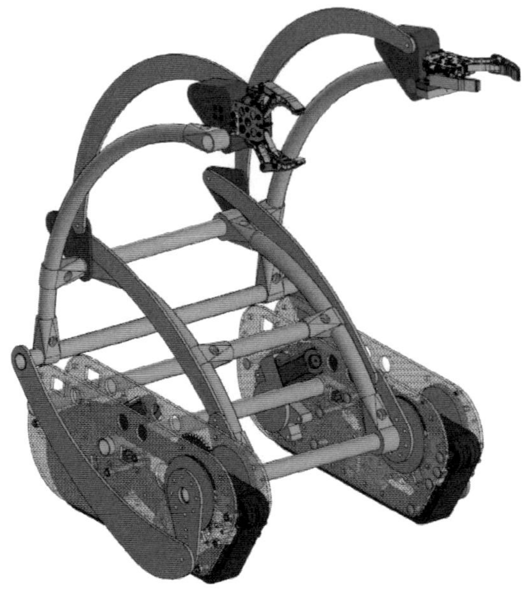

Fig. 3. 3D model of the upgraded design of the AMUR-307 service robot

To install objects with the manipulator's working tool for subsequent transportation, a table-cargo platform was added to the design. It contains another parallelogram mechanism, indicated in red in Figure 4. Thanks to this mechanism, the table-cargo platform remains in a horizontal position regardless of the manipulator's position. The inclined sides of the parallelogram form the lower link of the manipulator and the rear lever.

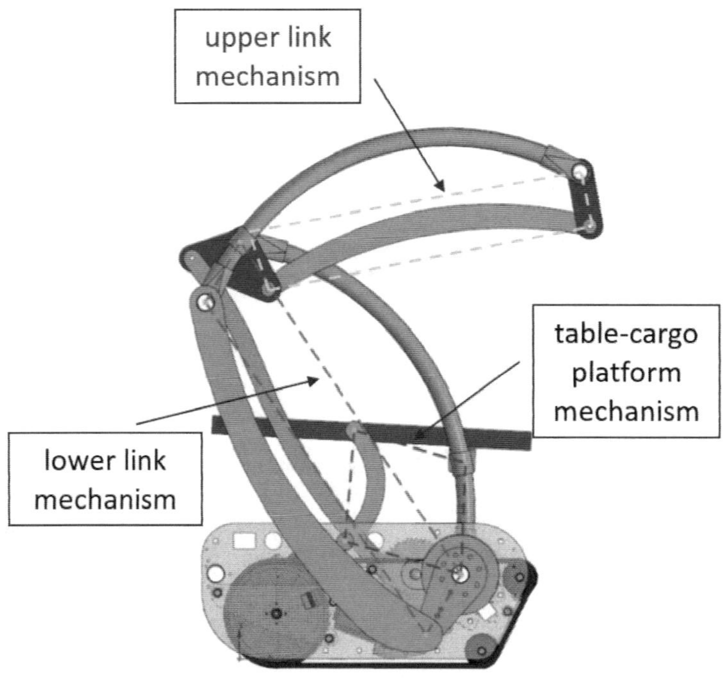

Fig. 4. Parallelogram mechanisms in the upgraded design of the AMUR-307 service robot

The movement of the upgraded three-dimensional model of the manipulator was analyzed using the KOMPAS 3D modeling software, and the extreme points of the manipulator's workspace were determined. By connecting them, the workspace area was constructed. The resulting kinematic diagram of the manipulator's links movements was constructed is shown in Figure 5.

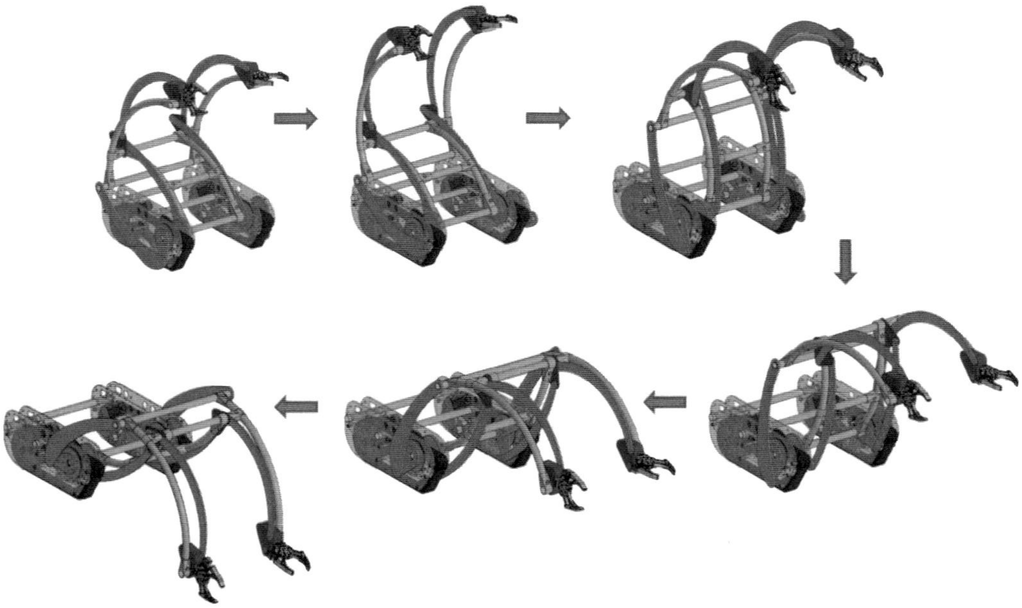

Fig. 5. Kinematic diagram of the manipulator's links movements of the AMUR-307 service robot

The drive of the lower link has the ability to rotate 360 degrees. This rotation mode allows the robot to perform stepping movements, which can be an additional function for overcoming obstacles during the movement of the crawler chassis. Figure 6 shows a visualization of the workspace of the manipulator obtained using animation of a 3D model.

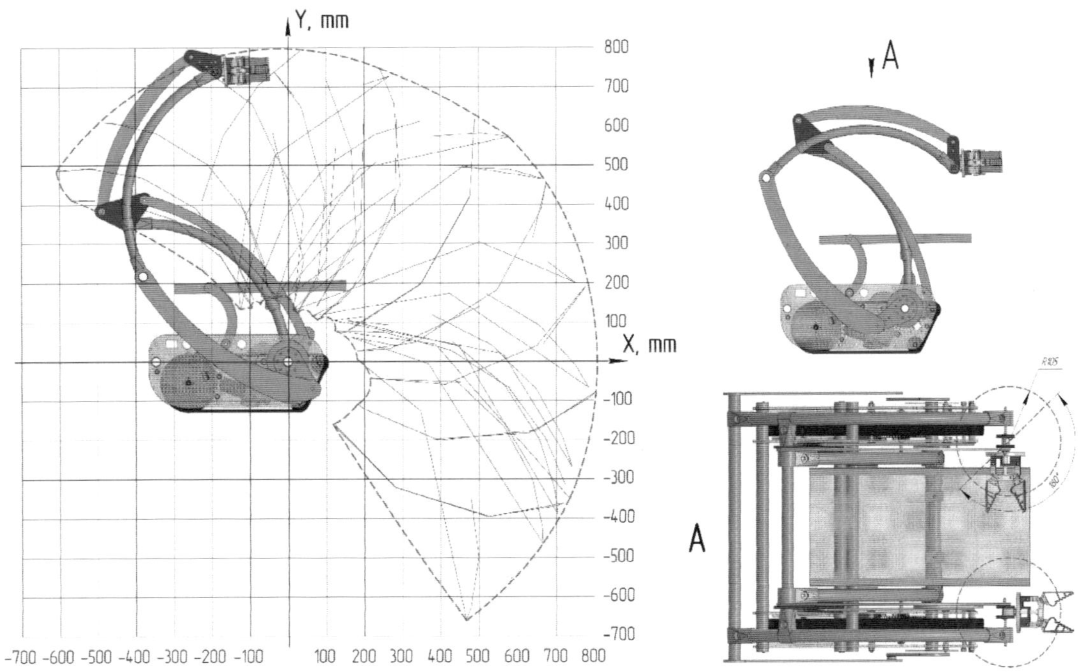

Fig. 6. Workspace of the AMUR-307 service robot manipulator

The obtained workspace of the manipulator is sufficient for performing the necessary technological operations by the AMUR-307 service robot.

4. Research of the manipulator drives

The existing design of the AMUR-307 service robot manipulator included worm gear motor connected to two-stage spur gear as the drives of the manipulator links (see Figure 7).

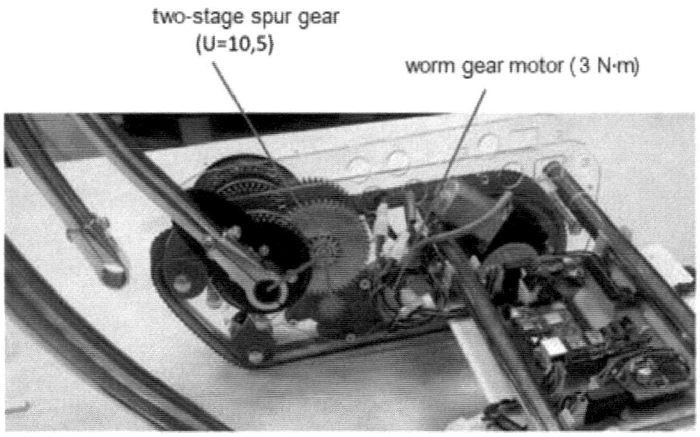

Fig. 7. Design of the drive for the lower link of the AMUR-307 robot

The nominal torque of the worm gear motor is 3 N·m. The two-stage spur gear has a gear ratio U = 10.5. With these drive parameters, without taking into account the safety factor, the total torque of the entire drive is equal to:

$$M = U \cdot M(w.g.) = 30.5 \text{ N} \cdot \text{m}$$

The required torque for one of the connected manipulators with a load on the entire manipulator of 7 kg is 84.5 N·m. With the main parameters of the manipulator, where the safety factor is 1.3, the load mass is 3.5 kg, the manipulator mass is 3 kg, and the length of the first and second links of the manipulator is 0.4 m and the length of the manipulator's working tool is 0.2 m. This drive is not powerful enough for the declared characteristics of the AMUR-307 service robot. Taking into account the conclusions obtained from earlier publications [1] and [2], it was decided to replace the two-stage spur gear with a modified cyclo gear box with a gear ratio of U = 29. This drive has a torque of 87 N·m, more compact dimensions, and a lower mass. The same manipulator is also suitable for installation on a larger crawler robot (120 kg) [8], as well as on an underwater walking apparatus [9], [3], [4], [5], [6] Fig. 8. The control and planning of the trajectories of such robots is carried out on the basis of earlier developments [10], [11], [12], [13], [14], [15].

Fig. 8. Crawler and underwater walking robots

The next step we plan to manufacture a model of a modified manipulator using the use of additive technologies, such as 3D printing, laser cutting, CNC milling. To do this, it is necessary to perform strength calculations of the structure and the development of the gripper. The study was partially funded by the grant from the Russian Science Foundation № 22-21-20115 and with KIAM support.

5. Conclusion

A three-dimensional model of the modernized manipulator was built, which made it possible to determine its working area. The obtained workspace of the manipulator is sufficient for performing the necessary technological operations by the AMUR-307 service robot. The upgraded kinematics and design of the manipulators made it possible to significantly improve such indicators as the load capacity of the manipulators - up to 30-40% of the total weight of the robot. The payload capacity has increased due to all the joint drives being installed in the robot's base.

The use of cyclo gear box instead of a gearbox with a two-stage spur gear has allowed for an increase in the gear ratio due to its smaller size. It became possible to capture large-sized objects with two grips at the same time for further loading onto a built-in table-cargo platform.

6. References

[1] Kudryavcev V.N. (1966). Epicyclic gears. Mashinostroenie Publishers, Leningrad branch. 307 p. Kudryavcev V.N. Planetarny`e peredachi / – L.: Mashinostroenie, 1966. – 307 s. – In Russian.

[2] M.D. Solovyeva, V.D. Shchepin. Improving the technological process of manufacturing cyclo gearboxes using laser processing. Scientific progress - creativity of young people: materials of the XI international youth scientific conference on natural and technical disciplines (Yoshkar-Ola, April 22-23, 2016): 4 parts redkol.: D. V. Ivanov [and etc.]. – Yoshkar-Ola: Volga State University of Technology, 2016. – P. 2. – 264 p. / M.D. Solovyeva, V.D. Shchepin / Sovershenstvovanie tekhnologicheskogo protsessa izgotovleniya planetarno-tsevochnykh reduktorov (PCR) s primeneniem metodov lezviynoy obrabotki // Nauchnomu progressu – tvorchestvo molodykh: materialy XI mezhdunarodnoy molodezhnoy nauchnoy konferentsii po estestvennonauchnym i tekhnicheskim distsiplinam (Yoshkar-Ola, aprel 22-23, 2016): 4 chasti redkol.: D. V. Ivanov [i dr.]. – Yoshkar-Ola: Povolzhskij gosudarstvennyj tekhnologicheskij universitet, 2016. – Ch. 2. – 264 s. – In Russian.

[3] Pryanichnikov V.E., Chernyshev V.V., Arykantsev V.V., Aryskin A.A., Eprikov S.R., Ksenzenko A.Ya., Petrakov M.S. (2018). Enhancing the Functionality of the Groups of Autonomous Underwater Robots, Proceedings of the 29th DAAAM International Symposium, pp.1319-1325, B. Katalinic (Ed.), Published by DAAAM International, ISBN 978-3-902734-20-4, ISSN 1726-9679, Vienna, Austria DOI: 10.2507/29th.daaam.proceedings.190.

[4] V.V. Chernyshev, V.V. Arykantsev, A.E. Gavrilov et al, Design and underwater tests of subsea walking hexapod MAK-1, in Proc. of the ASME 2016 35th Int. Conf. on Ocean, Offshore and Arctic Engineering (OMAE 2016), (Busan, Korea, 2016).

[5] V.V. Chernyshev, V.V. Arykantsev, Ya.V. Kalinin, «Underwater Tests of The Walking Robot MAK1», Human-Centric Robotics. Proceedings of the 20th International Conference on CLAWAR 2017, Professional Engineering Publishing, 2017, pp. 571-578.

[6] Bogdanovich A.V., Kirsanov K.B., Khelemendik R.V., Pryanichnikov V.E., Prysev E.A. (2018). Development of Technology for Manufacturing Hardware and Intelligent Levels of the Service Autonomous Mobile Robot AMUR-307, Proceedings of the 29th DAAAM International Symposium, pp.1313-1318, B. Katalinic (Ed.), Published by DAAAM International, ISBN 978-3-902734-20-4, ISSN 1726-9679, Vienna, Austria DOI: 10.2507/29th.daaam.proceedings.189

[7] S.Y. Yoo, B.H. Jun, H. Shim, Design of static gait algorithm for hexapod subsea walking robot: Crabster, Transactions of the Korean Society of Mechanical Engineers A, Vol. 38, Issue 9, (2014), pp. 989-997.

[8] Chernyshev, V.V. Traction properties of walking machines on underwater soils with a low bearing ability / V.V. Chernyshev, A.E. Gavrilov // Minerals of the Ocean – 7 & Deep-Sea Minerals and Mining – 4: abstracts of Int. Conf. / VNIIOkeangeologia. St. Petersburg, 2014. P.21-24.

[9] Chernyshev V.V. Design and underwater tests of subsea walking hexapod MAK-1 / V.V. Chernyshev, V.V. Arykantsev, A.E. Gavrilov et al. // Proc. of the ASME 2016 35th Int.Conf. on Ocean, Offshore and Arctic Engineering OMAE 2016, Busan, 2016. – 9 p.

[10] Davydov D.V., Eprikov S.R., Kirsanov K.B., Pryanichnikov V.E. (2017). Service Robots Integrating Software and Remote Reprogramming, Proceedings of the 28th DAAAM International Symposium, pp.1234-1240, B. Katalinic (Ed.), Published by DAAAM International, ISBN 978-3-902734-11-2, ISSN 1726-9679, Vienna, Austria DOI: 10.2507/28th.daaam.proceedings.172.

[11] Pryanichnikov V. E. Razrabotka distancionnyh sensorno-upravlyayushchih sistem, ispol'zuyushchih svyortki izobrazhenij / Zapis' i vosproizvedenie ob"yomnyh izobrazhenij v kinematografe, nauke, obrazovanii, media i v drugih oblastyah: XV Mezhdunarodnaya nauchno-prakticheskaya konferenciya, Moskva, 3–5 aprelya 2023 g.: Materialy i doklady // pod obshchej redakciej O. N. Raeva. — Moskva : IPP «KUNA», 2023. — 347 p., pp.87-101, – In Russian. ISBN 978-5-98547-144-1, https://www.rsuh.ru/education/minot/conf_3d-2023.php

[12] Programmno-upravlyayushchaya obolochka dlya servisnyh robotov s parallel'nymi vychisleniyami / R.B. Tarasov, O.I. Davydov, V.E. Pryanichnikov [i dr.] // Robototekhnika i tekhnicheskaya kibernetika. – 2021. – T. 9. – № 2. – pp. 99-105. – In Russian. – DOI 10.31776/RTCJ.9203. – EDN EUOQJP.

[13] A. Nüchter, J. Hertzberg. Towards semantic maps for mobile robots, Robotics and Autonomous Systems, Volume 56, Issue 11, November, 2008 , pp 915-926.

[14] Pryanichnikov V.E., Chernyshev V.V., Arykancev V.V., Aryskin A.A, Ksenzenko A.YA., Petrakov M.S., Travushkin A.S., Eprikov S R. Supervizornoe upravlenie shagayushchej podvodnoj bazoj s avtonomnymi robotami-sputnikami. Intellektual'nye adaptivnye roboty // Informacionno-izmeritel'nye i upravlyayushchie sistemy. 2019. T. 16. № 1–2. pp. 49–56. – In Russian. ISSN 2070-0814.

[15] Chernyshev V. V., Pryanichnikov V. E., Arykantsev V. V. Algorithms of Self-control of the Underwater Walking Device According to Information on a Collision of Feet with an Unorganized Support Surface. In: Yuschenko A. (Eds) Modern Problems of Robotics. MPoR2020. Communications in Computer and Information Science, vol 1426. Springer, Cham. DOI https://doi.org/10.1007/978-3-030-88458-1_14, Print ISBN978-3-030-88457-4 Online ISBN978-3-030-88458-1.

DOI: 10.2507/34th.daaam.proceedings.044

INVESTIGATION OF THE MOMENT ON THE CONTROL BODY OF A SYNCHRONISED ADJUSTABLE SWASH PLATES DOUBLE AXIAL PISTON HYDRAULIC MACHINE

Stazhkov Sergey, Elchinsky Victor, Kuzmin Anton & Feshin Boris

This Publication has to be referred as: Stazhkov, S[ergey]; Elchinsky, V[ictor]; Kuzmin, A[nton] & Feshin, B[oris] (2023). Investigation of the Moment on the Control Body of a Synchronised Adjustable Swash Plates Double Axial Piston Hydraulic Machine, Proceedings of the 34th DAAAM International Symposium, pp.0330-0334, B. Katalinic (Ed.), Published by DAAAM International, ISBN 978-3-902734-41-9, ISSN 1726-9679, Vienna, Austria
DOI: 10.2507/34th.daaam.proceedings.044

Abstract

In modern volumetric hydraulic drives, swash plate axial-piston hydraulic machines are widely used, which is due to their high-energy consumption and simplicity of design. A significant disadvantage of hydraulic machines of this type is a large insensitivity zone. To solve this problem, a design of a synchronised adjustable swash plates double axial piston hydraulic machine with the possibility of synchronous rotation of swash plates in opposite directions was proposed. The purpose of this work is to research the moment on the control body of a synchronised adjustable swash plates double axial piston hydraulic machine and compare it with the moment on the swash plate of a single-block swash plate axial piston hydraulic machine.

Keywords: Hydraulic machines; Axial piston machines; Synchronised adjustable double swash plates axial piston hydraulic machine; Flow pulsations; Control system; Volumetric hydraulic drive.

1. Object of investigation

The object of investigation in this work is a synchronised adjustable swash plates double axial piston hydraulic machine, which refers to a volumetric hydraulic drive, or rather to a type of swash plate axial piston hydraulic machine. Hydraulic machines of this type have high weight and size characteristics, which simplifies the layout in a limited space of machines [1]. The constructive difference between the synchronized adjustable swash plates double axial piston hydraulic machine and the single-block swash plate axial piston hydraulic machine consists in the presence of two groups of modified piston mechanisms [2], which rely on oppositely arranged swash plates[3].

The synchronised adjustable swash plates double axial piston hydraulic machine regulation is carried out by means of synchronous rotation of the swash plates around the axis passing through the center of the shaft of the hydraulic machine, at a given angle in opposite directions. Figure 1a on the left shows the synchronised adjustable swash plates double axial piston hydraulic machine [4] with the arrangement of swash plates that provide either the maximum angular speed of rotation of the shaft 5 (in motor mode), or the maximum flow of working fluid in pumping mode.

Figure 1b on the right shows a hydraulic machine with an arrangement of swash plates that provides zero angular velocity (in motor mode) or zero flow of working fluid (in pumping mode).

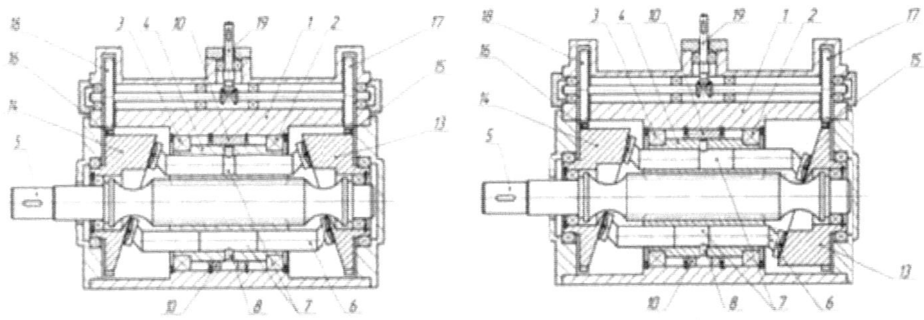

a b

Fig. 1. Synchronised adjustable swash plates double axial piston hydraulic machine

2. The effect of flow pulsation at the moment on the control body of the hydraulic machine

The flow rate in the synchronised adjustable swash plates double axial piston hydraulic machine is regulated by synchronously changing the angle of rotation of the swash plates in opposite directions and is calculated by the formula:

$$Q_s = \omega * F * R * tg\beta * \sum_{i=1}^{n} \left[\sin\left(\alpha + \frac{2\pi}{n}(i-1) + \gamma\right) + \sin\left(\alpha + \frac{2\pi}{n}(i-1) - \gamma\right) \right]$$
$$\alpha \in [0; \pi] \tag{1}$$

where ω is the speed of rotation of the hydraulic machine shaft, rad/s,
F – the area of the piston, m^2,
R – the radius of the cylinder block, m,
β – the angle of inclination of the swash plates, radians,
α – the angle of rotation of the cylinder block, radians,
γ – the angle of rotation of the swash plates, radians,
n – the number of pistons,
i – the piston number.

Figure 2 shows a diagram of the flow rate change Q_s at the speed of rotation of the shaft 104.72 rad/s for synchronised adjustable swash plates double axial piston hydraulic machine depending on the synchronous angle of rotation γ of the swash plates in opposite directions, where the maximum angle is 1.5708 radians, the minimum is 0, and the rotation of the cylinder block α, where the minimum is 0, the maximum is 0.698 radians, so as with n=9, the angle between the pistons is 0.698 radians.

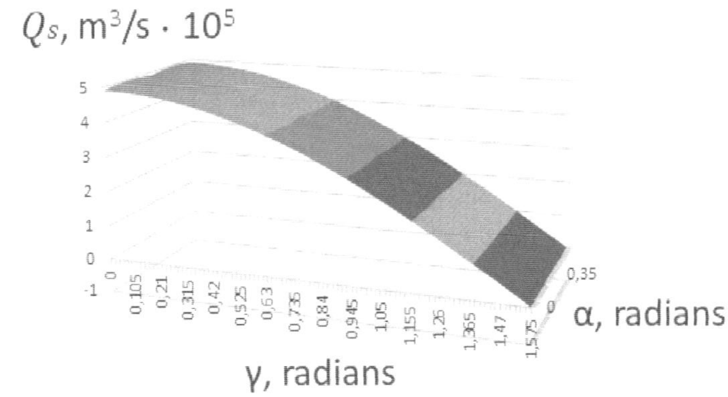

Fig. 2. Flow pulsation in a synchronised adjustable swash plates double axial piston hydraulic machine

Based on the schedule and the calculations made, the flow pulsation for synchronised adjustable swash plates double axial piston hydraulic machine is 1.5192%. This means that for further force analysis of the synchronised adjustable swash plates double axial piston hydraulic machine control body, when calculating the moment on the control roller, this pulsation can be neglected [5] and the pressure can be taken as a constant value [6].

3. Calculation of the moment on the control body of the synchronised adjustable swash plates double axial piston hydraulic machine

With the arrangement of swash plates providing zero angular velocity (in motor mode) or zero flow rate of the working fluid (in pumping mode), for calculation we assume that the oppositely arranged swash plates rotate in one direction, since the moments created by the piston mechanisms acting on the washers will try to tip them in the direction, shown in figure 3:

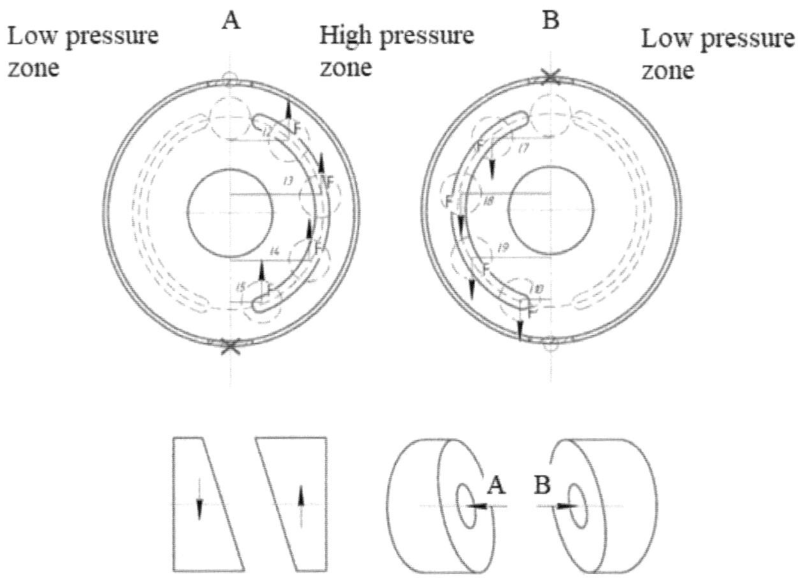

Fig. 3. Swash plates at zero angle position

The moments transmitted by the swash plates to the cylindrical gears, and then to the conical gears, will cause them to rotate in one direction. Then two moments will act on the synchronised adjustable swash plates axial piston hydraulic machine control body at once, which will turn the conical gear of the control body in opposite directions. This means that one moment counteracts the other. Then the total moment acting on the control body will be determined by the difference between the moments of the swash plates. At the same time, depending on the sign of the common moment, the direction of rotation of the control body will change. Then, to find the full moment of the control body of the control system of a synchronised adjustable swash plates double axial piston hydraulic machine, we use the formula:

$$\sum M = p * F * R * tg\beta * \sum_{i=1}^{n} \left[sin\left(\alpha + \frac{2\pi}{n}(i-1) + \gamma\right) - sin\left(\alpha + \frac{2\pi}{n}(i-1) - \gamma\right)\right] \quad (2)$$
$$\alpha \in [0; \pi]$$

where p is the pressure, MPa,
F – the area of the piston, m^2,
R – the radius of the cylinder block, m,
β – the angle of inclination of the swash plates, radians,
α – the angle of rotation of the cylinder block, radians,
γ – the angle of rotation of the swash plates, radians,
n – the number of pistons,
i – the piston number.

Figure 4 shows a diagram of the change in the total torque M on the control body during the mutual rotation of the swash plates and the rotation of the cylinder block:

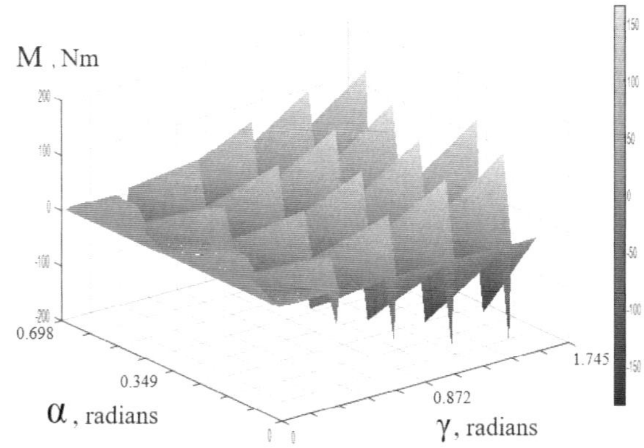

Fig. 4. Moment on the synchronised adjustable swash plates double axial piston hydraulic machine control body

It can be seen from the graph that with an increase in the angle of rotation of the swash plates, the moment on the synchronised adjustable swash plates double axial piston hydraulic machine control body increases.

4. Comparative analysis of the control moment of a single-block swash plate axial piston hydraulic machine and a synchronised adjustable swash plates double axial piston hydraulic machine

For a comparative analysis, it is necessary to calculate the total control moment of a single-block swash plate axial piston hydraulic machine [7]. Formula for calculating the total moment of force on a swash plate [8]:

$$M_s = p * F * \frac{R}{cos^2\gamma} * \sum_{i=1}^{n} \left[cos\left(\alpha + \frac{2\pi}{n}(i-1)\right)\right] \quad \alpha \in [0; \pi] \tag{3}$$

where p is the pressure, MPa,
F – the area of the piston, m^2,
R – the radius of the cylinder block, m,
α – the angle of rotation of the cylinder block, radians,
γ – the angle of inclination of the swash plate, radians,
n – the number of pistons,
i – the piston number.

Figure 5 shows a comparison of the control moments under the same conditions and the sizes of piston mechanisms for the single-block swash plate axial piston hydraulic machine and synchronised adjustable swash plates double axial piston hydraulic machine:

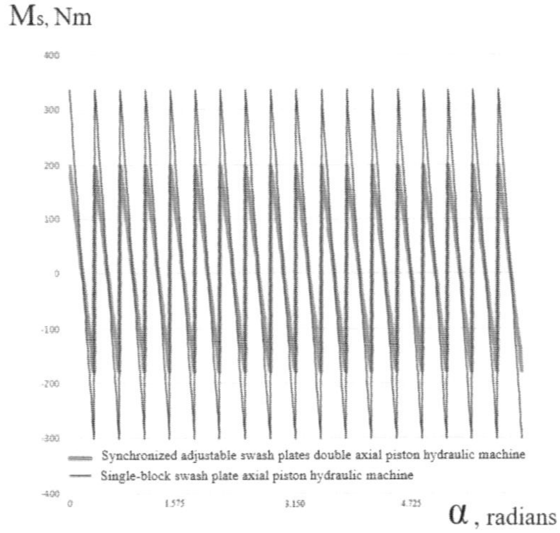

Fig. 5. The moment on the control body at the maximum angle of regulation

According to the results of calculations – for a single block swash plate axial piston hydraulic machine, the maximum moment is $M_s = 336.4$ Nm, for synchronised adjustable swash plates double axial piston hydraulic machine $M_s = 197.751$ Nm.

5. Conclusion

This investigation presents the dependence for calculating the flow rate and the moment on the control body of a synchronised adjustable swash plates double axial piston hydraulic machine. After the calculations, it was found that the pulsation of the flow rate in the synchronised adjustable swash plates double axial piston hydraulic machine is 1.5192%, therefore, for further calculations, this pulsation can be neglected. Calculations of the total moment of control body for single-block swash plate axial piston hydraulic machine and synchronised adjustable swash plates double axial piston hydraulic machine are carried out.

Force analysis showed that the total moment on the control body of the synchronised adjustable swash plates double axial piston hydraulic machine repeats the form of a single-block swash plate axial piston hydraulic machine, but has a smaller amplitude of oscillations. The moment on the control body of the synchronised adjustable swash plates double axial piston hydraulic machine is 1.7 times less than the control moment of a single-block swash plate axial piston hydraulic machine. Reducing the torque on the control body will reduce the dead zone of the hydraulic machine, and, consequently, increase the accuracy of the hydraulic drive control as a whole. In the future, our team plans to make a model of the dynamics of the control mechanism synchronised adjustable swash plates double axial piston hydraulic machine.

6. References

[1] V.A. Petrov (1988). Hydraulic volume transmissions of self–propelled machines. M.: Mashinostroenie.

[2] Kuzmin, A[nton]; Popov, V[alery] & Stazhkov, S[ergey] (2016). Advanced Axial Piston Swash Plate Pump Parameters Recommendations, Proceedings of the 27th DAAAM International Symposium, pp.0556-0561, B. Katalinic (Ed.), Published by DAAAM International, ISBN 978-3-902734-08-2, ISSN 1726-9679, Vienna, Austria

[3] S. Stazhkov, A. Kuzmin, V. Elchinskiy, N. Yakovenko (2020). Tribological tests of the improved piston mechanism of the axial piston hydraulic machine. IOP Conference Series: Materials Science and Engineering: 15, Nizhny Tagil, 18–19 June. DOI 10.1088/1757-899X/966/1/012120

[4] S. O. Zakharov, V. S. Yelchinsky, A. O. Kuzmin, S. M. Probationkov (2021). Synchronised adjustable swash plates axial piston hydraulic machine. Automation, mechatronics, information technologies: Materials of the XI International Scientific and Technical Internet Conference of Young Scientists, pp. 71-73, Omsk, May 18-19. Omsk State Technical University.

[5] V.N.Prokofiev, Yu.A.Danilov, L.N.Kondakov [et al.] (1969). Axial piston adjustable hydraulic drive, under the general editorship of V.N.Prokofiev. M.: Mechanical Engineering.

[6] V.N. Prokofiev (1976). High-pressure hydraulic drive. Bulletin of Mechanical Engineering, № 12.

[7] T.M. Bashta (1974). Volumetric pumps and hydraulic motors, hydraulic systems. M.: Mechanical Engineering.

[8] A.N. Dunaev (1985). Axial piston hydraulic machines with a swash plate. Manual. M.: MEI.

DOI: 10.2507/34th.daaam.proceedings.045

CHALLENGES IN INTEGRATING STATIC CODE ANALYSIS INTO DOMAIN-SPECIFIC LANGUAGE

Danilo Nikolic, Andjela Todoric, Dusanka Dakic,
Teodora Vuckovic & Darko Stefanovic

This Publication has to be referred as: Nikolic, D[anilo]; Todoric, A[ndjela]; Dakic, D[usanka]; Vuckovic, T[eodora] & Stefanovic, D[arko] (2023). Challenges in Integrating Static Code Analysis into Domain-Specific Language, Proceedings of the 34th DAAAM International Symposium, pp.0335-0340, B. Katalinic (Ed.), Published by DAAAM International, ISBN 978-3-902734-41-9, ISSN 1726-9679, Vienna, Austria
DOI: 10.2507/34th.daaam.proceedings.045

Abstract

Domain-specific languages (DSLs) have become increasingly popular for addressing specific domain problems in industry and academia. However, implementing static code analysis tools for DSLs presents unique challenges due to the evolving nature of the field. This paper explores major hurdles in implementing static code analysis tools for DSLs, including customization needs, limited tooling support, and complexity and expertise requirements. It emphasizes the significance of overcoming these hurdles to successfully integrate static code analysis tools into DSLs, enhancing code analysis and improving quality in DSL-based projects.

Keywords: domain-specific languages; static code analysis; integration; implementation challenges;

1. Introduction

In computer science and software engineering, domain-specific languages (DSLs) represent an important field of research. A DSL is a language adapted to a specific application domain and offers appropriate notations and abstractions [1]. Compared to general-purpose languages (GPLs), DSLs are more expressive and easier to use, increasing productivity and decreasing maintenance costs [1].

Providing the ability to assess the quality of DSL source code artifacts using well-known static analysis techniques can be an additional factor contributing to the successful adoption of DSLs [2]. Static program analysis enables the evaluation of software quality characteristics and the detection of various code smells and potential errors by using automatic tools for source code analysis [2]. GPLs can be used to build computer software with high-quality features and flexibility. Furthermore, DSLs offer ease of use and statement writing, with users not being required to have extensive domain knowledge.

Compared to GPLs, DSLs have a small community of software developers as well as domain experts. By using a DSL, domain experts can be more involved in developing a solution to a problem. They can more easily understand, check the code, and what impact a particular change will have [3].

Usability is one of the crucial attributes in language evaluation. DSL usability evaluation can be used in the language design process, which can significantly affect user productivity [4]. This paper is concerned with identifying challenges that may arise when integrating static code analysis into domain-specific languages. In order to improve the quality of DSLs, it is important to identify potential problems, thus enabling more efficient static analysis of the code. Solving the challenges leads to more reliable software solutions that rely on DSLs.

Understanding the challenges of integrating static code analysis into DSLs is crucial for efficient development and quality assurance. It enables developers to anticipate obstacles, allocate resources effectively, and ensure compliance with domain-specific rules and standards. By identifying these challenges, tool developers and researchers can develop specialized analysis tools and frameworks, driving the adoption and usage of DSLs in various domains. Stakeholders, including developers, quality assurance teams, tool developers, researchers, and organizations benefit from this understanding, leading to improved development processes, code quality, and tooling support in DSL-based projects.

Previous research in the field has explored the implementation and evaluation of various static code analysis tools on GPLs. These studies have investigated strategies for effective deployment, aiming to optimize code quality and identify vulnerabilities in GPL-based projects [5], [6], [8]. As a first step in this research, a systematic review of the literature in the field of DSL, as well as a static analysis of the code, was organized. In this way, relying on the literature in this field, various challenges in using static code analysis in integration with domain-specific languages are presented and analysed. The descriptive analysis also includes general purpose language, in order to adequately present the difference between GPL and DSL. The aim of this paper is to show aspects of the integration of static code analysis in domain-specific languages, and thus cover all the differences between GPL and DSL.

The remainder of the paper is structured as follows. The theoretical foundations of DSL and static code analysis are presented in Section 2. Section 3 reviews related works in this area. The following are the challenges of integrating static code analysis into DSL, presented in Section 4. Finally, Section 5 concludes the paper.

2. Theoretical foundations

This section provides a theoretical background on domain-specific languages and static code analysis.

2.1. Domain-specific languages (DSL)

The concept of DSL is based on the idea that a programming language can be designed specifically for a specific domain [9]. The use of DSLs is diverse. In addition to documenting the requirements and behaviour of a single domain, it can also be used to generate programs for the addressed domain [10]. By focusing on particular domain needs, DSLs can be more efficient and easier to use than GPLs.

A DSL development process includes an assessment of the effectiveness, usability, and suitability of the language for its purpose [4]. By using DSL, it is possible to hide implementation details and improve code readability [9]. Compared to GPLs, DSLs can offer significant gains in expressiveness and ease of use [9].

In software development, DSLs can significantly increase productivity and quality [11]. Also, since DSLs are explicitly designed for a specific domain, they can provide higher precision and accuracy than general-purpose languages. On the other hand, GPLs are designed to be applicable to a wide range of domains and applications [12]. For this reason, GPLs can be more flexible and adaptable to changing demands.

2.2. Static code analysis

With static code analysis, the source code can be analysed, and in this way, various defects in the program can be identified without its execution [8]. Static code analysis aims to detect problems early in development before the code is deployed. One significant advantage is their ability to identify the root cause of security issues rather than merely indicating symptoms [7]. This is particularly vital for ensuring that defects are effectively resolved.

Among the features offered by static code analysis, the most valuable one is the ability to identify common programming defects. Similar to how antivirus software detects viruses, static code analysis tools scan source code based on a specific set of patterns or rules [6]. Practitioners can use more advanced tools to add new rules to a set of predefined ones. However, it is important to note that if the patterns or rules do not specify a particular behaviour, the tools may not be able to detect corresponding errors.

During the analysis process, false positive and false negative errors may occur [13], [14]. A false positive error occurs if a static code analysis tool detects a problem that is not actually a problem in the code [14]. A large number of false positive errors can cause operational difficulties. In the analysis process, false positive errors are undesirable, but from a security perspective, false negative errors are a bigger problem. A false negative error occurs when it is impossible to detect a real problem in the code [14]. It implies that the source code still contains defects, even after the static analysis procedure is completed. As these errors can be problematic, more tools and methods should be used to cover more potential problems.

3. Related work

In recent years, DSL has become increasingly popular among professionals in various sectors and has more interest in the industry and literature. Also, research on the integration of static code analysis into DSL is increasingly mentioned in multiple works [2], [13], [14], [15], [16], [17], [18]. Despite the large number of existing DSLs, a lot of research is based on defining a new way of approaching and building DSLs [13], [15], [20].

In the paper [9], the authors believe that using DSL instead of GPL makes the code more readable, the mapping between the model and the code becomes more apparent, and thus the development time is reduced. Also, another advantage over the GPL is that DSLs can be updated and extended more easily than entire programming languages, such as Java, C, C++, etc. [9]. Specifically, the authors in the paper [18] showed that using DSL for robot systems is better than GPL. The authors [1] explain the difference between understanding DSL and the GPL, as well as the advantages of DSL over GPL.

In [15], the authors presented that static code analysis tools usually combine different types of analysis and look for a specific set of patterns or rules in the source code. The analyses performed by these tools consist of several independent rules that users can select each time the tool is run. These tools cannot find errors if patterns or rules do not specify such behaviour.

Given that patterns are significant both in software engineering and domain engineering, the paper [10] presented that DSL development can be integrated with application development. Also, with the help of forms, a solution can be provided for some common domain engineering problems. To start using DSL agilely, developers need to apply the patterns in the context of their existing programming knowledge.

Since GPLs are not always the most suitable for solving problems, DSLs have a significant advantage. DSLs can work independently but can also be designed from scratch or by extending some basic language [19].

The integration of static code analysis into DSL has gained traction in recent years, as evidenced by the growing interest in the industry and literature. The advantages of DSL over GPL, such as improved code readability, clear model-code mapping, and easier extensibility, make it a promising avenue for implementing static code analysis tools in the context of DSLs.

4. Aspects in the integration of static code analysis in domain-specific languages

Domain-specific languages are used to reduce development time and thus allow developers to work at a higher level of abstraction. Static code analysis is used to analyse the source code without executing it, thus identifying potential bugs, security vulnerabilities, and other problems [2].

Based on a descriptive review of the literature in this area and previous work, the conclusions are presented in this section. Several authors addressed some of the presented problems [9], [10], [11], [15], [18], [4].

Understanding the distinction between GPLs and DSLs is essential because it helps developers and tooling experts recognize the unique characteristics and challenges associated with each type of language. When integrating static code analysis into DSLs, it becomes crucial to tailor the analysis techniques and tooling to the specific constructs, abstractions, and the constraints of DSL, rather than relying solely on existing tools built for GPLs.

To integrate static code analysis into domain-specific languages in the best possible way, it is necessary to see the differences between general-purpose and domain-specific languages, as shown in Table 1.

	Domain-specific languages	**General-purpose languages**	**Source**
Development time	Save time during development	Long development time	[11]
Complexity	DSL syntax and semantics aid ease of parsing	A large, diverse codebase can make analysis difficult	[4], [10]
Identification of potential errors	The declarative approach describes the desired result	Lack of specialized functions for a specific domain	[10]
Community of experts	Lack of expertise among developers due to specificity	The larger community of developers and domain experts to develop tools	[2], [4]
Standardization	Lack of standardization can hinder the development of analysis tools	Established standardization for easier development of analysis tools	[9], [18]
Specificity	Precisely identifies problems due to specificity	Wide range of applications designed for flexibility and versatility	[13], [15], [18]

Table 1. Limitations and differences between DSL and GPL

Significant benefits can be seen by integrating static code analysis with domain-specific languages. One advantage is saving time during program development using DSL because errors and potential problems can be noticed much earlier in the process itself [11]. Therefore, removing them is much simpler. Integrating static code analysis with DSL can improve the quality of DSL by maintaining good design practices and principles. On the other hand, the time during program development using the GPL is significantly longer. As a result, it takes more time to debug a program developed using GPL.

Defining the syntax and semantics of the language facilitates the analysis of DSL itself. When designing the syntax and semantics, programmers can make DSL relatively easy to interpret, thus facilitating the identification of potential errors and problems through static code analysis [4]. For example, a language could be designed to facilitate parsing or to ensure that certain types of errors are impossible to express in the language itself. This could mean designing the language to facilitate the analysis or using a type system that catches common compile-time errors. The language could be designed to ensure that only valid operations and data types are used to help catch bugs early in development, thereby reducing the effort and time needed to fix those same bugs later. From Table 1 one can see the complexity of the language during its analysis. If the language has a large and diverse code base, the complexity may be greater, which is a characteristic of the GPL.

The declarative approach is a programming paradigm that emphasizes the logic and rules of the problem domain. It focuses on the desired result rather than the steps required to achieve it. With a declarative approach, a DSL can be easily analysed. In this way, identifying potential errors can be facilitated. Certainly, with the integration of static code analysis in DSL, a language is created that is more reliable, easy to use and maintain.

Another approach is to provide tools and frameworks that can enable the analysis of DSL code [10]. In this way, tools can be included to visualize and explore the structure of code and analyse the code for some common patterns and problems that may arise. Providing such tools and frameworks allows potential errors and issues to be more easily identified. In the case of complex DSLs that have many interdependent components, visualization tools can help to understand the structure and flow of DSL code and thus enable the identification of potential inefficiencies. Likewise, code analysis tools can identify common patterns and problems in DSL code, such as code smells or potential performance issues. Using tools and frameworks that support static code analysis can effectively improve the quality and reliability of DSLs, reducing the time and effort needed to debug and solve problems and making working with them more accessible. Combining these tools with a well-designed DSL can create a powerful and efficient tool for solving problems within a specific domain or application area.

The expertise and knowledge of programmers and researchers in that field are needed to analyse domain-specific languages effectively. Given that the area of static code analysis is still relatively new, a certain lack of expertise and knowledge can manifest itself in different ways. Because of language specificity, it can be difficult to develop practical analysis tools that can accurately identify problems.

In Table 1, the difference in standardization between DSL and GPL can be noticed. Established standardization can facilitate the development of analysis tools. But on the other hand, the lack of standardization or best practices for parsing domain-specific languages can challenge the development and implementation of efficient parsing algorithms that can be adopted and used [9].

General-purpose languages can be used for a wide variety of applications or tasks and are designed to be flexible and versatile. For this reason, they also lack some unique domain-specific features and constructs often found in domain-specific languages. These features may include domain-specific vocabulary, syntax, and semantics that reflect the specific needs and requirements of a particular application domain. This would be one of the challenges when analysing DSLs because the unique features of DSLs can make it difficult for static code analysis tools to accurately and reliably analyse code written in these languages. In the event that a DSL has domain-specific constructs and semantics that are not supported by existing analysis tools, then it can be quite challenging to identify potential problems and errors through static code analysis alone. It can lead to false positives or negatives, making it difficult to effectively use static code analysis to identify potential problems in DSL code. The best solution to these challenges is to develop custom analysis tools and frameworks designed to handle a particular DSL's unique features and constructs.

This may also include the development of custom parsers and parsing algorithms optimized for the specific syntax and grammar of the language, as well as incorporating domain-specific knowledge and heuristics into the parsing process [18]. To more easily identify the specific features and constructions of the language that are most important for analysis and to develop customized analysis tools that adapt to these needs, working with domain experts can be of significant importance. Aiming to improve the quality and reliability of DSL code, the static analysis of the code must be used efficiently [2]. Effective use can be achieved through tools for static code analysis that are designed for DSL, more precisely by integrating them into the development process itself. This can help to detect potential problems at an early stage of development.

Many existing static code analysis tools are designed and optimized for general-purpose languages such as C++, Java or Python. Adapting existing tools to work with domain-specific languages faces various challenges and may require a deep understanding of both DSLs and analysis tools. This can require significant expertise in language design, decomposition, and static analysis techniques. But it is also necessary to understand the domain and the specific requirements of DSL. The best solution is to develop new analysis tools for working with DSLs, but also existing analytical tools for working with DSLs can be modified or extended, which is also discussed in papers [13], [15].

Adapting existing tools can be much more challenging than creating a new one because it requires the appropriate expertise and approach. Still, in this way, the quality and reliability of DSL code can be obtained. Overall, based on the literature review and ongoing research, integrating static code analysis into domain-specific languages can improve the quality and reliability of DSLs and make it easier for developers to work with them. Previously mentioned is an area of ongoing research and development, and much remains to be explored regarding how best to integrate analysis techniques into DSLs.

5. Conclusion

This paper presents static code analysis in domain-specific languages to improve their quality and enable more efficient static code analysis. Various challenges and obstacles are associated with implementing static code analysis tools in domain-specific languages. One of the identified problems that can arise when integrating static code analysis into domain-specific languages is that they are not originally designed to support static code analysis and lack the basic features and constructs needed for efficient analysis. Consequently, existing static code analysis tools may be difficult or impossible to apply to such languages.

In addition, a lack of knowledge and expertise in static code analysis for domain-specific languages may exist since this field is still relatively new. The unique features and constructs of domain-specific languages can also present challenges because they may not be found in general-purpose languages, thus making it difficult for static code analysis tools to accurately and reliably analyse code written in these languages.

On the other hand, adapting existing static code analysis tools to work with domain-specific languages is another challenge since these tools are generally optimized for general-purpose languages. Therefore, modifying or extending tools to support domain-specific languages can require considerable effort and expertise.

Finally, implementing static code analysis tools in domain-specific languages requires overcoming various technical, knowledge and innovation challenges. Despite these difficulties, exploring the potential for incorporating static code analysis into domain-specific languages is crucial to improving their quality and supporting more efficient static analysis.

Future works could explore a concrete tool that would enable a static code analysis into domain-specific languages and thus overcome all the challenges identified in this paper. Additionally, future work could involve identifying ways of designing new and current DSLs to include the implementation of static code analysis tools easily. Through continued innovation and exploration, we can unlock the full potential of DSLs and enable their effective use in diverse application domains.

6. References

[1] Kosar, T.; Oliveira, N.; Mernik, M.; Varanda-Pereira, M. J., Črepinšek, M.; Cruz, D. & Henriques, P. R. (2010). Comparing General-Purpose and Domain-Specific Language: An Empirical Study. ComSIS Vol. 7, No.2, Special Issue, April 2010, pp. 247-264. https://doi.org/10.2298/CSIS1002247K

[2] Ruiz-Rube, I.; Person, T.; Dodero, J. M.; Mota, J. M. & Sanchez-Jara, J. M. (2019). Applying static code analysis for domain-specific languages. Springer-Verlag GmbH Germany 2019. https://doi.org/10.1007/s10270-019-00729-w

[3] Erazo-Garzon, L.; Cedillo, P.; Rossi, G. & Moyano, J. (2022). A Domain-Specific Language for Modeling IoT System Architectures That Support Monitoring. IEEE Access, Volume 10. https://doi.org/10.1109/ACCESS.2022.3181166

[4] Barišić, A.; Amaral, V.; Goulao, M. & Barroca, B. (2011). Quality in Use of Domain-Specific Languages: a Case Study. Proceedings of the 3rd ACM SIGPLAN workshop on Evaluation and usability of programming languages and tools, 65-67. https://doi.org/10.1145/2089155.2089170

[5] Stefanović, D.; Nikolić, D.; Havzi, S.; Lolić, T. & Dakić, D. (2021). Identification of strategies over tools for static code analysis. 9th International Conference on Engineering and Technology (ICET-2021), May 2021, Thailand. https://doi.org/10.1088/1757-899X/1163/1/012012

[6] Nikolić, D.; Havzi, S.; Dakić, D.; Lolić, T. & Stefanović, D. (2021). Evaluation of Strategies over Static Code Analysis Tools. Proceedings of the 32nd DAAAM International Symposium, Vienna, Austria. https://doi.org/10.2507/32nd.daaam.proceedings.069

[7] Nikolić, D.; Stefanović, D.; Dakić, D.; Sladojević, S. & Ristić, S. (2021). Analysis of the Tools for Static Code Analysis. 20th International Symposium INFOTEH-JAHORINA, March 2021, Jahorina. https://doi.org/10.1109/INFOTEH51037.2021.9400688

[8] Stefanović, D.; Nikolić, D.; Dakić, D.; Spasojević, I. & Ristić, S. (2020). Static Code Analysis Tools: A Systematic Literature Review. Proceedings of the 31st DAAAM International Symposium, Vienna, Austria. https://doi.org/10.2507/31st.daaam.proceedings.078

[9] Miller, A.; Han, J.,& Hybinette, M. (2010). Using domain-specific language for modeling and simulation: Scalation as a case study. Proceedings of the 2010 Winter Simulation Conference. https://doi.org/10.1109/WSC.2010.5679113

[10] Gunther, S.; Haupt, M. & Splieth, M. (2010). Agile Engineering of Internal Domain-Specific Languages with Dynamic Programming Languages. 2010 Fifth International Conference on Software Engineering Advances. https://doi.org/10.1109/ICSEA.2010.32

[11] Van den Bos, J. & van der Storm, T. (2013). A Case Study in Evidence-Based DSL Evolution. ECMFA 2013, Lecture Notes in Computer Science, vol 7949. Springer, Berlin, Heidelberg. https://doi.org/10.1007/978-3-642-39013-5_15

[12] Hrnčić, D.; Mernik, M. & Bryant, B. R. (2011). Embedding DSLs into GPLs: A grammatical inference approach. ISSN 1392-124X Information Technology and Control, 2011, Vol. 40, No. 4. http://dx.doi.org/10.5755/j01.itc.40.4.980

[13] Marcilio, D.; Furia, C. A.; Bonifacio, R. & Pinto, G. (2019). Automatically Generating Fix Suggestions in Response to Static Code Analysis Warnings. 2019 19th International Working Conference on Source Code Analysis and Manipulation (SCAM). https://doi.org/10.1109/SCAM.2019.00013

[14] Schiewe, M.; Curtis, J.; Bushong, V. & Cerny, T. (2022). Advancing Static Code Analysis With Language-Agnostic Component Identification. IEEE Access, Volume 10. https://doi.org/10.1109/ACCESS.2022.3160485

[15] Bodden, E. (2018). Self-adaptive static analysis. 2018 ACM/IEEE 40th International Conference on Software Engineering: New Ideas and Emerging Results. https://doi.org/10.1145/3183399.3183401

[16] Palomba, F.; Bavota, G.; Di Penta, M.; Oliveto, R.; De Lucia, A. & Poshyvanyk, D. (2013). Detecting bad smells in source code using change history information. 2013 28th IEEE/ACM International Conference on Automated Software Engineering (ASE). https://doi.org/10.1109/ASE.2013.6693086

[17] Kuruppu, T.; Tharmaseelan, J.; Silva, C.; Samaratunge Arachchillage, U. S.; Manathunga, K.; Reyal, S.; Kodagoda, N. & Jayalath, T. (2021). Source Code based Approaches to Automate Marking in Programming Assignments. Proceedings of the 13th International Conference on Computer Supported Education, Volume 1, 291-298. https://doi.org/10.5220/0010400502910298

[18] Mandal, A.; Mohan, D.; Jetley, R.; Nair, S. & D'Souza, M. (2018). A Generic Static Analysis Framework for Domain-specific Languages. 2018 IEEE 23rd International Conference on Emerging Technologies and Factory Automation (ETFA). https://doi.org/10.1109/ETFA.2018.8502576

[19] Mohagheghi, P. & Haugen, Ø. (2010). Evaluating Domain-Specific Modelling Solutions. Advances in Conceptual Modeling – Applications and Challenges, ER 2010, Lecture Notes in Computer Science, vol 6413, Springer, Berlin, Heidelberg. https://doi.org/10.1007/978-3-642-16385-2_27

[20] Ton, L. P. & Truong, T. M. (2015). Linking Rules and Conceptual Model in a Domain Specific Language. 2015 International Conference on Advanced Computing and Applications (ACOMP). https://doi.org/10.1109/ACOMP.2015.25

DOI: 10.2507/34th.daaam.proceedings.046

EFFECTIVENESS OF THE USE OF PASSENGER RAIL TRANSPORT IN THE TERRITORY OF THE CZECH REPUBLIC

Adam Malatinsky & Martin Hromada

This Publication has to be referred as: Malatinsky, A[dam] & Hromada, M[artin] (2023). Effectiveness of the Use of Passenger Rail Transport in the Territory of the Czech Republic, Proceedings of the 34th DAAAM International Symposium, pp.0341-0346, B. Katalinic (Ed.), Published by DAAAM International, ISBN 978-3-902734-41-9, ISSN 1726-9679, Vienna, Austria
DOI: 10.2507/34th.daaam.proceedings.046

Abstract

Passenger transportation began to be used already in ancient times. Among the first necessary means of transport were the use of wooden transport carts in Mesopotamia and later the use of ships and sailing ships in Egypt. Subsequently, other types of transport began to develop gradually. Until today, the method of transporting people has been constantly transformed to its current form. The article deals with rail transport on the territory of the Czech Republic. Specifically, it is about determining the effectiveness of passenger rail transport for individual regions based on the analysis. The results will determine the most efficient and, at the same time, the least efficient use of rail transport for individual regions in the territory of the Czech Republic. Determining efficiency within this context is appropriate for the regions themselves and their solutions to the traffic situation in the future. Regions with lower efficiency in transport use should create suitable conditions for improving the given issue.

Keywords: Transport; Rail Transport; Effectiveness; Analysis

1. Introduction

Humanity uses various means of transport for faster transportation from one place to another [1]. One of them is rail transport [2]. The article deals with the effectiveness of the use of passenger rail transport in the territory of the Czech Republic. The article first describes transportation in general. Subsequently, rail transport in the Czech Republic is described. This part is followed by a chapter regarding the analysis of passenger rail transport in the Czech Republic, where individual coefficients are gradually evaluated to determine efficiency. The last chapter deals with the efficiency of passenger rail transport, where four groups are created according to the values of the coefficients.

The result of solving the traffic situation is not only the transportation of people or things from place A to place B. [3] It is also about the efficient use of transport. Among the effective methods of transport is the solution to the traffic situation so that there are no queues, that passengers do not have long downtimes, that all types of transport communication are used effectively, that the traffic of means of transport is accelerated and many other ways. [4] The result of solving the traffic situation is not only the transportation of people or things from place A to place B. It is also about the efficient use of transport. Among the effective methods of transport is the solution to the traffic situation so that there are no queues, that passengers do not have long downtimes, that all types of transport communication are used effectively, that the traffic of means of transport is accelerated and many other ways.

2. Transport

Currently, transport is significantly developed [5]. Transport can be divided into four classifications. The first transport division is according to space, divided into land, water and air. Land transport can be road, rail and non-motorized (pedestrian and bicycle). Water transport is divided into inland, coastal and maritime. The second classification is according to the object and method of transport, which can be passenger or cargo. Personal transport can be divided into individual or collective. Freight transport is public or private. The third classification of traffic distribution is according to the territorial distribution of transport needs. This classification is divided into urban, local, national, regional or international. The last classification of the distribution of traffic is according to the relationship between the source and the traffic destinations concerning the given territory. This category includes internal, external or transit transport. [6]. Transport in the Czech Republic is primarily divided into road, rail, water and air. Road transport includes cars, buses, motorcycles, trolleybuses and bicycles. Rail transport includes trains, trams, subways or cable cars. In the Czech Republic, highways, roads for motor vehicles and other roads are designated for road transport travel. [7]

3. Railway transport in the Czech Republic

The railway network in the Czech Republic in 2022 consisted of 9,355 kilometres of state railway lines and over 100 kilometres of regional railway lines. The railway administration ensures the performance of the function of railway ownership. The structure of railway lines, according to the number of lines, is divided into single-track (7,287 km), double-track (2,003 km) and multi-track (65 km). The total length of electrified lines is 3,215 kilometres. There are 6,733 railway bridges with a length of over 156 kilometres and 169 tunnels with a length of fewer than 56 kilometres in the entire territory of the Czech Republic. There are 1,062 railway stations, and 1,546 railway stops in the Czech Republic as part of railway transport. [8] [9]

4. Analysis of passenger railway transport in the Czech Republic

This chapter consists of 4 parts, where the resulting coefficient is evaluated from each part, which serves to evaluate the efficiency of rail transport in the Czech Republic.

4.1 Number of passengers depending on the number of inhabitants

Specific values can be found in the following tables. Table 1 describes the number of passengers in the region based on the number of departures from the region, arrivals from the region and the number of passengers within the region (internal).

Region	Number of passengers (in thousands)			
	Departures	Arrivals	Internal	SUM
Prague	19 018	18 773	11 894	49 685
Central-Bohemian	12 593	12 758	6 528	31 878
South-Bohemian	1 466	1 467	4 008	6 941
Pilsen	1 587	1 630	7 280	10 497
Karlovy Vary	559	574	2 502	3 635
Usti nad Labem	1 659	1 714	7 931	11 303
Liberec	1 145	1 182	5 292	7 619
Hradec Kralove	2 077	2 160	6 036	10 273
Pardubice	3 372	3 303	5 455	12 130
Vysocina	1 086	1 134	2 399	4 619
South-Moravian	3 016	2 980	24 444	30 441
Olomouc	3 159	3 126	7 527	13 812
Moravian-Silesian	2 377	2 312	16 915	21 603
Zlin	1 517	1 517	3 915	6 949
SUM	54 630	54 630	112 125	221 384

Table 1. Number of passengers in regions [10, 11, modified]

The first coefficient (C_1) is determined based on the number of passengers per 1000 region inhabitants. The resulting coefficient value is based on the descending order of the number of passengers per 1000 inhabitants of the region. Specific values can be found in the following table.

Region	Population (in thousands)	Number of passengers per 1 000 inhabitants	C_1
Prague	1 275	39,0	1
Central-Bohemian	1 387	23,0	4
South-Bohemian	637	10,9	13
Pilsen	579	18,1	8
Karlovy Vary	253	14,4	10
Usti nad Labem	799	14,1	11
Liberec	438	17,4	9
Hradec Kralove	543	18,9	6
Pardubice	515	23,6	3
Vysocina	504	9,2	14
South-Moravian	1 185	25,7	2
Olomouc	635	21,8	5
Moravian-Silesian	1 178	18,3	7
Zlin	572	12,1	12
SUM	10 498	21,1	-

Table 2. First coefficient [10, modified]

4.2 Number of passengers depending on the size of the region

The second coefficient (C_2) is determined based on the region's number of passengers per 1 square kilometre. The resulting coefficient value is based on the descending order of the number of passengers per square kilometre of the region. Specific values can be found in the following table.

Region	Area (square kilometres)	Number of passengers per 1 square kilometre	C_2
Prague	496	100,13	1
Central-Bohemian	11 929	2,67	5
South-Bohemian	10 058	0,69	13
Pilsen	7 649	1,37	11
Karlovy Vary	3 310	1,10	12
Usti nad Labem	5 339	2,12	9
Liberec	3 163	2,41	7
Hradec Kralove	4 759	2,16	8
Pardubice	4 519	2,68	4
Vysocina	6 796	0,68	14
South-Moravian	7 188	4,24	2
Olomouc	5 272	2,62	6
Moravian-Silesian	5 431	3,98	3
Zlin	3 963	1,75	10
SUM	79 871	2,77	-

Table 3. Second coefficient [10, modified]

4.3 Number of passengers depending on the operation length of the railway line

The third coefficient (C_3) is determined based on the number of passengers per 1 km of line. The resulting value of the coefficient is based on the descending order of the number of passengers per 1 km of line. Specific values can be found in the following table.

Region	Length of railway line	Number of passengers per 1 kilometre of line	C_3
Prague	212	234	1
Central-Bohemian	1450	22	5
South-Bohemian	956	7	14
Pilsen	725	14	10
Karlovy Vary	443	8	12
Usti nad Labem	1079	10	11
Liberec	454	17	8
Hradec Kralove	667	15	9
Pardubice	581	21	6
Vysocina	577	8	13
South-Moravian	754	40	2
Olomouc	622	22	4
Moravian-Silesian	644	34	3
Zlin	357	19	7
SUM	9521	473,3	x

Table 4. Third coefficient [10, modified]

4.4 Area based on the operation length of the railway line

The fourth coefficient (C_4) is determined based on the number of square kilometres of the region and 1 kilometre of operational length of the railway line. The resulting coefficient value is based on the ascending order of the square kilometres per 1 kilometre of the operating length of the railway line. Specific values can be found in the following table.

Region	Region area per 1 kilometre of line	C_4
Prague	2,3	1
Central-Bohemian	8,2	7
South-Bohemian	10,5	11
Pilsen	10,5	12
Karlovy Vary	7,5	5
Usti nad Labem	4,9	2
Liberec	7,0	3
Hradec Kralove	7,1	4
Pardubice	7,8	6
Vysocina	11,8	14
South-Moravian	9,5	10
Olomouc	8,5	9
Moravian-Silesian	8,4	8
Zlin	11,1	13
SUM	115,2	x

Table 5. Fourth coefficient [10, modified]

5. Efficiency of passenger railway transport in the Czech Republic

The efficiency of railway transport in the Czech Republic is determined for each region of the Czech Republic separately. The resulting efficiency is evaluated based on the sum of the individual four coefficients. The minimum value can be four, and the maximum is 56. The resulting efficiency value is shown in the following table with the respective regions.

Region	Coefficients				SUM
	C_1	C_2	C_3	C_4	
Prague	1	1	1	1	4
Central-Bohemian	4	5	5	7	21
South-Bohemian	13	13	14	11	51
Pilsen	8	11	10	12	41
Karlovy Vary	10	12	12	5	39
Usti nad Labem	11	9	11	2	33
Liberec	9	7	8	3	27
Hradec Kralove	6	8	9	4	27
Pardubice	3	4	6	6	19
Vysocina	14	14	13	14	55
South-Moravian	2	2	2	10	16
Olomouc	5	6	4	9	24
Moravian-Silesian	7	3	3	8	21
Zlín	12	10	7	13	42

Table 6. Sum of coefficients [author]

The efficiency of rail transport in the regions is evaluated based on the sum of all coefficients. Efficiency is divided into four groups according to the corresponding number of points: very high efficiency, high efficiency, medium efficiency, and low efficiency. This division primarily has a comparative method for individual regions in the Czech Republic.

Efficiency	Very high	High	Medium	Low
Point spread	4 – 16	17 – 29	30 – 42	43 - 56
Regions	Prague South-Moravian	Pardubice Central-Bohemian Moravian-Silesian Olomouc Hradec Kralove Liberec	Usti nad Labem Karlovy Vary Pilsen Zlin	South-Bohemian Vysocina

Table 7. Efficiency of regions [author]

6. Conclusion

Nowadays, transport is an inherent method of transportation between places. The article dealt with the effectiveness of the use of rail transport in individual regions in the Czech Republic. The first chapter described transport in general. The following part dealt with transport on the territory of the Czech Republic. The next chapter dealt with the analysis of rail transport itself, where individual methods of evaluating coefficients were described. The last chapter dealt with the efficiency of rail transport in the Czech Republic, which divided the efficiency into four groups: very high efficiency, high efficiency, medium efficiency, and low efficiency. The Prague region and the South Moravian region belong to very high efficiency. High efficiency includes the Pardubice region, the Central Bohemian region, the Moravian-Silesian region, the Olomouc region, the Hradec Kralove region and the Liberec region. Medium efficiency includes the Usti nad Labem Region, the Karlovy Vary Region, the Pilsen Region and the Zlin Region. Low efficiency includes the South-Bohemian region and the Vysocina region. Based on the results, it is clear that more emphasis on using rail transport should be for medium and low-efficiency groups.

The results of the article and the evaluation of the efficiency of rail transport can positively impact other types of transport and their use. For example, in regions with a higher efficiency of rail transport, it should not hurt other types of transport. In regions with a lower efficiency in rail transport, there is a higher probability of inefficient use of other types of transport (for example, traffic jams, overcrowded buses, etc.). It follows from this that it is a direct ratio of the use of individual transport in the region. As the efficiency of one type of transport increases, the efficiency of other increases proportionally.

As part of the negative impact, it could be the need to create new transport lines and possibly purchase additional means of transport. Moreover, these activities increase the costs of the transport company. The limit of the article is data from 2022, as the yearbook is constantly updated at the end of the year. In the future, it is advisable to create efficiency for other types of transport and evaluate and compare them.

7. Acknowledgments

This research was based on the support of the Internal Grant Agency of Tomas Bata University in Zlín, IGA/FAI/2023/001 and The Department of Security Engineering, Faculty of Applied Informatics.

8. References

[1] Shiler, V.; Shiler, A. & Galiev, I. (2016). New Wheelset Design for Rail Transport, Proceedings of the 27th DAAAM International Symposium, pp.0896-0900, B. Katalinic (Ed.), Published by DAAAM International, ISBN 978-3-902734-08-2, ISSN 1726-9679, Vienna, Austria DOI: 10.2507/27th.daaam.proceedings.129

[2] Kotkova, B. & Hromada, M. (2021). Benefits and Risks of Multimodal Transport with a Main Focus on the EU, Proceedings of the 32nd DAAAM International Symposium, pp.0306-0310, B. Katalinic (Ed.), Published by DAAAM International, ISBN 978-3-902734-33-4, ISSN 1726-9679, Vienna, Austria DOI: 10.2507/32nd.daaam.proceedings.044

[3] Sharma, B., Pellegrini, P., Rodriguez, J. et al. A review of passenger-oriented railway rescheduling approaches. Eur. Transp. Res. Rev. 15, 14 (2023). DOI: https://doi.org/10.1186/s12544-023-00587-0

[4] Kristoffersson, I., Berglund, S. Modelling connection trips to long-distance travel. Eur. Transp. Res. Rev. 14, 30 (2022). DOI: https://doi.org/10.1186/s12544-022-00556-z

[5] Coruzzolo, A. M.; Lolli, F.; Gennari, F.; Marinello, S. & Gamberini, R. (2021). Drivers for Selecting the Type of Transport in the Modal Shift: Cost, Transit Time and Environmental Impact Analysis, Proceedings of the 32nd DAAAM International Symposium, pp.0423-0432, B. Katalinic (Ed.), Published by DAAAM International, ISBN 978-3-902734-33-4, ISSN 1726-9679, Vienna, Austria DOI: 10.2507/32nd.daaam.proceedings.062

[6] Adamec, V. (2007): Transport, health and environment. Grada Publishing, a.s., Prague. ISBN: 9788024721569. Prague

[7] Papouskova, K. (2021) The Concept of Railway Transport in the Czech Republic [online]. Transportation Research Procedia, 53, 154-158. ISSN 2352-1465. DOI: https://doi.org/10.1016/j.trpro.2021.02.020

[8] Basic characteristics of the railway network. Railway administration [online]. Available from: https://www.spravazeleznic.cz/o-nas/vse-o-sprave-zeleznic/zeleznice-cr/zeleznicni-sit-v-cr

[9] Info about transport in the Czech Republic. CZREGION [online]. Available from: https://www.czregion.cz/info-o-doprave

[10] Passenger transport by rail. Transport yearbook 2022 [online]. 2022. Available from: https://www.sydos.cz/cs/rocenka-2022/rocenka/htm_cz/cz22_512000.html

[11] Comparison of the region with the Czech Republic. Regional information service [online]. Available from: https://www.risy.cz/cs/krajske-ris/pardubicky-kraj/kraj

DOI: 10.2507/34th.daaam.proceedings.047

PROPOSAL OF THE APPLICATION OF SELECTED RISK ANALYSES FOR FIRE PROTECTION

Adam Malatinsky & Martin Hromada

This Publication has to be referred as: Malatinsky, A[dam] & Hromada, M[artin] (2023). Proposal of the Application of Selected Risk Analyses for Fire Protection, Proceedings of the 34th DAAAM International Symposium, pp.0347-0352, B. Katalinic (Ed.), Published by DAAAM International, ISBN 978-3-902734-41-9, ISSN 1726-9679, Vienna, Austria DOI: 10.2507/34th.daaam.proceedings.047

Abstract

Currently, there are many professional quantitative or qualitative analyses for determining the risk level for different types of security. The article focuses on the proposal and characteristics of selected risk analyses, which are intended for several types of security. For each selected analysis, a method designed for the field of fire protection is proposed. The result of the article is a proposal for applying methods that determine specific forms of risk analysis within the fire protection framework to minimize the occurrence of a threat and thus prevent damage to the asset of the given object. Applying fire protection methods in the future creates additional possibilities for determining the degree of fire threats, their analysis, identification, implementation, evaluation and comparison. The main result of the article is to show readers examples of ways to determine risks in fire protection for possible future research.

Keywords: Risk Analyses; Fire Protection; Fire Safety; Methods; Safety

1. Introduction

The threat as a factor hurts the development of society. Therefore, it is necessary to deal intensively with the threats. A risk analysis is suitable for determining the threat. Risk analysis is one of the most widely used commercial or public sector methods. [1] Each risk analysis is determined differently, but each has one main goal - to determine the level of risk in a given area. One of the researched areas of risk analysis is fire protection.

Fire protection, as one of the oldest types of security, has constantly evolved to its current form. Nowadays, fire protection is found in almost every industry. According to the relevant legislation, every production and non-production facility must take fire measures to minimize threat risk. The article deals with the proposal for applying selected risk analyses for fire protection. First, fire protection, in general, is described. Subsequently, selected individual risk analyses are characterized, from which the next chapter proposes examples of the application of these analyses and methods in the field of fire protection. These are only selected risk analyses. That is, there are many more analyses for determining fire risk. The benefit of the article is a suggestion for experts in fire protection on how to analyse fire risks in buildings. [2], [3], [4]

2. Fire Protection

The fire was and still is one of the primary threats that can occur anywhere and anytime worldwide. Currently, various organizations deal with fire protection, and the legal framework solves it through laws, decrees, regulations, technical standards and the like. Fire, as well as other threats, harms our society. Therefore, it is necessary to deal with this problem and then create and apply means for its elimination to protect people, animals and property. [5], [6]

3. Characteristics of risk analyses

In this chapter, the individual risk analysis methods that will be proposed for application to issues in the field of fire protection are characterized. Specifically, it is about: Checklist, Safety Audit, What-If Analysis, Fishbone Diagram, Preliminary Hazard Analysis, Process Quantitative Risk Analysis, Hazard Operation Study, Event Tree Analysis, Failure Mode and Effect Analysis, Fault Tree Analysis, Human Reliability Analysis, Relative Ranking, Causes and Consequences Analysis, Probabilistic Safety Assessment, Bow Tie, SWOT analysis and KARS analysis. [7], [8]

4. Applying risk analyses on fire protection

Within this chapter, possible proposals for fire protection are applied to selected risk analyses (from Chapter 3). The proposed solutions and determination of selected risk analyses within the fire protection framework are based on the author's expert estimate.

4.1 Checklist

The checklist is based on a systematic check of the fulfilment of predetermined conditions and measures. Control questions are usually generated based on a list of characteristics of the monitored system or activities related to the system and potential impacts, failure of system elements and damages. Their structure can vary from a simple list to a complex form that allows the inclusion of different relative importance of a parameter (weight) within a given set. [7]

This method can be used in several ways. One way of using this method is to create a checklist for the execution of fire protection repression. Specifically, it is a list of the procedures of the fire brigade members from the receipt of the report from the operation centre of the occurrence of an emergency until its liquidation. For example, a list of house fire disposal procedures from the exit of the fire protection unit from the fire station to its return. These lists would be appropriate to apply for various fire protection emergencies.

4.2 Safety audit

A safety audit identifies risky situations and proposes measures to increase safety. The method represents a procedure for searching for a potentially possible accident or operational problem that may appear in the system under consideration. [7]. A safety audit within fire protection would deal primarily with fire prevention and its control. The safety audit would be performed by a professionally qualified person in the field of fire protection, a fire protection technician or a fire protection preventive officer. Another entity performing a safety audit would be the state fire watchdog. A license to perform would be required to perform a security audit. A penalty would be assessed to the entity for violating the rules.

4.3 What – If Analysis

What-if analysis is a procedure for finding possible impacts of selected operational situations. It is a spontaneous discussion and search for ideas in which a group of experienced people familiar with the process ask questions or express thoughts about possible accidents. It is not an internally structured technique like some others. Instead, it requires the analyst to adapt the basic concept of saving to a particular purpose. [7]. The design of the application of this analysis is the gradual creation of scenarios in the event of a fire and its consequences. Each consequence would create other consequences until the last step, where the resulting consequences would be the fire threat. The result of the method would be a retrospective comparison of which causes of the fire are the most significant and subsequently deal with them and create a safe situation as a precaution. For example, in the event of a fire, how will the fire department behave if there is an explosion or leakage of a dangerous substance.

4.4 Fishbone diagram

The method helps to model and structure a process or identify possible causes of a problem. The purpose is to determine the most likely cause of the problem being addressed. [7] In this case, it is about determining the root cause, the gradual increase in the consequences of which can cause damage to lives or property. For example, a fire broke out in the building, and the root cause was overheating of machines and equipment.

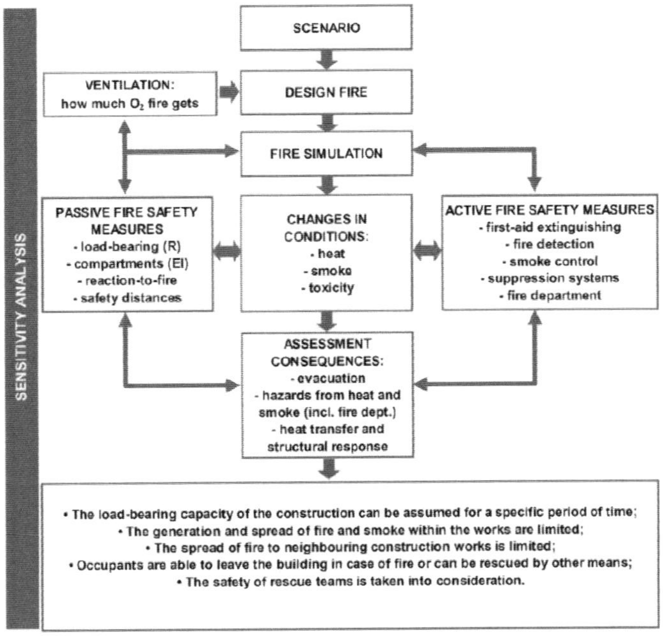

Fig. 1. Example of What – If Analysis [8]

4.5 Preliminary Hazard Analysis (PHA)

It is also called the quantification of risk sources. It is a procedure for searching for dangerous conditions and emergencies, their causes and effects, and classifying them according to predetermined criteria. [7]. This analysis would create groups from the least significant risk to the most significant. Causes would be assigned to each group, and the impacts of extraordinary events would be evaluated. For example, among the less significant risks in fire protection is the non-functioning of fire extinguishers, and among the more significant risks is improper smoke ventilation during a fire.

4.6 Process Quantitative Risk Analysis (QRA)

A comprehensive and systematic approach to predict the frequency and impact of accidents for equipment or system operation. [7]. This method would only deal with the preventive part of fire protection, where preventive measures and related activities would be comprehensively described. For example, they are solving strategic measures for the fire safety of buildings and updating legislation and technical standards.

4.7 Hazard Operation Study (HAZOP)

A procedure based on a probabilistic assessment of threats and the resulting risks. [7], [9]. This procedure would create coefficients with a certain weight and significance, from which the degree of risk probability and the possible emergence of a threat would be determined. For example, the coefficient of fire, smoke, explosion, and dangerous substance leakage would be evaluated.

4.8 Event Tree Analysis (ETA)

A procedure following the progress of the process from the initiating event through the construction of the event based on favourable and unfavourable possibilities. [7]. The method would compare the process of the preventive and repressive parts of fire protection in a favourable and, at the same time, unfavourable situation. For example, a fire incident was eliminated in a short period without harm to health, and at the same time, a fire incident lasting at least a week with a massive smoke leak, the necessary evacuation of the population.

4.9 Failure Mode and Effect Analysis (FMEA)

A procedure based on analysing failure modes and their consequences enables the search for impacts and causes based on systematic and structured equipment failures. [7]. The analysis would evaluate and create analyses of the reasons for the emergence of fire threats, where the main initiator would be the failure of the human factor, technical or natural. For example, the fire was caused by an employee smoking at the workplace, and a discarded butt caused the fire.

4.10 Fault Tree Analysis (FTA)

A procedure based on a systematic retrospective analysis of events using a chain of causes that can lead to a selected peak event. [7]. The analysis method is similar to the FMEA method. In this case, it describes the cause and the peak event. For example, failure to comply with fire safety measures resulted in a massive fire, which resulted in the loss of life of 5 employees. It compares the root cause and the worst impact of the given event.

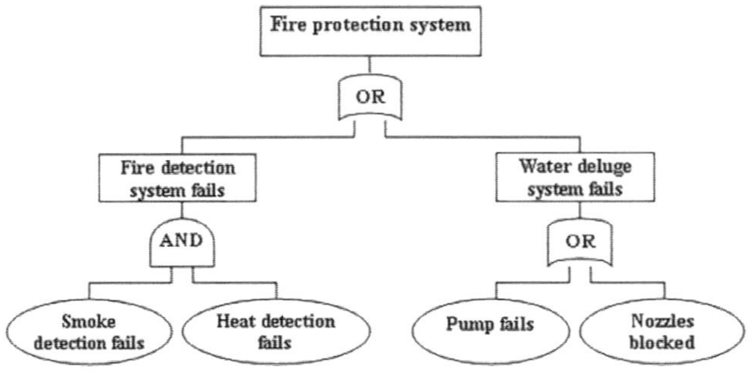

Fig. 2. Example of FTA [10]

4.11 Human Reliability Analysis (HRA)

Procedure for assessing the influence of the human factor on the occurrence of natural disasters, accidents, breakdowns, attacks, or some of their effects. [7]. In this case, it is an analysis of anthropogenic factors. That is, the cause of the emergency is the human factor. For example, a person was responsible for the threat of leaking a dangerous substance by not following safety regulations and improperly handling a dangerous substance.

4.12 Relative Ranking (RR)

An analytical strategy that allows analysts to compare the characteristics of several processes or activities and determine whether those processes or activities have sufficiently hazardous characteristics to warrant further, more detailed study by analysts. [7]. Fire protection analysts would have the primary say in this strategy. Specifically, it would be professionally qualified persons, fire protection technicians and fire protection preventionists, who all evaluate various processes and deal with a more detailed study strategy in the field of fire protection. For example, they are looking for new ways to extinguish fire using other means and equipment used today.

4.13 Causes and Consequences Analysis (CCA)

A mixture of fault tree analysis and event tree analysis. The resulting diagram of the analysis shows the relationships between the end states of the unacceptable impacts and the root causes. [7]. This analysis uses the ETA and FTA methods, resulting in an impact and cause analysis diagram. The analysis compares the ideal development of an extraordinary event with the worst possible way. At the same time, another determining factor would be monitoring root causes with the worst possible way of developing an extraordinary event. For example, the analysis follows the development of a fire with a high-speed extinguishing method and, simultaneously, with a prolonged fire extinguishing method. At the same time, this process includes looking for the root cause of the fire and evaluating the worst possible scenario of the fire.

4.14 Probabilistic Safety Assessment (PSA)

The method determines the contributions of individual vulnerable parts to the overall vulnerability of the entire system. [7]. The method primarily focuses on system vulnerabilities. As part of fire protection, it looks for the most severe vulnerabilities of the object, which can subsequently cause damage. For example, it evaluates the material from which the object is built and evaluates the fire resistance with the relevant characteristics of the material.

4.15 Bow tie

Risk visualization expresses a clear distinction between proactive and reactive risk management. [7]. This method graphically visualizes risk, especially the difference between proactive and reactive risk management. As part of fire protection for proactive risk management, it creates various safety precautions, increases frequent building checks, and adds more fire detectors and alarms. As part of reactive risk management, it evaluates how devices react reactively to the occurrence of fire and how they minimize and eliminate it. For example, it may be the risk of using sprinklers or heat and smoke extraction devices.

4.16 SWOT analysis

Analysis of strengths, weaknesses, opportunities and threats. The result of the analysis is also the evaluation of the resulting strategy, which can be offensive, defensive, liquidation or alliance. [11]. The analysis defines the object's strengths, weaknesses, opportunities and threats regarding fire protection. Weight and significance will be assigned to each part, where the most significant parts will be evaluated. An appropriate strategy will then be created from them. Analysts in the field of fire protection will further deal with the strategy and create appropriate preventive and punitive measures.

4.17 KARS analysis

Qualitative analysis using risk correlation. The analysis results in the division of risks into primary, secondary and tertiary. [11]. Based on the correlation of fire risks from the given list, the KARS analysis evaluates individual risks through the coefficient of activity and passivity. It divides them according to significance into primary, secondary and tertiary. For example, the primary risk of an object is fire, leakage of a dangerous substance or explosion. Secondary risks include overheating machines and equipment, blocked escape exits and insufficiently trained personnel. Tertiary risk includes an insufficient number of fire extinguishers in the workplace.

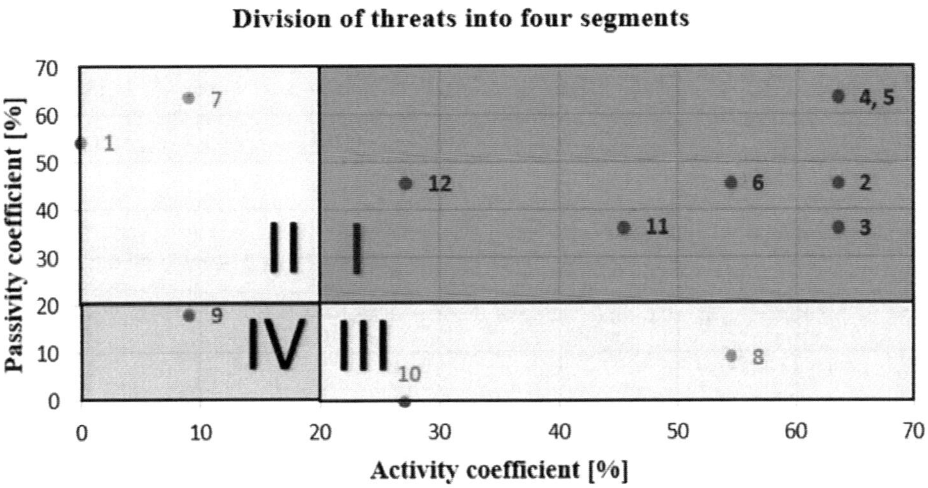

Fig. 3. Example of KARS analysis [11]

The chapter dealt with the characteristics of risk analyses, followed by their possible application proposal. For each risk analysis, a fire protection risk assessment method is proposed.

5. Conclusion

Risk analysis is one of the most important methods of determining the safety of an object. The object's risks also include risks in fire protection. The article dealt with the characteristics of selected risk analyses, followed by a proposal for a possible application in fire protection. Firstly, the chapter describes fire protection in general. The following chapter deals with the general definition and characteristics of 17 risk analyses, where a possible proposal for applying these risk analyses to the field of fire protection is presented, and specific examples for each analysis are also given. These suggestions suit all those who deal with fire protection, risk analysis, prevention experts and other interested people as a way of possibly implementing risk analysis to minimize them. In order to compare results from other sources where risk analysis is applied, the result of the article is also a brief overview of selected risk analyses.

The limit of the article is the use of only selected risk analyses, which are usually evaluated separately or differently. In the future, it is appropriate to apply these risk analyses to fire protection and to evaluate the probabilities and rates of occurrence of risks and thereby minimize them, whether from the preventive, proactive or repressive part of fire protection.

6. Acknowledgments

This research was based on the support of the Internal Grant Agency of Tomas Bata University in Zlín, IGA/FAI/2023/001 and The Department of Security Engineering, Faculty of Applied Informatics.

7. References

[1] Nolan, D. P. (2019). Chapter 7 – Risk Analysis [online]. Handbook of Fire and Explosion Protection Engineering Principles for Oil, Gas, Chemical, and Related Facilities (Fourth Edition), 151-168. ISBN 9780128160022. DOI: https://doi.org/10.1016/B978-0-12-816002-2.00007-6.

[2] Cajkova, N. & Dzermansky, M. (2022). Risk Analysis of the Post Office, Proceedings of the 33rd DAAAM International Symposium, pp.0352-0357, B. Katalinic (Ed.), Published by DAAAM International, ISBN 978-3-902734-36-5, ISSN 1726-9679, Vienna, Austria. DOI: 10.2507/33rd.daaam.proceedings.049.

[3] Cajkova, N. & Dzermansky, M. (2022). Risk Analysis of a Gas Station, Proceedings of the 33rd DAAAM International Symposium, pp.0358-0364, B. Katalinic (Ed.), Published by DAAAM International, ISBN 978-3-902734-36-5, ISSN 1726-9679, Vienna, Austria. DOI: 10.2507/33rd.daaam.proceedings.050.

[4] Nambisan, P. (2017) Chapter 10 - Risk Analysis [online]. An Introduction to Ethical, Safety and Intellectual Property Rights Issues in Biotechnology, Academic Press, 233-252. ISBN 9780128092316. DOI: https://doi.org/10.1016/B978-0-12-809231-6.00010-7.

[5] Malatinsky, A. & Hromada, M. (2022). Preventive and Repressive Measures for Fire Safety, Proceedings of the 33rd DAAAM International Symposium, pp.0397-0403, B. Katalinic (Ed.), Published by DAAAM International, ISBN 978-3-902734-36-5, ISSN 1726-9679, Vienna, Austria. DOI: 10.2507/33rd.daaam.proceedings.056.

[6] Xin, J. & Huang, C. (2013). Fire risk analysis of residential buildings based on scenario clusters and its application in fire risk management [online]. Fire Safety Journal, 2013, 62, 72-78. ISSN 0379-7112. DOI: https://doi.org/10.1016/j.firesaf.2013.09.022.

[7] Kratochvil, V., Navarova, S. & Kratochvil, M. (2021). Fire safety equipment in buildings: A brief encyclopedia for fire protection units, fire prevention and the professional public. Second revision edition. Prague: Association of Fire and Safety Engineering. ISBN 978-80-7385-238-2.

[8] Hietaniemi, J. & Mikkola, E. (2010). Design Fires for Fire Safety Engineering [online]. ISBN 978-951-38-7479-7. ISSN 1459-7683.

[9] Crawley, F. & Tyler, B. Chapter 10 - Advanced Aspects of HAZOP Study [online]. 2015, 53, 62-76. ISBN 9780323394604. DOI: https://doi.org/10.1016/B978-0-323-39460-4.00010-4

[10] Mahapatra, G. S. & Mahapatra, B. S. (2010). Intuitionistic Fuzzy Fault Tree Analysis Using Intuitionistic Fuzzy Numbers. International Mathematical Forum (21), 1015-1024.

[11] Malatinsky, A. & Hromada, M. (2022) Evaluation of the most important fire threats of the building," 2022 IEEE International Carnahan Conference on Security Technology (ICCST), Valeč u Hrotovic, Czech Republic, 2022, pp. 1-6, DOI: 10.1109/ICCST52959.2022.9896486.

fDOI: 10.2507/34th.daaam.proceedings.048

ANALYTICAL CALCULATION AND EXPERIMENTAL MEASUREMENT COMPARISON OF BOLTED JOINT WITH INSERT FOR COMPOSITE SANDWICH PANEL

Katerina Kasova & Martin Melichar

This Publication has to be referred as: Kasova, K[aterina] & Melichar, M[artin] (2023). Analytic Calculation and Experimental Measurement Comparison of Bolted Join with Insert for Composite Sandwich Panel, Proceedings of the 34th DAAAM International Symposium, pp.0353-0360, B. Katalinic (Ed.), Published by DAAAM International, ISBN 978-3-902734-41-9, ISSN 1726-9679, Vienna, Austria
DOI: 10.2507/34th.daaam.proceedings.048

Abstract

This paper addresses the design of a structural joint involving a composite sandwich panel. In the context of composite sandwich panels, ensuring a robust connection between the panel and other structural elements is essential. Typically, these components are secured to the composite panel through bolted joint, making the bolted joint between the composite panel and the structural structure the focus of this paper. The primary focus here is on the design methodology of a pass-through insert with a hot bonded joint.

Keywords: Composite sandwich panel; Bolted joint; Experimental test; Insert; Carbon fibre.

1. Introduction

For composite sandwich panels, ensuring the connection between the panel itself and other structural elements is crucial. These components are primarily attached to the composite panel through bolts. The mechanics of composite joints constitute a complex discipline, and a dedicated study is necessary for designing specific joints. The central research question revolves around designing a bolted joint with a sandwich composite panel for selected material combinations. This study delves into the intricate world of designing and assessing the integrity of these connections, which play a pivotal role in aerospace, construction, and various high-performance applications. While analytical calculations serve as a fundamental tool for initial design and estimation, their accuracy can be limited, particularly for complex material combinations. The incorporation of experimental measurements becomes essential to validate and refine these designs, ensuring they meet rigorous safety and performance standards. [1], [2].

By gaining a better understanding of the strengths and interplay between these two approaches, the main goal is to enhance the reliability and precision of bolted joints in composite sandwich panels. The insights gleaned from this analysis hold the promise of advancing engineering practices and, in the long run, contributing to the development of safer and more efficient designs across various industries. [2], [3]. A fundamental challenge with bolted joints involving composite sandwich panels is the compressive strength of the panel core. When a bolted joint is preloaded, the core can locally collapse, leading to joint failure (figure 1).

Fig. 1. Local core collapse in the bolted joint of a composite sandwich panel

Therefore, such joints are inadequate without additional reinforcing elements. In place of a bolted joint, an insert (Fig. 2) with high compressive strength must replace the core. Various materials are available for the insert, which must meet several essential criteria, including temperature resistance during panel curing, adequate compressive strength for preloading the joint, good adhesive properties, and more [4], [5], [6].

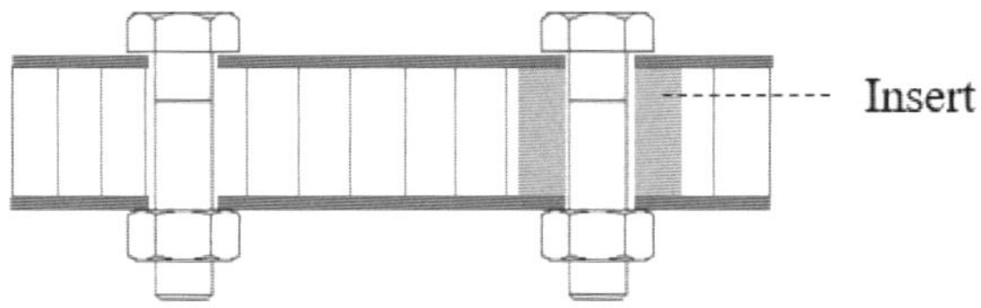

Fig. 2. Left bolted joint without insert; right bolted joint with insert

Inserts are categorized into two main types, depending on their attachment method within the sandwich panel:

- Hot-glued inserts are produced concurrently with the panel, with the skin/insert and core/insert bonded during the curing cycle of the skin.
- Cold-bonded inserts involve additional bonding of the insert after panel fabrication.

Other critical factors include whether the insert is pass-through or non-pass-through and the adhesive surface of the insert. Some standard insert types are illustrated in figure 3.

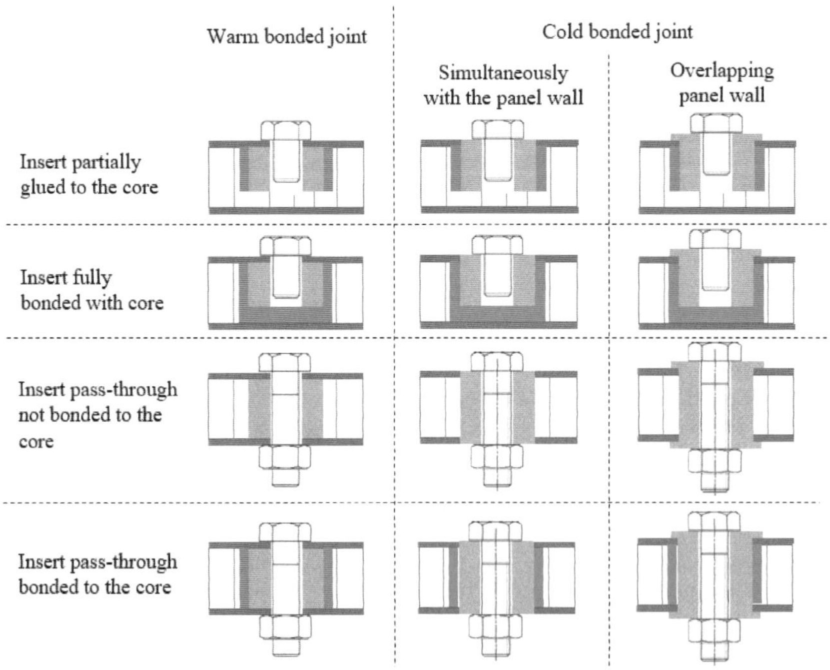

Fig. 3. Possible types of inserts and their method of insertion into the sandwich panel [7]

The design of a bolted joint with an insert is a highly intricate task influenced by various factors, including the materials of the sandwich panel, insert type and size, adhesive used, stress type, joint reliability, and more. Detailed guidelines for designing a bolted joint with a bonded insert are provided in the publication "Space Engineering – Insert design handbook from the European Cooperation for space standardization" [8]. Another publication, "Validity check of an analytical dimensioning approach for potted insert load introductions in honeycomb sandwich panels" [6], compares analytical calculations with experimentally determined values for the design of a pass-through joint with an insert, revealing significant discrepancies. This underscores the necessity of supporting the fundamental design with experimental measurements for a given material combination. The knowledge derived from these publications was applied in designing the bolted connection with an insert, as discussed in the following section.

2. Analytical calculation of bolted join with insert

Given the complexity of this structural aspect, this paper exclusively focuses on the design methodology of the pass-through with a hot-glue bonded joint. This calculation deals with two design states for uniaxial tensile loads conditions:

1. Maximum stiffness condition, where there is no permanent deformation or composite panel failure.

The analytical calculation for the maximum allowable tensile load of the insert (equation 1) applies under several assumptions, such as neglecting the flexibility of inserts and adhesive, the elastic modulus of the skin being significantly greater than that of the core; $Ef \ll Ec$, the skin thickness t is several times smaller than the core thickness c; $t \ll c$, the distance between the edge of the insert and the embedded part of the panel is uniform (it is a circular sample). [9],[10]. Based on the aforementioned assumptions and the ECSS publication [7], the highest shear stress occurs at the insert/adhesive/core interface, as depicted in the figure 4.

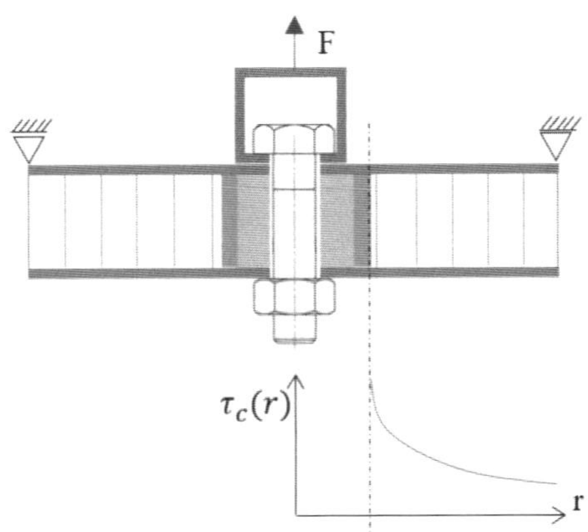

Fig. 4. Shear stress distribution in the panel section with insert

$$F_{cri} = 2\pi b_i d \tau_c \quad (1)$$

where F_{cri} is the critical load at which a failure occurs at the boundary between the insert and the core (N), b_i is the diameter of the insert including the adhesive layer (m), d is the mean distance of the skin (m), τ_c is the allowable shear load of the core (MPa).

2. Maximum tensile strength, where there is permanent deformation or composite panel failure (figure 5).

Shear failure of the skin occurs when the allowable shear stress of the skin is exceeded, as described in equation 2. The experimental measurement value of F_{kri-s} was lower than the calculated value because there is no ideal simultaneous failure of the outer and inner skin, as explained in equation 2.

$$F_{cri-s} = (\pi b_i t + \pi b_p t)\tau_{fmz} \quad (2)$$

where F_{cri-s} is the critical load at which the shear failure of the skin occurs when loading the bolted connection with the insert in tension (N), b_i is the diameter of the insert including the adhesive layer (m), b_p is the diameter of the washer (m), τ_{fmz} is the shear strength of the skin perpendicular to the plane of the skin (MPa).

The skin shear strength perpendicular to the plane of the skin is not a catalogue value and must be determined experimentally.

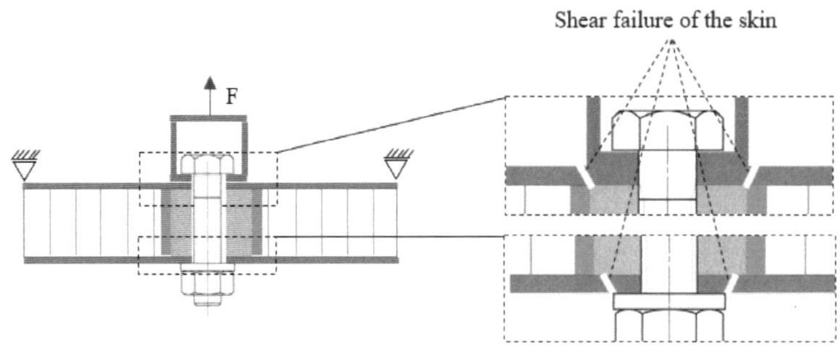

Fig. 5. Shear failure of the skin

3. Experimental measurement of skin shear strength

Experimental measurement of the shear strength of the skin in the direction perpendicular to the plane of the skin was conducted on 12 test specimens using various material combinations. Carbon fabric Toray 3K 200 Twill 2x2 was used for the skin material, while two different cores with different cell sizes were utilized for the core material. The 100x100 mm samples were produced via a two-cycle curing process in an autoclave, with IMP380FHC Black adhesive film used for the adhesive bond between the skin and core 250 g.m^{-2}.

Specimen	Material of skin	No. of layers in skin	Core	Core cell size (mm)	Panel height (mm)
Shear_P[4]_C[3,2/72]	Toray 3K 200 Twill 2x2	4	PAMG-XR-4.5-1/8-10-P-5056	3.2	21.8
Shear_P[4]_C[4.8/32]			PAMG-XR-2.0-3/16-07-P-5056	4.8	21.6
Shear_P[5]_C[3,2/72]		5	PAMG-XR-4.5-1/8-10-P-5056	3.2	22.2
Shear_P[5]_C[4.8/32]			PAMG-XR-2.0-3/16-07-P-5056	4.8	22.0

Table 1. List of test specimens for determining the shear strength of the skin

The shear strength tests were carried out under quasi-static loading at a speed of 5 mm.min^{-1} on a Zwick/Roell Z050 test rig, using a 25 mm push-through mandrel. A schematic of the test specimen arrangement is shown in the figure 6. The maximum shear strength of the skin perpendicular to the skin plane was determined from the maximum achieved load at the failure of the outer skin layer, as defined by equation 3.

$$\tau_{fmz} = \frac{F_{s-max}}{\pi d_{push} t} \tag{3}$$

where F_{s-max} is the maximum measured load during the test (N), d_{push} is the diameter of the shear mandrel (m), t is the thickness of the skin (m), τ_{fmz} is the shear strength of the skin perpendicular to the plane of the skin (MPa).

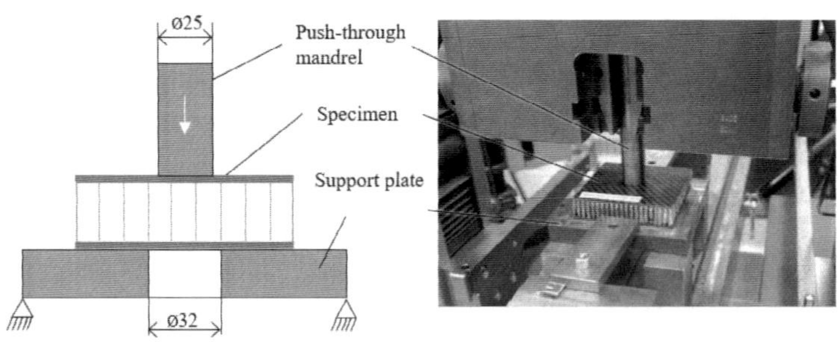

Fig. 6. Schematic of the specimen placement during the skin shear strength test

The results of the measurements were recorded in table 1. For each group of samples, the average value of the maximum load and the shear strength τ_{fmz} were calculated. Deviation in single specimen measurements is within 5% in most cases, except for Shear_P[4]_C[4.8/32], where the deviation is 7.3 %. It was evident from the measured values that the limit shear strength is significantly influenced by the core's stiffness, which provides support for the skin during testing.

sSpecimen	Maximum measured load (N)	Average maximum load (N)	Deviation of measured values from the average (%)	Thickness of the skin (mm)	Ultimate shear strength (MPa)
Shear_P[4]_C[3,2/72]_S[1]	6118.9				
Shear _P[4]_C[3,2/72]_S[2]	6011.4	6101.8	2.7	1.04	**74.70**
Shear _P[4]_C[3,2/72]_S[3]	6175.0				
Shear _P[4]_C[4,8/32]_S[1]	4968.6				
Shear _P[4]_C[4,8/32]_S[2]	4618.1	4787.6	7.3	1.04	**58.61**
Shear _P[4]_C[4,8/32]_S[3]	4776.1				
Shear _P[5]_C[3,2/72]_S[1]	7171.1				
Shear _P[5]_C[3,2/72]_S[2]	7414.2	7243.6	3.7	1.3	**70.94**
Shear _P[5]_C[3,2/72]_S[3]	7145.4				
Shear _P[5]_C[4,8/32]_S[1]	5435.8				
Shear _P[5]_C[4,8/32]_S[2]	5452.7	5493.3	2.8	1.3	**53.80**
Shear _P[5]_C[4,8/32]_S[3]	5591.4				

Table 2. Resulting measured values with specified shear strength

4. Experimental measurement of the strength of a pass-through bolted joint with insert

For the experimental measurement of the strength of the pass-through bolted joint with the insert, twelve test specimens were created. These specimens were identical to the ones used for shear strength measurements as shown in table 1. The sample dimensions were 120x120 mm, produced through a two-cycle curing process in an autoclave. IMP380FHC Black adhesive film was used to bond the skin and core, with an adhesive mass of 250 g.m^{-2}. The inserts were made of TECAPEEK CF30, with outer dimensions of 20 mm, height of 20 mm, and an inner diameter of 8 mm. Strength tests for the bolted joint with inserts were conducted under quasi-static loading conditions at a speed of 1 mm.min^{-1} using a Zwick/Roell test rig Z050. The tests followed the ECSS standard as per ECSS-E-HB-32-22A. The schematic of specimen placement during the tensile strength test is presented in figure 7.

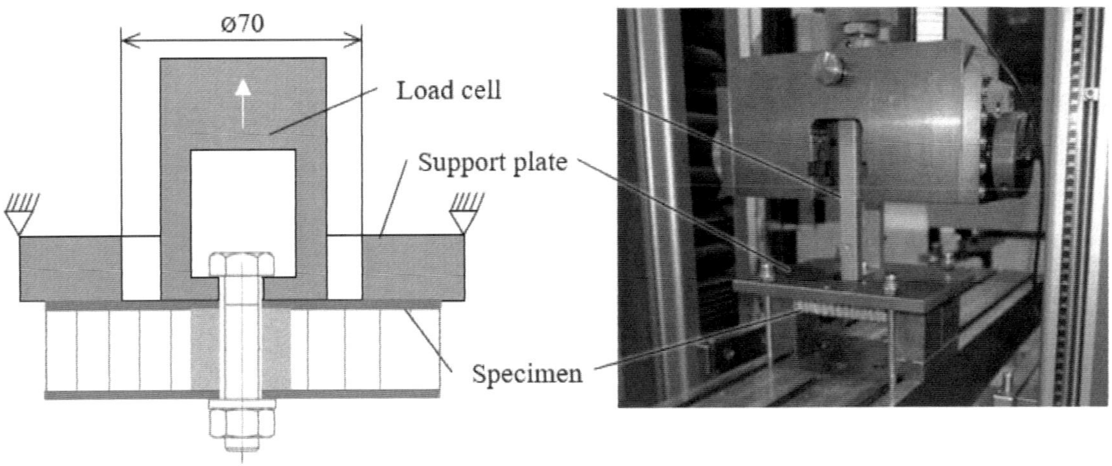

Fig. 7. Schematic of specimen placement during the tensile strength test of the insert

The recorded experimental measurement of the strength of the insert bolted connection for specimens Insert_P[4]_C[3,2/72] is shown in Graph 1. To determine the maximum stiffness, the dF/dy curve was plotted on graph 1. The measured force values at maximum stiffness for these specimens were approximately 60 % to 70 % of the proposed value of F_{cri} based on equation 1.

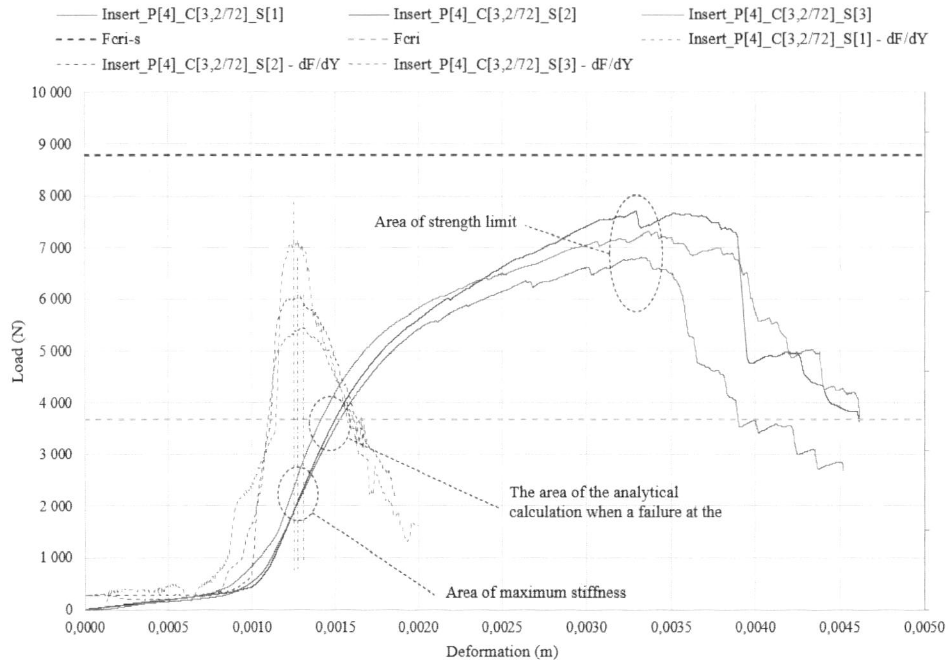

Graph 1. Experimental measurement of the strength of a bolted joint with an insert

The measured values are presented in table 3. The deviation of the measured values of the individual sample groups from the calculated average ranged from 5.9 % to 23.9 %. The term "uncertainty" refers to measurement deviation. [11] The deviation of the average critical load at the interface from the analytical calculation ranged from -17.2 % to -2.4 %. Using this deviation from the analytical calculation and the deviation from the measurement, a value was established for $F_{ckri\text{-}s0.7}$ according to equation 4.

$$F_{cri-s0,7} = 0,7\ F_{cri-s} \tag{4}$$

where $F_{cri-s0,7}$ is 70 % of the value F_{cri-s} (N), F_{cri-s} is the critical load at which the shear failure of the skin occurs when loading the bolted connection with the insert in tension (N).

Specimen	**Critical loads at ultimate strength (N)**	**Average critical load at ultimate strength (N)**	**Deviation of measured values from the average (%)**	F_{cri-s} **(N)**	**Deviation F_{cri-s} from the average measured critical load (N)**	**Deviation F_{cri-s} from the average measured critical load (%)**
Insert_P[4]_C[3,2/72]_S[1]	6808.7					
Insert_P[4]_C[3,2/72]_S[2]	7714.2	7277.8	12.4	8786	- 1508.2	-17.2
Insert_P[4]_C[3,2/72]_S[3]	7310.4					
Insert_P[4]_C[4,8/32]_S[1]	6838.7					
Insert_P[4]_C[4,8/32]_S[2]	7270.6	6725.9	17.9	6894	- 168.1	-2.4
Insert_P[4]_C[4,8/32]_S[3]	6068.3					
Insert_P[5]_C[3,2/72]_S[1]	9169.9					
Insert_P[5]_C[3,2/72]_S[2]	9179.2	8999.6	5.9	10430	- 1430.4	-13.7
Insert_P[5]_C[3,2/72]_S[3]	8649.7					
Insert_P[5]_C[4,8/32]_S[1]	6571.9					
Insert_P[5]_C[4,8/32]_S[2]	8286.0	7182.5	23.9	7910	- 727.5	-9.2
Insert_P[5]_C[4,8/32]_S[3]	6689.7					

Table 3. Final measured values of the critical load at the ultimate strength with deviation from the analytical calculation

The values of the specimens with the insert at maximum stiffness are provided in table 4. The deviation of the measured load concerning the load F_{cri} when failure occurs at the core-insert boundary was calculated, with deviations reaching up to 49.2 %.

Specimen	Load at maximum stiffness (N)	Average load at maximum stiffness (N)	F_{cri} (N)	Deviation F_{kri} from average measured load at maximum stiffness (N)	Deviation F_{kri} from average measured load at maximum stiffness (%)
Insert_P[4]_C[3,2/72]_S[1]	2261.3	2336.4	3648.3	-1342.0	-36.5
Insert_P[4]_C[3,2/72]_S[2]	2149.7				
Insert_P[4]_C[3,2/72]_S[3]	2598.0				
Insert_P[4]_C[4,8/32]_S[1]	952.4	981.3	872	102.9	11.7
Insert_P[4]_C[4,8/32]_S[2]	1061.1				
Insert_P[4]_C[4,8/32]_S[3]	930.4				
Insert_P[5]_C[3,2/72]_S[1]	1912.8	1879.5	3676.6	-1823.7	-49.2
Insert_P[5]_C[3,2/72]_S[2]	1685.3				
Insert_P[5]_C[3,2/72]_S[3]	2040.4				
Insert_P[5]_C[4,8/32]_S[1]	978.5	971.6	880.1	87.2	9.9
Insert_P[5]_C[4,8/32]_S[2]	1015.4				
Insert_P[5]_C[4,8/32]_S[3]	921.0				

Table 4. Final measured load values at maximum stiffness with deviation from the analytical calculation

5. Conclusion

This paper discusses the comparison of analytic calculations and experimental measurements of bolted joints with inserts in composite sandwich panels. The design of a bolted joint is a complex process influenced by various factors. To calculate the bolted joint with an insert, the shear strength rating of the skin was determined through experimental measurements, and these values are presented in table 2. Experimental measurements were also conducted to assess the strength of the bolted joint with the insert perpendicular to the plane of the skin. The load values of specimens with insert at maximum stiffness are shown in Table 4. The deviation of the measured load with respect to the load F_{cri} when failure occurs at the core-insert boundary was calculated. This deviation amounts to a total of 49.2 %. This is due to the insufficient understanding of the deformation of the core at the boundary with the insert and the difficulty in determining the first core and skin failure. The resulting deviation was directly dependent on the shear stiffness of the core, with respect to the measured parameters.

Based on the presented results, it is evident that relying solely on analytical calculations is insufficient when designing bolted joint with inserts. For the optimal design of bolted joint with composite sandwich panels, it is imperative to incorporate experimental measurements of tensile properties and the specific insert joint. Without this experimental foundation, the precision of design based solely on analytical calculations would not meet the required standards. The insights presented in this paper have led to a better understanding of the issues surrounding bolted joint with composite sandwich panels for specific material combinations. These findings enable the design of bolted joint for these specific material combinations with greater accuracy than could be achieved through analytical calculations alone. In the future, these insights will be leveraged for the design of specific bolted joint with composite sandwich panels, followed by a retrospective verification through experimental measurements.

6. Acknowledgments

The article contribution has been prepared under project SGS-2022-007.

7. References

[1] Rana S. & Fangueiro R. (2016). Advanced Composite Materials for Aerospace Engineering: Processing, Properties and Applications. Woodhead Publishing, ISBN 978-0-08-100054-0, UK.

[2] Zenkert D. (1995). Introduction to Sandwich Construction. Engineering Materials Advisory Services Ltd., ISBN 978-0-947817-77-0, UK.

[3] Barbero E. J. (2010). Introduction to Composite Materials Design, Second Edition, 2 edition, CRC Press, ISBN 978-1-4200-7915-9, UK.

[4] Elmarakbi A. (2013). Advanced Composite Materials for Automotive Applications: Structural Integrity and Crashworthiness, Edición: 1, ISBN13:978-1-11842-36-8, Chichester, West Sussex, United Kingdom: Wiley.

[5] Bitzer T. N. (1997). Honeycomb Technology: Materials, Design, Manufacturing, Applications and Testing, Springer ISBN 978-9-401158-56-5, Netherlands.

[6] Zweben C. H. & Beaumont P. (2017). Comprehensive Composite Materials II, Elsevier, ISBN 978-0-081005-33-0, Netherlands.

[7] Wolff J.; Brysch M. & Hühne C. (2018). Check of an analytical dimensioning approach for potted insert load introductions in honeycomb sandwich panels, Compos. Struct., Available from: https://www.sciencedirect.com /science/article/abs/pii, Accessed on: 2022-08-13.

[8] https://ecss.nl/hbstms/ecss-e-hb-32-22a-insert-design-handbook (2011). ECSS E HB 32 22A Insert Design Handbook. ESA Requirements and Standards Division, Noordwijk, Netherlands, Accessed on: 2023-06-08.

[9] Campbell F. C. (2010). Structural Composite Materials, New ed. edition. Materials Park: ASM International, ISBN 9781615031405, Ohio.

[10] Jingsi W.; Badu O. A.; Yonchen T. & George A. R. (2014). Design, Analysis, and Simulation of an Automotive Carbon Fiber Monocoque Chassis. SAE Int. J. Passeng. Cars - Mech. Syst. 7(2):838-861, Available from: https://doi.org/10.4271/2014-01-1052, Accessed on: 2022-10-9.

[11] Kubatova, D. & Melichar, M. (2018). Uncertainty of Surface Measurement, Proceedings of the 29th DAAAM International Symposium, pp.1239-1248, B. Katalinic (Ed.), Published by DAAAM International, ISBN 978-3-902734-20-4, ISSN 1726-9679, Vienna, Austria DOI: 10.2507/29th.daaam.proceedings.179

DOI: 10.2507/34th.daaam.proceedings.049

AERODYNAMIC COLLECTION EFFICIENCY: NUMERICAL SIMULATION AND ANALYSIS.

Mohammad Yousef Hdaib, Anas F A Elbarghthi, Musaddaq Azeem, Mohamed Abdelkader, Jiri Ragulik & Vaclav Dvorak

This Publication has to be referred as: Hdaib, M[ohammad] Y[ousef]; Elbarghthi, A[nas]; Azeem, M[usaddaq]; Abdelkader, M[ohamed]; Ragulik, J[iri] & Dvorak, V[aclav] (2023). Aerodynamic Collection Efficiency: Numerical Simulation and Analysis, Proceedings of the 34th DAAAM International Symposium, pp.0361-0364, B. Katalinic (Ed.), Published by DAAAM International, ISBN 978-3-902734-41-9, ISSN 1726-9679, Vienna, Austria
DOI: 10.2507/34th.daaam.proceedings.049

Abstract

The shortage of clean water resources is a common problem throughout the globe, which leads to an increasing interest in fog collection technologies. Therefore, finding new methods to optimize collection efficiency is essential. In this study, the aerodynamic efficiency of a fog collector was investigated numerically. A 3D Computational Fluid Dynamic (CFD) model was built based on the discrete phase modeling and Darcy-Forchheimer relation for Porous media. The Darcy-Forchheimer relation was fitted experimentally to find the porous media parameters, and the model was validated using results from the literature. Moreover, the velocity and pressure distribution around the fog collector were simulated and compared. The results revealed good agreement with experimental data with a high-pressure drop of 150 % and a velocity drop of 100 %.

Keywords: Fog collector; aerodynamic efficiency; CFD; Porous media.

1. Introduction

A fog collector is a device designed to extract clean water from fog droplets. It comprises three main components: mesh, gutter, and storage. The process of water harvesting occurs when droplets intersect the mesh and are directed to the storage via the gutter [1]. The performance of fog collection is assessed using the total collection efficiency, calculated as the product of three different efficiencies: aerodynamic collection efficiency, deposition efficiency, and drainage efficiency. This study focuses on the collector's resistance to airflow, represented by the aerodynamic collection efficiency (η_{ACE}). The aerodynamic collection efficiency indicates the proportion of water droplets in the upstream flow that could potentially impact the mesh fibers [2]. Several studies have developed theoretical models for aerodynamic collection efficiency [2]–[4]. In numerical simulations, porous media models have been employed to replace the mesh fibers [5]. This approach significantly reduces computation time by avoiding the structural complexities associated with mesh geometry. Carvajal et al. [5] conducted a large-scale simulation of a fog collector using the Darcy-Forchheimer porous medium model.

They fitted experimental data from Bresci et al. [6] to the Darcy-Forchheimer model and applied discrete phase modeling to simulate water droplets in the flowing fog. Their results demonstrated good agreement with those obtained by Bresci et al. [6]. Recently, several studies have aimed to enhance the aerodynamic collection efficiency of fog collectors. For example, Azeem et al. [7] introduced a harp fog collector design with parallel filaments in the mesh, reducing solidity and minimizing pressure drop. Damak et al. [8] proposed a method involving electrical forces that could surpass aerodynamic drag forces by utilizing different electrical charges for the droplets and the mesh. Both of these methods were combined, as demonstrated by Sharifvaghefi et al. [9], resulting in a collection efficiency of 82%.

CFD analysis has been utilized in various engineering applications to enhance the aerodynamic behavior of vehicles. [10], [11]. In this study a 3D CFD wind tunnel was used to simulate the aerodynamic collection efficiency of a fog collector. Firstly, the Darcy-Frochheimer model was fitted based on the experiment done by Bresci et al. [6]. Then the acquired Parameters: permeability (α), and the inertial resistance factor (C_2) were used with the discrete phase modeling on Ansys fluent to simulate the velocity and pressure fields around the fog collector. Finally, the results were compared with experimental data from the literature.

2. Theory

The physical model was created based on a numerical wind tunnel approach proposed by Carvajal et al. [5]. The realizable $k - \varepsilon$ model was used because it predicts the flow with high pressure gradients and circulations accurately [5]. The following assumptions were used:

- The flow is considered turbulent, steady, and incompressible.
- The system of differential equations was solved with time-independent properties (steady).
- Negligible heat transfer between the fog and collector was considered.

For the mentioned assumptions, the set of differential equations to be solved numerically is reduced to the following forms:

- Continuity:

$$\frac{\partial u_i}{\partial x_i} = 0 \tag{1}$$

- Momentum:

$$u_j \frac{\partial u_i}{\partial x_j} = -\frac{1}{\rho}\frac{\partial P}{\partial x_i} + \frac{\partial}{\partial x_i}\left[(v + v_t)\frac{\partial u_i}{\partial x_j}\right] + S_D \tag{2}$$

Where v and v_t are the kinematic and eddy kinematic viscosities, respectively. While The last term S_D represents the darcy-Frochhiemer equation which takes the form in the x-direction (flow direction) as:

$$(S_D)_x = -\rho\left(\frac{v}{\alpha}u_x + C_2\frac{|u|u_x}{2}\right) \tag{3}$$

Where α, and C_2 are the before-mentioned porous medium constants.

- Turbulent Kinetic energy (k):

$$u_j \frac{\partial k}{\partial x_j} = \frac{\partial}{\partial x_i}\left[(v + v_t)\frac{\partial k}{\partial x_j}\right] + G_k - \varepsilon \tag{4}$$

Where G_k represents the turbulence due to the mean velocity gradients.

- Rate of viscous dissipation (ε):

$$u_j \frac{\partial \varepsilon}{\partial x_j} = \frac{\partial}{\partial x_i}\left[(v + v_t)\frac{\partial \varepsilon}{\partial x_j}\right] + C_1 S\varepsilon - C_2\frac{\varepsilon^2}{k + \sqrt{v\varepsilon}} \tag{5}$$

Where ρ is the air density, and S is the mean rate-of-strain tensor. A specified number of particles were dispersed on a surface at a distance of 2.5h downstream of the collector, where h is the collector height and the surface is parallel to the fog collector.

The dispersed particles must be on a surface parallel to the fog collector and have the same line of action. Then the aerodynamic efficiency was calculated as shown in equation (6), Where N represents the number of particles, and s represents the solidity.

$$\eta_{ACE} = \frac{N_{collected}}{N_{released}} s \tag{6}$$

3. Results

The dimensions of the simulated collector are 12 meters in width and 4 meters in height. The wind tunnel maintains an undisturbed velocity at a constant 7 m/s. The fog collector was substituted with a porous jump, and this jump exhibits a pressure drop as indicated below:

$$(P1 - P2) = \Delta n\rho \left(\frac{\upsilon u_2}{\alpha} + C_2 \frac{u_2^2}{2}\right) \tag{7}$$

$P1$ and $P2$ represent the pressure before and after the fog collector, respectively. Δn stands for the thickness of the collector, and u_2 corresponds to the downstream velocity. Initially, the created model was validated using experimental data. Figure (1) Depicts the vertical velocity profiles downstream of the fog collector. The x-axis on the graph represents the dimensionless velocity (u/u_∞), whereas the y-axis indicates the dimensionless length (y/h), with "y" signifying the vertical distance from the ground and "h" representing the collector height (in the case of Bresci, h=4 meters). According to Figure (1.a), the results show good agreement with Bresci's dataset at a distance of 10 meters downstream of the collector ($x = 10\ meters$). The discrepancies can be attributed to the substantial simplifications applied to the model. These simplifications include considering the porous jump as a flat surface, despite its actual behavior of bending due to accumulated droplets and its weight. Moreover, the model assumes a constant pressure drop, whereas in reality, the pressure drop increases as droplets accumulate. Figure (1.b) depicts the evolving velocity profiles at distances exceeding 10 meters, illustrating a smooth flow development.

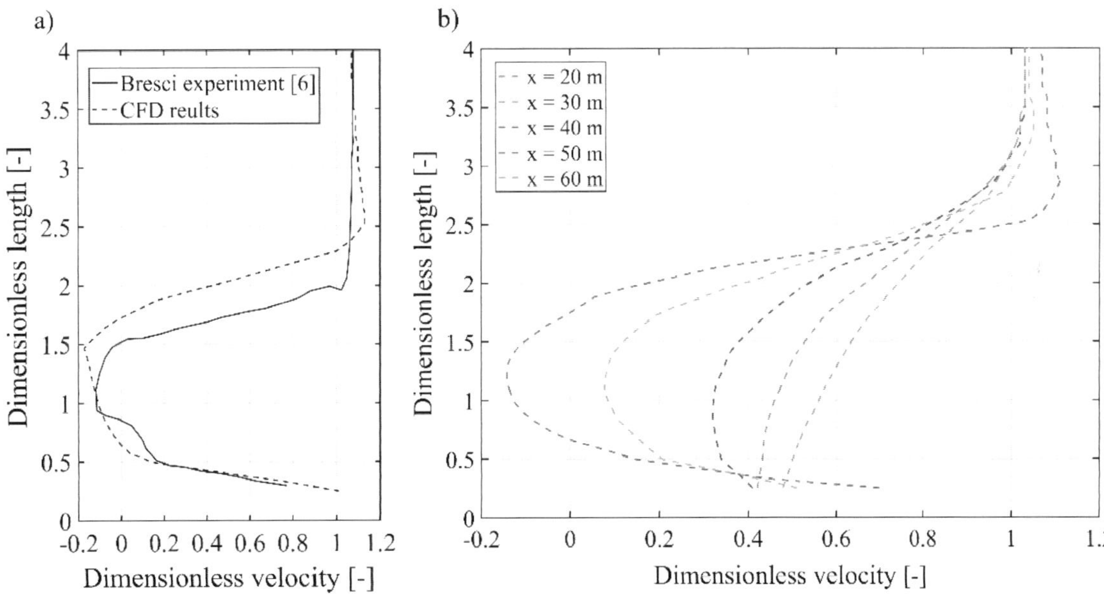

Fig. 1. Depicts vertical velocity profiles downstream of the fog collector. a) The validation using Bresci's dataset at $x = 10\ meters$, b) Vertical velocity profiles downstream of the collector at distances $x = 20, 30, 40, 50$ and $60\ meters$.

Figure (2) illustrates the difference between the distributed and undistributed flows around the fog collector for the top and side views. The results show good agreement with the CFD investigation carried out by Carvajal et al. [5]. According to Figure (2.a), the fog collector exhibits a significant pressure drop of 150 % due to its relatively high solidity (approximately 78%). It's important to recognize that this solidity increases as droplets accumulate on the collector leading to an even higher pressure drop. Based on the achieved results, only 42% of the released particles were collected by the porous jump leading to an aerodynamic efficiency of 33%. As shown in Figure (2.b), the high-pressure drop disrupts the flow influencing particles to pass by without being collected.

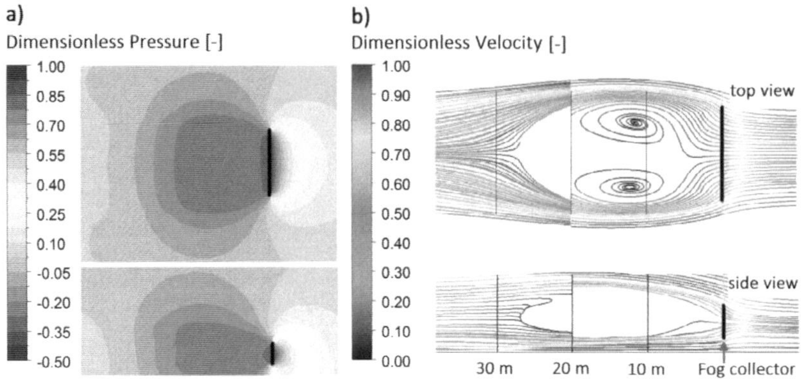

Fig. 2. Aerodynamic behavior around the fog collector. a) Dimensionless velocity streamlines, b) Dimensionless Pressure distribution $\left(\frac{P}{0.5\rho U_\infty^2}\right)$.

4. Conclusion

In this study, the aerodynamic collection efficiency of a fog collector was computed numerically. Initially, the Darcy-Forchheimer relation for Porous media was fitted to experimental data to find the porous media constants. Then, the Ansys fluent was used to perform the CFD simulation. The CFD results showed a high-pressure drop of 150 % and a velocity drop of 100 %. For future analysis, more consideration shall be considered regarding the porous media model because a lot of the collector structural parameters are fitted solely by three constants. Moreover, a transient simulation can be considered to investigate the change in pressure drop while the droplets accumulate on the collector.

5. Acknowledgments

This work was supported by the Student Grant Competition of the Technical University of Liberec under project No. SGS-2023-5323.

6. References

[1] M. Azeem, M. T. Noman, J. Wiener, M. Petru, and P. Louda, "Structural design of efficient fog collectors: A review," Environ. Technol. Innov., vol. 20, p. 101169, 2020, doi: 10.1016/j.eti.2020.101169.

[2] J. de D. Rivera, "Aerodynamic collection efficiency of fog water collectors," Atmos. Res., vol. 102, no. 3, pp. 335–342, 2011, doi: 10.1016/j.atmosres.2011.08.005.

[3] K. Steiros and M. Hultmark, "Drag on flat plates of arbitrary porosity," J. Fluid Mech., vol. 853, pp. 853R31-853R311, 2018, doi: 10.1017/jfm.2018.621.

[4] J.-K. Koo and D. F. James, "Fluid flow around and through a screen," J. Fluid Mech., vol. 60, no. 3, pp. 513–538, 1973, doi: 10.1017/S0022112073000327.

[5] D. Carvajal, L. Silva-Llanca, D. Larraguibel, and B. González, "On the aerodynamic fog collection efficiency of fog water collectors via three-dimensional numerical simulations," Atmos. Res., vol. 245, no. April, p. 105123, 2020, doi: 10.1016/j.atmosres.2020.105123.

[6] E. Bresci, "Wake characterization downstream of a fog collector," Atmos. Res., vol. 64, no. 1–4, pp. 217–225, Sep. 2002, doi: 10.1016/S0169-8095(02)00093-5.

[7] M. Azeem et al., "Optimal Design of Multilayer Fog Collectors," ACS Appl. Mater. Interfaces, vol. 12, no. 6, pp. 7736–7743, 2020, doi: 10.1021/acsami.9b19727.

[8] M. Damak and K. K. Varanasi, "A P P L I E D S C I E N C E S A N D E N G I N E E R I N G Electrostatically driven fog collection using space charge injection," no. D, pp. 1–9, 2018, [Online]. Available: http://advances.sciencemag.org/.

[9] S. Sharifvaghefi and H. Kazerooni, "Fog harvesting: combination and comparison of different methods to maximize the collection efficiency," SN Appl. Sci., vol. 3, no. 4, pp. 1–11, 2021, doi: 10.1007/s42452-021-04518-3.

[10] K. Tuzharov, S. Iliev, and D. Gunev, "Aerodynamic study of the environmental vehicle model with the flow simulation CFD program," Ann. DAAAM Proc. Int. DAAAM Symp., vol. 29, no. 1, pp. 0495–0504, 2018, doi: 10.2507/29th.daaam.proceedings.072.

[11] Z. Čarija, I. Lučin, B. Lučin, and L. Grbčić, "Investigation of numerical simulation parameters on fluid flow around trash-racks," Ann. DAAAM Proc. Int. DAAAM Symp., vol. 29, no. 1, pp. 1046–1052, 2018, doi: 10.2507/29th.daaam.proceedings.149.

DOI: 10.2507/34th.daaam.proceedings.050

ENHANCING HYDROPHOBIC PROPERTIES THROUGH LASER-INDUCED SURFACE STRUCTURING

Simon Syrovatka, Pavel Kozmin & Ondrej Marsalek

This Publication has to be referred as: Syrovatka, S[imon]; Kozmin, P[avel] & Marsalek, O[ndrej] (2023). Enhancing Hydrophobic Properties Through Laser-Induced Surface Structuring, Proceedings of the 34th DAAAM International Symposium, pp.0365-0370, B. Katalinic (Ed.), Published by DAAAM International, ISBN 978-3-902734-41-9, ISSN 1726-9679, Vienna, Austria
DOI: 10.2507/34th.daaam.proceedings.050

Abstract

This paper explores the use of lasers with a focus on surface engineering. Its primary aim is to generate surface structures capable of influencing the characteristics of the material that is tested. Mainly the wettability of the surface, more particularly the enhancement of hydrophobic properties. A laser machine with ultrashort pulses was used to produce hydrophobic surface structures. This paper concludes by evaluating the surface structures and describing the aspects that influence the genesis and proposed functions of these surface structures.

Keywords: wettability; laser; surface structure; contact angle; hydrophobic structure

1. Introduction

Laser machining is beginning to be a more and more common type of machining process. It can be divided into a lot of groups such as laser welding, cutting, machining, surface structuring, etc [1], [2], [3]. Each of these groups of technologies is very interesting and has already been discussed in academic works. This article focuses especially on surface structuring, and the main goal of this article is to describe possibilities of laser surface structuring which could lead to the change of its wettability.

Creating surface structures using a laser can lead to a modification of the inherent properties of the given surface. Laser structuring allows for the alteration of factors beside the surface wettability such as surface roughness, optical properties, tribology and so on. Wettability is the ability of a liquid to adhere to the surface of solid substances. This capability depends on the surface structure of the solid material, which affects the size of the contact angle between the liquid and the solid surface. Surfaces can be categorized in terms of wettability as hydrophilic, hydrophobic, and superhydrophobic– the differences between these types of surfaces are illustrated in Fig. 1, based on the magnitude of the contact angle, which can be measured using numerous methods.

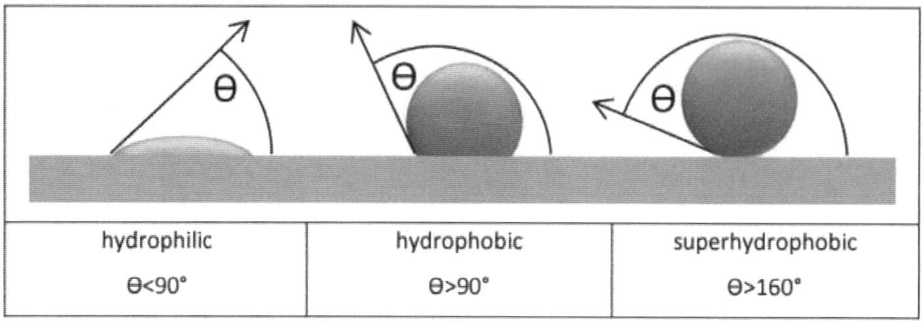

Fig. 1. Categorization of surfaces according to contact angle [4]

Contact angles could be measured by multiple methods of measuring. These methods include for example Young-Laplace method, the circle and polynomial fitting, the mask method, etc [5]. In the available academic publications, there are two main views on the designing of hydrophobic surface structures. The first one consists of known geometrical shapes such as cubes, dots, and grooves [4], [5]. These elementary shapes are often combined with special structures called LIPSS (laser-induced periodical surface structures).

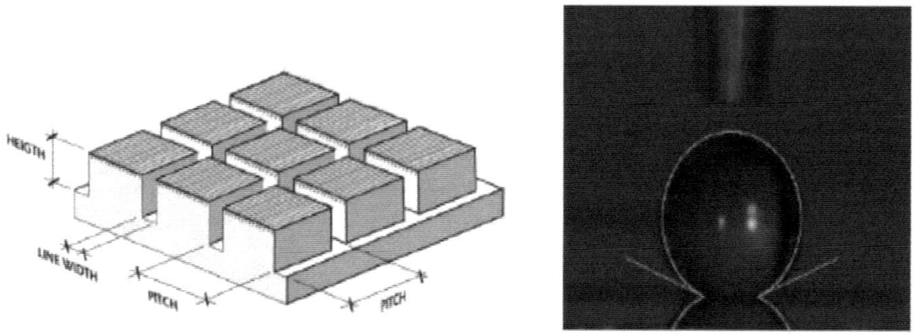

Fig. 2. Metrix structure with LIPSS and measured contact angle of this structure [7]

The second view of designing surface structures is inspired by nature. The shapes are more loose and whole structures are more complex. The best example of a hydrophobic surface structure is the surface of the lotus leaf.

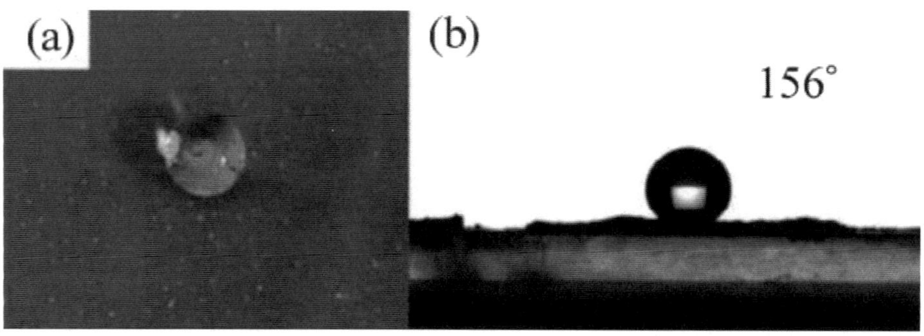

Fig. 3. The water droplet on the lotus leaf (a), the contact angle of the droplet [8]

Controlled wettability of the surface can play a crucial role in engineering in the development of new functional surfaces, the design of innovative machining technologies involving cutting fluids, or the production of workpieces with precisely defined surface properties. For instance, in the aviation industry, a significant challenge is the formation of ice, which detrimentally affects the operations and flight safety of aircraft and helicopters [6]. The complexity of surface structure laser fabrication and their overall effect on surface wettability are investigated in the experiment of this article.

2. Experiment

As the tested material the austenitic chromium-nickel steel stabilized with titanium (X6CrNiTi18-10) was used. This material holds significant prominence in numerous industrial domains that concentrate on laser surface structuring and its subsequent applications. The surfaces of the samples have been ground and the final roughness of the surface was approximately $R_a = 0,4$ µm this value of surface roughness was crucial for the rest of the experiment to measure only the effect of the manufactured surface structures on the wettability of the surface. Before the fabrication process the design of possibly hydrophobic structures was made.

1. Cube

This surface structure was modified to the 5 variants which dimensions can be seen in the picture bellow.

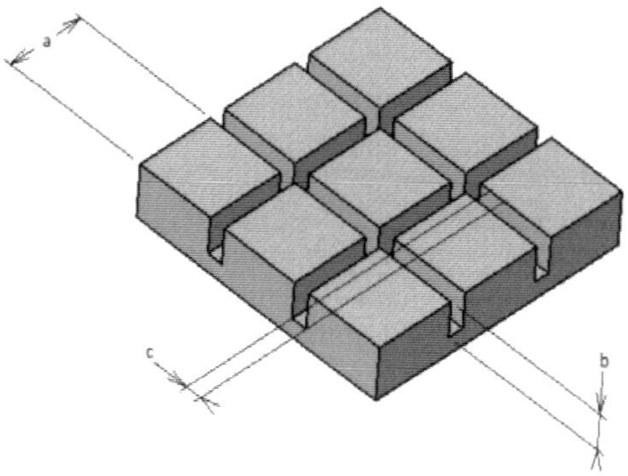

variant	a [µm]	b [µm]	c [µm]
I	50	5	20
II	50	10	20
III	50	10	10
IV	50	20	20
V	50	20	13

Fig. 4. Design and dimensions of structure Cube (a – side of a cube; b – depth; c – pitch)

2. Hole

This surface structure was modified to the 5 variants which specifications can be seen in the picture bellow.

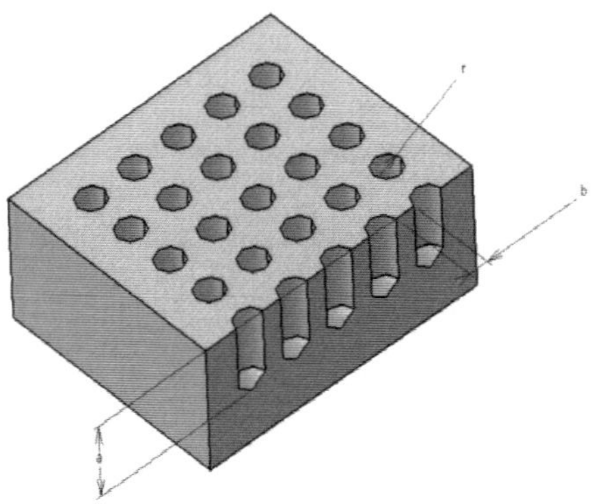

variant	r [µm]	a [µm]	b [µm]
1	10	12	20
2	10	12	10
3	15	12	20
4	15	12	10
5	15	12	5

Fig. 5. Design and dimensions of structure Hole (r – radius of a hole; a – depth; b – pitch)

Once the design of the individual structures was created, it was possible to proceed to their production. A laser system with ultrashort pulses was used for it, the parameters of which are:

- Wavelength = 532 nm
- Laser pulse length = 13 ps
- Repetition rate = 200 – 1000 kHz
- Average laser power < 12 W

It was necessary to used specific laser parameters for the manufacturing of all designed surface structures.

3. Results

After the production of the designed structures (it was produced 10 types of structures), each structure was scanned and measured by an optical microscope (Alicona Infinite Focus) to verify the fabrication of the correct dimension (Fig. 6 and Fig. 7).

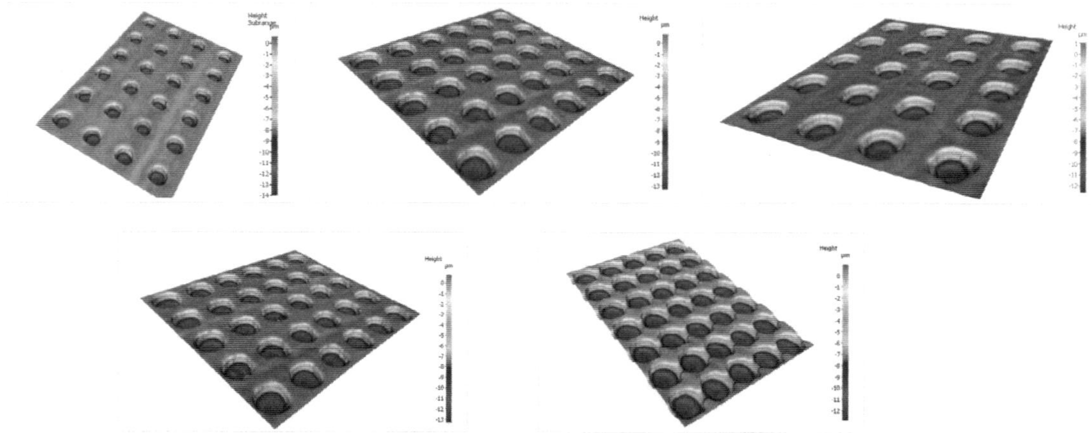

Fig. 6. Scans of fabricated surface structures type Hole (variant 1 – 5)

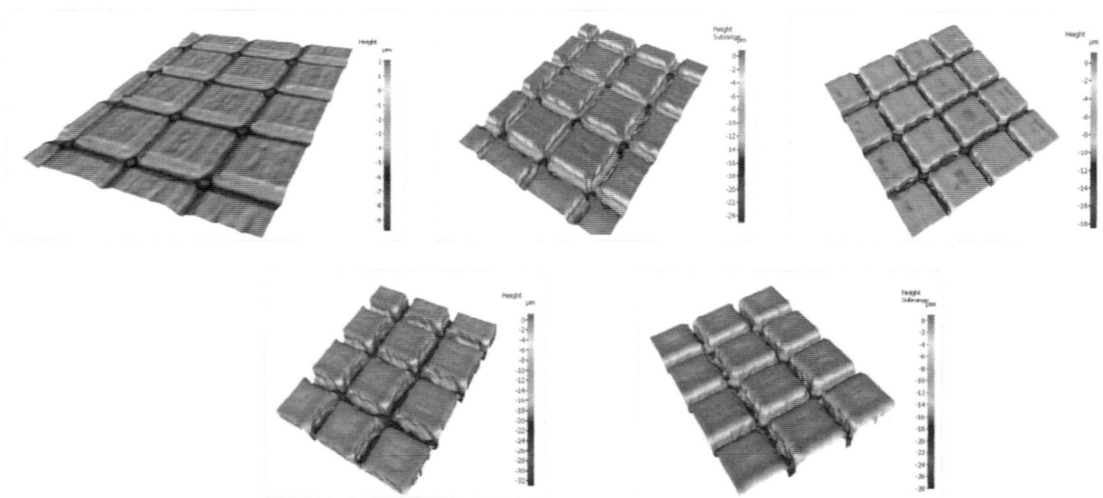

Fig. 7. Scans of fabricated surface structures type Cube (variant I – V)

All surface structures were produced with maximal dimensional deviation of 5 µm. The next step in the experimental process was to examine the influence of structures on the wettability of the surface (measurement of the contact angle). The method of Sessile drop (static contact angle for the surface) was used for the evaluation of the wettability. For this measurement, the Drop Shape Analyzer (DSA30E) was used. The actual test of wettability was conducted for each produced surface structure. Based on the research study, water droplets with a volume of 3.5 µl were chosen [6]. The wettability angles were measured twice for each sample – first immediately after landing on the created surface structure and secondly, 10 seconds after landing on the surface structure. The measurements took place approximately two weeks after the fabrication of the surface structures using the picosecond laser system.

Based on these measurements, it was found that the most hydrophobic structure from the type Hole was variant no. 5 and the least hydrophobic was variant no. 3. Regarding the surface structures type Cube the most hydrophobic was variant no. II and the least hydrophobic was the variant no. III.

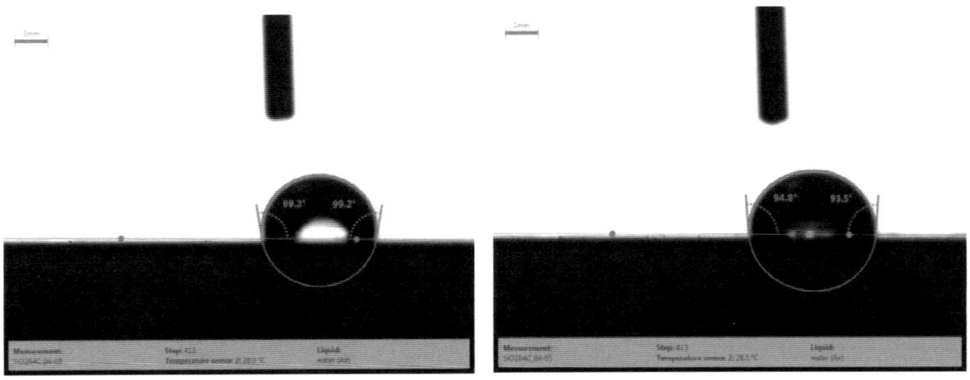

Fig. 8. The measured contact angle of surface structure Hole variant no. 5 (right – just after landing of the droplet on the structure, left – 10s after landing of the droplet on the structure)

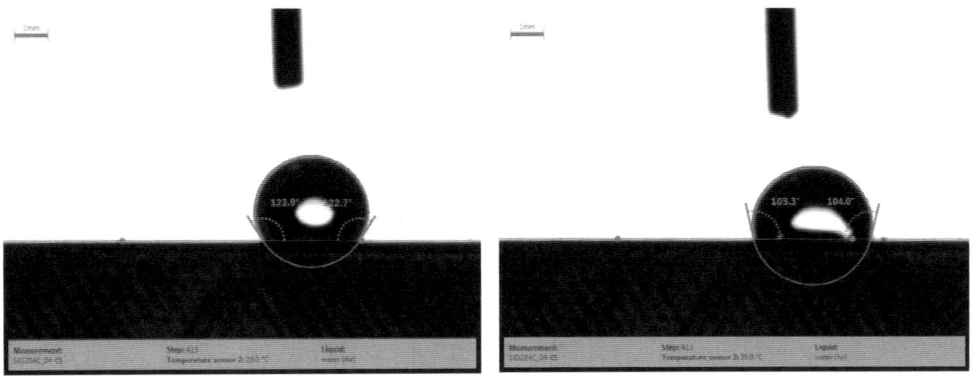

Fig. 9. The measured contact angle of surface structure Cube variant no. III (right – just after landing of the droplet on the structure, left – 10s after landing of the droplet on the structure)

4. Conclusion

Several surface structures whose purpose is to change the wettability of treated surfaces were designed based on published articles on a similar topic. These structures are the structures of recurring cubes and recurring holes. It was designed with five variations from each group. All 10 surface structures were manufactured using a picosecond laser system and their dimensional accuracy was verified using an optical microscope. After the fabrication, the contact angle of every structure was measured and based on which the wettability was determined. It was found that the contact angle of the most hydrophobic structure - type Hole was 99,2° (which is 3,8 degrees more than the non-treated surface). Regarding the structures - type Cube the biggest measured contact angle was 120,8° (which is 25,4 degrees more than the non-treated surface).

After all, it could be said, based on the experiments carried out, that the picosecond laser system is a useful tool for the fabrication of surface structures that have an influence on the surface property especially on the wettability of the surface. However, it is important to say that the wettability of a surface is not only affected by the morphology of the surface to determine the specific factors contributing to changes in surface wettability, but it would also be convenient to manufacture and test an even greater number of samples. Subsequently, the produced samples could be subjected to testing using an alternative method for measuring the contact angle of the liquid (such as the dynamic drop method) or employing a different measuring liquid (such as distilled water). This expanded scope of experimentation would enable a more comprehensive understanding of the underlying mechanisms affecting surface wettability. Also, for the next research work, it would be very interesting to manufacture and test the properties of some structures inspired by nature.

5. Acknowledgments

This article has been prepared under the project SGS-2022-007–Research and Development for Innovation in Engineering Technology–Machining Technology IV.

6. References

[1] Begic-hajdarevic, Derzija, Mugdim Pasic, Ahmet Cekic a Muhamed Mehmedovic (2016). Optimization of Process Parameters for Cut Quality in CO2 Laser Cutting Using Taguchi Method. In: Branko Katalinic, ed. DAAAM Proceedings [online]. 1. vyd. B.m.: DAAAM International Vienna, 2016 [vid. 2023-08-30], s. 0157–0164. ISBN 978-3-902734-08-2. DOI: 10.2507/27th.daaam.proceedings.024

[2] Topic, Angela, Branko Bauer, Zoran Kozuh a Nikola Knezovic (2016). Gas Composition Influence on the Microstructure and Geometry of Laser-Welded Joints in Duplex Stainless Steel. In: Branko Katalinic, ed. DAAAM Proceedings [online]. 1. vyd. B.m.: DAAAM International Vienna, 2016 [vid. 2023-08-30], s. 0734–0742. ISBN 978-3-902734-08-2. DOI: 10.2507/27th.daaam.proceedings.106

[3] Franz, Markus, Jens Bliedtner a Conrad Haupt (2014). Laser Metal Deposition Welding in the Field of Tool and Mould Making. Procedia Engineering [online]. 2014, 69, 237–240. ISSN 18777058. DOI: 10.1016/j.proeng.2014.02.227

[4] Conrad, Daniel a Lars Richter (2014). Ultra-short Pulse Laser Structuring of Molding Tools. Physics Procedia [online]. 2014, 56, 1041–1046. ISSN 18753892. DOI: 10.1016/j.phpro.2014.08.016

[5] Akbari, Raziyeh a Carlo Antonini (2021). Contact angle measurements: From existing methods to an open-source tool. Advances in Colloid and Interface Science [online]. 2021, 294, 102470. ISSN 00018686. DOI: 10.1016/j.cis.2021.102470

[6] Xing, Wei, Zu Li, Haiou Yang, Xinlin Li, Xinyun Wang a Ning Li (2019). Anti-icing aluminum alloy surface with multi-level micro-nano textures constructed by picosecond laser. Materials & Design [online]. 2019, 183, 108156. ISSN 02641275. DOI: 10.1016/j.matdes.2019.108156

[7] Li, Kewei, Nathaniel Myers, Grant Bishop, Yuxin Li a Xin Zhao (2022). Study of surface wettability on fused silica by ultrafast laser-induced micro/nano-surface structures. Journal of Manufacturing Processes [online]. 2022, 79, 177–184. ISSN 15266125. DOI: 10.1016/j.jmapro.2022.04.035

[8] Yang, Fuchao a Zhiguang Guo (2015). Characterization of Micro-Morphology and Wettability of Lotus Leaf, Waterlily Leaf and Biomimetic ZnO Surface. Journal of Bionic Engineering [online]. 2015, 12(1), 88–97. ISSN 1672-6529, 2543-2141. DOI:10.1016/S1672-6529(14)60103-7

DOI: 10.2507/34th.daaam.proceedings.051

INFORMATION WARFARE, MEDIA AND CIVILIANS

Miroslav Tomsu, Martin Dzermansky, Nikola Cajkova & Pavel Wanecki

This Publication has to be referred as: Tomsu, M[iroslav]; Dzermansky, M[artin]; Cajkova, N[ikola] & Wanecki, P[avel] (2023). Information Warfare, Media and Civilians, Proceedings of the 34th DAAAM International Symposium, pp.0371-0378, B. Katalinic (Ed.), Published by DAAAM International, ISBN 978-3-902734-41-9, ISSN 1726-9679, Vienna, Austria
DOI: 10.2507/34th.daaam.proceedings.051

Abstract

The progress in information and communication technologies in the transmission and sharing of data has fundamentally affected the armed forces, not only in terms of the development of information systems, control, and fire control systems but also in the conduct of war. The topic of the article is information warfare. The aim is to explain the meaning and concept of information warfare and to analyse the purpose of information warfare and its development in future security operations. The article first deals with the mutual relationship between information and military science and their concept of information. It continues with the characteristics of individual forms of information warfare, including the ways of conducting them. It explains the possible consequences of its action from information point of view. The benefit is considering the further possible development of some forms of information warfare and predicting probable implications in the event of their use.

Keywords: Information warfare; information operations; psychological operations; electronic warfare; intelligence operation.

1. Introduction

Contemporary society is increasingly approaching the ideas of the information society, and the main features are the predominance of work with information, interactivity, integration, and globalisation tendencies. From a technological point of view, an information society is a society with a high level of use of information and communication technologies based on computer technology and associated digitisation. However, the growing need for communication, sharing, and exchanging information characterises civilian and military institutions. The increasing dependence of military technology on electronic means that work with an unimaginable amount of information and the growing importance of the information itself has made information a lucrative target of combat action but, at the same time, a valuable weapon [1].

These changes are directly revolutionary from the point of view of information and from the point of view of conducting war. Unsurprisingly, there is more and more talk about the so-called information war nowadays. It has a lot to do with information science. The common subject of interest of information science and information warfare, which can be characterised in the broadest sense as „the joint deployment of available combat capabilities to influence, invalidate, dominate and disrupt enemy command capabilities, including automated means of decision-making; and at the same time to ensure the protection of all these own resources from the enemy [2], is indeed information.

In their studies, some authors mention the term information war in connection with a military conflict [3] [4] or the manifestation of the events of the last few years, [5] [6] but some authors also deal with it in the context of a socio-philosophical analysis of the processes associated with the formation of information culture among young people. [7] This article deals with information warfare in a military and general context with an emphasis on the role of the media and its impact on the civilian population.

2. Information environment

Information is a multifaceted term that every person encounters almost daily. It is a universal phenomenon, tightly connected not only with the daily life of man and society but also with living and non-living nature. From his point of view, its four basic meanings (are a psychophysiological phenomenon, numerical measure of disorganisation removal, potential information - sign-recorded and circulating data in technical devices, and expression of diversity in the objects and processes of nature). In the most general sense, information is information about the natural environment, its state, and the processes taking place [1].

2.1. Information from the perspective of information and military science

Information science primarily understands information as a message, communicable knowledge necessary for the recipient, or data that facilitates a choice between alternative decision-making options. [2] Vickery refers to information as knowledge or a type of knowledge. At the same time, he specifies command as what people know or think they know. In that case, new or modified knowledge is added or assimilated into the individual's „knowledge structure" [8].

In this context, Britz even believes that Vickery understands information from a social perspective as a process between the source (information) and the user (information). Information is, therefore, not only a part or product of a process but the process itself (from the creation of information to its use). This is, of course, one of the approaches to the concept of „information". User access can also be mentioned, which in the broadest sense means that information is what reduces the amount or degree of uncertainty. From the point of view of the content approach, what is communicated between people can be informed. According to Norbert Wiener's cybernetic concept, information is the content of what is exchanged with the external world when we adapt to it and act upon it with our adaptation. As Britz further states, ideological, knowledge, or anti-definitional approaches exist. Proponents of the latter oppose attempts to define the term information because it is said to have no practical meaning for information science [9].

2.2. Movement of information within military intelligence services

From the point of view of information, the military intelligence service can be considered an information institution within the armed forces. As with civilian information institutions, the information goes not only to the intelligence service – the intelligence service collects information and distributes it after processing and evaluation. The information provided is essential for the decision-making activities of the state [10].

The intelligence cycle is called the movement or chain of successive and interconnected activities during which information is obtained, processed, evaluated, and passed on to users. It comprises four phases: management, collection, processing, and dissemination, and it has a similar course to the classic information process. The names used for each step may vary, but these are synonymous terms. Even phase five is distinguished in American military regulations, such as the Doctrine for Intelligence Support of Joint Operations [11].

3. Information and military science

The difference in understanding information neither pits the two scientific fields against each other. On the contrary, they intermingle. Concerning its focus, military science, among other things, focuses on the military information environment, which is part of the global information environment and intertwines with the civilian sector. The military information environment is defined as an environment that is part of the worldwide information environment and is made up of information systems, friendly and hostile organisations, military and non-military, that support or significantly affect a specific military operation [12].

Military science is generally defined as „a system of knowledge about the nature and laws of war, the military preparation of the state and the armed forces for war, and the methods of conducting it." [13]. The primary subject of her research is armed struggle. In the broadest sense, information science can be defined as the science of information (physical, biological, cultural). „In a narrower sense, it is a science of an interdisciplinary nature dealing with the regularities of the processes of creation, processing, measurement, coding, storage, transformation, distribution and reception of information in society." [14]. The Explanatory Dictionary of Czech Terminology in Information Science and Librarianship further states that its goal is to secure and rationalise social information and communication processes. It is, therefore, clear that information science is not an isolated science but affects, among other things, the fields of technical sciences and information technologies (especially computer and communication technology), psychology, philosophy, etc. Military science also affects technical sciences, and, of course, information and communication technologies, philosophy, and psychology are also part of conducting war and commanding troops.

3.1. Information and the changing nature of wars

The introduction and use of modern weapons and the growing importance of electronics and computer technology have changed the character of war. Today, one of the most potent weapons is information. A completely new concept was born - information war (Information War - IW). Based on the generalisation of historical experience and experience from contemporary wars and armed conflicts, it is clear that the importance of information will continue to grow. It has always been, is, and will continue to be important in the future as well, the rule that whoever has information also has power and superiority. On the other hand, effectively using this information needs to be improved [15].

3.2. Information warfare

Information warfare is one of the means of hybrid warfare. We can define it as an activity aimed at gaining an advantage or superiority over an opponent in the management and use of communication and information channels or technology." Through the means of information, it takes place continuously from peacetime to wartime and then also during wartime. Previously, it was used only as a means of support. For example, in World War 1, leaflets were distributed among soldiers with calls to leave the resistance.

The main advantage of information warfare is a significant ratio of damage to the opponent compared to the percentage of costs incurred. Information warfare can be divided into offensive or defensive, depending on whether we protect information or try to influence it. This concept is often confused by the public with the term cyber war, which, according to the terminological dictionaries of the Ministry of the Interior, means: "The use of computers and the Internet to wage war in cyberspace. A set of large-scale, often politically or strategically motivated, related and mutually induced organised cyber-attacks and counter-attacks." Information warfare affects cyber warfare, but it also deals with other forms of action that intertwine with each other: command and control, intelligence operation, electronic warfare, psychological warfare, economic information war, diplomatic war, and, last but not least, a hacker war [16].

Christopher Bellamy of the Royal Military College of Science from Great Britain used an exciting definition in his analysis. According to him, the term information war could be a suitable designation for cyber warfare, provided that information is used to destroy, damage, or use the information of the adversary, information systems, and computer networks of the adversary. It also describes attacks against the mindset of commanders (influencing decision-making) - i.e., psychological warfare, deception, and mind-twisting. In this context, information war could be called a new term - psycho-cybernetic war, „psyberwar." In addition to these two areas, information warfare overlaps with warfare against command-and-control systems, electronic warfare, and information is crucial in deploying precision-guided munitions [17].

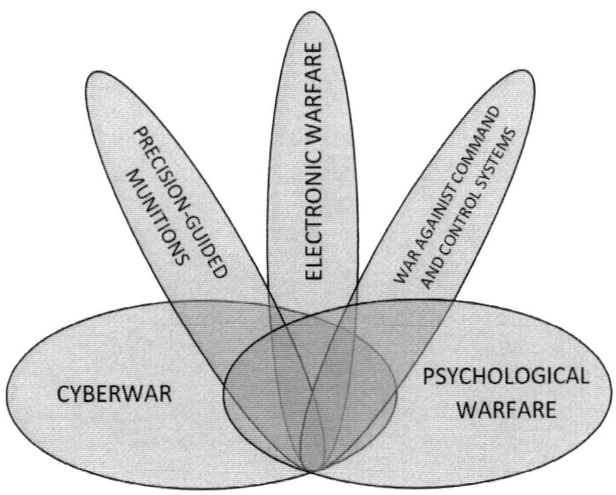

Fig. 1. Definition of the newly introduced term "psyberwar". [17]

4. Forms of information warfare

In recent years, in connection with the tactics of the great powers to abandon the complex techniques of conducting battles (about possible losses in the ranks of soldiers or the civilian population), the emphasis has shifted to the so-called soft forms of war. "This is why there has been a massive development of individual forms of conducting information warfare." Information warfare can be conducted by lethal or non-lethal means, brute force, or non-lethal means. It can also be divided into offensive or defensive - depending on whether the goal is to protect information or to act on it. However, such a division is crude, corresponding only to the primary division of combat activity into defence and attack [18].

Information warfare has yet to exist as a separate technique. It is made up of several forms that intertwine and complement each other. Currently, one can distinguish at least seven forms of information warfare:

4.1. Command-and-control warfare

Disarming a commander has been one of the goals for gaining dominance on the battlefield for centuries. However, the actual realisation of this plan has always been more difficult. However, the shift occurred at the moment of the creation of command centres. These are easily identifiable by basic features such as visible communication, computing equipment (associated with electromagnetic radiation), movements of printed documents and logistical material, and, most importantly, by atypical activities that are entirely different from the regular exercise of other units [18].

The purpose of the action is to destroy or disrupt the means of command and neutralise the control. This will interrupt the flow of information from the command structures to the executive element. These combat units will thus lose an overview of the overall situation. They will need to have, among other things, information on where to intervene, when to intervene, or how to intervene [19].

4.2. Intelligence-based warfare

The intelligence form of action in information warfare is discussed if intelligence is directly included in the combat activity - for example, in the process of selecting and allocating "targeting" targets or in determining whether the given means hit the target and the desired "Battle Damage Assessment" effect was met – and not only as an input to the command-and-control process. Therefore, including intelligence in combat is to obtain information to create a permanent and comprehensive picture of the battlefield. In addition, reporting also has the task of protecting data, i.e., preventing the enemy from failing to obtain information and form a complete picture of the battlefield, including the adversary's capabilities [19].

4.3. Electronic warfare

Electronic warfare is defined as „an activity focused on the use of the electromagnetic spectrum, which includes the search, interception, and identification of electromagnetic emissions, the use of electromagnetic energy, including directional radiation, to limit or prevent its use by the adversary against its means and ensuring their effective use“ [20].

4.4. Psychological warfare

Psychological operations are defined as „planned psychological activity in peace and war directed at an adversary, own or neutral persons to influence attitudes and behaviour, which affects the achievement of political and military objectives. They include psychological activity at the strategic level, psychological consolidation, and psychological activity on the battlefield“ [20].

But, according to Nastoupil, psychological warfare aims to influence the enemy's will without using weapons. From an information point of view, it is a planned and systematic distribution of information to various target groups. It must be carried out in a precisely defined period so that the announcement affects their awareness.

Psychological operations and propaganda are very close to each other. The boundary between the two activities needs to be clarified; some authors say these activities overlap. In general, peacetime psychological operations are based on presenting accurate information, or at least the truth as it is perceived. Propaganda, on the other hand, is perceived, according to the Oxford Guide to Military History authors, as the spread of lies (misleading information), and the target group is often one's population [21].

4.5. Economic information warfare

The economic information war was created by combining economic war and information war. Currently, two forms of it are distinguished: information blockade and information imperialism [22].

4.6. Hacker warfare

Information warfare is often simplistically considered a hacking war. In reality, however, the hacker war is one information war. Attacks are directed at specific systems (their structure and content) using technological shortcomings or "holes" in security measures. [36] The techniques or methods of hacker warfare vary according to the location from which the attacker launches the attack, the process of execution, and the expected extent of the damage. All of the hacker warfare techniques presented hereafter represent attacks against civilian targets. Attacks on civilian and, theoretically, military targets have common characteristics in a hacker war [22].

4.7. Cyberwarfare

Cyber warfare is considered a particular area of information warfare; it includes information terrorism, semantic attacks (intertwined with hacker warfare), simulated warfare, and Gibson warfare [22]. It is assumed that some of these forms will affect advanced military systems. Computer technology differs from other military assets because it is an integral part of all other assets used in modern armies; from this perspective, it is the critical component that many modern armies depend on, and potential enemies are well aware of this dependency. Cyber warfare has not yet been used against military assets, but it is increasingly being discussed, and its forms interest military experts [18].

5. Information warfare and civilians

In his famous 1921 essay on the importance of a free press, the long-time editor of the British Guardian, C. P. Scott, declared that comment is free, but facts are sacred. In quality independent journalism, it is good practice to verify facts from at least two independent sources and to separate clear facts from impressions and interpretations. Nevertheless, these principles cease to be the rule.

According to publicist and security analyst Jan Schneider, "An essential feature of current media events is the emphasis on creating a first impression. It is quickly passed off as a safe mainstream so as not to raise disturbing doubts or even questions of conscience. Political decisions are then built on this chimerical basis, which - put into practice - unfortunately already have irreversible material consequences. And all this happens even though the second impression, a closer examination of the original information, can show its shakiness, baselessness, or even falsity." [19]. On the other hand, some news sources still maintain a high ethical standard and rely on the truth of the facts they present to the public.

5.1. Making decisions based on desire and succumbing to fear

We don't necessarily have to lie if we want to distort reality to influence the recipient's thinking in a certain way. Yet, we have many tools to manipulate information to our advantage. One of the essential tools is the so-called agenda setting. Agenda-setting theory assumes that the emphasis and importance given to a specific issue by the media affects how the audience perceives and accepts the case.

The media have mechanisms that can determine what the audience is thinking about and which issues they perceive most important. The authors of this concept, Maxwell McCombs and Donald Shaw, concluded that the media set the public's agenda - they construct their reality and tell the people what to think about, what to notice, and what to consider necessary. Manipulation techniques in dealing with information that we can encounter in the media can be divided into several basic types:

- manipulation using purposeful selection of presented information (overwhelming with meaningless information, deliberate highlighting or ignoring of specific topics),
- sorting manipulation (this is the sorting of messages in a particular order when some messages are deliberately pushed aside and the main space is left for favoured messages - in the case of print, this is, for example, the front page),
- manipulation of emotions,
- manipulation of the selection of commentators (statements of witnesses or bystanders, which can have a significant influence on the overall tone of the report),
- manipulation of the context of communicated information (information taken out of context can have an opposite tone),
- manipulation by communicating incomprehensible messages,
- manipulation by editing – shortening and taking the statement out of context. [20]

6. Defence against information warfare

Psychology, as a scientific field, also has a dark side. Without realising it, we often become objects of psychological operations, the aim of which is to influence our attitudes, our opinions and our decisions. Psychological warfare methods are used in political, economic and hot conflicts. In a global society, it is a battle for territory and resources and for whom the world public will consider good and bad guys. It isn't easy to be utterly immune to psychological operations. However, as civilians in information warfare, we can defend ourselves with common sense, critical thinking, and knowledge of the manipulation techniques used against us. [21]. In addition to the criminalisation of the dissemination of hateful or alarmist messages, the main methods of defence against propaganda and information warfare are, in particular, the cultivation of the media and civil society and the responsible approach of the state.

6.1. Freedom and independence of the media

In the case of the media, it is about ensuring that the press is especially free and independent from the political sphere and that the market is rich and open to the whole spectrum of media. However, these media should have publicly declared and enforced standards of journalistic ethics and a transparent ownership structure.

There should be significant and independent public service media in the media market, sufficiently funded to follow current trends and compete with others. Although the media scene should be watched by a respected and impartial regulator that can effectively impose significant sanctions for breaking the rules - the media should be left with enough autonomy to set and enforce the laws through their journalistic organisations. The press should be assigned a credible disinformation rating. All media should be dedicated to uncovering misinformation or fake news and correcting it, as well as conducting impartial fact-checking when reporting the opinions and positions of actors in public affairs. Journalists and other media workers should be carefully and regularly trained on propaganda, disinformation and how to counter it.

6.2. Society Education

Expanding media and internet literacy through various educational programs is the most important among all citizens. Citizens should have the opportunity to participate with their opinions in a freely ongoing public debate and to get to know other nations and cultures, for example, through various international programs. Non-governmental organisations dealing with human rights and media freedom and quality should be created and supported in civil society to forge media control independent of the state and the journalistic form. States should allow such international non-governmental organisations to operate on their territory. For example, coercive actions should also come from the occasional company to dissuade advertisers from advertising in media that spread hateful materials or misinformation.

6.3. Steps taken by governments to prevent the spread of information warfare

In addition to implementing the measures mentioned above, governments should refrain from owning or exerting influence on the media and, of course, propaganda and spreading disinformation by any means. They should start dealing with the power of foreign propaganda and misinformation on their citizens; they should measure it and respond to it flexibly and effectively. From a position of authority, the state apparatus should also devote itself to refuting at least the most severe misinformation and fake news, and politicians should then speak out against propaganda and hateful opinions. At the same time, the government should also lead by example in improving its strategic communication – it should better explain the measures introduced and be generally transparent and open.

7. Weaknesses of information warfare

Weaknesses of information warfare: the meaning and importance of information generally will remain the same in the short term. However, the availability of information, the speed of its transmission and transmission, and its accuracy (accuracy of data obtained from advanced sensors) will increase. Dependence on advanced information technologies also has its weak side. Modern technology applies practically only to the ability to receive, share and disseminate information. There is an information war here, the weapons and targets of which are information and information technology. If properly conducted, it can interrupt the flow of information, shut down the communication of information systems, and even affect a person's decision-making process, the morale and mood of the population.

The intensity and amount of disinformation is increasing. Suppose most people agree that the official news is lying or that emotional contributions are deliberately made visible and amplified. In that case, this can lead to further problems. The constant impact of disinformation can lead to the fact that the population will claim that the constitutionally guaranteed freedom of speech is also that they, politicians, or the news media can knowingly publish false information or lie. Some see misinformation as part of the constitutionally guaranteed freedom of speech. The influence of disinformation can make the population feel that information warfare is just an excuse for Western governments to restrict freedom of speech and inconvenient media. [22]. Of course, information warfare is not limited to the population. It is fully applicable even in competition in a commercial environment. This area can also become a possible topic for further study and research.

8. Significance and development of information warfare in the early future

Information warfare plays a significant role in security operations as an integral part of them. It can be assumed that its position will be irreplaceable in the future as well; it will not lose its importance; on the contrary, it will increase. The speed of computing is growing dramatically while the size of computers is decreasing. In a study from the late 1990s, NASA's nanotechnology team states that soon, computers that will be millions of times more potent than the current ones can be made. [23]

Developed countries will thus dominate technologically on the battlefield, but paradoxically, their dependence on complex technologies will be a vulnerable point. One can, therefore, agree with Dunlevy's statement that "dependence on advanced technologies can be considered a potential Achilles' heel". [24]. No matter who is fighting on the battlefield, no matter what the image of the battlefield is (undersea depths, outer space, urban areas, mountainous terrain, virtual space of computer networks, etc.), future conflicts will be governed by the general principles of warfare, just like the current ones. Of these, the most important is that defence alone cannot be relied upon, for a "defensive war," as Welch points out, cannot lead to victory. [25]. How the information war will develop and what forms of information war will be used for this is a matter of imagination and consideration.

9. Conclusion

The progress in information and communication technologies in the transmission and sharing of data has fundamentally affected the armed forces in terms of the development of information systems, control and fire control systems, and the conduct of war. The meaning and importance of information in the military, in the general sense, will not change shortly. However, the accessibility of information, the speed of its transmission and transmission, and its accuracy (accuracy of data obtained from advanced sensors) will increase. Currently, when comparing the capacities of the armed forces and estimating the probable outcome of their confrontation, the factor of physical strength and speed (manoeuvrability) is considered in particular. In the future, it will probably be replaced by awareness - in the sense of situational preparedness, i.e., the ability to quickly deploy forces and resources to defend one's units or to strike at the enemy.

Dependence on advanced information technologies also has its weak side. Modern technology is practically only valid with the ability to receive, share, and spread information. This is where the information war occurs, the weapons and targets of which are information and information technology. If properly conducted, it can interrupt the flow of information, turn off weapons and information systems communications, and even affect the decision-making process of commanders, the units' morale, and the population's mood. Of course, information warfare is not limited to the military sector. It can also be fully applied in competition in a commercial environment. This area can also become a possible topic for further studies.

10. Acknowledgments

This article was supported by grant project IGA/FAI/2023/002, „Protection of the population, soft targets and the information environment", solved in 2023-2024.

11. References

[1] Cejpek, J. (2005). Information, communication and thinking: an introduction to information science, Nakladatelstvi Karolinum, ISBN 80-246-1037-X, Prague

[2] Ressler, M. (2006). Information science and librarianship: an explanatory dictionary of Czech terminology in the field of information science and librarianship, VSCHT Prague, ISBN 80-7080-599-4, Prague

[3] Li, Q., Liu, Q., Liu, S., Di, X., Chen, S., & Zhang, H. (2023). Influence of social bots in information warfare: A case study on @UAWeapons Twitter account in the context of Russia–Ukraine conflict. Communication and the Public, Vol. 8, No 2, pp. 54–80, DOI: 10.1177/20570473231166157

[4] Putkalets, L. A., & Vyshnevska, G. B. (2023). Regional mass media in the context of Russia's information and psychological war against Ukraine, In FUNCTIONING OF THE UKRAINIAN MEDIA DURING MARTIAL LAW IN UKRAINE, pp. 93-107, DOI: 10.30525/978-9934-26-270-8-6.

[5] Aro, J. (2016). The Cyberspace War: Propaganda and Trolling as Warfare Tools, European View, Vol. 15, No. 1, pp 121-132, ISSN 1781-6858, DOI: 10.1007/s12290-016-0395-5

[6] Deepa, K. (2006). Media, War, and Propaganda: Strategies of Information Management During the 2003 Iraq War, Communication and Critical/Cultural Studies, Vol. 3, No. 1, pp. 48-69, DOI: 10.1080/14791420500505650

[7] Mamasoliyevich, N. K. (2023). INFORMATION CULTURE: A NEW APPROACH. World Bulletin of Social Sciences, Vol. 19, pp. 129-134. Available: https://scholarexpress.net/index.php/wbss/article/view/2228

[8] Vickery, B. (2011). Information science – part 1, Available: http://www.lucis.me.uk/infosci1.htm

[9] Britz, J. (2006). A critical analysis of information poverty from a social justice perspective, Ph.D. Dissertation, University of Pretoria, Pretoria, South Africa

[10] Bucharova, L. (2006). Information sources in the field of state security, Diploma thesis, Faculty of Arts, Charles University, Prague, Czech Republic

[11] Doctrine for Intelligence Support to Joint Operations, (2000). Chiefs of Staff, Washington, Available: http://www.bits.de/NRANEU/others/jp-doctrine/jp2_0.pdf.

[12] Information Operations, (1996). Headquarters, Department of The Army, Washington, Available: http://www.iwar.org.uk/iwar/resources/usarmyio/fm100-6.pdf.

[13] Svoboda, O., et al. (1984). Concise Dictionary of the Military, Our army, Prague

[14] Ressler, M. (2006). Information science and librarianship: an explanatory dictionary of Czech terminology in the field of information science and librarianship, VSCHT Prague, ISBN 80-7080-599-4, Prague

[15] Piątkowski, K. (2003). Wojna w mediach, Nowa Technika Wojskowa, ISBN: 978-83-7395-246-1

[16] Kubesa, M. (2013). "Military Deception in the Information Age", Military perspectives, Vol. 22, pp. 160-164, DOI: 10.3849/2336-2995.22.2013.01.160-164

[17] Ch. Bellamy, Ch. (2001). "What is Information Warfare?" In: Managing the Revolution in Military Affairs, R. Matthews, J. Treddenick, (Ed.), pp. 234-252, Palgrave, ISBN 0333781899, Basingstoke

[18] Hodicky, J. (2007). "The Information Age in the Military," In the Information age, information society and military, pp. 76-102, MO CR-AVIS, ISBN 978-80-7278-379-3, Prague

[19] Schneider, J. (2016). Words do not fall into a void, ISBN 978-80-7376-468-5, Olympia, Velke Prilepy

[20] McCombs, M. E. & Shaw D., L. (2023). "The Agenda-Setting Function of Mass Media", The Public Opinion Quarterly, Vol. 36, No. 2, pp. 176–187. JSTOR http://www.jstor.org/stable/2747787.

[21] Lucas, E. & Pomeranzev, L. (2016). "Winning the Information War – Techniques and Counter-strategies to Russian Propaganda in Central and Eastern Europe: A Report by CEPA's Information Warfare Project in Partnership with the Legatum Institute", Center for European Policy Analysis, Available: https://cepa.ecms.pl/files/?id_plik=2715

[22] Tomsu, M. (2022). Reliable Information Sources in the Age of Propaganda, Proceedings of the 33rd DAAAM International Symposium, pp. 437-443, B. Katalinic (Ed.), Published by DAAAM International, ISBN 978-3-902734-36-5, ISSN 1726-9679, Vienna, Austria DOI: 10.2507/33rd.daaam.proceedings.061

[23] Globus, A.; Bailey, D.; Han, J.; Jaffe, R.; Levit, C.; Merkle, R. & Srivastava, D. (1998). NASA applications of molecular nanotechnology, The Journal of the British Interplanetary Society, Vol. 51, pp. 145-152. Available: http://www.nas.nasa.gov/News/Techreports/1997/PDF/nas-97-029.pdf

[24] Dunlevy, C.; Shimeall, T. & Williams P. (2001). Countering Cyber War, NATO Review, Director of Information and Press NATO, Vol. 49, pp. 16-19, Brussel

[25] Welch, D.; Buchheit, N. & Ruocco, A. (1999). Strike Back: Offensive Actions in Information Warfare, New Security Paradigms Workshop, pp. 47-52, Caledon Hills, Ontario, Canada Available: http://delivery.acm.org/10.1145/340000/335192/p47-welch.pdf?key1=335192&key2=9541386811&coll=&dl=&CFID=15151515&CFTOKEN=618461

DOI: 10.2507/34th.daaam.proceedings.052

SPEAR-PHISHING AND ITS SPECIFICS WITHIN SOCIAL ENGINEERING

Miroslav Tomsu, Veronika Rosiková, Petr Vojtek & Monika Hrabakova

This Publication has to be referred as: Tomsu, M[iroslav]; Rosikova, V[eronika]; Vojtek, P[etr] & Hrabakova, M[onika] (2023). Spear-phishing and Its Specifics Within Social Engineering, Proceedings of the 34th DAAAM International Symposium, pp.0379-0387, B. Katalinic (Ed.), Published by DAAAM International, ISBN 978-3-902734-41-9, ISSN 1726-9679, Vienna, Austria
DOI: 10.2507/34th.daaam.proceedings.052

Abstract

In recent years, we have observed an increasing emphasis on the digitisation of the environment, which leads to the fact that we also hear more and more loudly about the phishing issue, not only from security specialists and professional groups. Phishing is a social engineering attack consisting of deception and abuse of the human factor to break security and obtain sensitive information. The article's introduction provides a general view of social engineering and describes its principles, specifics, and forms of social engineering. It mainly emphasises the issue of spear-phishing. The article's primary goal is to solve a practical security problem and emphasises spear-phishing and the importance of digital security and the most vulnerable element - the user. The article is an introductory look at the issue of cyber security in the context of social engineering aimed at spear-phishing and the mapping of concrete measures against this threat.

Keywords: Security; social engineering; phishing; spear-phishing.

1. Introduction

We live in a time when the digital space increasingly penetrates our lives and often becomes the central scene of critical events, not only corporate ones. Digitisation is associated with emphasising information and the possible facilitation of life through modern technologies. Companies increasingly realise that information and its digital form are their greatest assets. [1] Thanks to them, individuals and companies can gain a competitive advantage or confidential information or make the right decision. [2] Therefore, we cannot be surprised that this area, as confirmed by the statistical data of global analytical companies, is becoming more and more attractive for cybercriminals, and it is not for nothing that there is talk of a new digital battlefield. [3]

And so, the well-known Moore's law of exponential growth applies to cybercrime and informatics. This only confirms that the rapid development of information technology does not only have positive effects. As information technology becomes an increasingly common part of our daily lives, cybercrime is on the rise, threatening individuals and companies capable of causing irreparable damage literally in minutes. We can hear about phishing as one of the most common attacks, which also benefited from the rapid move to the digital space. In recent times, thanks to many publicised problems and successful attacks using phishing techniques, we have observed an increased awareness and, above all, the interest of people and companies in a suitable solution for cyber security and the protection of their digital identity data. [4]

Many authors [5], [6], [7] try to describe spear-phishing as a method that cannot be prevented using traditional methods but emphasises the human factor, which is the most important in this problem. The issue of cyber security in the context of social engineering focused on spear-phishing is not only a technical problem, but it is primarily a problem of the human factor. The goal is to map concrete measures against this hidden threat and propose a possible solution.

2. Social engineering

Technological progress has been enormous in recent years of globalisation. Communication and part of our lives are increasingly moving into virtual space. Personal and sensitive information is often available online through various Internet services and social networks. Unfortunately, all information systems must adequately reflect this speed of change in society, and information is often protected poorly. The designs themselves become challenging to manage and highly vulnerable. Cybercrime is thus an ever-growing problem with an ever-increasing risk supported by the number of connected devices. Thanks to the current pandemic, we are observing a dramatic increase in the daily use of digital devices, internet traffic, entertainment technology, and various operational, corporate, and personal agendas. One of the vectors of possible cyber-attacks is social engineering, which uses the human tendency to trust and the relative ease of deceiving the human mind. The difference from other, primarily technical vectors of cyber-attacks is precisely that it does not matter so much on the technical level of the organisation, the robustness of the firewall, the quality of encryption, attack detection systems, and user authentication, but on the weakest link of security, the person himself. This reveals the difference between machines and men. People trust others more than machines and are guided by emotions and feelings. Social engineering attacks cannot be entirely prevented by software or hardware solutions. They can only be limited and detected. The basis is teaching individuals to recognise such attacks [8].

3. The human factor and psychology

As outlined above, the attacker, i.e., social engineering, relies on deception and abuse of human characteristics. His bad habits, stupidity, naivety, gullibility, and inattention often manifest in the so-called operational blindness, i.e., during repetitive and routine work. In the end, the victim does not even realise that he has become a victim and has given away some information that he did not have or did not want. This subchapter focuses on frequently abused human characteristics and the situations that allow abuse, although it does not cover all possible ones. In practice, it is common for the attacker to focus on several characteristics and circumstances simultaneously, thanks to which he creates the best possible position to manipulate the victim [9].

4. Phasis of a social engineering attack

We can rank the social engineer among the modern con artists of today's digital age. To exploit the weaknesses of the company's security policies, they use prepared manipulation scenarios that benefit from the human psychology described in the previous subsection or combine them with technical means. Individual social engineering attacks differ from each other, yet we can observe a characteristic pattern in them, forming the standard phases of an attack. Since the attacker's work is often iterative, we are discussing a cycle that can happen more than once within the same episode. The attacker deepens the obtained information as part of further iterations, completing the entire attack plan.

- **Information gathering** - the attacker begins by exploring open, freely accessible sources. This is information that the company or individual does not consider sensitive or exploitable and, therefore, does not protect or dispose of it securely. Under such information, we can imagine the company's website, organisational structure, contacts, marketing materials, public tenders, published job advertisements, employee lists, older company documents or guidelines, financial results, information from social networks or public whois databases, discarded memory media, printed or handwritten credentials, etc. This step of the cycle is often referred to as the most important.
- **Building relationships and trust** - based on the information gathered from the previous phase, the attacker considers all possible situations and prepares suitable scenarios of problems and questions that could arise and need to be answered so that he is always one step ahead of his victim. If possible, he chooses as a victim a person who does not realise the importance of the information he is working with. He pretends to be someone else and gradually builds mutual relations and trust with the victim. This is a time-consuming phase in which the attacker must be careful not to arouse suspicion in the victim. Still, on the other hand, he needs to obtain the desired information from the unsuspecting victim. Suppose the exploitation of the victim takes place in the workplace. In that case, it is essential to use a communication style appropriate to the given company, its jargon, commonly used expressions and terms, correct names of employees, addresses, etc.
- **Abuse of trust** - an attacker abuses the trust and relationship gained to get the victim to provide sensitive information or conduct a security incident. It is often a form of cooperation, both on the part of the victim and the attacker who asks for something, but the victim can also ask the attacker for cooperation in good faith.
- **Exploiting information** is the last stage of the cycle, where the attacker receives the desired information and uses it to achieve the desired result of the attack without leaving any evidence. Suppose the data obtained is only a partial step closer to the goal. In that case, the attacker returns to the beginning of the cycle and repeats it until he reaches the final destination [10] [11].

5. Definition of the principle of phishing

The word phishing was created by combining the words fishing and phreaking. This is a slight corruption of the English phrase fishing, meaning fishing, and it is here that we can observe an analogy with a fisherman who throws a hook with bait and hopes that a fish will catch it. Similarly, the attacker expects the victim to fall for the irresistible temptation he has created and thus obtain the desired data. The initial substitution of the letter f for ph comes from the tradition of hacker slang. It originates in the word phreaking, one of the first forms of hacking and targeted free calls through the US telecommunications network [12]. In general, phishing has been defined as "a type of computer attack that communicates socially crafted messages to people through electronic communication channels to persuade them to perform a specific action in favour of the attacker“ [13]. The essence of phishing, which is classified as a semantic attack that uses primarily human, not a system, vulnerabilities, is the attempt to obtain sensitive information from an unsuspecting victim that the attacker can then misuse, such as login credentials, credit card information, and other personal information leading to digital theft identity, for later misuse or in combination with ransomware-type attacks (according to (Gupta et al. 2017a), ransomware is attached to 9 out of 10 phishing e-mails) to block the user's access to their data and subsequently demand a ransom for their decryption [14].

6. Types of phishing attack

We will find many ways and views of dividing phishing attacks. The article compiles an overview of the currently preferred types of phishing:

6.1. Spear phishing

A targeted attack against a specific organisation or individual. An email attack is personalised by gathering all available information about the victim and tailoring the content. Such an attack is more demanding to plan, but it tends to be highly successful due to the difficulty of detection [15].

6.2. Whaling phishing

Attacks that target high-ranking people in organisations with privileged access to data or finances. The effort involved tends to be more significant but brings higher profits [16].

6.3. Business Email Compromise (BEC) Phishing

This attack also targets executives. He is not trying to deceive them but to impersonate them. After gaining the worker's awareness, the attacker sends a highly persuasive business email to get the subordinate or partner to open the link or attachment. In 2018, the total loss from reported FBI attacks was approximately $1.2 billion (FBI 2019), the most significant loss recorded for any cybercrime [15].

6.4. Social media phishing

Attackers easily create fake profiles that allow them to access the victim's published data or attack them via private messages. He may also post phishing links on his wall, on official sites, or in groups. A study (Stern 2014) shows as early as 2014 that 22% of phishing scams on the web target Facebook [17].

6.5. Instant Messaging (IM)

Attacks are not only based on links in the text but also use images, gifs, and files that can be included in the conversation. In addition to text, voice and video chat are also supported. Attacks can thus take place in real-time and use multiple attack techniques [14].

6.6. Vishing / IVR (Interactive Voice Response) phishing

Represents attacks through a phone call to manipulate people into providing their sensitive information for verification. They can be done using automated dialling and telemarketing. People generally find voice communication more trustworthy and are used to it. In addition, the older generation prefers the phone, which is also easier to handle [15].

6.7. Smishing

Smishing uses SMS and MMS messages. The attacker usually impersonates a trusted institution such as a bank. The message contains malicious code or links to a fraudulent website [18].

6.8. QRishing

QR codes are used to automate the collection and transmission of information. They are mainly used on mobile devices. The danger is that the user needs to learn what the QR code refers to or contains. It can lead to a fraudulent address or a link to download an infected file without the victim's knowledge. Some QR code readers will present a link for review before opening it. Using address shortening still makes it difficult for users to verify the link's legitimacy [18].

6.9. Sound Squatting

They exploit the voice interface (VUI) to exploit homonyms and input errors. Virtual assistants use keywords to open applications. An attacker registers a fake app with a voice keyword similar to the original app. The virtual assistant will open a phoney app when the user calls a specific app. Once opened, this fake app steals the user's sensitive information or performs other malicious activities [19].

6.10. Fake Phishing Websites

Visually mimic legitimate websites to obtain sensitive information from users. Links to them include e-mails and advertising banners on other legitimate websites, social networks, or video portals such as YouTube. People are generally less cautious when browsing the web [17].

6.11. Wiphishing (Wi-Fi)

Attackers use publicly available networks. A typical form is abusing the login portal or passing off a fake access point as a real one (same SSID network name) and capturing network traffic. A created hotspot from a phone or laptop is often enough for an attack [18].

6.12. Malvertizing

The attacker inserts malware into the ad, activated when clicked. These ads can be placed on legitimate websites without any effort. The victim usually does not consider advertising on a reputable website dangerous. Only some people realise the shortcomings of verifying ads before they are deployed. Thanks to information from websites hosting ads, it is possible to target specific groups [20].

6.13. Typo squatting

Targets a user's typographical error when entering a URL. An attacker on similar domain names creates visually similar websites that they use to steal sensitive information or do other malicious activities [18].

6.14. GUI Squatting

An attack targeting mobile devices. The automated method generates high-quality phishing copies of simple applications that bypass existing phishing prevention and detection techniques and steal user credentials [21].

7. Specifics of spear-phishing

While it is typical for traditional phishing to send out many e-mails, often to the wrong addresses, spear phishing is the opposite. In such a case, the attacker sends an e-mail to a specific person, usually an employee in a particular position or a high-ranking company manager. Such mail addressed to a specific person can be challenging to catch with an anti-phishing filter, mainly because it does not appear in large numbers and shows no signs typical of phishing mail. Please understand that neither the sender's nor the SMTP server's address is on any blocklist, and there are no links to the Internet from the e-mail.

In addition, if such an e-mail is sent to the victim directly from the environment of the given company, the latter usually does not even suspect that it could be phishing, even though they have been instructed in this direction. How could she, when the mail came from a person she knows well, there are no links from the mail to anywhere on the Internet, she is not required to enter any data, and the mail completely corresponds to the company's customs in terms of form and content. At this point, it must be emphasised that the attacker pays considerable attention to preparing the e-mail. For this purpose, the attacker is quick to familiarise himself in detail with the company culture, organisational structure, and conditions that prevail in the given company. Therefore, the correct language is used, and we do not find any grammatical errors or stylistic flaws in the text of the e-mail. Such a carefully prepared e-mail containing an attachment with malicious code is used as part of an APT attack when an unsuspecting victim opens the e-mail, clicks on the attached attachment, and, together with the associated application, launches malicious code that exploits some (zero-day) vulnerabilities to successfully load into memory and ensure that it runs even after a computer restart.

In the case of spear phishing, when the attacker's goal is to catch a big fish, if we continue to stick to the above comparison, it is not possible to underestimate its intelligence, so the attacker must first thoroughly explore the territory in which he decides to hunt and arm himself with a harpoon (in English spear, hence spear phishing) because they won't catch a big fish on some smelly bait [22].

Phishing	**Spear phishing**
Mail is sent to a large number of addresses.	Mail is sent to only one recipient.
The email contains a link to the Internet.	Mail does not have a connection to the Internet.
The mail will come from a person or company you know.	The mail will come from a person or company you know.
Specific data is required.	No data entry is required.
The mail contains (unintentional) errors.	Mail does not contain (unintentional) errors.
Mail usually does not contain any attachments.	Mail usually includes an extension.
The goal is to obtain login or personal information.	The goal is to get sensitive information that is the subject of intellectual property.

Table 1. Comparison of classic phishing and spear-phishing [22].

8. The most common examples of spear-phishing

8.1. E-mail

Spear-phishing can be very difficult to detect if done correctly. Emails come from a person the person knows and expects a response from. It is especially tragic in the case of native Microsoft clients. When a name is entered in the "from" field that matches the name of a person who works in your organisation, it will pull up his photo and display his status and other information. In that case, the user won't check the header to see what's in the "envelope-from" field. In most cases, sender users quickly follow what is listed in the "From" field. But that's not entirely true. The problem is that the email has two "from" areas. One is called "envelope-from," which contains the sender's actual address, and the other, called "from", anyone can write whatever they want.

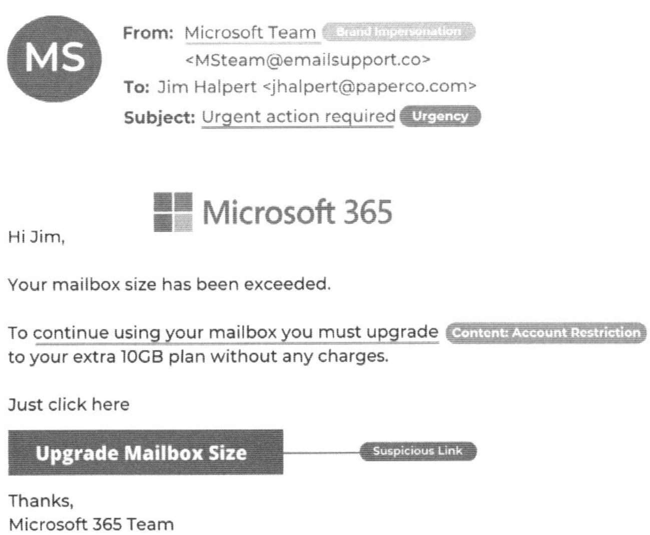

Fig. 1. A typical example of spear-phishing [27].

8.2. O365

Recently, there has been an increased number of attacks on employees of organisations that have moved their communications to O365. After all, thanks to O365, Microsoft has become the most popular brand among attackers who rely on phishing, and it is not surprising that the number of companies and users in O365 is growing at a rocket pace. An attacker can then attack practically anyone from a compromised account. The attack comes in several phases. It is a so-called multi-phase attack, where there is first classic phishing and then spear phishing on employees of a specific organisation using the Microsoft O365 cloud.

First, the attacker uses phishing to obtain the login data for the account of any employee in the given organisation. From this account, he contacts a specific person through spear phishing. Given that the e-mail comes from someone she knows, she does not suspect it could be an attack [23].

8.3. EML attachments

In other cases, the facts of how attachments are handled were used. They are scanned for malicious code, but if the branch is a file with the EML extension, a standard format for saving e-mails in a file represented by an envelope icon, it already has the same check as in an e-mail not in progress. And since the EML file can be easily created and edited in any text editor, it is possible to modify it so that both the sender and recipient fields contain the addresses you need. That is lessons from the sender's or recipient's organisation, which the recipient knows well [24].

8.4. Pictures

As part of the spread of SPAM, the text has been replaced with an image for years to bypass various filters. Here, a newly appeared image was created due to saving text in which homomorphic characters were intentionally used. Such a text is complicated to read with an OCR program and interpret correctly because it needs to make sense in any language during simple machine processing. A person can easily read it, but artificial intelligence is at a lower level to deal with it [24].

9. Defence against spear-phishing

While an attacker only needs a single click from the victim to open an infected attachment, spear-phishing defences consist of four broad, interconnected areas that combine technical, process, and human elements.

9.1. Regular users

For regular users, it is essential to follow several safety principles.

- Do not allow macros in programs - (especially Microsoft Office). If macros are required, verify the document's origin with its sender before enabling them. Do not unthinkingly open attachments and links in e-mails, especially from unknown senders.
- Check the e-mail address in case of urgent or unusual requests (payment of an invoice, request for confidential information).
- In case of uncertainty or suspicion of harmful e-mail, contact the IT department or other responsible workplace.
- Limit the sharing of employment information on social media and avoid sharing information about company hierarchy, security, and administrative processes.

9.2. Network administrators

Limit the attacker's access to the user. In their spear-phishing e-mails, more sophisticated attackers falsify their addresses to look like addresses within the organisation (spoofing). This technique can be made more difficult by using so-called anti-spoofing tools. These are DMARC (Domain Message Authentication Reporting and Conformance)12 technology, which verifies the sender's domain, and DKIM (DomainKeys Identified Mail)13 and SPF (Sender Policy Framework)14 technologies. These technologies evaluate which e-mail reaches the user and which ends up in spam. The attackers' chances of success can also be reduced by limiting publicly available information about the organisation. This is, for example, the organisation's exact organisational structure or the management's biographies. Users should also carefully consider what and with whom they share online. In this direction, individual organisations can inform users about the options for privacy settings on social networks, allowing users to limit the circle of people who can see the content on their profile. Restrictions on information sharing also apply to employment details, especially in critical infrastructure, government offices, and strategic enterprises, which are particularly interesting to sophisticated APT groups. The organisation should also be aware of what information is shared about it by third parties (partners, suppliers).

They are protecting the organisation from the effects of successful spear-phishing attacks.

If early threat detection fails, the consequences of a successful attack must be minimised. Here are universal cybersecurity principles that are not limited to spear-phishing:

- The organisation should use anti-virus software;
- Systems should have legal and supported software installed and updated to the latest versions;
- Administrator accounts should be limited based on necessary needs and should not be used for e-mail communication or Internet surfing;
- Limitation of macros for Microsoft Office;

- Using a two-factor authentication password manager (prevents 99.9% of account theft attempts);15
- Create regular data backups in case of data compromise or loss.
- Periodically assess whether the user needs all the access rights he has;
- Continuously delete the accounts of people who no longer work in the organisation.
- Rapid response to incidents.

Help users identify and report spear-phishing e-mails.

This step aims to create an environment where employees are unafraid to report suspicious e-mails. For this, it is necessary to clearly define within the organisation the person or persons to whom such e-mails can be written and how) and to inform employees about this. Given the ever-increasing sophistication of spear-phishing, it is unrealistic to expect users to identify 100% of malicious e-mails. Still, even minimal training and familiarisation with the reality of the threat can help reduce the risk [25].

10. Frequent reasons for failure of existing protection

The e-mail comes from someone the person knows and expects a response from. However, the attacker only sometimes succeeds in hacking the sender's e-mail account and, therefore, has to spoof the sender's address, which increases the chance of detection. You can quickly tell who the email came from by looking at what is written in the "from" field. But that's not entirely true. The problem is that the email has two "from" areas. One is called "envelope-from", which contains the sender's actual address, and the other, called "from", anyone can write whatever they want.

```
2859:2890:2902:2909:2933:2937:2939:2942:2945:2947:2951:29
4:3934:3936:3938:3941:3944:3947:3950:3953:3956:3959:4042:
388:9392:10010:10049:10175:11026:11233:11473:11529:11651:
80,0,RBL:122.201.87.195-irl.urbl.hostedemail.com-127.0.0.
bulk,SPF:fn,MSBL:0,DNSBL:neutral,Custom_rules:0:1:0
X-HE-Tag: chain43_902aac0ff3e60
X-Filterd-Recvd-Size: 3501
Received: from scuderia.turboservers.com.au (unknown [122
    by imf05.hostedemail.com (Postfix) with ESMTP
    for <jonathan@askwinters.com>; Wed,  8 Apr 2015 11:19
Received: from nobody by scuderia.turboservers.com.au wit
    (envelope-from <nobody@scuderia.turboservers.com.au>)
    id 1Yfo1N-0005mn-RS
    for jonathan@askwinters.com; Wed, 08 Apr 2015 21:19:4
To: jonathan@askwinters.com
Subject: About your last Transaction
```

Fig. 2. Spear Phishing Attack Example Raw Source [28]

The e-mail client takes the sender's address from this field, which it then displays to you. The check could be done by SPF (Sender Policy Framework), which ensures that it is not possible to receive e-mail from a mail server that is not authorised to send e-mail for the given domain. But in this case, it doesn't matter because SPF only checks what is stated in the "envelope-from", not what is said in the "from" field. If the attacker tried to change the value in "envelope-from" as well, then SPF would work, but why would the attacker do that when he needs to change the "from" field, which SPF doesn't check? And who would think to display the header of each email and see if something else is written in "envelope-from" than in "from".

A possible protection would be DKIM (DomainKeys Identified Mail). Still, the problem here is that this protection only creates a hash of the e-mail or only a specific part of it, which is encrypted with the private key that it has right only a person in the given domain. During the check, it was subsequently found that the given e-mail was protected using DKIM, so a query was made for the given environment. This obtains the public key and thus decrypts the message and obtains its hash. This is then compared with the hash calculated on the side of the message's recipient. If it fits, then the e-mail came from the given domain.

The only possible solution is DMARC (Domain-Based Message Authentication, Reporting, and Conformance), which compares the value given in the "from" and "envelope-from" fields, and it is possible to set how to handle an email that has a different address provided in both areas. In practice, however, it is used sparingly because, in many cases, these fields are other, even for legitimate reasons [26].

11. Conclusion

This article aimed to outline how easy it is for an attacker to create a successful phishing campaign and how, if at all, the recipient of such an e-mail can verify its authenticity and discover that it is fake before even clicking on a link in the e-mail. He will be redirected to a fraudulent site where he will be asked to enter his login information.

Company security does not begin and end with the IT professional or IT department. Cybercriminals often target the weakest link, so every employee must be up-to-date on the latest threats and tactics used by cybercriminals to help keep the business safe and protected from cyber threats that can have potentially devastating effects. It can be expected that the user will soon be the most vulnerable link in all systems. For that reason, there are reasons to emphasise educational programs and users' critical approach to their cyber security is essential, especially concerning the expected future development.

Spear-phishing attempts are usually not by random hackers but organised criminals seeking money, trade secrets, or other information who have vetted their target well to appear trustworthy. The apparent sender of spear-phishing is often one of the company's senior employees. A request in an e-mail is usually urgent and requires an immediate response. At the same time, it is more than likely that attackers in cyberspace will continue to refine the fraudulent techniques they will use. The opposite of this development can be, for example, integrating all social and age categories of users into the continuous education system. Healthy prevention, which unfortunately is often underestimated by many users, is to refrain from posting too much personal information in cyberspace because this is the most valuable food source for hackers who work with the spear-phishing method.

Further research aims to raise awareness of the dangers of phishing and teach users how to detect real phishing using software tools. Focus only on preparing phishing content and monitoring user reactions. The goal will also be to focus on preparing phishing content, monitoring user reactions, and evaluating the activities of individual recipients on the relevant phishing pages.

12. References

[1] Tomsu, M. (2022). Reliable Information Sources in the Age of Propaganda, Proceedings of the 33rd DAAAM International Symposium, pp.0437-0443, B. Katalinic (Ed.), Published by DAAAM International, ISBN 978-3-902734-36-5, ISSN 1726-9679, Vienna, Austria DOI: 10.2507/33rd.daaam.proceedings.061

[2] Shao, J., Zhang, T., Wang, H., & Tian, Y. (2022). Corporate social responsibility and consumer emotional marketing in significant data era: a mini literature review. Frontiers in Psychology, Vol. 13, DOI: 10.3389/fpsyg.2022.919601.

[3] McGuire, M., & Dowling, S. (2013). Cybercrime: A review of the evidence. Summary of key findings and implications. Home Office Research report, 75, United Kingdom.

[4] Salahdine, F., & Kaabouch, N. (2019). Social engineering attacks: A survey. Future Internet, Vol. 11, No. 4, DOI: 10.3390/fi11040089

[5] Bullee, J. W., Montoya, L., Junger, M., & Hartel, P. (2017). Spear phishing in organisations explained. Information & Computer Security, Vol. 25, No. 5, pp. 593-613, ISSN: 2056-4961.

[6] Stembert, N., Padmos, A., Bargh, M. S., Choenni, S., & Jansen, F. (2015, September). A study of preventing email (spear) phishing by enabling human intelligence. In 2015 European intelligence and security informatics conference, pp. 113-120, DOI: 10.1109/EISIC.2015.38.

[7] Butavicius, M., Parsons, K., Pattinson, M., & McCormac, A. (2016). Breaching the human firewall: Social engineering in phishing and spear-phishing emails. In Australasian Conference on Information Systems 2015, DOI: 10.48550/arXiv.1606.00887

[8] Salahdine, F. & Kaabouch, N. (2019). "Social Engineering Attacks: A Survey", Future Internet 11, No. 4, DOI: 10.3390/fi11040089

[9] Fan, W.; Lwakatare, K. & Rong, R. (2017), "Social Engineering: I-E based Model of Human Weakness for Attack and Defense Investigations", International Journal of Computer Network and Information Security (IJCNIS), Vol. 9, No. 1, pp. 1-11, DOI: 10.5815/ijcnis.2017.01

[10] Mitnick, K.; Simon, W. & Vasta, L. (2003) The Art of Deception: The World's Most Famous Hacker, Gliwice, Helion, ISBN 978-0764542800

[11] Jamil, A.; Asif, K.; Ghulam, Z.; Nazir, M. K.; Mudassar, A. S. & Ashraf, R. (2018). "MPMPA: A Mitigation and Prevention Model for Social Engineering Based Phishing attacks on Facebook," IEEE International Conference on Big Data (Big Data), pp. 5040-5048, Seattle, WA, USA, DOI: 10.1109/BigData.2018.8622505

[12] Fruhlinger, J. (2020) What is phishing? How this cyber-attack works and how to prevent it, CSO Online, 2020, Available: https://www.csoonline.com/article/2117843/what-is-phishing-how-this-cyber-attack-works-and-how-to-prevent-it.html.

[13] Cui, Q.; Jourdan, G. V.; Bochmann, G. V.; Couturier, R., I. & Onut, V. (2017). "Tracking Phishing Attacks Over Time", In Proceedings of the 26th International Conference on World Wide Web (WWW '17), International World Wide Web Conferences Steering Committee, pp. 667–676, Republic and Canton of Geneva, DOI: 10.1145/3038912.3052654

[14] Aleroud, A. & Zhou, L. (2017). "Phishing environments, techniques, and countermeasures: A Survey", Computers & Security, Vol. 68, pp 160-196, DOI: 10.1016/j.cose.2017.04.006

[15] Salahdine, F. & Kaabouch, N. (2019). "Social Engineering Attacks: A Survey", Future Internet 11, No. 4, DOI: 10.3390/fi11040089

[16] Fruhlinger, J. (2020) What is phishing? How this cyber-attack works and how to prevent it, CSO Online, 2020, [online] Available: https://www.csoonline.com/article/2117843/what-is-phishing-how-this-cyber-attack-works-and-how-to-prevent-it.html.

[17] Gupta, B. B.; Arachchilage, N. A. G. & Psannis K. E. (2018) "Defending against phishing attacks: taxonomy of methods, current issues and future directions", Telecommun Syst, Vol. 67, 2018, pp. 247–267, DOI: 10.1007/s11235-017-0334-z
[18] Alabdan, R. (2020). "Phishing Attacks Survey: Types, Vectors, and Technical Approaches", Future Internet, Vol. 12, No. 10, DOI 10.3390/fi12100168
[19] Shashank, K.; Rakesh, D. R.; Vaibhav, S. N. & Balkrishna E. N. (2020). "Applications of industry 4.0 to overcome the COVID-19 operational challenges," Diabetes & Metabolic Syndrome: Clinical Research & Reviews, Vol. 14, Is. 5, pp. 1283-1289, DIO 10.1016/j.dsx.2020.07.010
[20] Nagunwa, T. (2014) "Behind Identity Theft and Fraud in Cyberspace: The Current Landscape of Phishing Vectors," International Journal of Cyber-Security and Digital Forensics, Vol. 3, pp. 72–83. DOI 10.17781/P001287
[21] Chen, S.; Fan, L.; Chen, C.; Xue, M.; Liu, Y. & Xu, L. (2021). "GUI-Squatting Attack: Automated Generation of Android Phishing Apps," in IEEE Transactions on Dependable and Secure Computing, Vol. 18, No. 6, pp. 2551-2568, DOI 10.1109/TDSC.2019.2956035
[22] Butavicius, M.; Parsons, K.; Pattinson, M. & McCormac A. (2016) "Breaching the Human Firewall: Social engineering in Phishing and Spear-Phishing Emails," Available: http://arxiv.org/abs/1606.00887
[23] Dewan, P.; Kashyap, A. & Kumaraguru, P. (2014). "Analyzing Social and stylometric features to identify spear phishing emails," 2014 APWG Symposium on Electronic Crime Research (eCrime), Birmingham, AL, USA, pp. 1-13, DOI 10.1109/ECRIME.2014.6963160
[24] Bullee, J. W.; Montoya L.; Junger M. & Hartel P. (2017) "Spear phishing in organisations explained", Information and Computer Security, Vol. 25 No. 5, pp. 593-613, DOI 10.1108/ICS-03-2017-0009
[25] National Cyber and Information Security Agency, (2020). "Tailor-made phishing emails or social media messages: Spear-phishing and how to protect yourself from it", Available: https://www.nukib.cz/download/publikace/doporuceni/Doporuceni_spear_phishing_2.0.pdf
[26] Smejkal, V. (2015), Cybercrime, Ales Cenek, Plzen, ISBN 978-80-7380-849-5
[27] Gilmer, J. (2021) What does spear-phishing look like?, Available from: https://www.meshsecurity.io/spear-phishing
[28] Winters, J. N. (2015) Spear Phishing – Increasingly Common, Available: https://askwinters.com/2015/04/08/spear-phishing-increasingly-common

DOI: 10.2507/34th.daaam.proceedings.053

INFORMATION ECOLOGY, INFORMATION HYGIENE, AND ASSESSMENT OF INFORMATION VULNERABILITIES IN ORGANIZATION

Miroslav Tomsu, Jan Valouch, Jiri Vojtesek

This Publication has to be referred to as: Tomsu, M[iroslav]; Valouch, J[an] & Vojtesek, J[iri] (2023). Information Ecology, Information Hygiene, and Assessment of Information Vulnerabilities in Organization, Proceedings of the 34th DAAAM International Symposium, pp.0388-0394, B. Katalinic (Ed.), Published by DAAAM International, ISBN 978-3-902734-41-9, ISSN 1726-9679, Vienna, Austria
DOI: 10.2507/34th.daaam.proceedings.053

Abstract

The article tries to approach information hygiene as a tool that will make it possible to achieve higher efficiency in obtaining information through an appropriate knowledge and work regime arrangement. To put it into context, he points to the current state of society. In particular, he is concerned with the excess of information and the resulting problem with information vulnerabilities in the organisation, demonstrating the necessity of changing information habits. It deals with information hygiene mainly from the point of view of implementation by the user himself. It points out some negative phenomena that can occur when working with information. The article attempts to outline a new method for evaluating the information environment in an organisation from the perspective of information science. The article further deals with recommendations to organise the work regime better.

Keywords: Information ecology; information hygiene; information vulnerability; information.

1. Introduction

One of the essential entities is organisations and their information environment. The vulnerability of the information environment has become a topical issue in recent years. The prerequisite for ensuring a clean information environment is the identification of vulnerabilities in the information environment based on the given setting of the organisation itself. The information environment is an integral part of the social environment. Its structure, determined by the basic social levels, consists of the human factor, which is the originator, mediator, and user of the communication process, as well as the information and communication processes themselves, within which specific information goals are realised, certain information flows are ensured, and various information products are created. All this is provided by one particular material-technical base: information institutions, libraries, information technology and computer technology. [1]. The field of technology is vast. It contains many types of technology that enable human activities that were once impossible. However, with the rapid onset of technology, the volume of information that a person has to process also increases.

The rapid increase in information is called the "information explosion" and is described as "an extreme increase in the supply of available information." [2]. Davenport [3] evaluated a new way of looking at information management in a changing world that considers the information environment, which he called information ecology. According to him, "an information ecologist can mobilise not only architectural designs and information technology, but also information strategy, policy, behaviour, support staff, and work processes to create a better information environment.

A vital premise of the information ecology, defined as "the complete information environment," is that organisations must focus beyond "mechanical engineering" beyond information technology. Evaluation is an essential tool for improving management. Through organisational evaluation - commonly known as evaluation - the organisation's effectiveness is measured in terms of its functioning, problems, and results in terms of behaviour and social system - it evaluates economic impacts, impact evaluation, essential evaluation, analytical evaluation, and operational evaluation. [4] The heuristic method evaluates the usability of the user interface of the system without user involvement - the evaluation of the information system means the evaluation of performance in the areas of hardware, software, computer networks, data, and human resources [5], the evaluation of text documents and information resources assesses the extracted structured sources of information that are collected in text documents, and information about sources that indicate their reliability and relationships between them [6], website evaluations [7], among others.

The search for literary sources brings us an evaluation of organisations from a complex point of view, information systems, text documents, and information sources, which somehow evaluates the information environment but does not focus more closely on the evaluation from the point of view of information science of our described vulnerabilities - information overload,

The search for literary sources brings us an evaluation of organisations from a complex point of view, information systems, text documents, and information sources, which somehow evaluates the information environment but does not focus more closely on the evaluation from the point of view of information science of our described vulnerabilities - information overload, information smog, information waste, information fragmentation. Such an approach still needs to be improved. This often-neglected area can result in poorly done work, the breakdown of the organisation's processes, or significant financial losses. Based on the mentioned and other shortcomings of the current state of the problem, the need for a conceptual assessment of the vulnerability of the organisation's information environment was evaluated.

2. Information ecology

The German philosopher Rafael Capurro was the first author to deal with information ecology in a broad (mainly political) context. In his view, information ecology has a social, historical, and linguistic dimension. Information ecology takes care of and protects the information environment of the organisation and the system of interconnected social, cultural and political subsystems. The information ecology approach in an organisation affects what information will be produced, stored and accessed. It focuses primarily on the human element and not only on technology. Continuous development and changes in the organisation's environment are also considered. [3]

2.1. Knowledge society

Today, the knowledge society is an internationally recognised and promoted concept, according to which the further development and improvement of society must be supported by the fact that all people are educated throughout their lives; lifelong learning becomes part of the way of life. Let's look at the historical experience with the practical application of knowledge and information to make technical, economic and social decisions. The example of the decision-making of the mentioned King Solomon can illustrate the fact that such decision-making and management was and is considered wise, where not only information and knowledge contribute to the probability of a correct decision, but also actual emotional agencies such as feeling love, hope, faith, desires, motivation, common good, etc.

Therefore, the ability to harmonise informational and emotional agencies and select them for vital decisions about one's behaviour and the behaviour of surrounding systems is needed. Suppose we do not behave arbitrarily towards the physical environment of the world around us. In that case, we should also set certain limits and rules for maintaining the purity of the information environment. We live within the framework reserved by our European culture. It includes caring for the soul of the individual and our living space, and now it should also be caring for the quality of the information environment. [8]

2.2. Virtual smog

It comes to the words of the German philosopher Martin Heidegger, who observed that technology shortens distances but does not bring new closeness. The younger generation, in particular, lives partly in the virtual world, which is formed in an information environment rich in valuable components and polluted by dangerous information smog. There is talk of the need for a new kind of literacy that will eliminate the problem of "digital dementia".

We can consider the elements operating in the information environment as agencies. It is the term of the world-renowned professor Miloslav Jokl, who used it to denote the components of the microclimate, influencing its quality. Agencies have their sources, fields of transmission and effects, which in the information environment correspond to, for example, sources of news, social networks and the impact on users.

The influence of agencies is understandably increasing due to the rapid development of information and communication technologies. This is because telecommunications and media data formats converge, in which information agencies have unprecedented influence. It is vital to try to maintain the quality of these elements so that they serve people healthily. [9]

2.3. Information agency

In the information environment, we encounter agencies at different hierarchical levels. Essential elements at the lowest level are signals in the form of elementary information constructs. Data can already form a message at the next level, carrying information. An even higher level is knowledge. A separate area then consists of emotional agencies.

The individual levels deserve a more detailed explanation. A signal is a change or a sequence of changes in the states of state variables in the information environment. According to the physical nature of agency, we are talking about acoustic, optical, mechanical, electrical, chemical and other signals used by living beings or machines to process data, information and knowledge.

Data and databases represent sets of interpreted and uninterpreted data about the state of objects. This means that the possibility of emerging different interpretation models over the same collection of shared data can already be expected at this level. The concept of information is usually related to removing uncertainty (or increasing the orderliness of the system). Therefore, we can understand information, for example, as a message that brings the possibility of reducing uncertainty in the description of states and ongoing processes and a stimulus to changes in order in real-world systems or consciousness.

For example, we can state that a high level of information is contained in a cathedral building with an astonishing arrangement of elements and architectural signals. An example of the highest level of data is the organisation of the human genome, the manifestations of viability and the growth of the order of living organisms. Report usually answers questions: when, where, who, what, in what relationship, how long, how high, etc. [9]

2.4. Knowledge and emotion

A particular category of information agency is knowledge that answers more complex questions: how, why, with what perspective, in what causal relationships, where is the cause, what will be the consequence, how to formulate models of objects and processes on things, etc. It is the ability to assign, sort and filter data, data and information displays of probable states of items and their state transitions.

It is a question of whether any of the scientists in the field of computer science today have seen a computer cry. Perhaps such a question sounds naive, but it doesn't have to be so innocent in artificial intelligence. In any case, living beings sharing the information environment of the near and far surroundings show changes in the state of their consciousness and changes in the conditions of characteristic state variables and their arrangement in response to some types of agencies from the psychosocial component of the information environment.

Emotional agencies often stimulate support for the function of knowledge systems, especially in the stages of interpretation, which are essential for the intelligent design to make decisions about further changes in its state. An illustrative example is how King Solomon makes decisions in the well-known biblical story when two women argue over a child. Solomon wants to find out who the birth mother is, so he orders the child to be cut into two equal parts, after which the birth mother asks that the child's life be preserved even at the cost of giving it to another woman. The birth mother's emotions lead to solving the puzzle. [10]

2.5. Information hygiene

While abroad, in connection with the issues of information overload, internet addiction and security, they are spoken of as separate, interconnected areas to varying degrees. In the Czech and Slovak environments, the umbrella concept of information hygiene is gradually adopted for these topics. This term is a set of preventive measures enabling efficient work with information and preventing information overload. Information hygiene also includes the prevention of a situation where, on the other hand, we do not receive some information, i.e., the ability to search for information is part of information hygiene. [11]

2.6. Impact of reality

In 2018, the International Data Corporation predicted that the world's data volume would grow to 175 zettabytes by 2025. About five quintillion bytes of data are generated every day. The data the world will create in the next three years will be more than all the data made in the last 30 years. Interestingly, only about 0.5 per cent of all data generated is analysed and used. Our media consumption across all channels and formats is estimated to take more than 13 hours daily. Losses that are a direct or indirect result of information overload are estimated to be approximately $650 billion annually. It's the cost of interrupting productivity. In a company with 1,000 employees, all workers waste about 655,000 hours per year searching for information for their work. [12]

3. Vulnerabilities in an unhealthy information environment

Security plays a significant role in society. One of the critical elements of an organisation is its information environment. In an organisation, security ensures the protection of conditions so that it can perform its intended function. A prerequisite for ensuring safety is the identification of vulnerabilities in the information environment relative to the security environment and the organisation itself. We can talk here about a new concept - information vulnerability - as a vulnerable point, weakness or unintentional imperfection that can be exploited to threaten to cause damage or loss. Vulnerability itself is not the cause of harm. A vulnerability is merely a condition allowing a threat to affect assets. Vulnerabilities arising from various sources must be taken into account.

3.1. Information overload

Decision-makers need more cognitive ability to process information. Information overload occurs precisely when the amount of information exceeds this capacity. If informants are calm, the quality of their decision-making may improve. The problem of information overload arose during the Renaissance and the Industrial Revolution. However, with the information age and access to automated and efficient data collection that provides more information, the severity of this problem is increasing.

3.2. Information multitasking

Some people take great pride in their "multitasking" abilities. Especially women, who are said to be successful because the two hemispheres of their brain are more closely connected. If a person is constantly bombarded with information from different sources simultaneously, he cannot properly concentrate on any of them. It also affects short- and long-term memory and flexibility in solving new problems. Statistically, such people change jobs less often, even if they feel bored in their current job and it does not fulfil their career or financial dreams.

3.3. Inefficient way of working with information

The human brain works with information non-linearly, using connections and the ability to see links in the larger whole. Being able to process accurately and quickly find a large amount of information at any time is an increasingly important skill in the work of all executives. The most important thing is to be able to orient yourself, to choose an excellent source of information, and to satisfy the needs related to the performance of the task or the solution of the assigned task.

3.4. Little motivation to process information

A feeling of threat causes low motivation or demotivation, it manifests itself mainly by focusing on one's problems, limited concentration on receiving new information, on performance, inability to think creatively and find solutions, searching for an "escape" from the environment of receiving, processing and remembering information.

3.5. Information fragmentation

The fragmentation of information that only provides a partial context of events and their interrelationships, generated by many different sources of information, leads to a fragmented perception of reality for many people. This may be the reason for their childish view of reality. Especially nowadays, this damaging phenomenon is more visible and noticeable.

3.6. Chaotic arrangement of information

The problem is when the worker gets lost in the flood of documents and information. They need help finding the content they need or which document version is correct—the absence of knowledge when required.

3.7. Verification of information

Much information appears in the information space. They are attacking us from all sides, and it is challenging to navigate them. Some information has a subliminal effect on us without us being aware. An important people skill is to be able to resist influences and to be able to filter out information. In other words, be able to distinguish between those who are severe and those whose intent is to manipulate.

3.8. Unavailability of information

The absence of information can cause stress and uncertainty and create room for speculation.

3.9. Misunderstanding of information

Misunderstanding can be understood as something often unelected, something we need to be aware of. We have limited means of language and communication to convey what we think and imagine. The transmission of information always involves communication noise. An important aspect is also the information literacy of the individual.

3.10. Communication barrier

In the communication process (i.e., mutual sharing - communicating/receiving information), there are sometimes unavoidable obstacles, i.e., communication barriers. They interrupt successful communication. Communication barriers are as varied and enduring as human communication itself. In some situations, the blocks may be so severe as to cause armed conflict.

4. Method of quantifying and evaluating information vulnerability

Vulnerabilities arise due to undesirable information effects (information overload, information smog, information waste, information fragmentation, information credibility) on people in the organisation. Weaknesses or unintentional imperfections weaken or limit a person's quality of work and lead to damage or loss. Vulnerabilities arising from various sources must also be taken into account.

To reach the primary goal, you will need to complete the objectives using the following methods:

4.1. Detection of vulnerabilities:

- Input information about the organisation;
- Questionnaire survey (qualitative assessment) - deduction of specific vulnerabilities;
- Questionnaire survey (quantitative assessment) - information overload;
- Checklist – detection of deficiencies in the information environment;
- Observation - evaluation of the situation at the detected critical point;
- Creating a list of unnecessary information.

4.2. Compilation of a catalogue of information vulnerabilities, verbal evaluation and quantification

vulnerability name	vulnerability description	degree of worker influence	user impact rate	frequency of appearance	level of threat	level of threat
insufficient informational education	The organisation needs to organise additional training and courses on using digital technologies for its employees. Employees depend on their options, the time to complete a task increases, and productivity in other tasks decreases.	50 %	10 %	medium	1	20 %
chaotic arrangement of information/information sources	The worker has no order in his documents and cannot organise work with information/documents. Time increases when searching for information/documents.	50 %	40 %	medium	2	30 %
information overload	The worker is overwhelmed with information when serving multiple users simultaneously (personal, telephone, e-mail contact).	60 %	80 %	high	2	40 %
unavailability of information (missing orientation/information system)	The organisation's absence of an information/orientation system causes complications, especially for users from the public, who need to receive the correct information in time.	10 %	80 %	high	3	80
...	...	...	...	...	...	

Table 1. Sample catalogue of information vulnerabilities.

The catalogue of information vulnerabilities will describe not only vulnerabilities but also the degree of threat to the employee of the organisation and the degree of danger to the organisation's user (if it is an organisation that works with users / the public). These data will be determined by an expert opinion based on the information found from the local investigation. The overall degree of threat given in percentages will also be provided by an expert assessment based on the already assessed effects of the worker and the user and the basis of information from the local investigation. The resulting coefficient $\boldsymbol{n}$ for the organisation's vulnerability will be determined based on partial vulnerabilities n_1, n_2, n_3, ..., n_x. The following formula will be used to calculate the overall result:

$$n = \frac{n_1 + n_2 + n_3 + n_x}{n \times 2} \times 100$$

4.3. Compilation of evaluation criteria for the evaluation of organisational vulnerabilities

For the expert assessment, the evaluation method on a scale of 1-4 will be chosen, and vulnerability values, percentage expression of the problem, and finally, verbal expression will also be assigned to this scale.

Degree	Abbreviation	Vulnerability	Expression in %	
			from	to
1	L	low	0 %	25 %
2	M	medium	26 %	50 %
3	H	high	51 %	75 %
4	C	critical	76 %	100 %

Table 2. Degrees of vulnerability and expression of exposure. [13]

Degree	Verbal expression
1	The vulnerability does not represent a particular security problem for the organisation. The operation of the organisation is not threatened. Special measures should be kept secret. They can be consciously accepted.
2	Vulnerability represents a security problem for the organisation. The organisation's operation is not significantly threatened, but the issue must be addressed and solved.
3	Vulnerability represents a security problem for the organisation. The organisation's operation can be threatened. The issue needs to be solved and given long-term attention.
4	Vulnerability represents an acute security problem for the operation of the organisation. The function of the organisation is threatened. The issue must be resolved and eliminated.

Table 3. Degrees of vulnerability and verbal expression. [13]

4.4. Catalogue vulnerability assessment and overall assessment

The result of the coefficient for calculating the vulnerability of the information environment in the organisation will always contain a verbal evaluation and possible recommendations for improving the organisation's operation. Because it is always an environment where the main point is the person with their personality aspects – cognitive and social, it is necessary that the result, which objectively does not correspond to the current state of the information environment in the organisation, is corrected by an expert view with justification.

5. Principles of information hygiene environment

When our brain is overloaded, we make the mistakes above, get stuck in the middle of the thought process, ignore our surroundings, and forget important things on the agenda. Among other things, our ability to plan or deal with emotions can also be reduced. It can directly or indirectly lead to a deterioration in our overall well-being and performance at work. It can also directly lead to job error rates and threaten to cause financial damage to the organisation. From a human perspective, this could lead to chronic information syndrome or a total psychological breakdown. In the event of discovering or indicating any of the listed information vulnerabilities, it is necessary to change the style of work or life and apply the principles of an information-hygienic environment.

5.1. Create a list or outline of all tasks

To grasp and better plan tasks, you need to set priorities. That's precisely what these lists are for.

5.2. Let go of non-essential tasks

If you don't want to give them up, it's a good idea to write down the tasks you completed and the time you spent on them. It will help to place value on the work and realise that the time spent on these tasks stays the same. In the future, it will also improve the ability to plan.

5.3. Setting specific short-term goals

It is most effective to divide the task into short-term intervals, e.g., after 15 to 20 minutes.

5.4. Stop multitasking

Better to do only one or more tasks at a time, such as answering an email while sitting in a meeting. It will be more efficient if one does the tasks in sequence, one job at a time.

5.5. Don't be interrupted at work

There will be an increase in performance, and stress will not appear. It's better to check your inbox at intervals, such as once an hour or every 30 minutes, if you're worried about not responding quickly enough. It is a good idea to avoid constantly picking up new e-mails. Limiting visual and audio alerts such as pop-ups and notifications is advisable. Instead, set specific times to check them manually.

5.6. Tailor your breaks to the task at hand

If a person wants to regulate his information load effectively, it is advisable to take breaks that are different from work, i.e., they activate other parts of the brain. If one spends the day running a workshop and talking to people, a break is to read, listen to soothing music, or go for a walk. If, on the other hand, you sit at a computer all day, your break should be active. Move your body and socialise.

6. Conclusion

This article can be a first step towards a descriptive, conceptual analysis of the current state of information behaviour. Existing definitions and issues in the information ecology field and the relationship between these fields were analysed, and a brief body of knowledge was created for ongoing research. Many factors influence the collection and use of information, and other activities, such as information avoidance and manipulation, also change the role of chance in information gathering. Sufficient attention still needs to be paid to information barriers that may arise and changes in information needs. Quantifying and evaluating vulnerabilities is the basis for further research of various information activities in the organisation. This article discussed only selected main chapters of the field of information ecology, following on from psychology and providing a basis for further research. The benefit lies in evaluating the vulnerabilities of the information environment in organisations to increase the quality of information processes and information flows to minimise weak points and subsequent damages and losses (material and financial) for organisations.

7. References

[1] Sedlackova, B. (2015). Introduction to Information Science, Silesian University in Opava, Opava, ISBN 9788075101457

[2] Rudd, M., J. & Rudd, J. (1986). "The Impact of the Information Explosion on Library Users: Overload or Opportunity?", Journal of Academic Librarianship, Vol. 12, No 5, pp 304-308, ISSN 0099-1333

[3] Davenport, T. H. & Prusak, L. (1997). Information Ecology: Mastering the Information and Knowledge Environment, Oxford University Press, New York, ISBN 0195111680

[4] Lawler, E. E.; Nadler, D. A. & Cammann, C. C. (1980). "Organizational assessment: perspectives on the measurement of organisational behaviour and the quality of work life", In Measuring and assessing organisations, Wiley-Interscience, pp. 624-658, ISBN 978-0471048329

[5] Yusof, M. M.; Papazafeiropoulou, A. Paul R. J., Kuljis, J. & Stergioulas, L. K. (2008). "Investigating evaluation frameworks for health information systems" In International Journal of Medical Informatics, Vol. 77, No. 6, pp. 377-385, DOI 10.1016/J.IJMEDINF.2007.08.004.

[6] Mandalios, J. (2013). RADAR: "An approach for helping students evaluate Internet sources", Journal of Information Science, Vo. 39, No. 4, pp. 470-478, DOI: 10.1177/0165551513478889

[7] Olsina, L.; Godoy, D.; Lafuente, G. & Rossi, G. (1991). "Assessing the quality of academic websites: A case study", New Review in Hypermedia and Multimedia, Vol. 5, No 1, pp. 81-103, DOI 10.1080/13614569908914709

[8] Moos, P. & Malinovsky, V. (2008). Information Systems and Technologies, CVUT in Prague, Prague, ISBN 9788001040645

[9] Jokl, M. V. (1989): Microenvironment: The Theory and Practice of Indoor Climate, Charles C Thomas Publisher, Ltd., Springfield, ISBN 978-0398054359

[10] Bruner, J. S. (1973). Beyond the Information Given: Studies in the Psychology of Knowing, W.W. Norton, New York, USA, ISBN 978-0393093636

[11] Loukas, G.; Murugesan, S. & Andriole, S. J. (2022) "Information Hygiene: The Fight Against the Misinformation Infodemic", In IT Professional, Vol. 24, No. 2, pp. 16-18, DOI 10.1109/MITP.2022.3163007

[12] Rydning, J. (2022). Worldwide IDC Global DataSphere Forecast, 2022–2026: Enterprise Organizations Driving Most of the Data Growth, IDC Corporate, Available: https://www.idc.com/getdoc.jsp?containerId=US49018922

[13] Tomsu, M. (2022). Principles of Crisis Communication and Information Vulnerabilities in Crisis Management, Proceedings of the 33rd DAAAM International Symposium, pp.0429-0436, B. Katalinic (Ed.), Published by DAAAM International, ISBN 978-3-902734-36-5, ISSN 1726-9679, Vienna, Austria DOI: 10.2507/33rd.daaam.proceedings.060

DOI: 10.2507/34th.daaam.proceedings.054

IMPLEMENTATION OF TRAJECTORY PLANNING FOR THE MINIATURISED INDUSTRIAL ROBOTS OF THE MOROBOT PLATFORM

Claudia Holzgethan, Ali Aburaia, Kemajl Stuja & Mohamed Aburaia

This Publication has to be referred as: Holzgethan, C[laudia]; Aburaia, A[li]; Stuja, K[emajl] & Aburaia, M[ohamed] (2023). Implementation of Trajectory Planning for the Miniaturised Industrial Robots of the Morobot Platform, Proceedings of the 34th DAAAM International Symposium, pp.0395-0409, B. Katalinic (Ed.), Published by DAAAM International, ISBN 978-3-902734-41-9, ISSN 1726-9679, Vienna, Austria
DOI: 10.2507/34th.daaam.proceedings.054

Abstract

The International Federation of Robotics (IFR) has reported a 200% increase of installed industrial robots over the last decade, with further growth projected. To prevent a gap between supply and demand for a qualified workforce, Robotics in Education supports students with understanding robotics and sparking interest in technology. For a rapidly growing industry with new technologies and solutions emerging almost daily, the skills of and practice in critical comparison of methods and creative problem-solving are essential. The morobot platform of the University of Applied Sciences Technikum-Wien manufactures 3D-printed miniaturized industrial robots that enable cost-effective and easy access to knowledge in the robotics domain. However, the current implementation of the robot control leads to jerky movement and provides no possibility of creating a geometric path. In this paper, the platform is enhanced with trajectory planning to achieve smooth robot motions and implement continuous point-to-point and linear movements. Suitable velocity profiles are computed to achieve a desired motion. Given the high minimum velocity of the morobot motors, third- and fifth-order polynomials are compared regarding the smoothness of the resulting motion and the absolute positioning accuracy. Both methods achieved the desired smoothness. However, third-order polynomials provide a higher positioning accuracy, and have therefore been implemented for continuous point-to-point movement. For the linear movement, intermediate points are calculated to approximate a straight line between the start and goal robot position. Due to the high minimum velocity of the motors and the close proximity of the intermediate points, trajectory planning cannot resolve the jerky motion. Motors with lower minimum velocity should be identified for a future project phase.

Keywords: trajectory planning; third-order polynomials; fifth-order polynomials; geometric path; industrial robotics

1. Introduction

Be it industrial manufacturing or assistive technologies [1] robots are expected to play an increasingly important role in society, leading to a radical transformation in skill requirements. Robotics in Education focuses on strengthening the learning skills of the future workforce and motivating students for topics in Science, Technology, Engineering and Mathematics (STEM) [2], [3].

At the University of Applied Sciences Technikum Wien, the morobot platform was created. It focuses on manufacturing 3D-printed miniaturised industrial robots that enable inexpensive and easy access to knowledge in the robotic domain. The platform helps students to learn and understand aspects of industrial robotics, including the composition of different kinematic chains and their strengths and weaknesses, the computation of forward and inverse kinematics and the control of robots via inverse kinematics. However, the implementation of the platform comes with limitations regarding the control of the robot's geometric path and the smoothness of the motion. The current motor control leads to abrupt and jerky movements that induce mechanical stress on the robots' kinematic chains. Furthermore, it is only possible to define goal positions for the robot's tool centre point (TCP) that are approached in a discontinuous point-to-point movement. No continuous geometric path of the TCP can be composed.

To enhance the platform, this paper focuses on employing and comparing third- and fifth-order polynomials as trajectory planning methods for the miniaturised industrial robots. The goal is to minimise jerk and consequently create smooth and continuous movements, whilst maintaining a high positioning accuracy. Furthermore, continuous point-to-point and linear movements are implemented to enable the construction of geometric paths.

The paper is organised as followed: Section 2 gives an introduction to trajectory planning and explores different methods of creating and optimising trajectories. In section 3 the utilized hardware is presented, the limitations of the current implementation are discussed and the goals for this project are defined. Section 4 focuses on the motor limitations and the theoretical as well as practical realisation of trajectory planning and geometric paths for the miniaturised robots of the morobot platform. The paper closes with the discussion of the results in section 5 and the conclusion and suggestions for future projects in section 6.

2. State of the Art

In industrial robotics, a distinction is made between path planning and trajectory planning. Path planning mainly focuses on computing a geometric path along which the TCP should move [4]. Trajectory planning on the other hand deals with computing motion profiles, i.e. time sequences of positions, velocities, accelerations and higher derivatives of either the TCP or the robot's joint values [5]. The aim is to create desired motions, like smooth movements, by employing suitable displacement functions.

There are various approaches to generate trajectories for industrial robots. Bai et al. [6] use a five-segment interpolation function based on s-curves to create smooth displacement and velocity profiles for an ABB IRB 1200 industrial robot. S-curves are also employed in [7] and [8]. In the former, the s-curve trajectory is compared to a cubic spline trajectory with respect to execution time and jerk. The results show that the s-curve produces a faster and smoother trajectory than the cubic spline [7]. In [8], the authors use a sigmoid function to generate a desired jerk profile that enables an infinite order of continuity of the motion. This has the effect of minimising vibrations and increasing positioning accuracy. Zheng et al. [9] on the other hand utilise fifth-order polynomials and interpolation to compute a desired trajectory for a 6 degrees of freedom robot manipulator. Parikh and Dave [10] compare high order polynomials to parabolic and cubic functions. The results show that polynomials of order seven and nine generate smoother trajectories than the lower order polynomials but also lead to higher peaks of velocity and acceleration values. In [11] and [12] cubic splines and fifth-order B-splines are used to generate desired motion profiles.

Many papers also focus on optimising computed trajectories. Common optimisation criteria according to literature are minimum execution time [13], minimum energy consumption [14], and minimum jerk [7]. Often multiple of these criteria are combined to generate the optimal trajectory [12], [15].

2.1. Minimum execution time

Increasing productivity is a major aspect when it comes to working economically. One way of achieving this is by minimising the execution time of the robot's task [16], [15]. Li and Wang [13] use a cubic polynomial curve to connect the robot's path points and generate a smooth trajectory. Then a genetic algorithm is applied to optimise the execution time of the robot's joints. The results show that the robot reaches the desired poses and follows the desired trajectory in the shortest time.

2.2. Minimum energy consumption

The criterion of minimum energy consumption does not only focus on saving energy, which is beneficial as it reduces costs, but also on minimising the mechanical stress on the robot's actuators [16], [17]. Chen et al. [14] minimised the energy consumption for a bottle grasping task by applying a gaussian quadrature optimisation method.

2.3. Minimum Jerk

Jerk is defined as the time derivative of the acceleration, i.e. the third time derivative of the position. It indicates how quickly the acceleration of an object changes with respect to time. In general, jerk is undesirable in industrial robotics as it leads to an unnatural motion and increases the wear of the actuators and the robot's structure.

The minimisation of jerk does not only lead to a smoother execution of the robot's trajectory and reduction of vibration and wear but also reduces trajectory tracking errors [15]–[17]. Devi et al. [7] developed an algorithm to generate a smooth trajectory with minimum jerk. First the inverse kinematics of a 6-DOF PUMA-560 robot is solved with the help of an artificial neural network. The generated joint values are than fed into an s-curve equation to produce a smooth trajectory with minimum jerk and time.

2.4. Multi-criteria

Often multiple criteria are decisive to obtain an optimal trajectory. Therefore, many papers focus on multi-criteria optimisation. In [12], the contradictory criteria minimum time, minimum acceleration and minimum jerk are combined to produce an optimal trajectory for a glass-handing robot. The authors use a minimised objective function to find an optimal balance between the three criteria and fifth-order B-splines to produce an optimal trajectory. Rout et al. [11] face a similar challenge. Here, the contradictory criteria time, jerk and torque are combined to generate a smooth trajectory with minimum execution time. The best trade-off between the criteria is found by combining the Non-Dominated Sorting Genetic Algorithm and the Nelder-Mead simplex method. The output of the combined algorithms was used to compute the joint angles for the desired cubic spline trajectory path.

Chiddarwar and Babu [15] developed a method to achieve an optimal trajectory in terms of minimising the execution time, energy, jerk and acceleration, whilst accomplishing maximum manipulability, i.e. maintaining all of the robots degrees of freedom at any time. Here a trigonometric spline was used to represent the desired trajectory. The authors then used sequential quadratic programming and genetic algorithm to optimise the trajectory according to the optimisation criteria.

3. Problem Description

In this paper, trajectory planning is applied on the miniature robots of the morobot platform of the UAS Technikum-Wien. For a better understanding of the current limitations of the platform, this section starts with an overview of the available robots and the motors that are used. Based on the knowledge of the employed hardware, the limitations of the current robot control and the goals for this project are discussed.

3.1. Hardware

The morobot platform consists of several miniaturised industrial robots with different kinematic chains, including both parallel and serial structures. Fig. 1 shows an overview of the available miniaturised industrial robots.

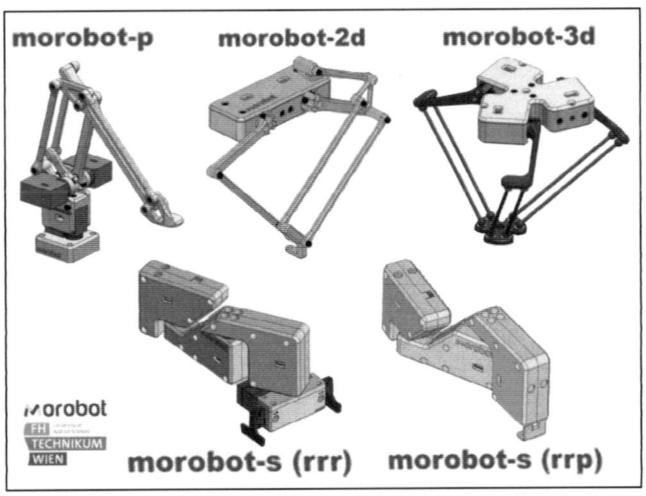

Fig. 1. Overview of the miniaturised industrial robots of the morobot platform (top row: palletising robot, two-dimensional delta robot, three-dimensional delta robot | bottom row: RRR and RRP SCARA)

The top row in the image shows, from left to right, the parallel structures of a palletising robot, a two-dimensional and a three-dimensional delta robot. The bottom row shows two versions of a SCARA robot, one with three rotational axes (RRR) and one with two rotational and one translational axis (RRP). Each of the robots takes up an approximate size of a DIN A5 sheet of paper. For this project, the robots are controlled with an Arduino Mega 2560. The motors that are used in the joints of the miniaturised robots are Smart Servos by Makeblock [18].

They dispose of an open-source programming library, which allows to both control the motors and receive information, like the current position, velocity or internal temperature. There are two main groups of commands with which the motors can be controlled:

A) Commands where the motor automatically moves to or by a defined angle and
B) Commands where a fixed motor velocity is set

In case of group A), the Makeblock Smart Servos use an internal velocity regulation to ensure that the target position is reached. On reaching the target position, the motor gets stopped. A continuous motion through via-points is not possible with these commands. In case of group B), a fixed velocity in form of a PWM (pulse width modulation) value gets sent to the motor. The motor then moves with this velocity until it is set to zero again.

3.2. Limitations of the implementation

The problems at this paper's start state are two-fold. First off, the robots can only be controlled by defining target poses for their tool centre point. Inverse kinematics is used to calculate the corresponding joint values and motor angles for the target pose. However, the geometric path, i.e. the path that the TCP follows to reach the target pose, cannot be defined by the user.

Secondly, the calculated goal angles are sent directly to the motors via commands of group A). As described in Section 3.1 above, with these commands, the motors always stop once a target position is reached. Therefore, only a discontinuous point-to-point movement is possible and no continuous geometric path of the robot's end-effector can be composed. Furthermore, the motor's internal velocity regulation leads to abrupt and jerky movements that induce mechanical stress on the robot's kinematic chain. This jerky movement is exemplified in Fig. 2. The image shows the motion of the TCP of the RRR SCARA when moving from a start position to a goal position. The jerk in the motion can be observed by the jumps in the red line and is magnified in the two positions a) and b).

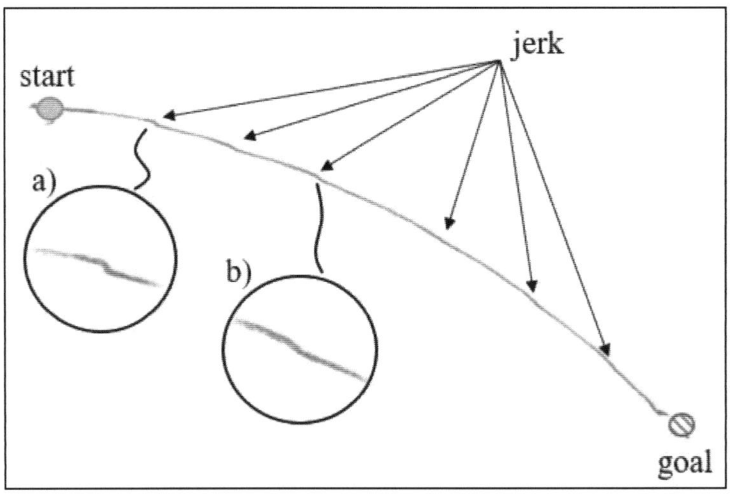

Fig. 2. Exemplary jerky movement of the TCP of the RRR SCARA with the current implementation of the robot control. The red line shows the motion of the TCP. The jerk is indicated by arrows and magnified at the two positions a) and b).

The problems of the current implementation can be summarised as follows:

1. The motor control leads to jerky and discontinuous movements
2. No geometric path can be defined by the user

The aim of this paper is to apply and compare two trajectory planning methods on the RRR SCARA of the morobot platform, with the goal of creating smooth and continuous motions with high position accuracy. Furthermore, the trajectory planning methods shall be used to implement a continuous point-to-point and linear movement of the robot's TCP.

4. Methods

The trajectory planning methods that are compared in this paper are third- and fifth-order polynomials. For a better understanding of the reasons that these methods were chosen, it is first necessary to examine the limitations of the employed motors.

4.1. Motor limitation – minimum velocity

As described in section 3.1, in addition to the commands for setting a target pose, the utilised motors also compose of commands with which a velocity can be set via a PWM value. This is important, as it allows to compute time sequences of velocities via trajectory planning and sending these velocities to the motors. Experiments, that were conducted on the three Smart Servos of the RRR SCARA showed that the minimum PWM value at which the motors rotate reliably is 8. To determine the actual velocity at the minimum PWM value, each of the motors was rotated by 50 degrees and the travel time was measured. The experiment was conducted 20 times per motor and the average travel time was calculated. The results are shown in Table 1.

	Motor 1	**Motor 2**	**Motor 3**
Average time [milliseconds]	1 923	1 826	1 812
Standard deviation [milliseconds]	8.82	10.15	16.10
Velocity [°/sec]	26.00	27.37	27.59

Table 1. Average time that is needed to rotate a motor by 50 degrees with the minimum PWM value of 8, taken from 20 measurements each

The minimum average time could be observed at motor 3 with a value of 1 812 [msec] and a standard deviation of 16.10 [msec]. After dividing the travelled distance of 50 degrees by this average travel time, a minimum velocity of 27.59 [°/sec] or 4.6 [rpm] can be obtained. To put the minimum velocity in perspective, an example can be considered. Fig. 3 shows the SCARA robot in the two maximum displacement positions of the first joint.

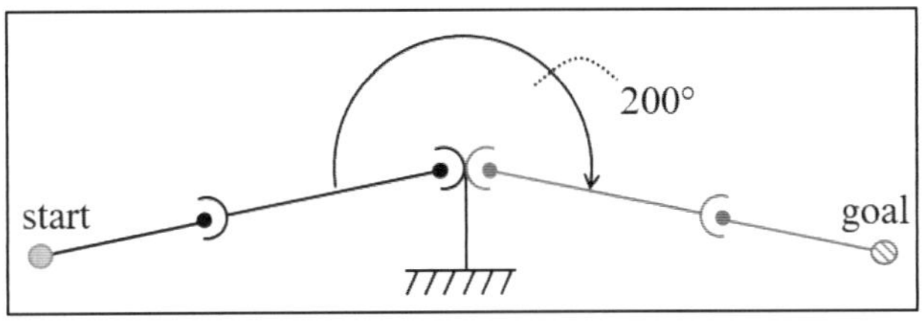

Fig. 3. Poses of the SCARA robot, when the first joint is moved to its minimum (start) and maximum (goal) angle

In total the first joint can be moved by 200 [°]. Dividing this value through the minimum velocity of the motors leads to a maximum travel time of 7.25 [sec]. This means that the first joint of the robot is not allowed to take more than 7.25 seconds to traverse its entire working area. Otherwise, the corresponding PWM value, that is sent to the motor, would be lower than the minimum value of 8 and the robot would not move reliably. In addition to the maximum distance of a joint, the minimum distance can also be considered. In case of the Makeblock Smart Servos the minimum distance that can be travelled at a time equals 1 [°]. Divided by the minimum velocity, a maximum travel time of 0.036 [sec] or 36 [msec] can be obtained.

4.2. Trajectory planning

As mentioned before, trajectory planning focuses on computing desired motion profiles. It can have different goals like minimising the jerk, the energy consumption or the execution time. In this paper, the goal is to create smooth and continuous movements for the miniaturised robots of the morobot platform and therefore reducing their jerk. In robotics and automation in general, a multitude of different trajectory planning methods exits to achieve such a motion [19]. Ranging from polynomial and trigonometric to spline functions, there are different levels of complexity that can lead to varying results of the computed trajectory. In case of the miniaturised robots, two main aspects must be considered when choosing suitable trajectory planning methods. First off, to achieve smooth motions, the first- and second-order derivative of the trajectory function must be continuous [5]. Secondly, due to the high minimum velocity of the motors and the need to calculate the trajectory function at runtime, the computational effort of the employed methods must be as low as possible.

Under these aspects third- and fifth-order polynomials were chosen to be applied and compared. Although third-order polynomials are not continuous in their second derivative and can therefore not guarantee a jerk-free motion, they prove to be promising since their computational effort is low compared to other methods. Fifth-order polynomials do dispose of continuous first and second order derivatives, however, more equations must be computed at runtime.

Trajectory planning can either be conducted for a robot's TCP in the robot coordinate space or for the individual robot joints in the joint space [5]. In this paper, the trajectories in the joint space are examined. In the following sections, the joint trajectory functions will be denoted as $q(t)$, $v(t)$ and $a(t)$ for the position, velocity and acceleration. It shall be noted that these expressions describe the trajectory of one individual joint. In practice, these functions must be computed for all *j* robot joints.

4.2.1 Third-order polynomials

Polynomials of degree three, also commonly known as cubic functions, and their first derivative can be expressed by the following equations (1) and (2) [5]:

$$q(t) = c_0 + c_1 t + c_2 t^2 + c_3 t^3 \tag{1}$$

$$v(t) = c_1 + 2c_2 t + 3c_3 t^2 \tag{2}$$

,where $q(t)$ and $v(t)$ are the joint value and joint velocity depending on time, t is the time that has elapsed since the start ($t = t - t_0$, with $t_0 = 0$) and c_0 to c_3 are unknown coefficients. Since there are four unknown coefficients, four constraints must be defined. These can for example be the start and final joint value and the velocity at the start and end of the motion:

At $t_0 = 0$:	At $t_f = t_1$:
$q(0) = q_0$	$q(t_1) = q_1$
$v(0) = v_0$	$v(t_1) = v_1$

By applying the conditions in the equations (1) and (2), the coefficients can be calculated as follows (3)-(6):

$$c_0 = \theta_0 \tag{3}$$

$$c_1 = v_0 \tag{4}$$

$$c_2 = \frac{3(q_1 - q_0) - (2v_0 + v_1)t_1}{t_1^2} \tag{5}$$

$$c_3 = \frac{-2(q_1 - q_0) + (v_0 + v_1)t_1}{t_1^3} \tag{6}$$

The robot's joints will be controlled by setting desired velocities depending on time. Therefore, the equations (4), (5), (6) and (2) must be solved during runtime to determine the corresponding velocity profiles of the motors.

4.2.2 Fifth-order polynomials

Similar to third-order polynomials, fifth-order polynomials can be described by the following equations (7)-(9) [5]:

$$q(t) = c_0 + c_1 t + c_2 t^2 + c_3 t^3 + c_4 t^4 + c_5 t^5 \tag{7}$$

$$v(t) = c_1 + 2c_2 t + 3c_3 t^2 + 4c_4 t^3 + 5c_5 t^4 \tag{8}$$

$$a(t) = 2c_2 + 6c_3 t + 12c_4 t^2 + 20c_5 t^3 \tag{9}$$

Here, $q(t)$, $v(t)$ and $a(t)$ describe the position, velocity and acceleration of the joint as functions of time. The variable t, again, stands for the time that has elapsed since the start of the motion with $t = t - t_0$ and $t_0 = 0$ and c_0 to c_5 are unknown coefficients. In contrary to third-order polynomials, six constraints must be defined to solve the unknown coefficients. In addition to the initial and final position and velocity of the joint, also the start and final accelerations can be specified:

At $t_0 = 0$:	At $t_f = t_1$:
$q(0) = q_0$	$q(t_1) = q_1$
$v(0) = v_0$	$v(t_1) = v_1$
$a(0) = a_0$	$a(t_1) = a_1$

With these conditions, the equations (7)-(9) can be converted as follows, to determine c_0 to c_5 (10)-(15):

$$c_0 = \theta_0 \tag{10}$$

$$c_1 = v_0 \tag{11}$$

$$c_2 = \frac{1}{2}a_0 \tag{12}$$

$$c_3 = \frac{20(q_1 - q_0) - (12v_0 + 8v_1)t_1 - (3a_0 - a_1)t_1^2}{2t_1^3} \tag{13}$$

$$c_4 = \frac{30(q_0 - q_1) + (16v_0 + 14v_1)t_1 + (3a_0 - 2a_1)t_1^2}{2t_1^4} \tag{14}$$

$$c_5 = \frac{12(q_1 - q_0) - 6(v_0 + v_1)t_1 - (a_0 - a_1)t_1^2}{2t_1^5} \tag{15}$$

As can be seen in the equations, the changes in position, velocity and accelerations are multiplied by higher values than in third-order polynomials. Therefore, even if the same constraints are applied, the coefficients c_n will have bigger values in fifth-order polynomials than in third-order polynomials. This leads to higher peaks in the velocity and acceleration. In case of fifth-order polynomials, six equations ((8),(11)-(15)) must be solved at runtime to calculate corresponding velocities depending on time.

4.3. Geometric path

In robotics it is often desired that the end-effector follows a specific geometric path to fulfil its tasks and avoid obstacles. Such a geometric path can be achieved by, for example, implementing and combining basic movements, like moving arbitrarily between points (point-to-point movement), following a line (linear movement) or following a circular arc (circular movement). In this paper, the previously described trajectory planning methods are utilised to implement a continuous point-to-point and linear movement.

4.3.1 Point to point movement

In a point-to-point movement a start and goal pose are specified in the robot coordinate system. The path that the TCP follows when moving from the start to the goal pose is unknown. Therefore, this type of movement is effective if no exact path is needed, like in pick-and-place or palletizing tasks. By combining multiple point-to-point movements, it is possible to avoid obstacles and approximate a desired geometric path. In this paper, the movement between a start and goal point is accomplished by calculating velocity profiles for each of the robot's joints with the described trajectory planning methods. The calculated velocity profiles are approximated by sending the corresponding velocities depending on the time, to the motors at different time stamps during the movement.

Furthermore, in order to achieve a smooth motion, the robot's joints are synchronised so that they start and finish their movement at the same time. Fig. 4 shows an overview of the main process steps that are used to achieve the point-to-point movement for the miniaturised robots.

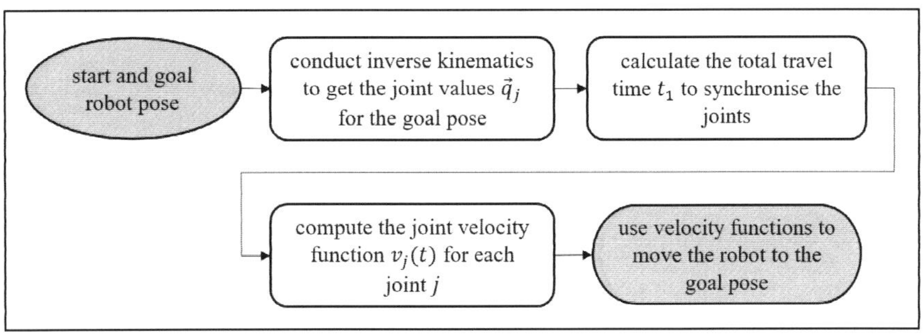

Fig. 4. Process steps for the point-to-point movement

The process begins with a start and goal TCP pose, whereby the start pose corresponds to the current pose of the robot. Inverse kinematics is used to calculate the joint values for the goal pose. Next, the total travel time t_1 is calculated. To achieve synchronisation, this travel time is the same for all joints. The velocity function $v_j(t)$ is computed for each joint, taking in consideration the start and goal pose, as well as the total travel time t_1. Depending on the employed trajectory planning method, also the start and end velocities and accelerations are required. Finally, the computed velocity functions are employed to move the robot from the start to the goal pose. For a better understanding, the synchronisation and motor control are discussed in more detail below.

A. Synchronisation

The synchronisation of the joints is achieved by computing a total travel time that is the same for all the robot's joints. This ensures that, if the motors are started at the same time, they also reach their target position at the same time. Since the employed Smart Servos have a high minimum velocity, the calculation of the total travel time is based on the fastest motor, i.e. the motor that travels the smallest distance to reach its goal position. If this motor were to move with the minimum velocity, it would need n seconds to reach its goal position. To achieve synchronisation, the other motors must reach their target position at the same time. Therefore, the total travel time is calculated as follows (16):

$$t_{1,max} = \frac{distance_{min}}{velocity_{min}} \tag{16}$$

,where $t_{1,max}$ describes the total travel time of the joints, $distance_{min}$ describes the minimum distance that has to be travelled and $velocity_{min}$ describes the minimum velocity of the Smart Servos. As indicated in the equation the corresponding travel time $t_{1,max}$ is the maximum total time that can be used in the velocity functions. Otherwise, velocities smaller than the minimum velocity might be computed and it cannot be guaranteed that all motors move reliably.

B. Motor control

Based on the determined total travel time, the start and goal joint values, velocities and accelerations, the velocity functions are calculated as described in section 4.2, with either third or fifth-order polynomials. Fig. 5 shows an example of such a velocity profile of a joint with a travel distance of 10 [°], a total travel time of $t_1 = 0.3$ [sec] and a start and goal velocity of $v_0 = v_1 = 25$ [°/sec]. A third-order polynomial was used to compute this exemplary motion profile.

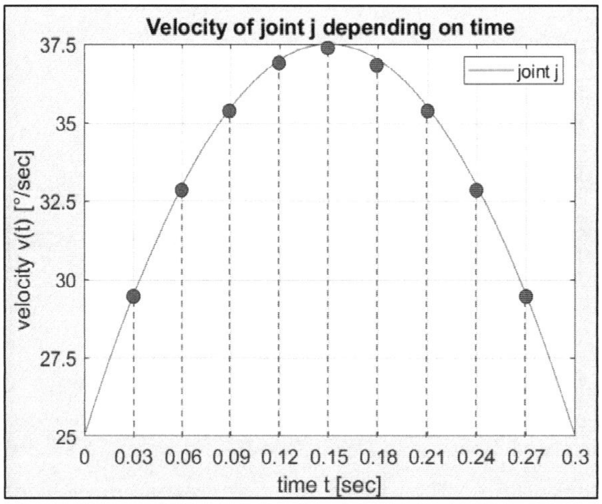

Fig. 5. Exemplary velocity profile with a total travel time of $t_1 = 0.3$ $[sec]$. The dots indicate the velocities that are sent to the motors at different time stamps

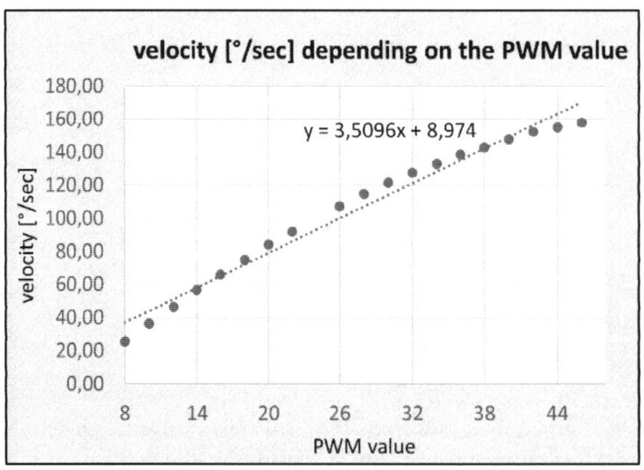

Fig. 6. Motor velocity in [°/sec] in dependence of the PWM value

Note, that for the start and goal velocity a value unequal to zero was chosen. Instead, these parameters were set to a value slightly below the minimum velocity. As a result, even though the motors start from and end in standstill, no values smaller than the minimum velocity are computed. In order to approximate the velocity profile, the velocity is calculated at several time stamps. This is indicated by the dots in the graph. Since in trajectory planning, functions of time are computed, the resulting velocities have the unit [°/sec]. To set the motor velocities, these velocity values must be converted to PWM values. In order to find the correlation between the motor velocity in [°/sec] and the PWM value, the velocity was measured for different PWM values and linear regression was applied. This is illustrated in Fig. 6 for positive velocities. As can be seen in the graph, a correlation of $y = 3.5096x + 8.974$ can be obtained. In this context y describes the velocity in [°/sec] and x the PWM value. Therefore, the PWM value can be computed as follows (17):

$$x = \frac{y - 8.974}{3.5096} \tag{17}$$

For a complete robot movement, this correlation must also be computed for negative velocities. To summarize, the motor control process is as follows:

1) Calculate the coefficients c_n, based on the total travel time, start and goal position, velocity and acceleration
2) Divide the total travel time t_1 in k parts
3) For each time stamp k:
   a) Use the coefficients c_n to calculate the velocity in [°/sec]
   b) Convert the velocity into a PWM value
   c) Send the PWM value to the motor

This process must be conducted for each of the j motors.

4.3.2 Continuous point-to-point movement

In the previous section, the fundamentals and operating principle of the point-to-point movement between a start and goal pose were described. However, in some cases, it is desired, that the TCP moves through multiple points in a continuous motion to approximate a geometric path. This can be achieved by defining start and goal velocities for intermediate points that are unequal to zero. Heuristic rules are applied as defined in equation (18) and (19), to determine optimal intermediate velocity [19].

$$v_k = \begin{cases} 0 & sign(d_k) \neq sign(d_{k+1}) \\ \frac{1}{2}(d_k + d_{k+1}) & sign(d_k) = sign(d_{k+1}) \end{cases} \tag{18}$$

$$d_k = \frac{q_k - q_{k-1}}{t_k - t_{k-1}} \tag{19}$$

In the equations, v_k describes the calculated intermediate velocity, q_k the position and t_k the current time at point k. With these rules, the intermediate velocity is set to zero if the motor changes directions and a value unequal to zero if the direction is maintained. An example of a corresponding position and velocity profile through intermediate points is depicted in Fig. 7

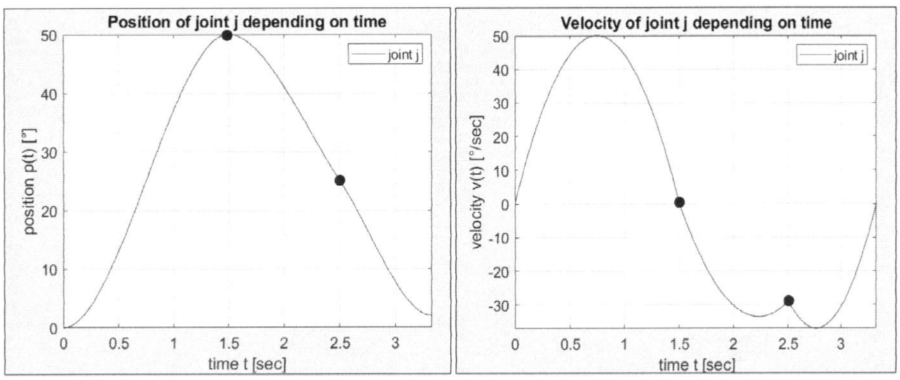

Fig. 7. Exemplary position (left) and velocity (right) profiles for a continuous point-to-point movement. The intermediate points are indicated by the dots at t=1.5 [sec] and t=2.5 [sec].

In the graphs, the intermediate points are located at the time stamps t=1.5 [sec] and t=2.5 [sec], as indicated by the dots. For a better visibility the velocity at the start, end and reversal point were set to zero. In practice, these parameters would, again, be set to a value slightly below the minimum velocity. In this example, the position and velocity profiles were calculated with third-order polynomials.

4.3.3 Linear movement

In a linear movement the robot's TCP follows a line to move from a start pose to a goal pose. This type of movement is suitable when an exact path must be maintained to fulfil a task, as it might be the case in assembly, cluing or painting processes. The implementation of a linear movement is comparable to a continuous point-to-point movement. However, in a linear movement, the intermediate points are not defined by the user, but must be calculated by the control system. The main process steps for the implemented linear movement are depicted in Fig. 8.

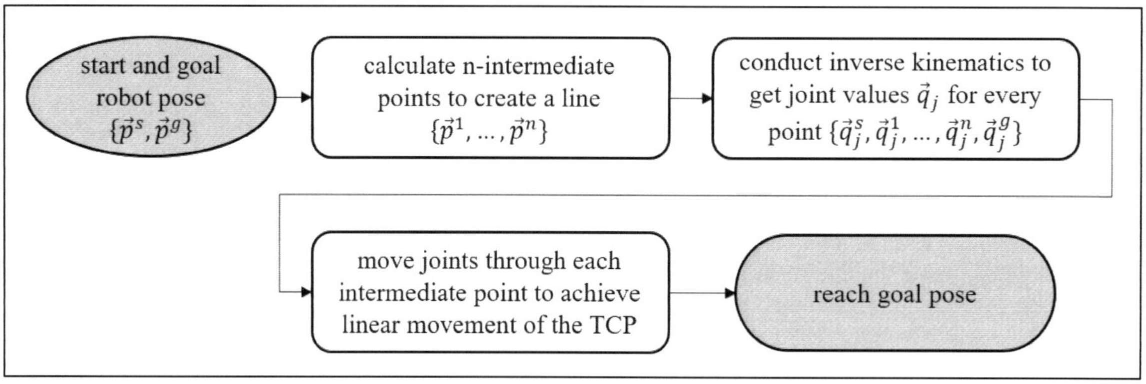

Fig. 8. Process steps for the linear movement

The process starts with a user-defined goal pose for the robot's TCP. The start pose corresponds to the current robot pose. In the first step, n intermediate points are computed to approximate a line. These points are then converted into the joint space by applying inverse kinematics. In order to make the TCP follow the approximated line, the robot must be controlled in a way that all joints pass through their intermediate joint positions simultaneously.

A. Computation of intermediate points

In order to compute appropriate intermediate points, first, a line is placed between the start and goal position. Since, depending on the robot structure, the line can be three-dimensional, it is described with vectors as expressed in equation (20).

$$\vec{x}(s) = \vec{a} + s \cdot \vec{m} \qquad (20)$$

$$\begin{pmatrix} x \\ y \\ z \end{pmatrix} = \begin{pmatrix} p_x^s \\ p_y^s \\ p_z^s \end{pmatrix} + s \cdot \begin{pmatrix} p_x^g - p_x^s \\ p_y^g - p_y^s \\ p_z^g - p_z^s \end{pmatrix}$$

In the equation, $\vec{x}(s)$ stands for the x, y and z coordinate of any point on the line, $\vec{a}$ is described by the start point $\vec{p}^s$ and $\vec{m}$ by the difference between the start point $\vec{p}^s$ and end point $\vec{p}^g$. With this definition, the parameter s has an interval of $[0,1]$. Therefore, an intermediate point s_i is calculated by dividing through the desired number of points n and multiplying the result with i (21).

$$s_i = \frac{1}{n} \cdot i \qquad (21)$$

An example of such a three-dimensional line is visualized in Fig. 9. The start point has the coordinates $\vec{p}^s = (4, 5, 2)^T$ and the end point $\vec{p}^g = (10, 20, 5)^T$. Eight intermediate points have been calculated with the equations (20) and (21) and are indicated in the graph by blue dots. The different coloured dots at the start and end of the line show the start and end point.

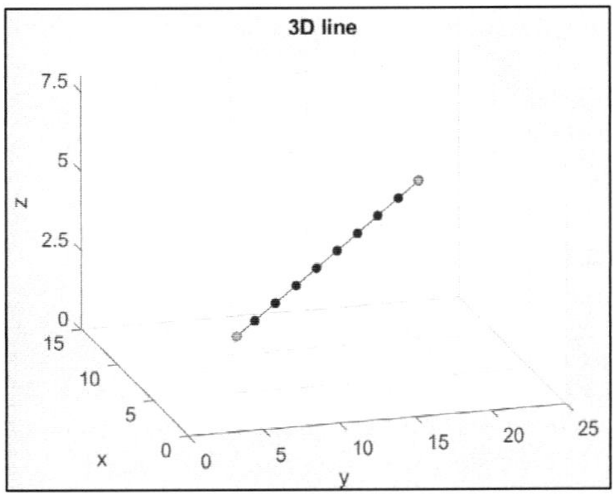

Fig. 9. Example of a three-dimensional line with intermediate points indicated by blue dots. (start point: $\vec{p}^s = (4,5,2)^T$, end point $\vec{p}^g = (10,20,5)^T$, number of intermediate point $n = 8$)

The optimal number of points n depends on the length of line and the desired resolution and is calculated as follows (22):

$$n = \frac{length\_of_{line}[mm]}{resolution\ [mm]} \tag{22}$$

B. Motor Control

After calculating and transforming the intermediate points in the joint space, appropriate motor commands need to be defined to create the correct TCP movement. To ensure that the TCP moves through the intermediate points and therefore follows the calculated line, all joints need to pass through their intermediate joint positions at the same time. This behaviour can be achieved by synchronising the motors in between the intermediate points, similar to the point-to-point movement. However, since for the approximation of a line, the intermediate points are required to be close together, the distances that need to be traversed by the motors in between points are small, i.e. often an individual joint must only move by 1 [°]. As described in the motor limitations in section 4.1, this means that in these cases the total travel time t_1 would assume a value of only 0.036 [sec]. To achieve synchronisation all motors would have to finish their movement in these 0.036 [sec], even if they have to travel higher distances like e.g. 5 [°]. This means that very high velocities are required for the motors, e.g. 139 [°/sec]. However, the motors are physically limited in their acceleration and deceleration and therefore, cannot guarantee to stop their movement in such a short time. This leads to high overshoot and makes following a line with motor synchronisation impossible. Fig. 10 shows an example of the resulting TCP movement with this approach. The line between the start and end point has a length of 123.76 [mm]. By defining a resolution of 5 [mm], in total 24 intermediate points have been computed. The movement between the intermediate points was achieved via a continuous point-to-point movement as described in section 4.3.2. The motion was computed with third-order polynomials. In the image, the actual TCP movement is depicted in blue, the desired linear movement is indicated by a dotted line. As can be seen, with this approach, the TCP has an extreme overshoot and its movement is far away from representing a straight line.

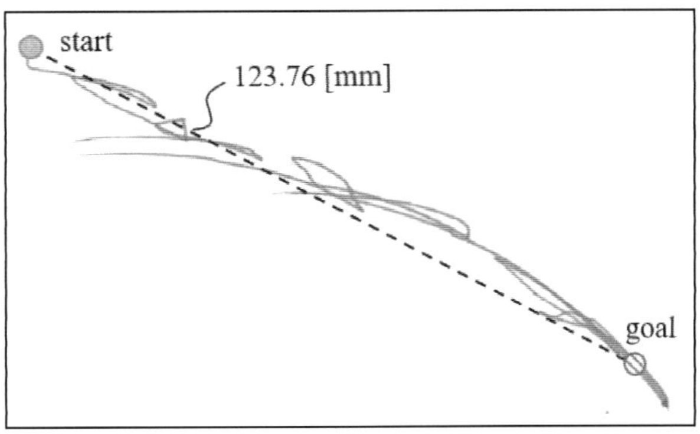

Fig. 10. Linear movement with motor synchronization. The actual TCP movement is illustrated in blue, the desired movement with a dotted line. Total length of line: 123.76 [mm], desired resolution: 5 [mm], number of intermediate points: 24

A different approach is needed to implement the linear movement of the miniaturised robots. For this, to maintain the short travel distances and the resulting short travel times between points, the motors are all set to the minimum velocity. The integrated Smart Servo functions are used to frequently check the current motor position. Once a motor has reached its target position, it is stopped. To achieve continuity, it is determined which motor must travel the greatest distance in between two adjacent points. Once this motor reaches a distance of 1 to 2 [°] from its target position, the joint travel distances to next TCP point are calculated and the motors are, again, set to the minimum velocity.

5. Results

Based on the theoretical knowledge of the applied trajectory planning methods and the practical implementation of the geometric paths, this section focuses on analysing and discussing the results. In particular, the smoothness of the TCP movement and the absolute positioning accuracy of the trajectory planning methods and geometric paths are examined.

5.1. Comparison of third- and fifth-order polynomials

To choose the optimal trajectory planning method, first the results of third- and fifth-order polynomials are compared. Fig. 11 shows the TCP movement of the RRR SCARA when applying the respective methods and moving between a start and goal pose.

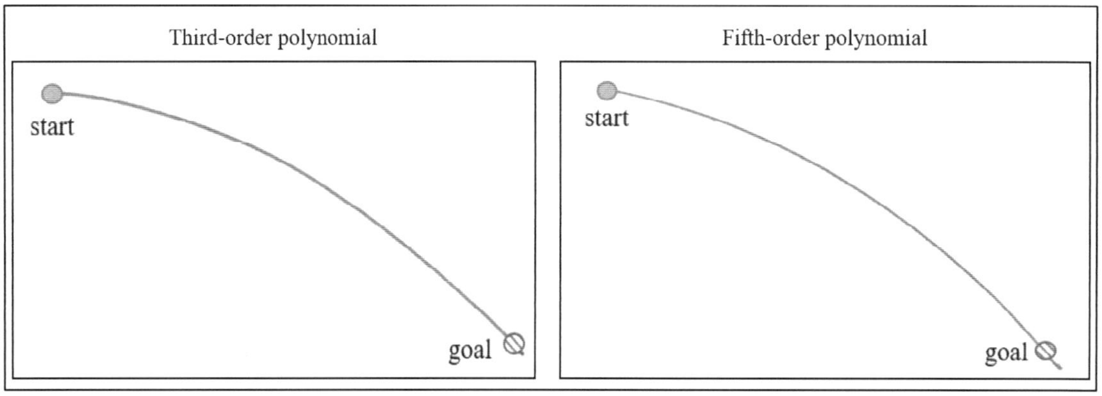

Fig. 11. TCP movement of the RRR SCARA with third-order polynomials (left) and fifth-order polynomials (right)

For both examples the robot started from the home position and was sent to the same goal position. As can be seen in the images, in both cases, the movement of the TCP is a lot smoother than in the start state (see Fig. 2). No jumps, in the form of jerk, are visible in the recorded movements. Therefore, both third- and fifth-order polynomials satisfy the criterion of smoothness. In terms of the accuracy, varying results could be observed. The robot was started from and sent to different positions and the absolute positioning error in x and y as well as the angular error of the motors were calculated. Due to the low repetitive accuracy of the Smart Servos, the actual TCP positions were computed from the joint positions via forward kinematics. Table 2 summarises the minimum, maximum and average positioning errors as well as the median of five individual point-to-point movements.

Method	Min [mm]	Max [mm]	Average [mm]	Median [mm]
3rd-order polynomial	0.55	29.4	10.26	5.05
5th-order polynomial	0.95	31.81	11.91	11.17

Table 2. Minimum, maximum and average absolute positioning error and median of five individual point-to-point movements for both third- and fifth-order polynomials

As can be seen, with third-order polynomials, smaller errors can be achieved in all four columns. The slightly higher positioning errors of fifth-order polynomials can be explained by the fact that with this method overall higher velocity values are computed. Fig. 12 shows the velocity profiles of all three joints of the RRR SCARA for a point-to-point movement as computed by third- and fifth-order polynomials. In the graphs, the synchronisation of the motors is visible, as their motion starts and ends at the same time. In the depicted example, the second joint of the robot travels the shortest distance and consequently the velocity profile stays close to the minimum velocity.

In both third- and fifth-order polynomials this joint reaches its exact target position of 18 [°]. The first joint has to move by the furthest distance of 53 [°]. In third-order polynomials, the velocity of this joint peaks at a value of about 100 [°/sec]. In fifth-order polynomials a maximum of about 120 [°/sec] is calculated. Both velocity values are so high, that the motor overshoots and misses its target position. However, since the third-order polynomial in generally produce smaller velocity values, it achieves an overall higher positioning accuracy.

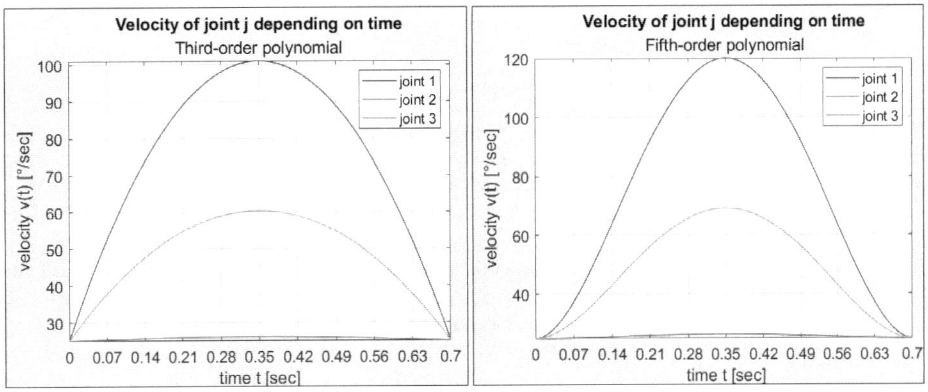

Fig. 12. Velocity profiles of all three RRR SCARA joints for a point-to-point movement, calculated with third-order polynomials (left) and fifth-order polynomials (right)

The variation of the absolute positioning errors of the individual trajectory planning methods can be traced back to the synchronisation of the motors. The greater the differences in distances that the individual motors have to travel, the higher the velocity peaks. High velocity peaks lead to high overshoot and consequently low positioning accuracy. In overall, third-order polynomials prove to be more suitable for the employed Smart Servos. The method achieves a better positioning accuracy than fifth-order polynomials whilst showing the same path smoothness. The fact, that the second derivative of third-order polynomials is not continuous, does not have visible effects on the motion of the robot.

5.2. Continuous point-to-point movement

Third-order polynomials were chosen to implement a continuous point-to-point movement for the RRR SCARA. The smoothness of the geometric path is the same as in the discontinuous point-to-point movement. However, differences in the accuracy can be observed. Experiments revealed that the angular positioning accuracy of the motors is higher at intermediate points where a change of direction occurs and the velocity is set to zero. These reversal points show an average error of 0 to 2 [°], whereas the average angular error at points with an intermediate velocity unequal to zero lies between 3 and 11 [°]. For the experiments the TCP of the RRR SCARA was moved five times through three intermediate points and the angular positioning error of the individual joints where determined. Overall, an absolute positioning error of the TCP of 12 [mm] in x and 4 [mm] in y could be obtained.

5.3. Linear movement

In case of the linear movement, due to the limitations of the minimum velocity, no smooth geometric path of the robot's TCP can be obtained. Fig. 13 shows the resulting TCP movement of the RRR SCARA with the implemented approach of the linear movement where all motors move with the minimum velocity. For the movement, the same constraints for the start and goal pose as well as the number of intermediate points as in Fig. 10 in section 4.3.3 were chosen.

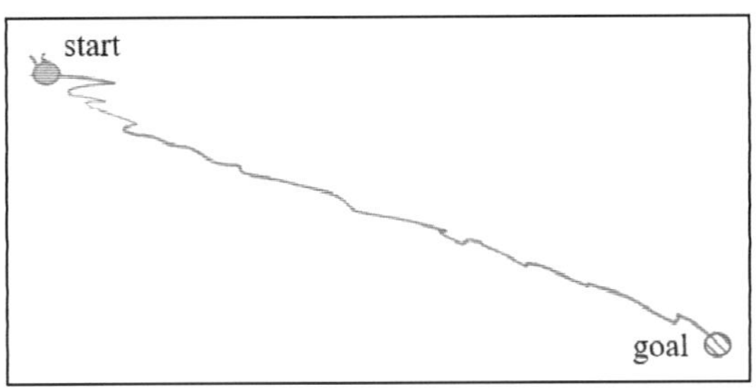

Fig. 13. TCP movement of the RRR SCARA with the implemented linear movement

As can be seen, with the implemented approach, the TCP movement does resemble a better line than in the continuous point-to-point movement in Fig. 10. However, since some of the motors are constantly stopped in between the intermediate points, there are a lot of jumps in the resulting TCP path. In the actual movement, these jumps correspond to jerk. However, the employed implementation of the linear movement leads to a high positioning accuracy. After five experiments with different start and goal positions a median of the positioning error of 0.85 [mm] was obtained.

6. Conclusion and outlook

The morobot platform focuses on producing miniaturised industrial robots for educational purposes. However, the original implementation of these robots came with the limitation of jerky movement and no possibility to create a continuous geometric path. In this paper, trajectory planning was applied to enhance the morobot platform and achieve smooth motions as well as implement a continuous point-to-point and linear movement. Due to the high minimum velocity of the employed motors in the robot's joints, third- and fifth-order polynomials were chosen to be compared. The methods are used to calculate velocities in dependence of time that are sent to the motors to achieve a desired movement of the TCP. To ensure smooth motions, motor synchronisation is employed by defining an overall total travel time for all motors. The results showed that third-order polynomials achieve a higher absolute positioning accuracy whilst producing the same smooth motions as fifth-order polynomials. Therefore, third-order polynomials were chosen to implement the continuous point-to-point movement. Compared to the start state, the movement of the robot is a lot smoother as there is no jerk visible anymore. The positioning error of the developed motor control ranges from 0.5 [mm] up to 30 [mm]. It can be explained by high velocity peaks, that are caused by the already high minimum velocity, and result in overshoot.

Additionally, to the continuous point-to-point movement, also a linear movement was implemented. Because of the high minimum velocity, the use of trajectory planning and motor synchronisation was not possible. Therefore, only a very jerky movement could be achieved. Both the high positioning error as well as the jerky linear movement might be solved by utilizing motors with a smaller minimum velocity. It would lead to smaller velocity peaks and consequently less overshoot and a higher accuracy. A smaller velocity would also allow to use trajectory planning to create smooth motions for the linear movement.

7. References

[1] Cooper, S., Di Fava, A., Vivas, C., Marchionni, L., Ferro, F. (2020). ARI: the Social Assistive Robot and Companion. In: 29th IEEE International Conference on Robot and Human Interactive Communication (RO-MAN), 31 August – 4 September 2020, Naples. Naples, Italy: IEEE, pp. 745-751.

[2] Ahmed, H., La, H., (2019). Education-Robotics Symbiosis: An Evaluation of Challenges and Proposed Recommendations. In: 2019 IEEE Integrated STEM Education Conference (ISEC), 16 March 2019, Princeton. Princeton, NJ, USA: IEEE, pp. 222-229.

[3] Curto, B., Moreno, C., (2016). Robotics in Education. Journal of Intelligent & Robotic Systems. 81(1), pp.3-4.

[4] Abu-Dakka, F., Rubio, F., Valero, F., Mata, V., (2013). Evolutionary indirect approach to solving trajectory planning problem for industrial robots operating in workspaces with obstacles. European Journal of Mechanics – A/Solids. 2013(42), pp. 210-218.

[5] Pratihar, D., (2017). Fundamentals of Robotics. Oxford: Alpha Science International Ltd.

[6] Bai, Y., Yang, X., Huang, W., (2020). Research on trajectory optimization of handling robot based on ABB. In: 2020 3rd World Conference on Mechanical Engineering and Intelligent Manufacturing (WCMEIM), 4-6 December 2020, Shanghai. Shanghai, China: IEEE, pp. 109-113.

[7] Devi, M., Prakash, C., Jadhav, P., Hebbar, P., Mohsin, M., Shashank, S., (2021). Minimum Jerk Trajectory Planning of PUMA560 with Intelligent Computation using ANN. In: 6th International Conference on Inventive Computation Technologies (ICICT), 20-22 January 2021, Coimbatore. Coimbatore, India: IEEE, pp. 544-550.

[8] Fang, Y., Hu, J., Liu, W., Shao, Q., Qi, J., Peng, Y., (2019). Smooth and time-optimal S-curve trajectory planning for automated robots and machines. Mechanism and Machine Theory. 2019(137), pp. 137-153.

[9] Zheng, X., Zheng, Y., Shuai, Y., Yang, J., Yang, S., Tian, Y., (2019). Kinematics analysis and trajectory planning of 6-DOF robot. In: 2019 IEEE 3rd Information Technology, Networking, Electronic and Automation Control Conference (ITNEC), 15-17 March, Chengdu. Chengdu, China: IEEE, pp. 1749-1754.

[10] Parikh, P., Dave, J., (2020). Trajectory Planning for the Five Degree of Freedom Feeding Robot Using Septic and Nonic Functions. International Journal of Mechanical Engineering and Robotics Research. 9(7), pp. 1043-1050.

[11] Rout, A., Mohanta, G., Gunji, B., Deepak, B., Biswal, B., (2019). Optimal time-jerk-torque trajectory planning of industrial robot under kinematic and dynamic constraints. In: 9th Annual Information Technology, Electromechanical Engineering and Microelectronics Conference (IEMECON), 13-15 March 2019, Jaipur. Jaipur, India: IEEE, pp. 36-42.

[12] Duan, H., Zhang, R., Yu, F., Gao, G., Chen, Y., (2016). Optimal Trajectory Planning for Glass-Handing Robot Based on Execution Time Acceleration and Jerk. Journal of Robotics. Vol. 2016, pp. 1-9.

[13] Li, G., Wang, Y., (2019). Industrial Robot Optimal Time Trajectory Planning Based on Genetic Algorithm. In: 2019 IEEE International Conference on Mechatronics and Automation (ICMA), 4-7 August 2019, Tianjin. Tianjin, China: IEEE, pp. 136-140.

[14] Chen, X., Zhang, W., Shi, Y., Fan, D., Zhang, T., Liu, G., Li, Q., Huang, Q., (2017). An energy optimization based planning approach for moving bottle grasping task using a seven-DoF robotic arm. In: 2017 IEEE International Conference on Mechatronics and Automation (ICMA), 6-9 August, 2017, Takamatsu. Takamatsu, Japan: IEEE, pp. 833-839.

[15] Chiddarwar, S., Babu, N., (2012). Optimal trajectory planning for industrial robot along a specified path with payload constraint using trigonometric splines. International Journal of Automation and Control. 6(1), pp. 39-65.

[16] Gasparetto, A., Boscariol, P., Lanzutti, A., Vidoni, R., (2012). Trajectory Planning in Robotics. Mathematics in Computer Science. 6(3), pp. 269-279.

[17] Ratiu, M., Prichici, M., (2017). Industrial robot trajectory optimization- a review. In: Annual Session of Scientific Papers IMT ORADEA 2017, 27-29 May (2017), Sanmartin. Sanmartin, Romania: EDP Sciences, pp. 54-59.

[18] Makeblock, (2021). Smart Servo MS-12A [Datasheet] Available at: https://www.manualshelf.com/manual/makeblock/950 80/data-sheet-english/page-5.html [Accessed on 14 December 2021]

[19] Biagiotti, L., Melchiorri, C., (2008). Trajectory Planning for Automatic Machines and Robots. Berlin: Springer Science & Business Media.

DOI: 10.2507/34th.daaam.proceedings.055

ENERGY CONSUMPTION IN INJECTION MOULDING: A CASE STUDY

Carlos Mario Flores Lazo & Edwing Isaac Rosales Molina

This Publication has to be referred as: Flores Lazo, C[arlos] M[ario] & Rosales Molina, E[dwing] I[saac] (2023). Energy Consumption in Injection Moulding: A Case Study, Proceedings of the 34th DAAAM International Symposium, pp.0410-0417, B. Katalinic (Ed.), Published by DAAAM International, ISBN 978-3-902734-41-9, ISSN 1726-9679, Vienna, Austria
DOI: 10.2507/34th.daaam.proceedings.055

Abstract

Energy consumption is a key parameter to determine manufacturing process efficiency and sustainability, therefore, analysing plastic injection moulding from an energy perspective is fundamental. In this context, this study presents the analysis on an industrial case to produce shoe soles. The daily injection process was monitored with one power quality analyser and the process was observed at the industrial site. Power data were examined in contrast with the PVC products generated by the rotary table injection machine. It was determined that the production method used was a key factor affecting energy consumption, resulting in a non-productive energy as high as 60%. Furthermore, an important amount of energy was not being used in value added activities. The net energy required to produce a product was found to be approximately 72 Wh for a juvenile sole and 160 Wh for an adult sole. Nevertheless, the energy per product increased to 128 Wh for a juvenile sole when the energy of a cycle (a table revolution) was distributed over the generated products. This analysis has provided the company a first approximation to the amount of non-productive energy in a manufacturing cycle.

Keywords: injection moulding; energy consumption; PVC; non-productive energy.

1. Introduction

Currently, there is a search for production processes that are economically feasible and sustainable. Sustainability of a process can be influenced by multiple factors, with energy consumption being a significant variable [1], [2]. In a low-volume manufacturing environment, it is critical for companies to optimise resource utilisation and minimise energy losses. In this regard, [3] suggests that the implementation of lean methodology can be an effective solution to improve operational efficiency and reduce costs associated with inefficient use of resources. The authors proposed a methodology that seeks to identify and eliminate energy inefficient processes, as well as eliminate non-value adding activities (e. g., non-productive energy consumption) with the aim of improving the overall efficiency of the manufacturing process. A close look at the process is made by [4], which analyses the optimisation of the injection process to reduce energy consumption without reducing product quality. The impacts of cooling time, screw rotational speed, mould temperature and nozzle temperature were studied.

The study revealed that the duration of cooling had the most significant effect on energy usage. Specifically, the examination showed that nearly 50% of the total energy consumed was attributed to the chiller. Maintaining the parameters at their minimum values facilitated a decrease in energy using without significantly impacting quality. A more comprehensive examination of energy usage is available in [5]. The authors examine each stage of the process, from injection to product ejection. According to their findings, the operations that consume the most energy are plasticising and cooling. These findings are in agreement to a certain degree with the outcomes found in [4]. One of the concrete proposals of this study [5] is to reduce the cooling time as much as possible without compromising the quality of the product.

In [6] the configuration of the injection moulding machine was analysed. The study assesses energy consumption based on the power unit used, indicating that choosing the appropriate power unit, at the time of purchase, can significantly decrease energy usage. The limitation of the implicit approach is that it is only suitable during machinery acquisition periods. Nevertheless, it is still crucial to articulate this viewpoint clearly.

A broader look at energy consumption in plastic injection moulding processes is given in [7]. The authors evaluate the plastic injection moulding manufacturing practices of several companies and identify both technical and managerial areas for improvement. Notably, it was found that the measurement and control of specific energy consumption is a weakness within all the companies under examination. This study also compares different machines with respect to specific energy consumption (SEC) and finds that it is depends on the injected mass, the machine output (productivity), and each mould-machine-material combination.

The previous ideas were considered when conducting an analysis of a plastic injection moulding process. Energy and power consumption were monitored during daily operations to gain insight into its behaviour. This study enables the association of energy consumption values with the manufactured component, a practice currently not implemented in the company. Additionally, this study enables the identification of components that can be targeted for implementing energy-efficient measures.

2. Experimental Procedure

In this case study, energy consumption was measured in an injection moulding process to comprehend its behaviour in daily operation. A rotary table injection machine (Desma Werke #1 and #3) was used to analyse, Fig. 1. Both rotary tables have spaces to fit 10 moulds. In this study, a complete rotation of the table was defined as a production cycle, and products were injected one at a time, resulting in PVC soles as the final product.

Fig. 1. Desma Werke injection moulding machine

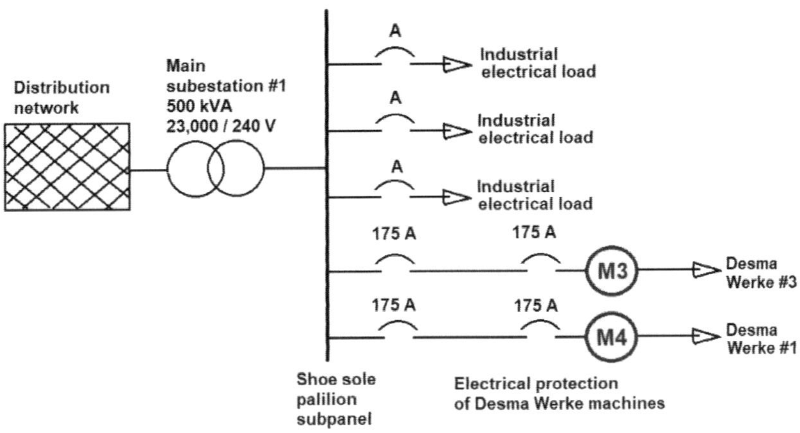

Fig. 2. Single-line diagram of the sole manufacturing pavilion.

To measure energy consumption and active power, a power analyser (Fluke 438-II) with an acquisition rate of 250 ms was installed in position M3 and M4, Fig. 2 but in different moment of time. The energy measurement procedure was defined based on the algorithm proposed in [8]. A general measurement was conducted, with the base case being the measurement of injection cycles executed in one rotation of the table. Power profiles were created and analysed to identify power peaks and determine the energy required for product generation.

The case study was approached in two phases. The first phase focused on measuring the energy consumption and active power of the process with position M3 of the analyser, Fig. 2. and Fig. 3.a. In the second phase with position M4 of the analyser, the production of the observed period was also recorded, Fig. 2. and Fig. 3.d. It was observed that during the cyclical operation of the equipment, not all ten spaces of the rotary table had moulds and some cases PVC was discarded, Fig. 4. This information was taken into account when analysing the productive and non-productive energy.

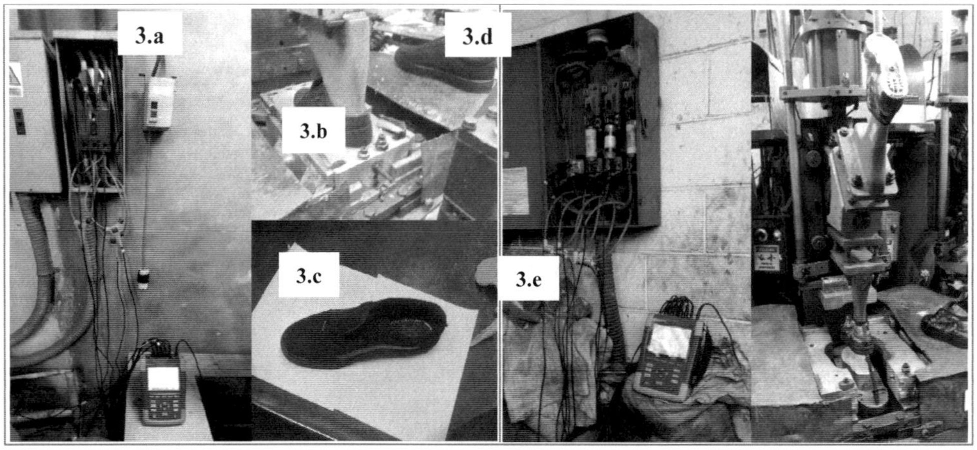

Fig. 3. a) installation of Fluke 438-II for measuring Desma Werke #3, b) moulding of adult sole in Desma Werke #3, c) final product of adult shoes, d) installation of Fluke 438-II for measuring Desma Werke #1, e) moulding of children sole in Desma Werke #1.

Fig. 4. PVC discard

The data for this study were collected in the plant during a daily production day, Desma Werke #3, under normal working conditions in the company, Fig. 3. b, c. In the second phase, Desma Werke #1, the production of the observed period was closely monitored, and the number of products produced during the measurement period was recorded, Fig. 3.e. This comprehensive data collection allowed for a thorough analysis of the energy consumption and production efficiency of the injection moulding process. Overall, the methodology used in this case study provided a systematic approach to measure and analyse the energy consumption of the injection moulding process, Table 1. By considering both the energy consumption and production data, a comprehensive evaluation of the process's energy efficiency was obtained.

Date of visit	Machine measured	Electric variable	Rate of measurement	Other measurement
June 30th	Desma Werke #3	• Energy consumption, kWh • Active power, kW	250 ms	-
August 16th	Desma Werke #1	• Energy consumption, kWh • Active power, kW	250 ms	Video to record the production of soles and PVC discard.

Table 1. Methodology of measurements of energy consumption and active power.

3. Results and discussion

3.1. Phase 1: Power and Energy Consumption

Power monitoring allows observing, as expected, a cyclical behaviour, Fig. 5. It was observed that the highest power peak was required at the time the plastic was injected into the mould. Therefore, each injection is shown on the graph as a power peak. The power profile also shows other peculiarities. The baseline machine power consumption is around 20 kW. This baseline appears to rise to 80kW, but the change is caused by the activation of the chiller used to cool the mould. This suggests that attention should be paid to other elements in addition to the injection moulding machine, as they significantly increase the power demand. This observation is consistent with what [4] discovered concerning the influence of the chiller. However, a more detailed analysis is required to assess the influence of this element in this case. The baseline value found is a first approximation that can be taken into account by the company when determining the costs linked to non-productive time.

The Fig. 5. shows regions where no power peaks are observed (around 1280 and 1420 s). This is due to the fact that no material injection was performed. These areas appear in a non-systematic way as the operator determined, for example, whether an out-of-mould injection was necessary to prevent the material from overheating. This highlighted the necessity to monitor whether a product was being produced in order to determine the consumption of non-productive energy. Therefore, recording was employed as a tool for process monitoring in the second phase.

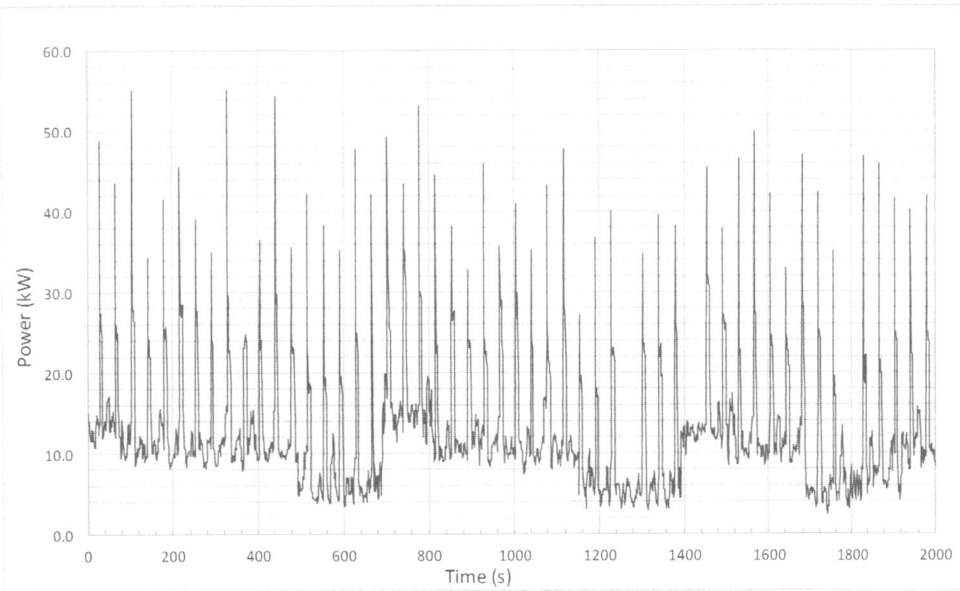

Fig. 5. Active power profile in Desma Werke #3

The energy consumed is shown in Fig. 6., presented in two formats: as energy consumed per second and as the cumulative value of these amounts. As anticipated, the energy curve for every second displays a corresponding pattern to the power graph, with the latter showing an upward trend. A comprehensive analysis of injecting a product enables an assessment of energy consumption. In this context, Fig. 7. has provided valuable insights. Consider the four injections that were carried out. The energy required for each injection was approximately 174.2, 164.7, 140.3 and 161.1 Wh. The corresponding average was 160.1 Wh. For each product, the energy consumptions corresponding to the abrupt power increases due to the injection represented 62.3, 76.0, 64.4 and 71.6 Wh. In terms of percentage, each product required injection of between 36% and 46% of the energy, with an average of 43%.

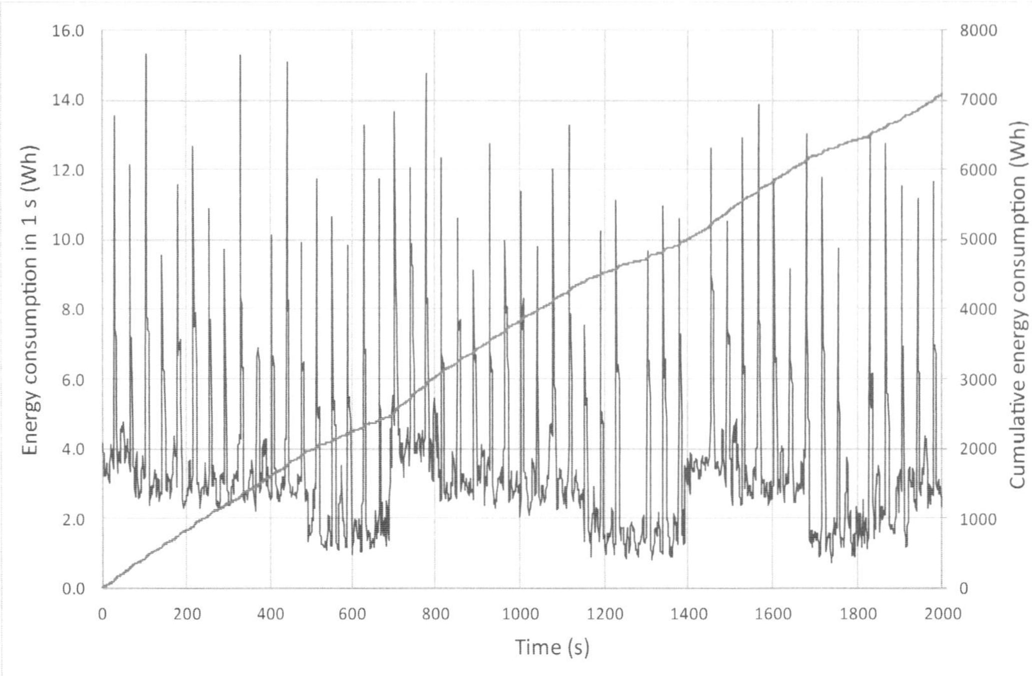

Fig. 6. Energy consumption in Desma Werke #3.

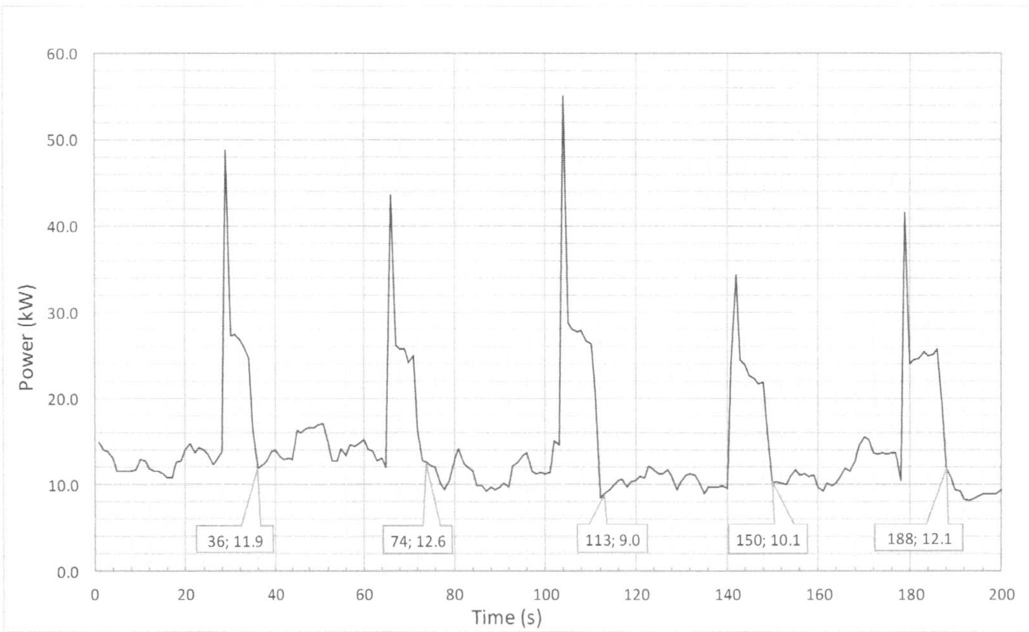

Fig. 7. Detail of the power profile in 5 injections in Desma Werke #3. For each ordered pair, time and power are indicated.

3.2. Phase 2: Power Consumption and Production

The power profile for phase two is displayed in Fig. 8., revealing six peaks that correspond to six injections. Only the required sole sizes are injected according to production plans, although the rotary table, which can hold ten moulds, must complete its rotation, resulting in higher energy consumption. This information highlights the necessity of considering the work method when examining energy consumption.

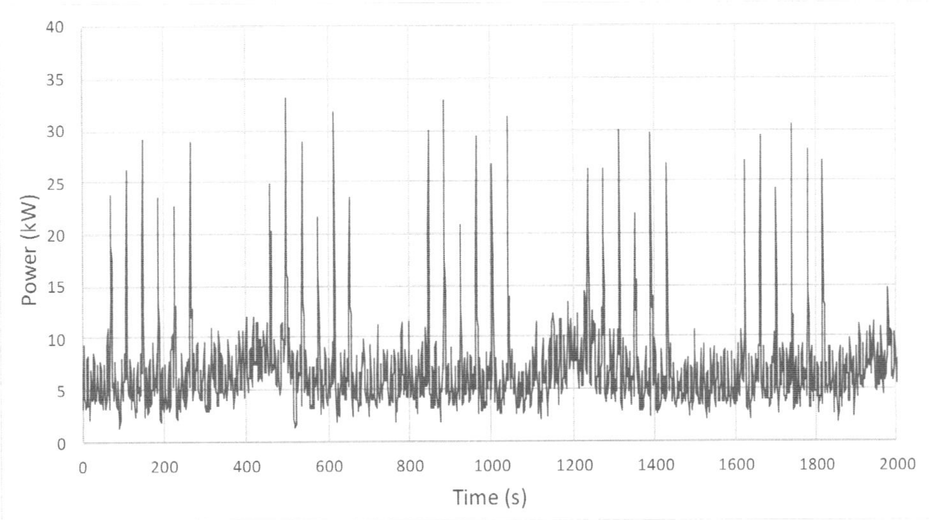

Fig. 8. Active power profile in Desma Werke #1.

Take into consideration a set of six injections, which require approximately 69, 69, 70, 74, 73, and 88 Wh of energy respectively, with an average overall of 73.83 Wh. The abrupt power increase from each injection consumed between 16 and 33 Wh of energy. Each material injection required between 23.1% and 37.5% of the total energy consumption for injection, with an overall average of 34.35%. As mentioned before, in addition to energy monitoring, it is necessary to identify whether the energy is productive or not. In this respect, the observed cycles behave as shown in Fig. 9., only four of the injections are generating product. In each cycle, two injections are made to discard material.

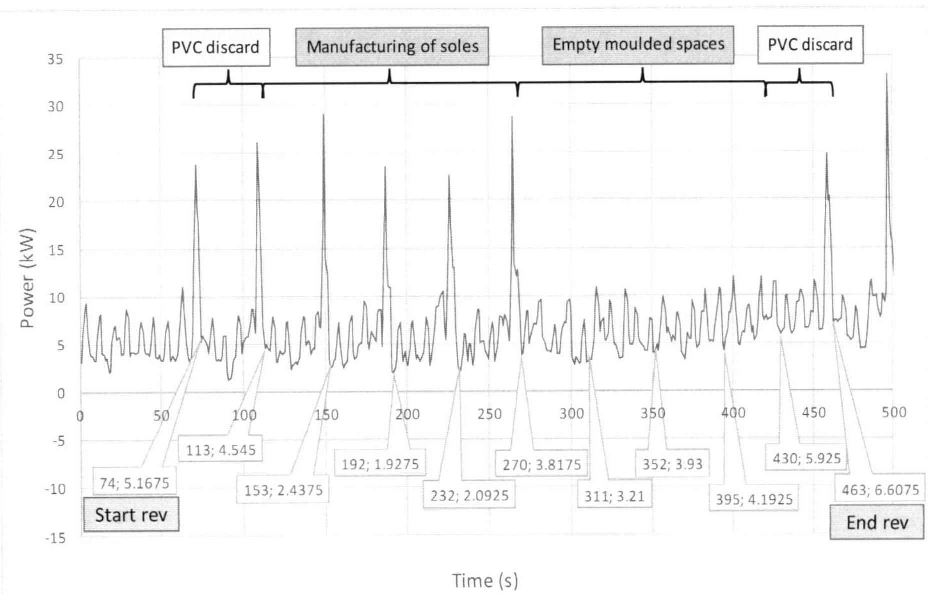

Fig. 9. Detail of the power profile in one revolution in Desma Werke #1. For each ordered pair, time and power are indicated.

The diagram in Fig. 9., is one of a series of 15 cycles that were observed (with number 16 excluded due to incompleteness). Most cycles follow a similar sequence as shown in Fig. 10., albeit with some deviations such as in the initial phase. Therefore, this underscores the importance of recording the working method and product generation during energy consumption measurements. For this case study, based on the previous idea, approximately 40% of the energy can be deemed productive. The above aligns with the suggestions outlined in [9], and identifies scheduling production as another issue to consider in scenarios with low volume. Furthermore, it is recommended to consider scheduling production in low-volume scenarios to reduce energy and resource losses and optimize resource allocation.

A careful analysis of challenges and opportunities associated with order scheduling is necessary for this purpose. Additionally, efficient planning can minimize downtime, thus enhancing energy efficiency and profitability of production.

Revolution	space moulds										Pattern
	1	2	3	4	5	6	7	8	9	10	
1	a	a	b	b	c	c	c	c	a	a	Regular
2	a	a	b	b	c	c	c	c	a	a	
3	a	a	b	b	c	c	c	c	a	a	
4	a	a	b	b	c	c	c	c	a	a	
5	a	a	b	b	c	c	c	c	a	a	
6	a	a	b	b	c	c	c	c	a	a	
7	a	a	b	b	c	c	c	c	a	a	
8	a	a	b	b	c	c	c	c	a	a	
9	a	a	b	b	c	c	c	c	a	a	
10	a	a	b	b	c	c	c	c	a	a	
11	a	a	b	b	c	c	c	c	a	a	
12	a	a	b	b	c	c	c	b	a	a	Irregular
13	a	a	b	b	c	c	b	b	a	a	
14	a	a	b	b	c	c	a	b	a	a	
15	a	a	c	c	c	b	c	b	c	a	
16	b	c	a	X							

Color description	
b	PVC discard
c	Manufacturing of soles
a	Empty moulded spaces
X	End of production
	Sample

Fig. 10. Working method in Desma Werke #1 for the period measured.

4. Conclusions

This study presents the analysis on an industrial case to produce shoe soles from an energy perspective. The daily injection process was monitored with one power quality analyser and the process was observed at the industrial site. Power data were examined in contrast with the PVC products generated by the rotary table injection machine. The analysis of the experimental results leads to the following conclusions:

1. When assessing energy efficiency, it is crucial to consider not only energy consumption but also the production process method. This involves calculating the ratio between the total energy used and the energy utilised to manufacture goods.
2. In this case study the net energy consumption was between 74 and 160 Wh, depending on the product. These values should be interpreted within the context of the entire production cycle since the working method may result in doubling the energy per product.
3. The production method used was a key factor affecting energy consumption, resulting in a non-productive energy as high as 60%.

In addition, this analysis has provided the company with a first approximation of the amount of non-productive energy in a production cycle. However, further studies should look more closely at the impact of specific equipment on consumption, the differences that occur between different products, the frequency with which non-energy-efficient working methods occur and their causes, among other things.

5. Acknowledgments

The authors would like to express their gratitude to Industrias Caricia, S. A. de C. V. and its team, particularly Eng. Héctor Peña, for their assistance and support.

6. References

[1] Pecas, P.; Götze, U.; Bravo, R.; Richter, F. & Ribeiro, I. (2019). Methodology for Selection and Application of Eco-Efficiency Indicators Fostering Decision-Making and Communication at Product Level—The Case of Molds for Injection Moulding, Advanced Applications in Manufacturing Engineering, pp. 1–52, DOI: 10.1016/B978-0-08-102414-0.00001-X.

[2] Cagno, E.; Accordini, D.; Trianni, A.; Katic, M.; Ferrari, N.; & Gambaro, F. (2022). Understanding the impacts of energy efficiency measures on a Company's operational performance: A new framework, Appl Energy, Vol., 328, DOI: 10.1016/j.apenergy.2022.120118.

[3] Karaulova, T.; Andronnikov, K.; Mahmood, K.; & Shevtshenko, E. (2019). Lean automation for low-volume manufacturing environment, Proceedings of the 30th DAAAM International Symposium, pp.0059-0068, B. Katalinic (Ed.), Published by DAAAM International, ISBN 978- 3-902734-22-8, ISSN 1726-9679, Vienna, Austria DOI: 10.2507/30th.daaam.proceedings.008

[4] Meekers, I.; Refalo, P. & Rochman, A. (2018). Analysis of Process Parameters affecting Energy Consumption in Plastic Injection Moulding, Procedia CIRP, Elsevier B. V., pp 342-347, DOI: 10.1016/j.procir.2017.11.042.

[5] Ávila-Cedillo, J.; Borja, V.; López-Parra, M. & Ramírez-Reivich, A. (2019). Energy consumption analysis of ABS plastic parts injected in a hybrid injection moulding machine, International Journal of Sustainable Engineering, Vol., 12, No. 2, pp. 115-122, DOI: 10.1080/19397038.2018.1521881.

[6] Liu, H.; Zhang, X.; Quan, L. & Zhang. H. (2020). Research on energy consumption of injection moulding machine driven by five different types of electro-hydraulic power units, J. Clean Prod., Vol. 242, DOI: 10.1016/j.jclepro.2019.118355.

[7] Mena-Ledezma, J. & Posada-Correa, C. (2023). Definition and evaluation of good manufacturing practices for plastic injection moulding, Energy Efficiency, Vol. 16, No. 4, DOI: 10.1007/s12053-023-10103-x.

[8] Flores Lazo, C. & Rosales Molina, E. (2022). Algorithm and criteria for the measurement of energy consumption in a CNC milling machine: a proposal, IEEE 40th Central America and Panama Convention (CONCAPAN), pp. 1–5. DOI: 10.1109/CONCAPAN48024.2022.9997715.

[9] Buchmeister, B.; Palcic, I. & Ojstersek, R. (2019). Manufacturing Scheduling Performance – a Case Study, Proceedings of the 30th DAAAM International Symposium, pp. 0020-0029, B. Katalinic (Ed.), Published by DAAAM International, ISBN 978-3-902734-22-8, ISSN 1726-9679, Vienna, Austria DOI: 10.2507/30th.daaam.proceedings.003

DOI: 10.2507/34th.daaam.proceedings.056

ASSESSMENT ON FLEXIBLE FIXTURE SOLUTION AND A NOVEL PIN ARRAY SOLUTION FOR AN INDUSTRIAL APPLICATION

Philipp Jelke, Ali Aburaia, Kemajl Stuja & Mohamed Aburaia

This Publication has to be referred as: Jelke, P[hilipp]; Aburaia, A[li]; Stuja, K[emajl] & Aburaia, M[ohammed] (2023). Assessment on Flexible Fixture Solution and a Novel Pin Array Solution for an Industrial Application, Proceedings of the 34th DAAAM International Symposium, pp.0418-0434, B. Katalinic (Ed.), Published by DAAAM International, ISBN 978-3-902734-41-9, ISSN 1726-9679, Vienna, Austria
DOI: 10.2507/34th.daaam.proceedings.056

Abstract

The demand for customisation in today's product assortment challenges the manufacturer to produce unique goods rapidly and economically. Whereas a robot arm performing welding/drilling can be programmed for multiple pieces/functions, the actual standard fixture holding the workpiece in place is not only inflexible, but also lacks the adaptability required to accommodate workpieces of different sizes and shapes. Furthermore, the industry demands a more cost-effective holding solution to secure the workpiece during machining, to reduce the typical 20% of the total system cost for dedicated fixtures. The problems lie in the necessity to lock all six Degrees of Freedom of an irregular and unknown shape, and to define a workpiece pose for automatic production. This paper describes the development and implementation of a flexible fixture that can adjust itself to any form and grasp any workpiece securely. The fixture consists of 112 pins embedded in a 100x200x80 mm base and prodding 60 mm from the top. In this fully automated system, a robot places the piece into the pin array, which closes itself automatically. Multiple revolving active pin arrays conform to the surface and clamp the part. a Pneumatically actuated pins combined with a brake enable the fixture to optimally adapt and support all pieces. The built-in revolting mechanism ensures an ideal connection between the pins and workpiece and provides the required holding force. Assessments of the prototype showed great potential for the method, but also demonstrated the need for high precision parts and fine control of the pins. The pin array presented great adaptability, and the rotating mechanism could significantly increase contact points. Therefore, a future project phase should build on and add to the valuable inside information gathered during this paper to refine the prototype.

Keywords: automated production; fixtures; manufacturing

1. Introduction

Due to changing consumer demand for more customisable and personalised goods, manufacturers adapted their portfolio to accommodate greater variety and shorter product life-cycle. Additionally, the rise in global competitors demands a lower unit cost, higher production quality and shorter lead times. Therefore, producers face the challenge of automating low-volume, high-mix production, which traditionally is expensive and inefficient [1,2]. One essential part, which still lacks an industrial-grade solution, is a flexible fixture of the workpiece.

Commonly used solutions are changing dedicated fixtures, which are specially made for each specific part, between production items. However, every dedicated solution can account for 10-20% of the total system cost due to the need for careful design and high work experience (Wang et al., 2002). Flexible fixtures would eliminate the effort of manufacturing multiple dedicated fixtures and enable the possibility of grasping any item, regardless of its form. This paper describes the development of such a flexible fixture relating to the Morobot [3] production at the Technikum Digital Factory at the UAS Technikum Wien. The aim is to compare different approaches to flexible fixtures currently in development in the science community and further develop the most promising method into a physical prototype. The main focus lies in the explore their uses and problems in an industrial environment. With the aid of the physical prototype, possible problems for a fully automated process should be explored and documented.

2. State of the Art

One essential but often overlooked part of any assembling, manufacturing, inspecting process is a fixture. The primary purpose of fixtures is to fixate, position and support the workpiece during manufacturing operations. Therefore, fixtures are essential to ensure high quality and low variability in parts and often reach 10-20% of the total system cost alone. Moreover, poor fixture design is the source of about 40% of the rejected part because of dimensioning errors [4]. Initially, fixtures were designed for one specific workpiece in mind, which are called dedicated fixtures. These can consequently only be used while producing this piece, which constricts the whole production process. On the other hand, dedicated fixtures also present some benefits, like high rigidity and low tolerances. Current development and production practices demand a new flexible approach to fixtures. Flexible fixtures aim to hold a wide range of workpieces, to adjust to the massive increase of CNC and robot-based manufacturing [1,5-8].

2.1. Purpose of Fixtures

As mentioned, the objectives of fixtures are to fixate, position and support a given workpiece. To achieve these objectives, three main functions must be fulfilled by fixtures, which are clarified in the following paragraphs [9]. In addition, a fixture can also provide various additional functionalities like fast loading and unloading of parts, constant disposal of chips and coolant, be durable and easy to operate by a machinist or automatic [5,8]

Locating: Locating in the context of fixtures describes the positioning of the workpiece inside a 3D workspace at a specific pose (position and orientation). This is necessary to machine part automatically, which depicts the majority of produced industrial goods. For such automatic processes, the parts must be placed at the exact same pose every time to ensure fast, reliable, and precise operations on a workpiece. Furthermore, the correct positioning of the part inside the machine directly influences the reject rate. For example, if a cut or drilling is off by a few millimetres, the part cannot be used in further assembly and must be discarded [5,8,9].

Holding: Holding describes the fixation of the part at the pose in the work area. The held item must not move during the process. If the workpiece moved during the processing, its pose would be incorrect, and the same problems mentioned before would occur. Most of the time, this is achieved by clamping the workpiece between multiple surfaces. The complexity of clamping lies in the supply of sufficient clamping force and in limiting the strength to prevent damage to the part. The necessary clamping power guarantees a rigid and stable fixture of the workpiece in a 3D area and prohibits the movement in the three directions (X, Y, Z) and rotations (RX, RY, RZ), and thereby removing all Degrees of Freedom (DoF). By providing too much force, the part could be damaged or deformed, which would lead to further problems, described in detail in the next paragraph [5,8,9].

Supporting: The last primary function is the support of the whole workpiece against external forces. The part must be supported to prevent the elastic and plastic deformation of the workpiece due to the clamping forces or the machining. This is especially important by thin or fragile workpieces, like Metal sheets or plastics. Errors due to deformation can range from wrong dimensions, tolerances or even damage to tools, and therefore must be minimised to reach a reliable workpiece quality [5,8,9].

2.2. Flexible Fixtures

The change in consumer demand and the growth of global competitors led the producer to increase product diversity, shorten product life cycle and lead-time, lowering unit cost in combination with higher quality and a need to innovate constantly. To be recognised on the global market, manufacturers must be flexible and automated, leading to strife for Flexible manufacturing systems (FMS). FMS enable manufacturers to produce a vast accumulation of different part, varying wildly from each other efficiently and effectively. FMS attain this variation by combining hardware and software components to adapt and react to a changing demand rapidly. The most common implementations of FMS are CNC Systems and robotic manufacturing.

2.3. Types of Fixtures

This paper aims to design and implement a Flexible Fixture System for a mobile mover in the Technikum Digital Factory at the UAS Technikum Wien.

It is compulsory to examine current development in this area and actual industrial applications to make an informed decision. The Categorisation in Fig. 1 is based on Bijan Shirinzadehs [10] grouping of Flexible Fixtures and is cumulated and curtailed. This results in the five main types, where four are described in the following paragraphs.

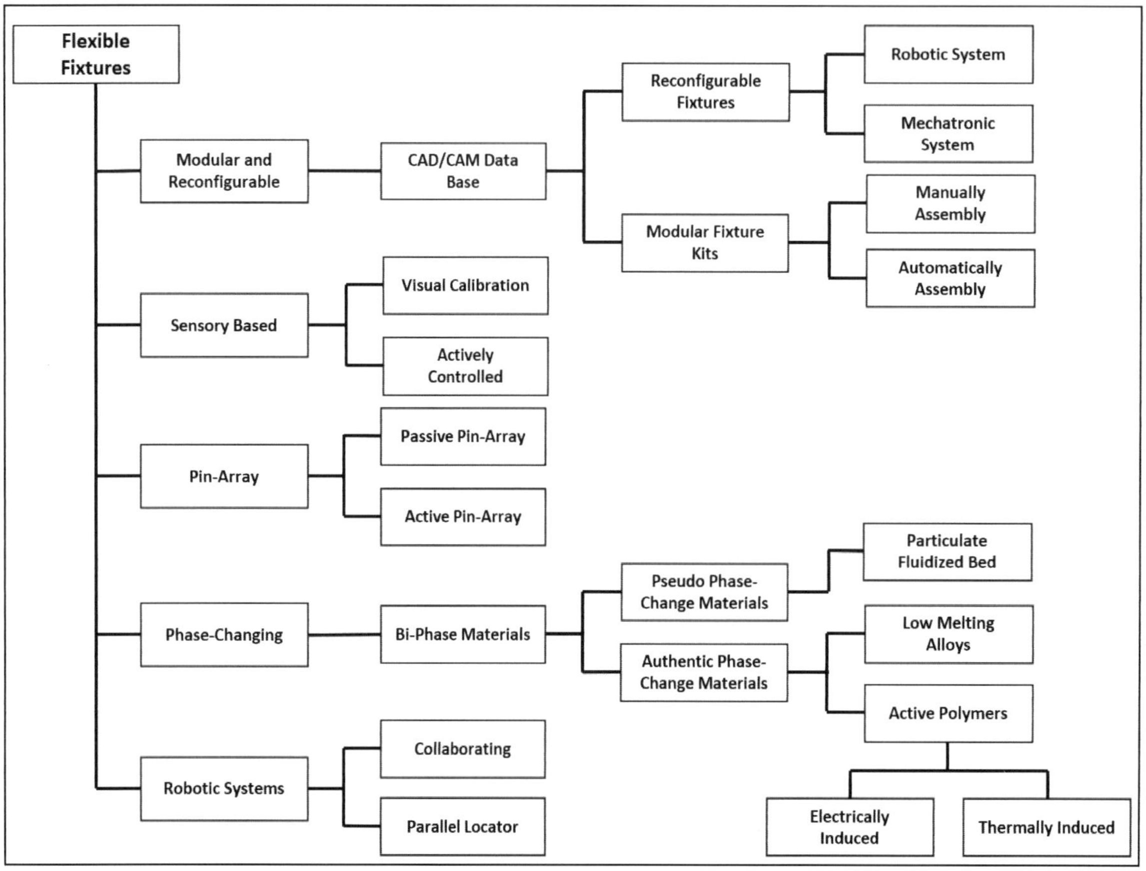

Fig. 1. Categorisation of Flexible Fixtures [10]

2.3.1. Modular and Reconfigurable

Modular and reconfigurable fixtures are descendants of classical systems, splitting the fixtures into multiple parts and making all of them interchangeable. The elements of such a system are grouped into three different types [11]:

Baseplate: A sturdy structure used to connect all parts of the fixture with a single mounting point. Most of the time, it is a plate with multiple parallel T-slots or a hole Matrix System. The baseplate defines the location of the workpiece by attaching the clamping elements at defined positions.

Locating and supporting elements: These elements are placed on the baseplate to align the workpiece onto the fixture and define the pose. In addition, they also support the workpiece during machining by transferring forces to the baseplate.

Clamping elements: Used to clamp the workpiece to the baseplate, thereby restricting the movement during manufacturing processes. The force can be transferred by friction or through a geometric form. They are also used to insert the workpiece into the fixture.

The position of the parts must be calculated or tested for every new part, which should be manufactured. There was a rapid development of an automatic process to design these (CAFD - Computer-aided fixture system [5]), to apply these in an industrial environment [12]. The elements of the modular fixture must be precisely placed on the baseplate before each workpiece change. This is highly skilled and labour-intensive work; therefore, it is often outsourced to a robot system. The robot gathers the needed elements from storage and places them on the baseplate to construct the fixture [10,11,13].

2.3.2. Sensory Based

Sensor-based fixtures are dedicated fixtures, which are enhanced by sensors and software. The information gathered by the sensors is used to compensate for any positioning errors, thereby reducing the fixture's needed accuracy. A "visual fixture" is generated by the computer vision system, which is applied to triangulate the pose of the workpiece in 6 DoF.

With the exact pose of the workpiece, the computer system adapts the manufacturing process to counteract the uncertain mechanical positioning of the fixture. A typical application of this technique is the insertion of windshields into the automotive body [10].

2.3.3. Phase-Changing

One promising idea, in recent years, is the utilisation of materials with phase-changing abilities to change the material from a fluid to a solid and back to a fluid again [14]. A central Obstruction of this technology is the limitation through mandatory functions. The material and process must not harm the workpiece, and the phase change must be simply controlled. The functionality of the fixture can be described in two steps. First, the workpiece is placed in the liquid form of the medium. By triggering the phase-change of the material, the workpiece gets secured in its pose and the manufacturing process can be started. There are currently two different types of phase-changing materials in use.
Authentic phase-change uses the temperature to undergo the phase change. Low melting allowed are primarily used in this category, but even this temperature change can introduce unwanted deformations in the workpiece.
Pseudo-phase-change utilises a particulate fluidised bed, which is a vessel filled with particles (e.g., sand). A controlled stream of air flows through a porous material at the bottom of the vessel. This stream loosens up the particulate material and enables the fluid-like behaviour of the material [10,14].

2.3.4. Pin-Array

Pin-array fixtures utilise the concepts of multiple mechanical contacts to secure a workpiece at a specific location. It consists of multiple pins, also called fixels or fingers, and a base block [15]. The pins can move freely in on direction normal to the base's surface. The workpiece is surrounded by multiple pins, which conforms to the geometry and thereby secures the pose in multiple DoF. Fig. 2 demonstrate the adaptation of a pin array to an irregular workpiece.

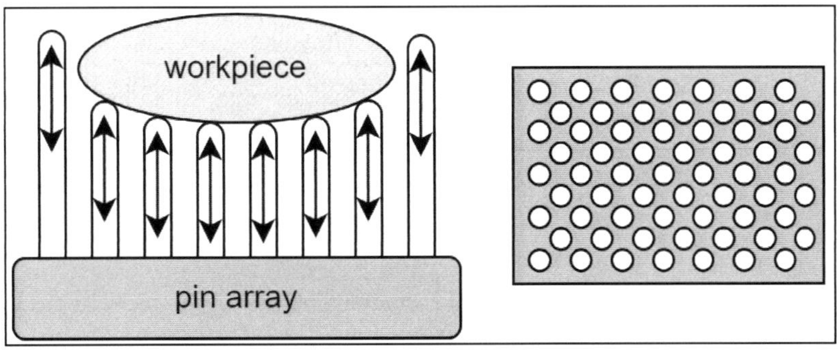

Fig. 2. Schematic of a pin array and representation of its functionality. The right side shows the pins adapt to a workpiece's irregular surface, and the left side is a view from above to demonstrate the pin positioning.

In addition, the workpiece is excellent support over its whole surface. The pin-array is used in two different forms in an industrial setting, a passive and active form. The passive form requires the workpiece to be pushed into the array since the pins cannot be actuated. An active pin-array fixture is capable of moving the pins in a defined direction. This motion is achieved by air or hydraulic pressure or a tiny electric actuator. A friction-based locking mechanism often holds the pins in place [15] to secure their position after adaptation.

Another emerging use of pin arrays are grippers for robots to grab any object and manipulate it. For example, a recent development by Shi [2] utilises a shape-memorable adaptive pin array to grasp any contour. The process consists of two parts. First, multiple objects are learned by pushing them inside a passive pin array, which holds its form through friction. The next step is the clamping part, where the pin rows alternatively move in different directions, which clamps the inserted workpiece. A similar principle is used by the Concentric Rotation Pin Array Gripper [16]. It also uses a passive pin array to conform to the surface, but a twisting motion introduces the clamping forces against multiple revolting parts, which house the pins. Due to the rotational movement, every direction of a workpiece can be grasped.

3. Problem Description

The standard fixture solutions, which are currently in use in an industrial environment, lack the adaptability needed to fulfil the new requirements, presenter to the manufacturer. These requirements consist of a need to produce a small batch size, down to one, to perfectly fulfil all consumer demands, thereby staying competitive in the global market of manufacturing and production. The need for such low batch sizes originates from the consumers' desire for a personalised product that must be customised during production. Moreover, the industry faces the challenge of implementing a cost-effective production process to avert substantial price increases, which the consumer would not accept.

In addition, if a product is custom-made, it cannot be produced in advance and shelved, thereby increasing the delivery time for the client drastically. Therefore, the production process must be as fast as possible to prevent enormous lead times. The necessary machining equipment is often already in use to achieve all these requirements. For example, CNC machinery can produce a variety of different products by just adjusting the program code without additional setup times. Moreover, robots execute multiple different tasks in an autonomous production line, like welding, drilling, and part handling, because they provide an adaptable and programable base framework. In combination, the machining side of a production line is often already suited to manufacture multiple parts without any additional production time and costs besides the initial programming. Whereas the machining part can be programmed for multiple pieces/functions, the existing standard fixture holding the workpiece in place is inflexible and lacks the adaptability required. The fixture fulfils the task to hold the workpiece in place during any kind of processing. Therefore, all the parts six degrees of freedom must be locked to prevent any movement, which would alter the workpiece's dimensions and render the process useless. Traditionally, every part, which can be manufactured at a production line, has its dedicated fixture to fulfil this role, but this is not an option with the need for adaptation and customizability. A dedicated fixture would present the manufacturer with massive additional costs and time for each part. Therefore, a fixture solution is needed to accommodate workpieces of different sizes and shapes. The problem with the standard dedicated fixtures is demonstrated on a vice in Fig. 3. When an irregular shaped object is placed in a fixture there can only be a limited amount of contact points, which highlighted as red dots in the image.

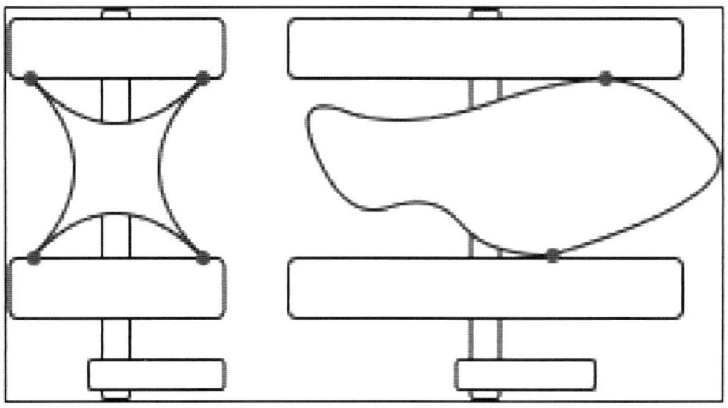

Fig. 3. Limited contact points resulting from an irregular shaped workpiece

Additionally, the fixture should support the part during machining to prevent deformation, ensuring a suitable workpiece. The limited contact points counteract this function and can even cause the part to move during machining and loosen itself, by rotating in the fixture, as seen in Fig. 4. This would lead to a rejects part or could even damage the production machine, which results in additional delays and costs. The third task of a fixture is the introduction of clamping forces to counteract the forces produced by the machining. However, with such a low contact point count, these areas of the workpiece are stressed extraordinary which can lead to damage.

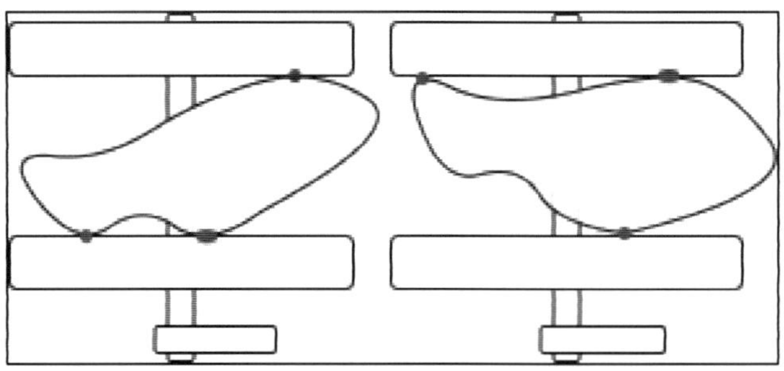

Fig. 4. Uncertain orientation of the workpiece, due to limited contact points between fixture and piece

4. Materials and Methods

The overarching idea was to develop a new flexible fixture for the use in a fully automated industrial setting. The digital factory of the UAS Technikum Wien provides a fully automatic production line of miniature robots in which this fixture is implemented in a future project, to be more precise. This paper provides the groundwork and conceptional design for such a flexible fixture. To achieve that, three promising concepts were compared in their versatility and scope of functions. An evaluation matrix was developed, and the solutions examined and assessed in detail. The best concept was further developed and adapted to the needed requirements.

Lastly, a prototype was manufactured and tested. Selective laser sintering (SLS) 3D printing was utilised to manufacture the prototype, which enables internal mechanical structures but limits the material selection to a thermoplastic polymer. The prototype consists of a subpart of the complete fixture and was only used to examine the developed fundamental concept. The focus during testing lay on the general fulfilment of the desired functionality and not on the performance of the fixture, due to the fact that this project provides the groundwork for a flexible fixture.

5. Practical Realisation

Despite their incredible benefits, flexible fixtures are not yet a widespread part of the industry. Nonetheless, some ideas and concepts could provide a solution to this problem. These notions often come from other industry fields and can be adapted to create a flexible fixture solution. The conceptualisation of the proposed solutions starts with examining and evaluating the most promising concepts. To compare different solutions, we first define what requirements a flexible fixture must fulfil to meet the needs of the industries. The three essential aspects are explained in detail in 1 and must be fulfilled by all fixtures, providing an excellent parameter for comparing these concepts. Next, different qualities and functionalities of a fixture are weighted as additional benefits for the solution and are explained in their corresponding sections. Six specific qualities of fixtures are evaluated to provide a quantified and comparable picture of each concept to quantify these aspects. These examined features are as follows:

- *Complexity*
- *Control*
- *Stress to the workpiece*
- *Position accuracy of the part*
- *Repeatability*
- *Automation*

These characteristics are a general matrix for flexible fixtures and not adjusted to the compared one. Therefore, additional concepts can be included in the comparison without any further adaptation or re-evaluation of the presenter concepts in future developments. This paper covers three concepts, presenting a unique approach to adaptation and flexibility. The concepts are a variation of a fractal vice (Section 5.1), a modular fixture system (Section 5.2) and a pin array (Section 5.3). All solutions are examined individually, using the before defined features. After ranking and comparing the concepts, the best idea is the basis for further developing a flexible fixture in Section 5.5.

5.1. Fractal Vice

A fractal vice is an early 20th-century evolution of the vice jaws, which enables a vice to hold irregularly shaped objects and was patented by Mantle & Co [17]. An excerpt of the patent, where the construction of the vice is demonstrated, is shown in Fig. 5.

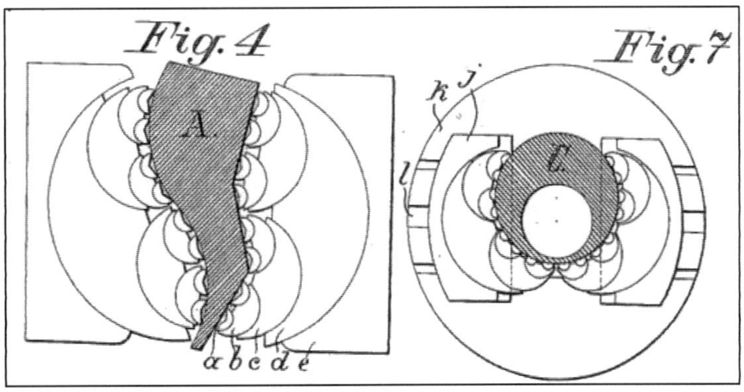

Fig. 5. Excerpt of the patent of the original fractal vice, demonstrating the adaptability of the jaws (Source: [17])

This technology was pushed into the background with the rise of mass production and standardisation, but it currently has an upswing in the 3D printing community. The vice utilises multiple semi-circular jaws to change its form and conform to the shape of the workpiece. To achieve this adaptation, the jaws are placed in a series of linked shrinking semi-circles, most of the time in four steps. The individual jaws can swivel about 45 degrees in each horizontal direction. This system is a passive mechanical chain and is only actively moved at the base of the jaw. The joints between the parts are sliding contact bearings. Therefore, they must meet specialised requirements, like surface quality and lubrication. Nonetheless, the mechanical system of the fractal vice is comparatively straightforward and proven.

The only attention must be given to the design of the joints between the part, due to their before mention needs, yet they are no hinders with today's manufacturing qualities. Fig. 6 shows the 3D printed prototype of the design to evaluate the system better.

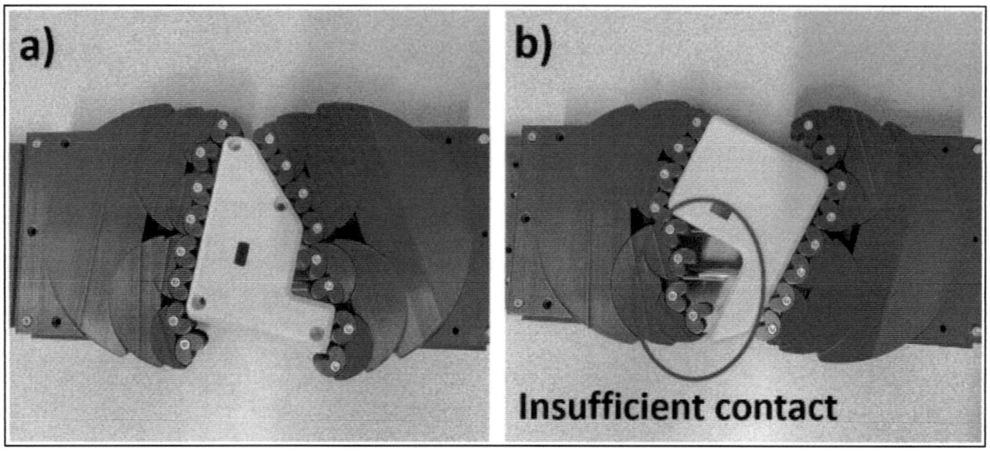

Fig. 6. 3D printed prototype clamping irregular shaped objects. a) Demonstrates a good grasp of the part and b) shows an insufficient contact due to the workpiece geometry

The control of the system is done through a single driveshaft, which can be powered manually, with a handle or automatically with a stepper motor and torque sensor. Therefore, the fractal vice qualifies for an autonomous industrial setting due to the low implementation requirements for the manufacturing plant. The production line must provide power and a digital I/O connection to control the vice. However, the fractal vice presents difficulties in an autonomous environment due to its poor position accuracy and repeatability. It is impossible to say how the part will align itself in the vice during the closing, thereby complicating further workpiece processing. Moreover, as seen in Fig. 6 b, it can occur that the part and vice do not form a sufficient contact and thereby limit the functionality of the vice. These problems can be limited by adjusting the jaw positioning and adding an additional jaw pair 90 degrees rotated, but not wholly prevented, which is needed in an autonomous industrial environment. Another drawback of the concept is the reliability of clamping forces to move the vice parts and conform to the workpiece. These might not be a problem with sturdy materials like metal or plastic but limit the concept's application range. The before-defined criteria assessment can be found in Table **1** in Section 5.4.

5.2. Modular Fixture

The most common solution currently used in an industrial environment is a modular fixture system. However, only specialised manufacturers offer these and often contribute high costs to the production line. Moreover, they do not wholly characterise as flexible fixtures but present a great middle-ground between total flexibility and precision. This section examen its potential as a flexible fixture and what disadvantages hindered its universal adoption. Section 1 describe the general functionality and construction of modular fixtures. Thus, this section concentrates on their implementation and functionalities in an autonomous industrial environment. As mentioned previously, the fixtures consist of multiple parts, like support, locator, clamps, and mounts attached on a baseplate, which has holes or grooves as a means of mounting the part to it. Consequently, the fixture must be assembled and reconfigured for every new part in the holding solution. A highly skilled worker often does this. However, this is not a suitable solution for autonomous production. Manufacturers often overcome the problem by storing the assembled and configurated fixtures separately and changing them with the workpieces. If a new part is designed, an additional fixture is configured, or a previously used one gets adopted. Another promising approach is to outsource the fixture assembly to a robot system. It can place the part precisely where they are needed, and the robot can prepare the following configuration of the fixture during machining of the current part. A modular fixture combined with a robotic reconfiguration station presents the manufacturer with some challenges during implementation into a production line. First, the fixture parts must be stored and accessible to the robot, occupying variable production space. Moreover, this process takes time and could present itself as a bottleneck of the production line. Last, the fixture configuration must be designed, which presents additional work. There are tools to generate these fixtures automatically, but these are often experimental and costly. These implementations challenges mean that the whole system, consisting of the fixture, the robot assembly station, and part storage, are taken into account in the assessment. The individual passive part reduces the complexity of the fixture substantially, but the addition of the assembly station in combination with a configuration tool makes it very time-consuming and error-prone. Likewise, the control of the fixe is the simplest one because it consists entirely out of passive components and is used as a dedicated fixture, but the robotic systems present a significant amount of work.

The modular fixture outperforms in stress of the workpiece, position accuracy and repeatability. If sufficiently well-built and assembled, it competes with dedicated fixtures and presents no problem in these categories at all. As mentioned before, the system can be fully automated, but only with substantial additional work, space, and cost.
Moreover, due to the fixture's design, it cannot accommodate every possible shape of the workpiece. The individual parts of the fixture are off-the-shelf, and the flexibility is only achieved by combining these parts. Therefore, the modular fixture can adjust itself only in small defined steps. The assessment of the criteria once more can be found in Section 5.4.

5.3. Pin Array

The last examined concept of a flexible fixture in this paper is the pin array. It has only a limited application in today's industrial production, but several projects are currently developing. This section only describes the variations and adaptations of pin arrays. The basic functionality and structure are described in section 2.3.4 in detail. Moreover, Fig. 1 in this section is used as a reference here. As it shows, a pin array consists of multiple movable pins. There are two main types of pin array, active and passive. Active pin arrays can actuate the pins and control them. The passive pin array relies on a spring system to push the pin into a before defined position and introduce a force into the workpiece. This type of pin array requires heavy parts to function properly and is sometimes used in engine block manufacturing. For that reason, an active pin array is evaluated in this scope. There are multiple possible variations of actuation, but to two most common ones are a pneumatic system and a linear electric motor. A pneumatic system presents itself as a suitable fit in an industrial environment due to its low implementation cost since nearly every production line has a pressure air system and the great potential power to area ratio at the pins. That means that each pin could introduce more power to the workpiece with a pneumatic system than with an electric one. A pneumatic system with many pins in an array presents a rather complex system to produce and control. Each pin must be treated as a cylinder, with small tolerances and sealing rings. Moreover, the piping inside a small footprint presents its own challenges. The control of the system is achieved through multiple valves and a controller. The pneumatic control and the multiple pins enable the system to be fine controlled and only strain the workpiece with minimal force. An aspect, which standard pin arrays could improve is the position accuracy and repeatability. It is dependent on the minimal distance between the individual pins, but a system can be implemented to additionally move the pins against each other to counteract this disadvantage. This system adds significant complexity but presents a near-ideal performance of the fixture system. If a pin array should be implemented into an autonomous environment, it can be done quickly by adding a robot to position the workpiece into the array. When the piece is placed at the correct location, the pin array locks its form, grasping and holding the workpiece ideal. Moreover, a pin array is arbitrary expandable without losing its ability to hold a small part.

5.4. Comparison

This section provides a direct comparison between the examined concepts (5.1-5.3), providing a solid base to make an educated decision on which solution should be used for further development. The categories the assessment is based on can be found in the first paragraph of section 5. "Complexity" describes the overall difficulty and expenditure of the mechanical and electrical design and manufacturing of a prototype. The scope and complexity of the control, including programming, number of actors and types of power supply, are considered in "Control". "Stress to the workpiece" evaluates the forces acting onto the workpiece during clamping, which are unnecessarily high or unwanted. These forces limit the potential to grasp delicate and soft workpieces. "Position accuracy of the part" and "Repeatability" have the same meaning as in robotics but are applied to the workpiece inside the fixture after clamping. The last metric evaluated is "Automation". It depicts the overall difficulty and expenditure to implement the fixture into a completely automated production line. Important factures here are the footprint, control interface, power supply and speed.

	Fractal Vice	Modular Fixture	Pin Array
Complexity	Average	Good	Fair
Control	Excellent	Poor	Good
Stress to the workpiece	Average	Excellent	Excellent
Position accuracy of the part	Poor	Excellent	Good
Repeatability	Poor	Excellent	Good
Automation	Excellent	Poor	Excellent

Table 1: Comparison of the three flexible fixture concepts

Table **1** lists the individual ratings of the systems split into the defined categories. A grading metric from 1 to 5 (Poor/Fair/Average/Good/Excellent) is used as a rating system. The fractal vice presented two significant problems in the scope of this paper. On the one hand, the aim is to design a holding solution for an autonomous industrial environment, demanding a high position accuracy and repeatability to function correctly. Moreover, a misaligned part can halt the whole production line, resulting in additional problems. On the other hand, the vice must be built with a specific dimension range of the workpiece in mind.

If the object is too small, the vice cannot adjust itself due to the size of its smallest semi-circular jaws. Additionally, when the part is too big, the vice cannot open wide enough to fit the part, and only a small area of the workpiece would be in contact with the fixture, limiting the force transmission and clamping capabilities. The modular fixture convinced with its performance but presents too many obstacles during implementation at a smaller scale. Moreover, the paper's aim is a fully adjustable fixture that can grapes any shape securely, requiring massive modifications and substantial changes to the existing and proven system.

The pin array could not outperform in any category but achieve a relatively good and consistent rating in all groups. Moreover, most aspects, which lowered the score can easily be rectified by minor adaptations and additions to the solution, as elaborated in section 5.3. Therefore, the pin array was chosen as the basic concept of the flexible fixture. The following sections describes the design, adaptation and additions needed for a fully automated flexible fixture.

5.5. Development of a Flexible Fixture

This section covers the development and implementation of the new approach to fixate a part in an industrial environment. The aim is to hold any shape and form securely in place, exert enough holding forces to process the workpiece and provide an automatic process of insertion and closing of the fixture to avoid any need for human interaction. To accomplish this, two new techniques are adapted and combined. These are an active pin array to conform and support and a rotating mechanism to hold and secure the workpiece, which are explained in detail in Section 3 and 5.5.2. In addition, the pneumatic and power supply system and control design are examined in their corresponding paragraphs 5.5.3 and 5.5.4. Finally, a smaller prototype of the complete flexible fixture system was made and tested based on the adaptations and new ideas. The results of these tests can be found in section 6.

5.5.1. Active Pin Array

As elaborated in the previous comparison between different concepts in section 5.4, a pin array was chosen for this fixture. Despite the excellent capabilities of classical pin arrays, some adaptations had to be made. The first significant addition to existing pin array concepts was the active manipulation of each pin. Existing concepts of pin array fixtures depend on springs and the pushing force of the workpiece to adjust their pins. This constricts the capability of the fixture to adapt to any form without unnecessary additional forces. Each spring exerts a specific force depending on its characteristic curve, which cannot be adjusted easily after installation, if at all. Therefore, a spring-based fixture can only accommodate a limited range of materials and workpieces without a spring change. If a process demands soft and hard materials handling, such a fixture is not suitable. Hence, the present solution allows adjusting the force exerted through the pin onto the workpiece, thereby making it possible to grasp a delicate soft piece without damaging it, directly after holding a rigid part during heavy machining like drilling or milling, which demand high holding forces. This great adaptability is achieved by replacing the spring with a pneumatic system. Each pin acts like a miniature cylinder and possesses the same functionality. Therefore, the force applied by the pin is linked to the pressure inside the cylinder. Consequently, by adjusting the pressure, the force on the workpiece can also be varied. To provide an even better control, the pins are designed as a double-acting cylinder. This means the pin can be extended and retracted by the controller without any external interactions. In addition, if both chambers of the cylinder are pressured, the position of the pin is locked. This comes from the fact that both sides of the piston have the same pressure and thereby exerting the same force on the piston in the opposite direction. Thus, the pin does not move in any way and even can absorb some force, corresponding to the internal pressure used.

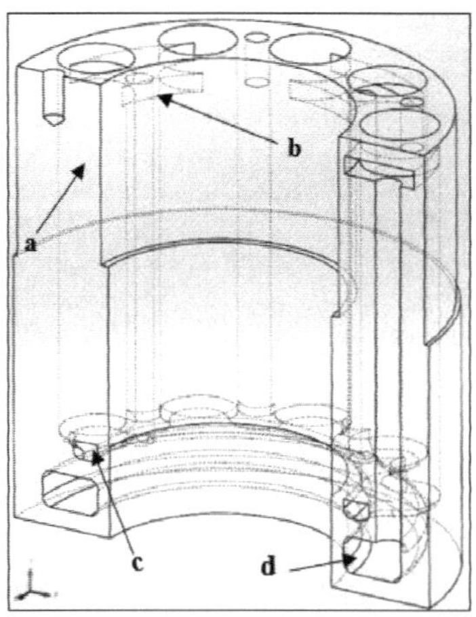

Fig. 7. Section view of the outer housing of a module. Visualisation of the internal air piping and double-acting cylinder construction. a: Cylinder tube, b: Piping and splitting for two top pressure chambers, c: Distribution for all bottom pressure chambers, d: Distribution for all top pressure chambers

Fig. 7 shows the design of these described features and how they were realised in this flexible fixture. This paper's functionality and design decisions are demonstrated on the base part of the outer housing and applied with minor adaptations to the inner base part, which can be seen in Fig. 10. The outer part houses ten cylinder tubes which are radial oriented, as seen in Fig. 7 a. They have a diameter of 8,4 mm, a length of 42 mm and are placed at a distance of 39 mm around the centre. In each tube sits a pin splitting the cylinder into two parts. These are the two pressure chambers needed for extension and retraction. Internal piping is needed, to actuate the pins by filling the chambers with air pressure. The top chamber is supplied with air by five internal tubes between the cylinders (Fig. 7 b). At the top of each tube, it splits up and supplies two top pressure chambers. All five internal tubes are connected through a distribution chamber at the bottom of the part, which can be seen in Fig. 7 d. The bottom chambers of each cylinder are also combined by a separate distribution chamber (Fig. 7 c). Therefore, all cylinder in the outer housing of each module work in unison, regarding extension and retraction of the pins. Hence only two pneumatic tube connections are needed per module, one for the top and one for the bottom chamber, which reduces complexity and piping effort inside the fixture without losing significant flexibility. That is because the position of the individual pins is not linked with this connection. The position of the pin is a combination of the pressure in the chambers and the connection point of the pin with the workpiece. During extension, the pressure of the bottom chamber is higher than in the top chamber, and the pin has no physical resistance, therefore moving up. As soon as the pin connects with the workpiece, the pressure inside the chamber rises to the set air pressure and exerts the force through the pin onto the workpiece. Since the maximum pressure reached is the set pressure, the pins can undergo this sequence without interfering with each other.

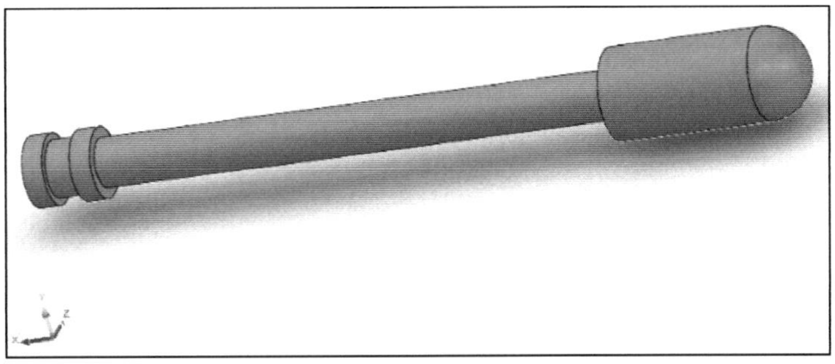

Fig. 8. Single pin, which is used multiple times in the flexible fixture. Additionally, the optional pinhead is also displayed

The pin, which is placed inside each cylinder tube, can be seen in Fig. 8. Every pin is made of two parts, the pin itself and an optional pinhead. The pin functions as the piston of the cylinder, where one side sticks out of the top and contacts the workpiece. It has a diameter of 4,5 mm at the rod and extends at the bottom to hold a gasket. For the gasket, an O-ring with an inner diameter of 4,5 mm, is used to seal the pin and the cylinder walls. This sealing also provides the separation of the top and bottom pressure chamber. The total length of the piston is 70 mm, which allows the pin to extend 60 mm. To ensure a smooth movement of the pin and prevent wedging, the pin is guided by the gasket at the bottom and an additional O-ring in the lid of the housing. This additional O-ring also provides the sealant between the outside and the top pressure chamber. To further prevent any cramming and provide a better connection between the pin and the workpiece, the top of the pin is rounded. Due to the physical space required by the double-acting cylinder and gaskets, the distance between the pins reaches a maximum of 7,5 mm. To reduce this distance, additional pinhead can be installed. These are 3D printed caps, which hold on top of the pins by friction. In this way, the distance can be reduced to 3,6 mm. If a passive movement of the pins is desired, or additional forces should be compensated, there is the possibility to add a spring in the bottom chamber as well as in the top chamber. However, in the majority of use cases, these springs are not needed.

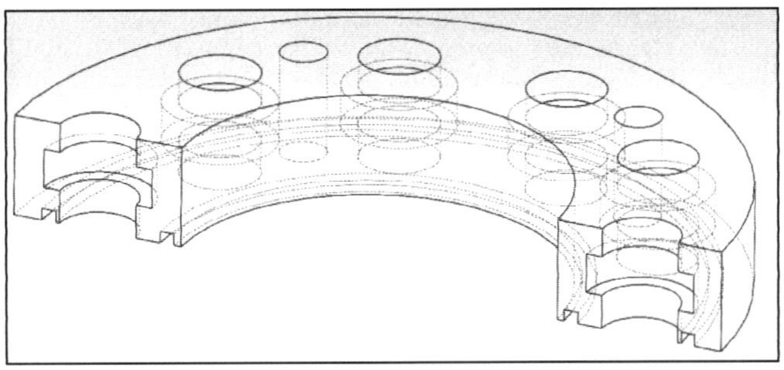

Fig. 9. Section view of the outer housing lid of a module

The final part of the outer housing is the lid, which can be seen in Fig. 9. It is made of a ring with an inner diameter of 27 mm, which surrounds the inner housing and an outer diameter of 51 mm, which is also the maximum dimensions of a module. Ten holes, one for each pin, encircles the centre corresponding to the ten cylinder tubes in the base. Additionally, these holes present a groove for the before mentioned O-rings. These serve the purpose of guidance for the pins and sealant for the upper chamber. Additional O-rings are placed between the lid and the base part to seal this connection and prevent air leakage. To mount the lid to the base, five M3 through-holes are placed rotational symmetrically to ensure an even mounting pressure, which is needed for the O-rings.

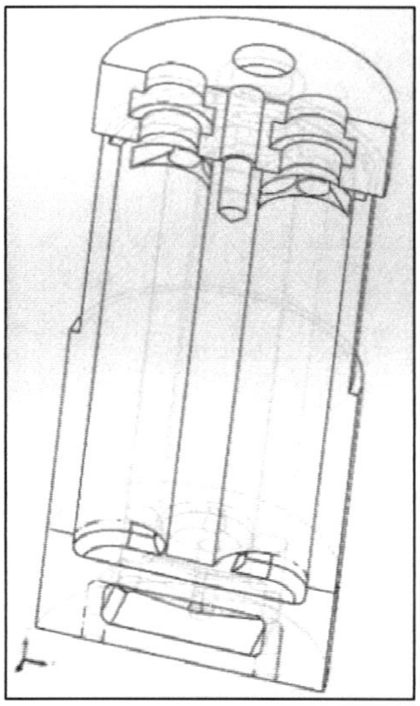

Fig. 10. Section view of the inner housing of a module. Depicts the base part and the lid, without the enclosed pins for better visibility

Fig. 10 presents a section view of the inner housing. The same design process was used, and many similarities to the outer housing are visible. The structure is the same: two distribution chambers at the bottom, a double-acting cylinder in the middle, and a lid with guidance and gaskets at the top. The connection for the top pressure chamber consists of four internal tubes, which are linked into all four top chambers together. A single M3 Screw in the middle is used to assemble the part. The inner housing is placed in the hole of the outer housing and connected with a barring, making it possible to move the inner part around its axis. Section 5.5.2 describes the rotating mechanism and its use in greater detail. A physical prototype of such a module was made and tested. The results and examination methods can be found in section 6. 3D printing was utilised for the prototype due to the complex internal geometrics and piping. The SLS (selective laser sintering) process presented itself suitable to print the parts. It provides relatively good accuracy and can produce small footprints reliable. This is needed because some walls require only 2-3 mm thickness, which is not feasible with a commonly used FDM (fused deposition modelling) printer.

5.5.2. Rotating Mechanism

This section describes the functionality of the rotating mechanism and what benefits it provides. The mechanism addresses the limited number of contact points between the workpiece with an irregular shape and the fixture. The issue is described in detail in section 3 and states the main difficulty with flexible fixtures. If just a static pin array is used, the number of contact points is very limited and often, only 1-3 contacts can be achieved, as seen in Fig. 11 on the left side. The number of contact points highly depends on the exact location of the part inside the pin array. It may even happen that no contact point occurs, which means the workpiece is not supported and moves slightly until a connection can be made. Moreover, a minimum number of contact points is needed to prevent any movement in the six degrees of freedom. Therefore, the number of contact points must be maximised at the precise location of the workpiece. To achieve this, the pins move in the direction of the workpiece to make contact. However, linear movement of the pin could result in special cases where no improvement is achieved. This is the situation when a narrow part is orientated parallel to the movement direction of the pins. As a result, the pins do not move in the direction of the workpiece and, therefore, cannot make additional contacts. The proposed solution is a rotating pin array. This means that two sets of rotationally oriented pins rotate in opposite directions, as seen in Fig. 11 on the right side. By doing that, it is assured that every line drawn through the fixture have at least 2 pairs of pins, which move together, clamping the line in the middle. Additionally, the position of the part inside the fixture is defined by its placement since the pins move onto the part. Thus, the number of contact points and position accuracy can be significantly increased, as seen in Fig. 11 by comparing the left and right sides.

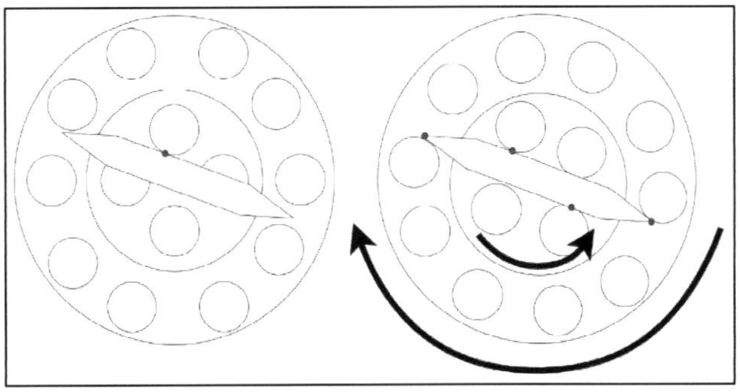

Fig. 11. Contact point between fixture and workpiece with (right) and without (left) rotating

The rotation of a module's inner and outer housing was realised with a pneumatic cylinder. The pneumatic system was chosen since the active pin array already uses an air supply, reducing system complexity. The installed actuator is a spring-loaded single-acting cylinder with a diameter of 4 mm and a 10 mm lift. Fig. 12 demonstrates the mechanism used to move the flexible fixture.

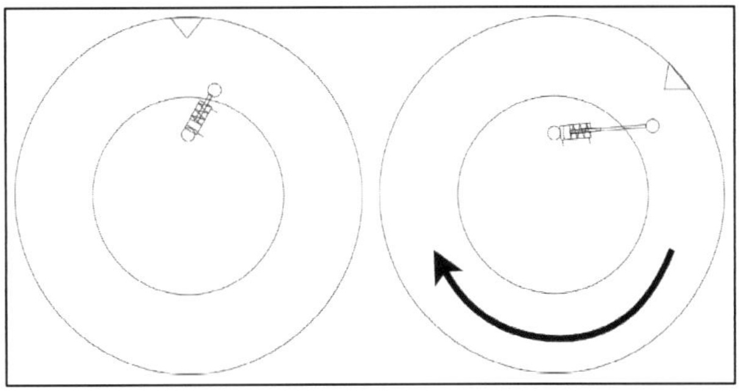

Fig. 12. Transforming a linear motion into a rotation motion to move the inner and outer housing against each other

The rotational movement in opposite directions is achieved by placing the cylinder between the two parts. If the cylinder extends, it pushes the components apart and introduces a force, which rotates the parts in opposite directions since neither the inner piece nor the other part is fixated. Because the movement of the cylinder would interfere with the air pressure fitting, it was lifted above those, and angel fittings were chosen. As a result, the cylinder is mounted on two stilts, which must be able to rotate, as seen in Fig. 13. This is necessary because, during expansion, the cylinder rotates, and the mounts must stay normal to it. Therefore, a friction bearing with two washers is a space-efficient and reasonable approach.

Fig. 13. Assembly of the rotation mechanism without piping to enhance visibility

5.5.3. Pneumatic System

The pneumatic system of the flexible fixture is the primary power source and responsible for all physical movement of the fixture. As elaborated in the previous sections, both the pins and the rotating mechanism are powered by air pressure. Therefore, an air pressure supply is needed to use the fixture. It is common in an industrial environment that a production line-wide air system is present. Thus, the flexible fixture system was designed to be connected to such a system with a single connection. All controls and routings are done inside the fixture, which eases the implementation into the production line significantly. The internal air system consists of a valve terminal, fittings, manifolds and tubes. A schematic of the layout is shown in Fig. 14. The valve terminal contains a distribution rail and five valves, which are controlled by 5V. The valves regulate the following parts individually: The top and bottom chamber of the other housing, the inner housing's top and bottom chamber, and the cylinder for the rotating mechanism. Each valve feeds a manifold, enabling a single valve to control multiple modules. For the fittings and tubing, a range of diameters from 4-6 mm are in use, depending on their expected throughput. The tubes which lead into a manifold are larger than the outgoing ones. The fittings are equipped with flow control, through which the extension and retraction of the pin can be fine-tuned.

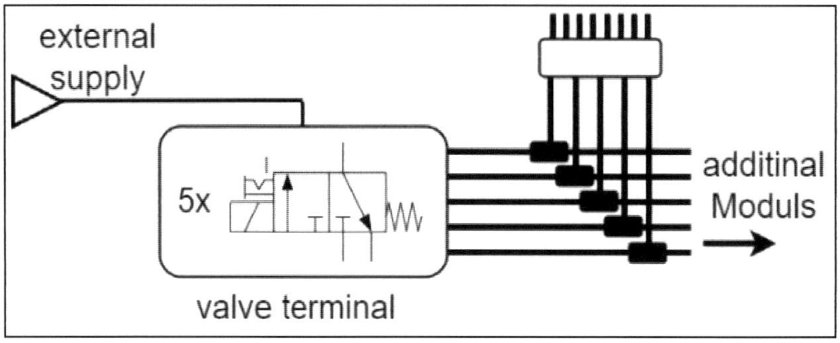

Fig. 14. Schematic of the internal air supply and piping in the base of the flexible fixture

5.5.4. System Control

This section discusses the control system of the fixture and what must be present for implementation into an autonomous process. Additionally, the individual steps performed by the fixture during clamping are listed and described. For the main process unit an Arduino Uno [18] is used, which is a microprocessor with an integrated I/O shield. It provides multiple 5V digital input/output ports and supports varies different serial transmission protocols. As mentioned in section 5.5.3, the pneumatic control is handled by a valve terminal. This terminal is controlled by the 5V digital outputs of the microcontroller, for each valve an individual output. The valves are normally closed (NC), therefore open, if a high signal is present on the corresponding pin. The microcontroller needs a supply of 7-14V to function property, which must be provides by the production line. Hence, the flexible fixture must only be supplied externally with pressured air and 7-14V power. The interaction between the manufacturing plant and the fixture is realised with a digital input, which triggers two different processes. "Low" state starts the opening process of the fixture to release the workpiece and hold the fixture open, while waiting for a new part. The "high" state signals the fixture to close itself and grasp the workpiece. This signal is provided by the plant controller, which also is responsible for the coordination with the placement of the workpiece inside the fixture.

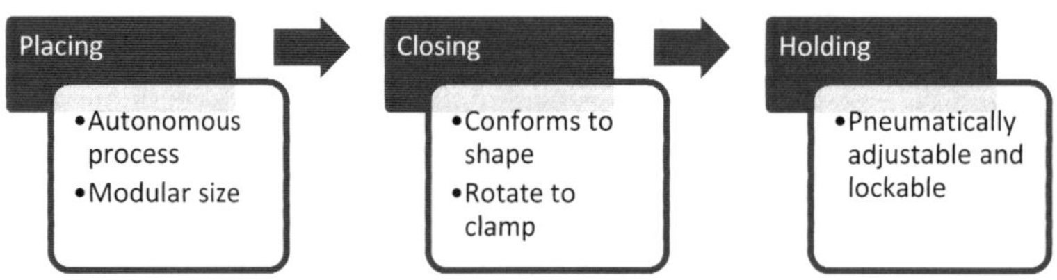

Fig. 15. Sequence of the complete clamping and holding procedure of the proposed flexible fixture

Fig. 15 visualises the steps the fixture undergoes to receive and hold a part. The opening process has the same steps but in reversed order. The first step is to place the workpiece inside the fixture. That means that, the pins are extended (bottom chamber pressurised) and a robot or similar mover places the workpiece at the exact position and orientation. By pressing the part into the fixture, some pins compress to make way to the part and conform to the surface of the part. In the next step, the rotating mechanism activities and rotates the fixture part against each other. In that way, a maximum number of pins contacts the part and exert a force to secure it in place. Last, the pressure in all chambers (top and bottom) rises and the workpiece can be released from the robot gripper.

The additional pressure strengthens the support of the underside of the workpiece, to achieve a higher absorption of manufacturing forces. With the pressurisation of the top chamber, the pins are locked into position, to prevent any unintentional forces.

6. Results

The main focus of this work was to evaluate the concept and create a knowledge base for a future project. Therefore, the fixture's performance was measured by the functionality and not by specific performance parameters. Fig. 16 shows the finished prototype of a single module inside a mounting for evaluation purposes. The complete fixture consists of eight modules inside a joint base, which also houses the tubing and control system. However, in this project, only a single module was manufactured and tested. The 14 pins can be seen in their retractive position in the upper part of the image. The whole fixture is mounted with a bearing in the middle to test the fixture without outside interferences. At the bottom, the four connection ports for the pressure chambers are noticeable, but the tubes were removed for better visibility. Moreover, the small cylinder, responsible for the rotational movement can be seen behind the connectors.

Fig. 16. Prototype of a single module of the proposed flexible fixture

The first part tested was the pin array and how it interacts with irregularly shaped objects. The assessment evaluates, the holding performance in regard of general functionality. Two pictures of an inserted part inside the fixture can be seen in Fig. 17. In Fig. 18, the adaptation to horizontal irregular shapes is demonstrated. The pins underneath the part extended until they contacted the workpiece and then stopped and supported. This functionality was desired and is one outstanding feature of the pin array. Therefore, the general functionality of the pin array was quite good, and the part could be held in place by the fixture. However, some significant problems occurred with this prototype and the manufacturing methods. The SLS printing procedure could not achieve the dimensional accuracy needed for such a small and precise tool. Moreover, the surface quality was not as desired, which led to additional friction inside the cylinders. This results in two noticeable downfalls. First, the pressurisation of both chambers (top and bottom) could not be accomplished due to the inaccuracies and the resulting leaky seals between the pins and the outside.

Furthermore, a higher pressure of 4 bar (58 psi) was needed to extend the pins, which resulted in additional forces acting on the part and a rapid extension. However, a satisfactory extension and retraction process is feasible with better sealing and flow regulators. Additionally, the inaccuracies and surface quality caused the pin to wedge and move horizontally inside the cylinder tube easily. A wedged pin can be seen in Fig. 17 in the left image in the middle. The pin leaned to the right, which led to fewer contact points between the fixture and workpiece. Moreover, wedged pins are often halted in a permanent position, resulting in a required loosening by hand or disassemble in a few major cases. This is unacceptable in an autonomous environment and must be addressed in a future project. Additionally, to the inaccuracies, the pins bend under load, which furthers any wedging and hinders the optimal performance of the fixture. The bending comes from the choice of material, which performs inadequately in such small dimensions. Bending can be observed in both images of Fig. 18, on the right, the outermost pin and on the left, the pins underneath the slant.

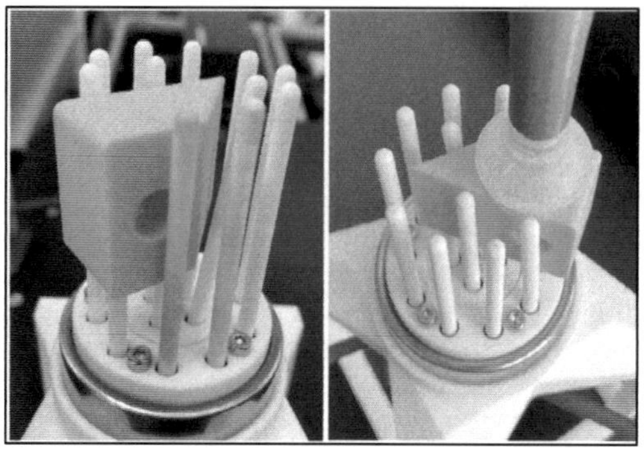

Fig. 17. Irregular part inside the fixture with extended pins. Right: Part is placed by a robot inside the fixture, and it closes around it

Fig. 18. Demonstration of the vertical support of irregularly shaped objects inside the fixture

The second area examined was the rotating mechanism and its performance regarding the workpiece's contact points and position accuracy. A demonstration of the clamping process with the rotating mechanism can be seen in Fig. 19. On the left side is the fixture with an embedded part before the rotating mechanism is activated. The markers show the two potential pins close to the workpiece but not touching it. Therefore, they did not contribute any grasping forces. On the right side, the fixture is shown after clamping the part. By comparing the marked areas, it can be seen that indeed two new contacts could be achieved. Therefore, the hypothesis that a rotating pin array can increase the number of contact points could be confirmed. Moreover, with additional modules, the contact point gain could be further expanded.

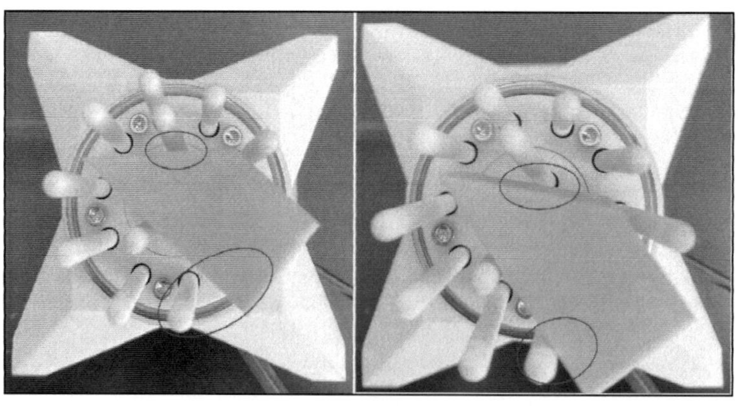

Fig. 19. Experiment to validate the rotation mechanism. The two additional contact points gained by the rotation are marked. Left: Before rotation; Right: After rotation, the workpiece is clamped

Nonetheless, the implemented methods grappled with the same problems as the pin array. The surface quality of the SLS print could be mostly circumvented using bearings between the inner and outer ring and the mounting rack, but the friction bearing used for the stilts are affected, as seen in Fig. 20. The additional friction in combination with the tendency of the material to bend resulted in sometimes in a deformation of the stilts, which hinders the smooth operation of the cylinder, and the linear motion could rotate the fixture with sufficient force.

Fig. 20. Close up of the bending of the stilts during rotation of the fixture

7. Conclusion and Outlook

This work tried to develop a flexible fixture and produce a groundwork for a follow-up project. These goals were fulfilled with the improvement for an active pin array combined with a rotating clamping mechanism and the assessments with the first prototype. The concepts functionality could be validated, and valuable knowledge of flexible fixtures gathered for the next iteration. However, the prototype also presented some significant obstacles and challenges, like tolerances and surface qualities. As the results state clearly, SLS 3D printing is not a suitable production method for such a small and precise part. The project's next step is to use other production means, like CNC-routing, to produce a better prototype to test the functionality purely without the before mentioned material limitations. For this purpose, the design must be adapted to accommodate the new production restrictions.

In this process, the flexible fixture can be enhanced with multiple features that provide better functionality and ensure a seamless implementation into a fully automatic environment. The Arduino was a suitable controller for a simple prototype, but a Linux based computer, like the raspberry pi, provides more versatility and better interfaces, which supports standard industrial protocols, and is therefore preferable for the implementation into production lines. In addition, a closer pin configuration combined with individual pin control would ensure a safer grip of the workpiece by the increased number of contact points and more nuanced control of the partially pushed pins. The rotating mechanism is an excellent method for introducing clamping force into a pin array, but the proposed mechanism showed some difficulties and should be improved in future projects to ensure a smooth operation. The pins would also benefit by guiding to accommodate the sideways clamping forces better. Lastly, the flexible fixture could be improved by changing the directional valves with electrically adaptable pressure units and sensors to enable more nuanced control and enhance the controllability significantly. The project revealed the great potential of flexible fixtured and provided the first draft of a possible solution. Despite the semi-successful prototype, the concept looks promising and should be further developed into a fully functional prototype for an industrial application.

8. References

[1] Bi, Z., M. & Zhang, W., J., 2001. Flexible fixture design and automation: Review, issues and future directions. International Journal of Production Research. 39(13), pp.2867-2894

[2] Shi, P., Hu, Z., Nagata, K., Wan, W., Domae, Y. & Harada, K., (2021). Development of a Shape-memorable Adaptive Pin Array Fixture. Advanced Robotics. 35(10), pp.591-602

[3] FH Technikum Wien, (2021). Morobot. [Online] Available at:<https://engine.ie.technikum-wien.at/webpage/minifab.html [Access 04.11.21]

[4] Wang, H., Yiming, R., Hua, L. & Price, S., (2002). Computer aided fixture design: Recent research and trends. Computer-Aided Design. 42, pp.1085-1094

[5] Price, S., 2009. A Study of Case Based Reasoning Applied to Welding Computer Aided Fixture Design. [Master Degree] Location: Worcester Polytechnic Institute.

[6] Lazarevic, M., (2018). Modelling and Simulation of a Novel Modular Fixture for a Flexible Manufacturing System. International Journal of Simulation Modelling. 18(1), pp.18-29

[7] Gothwal, S. & Raj, T., (2017). Different aspects in design and development of flexible fixtures: review and future directions, International Journal of Services and Operations Management, 26(3), pp.386 – 410

[8] Gameros, A., Lowth, S., Axinte, D., Nagy-Sochacki, A., Craig, O. & Siller, H.R., (2017). State-of-the-art in fixture systems for the manufacture and assembly of rigid components: A review. International Journal of Machine Tools and Manufacture. 123, pp.1-21

[9] Bakker, O., J., Papasathis, T., Ratchev, S. & Popov, A., (2013). Recent Research on Flexible Fixtures for Manufacturing Processes. Recent Patents on Mechanical Engineering. 6(2), pp.107-121

[10] Shirinzadeh, B., (2002). Flexible fixturing for workpiece positioning and constraining. Assembly Automation. 22(2), pp.112-120

[11] Dai , J. R., Nee , A. Y. C., Fuh, J. Y. H. & Kumar, A. S., (1997). An approach to automating modular fixture design and assembly. Proceedings of the Institution of Mechanical Engineers, Part B: Journal of Engineering Manufacture. 211, pp.509-521

[12] Hashemi, H., Shaharoun, A., M., & Izman, S., (2014). Fixture Designers Guidance: A Review of Recent Advanced Approaches. Jordan Journal of Mechanical and Industrial Engineering. 8(6), pp.377 – 384

[13] Hongjian, Y., Bing, L., Yang, W. & Ying, H., (2012). Conceptual design and workspace analysis of reconfigurable fixturing robots for sheet metal assembly. Assembly Automation. 32(3), pp.293 – 299

[14] Rong, Y., Tao, R. & Tang, S., (2000). Flexible Fixturing with Phase-Change Materials. Part 1. Experimental Study on Magnetorheological Fluids. The international Journal of Advanced Manufacturing Technology. 16, pp.822-829

[15] Hurtado, F. A., Melkote, S., 2002. A model for synthesis of the fixturing configuration in pin-array type flexible machining fixture. International Journal of Machine Tools and Manufacture. 42(7), pp.837-849

[16] Mo, A., Fu, H., Luo, C., & Zhang, W., 2018a. Concentric Rotation Pin Array Gripper for Universal Grasp. In: 2018 3rd International Conference on Advanced Robotics and Mechatronics (ICARM). IEEE, pp. 112-117

[17] Kunze, K. P., (1913). Device for obtaining intimate contact with, engaging, or clamping bodies of any shape. US Patent 1059545

[19] Arduino, (2022). UNO R3. [Online] Available at: https://docs.arduino.cc/hardware/uno-rev3 [Accessed 5 January 2022].

DOI: 10.2507/34th.daaam.proceedings.057

AUTOMATIC PRINT BED LEVELING FOR INDUSTRIAL ROBOT SYSTEMS

Clemens Hofbauer, Ali Aburaia, Kemajl Stuja & Mohamed Aburaia

This Publication has to be referred as: Hofbauer, C[lemens]; Aburaia, A[li]; Stuja, K[emajl] & Aburaia, M[ohamed] (2023). Automatic Print Bed Leveling for Industrial Robot Systems, Proceedings of the 34th DAAAM International Symposium, pp.0435-0447, B. Katalinic (Ed.), Published by DAAAM International, ISBN 978-3-902734-41-9, ISSN 1726-9679, Vienna, Austria
DOI: 10.2507/34th.daaam.proceedings.057

Abstract

Additive manufacturing with 3D-printers enables the creation of any kind of object by printing it layer by layer. However, the object size and complexity that can be reliably produced with standard industrial applications is limited by the 3D-printer's size and degrees of freedom. In contrast, industrial robots typically have larger work envelopes and more degrees of freedom. Combining a robotic system with a 3D-printer increases the possible print area and thus the opportunity to create bigger objects. However, a larger print surface becomes increasingly uneven because of mechanical constraints and deformation caused by heat, reducing the adhesion quality. Adhesion is required to achieve a high product quality, or even a successful print at all. The UAS Technikum Wien Digital Factory robotic printing system consists of an ABB IRB 2400 and a custom-built print bed. This allows the production of objects with a surface area of over 3 m^2 in a short time span, which allows for faster prototyping durations of different projects. However, the adhesion properties suffer severely. This report describes the implementation and evaluation of an automatic print bed leveling system, as well as the possibility to print on curved surfaces. A grid-based probing approach controlled by using a touch probe sensor with an appropriate mounting and offset calculation was designed to compensate for the disadvantage of the sizable print bed. This results in a higher percentage of even surface area and therefore to an increase in overall adhesion quality. For printing on curved surfaces, it is recommended to further expand the system using a collision check as well as an orientation interpolation.

Keywords: Additive Manufacturing; Auto Bed Leveling; 3D-Printing; Robotic System; Print Bed Leveling

1. Introduction

3D-printing or more precisely fused filament fabrication (FDM) has become a regular application in the industry. It is widely used to create different and mostly complex components in one step compared to the numerous steps that occur in conventional manufacturing processes. In order to create such objects, a fine-tuned printing system is needed. One of the most common problems that occur during these prints is the bad adhesion properties between the actual object and the print bed. The reason for this can be many things like temperature differences or bad filament. But most of the time the problem occurs because of a non-constant distance between the extruder and the print bed. This leads to objects warping or sliding around the bed during the print.

This effect is even more evident when printing with robotic systems. The problem with those systems is, that they usually don't level the bed before a print. Because the bed itself isn't perfectly even, the distance between the extruder and the actual bed changes constantly during the print. Even though the differences in distance are minimal, the actual result of the print quality depends highly on a constant distance during the first layer of the print.

The goal of this work is therefore the implementation of an automatic bed leveling system for the current application of the 3D-printing robot system situated in the digital factory of the UAS Technikum Wien. For this purpose, different sensors will be evaluated, compared and finally implemented in a modular bed leveling application that can be used for different robotic systems. Therefore, the current printing process will be expanded with the bed leveling system. More importantly a quality increase of the prints is to be expected. Furthermore, approaches for leveling with the system on curved surfaces are being examined. Chapter 2 examines the actual problems that occur on the current printing system. In chapter 3, the current state of the art for commercial and robotic leveling systems, including sensors and algorithms, as well as curved surface leveling is examined. Chapter 4 describes the methods used to implement the system, following with chapter 5 where the practical realization will be documented. The subsequent chapter will summaries the results with the conclusion giving a glimpse into the future and ongoing work about this topic.

2. Problem description

3D-printing is becoming a common technology in the industry and for private consumers. The applied technology that is utilized with those commonly used printing machines cannot easily be transferred to a robotic system. This spans from slicing an object with a slicer to converting the result into the actual robotic system. More importantly this applies to functionalities which are normally not used while printing with robotic systems, like using a bed leveling system.

2.1. Robotic system

Currently the actual robot system that is used in the Digital Factory of the UAS Technikum Wien for printing large objects consists of an ABB robot system with an appropriate extruder and customized heated print bed. More precisely the robot used is an ABB IRB 2400/16. The maximum weight of this robotic system lies at about 16 kg. One problem that is currently leading to an imprecise operation during printing is caused by the weight of the extruder contraption. The used extruder is called a Noztek Xcalibur and weighs with all the additional parts mounted to it about 20kg. This can lead to limitations [1] of the actual robot system that occur because of this overweight extruder. Another key characteristic of the robot is the position and path accuracy. These two values sit at 0,03 mm for the position and at 0,11 – 0,15 for the path accuracy. It is to mention that those two values are for the repeatability of the robot. Using the robot for points resulting from a sliced object applies the absolute accuracy, which has a value of around 0.3 mm [1]. The reason for the implementation of the bed leveling process is to counteract those limitations created by the robotic system. Above all, this has to be done on the first few layers of the print in order for it to get the best adhesion possible. Of course, it is not possible to fully reduce those limitations with just a bed leveling system, but the implementation will by all means improve the actual result. The current state of the robotic printing station can be seen in Fig. 1. The different parts of the system are highlighted respectively.

Fig. 1. Robotic system used for 3D-printing, positioned in the Digital Factory of the UAS Technikum Wien

2.2. Print bed and sensor systems

The robotic system has a customized heated print bed. This print bed consists of two big aluminum plates mounted together in order to form the actual print bed. On top of those plates are mounted heat pads. These are used to raise the bed temperature to about 60 °C during operation, which is the optimal temperature for printing with polylactide (PLA).

The actual surface on which it is printed on is realized with a glass plate on top of the heat pads. Glass usually has a very smooth surface texture, which is actually a good thing for 3D-printing. The main problem that can happen despite the smooth surface, is an actual tilt of the bed. This can occur because of many reasons. One could be an extension of the mounting structure because of the increase of the temperature. Another could be that the actual mounting cannot be done with such high precision, so that no tilt is occurring at all. All these reasons lead to a bad adhesion between the actual print and the surface of the bed. With an appropriate bed leveling system, the tilt can be balanced out using the results of the probe points.

Because the main print surface consists of glass, the sensors that can be used for leveling the bed are very limited. Typically used sensors like inductive ones, which are utilized by many common printers, cannot be applied for this reason. Instead, other solutions need to be considered. Therefore, material independent sensors should be the main focus for realization. One disadvantage that the material independent sensors provide is, that other sensor types that would be a better use for curved surface leveling, for example laser systems, cannot be applied because they rely on specific materials. Consequently, the sensors used for the planar leveling should also be able to perform curved surface operations.

2.3. Software implementation

The robotic system utilized for printing relies on three different programs used to realize the actual print. The first component applied consists of a PC based programmable logic controller used to control external components like the extruder, the material feeding process and the print bed heating process. The second system used is the actual slicer of the 3D-printed parts. For this case the Cura slicer is utilized. One problem that occurs using traditional slicers like Cura slicer is, that there are only two predefined print forms that can be applied. One is a cylindrical shape, the other one is a traditionally used rectangular cuboid. The dispute that occurs is, that the actual print bed is shaped in a special way according to the robot's workspace. For this matter the rectangular cuboid is utilized in the slicer. The size defined is far bigger than the actual print bed in order to cover it completely. The actual differences in size of the print area and slice area can be seen in Fig. 2. Additionally, the position where the height of the glass was teached in the system is visualized. The further away the position of the object is from the work object point, the more severe the error to the real print bed becomes.

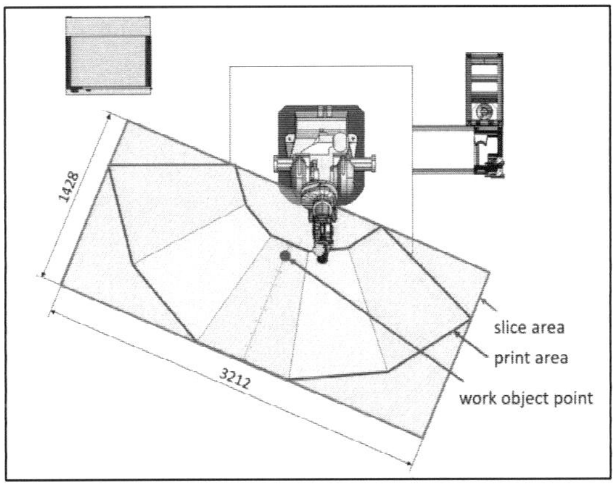

Fig. 2. Comparison of print area and slice area (top view of the system)

The actual resulting G-code gets then moved in the correct position using the third software called RoboDK. The problem that results is that the leveling process should only be done on the actual print surface. Therefore, the differences between the rectangular surface and the print area need to be considered.

2.4. Printing results of the current station

Although a lot of objects that have already been printed on the newly designed and extended print area turned out to be good results for printing with PLA in this order of magnitude, some shapes tend to create a few problems. The same problems that occur on standard printing machines at a smaller scale, also take place while printing with a robotic system. This is even more apparent, the bigger or longer the actual object is. In order to compare the prints with and without bed leveling, the same object is being examined. For this purpose, a long rectangular like shaped box was printed without the actual leveling applied. The resulting print has severe issues that occurred because of the material cooling down and leading to warping effects. This warping effect even led to a tear in the wall. The reason for this was the movement of the wall during the print resulting in wrong positions for the robot, where the filament was being extruded. The actual print can be seen in Fig. 3 which is split into a top and bottom view of the resulting object. The object has a dimension of 145 x 37 x 15 cm.

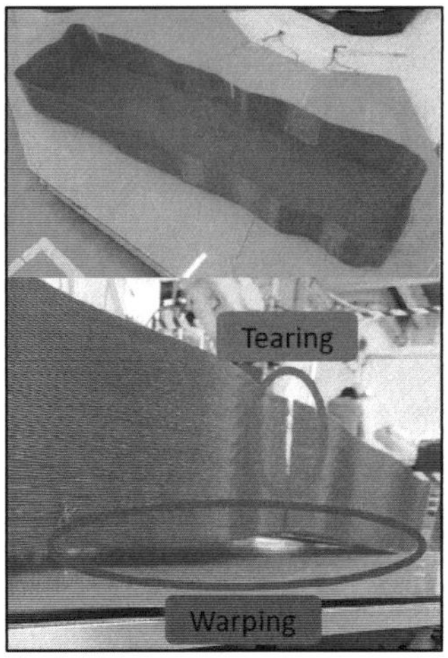

Fig. 3. Problems with current printing results on a relatively long object (top: aerial view of object; bottom: warping and tearing of the object)

3. State of the art

Currently mostly all of the commercial 3D-printers available use some sort of leveling system. This leveling of the print bed is either done manually or automatically on more expensive devices [2]. In either way the goal of the leveling process is to keep the nozzle of the extruder at a constant distance to the actual bed [3]. If the nozzle is too high or too low, different results may occur as seen in Fig. 4.

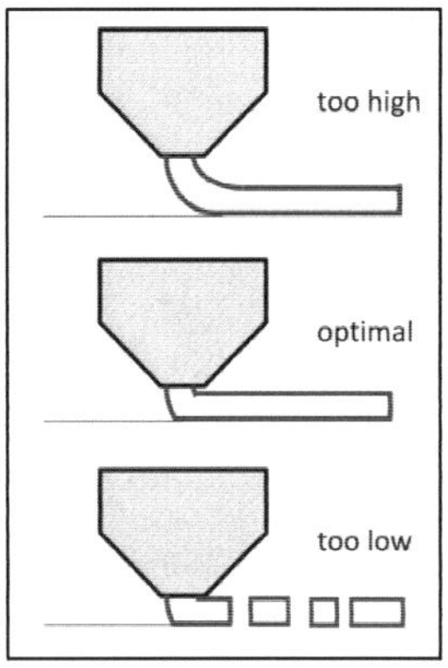

Fig. 4. Extruder nozzle distance to print bed (modified adopted from: [3])

3.1. Leveling processes and applied sensors

The manual leveling process on commercial 3D-printers is usually done using a sheet of paper. This paper is placed beneath the extruder and the actual print bed. The position between the nozzle and the bed is then reduced to a point where the paper is moving with a little bit of resistance from the pressure that occurs. This is usually the optimal distance between the nozzle and the print bed. This position is then saved and done on the next point [2]. Usually this is done on the four ends of the print bed. This manual process needs to be done every time a new print is about to be started. Due to the manual process a lot of errors can occur which leads to an imprecise method of leveling the print bed [4].

In case of an automatic print bed leveling system, a sensor is used to measure the exact distance to the print bed at a specific point [5]. There are many different sensors that can be used for this specific task. The most common sensor types are infrared, ultrasonic or proximity sensors like capacitive or inductive types. Even load or force sensitive resistor (FSR) sensors can be used for leveling a print bed [3]. Additionally, switch sensors can be applied which need to be set in front of the extruder to sense the bed and need to be removed afterwards in order to print properly. Inductive, infrared or capacitive sensors can only be used on specific bed materials. Ultrasonic and switch sensors can basically be used on any print bed surface. Furthermore, each sensor type has different results in precision which leads to a distinct quality of the actual print. Hence each sensor has its advantages and disadvantages for each application [4]. Most common 3D-printers use metal sheets as print beds.

This leads to preferable thermal properties and to the use of inductive sensors for the print bed leveling process. Big brands like for example Prusa use inductive sensors for their leveling process. Their specific inductive sensors, called PINDA, are commonly used in the 3D-printing market. The use of inductive sensors combines the advantage of precision and range. Therefore, the sensors can be positioned behind the extruder leading to no complications during the print [6].

3.2. Algorithms used for leveling print beds

Either manual or automatic print bed leveling processes rely on different algorithms which are applied to level the actual bed. The most basic algorithm typically used by common 3D-printers is a three-point leveling system. As the name applies, three different points are probed on the print bed. Thee points lead to a plane in a 3D space which can be further analyzed. This is necessary in order to get a rotational matrix which tries to describe the actual tilt of the bed as precise as possible [5]. Another typical algorithm is the planar grid approach. Hereby a grid is set up across the whole bed. The actual crossing points of the lines from the grid are the probe points for the algorithm. With the different probe points in place, a linear relation between the actual points can be achieved. This leads to more precise values depending on how fine the grid is set up [5].

A more precise application that is typically used for leveling the actual print bed is a bilinear approach. This approach is basically the same as the planar grid, but the relation between the different probe points is bilinear in this case. The precision is higher as compared to a planar approach. Hence the calculation amount of the resulting mesh is higher [5]. The goal of the algorithms is to keep the nozzle at a constant distance to the bed. If the bed has a tilted surface, then a normal print without leveling would lead to a too small distance on one side and a too large distance on the other side. The auto bed leveling algorithm changes the z-axis value of the extruder according to the position it is currently at. An example of some layers with and without leveling can be seen in Fig. 5 [7].

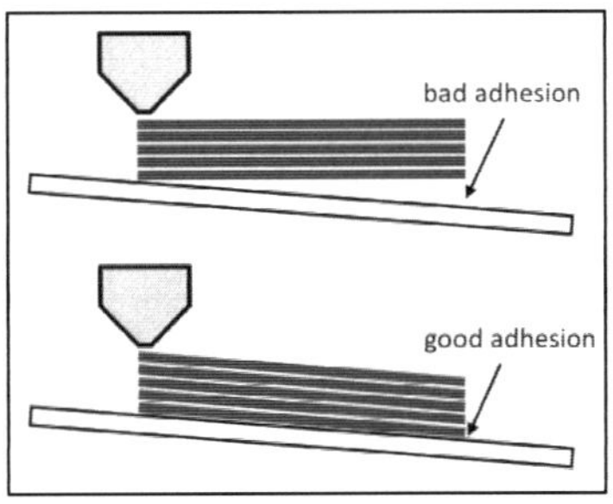

Fig. 5. Print with no bed leveling (top) and with bed leveling (bottom) (modified adopted from: [7])

3.3. Printing with robotic systems

Robotic systems are currently commonly used to act as a complete 3D-printer. Typically, 3D-printers use control units which generate the motion of the printer based on a G-code file. This G-code is generated by slicing the desired object using a so-called slicer-software. Robotic systems usually cannot use G-code information in order to move the robot directly [8]. To convert the G-code into an actual robot motion, different software options are currently applied. Big robot companies like ABB and KUKA created their own software which enables this sort of application. ABB enabled 3D-printing in their robot programming software called Robot Studio [9]. KUKA integrated their own slicer and use KUKA.CNC in order to generate the robot motion [10]. Additionally, KUKA implemented a software application called KUKA.TouchSense, which can be used to compensate irregularities on an actual workpiece. This application was created for arc welding but can actually be used for any application that requires a constant distance to a specific surface, like 3D-printing. For that purpose, a wide array of sensors can be applied to measure the distance to the object [11]. A more common used method for 3D-printing with robotic systems is the software called RoboDK. With this software it is possible to use a vast variety of robots to accomplish a 3D-printing operation. The objects can directly be sliced using an in-built slicer. Additionally, custom python scripts and plugins can be used to perform different movements based just on code ([8], [12]). In the implementations used by ABB and RoboDK, an in-built bed level system is not available. Therefore, these solutions rely on a print surface implemented using a CAD file of the print bed in the software ([9], [12]). Applying the solution offered by KUKA, an actual print bed leveling process can be achieved, using the additional application typically used for arc welding ([10], [11]).

3.4. Projection algorithms for curved surfaces

Additive manufacturing processes typically use planar slicers in order to slice the printed object in layers with a specific height. Therefore, a flat plane is used, where the object is placed and sliced [13]. Some advanced slicer algorithms even try to create different planes for different parts of the object as a means to reduce stair effects and support material during the printing process [14].

However, intending to print on a curved surface, the planar result needs to be projected using some sort of algorithm. This can either be done using just a projection or a lossless projection. Using just a projection, the actual plane gets distorted resulting in a warped object. Applying a more complex lossless projection, the result on the curved surface will resemble the actual object more precisely, with some errors still available. In an effort to achieve such a projection, the curved planes, or more precisely the triangles of the STL file, are converted into a 2D plane. Basically, the surface of the object gets unfolded. The actual print can then be projected on the 2D surface. Because of the resulting transformation matrix from the unfold process, the plane with the print can be transformed back to the actual curved surface, resulting in a lossless projection [15]. The actual process of this projection algorithms can be seen in Fig. 6.

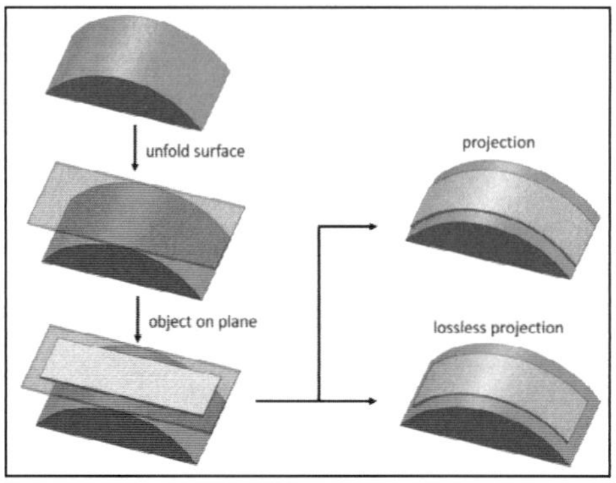

Fig. 6. Projection of a 2D object on a curved surface using normal projection and lossless projection (modified adopted from: [15])

With the usage of a projection algorithm, it is basically possible to project 2D data onto a 3D surface. The result of a sliced object are hundreds of planes stacked on top of each other. Therefore, a projection of those planes is theoretically possible. Any kind of 2D plane can be projected on a curved structure [16], which includes grid planes for leveling as well. In this case a 2D leveling grid plane algorithm could be projected on a curved object, leading to a leveling process for that kind of surfaces. The plane sliced object could then be adapted to the curved surface using the leveling results.

4. Materials and methods

In order to achieve a print bed leveling on such a scale, different methods need to be addressed. These include the specification of an appropriate sensing device, programing of the logic that can be executed on the robot and finally the adjustment of the G-code with the leveling result. Additionally, methods need to be examined in order to apply a print leveling on a curved surface.

4.1. Evaluation of applied sensors

To appropriately measure the distance to the print bed, a sensor has to be found, which enables this kind of application. Because the applied surface area of the print bed is a glass plane, the sensor selection is limited. Commonly used sensors like inductive types cannot be utilized in this situation. Therefore, other categories need to be examined. The two main sensors that come in mind when sensing glass are touch probes and ultrasonic sound devices. In both cases the accuracy of those sensors is a main criterion for the evaluation. As mentioned by [1] the absolute position accuracy of the ABB IRB 2400/16 lies at around 0,3 mm, while the repeatability has a value of 0,03 mm. In order to get reliable results with the sensor, an accuracy of lower than this amount would be perfect for the application. Ultrasonic sound sensors tend to have a quite unprecise accuracy most of the time. However, there are still products available, which have a precision in those small ranges required. The UM18 series of SICK has accuracy levels in a range that is still acceptable for the leveling application. For this case the UM18-2 Pro version was selected to be further examined. With a defined precision of about 0.069 mm, the sensor is just a little bit over the repeatability accuracy of the robot, which is acceptable in this case [17].

Furthermore, a touch probe sensor can be utilized to sense the glass plane. Commercial printers without a bed leveling sensor can often be enhanced with such a probe sensor in order to achieve an auto bed leveling result. In this case the probe sensor from ANTCLABS called BLTouch was chosen for further examination. This simple sensor can reach a precision of around 0.01 mm [18]. In this case the accuracy is way better compared to the ABB IRB 2400/16. In order to compare the two sensors and find out which of the two is more reliable for the operation, a small 3D-printed measuring station was used. In this station, the ultrasonic and probe sensor were installed and teached on a fixed distance. After that, a wall was slowly moved towards the sensor leading to an activation of the digital output. Subsequently the distance was measured and repeated numerous times. Using a normal distribution, the results were visualized and compared. The actual diagram can be seen in Fig. 7.

As expected, the touch probe provides the better reliability compared to the ultrasonic sensor on a fixed distance. Even though, the ultrasonic sensor has the advantage of sensing the bed at a safe distance, a more reliable and precise sensor is a more important factor in this case.

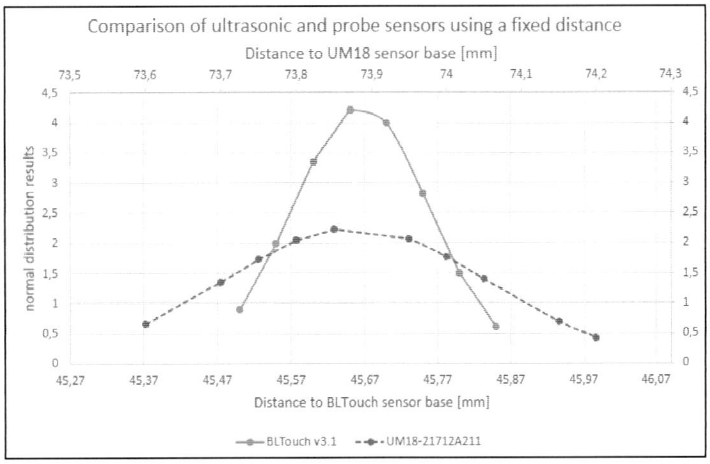

Fig. 7. Comparison of ultrasonic and probe sensors using a normal distribution

4.2. Probing algorithm

To properly achieve an automatic bed leveling process, an algorithm needs to be realized. As mentioned before, the print bed isn't shaped like a commercial printing area. Therefore, a simple level algorithm cannot be applied in this case. Hence the actual print area needs to be separated in smaller sections, on which the leveling can be achieved normally. In order to achieve that, a grid, the size of the slicer volume, needs to be pictured on the current print bed. This results in a lot of probe points depending on the fineness of the grid. However, some of the probe results are not on the actual print bed and need to be marked as invalid in this case. The remaining cross sections are therefore valid probe points and can be used for the actual leveling algorithm. Because the probing of such a huge grid will take quite a long time, the grid fineness needs to be a variable factor in order to reduce it. A finer grid will lead to a lot of points being measured using the robot. The bigger the grid distance becomes, the fewer points need to be probed, but the actual result will not be as precise. The basic concept of the probing algorithm with the specified grid can be seen in Fig. 8. It is also visible that an additional reference point needs to be set up in order to calculate the x and y positions of each probe point on the print bed.

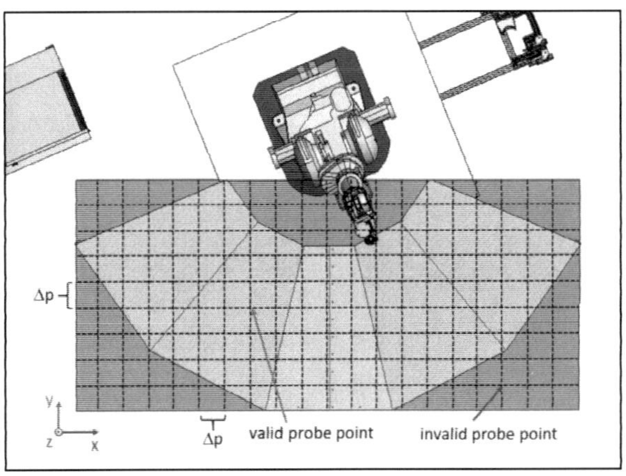

Fig. 8. Probing grid algorithm concept

4.3. Adjustment of the G-code

The result of the probing algorithm should be a lot of points with different z-values according to the sensor readings. Those results need then be applied on the actual G-code itself. Because each G-code line is a target for the robot to reach, those targets can then be adjusted according to the leveling results.

As mentioned, a grid gets projected above the print area. Each G-code target will sit somewhere in a square that results from the projection of the gird. Therefore, a bilinear approach [5] can be applied to edit the z-value of the appropriate G-code target. This approach needs to be repeated for every target that exists in the converted G-code. The result will be a leveled object on the print surface.

4.4. Leveling on a curved surface

Leveling 3D sliced objects onto a curved surface requires some additional problems that need to be solved. According to Guo et al. [15] any 3D surface can be flattened into a 2D plane. This approach then needs to be applied on the surface on which it is going to be printed on. Furthermore, the normal bed leveling algorithm needs to be expanded to handle curved surfaces. What this means is, that the different probe points which result on the flattened surface need to be projected back to the surface of the object as described by Luo & Tseng [16]. After that the probing needs to be achieved with the appropriate orientation of the nozzle resulting from the curvature. The position of the extruder and in this case also from the sensor needs to be orientated accordingly in order to not scratch or even destroy the surface or object. After that, the G-code, which was sliced on a plane, needs to be adapted with the new z-values resulting from the curved surface leveling. Each point between the projected grid points needs to be interpolated not just for the z-value but also for the orientation between the four neighboring grid points. Therefore, the grid on a curved object needs to be as fine as possible to reduce orientation errors. This complex procedure then needs to be programed and executed using RoboDK, which itself creates a whole new number of problems, resulting from the application programing interface (API) limitations.

5. Practical realization

The actual implementation of the leveling process requires different tasks to be solved. Those include mounting options, sensor readings, the RoboDK API implementation as well as handling the actual robot movements without damaging the surface of the print area. Additionally, the results of the leveling process need to be implemented in the current operation of the 3D-printing station without changing the procedure.

5.1. Sensor mounting and output reading

The BLTouch sensor consists of a base body and a pin that gets extracted to a probe position. If the pin gets pushed in, the sensor outputs a signal. The actual length of the pushed pin is 6.6 mm [18]. Therefore, the sensor needs to be mounted in such a position in which it can detect the surface before the nozzle touches the glass and retract behind the nozzle in order to not disturb the actual printing process. Because the maximum reach of the probing pin is such a small length, the mounting needs to be very precise. Additionally, the pin needs to be pushed in approximately 1 mm to send the signal. According to ANTCLABS [18] the sensor needs to be placed at around 2.3 to 4.3 mm behind the nozzle to achieve good results and not disturb the printing operation. Therefore, a mounting plate was constructed which can be fixed onto the Noztek Xcalibur assembly. The actual mounting solution can be seen in Fig. 9. It is also important to mention, that the touch distance and actual trigger distance are marked with planes. The green one marks the distance at which the probe pin touches the glass, the red one marks the actual triggering distance of the sensor.

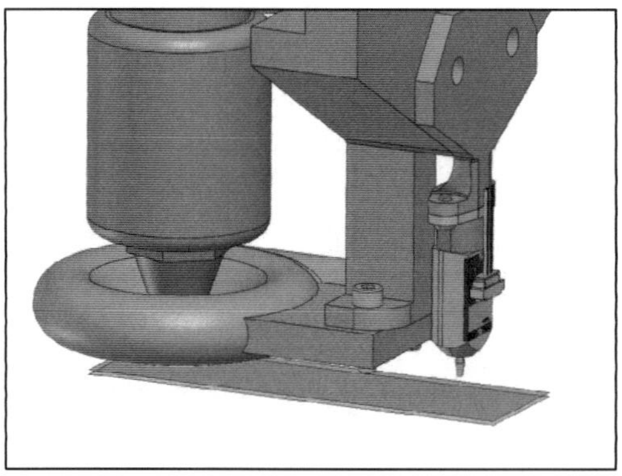

Fig. 9. Mounting point on the Noztek Xcalibur extruder using a 3D-printed plate

To get the real z-value from the sensor, a small calculation of the resulting offsets needs to be applied. To better understand the mounting procedure, a graphical interpretation can be seen in Fig. 10. Because the print bed surface isn't positioned perfectly even, it may occur that the teached in z-value of the bed is higher or lower than expected. Therefore, the subtraction of those offset measurements has to be done in order to receive the real z-value of the print bed at any given probe point.

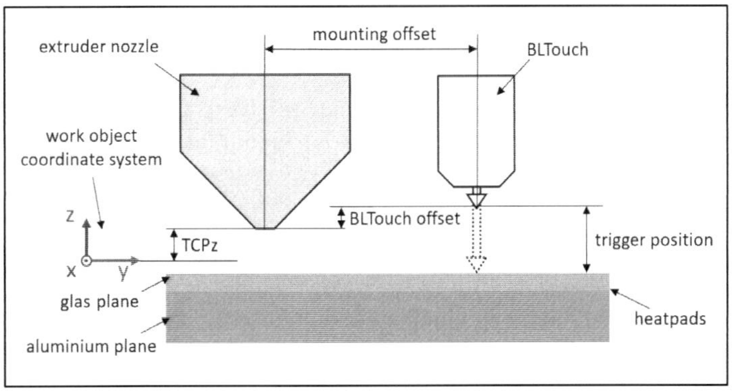

Fig. 10. Interpretation for the different offsets occurring with the mounting of the sensor

To actually receive the signal from the BLTouch, an Arduino board needs to be set up appropriately. Therefore, a small script was implemented, which sends a serial command when the sensor activates via USB to the Laptop on which RoboDK is running. The actual instruction to push the pin down and up again is achieved via PWM signals. Additionally, the script pulls the pin up as soon as the sensor triggers.

5.2. RoboDK API implementation

The actual leveling algorithm was implemented using the RoboDK C++ API. For that, the plugin which is already used for printing was utilized as a base for the implementation. The leveling process was programed in such a way, that the current plugin is not affected, if no leveling is applied. Therefore, an additional section was created in the interface for the user. The user has a few parameters which can be changed to perform the leveling process. These include setting the grid distance in mm as well as an offset to the border of the print surface. This is necessary because with a specific grid distance value it is possible to exactly place a probe point on the corner of the print bed, which would lead to bad results at that position. For this case, a fixed offset is already present, but can be changed according to the user.

With the parameters set, the actual leveling algorithm is set in motion. This is done using the "Calculate Leveling Mesh" button. On pressing the button, the leveling routine is called. The sequence starts with the mesh being created. Therefore, the grid distance is applied. With the mesh created, the actual probe points get calculated, depending on the offset defined. After that the leveling program gets created. This includes every probe point target on the print bed with the appropriate orientation to the robot as well as the offsets to the sensor and position on the surface of the bed. An example of the result visible in RoboDK can be seen in Fig. 11.

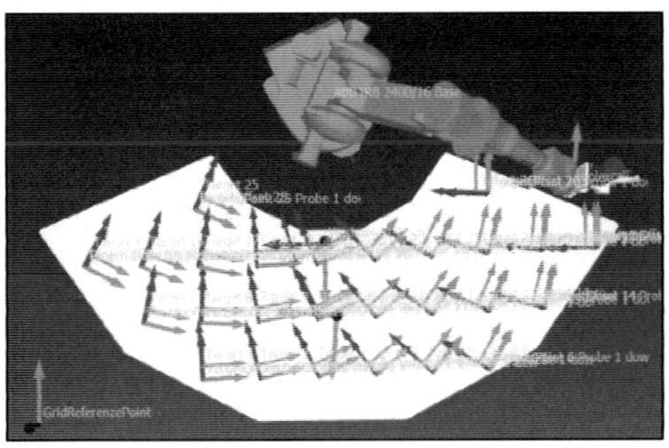

Fig. 11. Top view of RoboDK with the resulting probe points of the leveling algorithm using a grid distance of 300 mm with an offset of 100 mm

One disadvantage that is present when using the RoboDK's C++ API is, that code can only be executed once and not during the whole operation of a task. If the function, which is executed on the press, doesn't finish, or stays in an infinite loop, RoboDK will be unresponsive or even crash. Therefore, a solution has to be found, in order to read sensor inputs during the operation of the robot. For this case, RoboDK supports python scripts [12]. Python scripts can be executed via the button press and can run during the operation of the actual leveling program. As soon as the leveling begins, a custom python script is launched where a connection to the sensor is being established. If the sensor detects the ground while probing, the current tool center point (TCP) value is used for further calculations. One constraint that occurred during the implementation on the robot was, that the TCP value cannot be read during the operation of the robot. This is because of the RoboDK communication socket only supporting one connection to the robot at a time.

In order to overcome this problem, the simulation of RoboDK was used. TCP values can be accessed via the simulation even if the robot is currently executing a task. As a means to reduce simulation and physical TCP errors, the simulation was synchronized with the robot motion before each probing point. Therefore, the task was stopped before the next probe point, the simulation was synchronized with the real robot and after that, the task was continued. This results in the python script to receive the values from the simulation during the operation of the actual task being executed on the robot.

5.3. Applying leveling results

With the leveling results in place, it is possible to apply them to the actual G-code targets. If the user ticks the leveling checkbox, each target gets examined during the path planning algorithm for the robot. If the target is positioned between four neighboring points, a bilinear interpolation of the z-values is performed. This is done for every target resulting from the G-code. If the user positions the object way outside of the grid area, but still on the print bed, different edge cases are applied. Those basically include the outside area of each side from the grid. On such a case, the remaining points that are not in the grid get generated with a z-value of zero. Therefore, an interpolation to the teached in zero level is applied outside of the probe point area. This results in similar results as printing without the leveling in process.

6. Results

Following the implementation of the bed leveling system, different results can be examined. These include the deformation of the actual print bed, the appliance of the leveling results to the G-code and even the quality increase of the resulting objects printed with the system.

6.1. Print bed leveling results

Applying the implemented auto bed leveling system onto the real print bed, results in the actual deformation of the bed. Depending on the user defined values set in the interface, different results with an appropriate fineness are achieved. Even though a higher fineness results in a better leveled bed, the overall unevenness of the print bed is evident in every result. To better visualize the actual deformation, a leveling process was done using a grid distance of 150 mm which is a small step compared to the whole print bed. The results of the actual bed can be seen in Fig. 12. As seen in the resulting bed structure, an offset range of about 3 mm occurs from the lowest to the highest point. It is also clearly visible that the area where the print bed was teached in, in this case the middle section of the bed, results in a negative value to the actual position. This behavior occurs because the teaching of the bed cannot be done perfectly by a manual approach. The further the probe points are away from the teached zero point, the more severe the unevenness becomes. Printing an object on the far side of the print bed to the left or right without using the leveling system would result in the nozzle touching or even damaging the glass. Printing in those areas would result in severe failures of the object at those areas. A proper fist layer adhesion in those areas would not be possible using a non-leveled bed.

Comparing a finer grid to a rougher probing setup shows that the overall unevenness is the same. Even though fewer probe points were applied, the interpolation allows to adjust the areas where no probing was done at all. The rougher the user defined grid is, the faster the actual algorithm is executed. The resulting grid with a grid distance of 300 mm can also be seen in Fig. 12. The same result can be seen on an even rougher grid setup, where the offset was set at about 450 mm. The result can be seen in Fig. 12. This process took about 13 min to finish on the actual robot.

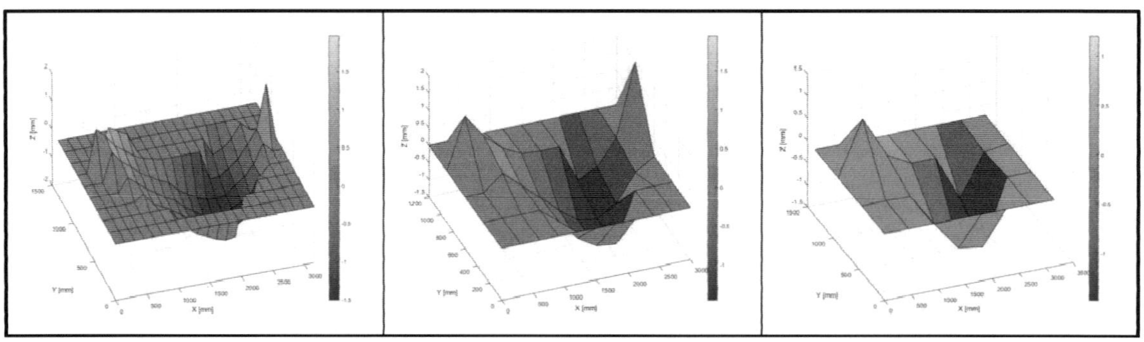

Fig. 12. Bed leveling results, 150 mm grid distance (left), 300 mm grid distance (middle), 450 mm grid distance (right)

Even though a finer grid setup takes longer to execute than a rougher one, probed grids can always be used as many times as needed and don't need to be redone every time before the print. However, executing a probe setup after a specific time, for example when the temperature in the room heats up during summer, will result in more precise values than using a few month-old results.

6.2. G-code results

If the user decides to apply the latest bed leveling results onto the G-code, the interpolation process begins on pressing the calculate path button. Even though the interpolation is done using a lot of coordinate system transformations, the actual time difference is not relevant. A geometry with a high number of G-code lines was used to measure the differences. Calculating the robot motion without using the bed leveling results in a time of 19 minutes and 41 seconds. Applying the same procedure but using the leveling system with integrated interpolation results in a time consumption of 20 minutes and 3 seconds. The overall random-access memory (RAM) usage was the same for both procedures. It is to mention that simple structures with less G-code lines take just a few minutes to complete.

In order to see the results in the simulation itself, arbitrary numbers were written into the leveling results. Those numbers were set at high values to see the G-code bend around the different probe points. In reality, those numbers are so small that there is no difference visible in the actual simulation representation. Examining the simulation results with the respective real values however shows that even small probing results adjust the G-code. The arbitrary G-code manipulation with the self-defined values can be seen in Fig. 13. In this case a simple circle structure was used to show the results. The curvature resulting from the interpolation of the different probe points is clearly visible.

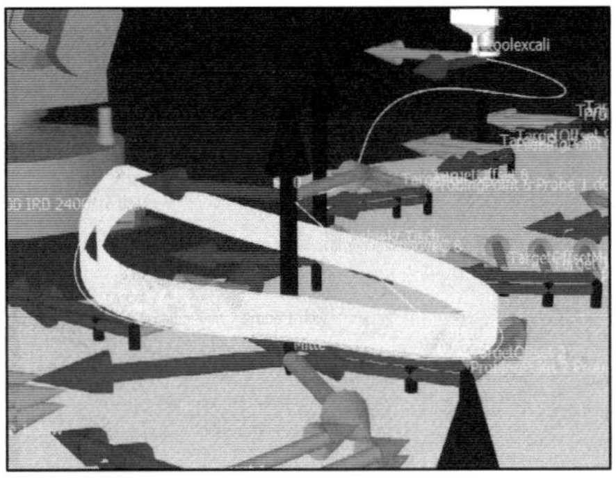

Fig. 13. Applying arbitrary probe results to the G-code in the simulation

6.3. Printing results using the bed leveling system

Printing with the enabled leveling system allows the extruder to keep an almost constant distance to the actual bed. The extruded material can therefore stick better to the glass surface. Without the bed leveling system active, the extruder drops the material onto the surface without applying pressure, resulting in grooves between the lines. In order to visualize this effect, a hexagonal object was printed on the same spot with and without the system active. The results can be seen in Fig. 14. The grooves or ripples occurring because of the high distance between the nozzle and the bed can be clearly seen. Keeping a constant distance with the leveling system, almost removes those ripples entirely. This in contrast increases the area of the object touching the bed resulting in a better adhesion property.

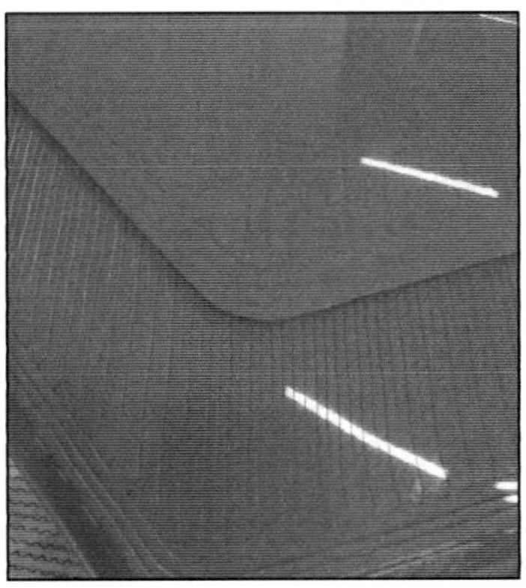

Fig. 14. Surface between object and print bed (top: with bed leveling system, bottom: without bed leveling system)

Furthermore, using the standard extruder settings and applying the leveling systems, results in too much material being extruded. This is because of the height difference that occurs during the operation without the leveling in place. The station was optimized to perform well without the leveling system and is therefore extruding more material than actually needed. Printing with a leveled object results in the material being expanded and pushed away from the extruder. Hence the rotational speed of the extruder needs to be set to an appropriate amount.

Fig. 15 shows the difference between some extruder properties applied during printing. Using the normal value of 57 1/min results in ripples on the surface of the object. These ripples can be 1 – 2 mm high, which reduces the quality of the print. Printing with almost half the speed of about 30 1/min results in a much smoother surface, without losing the actual stiffness of the object. This is because the extruder keeps the optimal distance to the surface, resulting in less material needed, compared to a higher gap. Different rotational speeds were tested ranging from 15 – 30 1/min.

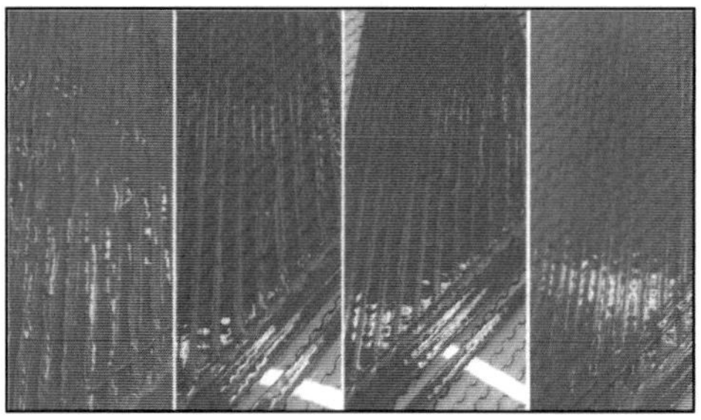

Fig. 15. Print quality with different extruder rotational speeds (from left to right: 1: 57 1/min [standard speed], 2: 30 1/min, 3: 20 1/min, 4: 15 1/min)

Additionally, the lines which follow the first layers were examined using two different rotational speed values. As in Fig. 16 depicted, printing with 15 1/min leads to the lines not sticking to each other whereas 25 1/min shows better results in this regard.

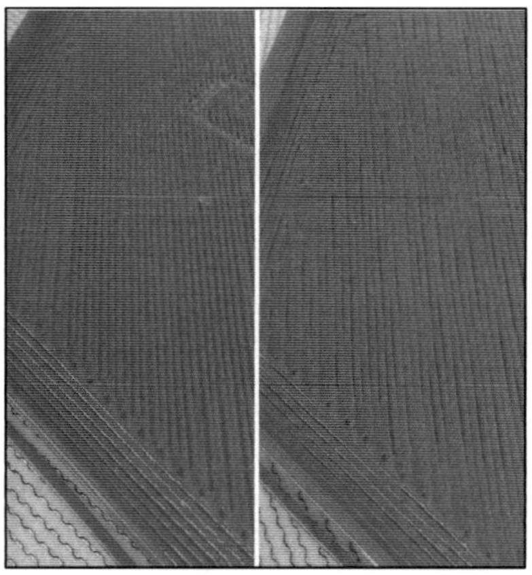

Fig. 16. Printing lines difference (left: 15 1/min, right: 25 1/min)

As examined from the printing tests, using 30 1/min results in an almost optimal printing quality and adhesion characteristic. Reducing the parameter allows for less ripples in the first layers but reduces the line thickness in the following ones. Printing with less than 15 1/min using the bed leveling system would result in too little material being extruded during the process. However, using the system requires overall less material in order to gain an almost equal result. Additionally, it has been examined that the layer height of a non-leveled print results at about 2 mm but should actually be 1.5 mm. This again is because of the gap between the extruder and the actual surface of the object. Comparing the layer height to a leveled object shows that they are at a constant 1.5 mm height throughout the object as it should be.

7. Conclusion and outlook

The implemented auto bed leveling system visualizes the uneven behavior of the customized print bed used in the Digital Factory of the UAS Technikum Wien. It allows the user to take advantage of the whole area which the bed provides, without sacrificing print quality because of deformations resulting from mechanical and thermal expansion properties. The resulting probe values clearly show the high amount of deformation happening at the sides of the print bed. Without a leveling system in place, printing in those areas would result in an unsuccessful print or even worse, to a damage on the glass on top of the actual bed. Printing with the system active has shown that the actual surface area between the object and the print bed increases. Therefore, a stronger adhesion between the print bed and the object can occur. Using the system allows the extruder to keep an almost constant distance to the print bed. Because of this, the actual amount of material that was used without the system needed to be reduced drastically in order to achieve optimal results. Hence a material decrease of about 40 % was possible without sacrificing in print quality.

Applying the leveling approach on curved surfaces would result in a similar approach. The leveling needs to be adjusted according to the actual tilt of the probe point on the curved object. Planar sliced objects can then be projected via the resulting mesh on the actual object itself. In order to achieve the required adhesion properties on an actual object, a leveling system is unavoidable. Even though the applied bed leveling system works for the defined print bed, applying it to curved objects still requires a lot of work to solve occurring problems. Additionally, ways to implement non-planar printing could be examined in future work. Furthermore, optimizations for different parameters like robot speed and extruder rotation could be examined resulting in an optimal operation during the print. The combination of a bed leveling system for plane and curved surfaces, as well as non-planar approaches during the print would result in a highly flexible station to create different kinds of objects for various applications.

8. References

[1] ABB, (2019). ROBOTICS IRB 2400 Industrial Robot. [Datasheet] Zurich, Switzerland: ABB. Available under: < https://new.abb.com/products/robotics/industrial-robots/irb -2400/irb-2400-data> [Access on 15.9.2021]

[2] Lianghua, Z. & Xinfeng, Z., (2019). 3D Printer Self-adaptive Levelling Algorithm and Structure Design. In: 4th International Conference on Mechanical, Control and Computer Engineering (ICMCCE), 25.-27. October, Hohhot. Hohot, China: ICMCCE, pp. 1054-1056.

[3] Huiyang, Y., (2019). Improving Desktop FDM 3D printer by User-Centered Design. [Masterthesis] Espoo: Aalto University Department of Design.

[4] Heras, E. S., Haro, F. B., De Augustin Del Burgo, J. M. & Islan Marcos, M. E., (2017). Plate auto-level system for fused deposition modelling (FDM) 3D printers. Rapid Prototyping Journal. 23(2), pp. 401-413.

[5] Hsieh C.-T., (2018). Developing Low Cost and Smart Laser-based Paper Cutter based on Open Source 3D Printer. In: IEEE International Conference on Advanced Manufacturing (IEEE ICAM), 16.-18. November, Yunlin. Yunlin, Taiwan: IEEE ICAM, pp. 109-112.

[6] Prusa, (2018). 3D Printing Handbook User Manual For 3D Printers -Original Prusa i3 MK3 KIT -Original Prusa i3 MK3. [Datasheet] Prague, Czech Republic: Prusa Research. Available under: <https://help.prusa3d.com/en/article/3d-printing-handbooks_125045> [Access on 5.9.2021]

[7] Stevenson, K., (2017). Wondering About 3D Print Auto-Leveling. [Online] Available under: <https://www.fabbaloo. com/2017/06/wondering-about-3d-print-auto-leveling> [Access on 5.9.2021]

[8] Pollak, M., Török, J., Zajac, J., Kocisko, M. & Teliskova, M. (2018). The Structural Design of 3D Print Head and Execution of Printing via the Robotic Arm ABB IRB 140. In: 5th International Conference on Industrial Engineering and Applications, 26.-28. April, Singapore. Singapore, Singapore: ICIEA, pp. 194-198.

[9] ABB, (2020). RobotStudio 3D Printing PowerPac Digital manufacturing with state-of-the-art 3D printing. [Datasheet] Zurich, Switzerland: ABB. Available under: < https://new.abb.com/products/robotics/de/applikationssoftware/3d-printing-powerpac> [Access on 7.9.2021]

[10] KUKA, (2021a). Additive manufacturing and industrial 3D printing. [Online] Available under: <https://www.kuka. com/en-at/products/process-technologies/3d-printing> [Access on 7.9.2021]

[11] KUKA, (2021b). KUKA.TouchSense. [Online] Available under: <https://www.kuka.com/en-at/products/robotics-systems/s oftware/application-software/kuka_touchsense> [Access on 28.9.2021]

[12] RoboDK, (2021). Robot 3D Printing Project. [Documentation] Available under: <https://robodk.com/doc/en/Robot-Machining-Robot-3D-Printing-Project.html> [Access on 7.9.2021]

[13] Shan, Y., Gan, D. & Mao, H. (2021). Curved Layer Slicing based on Isothermal Surface. In: 49th SME North American Manufacturing Research Conference, 21.-25. June, Cincinnati. Cincinnati, USA: NAMRC, pp. 484-491.

[14] Wu, C., Dai, C., Fang, G., Liu, Y-J. & Wang, C., (2020). General Support-Effective Decomposition for Multi-Directional 3-D Printing. Transactions on Automation Science and Engineering. 17(2), pp. 599-610.

[15] Guo, J., Liu, G., Zhang, G., Guan, Y. & Zhang, H., (2015). Projection algorithm for 3D laser marking. In: International Conference on Robotics and Biomimetics, 6.-9. December, Zhuhai. Zhuhai, China: ROBIO, pp. 2199-2204.

[16] Luo, R. C. & Tseng, P., (2017). Carving 2D image onto 3D curved surface using hybrid additive and subtractive 3D printing process. In: International Conference on Advanced Robotics and Intelligent Systems (ARIS), 6.-8. September, Taipei. Taipei, Taiwan: ARIS, pp. 40-45.

[17] SICK, (2021). Ultrasonic sensors UM18 / UM18-2 Pro. [Datasheet] Waldkirch, Germany: SICK. Available under: <https://www.sick.com/at/en/distance-sensors/ultrasonic-sensors/um18/um18-21712a211/p/p282744> [Access on 30.10.2021]

[18] ANTCLABS, (2021). BLTouch :: Auto Bed Leveling Sensor for 3D Printers. [Datasheet] Seoul, Republic of Korea: ANTCLABS. Available under: < https://www.antclabs.com/bltouch-v3> [Access on 30.10.2021]

DOI: 10.2507/34th.daaam.proceedings.058

INFLUENCE OF DIRECT METAL LASER SINTERING ON THE ULTRASONIC VELOCITY MEASUREMENTS

Maria Grozdanic & Morana Mihaljevic

This Publication has to be referred as: Grozdanic, M[aria] & Mihaljevic, M[orana] (2023). Influence of Direct Metal Laser Sintering on the Ultrasonic Velocity Measurements, Proceedings of the 34th DAAAM International Symposium, pp.0448-0454, B. Katalinic (Ed.), Published by DAAAM International, ISBN 978-3-902734-41-9, ISSN 1726-9679, Vienna, Austria
DOI: 10.2507/34th.daaam.proceedings.058

Abstract

As a result of additive manufacturing, where the part build-up is usually done layer by layer, parts have an anisotropic material behaviour, which directly affects the ultrasonic velocity. In ultrasonic testing, the ultrasonic velocity is one of the most important parameters to be adjusted to obtain the most accurate test results for the tested part, and it must be uniform throughout the whole volume of the part. This paper investigates whether the longitudinal wave velocity in a specimen, manufactured using the direct metal laser sintering process (DMLS), depends on the test direction. The ultrasonic velocity is determined in two perpendicular directions: the layer direction and the build-up direction. Detailed statistical data analysis is performed to determine whether there was a statistically significant difference between ultrasonic velocities in the same test direction (paired t-tests) and between the ultrasonic velocities at perpendicular directions (one-way ANOVA).

Keywords: ultrasonic velocity measurement; additively manufactured specimen; directional dependence; statistical analysis.

1. Introduction

Additive manufacturing (AM) is an advanced production technology that enables the production of objects from 3D model data through the process of joining materials, usually layer by layer. The use of non-destructive testing (NDT) on additively manufactured parts is increasingly common for determining various mechanical and physical properties of materials since they do not damage AM parts during testing [1]. Ultrasonic testing (UT) has found its application in the characterisation of AM parts as well [2], [3]. Material characterization using UT enables the determination of changes in acoustic parameters, which occur due to the propagation of ultrasonic pulses through the test specimen and their interaction on different microstructure. Commonly, UT is used to determine elasticity constants, microstructure, irregularities, and mechanical properties [4]. One of the relevant features for material characterisation is the ultrasonic velocity of the ultrasonic pulse [5], as an important acoustic parameter. The ultrasonic velocity is related to the elastic constants and density of the material, so it provides information about the mechanical, anisotropic, and elastic properties of the medium through which it passes. Several elastic properties of materials, such as Young's modulus of elasticity E, shear modulus G, and Poisson's ratio v, can be calculated with the ultrasonic velocities.

Honarvar et al. [1] made a review of ultrasonic testing applications in the material characterisation of AM parts. They stated that the main reasons for the effectiveness of the UT in the characterisation of AM parts are used additive manufacturing process and the capabilities of used ultrasonic technique.

1.1. Influential parameters of additively manufactured parts on the ultrasonic velocity

Before conducting ultrasonic testing, it is necessary to determine the ultrasonic velocity in the test specimen. The ultrasonic velocity is a characteristic of every material, and it results from the microstructure and metallurgical state of the material, as well as external influences such as temperature and stress. Consequently, a change in ultrasonic velocity in the same specimen can occur within the same material if the material is not a homogeneous structure [6]. Even though AM has many benefits, such as higher efficiency, producing highly optimized parts, flexibility in the development and production, production of prototypes, production of parts that are not impossible to make with traditional manufacturing processes, et cetera, AM also has disadvantages, such as anisotropy, heterogeneity, porosity, surface roughness. Porosity [7], mechanical properties of the material [8], and the degree of anisotropy [9] are highly dependent on process parameters, device settings, and the material in use. Castro-Sastre et al. [10] investigated the influence of used materials. They stated that properties of used material for AM, such as morphological shape and size of the powder particles, chemical composition, and existence of different phases, affect the manufacturing process and the properties of AM parts.

Porosity significantly impacts the properties of AM parts and their application. Pores are stress concentrators with a higher probability of initiating fracture and affecting mechanical properties; therefore, Cook et al. [11] investigated the effect of porosity on ultrasonic velocity. They concluded that the ultrasonic velocity waves decrease with an increase in the proportion of pores. To the same conclusion came Huang et al. [12]. The mechanical properties of components produced by additive manufacturing depend on the material and process parameters; therefore, the 3D printer must correctly adjust before production. The ultrasonic device used limits the settings of production parameters. Different parameters during the additive manufacturing process, such as layer thickness, production speed, laser power, deposition energy [13], and hatch spacing [7], affect the obtained specimen's mechanical properties, ultrasonic velocity, and microstructure. Javidrad et al. [9] investigated the effects of scanning speed, laser power, and the distance between the laser beam paths on the elastic constants of materials in different directions regarding the building direction. The elastic constants of the material were determined by measuring the ultrasonic velocity inside the specimens.

Anisotropic materials have direction-dependent properties. Anisotropy, in additively manufactured specimens, is related to the building direction, where the build-up direction is generally the weakest [14]. Nearly all additive manufacturing processes, due to the nature of building parts, typically material being added together layer by layer, have different properties in the build-up direction than those in the layer direction. Therefore, when producing specimens, it is necessary to take care of the orientation of the specimen regarding the building direction and the required function of the specimens to obtain desired values of its properties. Pandzic [15] investigated the influence of build orientation, layer height, and post-curing on AM part's elastic modulus and tensile strength. He concluded that different combinations of those three parameters lead to different observed property values. It is clear that because of anisotropy in the microstructure, anisotropic mechanical properties occur [16]. Therefore, the ultrasonic velocity can also differ significantly in different directions. Several authors investigated the dependence of the velocity of ultrasonic waves in different directions regarding the building direction in additively manufactured specimens. The specimens were made of different metal materials with different additive manufacturing processes and different process parameters. Lin et al. [17] measured the time-of-flight of ultrasonic waves and calculated the L-wave velocity in perpendicular directions in the specimens (in the build-up direction and the layer direction) on specimens of 316L stainless steel made by selective laser melting (SLM) process with different power laser (from 175 W to 250 W). They determined that the L-wave velocity in the layer direction is higher than in the build-up direction. Also, they found that increasing the laser power increases the density of the specimens. Consequently, with an increase in the density of the material, L-wave velocities increase. Kim et al. [7] investigated the influence of the distance between the laser beam paths on the phase velocity of specimens made of 316L stainless steel manufactured with the Laser Powder Bed Fusion process. They concluded that the phase velocity is always lower in the build-up direction, compared to the layer direction, for all distances between the paths of the laser beam. Sol et al. [18] investigated the potential anisotropy in additively manufactured AlSi10Mg specimens by the SLM process using the Pulse-Echo technique. Measuring and calculating the ultrasonic velocities in different directions, they concluded that anisotropy appears only in the transverse wave ultrasonic velocities, and L-wave velocities are almost the same in all directions. The success of ultrasonic testing depends on the adjustment of the ultrasonic device. Therefore, before conducting ultrasonic testing on AM specimens, it is necessary to determine the ultrasonic velocity value in the testing direction so could the ultrasonic device be adjusted.

2. Methods

The additively manufactured specimen, shown in Fig. 1, is made of EOS ToolSteel 1.2709 (X3NiCoMoTi 18-9-5) on the EOSINT M280 3D printer from EOS GmbH with Direct Metal Laser Sintering (DMLS) manufacturing process. The DMLS process is based on the melting and sintering fine metal powder using a laser beam, layer by layer, to create a specimen of the desired characteristics and dimensions. The dimensions of the specimen are (140×15×15) mm.

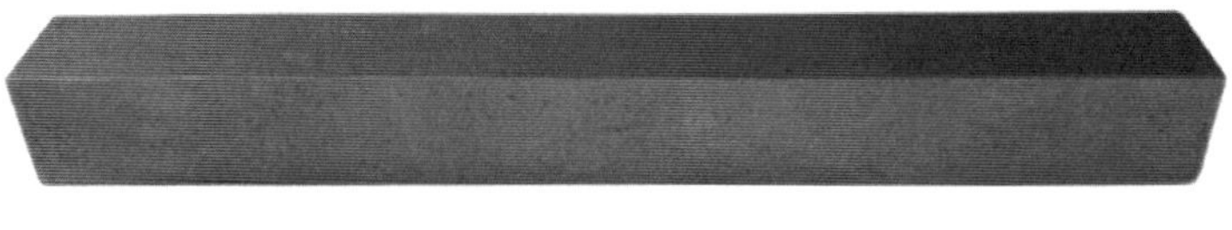

Fig. 1. Additively manufactured specimen

The time-of-flight (t_{TOF}) measurements in the experimental part of this work were carried out with the following ultrasonic system (Figure 2):

- Probe: K5N
- Device: Krautkrämer USM 36,
- Oscilloscope: LeCroy 9310AM
- Couplant: gel
- Technique: pulse overlap

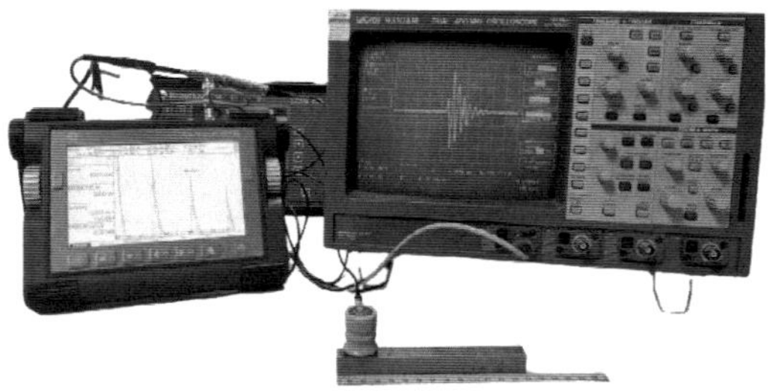

Fig. 2. Ultrasonic system

2.1. Ultrasonic velocity measurement

Ultrasonic velocity is determined by measuring the time-of-flight (t_{TOF}) and knowing an ultrasonic path (s). The ultrasonic path is the distance travelled, by the ultrasonic pulse, between two consecutive echoes from the back wall. It corresponds to the double thickness of the specimen. The velocity of ultrasonic longitudinal waves (v_L) calculates according to the equation:

$$v_{\mathrm{L}} = \frac{s}{t_{\mathrm{TOF}}} \tag{1}$$

Time-of-flight measurements were performed at two measurement places in the layer direction (x-axis) – M1 and M2, and at two measurement places in the build-up direction (z-axis) – M3 and M4. The measuring places are 20 mm from the edges of the specimen, and the central axis of the beam is 7.5 mm from both sides of the edges. The positions of the measuring places are marked in Figure 3.

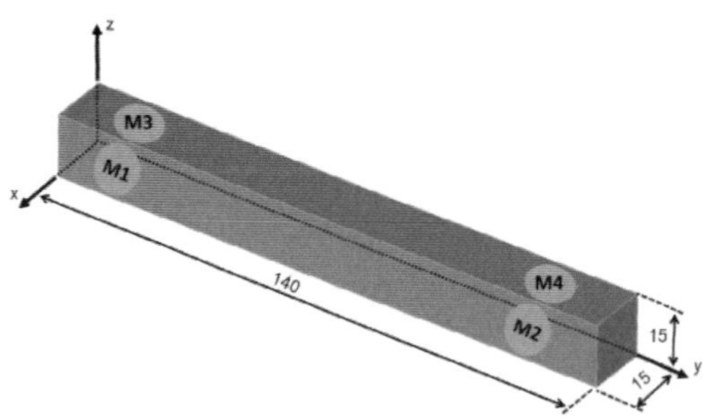

Fig. 3. Test axes and display of measurement places

The Pulse overlap technique is used. t_{TOF} is measured by overlapping ultrasonic pulses between echos from the back wall of the AM specimen. On the overlap between the first and the second echo from the back wall, five points were selected where t_{TOF} was measured. Measurements were repeated five times at the same measuring place. Therefore, 25 measurements were made at each measuring place, and the arithmetic means are presented in Table 1.

Measuring place	Time-of-flight, t_{TOF} µs
M1	5.3822
M2	5.3689
M3	5.3379
M4	5.3375

Table 1. Time-of-flight at measuring places

The thicknesses of the measuring places were measured with Mitutoyo LH-600 digital altimeters. At each measuring place, thickness measurements were performed three times, and the arithmetic means are shown in Table 2. Table 2 also presents ultrasonic paths.

Measuring place	Specimen thickness, d mm	Ultrasonic path, s mm
M1	15.0913	30.1825
M2	15.0570	30.1141
M3	14.9110	29.8221
M4	14.9126	29.8252

Table 2. Thickness and ultrasonic paths of the measuring places

3. Results and discussion

L-wave velocities are calculated using measurement results of t_{TOF} and specimen thickness d using (1). Type A standard measurement uncertainty was estimated according to the GUM method. The expanded measurement uncertainty is $U = 2.5$ m/s, with the coverage factor $k = 2$ and coverage probability $P = 95$ %. The results of the ultrasonic velocities with the expanded measurement uncertainty are expressed in Table 3.

Measuring place	L-wave velocity, v_L m/s
M1	5608 ± 2.5
M2	5609 ± 2.5
M3	5587 ± 2.5
M4	5588 ± 2.5

Table 3. Time-of-flight at measuring places

To determine whether there is a statistically significant difference between ultrasonic velocity in the same direction at two measuring places and different directions detailed statistical data analysis was performed using the Minitab software package. The results of the L-wave ultrasonic velocities are statistically processed and graphically presented below. The box plots provide a visual summary of the data and show how the values in the data spread out. Thus they were made before statistical analysis.

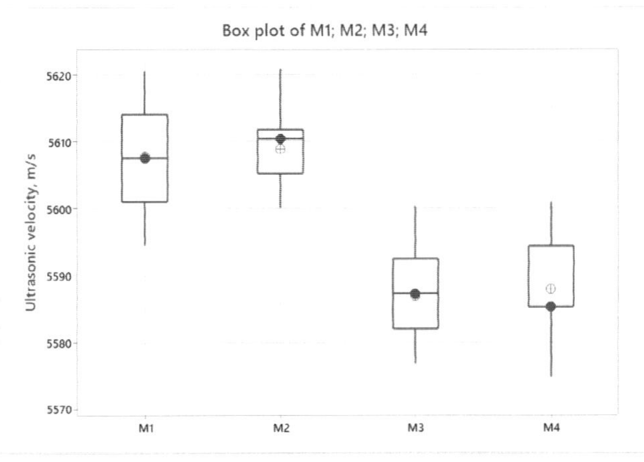

Fig. 4. Box plots for the ultrasonic velocities

Figure 4 shows the box plots for the ultrasonic velocities at all four measuring places (M1 – M4). The box plots show the minimum, the maximum, the lower and upper quartiles, the median (full circle), and the mean (circle with a cross). The graph shows that the measurement places in the same test directions (M1 and M2 in the direction of the x-axis; M3 and M4 in the z-axis direction) have similar velocities, while the difference is more significant in different directions. The difference in mean values of L-wave velocities between measurement places in the same direction is 1 m/s, while the differences in mean velocities in perpendicular directions and at the same ends of the specimen (comparison M1 and M3; comparison M2 and M4) are 21 m/s. For statistical analysis, it was necessary to check the normal distribution of the specimen data. Testing whether the data has a normal distribution was performed with the Anderson-Darling test.

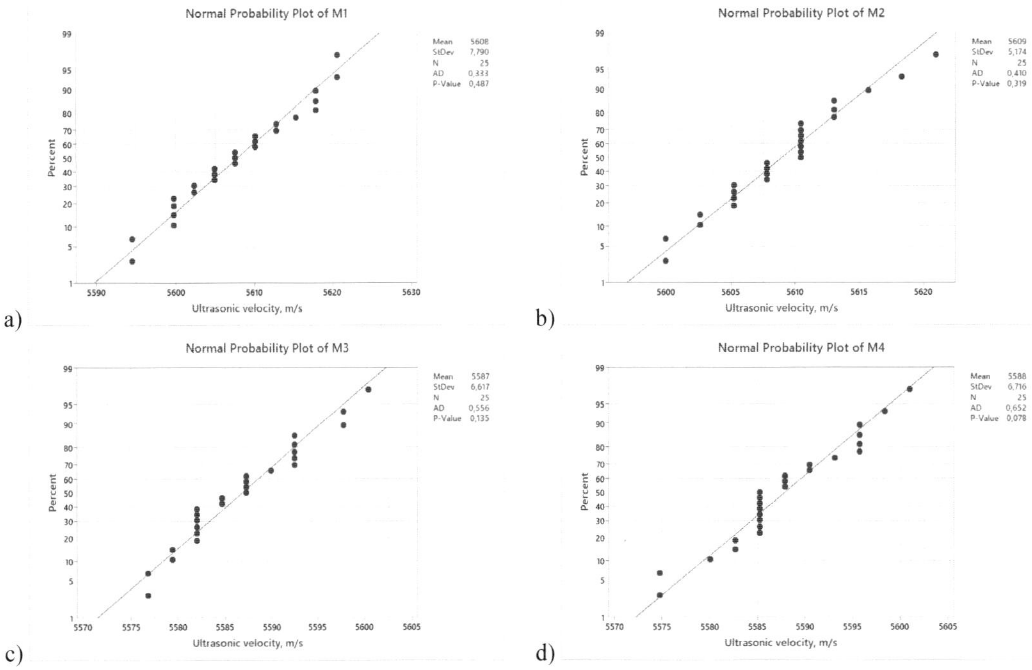

Fig. 5. The normal probability plots of ultrasonic velocities at measurement places: a) M1; b) M2; c) M3; d) M4

The Anderson–Darling test established the normal distribution of the results at all four measurement sites (Figure 5), as *P*-values at all four measuring places are higher than the significance level (*P*-value > 0.05). The null hypothesis (which assumed that results follow a normal distribution) is accepted with the error probability of $\alpha = 5\%$. To determine whether there is a statistically significant difference between the L-wave velocities means in the same direction, but at different measurement places (comparison M1 and M2; comparison M3 and M4), paired t-tests were performed. The results of the t-tests are presented in Table 4.

Measuring places	Mean	Standard deviation	Standard error of the mean	*t*-value	$t_{crit.}$	*P*-value
M1 and M2	-1.13	8.23	1.65	-0.69	2.06	0.498
M3 and M4	-1.01	11.87	2.37	-0.42	2.06	0.675

Table 4. Results of Paired t-tests

The estimate for the population mean difference in ultrasonic velocities between M1 and M2 is -1.13, with 95% confidence that the population mean difference is between -4.53 and 2.26. Comparing the obtained *P*-value, which is 0.498, to the significance level $\alpha = 0.05$, it is evident that the difference between the means is not statistically significant (*P*-value > α). The decision is to fail to reject the null hypothesis and to conclude that the difference between the means of ultrasonic velocities in the layer direction is not statistically significant. The same analysis is performed for the build-up direction. The estimate for the population mean difference in ultrasonic velocities between M3 and M4 is -1.01, with 95% confidence that the population mean difference is between -5.90 and 3.89. Since the *P*-value of 0.675 is higher than the significance level $\alpha = 0.05$, the difference between the means is not statistically significant. The decision is to fail to reject the null hypothesis and to conclude that the difference between the means of ultrasonic velocities in the build-up direction is not statistically significant. Based on the results of t-tests was concluded that the difference of 1 m/s in the same direction (the layer direction and the build-up direction) is not statistically significant.

To determine whether there is a statistically significant difference between the arithmetic means of the L-wave velocities at different measurement places, an analysis of variance with one variable factor (ANOVA) – the measurement place/the test direction - was performed. The ANOVA is summarized in Table 5.

Source of Variation	Sum of Squares	Degrees of Freedom	Mean Square	F_0	*P*-value
Measurement place	11113	3	3704.23	84.02	< 0.01
Error	4232	96	44.09		
Total	15345	99			

Table 5. ANOVA table

The *P*-value from Table 5 is compared to the significance level of 0.05 to assess the null hypothesis. The null hypothesis states that all the ultrasonic velocity means are equal. The alternative hypothesis states that not all means are equal. Since the *P*-value (the exact *P*-value is $9.50 \cdot 10^{-27}$) is less than the significance level, the null hypothesis is rejected and concluded that not all ultrasonic velocity means on all measuring places are equal. This result indicates that the test direction on AM specimen significantly affects ultrasonic velocity.

4. Conclusion

Longitudinal wave velocity in the additively manufactured specimen, made of X3NiCoMoTi 18-9-5 with the DMLS manufacturing process, depends on the test direction. The ultrasonic velocity in the layer direction (x-axis) is higher than the ultrasonic velocity in the build-up direction (z-axis). Differences between ultrasonic velocities on the same sides of the specimen (comparison M1 and M2; comparison M3 and M4) are 1 m/s, which indicates uniform ultrasonic velocity in the same direction. It is confirmed by performing paired t-tests, where obtained *P*-values (for the x-axis *P*-value is 0.498, and for the z-axis *P*-value is 0.675) are higher than the significance level of 0.05. On the contrary, the differences in velocities in perpendicular directions and at the same ends of the specimen (comparison M1 and M3; comparison M2 and M4) are 21 m/s. ANOVA analysis was conducted to determine the existence of a statistically significant difference between the arithmetic means of the L-wave velocities at measuring places. ANOVA analysis confirmed that the difference between L-wave velocities means in different test directions is statistically significant (P-value < 0.01) and that the test direction on AM specimen significantly affects ultrasonic velocity. Since the accuracy of the adjustment of the ultrasonic device affects the accuracy of the ultrasonic test results, a uniform ultrasonic velocity is mandatory throughout the test specimen. The reduction in the difference in ultrasonic velocities in different directions can be enhanced, by changing the microstructure, that is, by achieving a more uniform microstructure. To achieve a more uniform microstructure it is necessary to subject the specimen to heat treatment; therefore, the next step of the research is subjecting the specimen to heat treatment to achieve more uniform ultrasonic velocity throughout the whole volume of the specimen along all directions.

5. References

[1] Honarvar, F. & Varvani-Farahani, A. (2020). A review of ultrasonic testing applications in additive manufacturing: Defect evaluation, material characterization, and process control, Ultrasonics, Vol. 108, 106227, ISSN 1874-9968, DOI: 10.1016/j.ultras.2020.106227

[2] Park, S. H., Jhang, K. J., Yoon, H. S. & Sohn, H. (2021). Porosity Evaluation of Additive Manufactured Parts: Ultrasonic Testing and Eddy Current Testing, Journal of the Korean Society for Nondestructive Testing, Vol. 41, No. 1, pp. 1-10., ISSN 2287-402X, DOI: 10.7779/jksnt.2021.41.1.1

[3] Taheri, H., Shoaib, M., Koester, L. W., Bigelow, T. A., Collins, P. C. & Bond, L. J. (2017). Powder-based additive manufacturing - a review of types of defects, generation mechanisms, detection, property evaluation and metrology, International Journal of Additive and Subtractive Materials Manufacturing, Vol. 1, No. 2, pp. 172-209, ISSN 2057-4983, DOI: 10.1504/IJASMM.2017.10009247

[4] Pandey, D. K. & Pandey S. (2010). Ultrasonics: A Technique of Material Characterization, In: Acoustic Waves, Dissanayake, D., (Ed.), pp. 397-430, Intechopen, ISBN 978-953-51-4543-1

[5] Keran, Z., Mihaljevic, M., Runje, B. & Markucic, D. (2017). Ultrasonic testing of grain distortion direction in cold formed aluminium profile, Archives of Civil and Mechanical Engineering, Vol. 17, No. 2, pp. 375-381, ISSN 1644-9665, DOI: 10.1016/j.acme.2016.11.003

[6] Krstelj, V. (2003). Ultrasonic Testing - Selected Chapters, University of Zagrebu, Faculty of Mechanical Engineering and Naval Architecture, ISBN 953-6313-54-5, Zagreb, Croatia

[7] Kim, C., Yin, H. S., Shmatok, A., Prorok, B. C., Lou, X. Y. & Matlack, K. H. (2021). Ultrasonic nondestructive evaluation of laser powder bed fusion 316L stainless steel, Additive Manufacturing, Vol. 38, 101800, ISSN 2214-7810, DOI: 10.1016/j.addma.2020.101800

[8] Fayazfar, H., Salarian, M., Rogalsky, A., Sarker, D., Russo, P., Paserin, V. & Toyserkani, E. (2018). A critical review of powder-based additive manufacturing of ferrous alloys: Process parameters, microstructure and

mechanical properties, Materials & Design, Vol. 144, pp. 98-128, ISSN 0264-1275, DOI: 10.1016/j.matdes.2018.02.018

[9] Javidrad, H. R. & Salemi, S. (2020). Determination of elastic constants of additive manufactured Inconel 625 specimens using an ultrasonic technique, International Journal of Advanced Manufacturing Technology, Vol. 107, No. 11-12, pp. 4597-4607, ISSN 2456-4346, DOI: 10.1007/s00170-020-05321-x

[10] Castro-Sastre, A.; Fernandez-Abia, A. I.; Rodriguez-Gonzalez, P.; Martinez-Pellitero, S. & Barreiro, J. (2018). Characterization of materials Used in 3D-Printing Technology With Different Analysis Techniques, Proceedings of the 29th DAAAM International Symposium on Intelligent Manufacturing and Automation, Zadar, ISSN: 1726-9679, ISBN 978-3-902734-20-4, Katalinic, B. (Ed.), pp. 0947-0954, Published by DAAAM International, Vienna, DOI:10.2507/29th.daaam.proceedings.136

[11] Cook, O., Huang, N., Smithson, R., Kube, C., Beese, A. & Arguelles, A. (2022). Ultrasonic characterization of porosity in components made by binder jet additive manufacturing, Materials Evaluation, Vol. 80, No. 4, pp. 37-44, ISSN 255327, DOI: 10.32548/2022.me-04266

[12] Huang, N., Cook, O. J., Smithson, R. L. W., Kube, C. M., Arguelles, A. P. & Beese A. M. (2022). Use of ultrasound to identify microstructure-property relationships in 316 stainless steel fabricated with binder jet additive manufacturing, Additive Manufacturing, Vol. 51, 102591, ISSN 2214-8604, DOI: 10.1016/j.addma.2021.102591

[13] Pal S., Gubeljak, N., Hudak, R., Lojen, G., Rajtukova, V., Predan, J., Kokol, V. & Drstvensek, I. (2019). Tensile properties of selective laser melting products affected by building orientation and energy density, Materials Science and Engineering a-Structural Materials Properties Microstructure and Processing, Vol. 743, pp. 637-647, ISSN 0921-5093, DOI: 10.1016/j.msea.2018.11.130

[14] Frazier, W. E. (2014). Metal Additive Manufacturing: A Review, Journal of Materials Engineering and Performance, Vol. 23, pp. 1917-1928, ISSN 1544-1024, DOI: 10.1007/s11665-014-0958-z

[15] Pandzic, A[di] (2021). Influence of Layer Height, Build Orientation and Post Curing on Tensile Mechanical Properties of SLA 3D Printed Material, Proceedings of the 32nd DAAAM International Symposium, pp.0200-0208, B. Katalinic (Ed.), Published by DAAAM International, ISBN 978-3-902734-33-4, ISSN 1726-9679, Vienna, Austria, DOI: 10.2507/32nd.daaam.proceedings.030

[16] Kok, Y., Tan, X. P., Wang, P., Nai, M. L. S., Loh, N. H., Liu, E. & Tor, S. B. (2018). Anisotropy and heterogeneity of microstructure and mechanical properties in metal additive manufacturing: A critical review, Materials & Design, Vol. 139, pp. 565-586, ISSN 0264-1275, DOI: 10.1016/j.matdes.2017.11.021

[17] Lin, Y. Zou, D., Ye, G. & Liang J. (2021). Density Analysis of Additive Manufactured Metal Parts Using Laser Ultrasonics, Proceedings of 2021 3rd International Academic Exchange Conference on Science and Technology Innovation (IAECST 2021), Guangzhou, China, 9781665402682, pp. 953-956, Institute of Electrical and Electronics Engineers (IEEE), DOI: 10.1109/IAECST54258.2021

[18] Sol, T., Hayun, S., Noiman, D., Tiferet, E., Yeheskel, O. & Tevet, O. (2018). Nondestructive ultrasonic evaluation of additively manufactured AlSi10Mg samples, Additive Manufacturing, Vol. 22, pp. 700-707, ISSN 2214-8604, DOI: 10.1016/j.addma.2018.06.016

DOI: 10.2507/34th.daaam.proceedings.059

THE INFLUENCE OF DRAG FINISHING ON CARBIDE END MILLS MANUFACTURED IN DIFFERENT DIAMETERS

Ridvan Tenk & Anil Akdoğan

This Publication has to be referred as: Tenk, R[idvan] & Akdogan, A[nil] (2023). The Influence of Drag Finishing on Carbide End Mills Manufactured in Different Diameters, Proceedings of the 34th DAAAM International Symposium, pp.0455-0460, B. Katalinic (Ed.), Published by DAAAM International, ISBN 978-3-902734-41-9, ISSN 1726-9679, Vienna, Austria
DOI: 10.2507/34th.daaam.proceedings.059

Abstract

The purpose of this paper is to find out how different diameter milling tools are affected by drag finishing. In order to produce a high performance tool, it is not enough to determine the rake angle, clearance angle and other important geometries alone. In addition to these, carbide tools must be finished to improve the surfaces and cutting edges produced by grinding method. The most important benefit of additional finishing on cutting tools is the rounding of sharp corners and the smoothing of surfaces. In this study, experimental investigations are presented to analyse the effects of drag finishing operation on solid carbide milling cutter production. Solid carbide milling cutters with diameters of 6, 8, 10, 12 mm were machined by drag finishing and the cutting edge geometry and surface roughness were analysed by means of 3D optical measurement. The results of the finishing rate on the cutting tool due to the contact density between the workpiece and the abrasive medium are explained.

Keywords: Carbide end mill manufacturing; Drag finishing; Cutting edge preparation; Roughness; 3D optical measurement

1. Introduction

Solid carbide milling cutters have complex geometries that are difficult to produce, and it is important to finish the surfaces formed during grinding. Therefore, cutting edge preparation is an essential production process in tool manufacturing. Today, the machining industry needs high productivity and consistent quality of the tools used. These needs can be met by edge preparation applied to cutting tools [1]. Cutting edge preparation, in addition to increasing the mechanical strength, provides better wear resistance of the tool by reducing chipping and crack formation along the cutting edge. Besides, the adhesion ability of the coating to be applied to the cutting tools is increased by cutting edge preparation and the coating process can be facilitated [2], [3], [4]. The most important effect of finishing on cutting tools is the rounding of sharp corners and smoothing of surfaces. Cutting edge rounding is the process of blunting the cutting edges after grinding the tool. Cutting edge shape and cutting edge condition are very important for machining operations. Also, the quality of the workpiece surface is affected by the shape of the cutting edge [5]. As a result of rounding the cutting edges, the problem of wear and breakage of the tools is prevented, and longer-lasting tools can be obtained.

There are different surface treatment methods used for cutting edge preparation, such as brushing, laser machining and drag finishing [6]. Regardless of the process applied for surface finishing, the advantages of cutting edge preparation have been scientifically proven by many studies [7], [8]. The surface treatment method used in this study is the drag finishing method. In the drag finishing process, the cutting tools are fixed in the holders of the machine and dragged at high speed in a roational movement through the container filled with abrasive granules. The speed of the rotational movement creates a contact density between the cutting tool and the abrasive media. Finishing is done by removing chips from the cutting tool surface with contact density. In the literature, there are studies on speeds [9], time, immersion depth [10], and abrasive granules [11], which are factors affecting the finishing process. In a study [12], the relative motion between the workpiece and the abrasive media in drag finishing was modelled using the Discrete Element Method (DEM). In the study, the chip removal rate on the cutting tool is based on the contact density between the workpiece and the abrasive media. It is stated that the accuracy of the model is consistent with the experiments. As a result of the increased contact density between the cutting tool and the abrasive media, it is expected that the machining of the surfaces of tools with larger diameters will occur in a shorter time or at a higher rate. Is it really possible to machine larger diameter cutters in less time or at a higher rate in the same period? This study evaluates the finishing process and results of four different diameter milling cutters most commonly used in the machining industry.

2. Materials and Methods

The cemented carbide raw material with 10% Co content to be used in this study is UF10 grade raw material of IMC company and some of its properties are given in table 1.

Grade	Unit	UF10
ISO Range		K20-K50
WC + other carbides	%	90
Co	%	10
WC Grain Size		Submicron
Density	g/cm³	14.45

Table 1. Table example

The 4-flute end mills used in the study were produced on a 5-axis ANCA MX7 grinding machine. The cutting edges of carbide milling cutters manufactured in different diameters were improved with OTEC DF-3 drag finishing machine using OTEC HSC 1/300 abrasive granules. The abrasive material used in the study, HSC 1/300, is a mixture of 30% silicon carbide (SiC) with a grain diameter of dT ~ 200 µm and 70% walnut shell granules with a grain diameter of 0.8 mm ≤ dT ≤ 1.3 mm. The abrasive media of this mixture is suitable for the preparation of cutting edges of cutting tools made of cemented carbide [9].

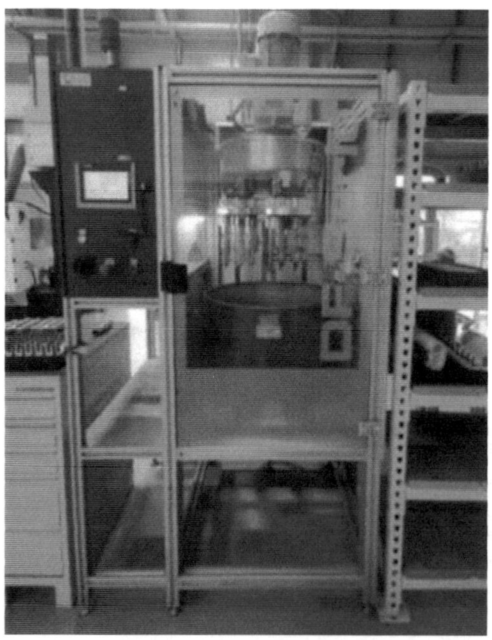

Fig. 1. OTEC DF-3 Drag finishing machine.

The finishing parameters, which were used in experiment is shown in the Table 2.

Rotor speed [min^{-1}]	Holder speed [min^{-1}]	Immersion depth [mm]
50	80	430

Table 2. Finishing parameters

Fig. 2. Alicona Edgemaster

The milling cutters, whose cutting edges were prepared by drag finishing, were measured with the Alicona Edgemaster measuring device. For a better understanding of how the cutting edge radius and surface roughness are measured using 3D optical measurement, see the figure below.

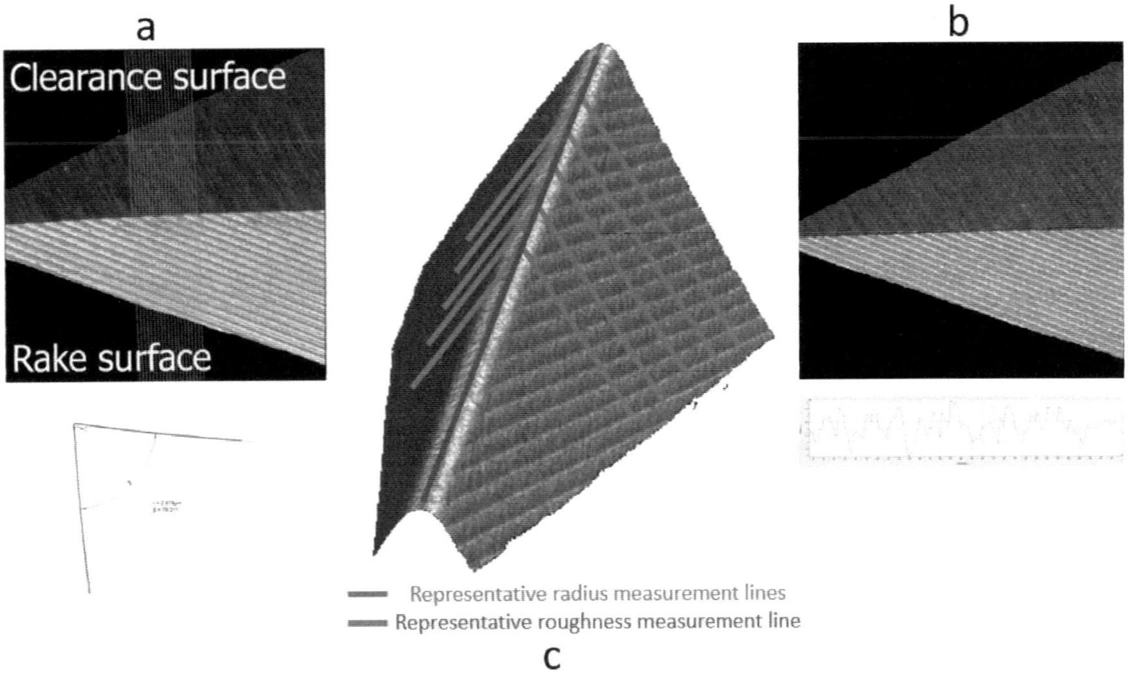

Fig. 3. Measurement lines

3. Results

In the experimental study, investigations are presented to analyse the effects of the drag finishing operation on solid carbide milling cutters manufactured in different diameters. For the investigations, carbide milling cutters with diameters of D=6, D=8, D=10, D=12 mm manufactured at TaeguTec Turkey were machined by drag finishing and the cutting edge rounding and surface roughness were analysed by means of 3D optical measurement. HSC 1/300 granule, the content of which was previously specified, was used as abrasive media. The cutting edge rounding limit of this granule is ER=20 µm and the surface roughness processing limits are in the range of Ra=0.3-0.4 µm.

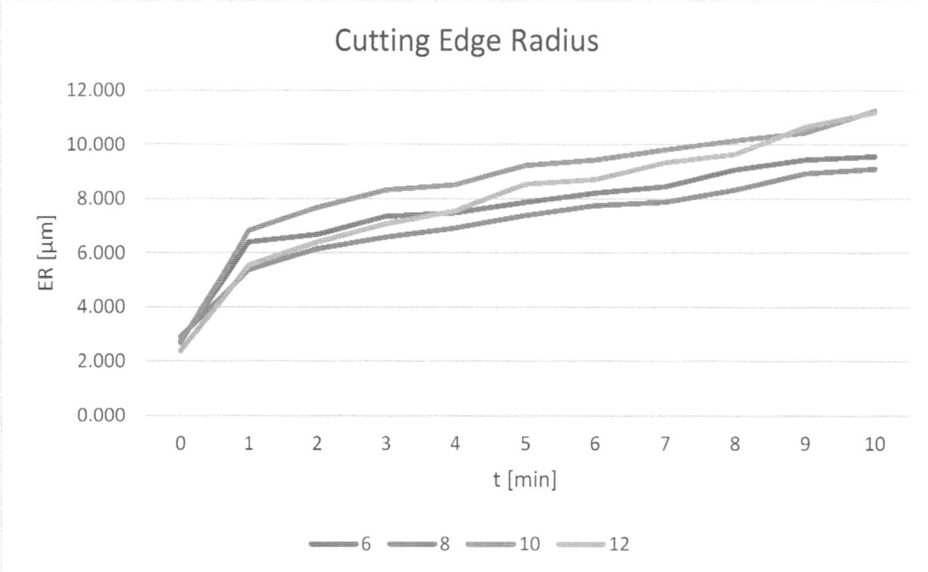

Fig. 4. Cutting edge radius, 10 minutes

The cutting edge radii of the milling cutters of different diameters after finishing are shown in Figure 4 for the first 10 minutes. The reason why this graph is given separately is to show the rapid change in the first minutes in more detail. Comparing Figures 4 and 5, it can be seen that the change observed in the first minutes is faster, while after a certain time the machining rate slows down. It can be seen that the cutting edges of the milling cutters with relatively larger diameters reach the rounding capability limit of the abrasive media, 20 µm, faster.

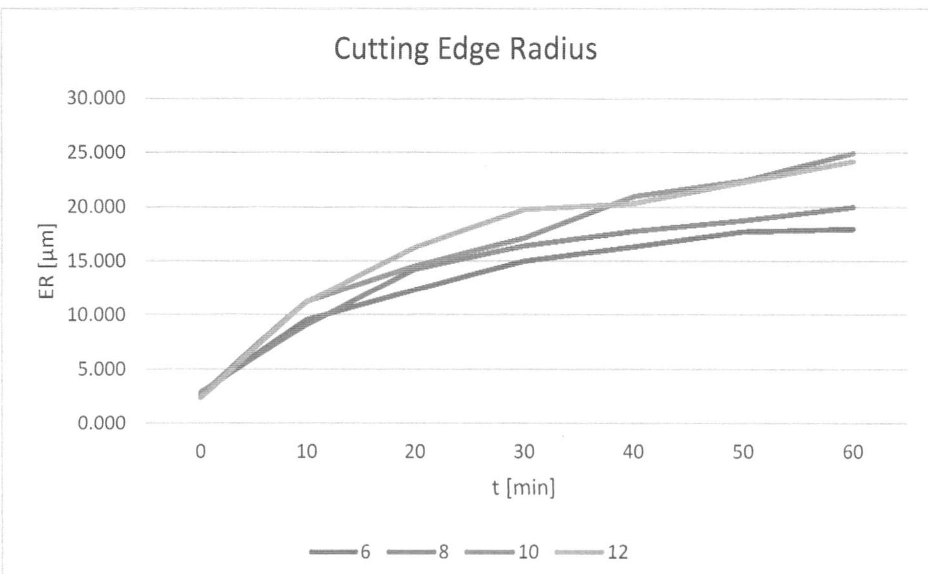

Fig. 5. Cutting edge radius, 60 minutes

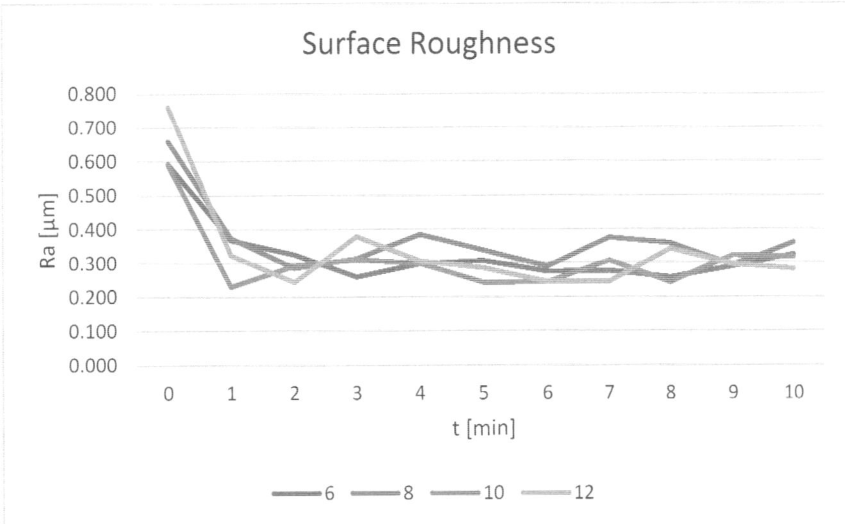

Fig. 6. Surface roughness, 10 minutes

In the graph in Fig. 6, which shows the roughness measurements in the first 10 minutes, it is seen that the change is faster in the first minutes, similar to the machining effect in the cutting edge radius graph. When the graph with longer measurements is analysed, it is seen that after 10 minutes, the change slows down and remains at 0.3-0.4 µm, which is the surface treatment limit.

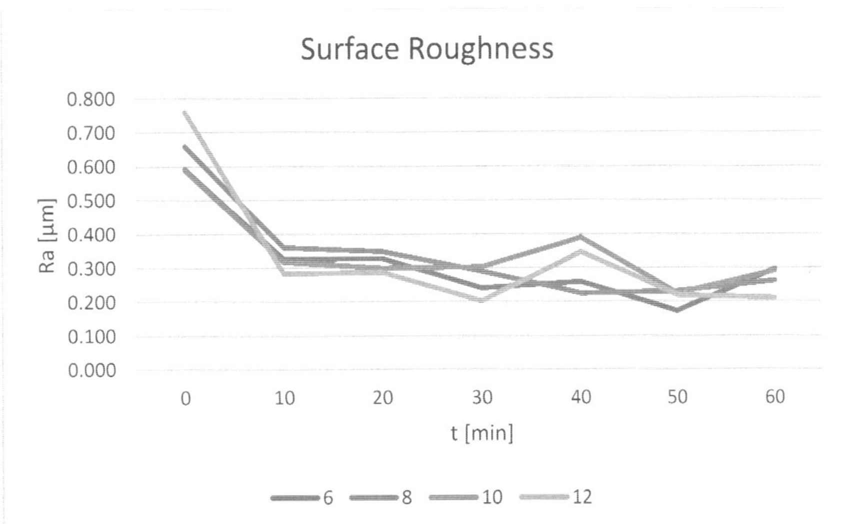

Fig. 7. Surface roughness, 60 minutes

Finally, below you can see the 3D image of a cutting edge before and after drag finishing.

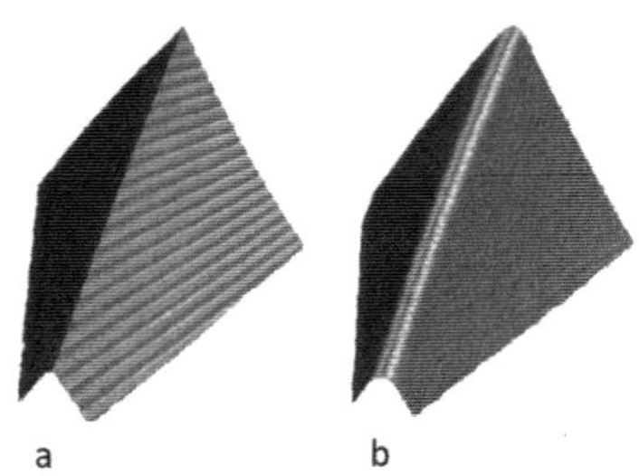

Fig. 8. a) Before drag finishing, b) After drag finishing

4. Conclusion

The focus of this study was to examine the effects of drag finishing on the cutting edge of a carbide milling cutter by performing 3D dimensional measurements. Therefore, an investigation was proposed in which carbide milling cutters of four different diameters would be machined on a drag finishing machine. 6, 8, 10, 12 mm diameter milling cutters with the same working parameters were machined on the drag finishing machine for surface treatment. After the drag finishing process, the differences between the cutting edges of the milling cutters of different diameters were determined. After a processing time of 60 minutes, the carbide milling cutter with a diameter of 6 mm could be rounded to approximately 18 µm. For 8 mm, this value was 20 µm, while the tools with 10 mm and 12 mm diameters exceeded 20 µm with the machining capability of the abrasive media. When the roughness measurements are examined, the improvement rate of the roughness in 10 mm and 12 mm milling cutters, which are larger diameters, is greater than that of 6 mm and 8 mm milling cutters. The reason for these two effects is that the contact density is higher due to the higher surface area of tools with higher diameters.

The limitation of the study is that only HSC 1/300 abrasive media was used. In future studies, the results that can be obtained from different types of abrasive granules can be compared. When previous studies in which HSC 1/300 abrasive granules used in the study were examined, it was seen that the results took a shorter time to reach the limit value. The condition of the abrasive media, which has not been mentioned in any previous study, may have caused this result. After working for certain hours, the abrasive granules must either be replaced completely, or the machining feature must be maintained by renewing the volume at certain periods. For example, the machining ability of the abrasive media is higher in the first hours of operation and the surface processing times are shorter. However, since the abrasion properties of granules that have been used for a long time are lost, surface treatment times are longer. One of the most important points that manufacturers should pay attention to is continuity in quality. Therefore, it is more suitable to replace some part of the abrasive media volume in certain working periods to ensure process continuity by achieving the same cutting edge rounding rate and surface quality in the same period.

In future studies, the effects of different finishing methods or different finishing medias on cutting tools in drag finishing can be investigated. Research can be conducted to examine the effects of surface-improved milling tools on the machining performance or the surface quality of the workpiece.

5. References

[1] Tönshoff, H.K. & Denkena, B. (2013). Basics of Cutting and Abrasive Processes, Springer, Berlin

[2] Bassett, E.; Köhler J. & Denkena, B. (2012). On the honed cutting edge and its side effects during orthogonal turning operations of AISI1045 with coated WC-Co inserts, CIRP Journal of Manufacturing Science and Technology. 5. 108–126.

[3] Wyen, C. & Wegener, K. (2010). Influence of cutting edge radius on cutting forces in machining titanium, CIRP Annals-Manufacturing Technology, 59, 93-96.

[4] Denkena, B.; Koehler, J. & Rehe, M. (2012). Influence of the honed cutting edge on tool wear and surface integrity in slot milling of 42CrMo4 steel", 5th CIRP Conf. on High Performance Cutting.

[5] Durakbasa, M.N.; Akdogan, A.; Vanli, A.S. & Bulutsuz, A.G. (2015). Optimization of end milling parameters and determination of the effects of edge profile for high surface quality of AISI H13 steel by using precise and fast measurements", Measurement, 68, 92-99.

[6] Denkena, B. & Biermann, D. (2014). Cutting edge geometries, CIRP Annals-Manufacturing Technology, 63, 631-653.

[7] Simna, V; Jurina, F; Vopat, T & Kuruc, M. (2019). Edge Preparation of Cutting Tools and it Impact on Cutting Forces During Milling, Proceedings of the 30th DAAAM International Symposium, pp.0764-0769.

[8] Denkena B. & Breidenstein, B. (2012). Residual stress distribution in PVD-coated carbide cutting tools-origin of cohesive damage, Tribology in Industry. 34. 158-165.

[9] Uhlmann, E.; Oberschmidt, D.; Kuche, Y. & Löwenstein, A. (2014). Cutting edge preparation of micro milling tools, Procedia Cirp, 14, 349-354.

[10] Hronek, O.; Zetek, M. & Baksa, T. (2017). The Immersion Depth Influences on Cutting Edge Radius During Drag Finishing, Proceedings of the 28th DAAAM International Symposium, pp.1030-1036.

[11] Dejin, L.; Wang, Y.; Yu, X.; Chen, H. & Gao, Y. (2021). Analysis of Abrasives On Cutting Edge Preparation By Drag Finishing, The International Journal of Advanced Manufacturing Technology, 119(5-6), 3583-3594.

[12] Uhlmann, E.; Eulitz, A. & Dethlefs, A. (2015). Discrete element modelling of drag finishing", Procedia Cirp, 31, 369-374.

DOI: 10.2507/34th.daaam.proceedings.060

AUTOMATIC TRANSFORMATION OF OPC UA SKILLS MODELS INTO KNOWLEDGE GRAPHS

Aleksandra Müller, Miray Özakkas, Oliver Petrovic,
Werner Herfs & Christian Brecher

This Publication has to be referred as: Müller, A[leksandra]; Ozakkas, M[iray]; Petrovic, O[liver]; Herfs, W[erner] & Brecher, C[hristian] (2023). Automatic Transformation of OPC UA Skills Models into Knowledge Graphs, Proceedings of the 34th DAAAM International Symposium, pp.0461-0469, B. Katalinic (Ed.), Published by DAAAM International, ISBN 978-3-902734-41-9, ISSN 1726-9679, Vienna, Austria
DOI: 10.2507/34th.daaam.proceedings.060

Abstract

As part of the Industry 4.0 initiative, OPC UA as a uniform standard for data and information exchange enables heterogeneous machine data to be transported and described so that other production participants can access it. In addition, control with the help of process-relevant capabilities, so-called OPC UA Skills, promises simple and flexible exchange and seamless integration of individual production participants and processes through uniform interfaces. In addition to interoperability, however, an Industry 4.0 network should be expandable by new information and provide possibilities for recognizing new connections of the existing data (reasoning). Information modelling based on the Semantic Web (e.g., by means of RDF) offers a solution for this. However, the already existing semantic OPC UA modelling on the shop floor is not directly interoperable with other semantic data models such as RDF/OWL. Furthermore, the OPC UA models lack query functionality, making them cumbersome to search. For these reasons, the focus of this thesis is on the transformation of flexible process control models based on OPC UA Skills into semantic RDF/OWL Knowledge Graphs. For this purpose, transformation rules are established, which are validated by means of application to a robot-based industrial application example.

Keywords: OPC UA Skills; semantic modelling; RDF; model transformation

1. Introduction

The goal of the Industry 4.0 research initiative is to enable globally networked, flexible and thus adaptable production [1]. In the vision of the Internet of Production (IoP) cluster of excellence, this requires the creation of a continuous network in which semantically adequate and context-dependent production data is available. One of the challenges here is the fact that production data is often heterogeneous and distributed. The OPC UA architecture as a uniform standard enables both a description of dynamic process data and a semantic modelling of the hardware and software structure of a production process with the help of the information models contained in the standard. In addition, the use of OPC UA enables cross-manufacturer data exchange between the production process participants [2]. In combination with the concept of a capability-based control with OPC UA Skills, a significant increase in the flexibility of the production process is made possible.

On the other hand, an IoP network should be expandable by new information and provide possibilities for recognising new connections of the existing data. Information modelling based on the Semantic Web is a suitable solution for this. However, the already existing semantic OPC UA modelling on the shop floor is not directly interoperable with other semantic data models such as RDF/OWL.

In this context, the integration of OPC UA with ontologies plays a decisive role. This allows the structures and information from the OPC UA information model to be transferred into an RDF model. In this way, with SPARQL as the query language, complex search queries can be made on RDF data in order to retrieve specific information easily. This facilitates the search and understanding of process data. By using OPC UA as the communication protocol and RDF as the data representation format data can be effectively managed, analysed and exchanged in the automation industry. There are already approaches that aim to transform OPC UA models into RDF, but these deal with the general methodology of OPC UA - RDF translation without considering the OPC UA Skills concepts. However, this is inevitable in order to map flexible control concepts in their fullness.

The focus of this thesis is on the transformation of OPC UA Skill information models into semantic RDF/OWL Knowledge Graphs. For this purpose, Section 2 presents existing approaches to the use of OPC UA Skills, semantic technologies in production and approaches to OPC UA - RDF translation. This is followed by a concept description of an OPC UA-RDF algorithm for the translation of control-relevant skills in Section 3. The algorithm is then prototypically implemented on an industrial robot-based demonstrator, described in Section 4. The thesis concludes with a summary and an outlook on further research possibilities.

2. State of the Art

2.1. OPC UA Skills

In industrial automation, OPC UA (Open Platform Communications Unified Architecture) has become the established standard for industrial communications, approved by RAMI 4.0 as uniform recommendation on the implementation of the communication layer. OPC UA specifies data exchange and enables semantic interoperability through more extensible information models. It provides a framework for object-oriented representation and exchange of data and information [3]. In addition, OPC UA introduces the concept of OPC UA Skills to enable flexible process control [4]. According to the Industry 4.0 platform, skills should be modelled as state machines, as this enables targeted monitoring and control of them [5]. The concept of OPC UA Skills allows complex processes to be decomposed into smaller units that can be executed in a specific sequence based on specific production requirements. This enables the adaptation of production systems for different processes, utilizing only the necessary functionalities for each process [6]. The implementation of OPC UA Skills allows precise control of the process flow, considering criteria such as production demand or energy efficiency. In this way, process control can be adapted to meet specific demands, leading to optimal performance. This requires an efficient analysis and processing of skills and process-related properties. Queries can be used to collect the relevant data to perform essential evaluations. However, there are currently no query tools available for extracting information from OPC UA documents [7].

2.2. Semantic Technologies in production

To model a knowledge base in a production environment, the available data needs to be extended by means of semantics [8]. Following the rules of Semantic Web, the data can be interpreted by a machine [9], establishing explicit relationships and creating an accessible knowledge repository for both machines and humans. For semantic enrichment of data, the Resource Description Framework (RDF) is used as a standardized syntax and core technology. To provide a consistent and declarative structure, the RDF Schema (RDFS) has been introduced to enable the semantic modelling of contexts. When merging many so-called RDF and RDFS statements, the resulting graph must be made readable to the machine. This is done through various serializations that specify the data format and thus the syntax [10]. The respective syntax can be used for the Ontology Web Language (OWL) as a data format and represents the essential OWL ontology. OWL enables the complex modelling of relations with additional features such as cardinalities or inverse relations [8]. Thus, ontologies using OWL create a suitable data modelling for complex relations and enable the intelligent derivation of knowledge [11]. This flexibility allows for adaptation and transferability to other data models. Additionally, SPARQL queries can be used to perform complex searches on RDF models to extract valuable insights and enable sophisticated data analysis.

Despite the potential benefits of semantic integration, the lack of standardization of Semantic Web Technologies presents challenges. There are some academic and industrial approaches to implement them in production, e.g. [12], [13], [14] and many more, but the vast majority of them lack standardization of the models. In this regard, the combination of OPC UA and Semantic Web Technologies can be promising to overcome the limitations of OPC UA for semantic descriptions in production environments. However, a conceivable parallel semantic modelling in the form of an OPC UA information model and an RDF graph, which are created independently of each other but model the same concept, proves to be unfavorable unless the additional effort of synchronizing both models can be automated.

2.3. Related work: OPC UA RDF Transformation

Previous studies have already investigated the integration of OPC UA with semantic modelling technologies. In [15] an approach for the integration of OPC UA into a Linked Data Environment is described. A new ontology is created specifically for the OPC UA information model and the built-in concepts of OPC UA are not used to their full extent. In [16], a mapping from OWL to OPC UA is proposed. However, this mapping does not consider every single concept of OPC UA, such as ReferenceType. In [17] it is pointed out that Schiekofer et al. in [7] presented the first complete formal mapping between OPC UA and RDF. Their work shows that a trivial mapping between OPC UA and OWL DL is not possible because the expressiveness of OWL DL is not sufficient to represent the entire OPC UA information model. Moreover, in [18] Steindl et al. proposed an alternative mapping from OPC UA to OWL by defining mapping rules to extract the required information from the OPC UA information model and use it in a domain-specific target ontology.

These previous studies lay the groundwork for developing an interoperable connection between OPC UA and Semantic Web Technologies. They emphasize the significance of a comprehensive mapping of OPC UA into RDF to ensure a better understanding and effective processing of OPC UA data. This integration of OPC UA with ontologies and knowledge graphs opens up great potential for automation technology.

While several approaches and works have explored combining OPC UA information models with Semantic Web Technologies, the predominant focus has been on converting entire OPC UA servers into RDF. Notably, there has been no concerted effort to specifically transform OPC UA Skill information into RDF. The challenge arises due to the specific structures and semantics of the OPC UA Skill concept. To address this, this thesis focuses on meeting the specific requirements associated with the OPC UA Skills. This involves a comprehensive mapping of the way the skills are organized, encompassing their hierarchical structures and properties, as well as their relationships into RDF. The objective is to establish a formal mapping between OPC UA and RDF that not only includes the data structures but also takes into account the functionalities and properties inherent in OPC UA Skills. This semantic representation facilitates efficient analysis and enables improved information retrieval.

3. Transformation concept

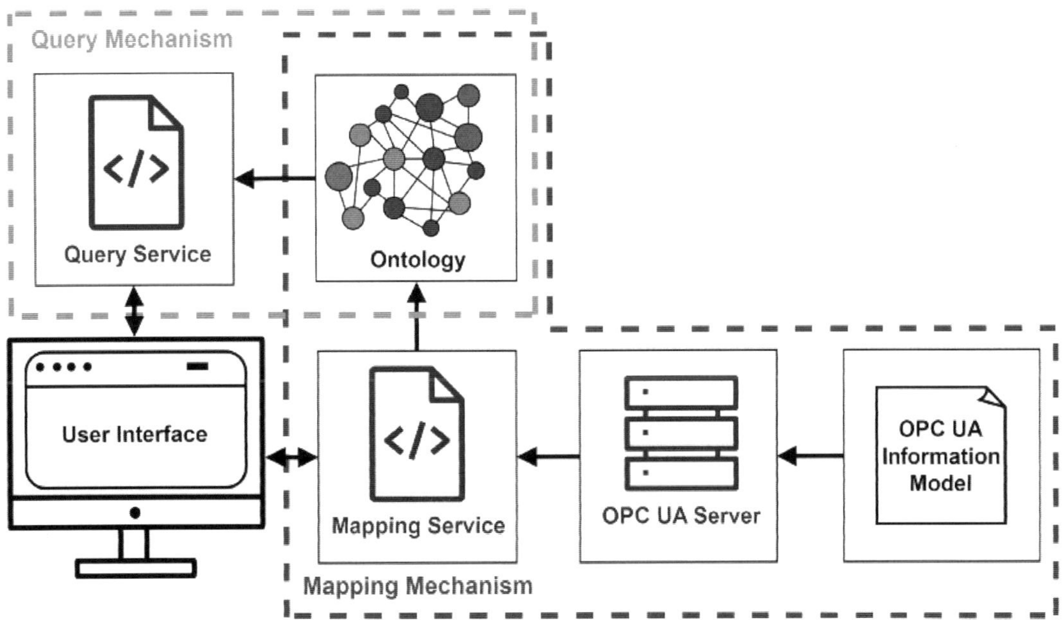

Fig. 1. Transformation service architecture

The proposed concept for transforming OPC UA information models into ontologies and its effective utilization introduces a system comprising two key mechanisms: the mapping mechanism and the query mechanism. These mechanisms are accessible through a user interface, offering a user-friendly and interactive approach. Within the user interface, the user can select the OPC UA server to be translated. The proposed concept allows the transformation of one or more OPC UA servers, each operating as a process participant. For this purpose, the corresponding information models must already be implemented and running on the OPC UA Servers, with the address space of each server being accessible through the OPC UA client. The mapping service processes the user's selection and transforms the selected OPC UA information model into an ontology. This involves establishing a connection with the OPC UA server hosting the desired information model and executing the transformation process based on the defined mapping rules, as shown in Table 1 (s. below).

Once the mapping process is complete, and the selected OPC UA information model has been successfully transformed into an ontology, the query mechanism comes into play. This allows users to interact with the created ontology and perform relevant queries. Here, a query service is employed that effectively searches the ontology using user-defined queries and returns the corresponding results. These results, obtained through the query mechanism, can be used for further analysis and applications. The overall system structure is shown in Figure 1.

3.1. Mapping Service

The Mapping Service enables the transformation of OPC UA information models into RDF models and ensures a consistent transfer of OPC UA skills into ontologies. By analyzing the structure and semantics of both models, equivalent concepts are identified, leading to the development of mapping rules. These rules govern the transformation process, ensuring accurate and meaningful representation of OPC UA data into RDF format.

OPC UA and RDF are two different technologies, each serving different purposes and applications. However, they have relevant similarities in the modelling methodology. The primary goal of OPC UA is to provide a standardized data exchange between different devices and systems in industrial environments. It provides information models to represent data in a structured way. On the other hand, RDF is designed for data representation and exchange on the Web. It aims to visualize the data in a graph-based, machine-readable format.

Although OPC UA and RDF are used in different areas, they have in common that they deal with structured data representations and exchange based on a graph-based structure composed of nodes and their connections. In both models, the connection between a *Source Node (Subject)* and a *Target Node (Object)* is established by a *Reference* (*Predicate*), as depicted in Figure 2. However, the specific XML format used by OPC UA for serialization is not directly compatible with RDF, making it difficult to directly map OPC UA information models to RDF [17]. Therefore, mapping rules must be developed to map the structure and semantics of OPC UA to RDF. The main mapping rule entails mapping OPC UA nodes as RDF resources and OPC UA references as RDF properties.

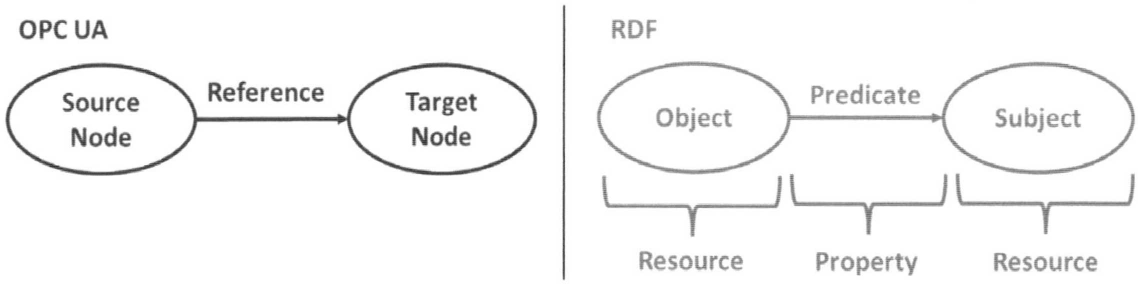

Fig. 2. *Target Node-Source Node* in OPC UA, *Object-Subject* in RDF

The exact transformation of OPC UA information models into RDF models depends on the specific requirements of the system. It should be noted that OPC UA and RDF use different semantics to define concepts. In the context of this work the goal is to map OPC UA Skills into ontologies while preserving the semantics and meaning of the skill data. However, the complexity of the OPC UA information models is a challenge. The complexity increases with the number of classes, instances of those classes and their relationships and attributes. The OPC UA Base Information Model is already extensive and contains numerous classes to create instances for system modeling [19]. Yet not all classes are used in every context. Therefore, the concept of mapping the relevant classes is developed. Relevant classes are those for which at least one instance has been created in the OPC UA information model.

By adopting this approach, only essential information is transformed from OPC UA to the ontology and the complexity of the resulting graph is minimized. This facilitates fast data processing and streamlines information retrieval. Table 1 presents the mapping rules in more detail, specifying how different OPC UA nodes are mapped to specific types of RDF resources and properties. The focus is primarily on the OPC UA nodes of *Object*, *ObjectType*, *Variable*, *VariableType*, *Method* and *ReferenceType* node classes, as they play the central role in representing OPC UA Skills. In the mapping process, a crucial decision is made to represent OPC UA references as RDF properties, while all other OPC UA nodes are mapped to resources, with one exception. While inheritance exists in OPC UA, the explicit relationship between inherited features and their source classes is not explicitly defined, as shown in Figure 3.

This lack of explicit relationship makes it difficult to identify and establish the connection between the inherited features and their respective source classes. To address this issue, an exception is made for the nodes belonging to the *Variable* node class which are instances of the *PropertyType* class. Because these nodes semantically represent properties, they are mapped as datatype properties in the RDF model. This mapping solves the problem by providing a clearer and more explicit representation of the relationship between nodes, their properties and their parent classes.

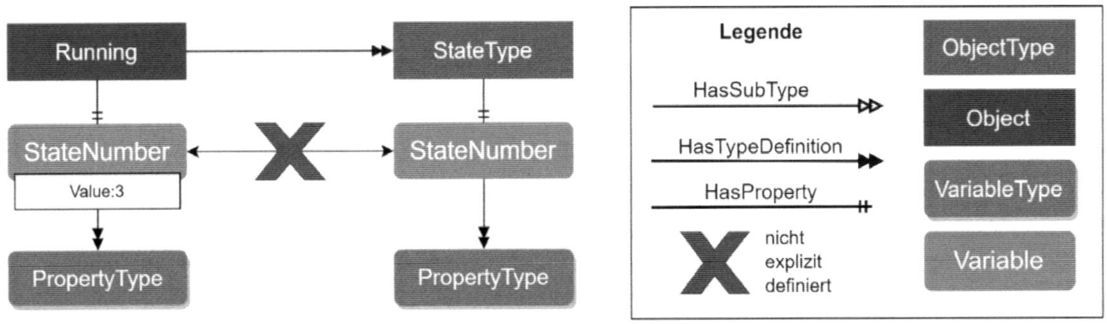

Fig. 3. Example of an object node and its property

3.2. Query Service

The query service is designed to provide a user-friendly interface without requiring the user to have in-depth knowledge of the ontology, OPC UA server or SPARQL. Users can select a skill from the OPC UA server, and the query service processes this selection and searches the ontology for the variables, objects and methods associated with the selected skill. The retrieved results are then presented to the user for further refinement of the selection. The query service dynamically adapts its search operations based on the user's selection and displays the relevant results.

4. Limitations

It is important to note that the proposed mapping approach focuses only on transforming OPC UA Skills into RDF. This selective approach is intended to ensure that the resulting knowledge graph contains only the essential data required for an efficient OPC UA Skill analysis, since the OPC UA information models are often complex and consist of numerous nodes and references. Consequently, it does not cover the entire OPC UA information model. Furthermore, the viability of the proposed transformation concept depends on the OPC UA server being built according to the OPC UA Skill structure. This conformance check requires manual verification, as the mapping process itself does not include server conformance assessment. In cases where the development of the server is incomplete or where the structure of skills deviates from the expected standard, the implementation of the transformation concept will require adjustments to align with the actual server structure.

5. Implementation

5.1. Use Case

The proposed concept is practically applied in a robotic sub-process of glass pane assembly. The sub-process consists of Cleaning, Handling, and Priming stations. In the first station, a KUKA iiwa robot is used to clean the working trajectory of a car window. In the Priming station a KUKA Agilus (KR 6) is used to apply glue to the cleaned shape of the glass. Each station is equipped with workpiece carriers for securely holding the glass pane and proximity sensors for precise positioning. The handling robot (UR 5) facilitates the transfer of the panes between the stations. Each station is assigned different skills to perform its specific tasks. Additionally, there is a SCADA (Supervisory Control and Data Acquisition) system that includes the compilation of the skills of all the individual units that are involved in the assembly process. Figure 4 illustrates an overview of the different systems and their corresponding OPC UA servers, showing the skills of each robot at its respective station.

Cleaner, Handling Robot and SCADA systems share the same OPC UA server structure, while there is a slight deviation in the Priming unit due to the limited import functionality of the individual information models in the TwinCAT software. Therefore, the implementation of the mapping algorithm for the standardized OPC UA Skills structure is covered first. This is followed by an explanation of the adjustments required to adapt the mapping algorithm to Primer.

5.2. Implementation of Mapping Service for Cleaner, Handler and SCADA

The OPC UA information models of the process units are embedded in their respective OPC UA servers. To access these information models, the connection to the OPC UA servers must be established first. Since each OPC UA process unit has its own dedicated OPC UA server, the mapping service is implemented in a separate document for each unit. The following section presents the implementation of the mapping service for a single server. The same implementation is used for all other servers, only the server URL and the name of the ontology document need to be adapted.

Rules

1. OPC UA nodes of the node class *ReferenceType* are defined as owl:ObjectProperty.

2. The references that have an *InverseName* are linked to their inverse *ObjectProperty* via owl:inverseOf.

3. Nodes of node classes *ObjectType* and *VariableType* are defined as owl:Class.

4. Nodes of node classes *ObjectType* and *VariableType* are defined as subclasses of their parent nodes using rdfs:subClassOf, to which they are connected through the *SubTypeOf* reference.

5. Nodes belonging to the node class *Object* are defined as instances of the type node to which they are linked by the *HasTypeDefiniton* reference.

6. Nodes belonging to the node class *Variable*, which are not instantiated from the *PropertyType* node, are defined as instances of the type node to which they are connected via *HasTypeDefinition* reference.

7. Nodes belonging to the *Variable* node class, which are connected to the *PropertyType* node via the reference *HasTypeDefinition*, are mapped as owl:DatatypeProperty.

8. The range of *DatatypeProperties* mapped from the nodes instantiated from *PropertyType* are set on the class they are associated with *HasProperty* reference.

9. The range of DatatypeProperties mapped from *PropertyType* nodes is set to the corresponding XSD data format depending on their OPC UA Datatype.

10. The value attributes of nodes belonging to *Variable* node class are mapped as owl:DatatypeProperty named hasValue.

11. The SourceTimestamp attributes of nodes that are not typed by *PropertyType* are mapped as owl:DatatypeProperty named hasSourceTimestamp.

12. The SourceTimestamp attributes of nodes typed by *PropertyType* are mapped as owl:DataTypeProperty named has[Name of Property]SourceTimestamp.

13. Node ids are used to determine URIs of Resources.

14. The *DisplayName* attribute is mapped as rdfs:label.

15. Attributes, except *DisplayName*, *Value* and *SourceTimestamp*, are mapped as owl:AnnotationProperty.

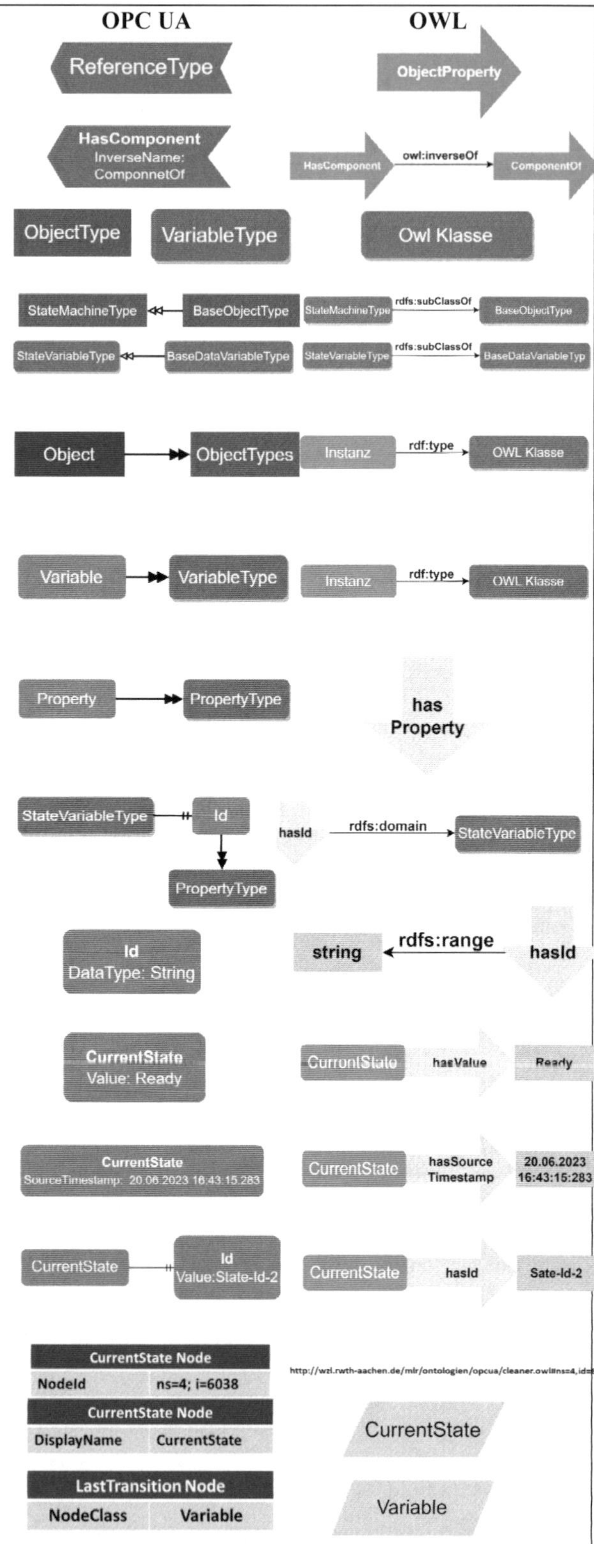

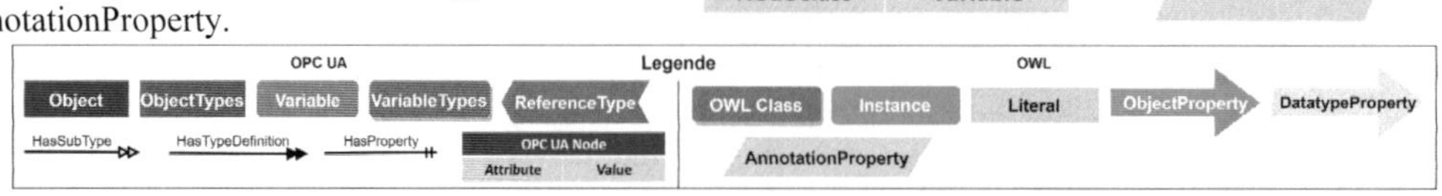

Table 1. Mapping Rules

The service uses the opcua-asyncio Python library to connect to the OPC UA server, allowing seamless navigation and access to the nodes, their properties and relationships. Initially, the OPC UA references are mapped to RDF, starting from the *References* node which serves as the primary reference node from which all other references inherit. Its *NodeId* (ns:0; i=31) is predefined within the mapping service and serves as a consistent starting point for mapping all OPC UA references. For this use case, the *NodeId* is constant across all servers and is unaffected by server restarts.

To map the primary reference node as an object property, the function create_object_property_for_reference() is implemented. This function creates an object property for the primary reference node and translates its attributes into OWL AnnotationProperties. Afterward, the get_reference_children() function is employed recursively to find all child references of the current node and map them as object properties in OWL. These child references are linked to their parent using rdfs:SubPropertyOf. This mapping approach captures the hierarchical structure of OPC UA references, by placing the subordinate references in a hierarchical relationship to the parent references. In addition, the attributes of the references such as *DisplayName*, *NodeId*, *NodeClass* are passed as AnnotationProperty in OWL. The get_reference_children() function also takes into account the *InverseName* of a reference and creates a corresponding object property for the inverse reference. This allows the representation of inverse relationships of the OPC UA references within the OWL ontology.

After successfully transforming all references, the mapping service proceeds to identify the skill nodes that are located under the *Objects* node with the *NodeId* (ns=0; id=85). To do this, the get_type_def() function is used to search for the node connected to the *Objects* node via the *HasTypeDefinition* reference. The node found represents the type class of the *Objects* node. Next, the get_super_class() function is used to recursively identify parent classes. These classes are then mapped as OWL classes using the create_class() function and linked to their subclasses using rdfs:subClassOf. Once the type class (e.g., *FolderType*) and parent classes (e.g., *BaseObjectType*) are mapped, the instances, such as *Objects*, of the corresponding class can be created using the create_instance() function. The create_instance() function then calls the get_referenced_nodes() function, to further capture the child nodes that are connected to the instance node by a specified reference. In this use case, the *Objects* node is connected to child nodes through the *Organizes* reference. Therefore, the Organizes reference is passed to the get_referenced_nodes() function to identify and map the child skill nodes to the ontology. To ensure that only the relevant nodes are included in the ontology and to omit other irrelevant nodes such as *Aliases*, *DeviceSet*, *DeviceTopology*, etc., only the nodes instantiated by the *StateMachineType* class are transformed into the ontology, as they represent the skill nodes. Then, the *HasComponent* reference is passed to the get_referenced_nodes() function, as skills are linked to their child variables and objects through this reference. These objects and variables of a skill can therefore be called as components. To map these components in OWL, the functions create_instance(), get_superclass() and get_referenced_node() are called. Through this recursive process, all relevant nodes, and their relationships for the OPC UA skills are effectively mapped to the ontology.

For value and timestamp attributes of a node, data type properties are created. The get_data_type_property() function finds nodes linked by the *HasProperty* reference and creates data type properties for them. It establishes the data tzpe property's domain based on the inherited type class. To set the range of each data type property, it maps the OPC UA data type of the property node's value to the appropriate XSD (XML Schema Definition) data format. Finally, RDF triples are created by connecting the instance with its value using newly created data type propert. The resulting ontology is then stored in XML format. Figure 5 provides an illustrative example of mapping OPC UA type classes and instances to RDF format.

5.3. Implementation of Mapping Service for Primer

The OPC UA information model of the Priming unit differs from other units in terms of node and relationship structure. The implementation must handle these differences. In the Priming unit, the skill nodes are located under the *MAIN* node and connected by the *HasComponent* reference, unlike other units where they are listed under the *Objects* node and connected by the *Organizes* reference. The variables and objects subordinate to the skill nodes remain unchanged. The Priming unit has an additional object node named *skill_function* that is subordinate to the *priming_skill* node. This skill node contains *Parameters* node linked to the other variables through the *HasProperty* reference. The previously used exception to transform only those nodes that are linked to their parents via the *Organizes* reference and are of type *StateMachineType* in the ontology, can therefore no longer be used for the Primer. This exception has to be removed from the implementation. The variables of the skills have no other properties, while the methods, such as *Start*, have properties like *InputArguments* and *OutputArguments*.

Therefore, a data type property is created for each method property and its domain is set to the method class. In addition, custom type classes derived from the BaseObjectType class are defined for the skills, states and state transitions, but these custom type classes are linked to the nodes, such as *PLC1*, which is instantiated from *FolderType*, through the *Organizes* reference. To maintain the hierarchical structure and avoid confusion, these folder nodes, *PLC1* and *BeckhoffControlTypes*, are excluded from the implementation, because they do not represent type classes and do not share inheritance relationships with their subordinates through *HasSubType* reference. The type classes for the skills (e.g., *clamping_skill*) are subclasses of *BaseObjectType*. This does not require any changes to the existing implementation, as it already supports the creation of OWL classes and subclasses.

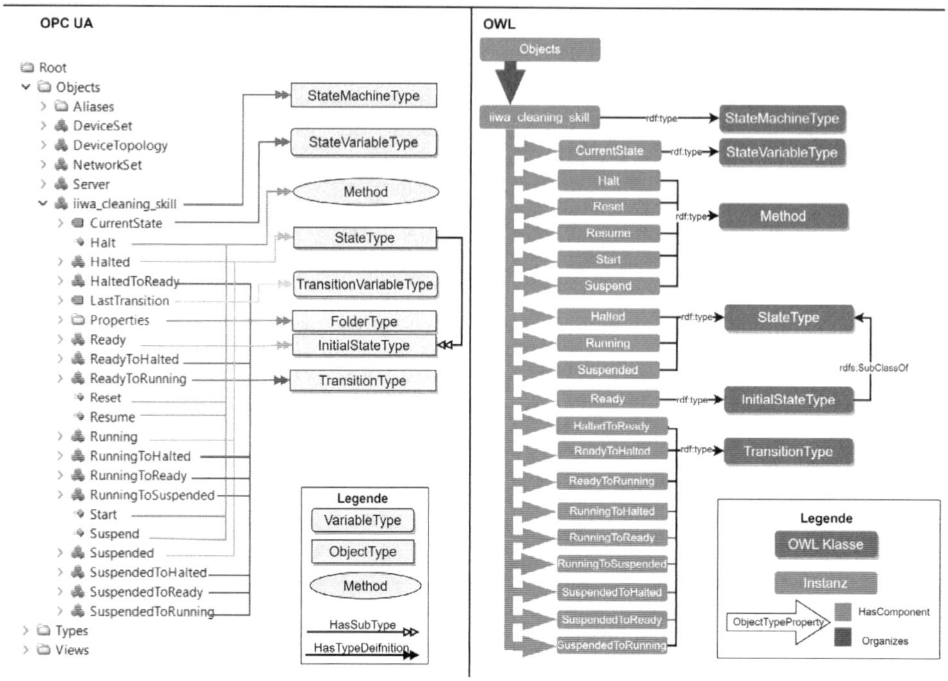

Fig. 4. Example of Mapping OPC UA Skills in OWL

5.4. Implementation of Query Service

The query service is implemented using the Python Dash library for an interactive and user-friendly interface with four buttons: *Start Transformation* to select an OPC UA server for ontology transformation and four additional *View Ontology* buttons to select the ontology post-transformation. Clicking the *Start Transformation* button executes the code to transform the chosen OPC UA server into the ontology. Once the transformation process is done, the user can select an ontology using corresponding *View Ontology* button. The interface directs users to a new page where the visualization of the associated ontologies is generated using the dash-cytoscape library. Additionally, two dropdown menus are integrated to allow the user to execute specific SPARQL queries and retrieve the desired values from the ontology.

6. Summary and outlook

In the context of this work, a concept for translating OPC UA information models into ontologies with a focus on OPC UA Skills was developed. This concept was implemented and validated using an industrial robot-based use case of glass pane completion. It has been shown that OPC UA information models containing the OPC UA Skills can be transformed into ontologies and that the desired information about their execution can be easily retrieved from the ontology by SPARQL queries. However, the developed concept requires a full compliance of the models to be translated with the OPC UA skill concept. The skill information models that are not created according to the OPC UA Skill standard should be treated differently. As in the case of the Priming unit application example, an analysis of the structure of the OPC UA information model is necessary to adapt the implementation of the concept to the different structure of the OPC UA information model, so that all required nodes are transferred to the ontology. Especially in the cases where user-defined type nodes are present in the OPC UA servers or different references between the nodes are used, additional mappings and contexts have to be considered. In future work, it could be advantageous to extend the translation system with a checking mechanism that can check whether the OPC UA server to be translated is built or complete according to the OPC UA standards. In the case that the OPC UA server is not complete, then the system should give suggestions to the user to complete the server structure or create the missing nodes automatically. Furthermore, the algorithm is to be used as part of the assistance system to be developed for knowledge-based control process reconfiguration, in the context of the scientific project described above.

7. Acknowledgments

Funded by the Deutsche Forschungsgemeinschaft (DFG, German Research Foundation) under Germany ´s Excellence Strategy – EXC-2023 Internet of Production – 390621612. Furthermore, a vivid exchange with Chair for Production Metrology and Quality Management at the WZL of the RWTH Aachen University (WZL-IMA), especially with Michael Zeng, is highly appreciated.

8. References

[1] B. Bajic, I. Cosic, B. Katalinic, S. Moraca, M. Lazarevic und A. Rikalovic (2019), Edge Computing vs. Cloud Computing: Challenges and Opportunities in Industry 4.0, 30TH DAAAM International Symposium on Intelligent Manufacturing and Automation, doi:10.2507/30th.daaam.proceedings.120

[2] M. Obdenbusch (2018), Referenzarchitektur für cloudbasiertes Condition Monitoring am Beispiel von Verpackungsmaschinen, Apprimus Wissenschaftsverlag, ISBN:978-3-86359-573-9, Aachen

[3] I. Reithner, M. Papa, B. Lueger, M. Ćato, S. Hollerer und R. Seemann (2020), Development and Implementation of a Secure Production Nretwork, 31ST DAAAM International Symposium on Intelligent Manufacturing and Automation, Vienna, Austria, doi: 10.2507/31st.daaam.proceedings.102

[4] S. Profanter, A. Perzylo, M. Rickert und A. Knoll (2021), A Generic Plug & Produce System Composed of Semantic OPC UA Skills, IEEE Open Journal of the Industrial Electronics Society, p. 128–141, doi:10.1109/OJIES.2021.3055461

[5] plattform-i40: Information Model for Capabilities, Skills & Services, Available: https://www.plattformi40.de/IP/Redaktion/EN/Downloads/Publikation/CapabilitiesSkillsServices.html.

[6] K. Dorofeev und A. Zoitl (2018), Skill-based Engineering Approach using OPC UA Programs, IEEE 16th International Conference on Industrial Informatics, doi: 10.1109/INDIN.2018.8471978.

[7] R. Schiekofer und M. Weyrich (2019), Querying OPC UA information models with SPARQL, 24th IEEE International Conference on Emerging Technologies and Factory Automation, p. 208–215, doi: 10.1109/ETFA.2019.8868246.

[8] A.Dengel (2012), Semantische Technologien: Gundlagen - Konzepte - Anwendungen, Spektrum Akademischer Verlag, ISBN: 978-3-8274-2663-5, Heidelberg.

[9] M. Merdan, A. Zoitl, G. Koppensteiner und F. Demmelmayr (2010), Semantische Technologien – Stand der Technik, Elektrotech. Inftech., p. 291–299, doi: 10.1007/s00502-010-0777-3.

[10] G. Saake, K. Uwe, A. Heuer (2018), Datenbanken-Konzepte und Sprachen, mitp Verlag, ISBN: 9783958457782

[11] A. Mueller, W. Jesse, S. Storms, W. Herfs und C. Brecher (2023), Ontology-based assistance system for control process reconfiguration of Robot-Based Applications, 4th Conference on Production Systems and Logics, Hannover, doi: 10.15488/13444.

[12] C. Brecher, M. Buchsbaum, A. Müller, K. Schilling, M. Obdenbusch, S. Staudacher and M. Chaikh Albasatineh (2021), Gaining IIoT insights by leveraging ontology-based modelling of raw data and Digital Shadows, 4th IEEE International Conference on Industrial Cyber-Physical Systems, doi: 10.1109/ICPS49255.2021.9468116

[13] J. Lipp und K. Schilling (2020), The semantic web in the Internet of Production. A strategic approach with use case examples, the Fourteenth International Conference on Advances in Semantic Processing, p. 68-72, ISBN: 978-1-61208-813-6

[14] S. Pieske, W. Herfs, M. Zenke, S. Storms und C. Brecher (2022), Semantic modeling of a cyber-physical biological production platform, 27th IEEE International Conference on Emerging Technologies and Factory Automation , Stuttgart, doi: 10.1109/ETFA52439.2022.9921616

[15] M. Graube, L. Urbas und J. Hladik (2016), Integrating industrial middleware in Linked Data collaboration networks, 21st International Conference on Emerging Technologies and Factory Automation , pp. 1-8, doi: 10.1109/ETFA.2016.7733710

[16] A. Bunte, O. Niagemann und B. Stein (2018), Integrating OWL Ontologies for Smart Services into AutumationML and OPC UA, 23rd International Conference on Emerging Technologies and Factory Automation, pp. 1383-1390, doi: 10.1109/ETFA.2018.8502593

[17] A. Weiss und S. Ihlenfeldt (2022) , Integration of OPC UA Information Models into Enterprise Knowledge Graphs, Journal of Machine Engineering, doi: 10.36897/jme/150008

[18] G. Steindl und W. Kastner (2021), Transforming OPC UA Information Models into Domain-Specific Ontologies, 4th IEEE International Conference on Industrial Cyber-Physical Systems, , pp. 191-196, doi: 10.1109/ICPS49255.2021.9468254

[19] M. Graube, S. Hensel, C. Iatrou und L. Urbas (2017), Information models in OPC UA and their advantages and disadvantages, 22nd IEEE International Conference on Emerging Technologies and Factory Automation, doi: 10.1109/ETFA.2017.8247691

DOI: 10.2507/34th.daaam.proceedings.061

INVESTIGATION OF IRREGULAR THERMAL BEHAVIOUR OF ALSI10MG COMPONENTS MANUFACTURED BY LPBF

Alexandra Morvayova, Maria Emanuela Palmieri,
Giuseppe Casalino & Luigi Tricarico

This Publication has to be referred as: Morvayova, A[lexandra]; Palmieri, M[aria] E[manuela]; Casalino, G[iuseppe] & Tricarico, L[uigi] (2023). Investigation of Irregular Thermal Behaviour of AlSi10Mg Components Manufactured by LPBF, Proceedings of the 34th DAAAM International Symposium, pp.0470-0474, B. Katalinic (Ed.), Published by DAAAM International, ISBN 978-3-902734-41-9, ISSN 1726-9679, Vienna, Austria
DOI: 10.2507/34th.daaam.proceedings.061

Abstract

The unique thermal history associated with the LPBF manufacturing process is closely linked to the distinct behaviour exhibited by parts manufactured via this method, which significantly deviates from that of parts produced via conventional techniques. Specifically, AlSi10Mg specimens manufactured via LPBF exhibit a heterogeneous microstructure, unequivocal high-strength properties, and a distinct response to thermal loading. This latter characteristic can be regarded as a crucial parameter in determining the potential applications of LPBF-manufactured AlSi10Mg specimens, as it indirectly indicates the microstructural alterations induced by thermal loading. In this study, we explore the thermal behaviour of AlSi10Mg cylinders produced via L-PBF through physical simulations. Three different thermal cycles were applied to the cylinders, and the imposed and actual temperatures were evaluated and compared. Subsequently, the obtained signal was filtered to eliminate the effects of the PID controller. The filtered signal was then subjected to a thorough analysis, with a focus on comparing the identified peaks with the distinct thermal behaviour previously reported in the literature for LPBF AlSi10Mg specimens. These identified peaks were subsequently linked to temperature-induced microstructural changes. The noteworthy correlations established in this investigation can be perceived as foundational knowledge for tailoring the heat treatment processes applied to LPBF-manufactured AlSi10Mg parts.

Keywords: thermal behaviour; physical simulation; LPBF; AlSi10Mg.

1. Introduction

Prominent progress in process reliability, productivity and product quality over past years can be perceived as the major reasons for the increasing popularity of LPBF processes in automotive and aerospace industries, especially in the production of parts with high strength-to-weight ratio and intricate designs [1], [2], [3]. The specific, and highly complex nature of this manufacturing technology however facilitates the strongly distinctive characteristics of the manufactured parts, that significantly differ from those ones of the same materials but processed conventionally. Currently, a high number of researchers describe the particular associations within the compounded process-structure-property interrelationship of the parts manufactured by LPBF, however, this area still needs a more thorough understanding [4], [5].

Notably, it is already generally recognised that a good part of the unique characteristics of LPBF parts can be associated with their solidification conditions with rapid cooling rates, reheating thermal cycles and overall complex thermal history, and their consequent fine-grained and heterogeneous microstructure. To contribute to the better comprehension of the threefold interrelationship underlying the LPBF manufacturing process, and to improve the overall applicability of this manufacturing technology in various industries, rigorous knowledge of the specific behaviour of components manufactured by LPBF is considered crucial [6], [7], [8].

The behaviour of LPBF-manufactured specimens under thermal loads can be perceived as an important parameter for their prospective application and can indirectly indicate the microstructural changes within the specimens caused by thermal loads. The lack of existing research studying the behaviour of LPBF-manufactured specimens under thermal load is reflected in the current practice of thermal treatment of LPBF-manufactured specimens, where rather conventional types of treatment are still applied. It should be noted that fundamental differences between the conventionally manufactured specimens, and specimens manufactured by LPBF make the traditional types of thermal treatment often no longer effective [9], [10].

The present study investigates the thermal behaviour of AlSi10Mg cylinders manufactured by L-PBF using physical simulations. The AlSi10Mg cylinders were subjected to 3 different thermal cycles, where imposed and the real temperature of the cylinders were compared. Subsequently, the received signal was filtered with the simulated influences of the PID controller. The filtered signal was then closely analysed, comparing the identified peaks with the unique thermal behaviour of LPBF AlSi10Mg specimens found in the literature. The identified peaks were further associated with temperature-induced microstructural changes.

2. Materials and methods

Three cylindrical specimens with a diameter of 12 millimetres and a total length of 120 millimetres (Fig. 1) were fabricated by Laser Powder Bed Fusion (L-PBF) using EOSINT M270 machine (EOS, Germany). The specimens were manufactured from gas-atomized virgin AlSi10Mg powder, supplied by EOS. The employed processing parameters are represented in Table 1. The specimens were positioned in a vertical direction, and a zigzag scanning strategy, with a rotation angle of 67° between consecutive layers was employed.

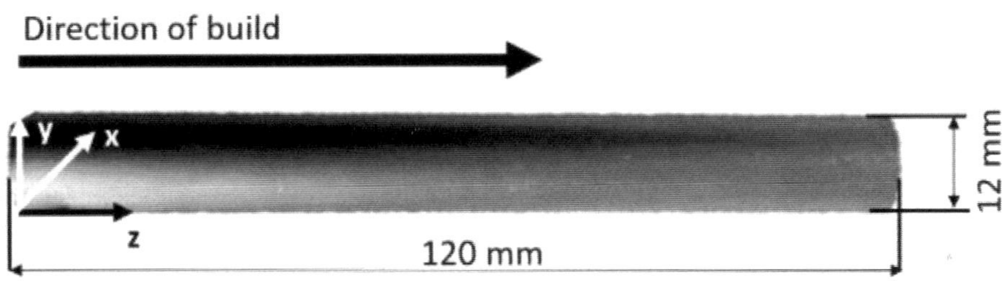

Fig. 1. LPBF manufactured AlSi10Mg cylinder

Laser power[W]	Scanning speed [mm/s]	Layer thickness [mm]	Hatch space [mm]
275	400	0,03	0,08

Table 1. L-PBF processing parameters

Three different types of thermal loads were applied using the Gleeble 3800 physical simulator, as outlined in Table 2. The thermal loads included a heating phase (from 20 °C to 500 °C), a holding phase (maintaining a constant temperature of 500 °C), and a cooling phase (from 500 °C to 20 °C). It is important to note that subsequent analysis revealed that the physical simulator was unable to achieve the desired cooling rate, thus the thermal behaviour during the cooling phase could not be evaluated. The physical simulation system employed two thermocouples: one for heating the specimen using Joule heating principles, and another to measure the instantaneous temperature of the specimen (Fig. 2).

Thermal load	Heating rate [°C/s]	Keeping phase time [s]	Cooling phase [°C/s]
TH1	10	60	10
TH2	7.5	80	7.5
TH3	6	240	6

Table 2. Imposed thermal loads

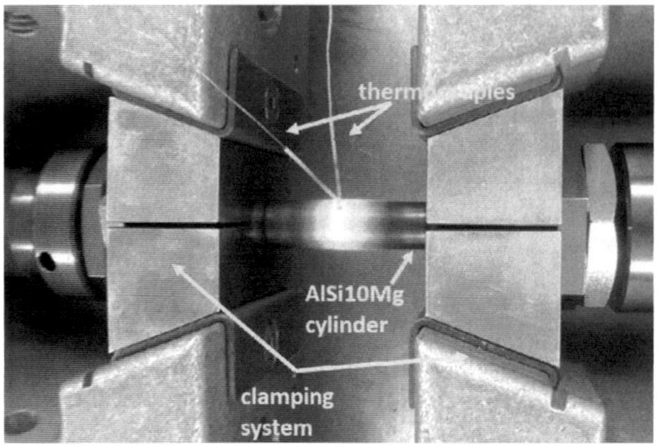

Fig. 2. AlSi10Mg cylinder during physical simulation

The measured (T_m) and imposed (T_i) temperatures were subsequently analysed, their difference (T_{diff}) for each time step was calculated (1), and the T_{diff} over the imposed temperature was plotted.

$$T_{diff} = T_m - T_i \tag{1}$$

3. Filtering of the signal

The signal of received temperature deviation, plotted over the real temperature was thereafter filtered from the influence of the PID controller, used for process control in the applied physical simulator. The influence of the PID controller for each case of applied thermal loads was simulated in Comsol Multiphysics, according to the standard PID control algorithm (2) [11], [12].

$$u(t) = u_{bias} + k_p[c_{set}(t) - c(t)] + k_i \int_0^t [c_{set}(t) - c(\tau)]d\tau - k_d \frac{d}{dt}c(t) \tag{2}$$

Where u_{bias} stands for bias of PID, k_p is the proportional gain, c_{set} is the reference value and k_d is a derivative gain. Furthermore, the filtering of the derivative was enabled. The PID control parameters were used according to the real characteristics of the PID controller applied in the physical simulator.

4. Results and discussions

Fig. 3 shows the comparison of raw, unfiltered signal of T_{diff}, plotted over the imposed temperature.

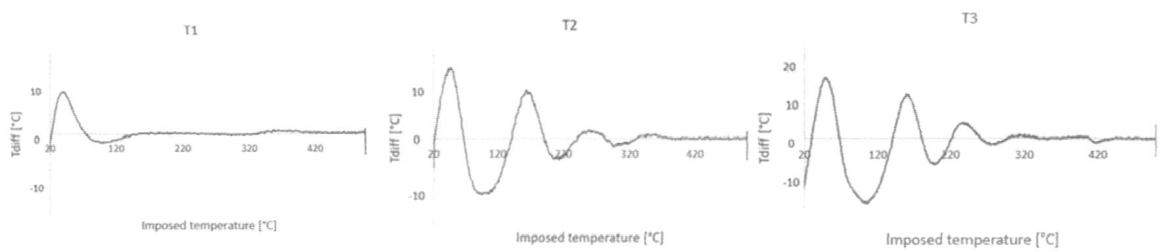

Fig. 3. Comparison of unfiltered T_{diff}, plotted over time in specimens under different types of thermal load

Significant differences between the plotted curves of different thermal loads were observed already in the unfiltered state. The cylinder under thermal load with the highest heating rates demonstrated the least irregularities, and as the heating rate increased, more irregularities were observed. All of the observed curves initially demonstrated similar irregularities, starting with a larger peak in a positive direction, immediately followed by a negative peak. It should be noted that the height of the first peak correlated with the initial deviation at Ti=20 °C. The deeper the initial deviation, the higher the initial peak, suggesting that it was attributed to the PID controlling compensating for the higher initial temperature of the specimen than was the temperature set.

The same applies to the second negative peak, where the lower initial deviation resulted in more shallow peaks. It should be also noted that lower heating rates (T2, T3) resulted in more peaks than thermal load with the higher heating rate (T1). Fig. 4 shows curves of Tdiff plotted over the imposed temperature of cylinders under different thermal loads, already filtered from the influence of the PID controller.

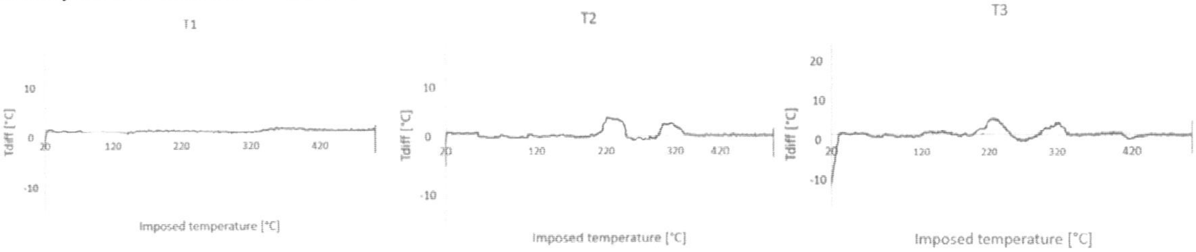

Fig. 4. Filtered signal of T_{diff} plotted over the imposed temperature for different heating rate

It should be noted that the number of irregularities identified in the signal correlated with the applied heating rate. The temperature of the specimen under the thermal load T1 mimicked the imposed temperature most accurately, resulting in the least number of irregularities. The irregularities between the temperature of the AlSi10Mg cylinder and the imposed temperature of the T1 load were located mainly until 120 °C, consisting of numerous very short peaks. Another irregularity in the analysed temperature was observed between 320°C and 400°C. The observed peak has a positive direction and prolonged, short, and doltish shape.

Two main irregularities between the real temperature of the AlSi10Mg cylinders and the temperature imposed in the T2 thermal load were observed. The first positive peak was identified between 210 °C and 280 °C, and the second positive peak was identified between 300 and 350 °C. Both peaks demonstrated similar shape characteristics, however, the first identified peak was slightly higher and thicker. It should be noted that the temperature of the cylinder between both identified peaks (280-300 °C) relatively closely followed the imposed temperature and thus no significant irregularities in this temperature range were identified.

Similar irregularities were identified between the temperature of the AlSi10Mg specimen and the imposed temperature of thermal load T3. The first positive peak was identified at a slightly lower temperature than previously, starting at around 190 °C. It is noteworthy that, unlike in the thermal load T2, the first and second peaks are seemingly connected with a shallow negative peak, occurring between 270 and 290 °C. The second positive peak was subsequently identified between 290 and 340 °C. An additional negative peak was identified between 410 and 440 °C.

The first peak, observed in cylinders under T2 and T3 thermal loads, has been, according to Van Cauwenbergh et al. [14] attributed to microstructural changes induced by thermal loading. These changes are primarily characterized by the precipitation of silicon (Si) particles from the supersaturated aluminium (Al) cells. The missing of the first peak during the analysis of the first thermal load indicates that the applied heating rate did not provide sufficient activation energy for this change to fully occur.

Similarly, the second peak, observed during all three thermal loads, is also linked to temperature-induced microstructural alterations. Within this temperature range, the microstructural changes involve both diffusional activity and spheroidization of the Si particles [13], [14], [15]. The third peak, observable during T3 thermal load can be attributed to the coalescence of the primary Si particles. It should be noted that this microstructural transformation requires a higher activation energy, and consequently, it only manifests at elevated temperatures with sufficiently low heating rates [16]. It should be noted that the exact position and shapes of the identified peaks also depend on the applied heating rate, correlating to the total thermal energy delivered to the specimen under the respective thermal load. Thermal loads with lower heating rates deliver the thermal energy sufficient to activate the respective microstructural change sooner, and thus the peak occurs at a slightly lower temperature [17].

5. Conclusions

The present study investigates the thermal behaviour of 3 AlSi10Mg cylinders under different thermal loads. The study includes filtering of the received signal from the influences of the PID controller using numerical simulation. Analysis of the filtered temperature of each L-PBF manufactured cylinder under the thermal loads has provided several conclusions.

- The PID controllers might substantially influence the difference between imposed and real temperature measured in the cylinder under thermal load. Thus, in order to unambiguously analyse the irregularities between the temperature of the cylinders and the imposed temperature, the influence of the PID controller must be considered.
- The behaviour of AlSi10Mg cylinders manufactured by L-PBF under thermal load is influenced by the applied heating rate, where lower heating rates result in more observable irregularities between the imposed and immediate temperature of the specimen. The applied heating rate also influences the total thermal energy delivered to the cylinder under thermal load and thus influences the exact position and shapes of the identified peaks.

- The first identified peak, observed between 210-280 °C in cylinders under T2 and T3 thermal loads can be attributed to the precipitation of Si particles from supersaturated Al cells.
- The second identified peak, observed between 300-400 °C in cylinders under all can be attributed to diffusional activity and spheroidization of Si particles.
- The third identified peak, occurring between 410-440 °C in a cylinder under T3 thermal load can be attributed to the coalescence of the primary Si particles.

Investigating the thermal behaviour of AlSi10Mg components manufactured using L-PBF under thermal load and establishing a correlation between the observed irregularities and the microstructural changes caused by temperature, has the potential to become a crucial element in identifying the complex process-structure-property relationship in L-PBF processes of the AlSi10Mg alloy. The distinguished correlations can be further applied as a base knowledge for tailoring the heat treatment for LPBF-manufactured AlSi10Mg parts.

6. References

[1] Abd-Elaziem, W., Elkatatny, S., Abd-Elrahim, A., Khedr, M., Abd El-baky, M. A., Hassan, M. A., Abu-Okail, M., Mohammed, M., Järvenpää, A., Allam, T., & Hamada, A. (2022). On the current research progress of metallic materials fabricated by laser powder bed fusion process: a review. Journal of Materials Research and Technology, 20, 681-707.

[2] Kroft, L.; Bicova, K[aterina] & Milsimerova, A[neta] (2020). Use of 3D Printing Technology in Teaching the Basics of Milling Technology, Proceedings of the 31st DAAAM International Symposium, pp.0323-0327, B. Katalinic (Ed.), Published by DAAAM International, ISBN 978-3-902734-29-7, ISSN 1726-9679, Vienna, Austria

[3] Casalino, G., Campanelli, S. L., Contuzzi, N., & Ludovico, A. D. (2015). Experimental investigation and statisticaloptimisation of the selective laser melting process of a maraging steel. Optics & Laser Technology, 65, 151-158.

[4] Eskandari Sabzi, H., & Rivera-Díaz-del-Castillo, P. E. J. (2020). Composition and process parameter dependence of yield strength in laser powder bed fusion alloys. Materials & Design, 195, 109024.

[5] Morvayová, A., Contuzzi, N., & Casalino, G. (2023). Numerical prediction of lack-of-fusion defects in selective laser melted AlSi10Mg alloy. Materials Science and Technology.

[6] Hyer, H., Zhou, L., Park, S., & Lee, J. (2020). Understanding the Laser Powder Bed Fusion of AlSi10Mg Alloy. Metallography, Microstructure, and Analysis, 9, 484-502.

[7] Caiazzo, F., Alfieri, V., & Casalino, G. (2020). On the Relevance of Volumetric Energy Density in the Investigation of Inconel 718 Laser Powder Bed Fusion. Materials (Basel, Switzerland), 13(3), 538.

[8] Roveda, I., Serrano-Munoz, I., Kromm, A., & Madia, M. (2022). Investigation of residual stresses and microstructure effects on the fatigue behaviour of a L-PBF AlSi10Mg alloy. Procedia Structural Integrity, 38, 564-571.

[9] Padovano, E., Badini, C., Pantarelli, A., Gili, F., & D'Aiuto, F. (2020). A comparative study of the effects of thermal treatments on AlSi10Mg produced by laser powder bed fusion. Journal of Alloys and Compounds, 831, 154822.

[10] Ghasemi, A., Fereiduni, E., Balbaa, M., Elbestawi, M., & Habibi, S. (2022). Unraveling the low thermal conductivity of the LPBF fabricated pure Al, AlSi12, and AlSi10Mg alloys through substrate preheating. Additive Manufacturing, 59(Part A), 103148. https://doi.org/10.1016/j.addma.2022.103148

[11] Armstrong, B., & Wade, B. (2000). Nonlinear PID control with partial state space knowledge: Damping with derivative. International Research, 18(8), 715-731.

[12] Cedro, L., & Wieczorkowski, K. (2019). Optimizing PID controller gains to model the performance of a quadcopter. Transportation Research Procedia, 40, 156-169.

[13] Van Cauwenbergh, P., Beckers, A., Lore, T., Hooreweder, B., & Vanmeensel, K. (2018). Heat Treatment Optimization via Thermo-Physical Characterization of AlSi7Mg and AlSi10Mg Manufactured by Laser Powder Bed Fusion (LPBF).

[14] Marola, S., Calignano, F., Manfredi, D., Ambrosio, E. P., & Biamino, S. (2018). A comparison of Selective Laser Melting with bulk rapid solidification of AlSi10Mg alloy. Journal of Alloys and Compounds, 742, 271-279.

[15] Van Cauwenbergh, P., Samaee, V., Thijs, L., Delroisse, G., & Kruth, J. P. (2021). Unravelling the multi-scale structure-property relationship of laser powder bed fusion processed and heat-treated AlSi10Mg. Scientific Reports, 11(1), 6423.

[16] Cabrini, M., Lorenzi, S., Testa, C., Manfredi, D., Lombardi, M., Aversa, A., Andreatta, F., Fedrizzi, L., Dekhtyar, Y., Sorokins, H., & Biamino, S. (2021). Effect of Heat Treatment on Microstructure and Selective Corrosion of LPBF-AlSi10Mg by Means of SKPFM and Exo-Electron Emission. Materials, 14(19), 5602.

[17] Megahed, S., Bühring, J., Duffe, T., Bach, A., Schröder, K.-U., & Schleifenbaum, J. H. (2022). Effect of Heat Treatment on Ductility and Precipitation Size of Additively Manufactured AlSi10Mg. Metals, 12(8), 1311.

DOI: 10.2507/34th.daaam.proceedings.062

OPTIMIZATION OF THE LOGISTIC "FILL RATE" KEY PERFORMANCE INDICATOR THROUGH THE APPLICATION OF THE DMAIC APPROACH

Samuele Marinello[1], Qian Zhao[*2], Antonio Maria Coruzzolo[2], Elia Balugani[2], Rita Gamberini[1,2] & Francesco Lolli[1,2]

[1] En&Tech Interdepartmental Center of the University of Modena and Reggio Emilia, Piazzale Europa 1, 42124 Reggio Emilia, Italy,
[2] Department of Sciences and Methods for Engineering, University of Modena and Reggio Emilia, Via Amendola 2, Padiglione Morselli, 42122 Reggio Emilia, Italy
* Corresponding author

This Publication has to be referred as: Marinello, S[amuele]; Zhao, Q[ian]; Coruzzolo, A[ntonio] M[aria]; Balugani, E[lia]; Gamberini, R[ita] & Lolli, F[rancesco] (2023). Optimization of the Logistic "Fill Rate" Key Performance Indicator Through the Application of the DMAIC Approach, Proceedings of the 34th DAAAM International Symposium, pp.0475-0484, B. Katalinic (Ed.), Published by DAAAM International, ISBN 978-3-902734-41-9, ISSN 1726-9679, Vienna, Austria
DOI: 10.2507/34th.daaam.proceedings.062

Abstract

Measuring and monitoring the performances of supply chains over time is a primary interest factor for companies. In this way, it is possible to determine the effectiveness and efficiency of strategies for being competitive in global markets, verify the achievement of the predetermined targets, and establish intervention and improvement measures. In this context, key performance indicators (KPIs) are widely used to measure the numerous activities performed across a supply chain. Numerous KPIs are available in the literature, and they are often customized by each user to make them more suitable for their reference context. This paper analyzes the logistic "fill rate" KPI that characterizes the shipping phase of goods by evaluating the fill rate of the transport unit used. A case study analyzes the fill rate indicator used by a multinational corporation that produces and markets food packaging. Through the DMAIC (Define, Measure, Analyze, Improve, and Control) approach, the criticalities of the current formulation of the index are highlighted, and a new model for calculating the index is proposed and applied experimentally at a plant in northern Italy.

Keywords: Key performance indicators (KPIs); Fill rate; Supply chain management; DMAIC approach

1. Introduction

The extremely dynamic global market has significantly increased competition between companies that must continually find the best ways to lower the price of products and continuously improve their characteristics [1]. The tools and approaches of supply chain management practices support improvements in organizational performances and the competitive advantages of companies in global markets [2], [3].

Hence, logistics are treated as part of supply chain management, ensuring that the various corporate functions involved in planning, implementing, and controlling the material and information flows occur efficiently and effectively [4]. This integration roles between the various business functions can be identified from the definition of logistics provided by the Council of Logistics Management. In particular, the logistic function can be distinguished in different macro-activities that correspond to different phases but are integrated [1], [5]. These are inbound logistics, stock management, outbound logistics, and return logistics. When so framed, the role of logistics for a company represents the link between internal operations and the surrounding ecosystem. The choices made in a certain area of the logistics activities impact all other areas, generating mutual conditioning [6]. Logistics management significantly affects the overall direct and indirect costs of a company that are necessary for moving products, from sourcing raw materials to delivering goods to customers, including all of the intermediate steps. Logistics have a strong impact on earnings [7]. These costs could be divided into numerous components, including: transport, warehousing, stock management, administration, packaging, and indirect logistics costs. Marotta et al. [8] divided the first four components into percentages and a temporal evolution. The incidence of logistic costs changes among companies in different industries since they pertain to various business areas (e.g., sales, purchasing, and manufacturing). Rantasila and Ojala [9] and Škerlič and Sokolovskij [10] quantified the share of sales revenues that is linked to logistics costs as being at least 6%, but they may also reach a share of 25% [7]. The evolution of the effectiveness and efficiency of the processes that characterize a supply chain (also from an economic point of view) can be monitored through performance measurement systems (PMSs). These are tools that integrate measures that support the decision-making process. This information is analyzed through appropriate key performance indicators (KPIs), which are financial and non-financial measurable values that estimate how effectively a company achieves its key business objectives [11]. These company data in which metrics, ratios, and percentages are conveyed form the basis of the decision-making process for planning improvement actions [12]. Organizations use KPIs at multiple levels to evaluate their success in reaching targets. KPIs analyze performance dimensions; in particular, asset management, time, cost, productivity, and quality of service [13]. Considering the strong impacts of logistics activities on the performance of a company (including economics), having suitable and adequate KPIs is a primary objective. This work describes a specific case study represented by a company that deals with producing and marketing packaging products. The paper describes the current formulation of the FR indicator. It suggests an optimized formulation that will increase the effectiveness of the logistics processes and the methods for monitoring performance.

2. Literature review

Organizations search for KPIs that are aligned with their specific targets. A widespread solution, especially by software vendors who have tried to include the greatest possible choice of KPIs in their products, is to favor the custom development of KPIs. In this study, a novel approach was developed for the automated prediction of relevant KPIs for organizations to favor the indicator customization process. The problem of the number of KPIs describing the performance of a supply chain was also addressed in a study by Brint et al. [14], where they highlighted difficulties in the collection and management of the many data necessary for the quantification of the indicators and the overlaps between the indicators. To overcome this problem, they used principal component analysis (PCA) to reduce the number of KPIs, followed by TOPSIS (technique for order performance by similarity to an ideal solution) for validating the results, reducing the initial dataset of their case study by 28 KPIs to a set of only 8 KPIs. The same problem has been described in numerous review papers available in the literature, aimed at both the study of the entire supply chain and, specifically, the aspects of logistics and transport. Karl et al. [15] analyzed the influence of non-financial KPIs on supply chain resilience and identified which ones had greater impacts as order and delivery lead times, on-time deliveries, supplier delivery efficiencies, and customer satisfaction. Maestrini et al. [16] summarized the reference context of supply chain performance measurement systems through a life cycle perspective. Qorri and Mujki [17] focused on measuring the sustainability performance of supply chains, supported using numerous multi-criteria decision-making methods.

Oriented to measures of logistics management, the studies by Domingues et al. [11] and Wudhikarn et al. [6] focused on reviewing third-party logistics providers and the tangible assets and intellectual capital (IC) of organizations, respectively. Several logistics indicators have been selected as those most used in the literature, such as on-time or out-of-date delivery performance, order fill rate (FR), inventory level and condition, delivery time or speed, and information and documentation accuracy. Larrea [18] evaluated several KPIs applied to logistics management during humanitarian operations. Domingues et al. [11] evaluated the representative indicators of third-party logistics providers through a multi-level approach (decision level, activities, and actors) and provided a comprehensive and innovative performance measurement framework. Dornhofer et al. [19] developed a framework of KPIs for evaluating the effectiveness and efficiency of current logistics processes through a lean perspective, evaluated in an automotive context. Dumitrache et al. [20] evaluated the KPIs applied in large Romanian transport companies to improve their fleet utilization, average daily trips, and transport capacities. Kucukaltan et al. [21] integrated the balanced scorecard (BSC) model and the Analytic Network Process (ANP) method to provide a novel way to evaluate logistics performance indicators from a logistician's perspective. The specific aspect of the reverse supply chain has also been analyzed using KPIs [22], [23]. Another aspect of considerable interest is using KPIs for sustainability assessments. Numerous aspects have been analyzed, including energy, emissions into the atmosphere, and the use of resources. Often, these assessments involve the three traditional pillars of sustainability: environment, economy, and society [24], [25], [26].

3. Methodology and case study

The methodology adopted in this study applied the principles of Lean Six Sigma as a combination of the speed inherent in the Lean production method with the statistical rigor of Six Sigma approaches. Specifically, the define, measure, analyze, improve, and control (DMAIC) approach represented the operational framework, which, through different tools, was used to fix problems with existing processes [27]. The description of phases that formed the Six Sigma DMAIC approach is reported in Table 1, identifying the objectives, activities, and main outputs that characterized this approach. The DMAIC approach was applied sequentially to the case study using different operational tools.

The case study described in this paper refers to a company that develops, transforms, and distributes packaging solutions such as packaging materials, machinery, and accessory products packaging. The multinational company operates in various countries around the world through numerous production plants and trading companies divided into clusters according to the geographical distribution. The case study supported the analysis of a specific plant located in Italy responsible for producing packaging materials and its marketing, especially to Italian customers. The logistics transport was the focus of this study.

The company monitors the deployment of logistics losses and costs through appropriate KPIs. It applies indicators to monitor the following aspects: logistics cost, inbound cost, outbound cost, warehouse cost, savings from co-loading, and fill rate. The goal of the distribution of the loss of the logistics costs was to control the end-to-end flow from the supplier to the customer, grouping the elements belonging to the same category and measuring their influence on inbound and outbound costs. For all the indicators used, the activities carried out at the plant considered in this paper guaranteed the achievement of the reference targets. The only exception was related to the FR, which reached a score below the reference target. This performance was confirmed for several past years. The FR was an outbound KPI included in the transportation losses that evaluated the space maximization of the transportation by a vector (truck for inland transportation and container for sea transportation). It was used for the inbound transportation of base materials and outbound transportation of the packaging materials and additional materials. The FR was the KPI target of the present study. The actual formula of the FR calculation of a shipment was based on the partial occupation of the additional material expressed in volume, added to the partial occupation of the packaging expressed in weight. Its formulation is reported in (1). This KPI calculation considered the additional material and packaging material separated because additional the materials grew in volume, thus decreasing the potential total "full" truck weight. In the case of the packaging materials, the maximum weight limitation of the truck or container was reached before the truck/container was filled up. For light additional materials, the truck or container was filled up before the weight limitation was reached. The FR losses considered the cost loss derived from the space not used in a vector for a shipment. It considered the actual shipment cost, and so it was evaluated on the contract prices with the carriers and the unplanned freight cost less extra charges such as cancellation, express charge, and waiting time. Its formulation is reported in (2).

DMAIC approach	Objectives	Activities	Possible tools	Outputs
Define the problem	Identification of the purpose of the work to clarify to everyone what improvements you want to make to the process under review	- Identification of team members - Selection the process under investigation - Definition of the project's objectives	- Benchmarking - SIPOC method - Gantt Charts - Project charter - Voice of Customer - Stakeholders' analysis - Quad-Chart - Multigenerational plans - ROIC analysis tools	- Project charter - Gantt chart - SIPOC map
Measure the baseline performance	Collection of useful data to characterize and analyze the objectives of the project	- Process flowcharts - Data collection - Data assessment	- Process flowcharts - Benchmarking - Value stream map - Gage reproducibility and repeatability - Pareto chart - Capability analysis	- Process Sigma value - First conclusion and/or observation on collected data
Analyze the root causes	Identification of the root causes of the business inefficiencies	- List of potential causes - Sort the causes	- Brainstorming - 5WHY technique - Modell expectation maximization - Diagram cause-effect - Hypothesis testing - Multi-vari chart - Root cause analysis (RCA) - Failure mode and effects	- - Identification of the gaps between the actual and goal performances, the causes, and the opportunities for improvement

			analysis (FMEA)	
Improve the process	Improvement in the process by determining potential solutions and ways to implement them	- Identification of possible improvement solutions - Analysis of improvement solutions	- Brainstorming - Expectation maximization model - Process flow improvement - Risk analysis - Lean tools - Simulation software - Mistake-proofing (Poka Yoke) - Prototyping	- Evaluate whether the solution is effective and financially viable
Control the improved process to prevent regression	Develop metrics that help leaders monitor and document continued success	- Process monitoring - Process analysis	- Statistical process control - Quality control plan - 5S - Mistake proofing (poka-yoke)	- Process control plan to mainstream gains

Table 1. Operational framework of the DMAIC approach

$$\text{FR KPI} = \frac{\textit{Shipment Weight}(PM, Strips, Lid\ Mat, Tab\ Mat, Film)(kg)}{\textit{Truck weight capacity}(kg)} + \frac{\textit{Shipment Volume}(Straws, Closures)(m^3)}{\textit{Truck Volume capacity}(m^3)} \quad (1)$$

$$\text{FR Losses} = [100\% - \text{Fill Rate of the vehicle}] \times \text{Shipment cost or Actual truck cost} - \text{Pro-rated full truck cost} \quad (2)$$

4. Results and discussion

This section describes the results obtained by applying each step of the DMAIC approach to the case study, and it provides some discussion.

4.1 Define the problem

To properly define the problem, the project charter was created, as reported in Table 2. This approach allowed us to characterize the main elements that defined the problem, identifying the objectives, the times, and the ways to proceed. Furthermore, the project team was created, involving members from different logistics areas to provide knowledge on the topic. Each was assigned a role in the project activities, such as manager, operational support, or action support.

Problem statement	Project scope	Objectives	Key process measurement
All the KPIs used by the company to assess the efficiency and effectiveness of the logistics losses and costs are consistent with the expected objectives, except for the FR, which has unsatisfactory performance	Understand the reasons for the failure to achieve the target set for the FR and establish the approach to achieve it	- Understanding the causes - Establish the necessary interventions - Monitor post-modification results	- FR target = 58% - Approximately 72% of the local shipping destinations and 28% of the export destinations - Approximately 1530 trucks/year
Timeline	**Deliverables**	**Leveraging opportunities**	**Project team**
D June M July A September I October–November C December	- Process data - Document plan - Reports and information	- Tools - Test results - KPIs - Improvements in operational management	- Director - Project coordinator - Logistics coordinator - Logistics export coordinator - Shipping team

Table 2. Project charter

4.2 Measure

To analyze the reference context, data about the FR and shipment details (cost and material quantities) were collected.

The maximum weight and volume restrictions and destination country were also analyzed.

Factories	Factory 1	Factory 2	Factory 3	Factory 4	Factory 5	Factory 6	Factory 7	Factory 8
FR Adjusted	80%	65%	65%	63%	59%	56%	55%	54%

Table 3. FR indices of the eight factories for the year 2019

We used benchmarking to analyze the performances of the FRs at different plants that formed the cluster to which the case study of this paper also belonged. Table 3 shows the annual value (for the year 2019) of the FRs for the eight cluster factories. Factory 8 had the poorest performance. Considering only the system that had the worst performance for the FR, the assessment of the "number of shipments" and "destination country" made it possible to isolate the highly frequented and lowly frequented destinations (Fig. 1). The highly frequented destinations were Spain, Portugal, Switzerland, France, Slovakia, Slovenia, Finland, and Germany. The lowly frequented destinations were Armenia, Bosnia and Herz, Bulgaria, China, Malaysia, Macedonia, Turkmenistan, and Uzbekistan.

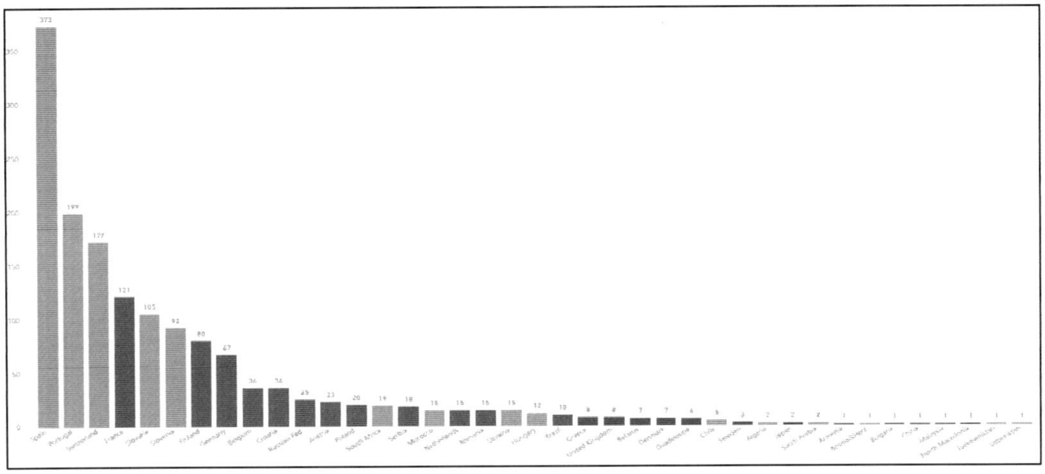

Fig. 1. FRs for the year 2019 per country (the numbers indicate the annual shipments to that country)

Fig. 1 indicates the destinations with an FR that met the company's target (in green) and those that did not reach the threshold (in red). The most popular destinations met the target (except for France, which was very close to the threshold), while the insufficient values refer to the countries served less frequently. It was interesting that those below the threshold were mainly European countries close to the plant that was the subject of this study, with an average number of 27 shipments per year. The most distant destinations (e.g., those outside of Europe) usually reached the FR target, and in any case, they were characterized by a limited number of shipments per year (approximately eight). This information allowed us to understand how, for distant shipments, more significant transport consolidation and optimization were always sought to reduce logistical costs. The second analysis concerned the composition of the shipments, verifying the frequency of the combined loads (packaging materials and additional materials). Only 4% of the total shipments to the market included additional materials in the order, while most of the loads (96%) were made up of packaging materials only. The conclusions were: first, separating the dimensions of the items' volumes and weights in the formula was not influential. The formula had its best scenario with 100% of the weight maximization plus (+) 100% of the volume maximization. However, this assumption did not reflect the reality because it was physically impossible to fix all of the trucks' spaces with the pallet combinations. Second, it was impossible to fill all the trucks' spaces because there were technical limits and constraints, such as the spaces next to the trucks' doorways, special palletization, and special requests, such as the prohibition of stacking. Third, the most frequent condition was that the items' weights for a single shipment influenced the FR.

4.3 Analize

Statistical tools and brainstorming were used to study the phenomenon. The 5WHY tool was the first method applied to delineate which variables were impactable (or not) in our study. This tool allowed us to analyze the emerging problems and explore their cause–effect relationships. From its application, four main elements that caused a low FR emerged (Table 4), providing useful indications that allowed for the evaluation of possible improvement interventions. In particular, the estimated time of arrival (ETA) comparability, the destination comparability, the unique delivery requests, and the numerosity of the items in a shipment and their space combinations were the aspects that were analyzed.

Problem: FR decreased due to a lack of delivery order consolidation				
Why	Why	Why	Why	Why
Completed the order with the same destination (or next) and different date ETAs	The customer did not accept the change to the ETA	It was assumed that the customer would not be satisfied	Lack of cooperation	Procedure not focused on the consolidation saving or there was no collaboration with the planning material
Problem: FR decreased due to not enough truck space for consolidation				
Why	Why	Why	Why	Why
Completed the order with the same destination (or next) and same date ETAs	Not enough space in the truck to merge the two orders	No collaboration between the CSRs during the DO planning	Satisfaction of the customer quantity requested	Procedure not focused on the consolidation saving or there was no collaboration with the planning material
Problem: FR decreased due to special palletization				
Why	Why	Why	Why	Why
There was more space booked in the truck to satisfy the special palletization or special request	It was assumed that it was possible to load 22 pallets in one truck regardless of the characteristics of the pallets	It was not possible for shipping to consider all of the special requests	There was no procedure defined	The formula did not take it into consideration
Problem: FR decreased due to not enough truck space and volume combination				
Why	Why	Why	Why	Why
There were items that could not fill the full truck space	The pallet configurations and reel weights	Each type of reel in the order had different weights and volumes	It was a problem of backpack optimization	We did not have the power to modify the items

Table 4. 5WHY tool

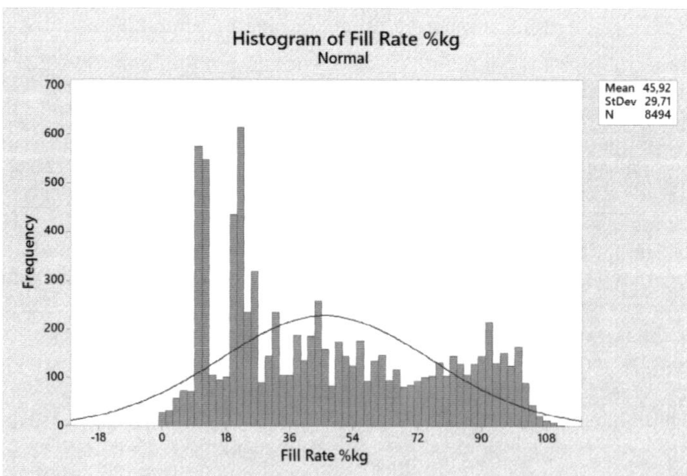

Fig. 2. FR Histogram of the plant analyzed in the case study

These elements were crucial for achieving merged transportation and, thus, improving the FR of a transport. The brainstorming showed that the main problems were caused by actions and decisions that did not directly depend on the logistics departments. Therefore, with these elements, it was assumed that to have a KPI that correctly evaluated the efficiency of the transport from the logistics side, it should not consider any elements that the logistics departments could not modify or those elements for which the logistics departments had no decision-making power.

Therefore, only data directly representative of the actions directly related to the logistics functions were considered. An analysis of the representative data of FRs considering all the shipments together is shown in Fig. 2, and it is evident that it was impossible to obtain a normal distribution of the index using the current formulation. This meant that the values were not concentrated around a single average value and grouping the data in any bell shape was impossible.

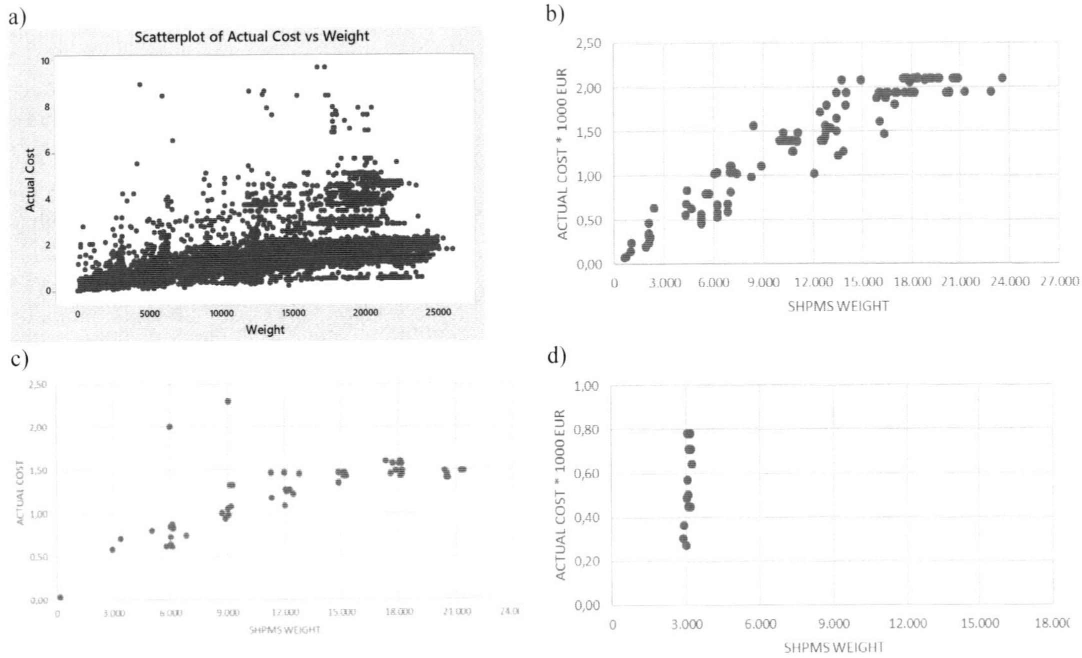

Fig. 3. Overall actual situation (cost vs weight): a) representation with all the distinctions without distinction; b) example of a destination where the trend was easily identifiable; c) example of a destination where the trend was identifiable, though with some anomaly; and d) example of a destination where the trend was unusual

Fig. 3 shows the scatter plots of the cost of shipping versus weight. Considering all the shipments together, Fig. 3a shows that there were no trends or behaviors because all the weights ranging up to 25 tons were exploited, and that was correct since we delivered the CSRs' orders that (most of the time) were randomly made without considering transport optimization. This result evidenced that every shipment's goods and weights were unpredictable. Instead, correlations were detectable if the shipments were divided by destination to cluster the data considering the 172 destinations where the analyzed site shipped. A total of 95% of the destinations are shown in 3b, 3c, and 3d, illustrating examples of destinations where the trends were easily identifiable, destinations where the trends were identifiable (though with some anomaly), and destinations with unusual trends, respectively. The costs increased with the weights of the loads in an almost constant trend up to a threshold, where the shipments were paid as full trucks loaded, regardless of the loads. The straight line had a slope that depended on a cost increase coefficient per pallet. A threshold represented the situation where the cost did not increase with the weight but remained constant within an error. There was not necessarily a point of discontinuity. The data analysis showed that 20% of the actual cost per shipment was larger than the full truck cost. This meant that these were cases with extra costs. A total of 7% of the shipments were paid as full trucks.

4.4 Improve

The improvement phase was divided into two main activities: developing a new formulation of the FR indicator and experimental application to the case study. The Expectation Maximization machine learning algorithm was applied to cluster the data for each destination. The algorithm allowed for defining a new formulation of the FR that was more representative of the analyzed context. A formulation capable of analyzing the most effective cost that each shipment had referred to the specific destination, and the weights of the goods were defined. In this way, the new FR was able to assess how much the real cost of shipping deviated from the best-expected cost. The new index was called pay weight. The developed model was based on general Bayes' rule, shown below in (3):

$$P(z|y,\theta)=\frac{P(y|z,\theta)\cdot P(z|\theta)}{P(y|\theta)}\Leftrightarrow P(y|\theta)=\frac{P(y|z,\theta)\cdot P(z|\theta)}{P(z|y,\theta)} \tag{3}$$

$$P(A_1 \mid B)=\frac{P(A_1 \cap B)}{P(B)}=\frac{P(B \mid A_1)\cdot P(A_1)}{P(B \mid A_1)\cdot P(A_1)+P(B \mid A_2)\cdot P(A_2)}=\frac{P(B \mid A_1)\cdot P(A_1)}{P(B \cap A_1)+P(B \cap A_2)}=\frac{P(B \mid A_1)\cdot P(A_1)}{P(B)} \rightarrow CVD \tag{4}$$

where A_1 and A_2 are the two models that the universe is divided into and B is the probability that the event observed belongs to model A_1 or to model A_2. Applying the model to the case study, its formulation became the following:

$$\begin{aligned} P(Model1 \mid y_i \cap x_i) &= \frac{P(Model1 \cap y_i \mid x_i)}{P(y_i \mid x_i)} = \frac{P(y_i \mid x_i \cap Model1)\cdot P(Model1 \mid x_i)}{P(x_i \cap Model1 \mid x_i)+P(y_i \cap Model1 \mid x_i)} \\ &= \frac{P(y_i \mid x_i \cap Model1)\cdot P(Model1 \mid x_i)}{P(y_i \mid x_i \cap Model1)\cdot P(Model1 \mid x_i)+P(y_i \mid x_i \cap Model1)\cdot P(Model2 \mid x_i)} \end{aligned} \tag{5}$$

$$\begin{aligned} P(Model2 \mid y_i \cap x_i) &= \frac{P(Model2 \cap y_i \mid x_i)}{P(y_i \mid x_i)} = \frac{P(y_i \mid x_i \cap Model2)\cdot P(Model2 \mid x_i)}{P(x_i \cap Model2 \mid x_i)+P(y_i \cap Model2 \mid x_i)} \\ &= \frac{P(y_i \mid x_i \cap Model2)\cdot P(Model2 \mid x_i)}{P(y_i \mid x_i \cap Model2)\cdot P(Model1 \mid x_i)+P(y_i \mid x_i \cap Model2)\cdot P(Model2 \mid x_i)} \end{aligned} \tag{6}$$

where *Model*1 is event 1 (the linear trend with a slope factor in the clustering), *Model*2 is event 2 (the linear trend without a slope factor in the clustering), *Model*1 ∩ *Model*2 is the universe, x_i is the weight of a shipment and an input (derived from the master data), y_i is the actual cost of a shipment and an input (derived from the master data), and, in general, P(*Model*1 | x_i) = P(*Model*1) and P(*Model*2 | x_i) = P(*Model*2).

The iterative algorithm calculated the probability that a data point x_i belonged to a specific cluster (*Model* 1 or *Model* 2) and converged to a local maximum. The model was also formulated through MATLAB code to make it operational. The results indicated a situation in which, first, the shipment had a good performance because its actual cost was equal to the prediction; second, the shipment had a good performance because its actual cost was lower than the expected; and third, the shipment had a bad performance because its actual cost was greater than expected. All the destinations were applied to evaluate the goodness of the new pay weight indicator.

4.5 Control

The calculation of the FR was conducted through both formulations (the initial and the new one) for some key destinations that were selected with a significant number of shipments. Table 5 reports the percentage of similarity between the evaluation results, intended as a percentage difference between +/- 5% between the results. The two formulations produced results that were not very similar in all destinations, with values between 18% and 44%. Only one destination in Spain had a high level of similarity that reached 70%. Graphically, Fig. 4 reports the trends for two destinations, i.e., number 1 in Fig. 4a and number 13 in Fig. 4b. With the new formulation, the level of the FR for the structure analyzed had achieved the performance objectives established by the company's management.

Destination	**Number of shipments**	**Similar results between the two formulations**
Destination 1: Spain	71	70%
Destination 2: Portugal	53	44%
Destination 3: Germany	131	28%
Destination 4: France	109	29%
Destination 5: Switzerland	349	23%
Destination 6: Spain	151	29%
Destination 7: Belgium	175	29%
Destination 8: Portugal	125	28%
Destination 9: Portugal	391	19%
Destination 10: Spain	342	22%
Destination 11: France	75	31%
Destination 12: Germany	262	19%
Destination 13: Belgium	109	18%
Destination 14: Spain	187	36%
Destination15: Spain	181	30%

Table 5. Project charter

a)

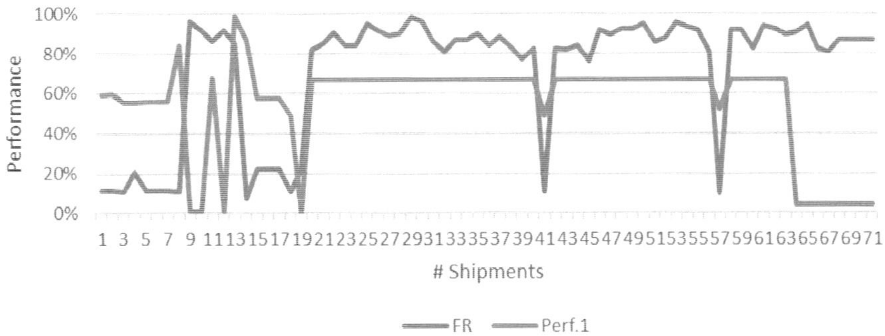

b)

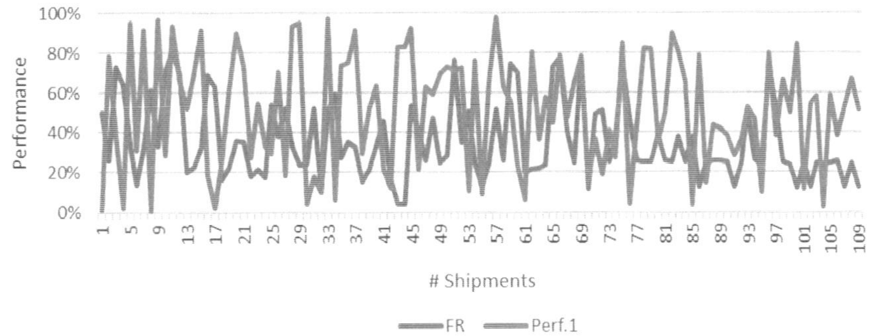

Fig. 4. Similarity trends between the two KPIs for two specific destinations

5. Conclusions

This paper proposes a novel approach for improving the understanding and control of logistics performances using optimized KPIs. The approach is an integral part of Lean Six Sigma, which is applied in production and process management as a method of improvement. With a specific case study, we evaluated its application to an improvement in the formulation of the FR applied to the monitoring of the performance of distribution logistics. The result was as follows: the company, with specific reference to one of its structures located in Italy, had improved the performance of the FR by reaching the reference targets set by the company and aligning the results with the performance of the other plants. The analysis showed that it proposed a formulation that was more suitable for understanding the phenomenon's complexity and overcoming the identified limitations. Based on the existing process used in Six Sigma, this paper improves the DMAIC approach. Due to the complexity of logistics, a new process was needed. Therefore, we can extend the define, measure, analyze, design, and verify (DMADV) approach to new processes in the future.

6. References

[1] Ristovska, N., Kozuharov, S., & Petkovski, V. (2017). The impact of logistics management practices on company's performance. International Journal of Academic Research in Accounting, Finance and Management Sciences, Vol. 7, No. 1, pp. 245-252. https://doi.org/10.6007/ijarafms/v7-i1/2649

[2] Gani, A. (2017). The logistics performance effect in international trade. The Asian Journal of Shipping and Logistics, Vol. 33, No. 4, pp. 279-288. https://doi.org/10.1016/j.ajsl.2017.12.012

[3] Ghoumrassi, A. & Tigu, G. (2018). The impact of the logistics management in customer satisfaction. In Proceedings of the International Conference on Business Excellence, Vol. 12, No. 1, pp. 407-415. https://doi.org/10.2478/picbe-2018-0036

[4] Omoruyi, O. (2018). Competitiveness through the integration of logistics activities in SMEs. Studia Universitatis Babes Bolyai-Oeconomica, Vol. 63, No. 3, pp. 15-32. https://doi.org/10.2478/subboec-2018-0012

[5] Lambert, D. M., & Burduroglu, R. (2000). Measuring and selling the value of logistics. The International Journal of Logistics Management, Vol. 11, No. 1, pp. 1-18. https://doi.org/10.1108/09574090010806038

[6] Wudhikarn, R., Chakpitak, N., & Neubert, G. (2018). A literature review on performance measures of logistics management: an intellectual capital perspective. International Journal of Production Research, Vol. 56, No. 13, pp. 4490-4520.

[7] Muha, R. (2019). An overview of the problematic issues in logistics cost management. Pomorstvo, Vol. 33, No.1, pp. 102-109. https://doi.org/10.31217/p.33.1.11

[8] Marotta, A., Studer, L., Marchionni, G., Ponti, M., Gandini, P., Agriesti, S., & Arena, M. (2018). Possible impacts of C-ITS on supply-chain logistics system. Transportation research procedia, Vol. 30, pp. 332-341. https://doi.org/10.1016/j.trpro.2018.09.036

[9] Rantasila, K., & Ojala, L. (2015). National-level logistics costs: an overview of extant research. International journal of logistics research and applications, Vol. 18, No. 4, pp. 313-324.

[10] Škerlič, S., & Sokolovskij, E. (2019). A model for managing logistics costs throughout a product's life cycle: a case study of a multinational manufacturing company. Transport, Vol. 34, No. 5, pp. 517-528. https://doi.org/10.3846/transport.2019.11080

[11] Domingues, M. L., Reis, V., & Macário, R. (2015). A comprehensive framework for measuring performance in a third-party logistics provider. Transportation Research Procedia, Vol. 10, pp. 662-672. https://doi.org/10.1016/j.trpro.2015.09.020

[12] Badawy, M., Abd El-Aziz, A. A., Idress, A. M., Hefny, H., & Hossam, S. (2016). A survey on exploring key performance indicators. Future Computing and Informatics Journal, Vol. 1, No. 1-2, pp. 47-52. https://doi.org/10.1016/j.fcij.2016.04.001

[13] Schmidt, M., Maier, J. T., & Härtel, L. (2019). Data based root cause analysis for improving logistic key performance indicators of a company's internal supply chain. Procedia CIRP, Vol. 86, pp. 276-281. https://doi.org/10.1016/j.procir.2020.01.023

[14] Brint, A., Genovese, A., Piccolo, C., & Taboada-Perez, G. J. (2021). Reducing data requirements when selecting key performance indicators for supply chain management: The case of a multinational automotive component manufacturer. International Journal of Production Economics, Vol. 233, pp. 107967. https://doi.org/10.1016/j.ijpe.2020.107967

[15] Karl, A. A., Micheluzzi, J., Leite, L. R., & Pereira, C. R. (2018). Supply chain resilience and key performance indicators: a systematic literature review. Production, pp. 28. https://doi.org/10.1590/0103-6513.20180020

[16] Maestrini, V., Luzzini, D., Maccarrone, P., & Caniato, F. (2017). Supply chain performance measurement systems: A systematic review and research agenda. International Journal of Production Economics, Vol.183, pp. 299-315. https://doi.org/10.1016/j.ijpe.2016.11.005

[17] Qorri, A., Mujkić, Z., & Kraslawski, A. (2018). A conceptual framework for measuring sustainability performance of supply chains. Journal of cleaner production, Vol. 189, pp. 570-584. https://doi.org/10.1016/j.jclepro.2018.04.073

[18] Larrea, O. (2013). Key performance indicators in humanitarian logistics in Colombia. IFAC Proceedings Volumes, Vol. 46, No. 24, pp. 211-216. https://doi.org/10.3182/20130911-3-BR-3021.00089

[19] Dörnhöfer, M., Schröder, F., & Günthner, W. A. (2016). Logistics performance measurement system for the automotive industry. Logistics Research, Vol. 9, pp. 1-26. https://doi.org/10.1007/s12159-016-0138-7

[20] Dumitrache, C., Kherbash, O., & Mocan, M. L. (2016). Improving key performance indicators in Romanian large transport companies. Procedia-Social and Behavioral Sciences, Vol. 221, pp. 211-217. https://doi.org/10.1016/j.sbspro.2016.05.108

[21] Kucukaltan, B., Irani, Z., & Aktas, E. (2016). A decision support model for identification and prioritization of key performance indicators in the logistics industry. Computers in Human Behavior, Vol. 65, pp. 346-358. https://doi.org/10.1016/j.chb.2016.08.045

[22] Butzer, S., Schötz, S., Petroschke, M., & Steinhilper, R. (2017). Development of a performance measurement system for international reverse supply chains. Procedia Cirp, Vol. 61, pp. 251-256. https://doi.org/10.1016/j.procir.2016.11.264

[23] Sangwan, K. S. (2017). Key activities, decision variables and performance indicators of reverse logistics. Procedia Cirp, Vol. 61, pp. 257-262. https://doi.org/10.1016/j.procir.2016.11.185

[24] Kurtagic, S. M. & Kadric, D (2022). Improvement of Airports Energy Efficiency, Case Study of Airport Sarajevo, Annals of DAAAM for 2022 & proceeding of the 33th International DAAAM Symposium, Published by DAAAM International, pp.133-142, https://doi.org/ 10.2507/33rd.daaam.proceedings.019

[25] Neri, A., Cagno, E., Lepri, M., & Trianni, A. (2021). A triple bottom line balanced set of key performance indicators to measure the sustainability performance of industrial supply chains. Sustainable Production and Consumption, Vol. 26, pp. 648-691. https://doi.org/10.1016/j.spc.2020.12.018

[26] Singh, C. S., Soni, G., & Badhotiya, G. K. (2019). Performance indicators for supply chain resilience: review and conceptual framework. Journal of Industrial Engineering International, Vol. 15, pp. 105-117. https://doi.org/10.1007/s40092-019-00322-2

[27] Kowalik, K. (2018). Six Sigma as a method of improving the quality of service process. Production Engineering Archives, Vol. 19, No. 19, pp. 10-15. https://doi.org/10.30657/pea.2018.19.03

DOI: 10.2507/34th.daaam.proceedings.063

CARGO DIVERSIFICATION OPPORTUNITIES AND METHODS IN PORTS UNDER THE INFLUENCE OF ECONOMIC CHANGE

Astrida Rijkure

This Publication has to be referred as: Rijkure, A[strida] (2023). Cargo Diversification Opportunities and Methods in Ports under the Influence of Economic Change, Proceedings of the 34th DAAAM International Symposium, pp.0485-0490, B. Katalinic (Ed.), Published by DAAAM International, ISBN 978-3-902734-41-9, ISSN 1726-9679, Vienna, Austria

DOI: 10.2507/34th.daaam.proceedings.063

Abstract

Current study provides evaluation of the processes for assessing the efficiency in the transport cargo structure process based on real-time research using the same technologies and methods as well as offering new methods for assessing cargo diversification. The aim of this study was to confirm the importance of timely estimation of possibilities to diversify the cargo structure in order to take respective measures before the specified risks occur in the multimodal transport system by developing a methodological proposal for monitoring and assessment of the efficiency of the cargo structure as well as for measures and comparative assessments. Methods and materials: The principal materials used for the studies are as follows: various sources of literature, e.g. scholars' articles, research papers and the reports of institutions. The following suitable qualitative research methods have been used: monographic; analysis and synthesis; grouping, logical and abstractive constructional etc.

Keywords: ports economic; cargo diversification; port efficiency; blue economy; diversification methodology

1. Introduction

This study investigates the impact of diversification-driven industrial transformation, adopted by port enterprises to combat slowing economic growth and domestic competition, on their operational performance. The study focuses on publicly listed port-related enterprises in corresponding port cities, considering their interaction. The study analyzes the impact of the industrial transformation and urban economic development level of port cities on three aspects of companies' operating performance: profitability, operating capacity, and development capability. Findings indicate that the relationship between the industrial transformation and operating capacity is nonlinear and inversely shaped, suggesting an increasing negative impact on profitability and development capability. The level of urban economic development within port cities enhances profitability and development capability of port enterprises, has minimal effect on their operating capacity, but positively regulates the relationship between the diversification-driven industrial transformation and operating performance. The port is one of the most important segments in the transport chain and one of the most important facilitators and providers of the economic development of countries. Each port is unique in terms of its geographical location, infrastructure and offered range of services, which are aimed at servicing the maximum number of ships and processing the maximum amount of cargo, as well as being competitive in the global market of logistical services.

In order to make an overall evaluation of a port, systems of port operation efficiency must be developed, because, if the efficiency of port operation is known, more efficient use of the resources at the disposal of the port can be promoted, thus raising the competitiveness of the port. The directions of cargo transit flow often change radically under the influence of contemporary economic processes as well as political conditions, therefore ports must be capable of diversifying their structure, by sustaining minimum financial loss.

Ports play an important role in the global supply chain through efficient logistics operations, and each port enterprise aims to increase economic value added through its unique competitive advantage [21]. Especially under the pressure of global competition, port authorities are looking for ways to improve their operational performance, which is an important indicator of the effectiveness of the activities the enterprises perform [20]. Scholars have conducted extensive research on the impact of diversification-driven industrial transformation strategies on business performance, but have arrived at dissimilar conclusions. This indicates that there is a complex relationship between a port's diversification-driven industrial transformation and its operating performance. Studies on port–port city relationships have proven that coordination between the two entities is conducive to achieving positive outcomes for both. However, to date, few scholars have analyzed the impact of port enterprises' diversification-driven industrial transformation on their operating performance from the perspective of port–port city relationships. As an important external environmental factor, the urban economic development level of port cities can provide essential synergy for port enterprises that implement diversification-driven industrial transformation strategies to improve performance. [19]

2. Materials and methods

The present study provides the evaluation of the processes that are used to evaluate the efficiency in the transport cargo structure process, based on the research studies performed in real time by using the same technologies and methods, as well as by offering new cargo diversification assessment methods. The objective of the study was to confirm how important prompt performance of potential calculations are, regarding the possibilities of diversifying the cargo structure, in order to ensure prompt implementation of appropriate measures, before the determined risks occur in a multimodal transport system, by determining the methodical proposal for the monitoring and assessment of the efficiency of cargo structure, as well as for the development of measures and comparative assessments.

3. Results and discussion

Ports are one of the most significant segments in the logistical chain of cargo and passenger transportation, which occupy an important role at economical, political and national level in a country. Each port is unique in terms of its geographical location, infrastructure and the range of services offered. The efficiency parameters of port operation will depend on the efficiency and scope of the use of port resources. Therefore, organisation of the economic activity of the port in the manner that balances the capacity of the port, with the existing and prospective flow of overland cargo and the development of infrastructure, considering the alternative directions of cargo transportation through other ports, is important to ensure the efficient operation of a port. Ports have a significant role in the transport industry and promotion of competitiveness and their potential in creation of workplaces and attracting investment, is huge. [9]

The port industry has a significant influence on the economy, which is expressed both as the promotion of employment and activity in the industry of ports (direct effect), the influence on the supply chain (indirect effect), as well as a wider influence on the economy (induced effect). Ports are business locations of companies providing a wide range of services – for instance, coal, container and other cargo reloading services. Costs and quality of services provided by ports are an important element of the business environment. Port costs can constitute a significant part of the total costs of logistical services. Ports considerably differ in terms of their productive capacity. Currently, the cargo structure significantly differs at different ports. Differences in capacity cause a significant drop in efficiency – routes become longer as a result of significant deviations in them, the transportation requires more time, the distances covered by sea and road last longer and, consequently, more transport emissions and transport overload occur, which harms residents and the economy.
Considering only tentative economic growth forecasts, the cargo volume at ports is expected to grow by 50% until 2030 and the rapidly growing sector of container traffic is expected to grow even faster. In order to operate under the conditions of forecasted growth, the decisions must be taken as early as today, while the effect of such decisions will be experienced after 5 to 10 years. If ports fail to act now, the economic growth will be endangered by the risk of overload and high external costs – especially in port cities and regions and port connections with larger mainland areas between them. [7]
The efficiency of ports differs considerably: not all ports are equally successful in their operations and the differences between ports that have adapted to the new logistical and economic conditions, and the ports that have failed to adapt have only increased over the last years. As a result of differences in productive capacity, traffic flows are being diverted considerably, which adversely affects the business opportunities at the ports that are operating well. This endangers the efficiency and sustainability of the transport network and the overall competitiveness of the economy.

Ports must modernise the provision of their services, improve connections between ports and ensure that all ports are capable of offering as high quality of services, as possible. This means that ports that are falling behind must gain on the ports that are developing faster. Transit cargoes dominate the cargo structure of several ports, which, in turn depend on the economic, as well as political development of other countries. Therefore, there is a risk that, as a result of economic or political decisions, these transit cargoes will be diverted to other ports with more favourable business conditions.

Previous research studies regarding terminal operators' strategies focused predominantly on the activities and strategies of container terminals that operate in large seaports, i.e., specialisation strategies. There are few studies addressing diversification strategies in business models followed by operators of multipurpose terminals, operating in multipurpose, small and medium ports or secondary ports [5] [17]. Secondary ports function in a highly competitive environment of large hub ports, also providing complementary functions for them [11]. This has a significant effect on the directions of strategic development of terminal operators in secondary ports. Compared to hub ports, secondary ports to a larger extent experience the effects of structural changes taking place in the world economy, maritime trade and transport, which are also stimulated by the increasingly strict climate policy [12] [13] [14]. The impact of those changes on the operations of seaports is manifested by the gradual decrease in trading of traditional bulk cargo groups (coal, ores) or general cargo. The competitiveness of this category of ports is also limited by their technical parameters, which prevent the ports from handling increasingly large vessels that are put in operation. Consequently, secondary ports are under strong pressure as they need to attract new cargo groups to replace those on the decline. This justifies implementation of diversification measures by terminal operators, focusing on the search for new cargo groups, new cargo flow directions and expanding the range of offered services. [16]

For this reason, ports must promptly evaluate the sustainability of their cargo structure and assess the possibilities of diversifying the cargo and service structure of the port. To perform a study of the influence of contamination processes of heat exchange equipment on reducing the quality of regulation, it is necessary to assess the quality of closed loop single-circuit regulation of the heat exchanger, taking into account the technological wear of the equipment and to simulate the change in its characteristics over time. [15]

Diversification is a business development strategy allowing a company to enter additional lines of business that are different from the current products, services and markets. In the current conditions of dynamic markets and strong competition, a successful instrument of risk management is to avoid focusing on a single product, service and/or their distribution to a single limited market. When implemented wisely it contributes to keeping the company stable even in hard times since the economic downturn usually occurs simultaneously in all sectors and all markets. Diversification of business activities brings competitive advantages allowing companies to reduce business risks. That is why it is a great tool for business development. However, its successful implementation requires profound knowledge and thorough preliminary assessment of the company and its environment. And, although sometimes diversification is difficult for the small companies, it can prove to be inevitable when their original markets become unviable.

Diversification, being a strategic approach, is the subject of extensive research aiming to examine its relation to the financial results of the market. In the majority of studies a comparison is done between results of related and unrelated diversification, although the distinctive line between them is still not clear. [18]

Depending on the direction of company diversification, the different types are:

- *Horizontal Diversification* - acquiring or developing new products or offering new services that could appeal to the company´s current customer groups. In this case the company relies on sales and technological relations to the existing product lines.
- *Vertical Diversification* - occurs when the company goes back to previous stages of its production cycle or moves forward to subsequent stages of the same cycle - production of raw materials or distribution of the final product.
- *Concentric Diversification* - enlarging the production portfolio by adding new products with the aim of fully utilising the potential of the existing technologies and marketing system. The concentric diversification can be a lot more financially efficient as a strategy, since the business may benefit from some synergies in this diversification model. It may enforce some investments related to modernizing or upgrading the existing processes or systems. This type of diversification is often used by small producers of consumer goods.
- *Heterogeneous (conglomerate) diversification* - is moving to new products or services that have no technological or commercial relation with current products, equipment, distribution channels, but which may appeal to new groups of customers. The major motive behind this kind of diversification is the high return on investments in the new industry. Furthermore, the decision to go for this kind of diversification can lead to additional opportunities indirectly related to further developing the main company business - access to new technologies, opportunities for strategic partnerships, etc.
- *Corporate Diversification* - involves production of unrelated but definitely profitable goods. It is often tied to large investments where there may also be high returns. [18]

Diversification of risks is required, since the expansion of the market, as well as customer range, makes the company or a market segment independent of the economic situation in the local market.

a – growing trend of cargo volume;
x – critical point of cargo volume growth;
S – the required mechanism of compensation for the fall in cargo volume;
Sd – cargo diversification.

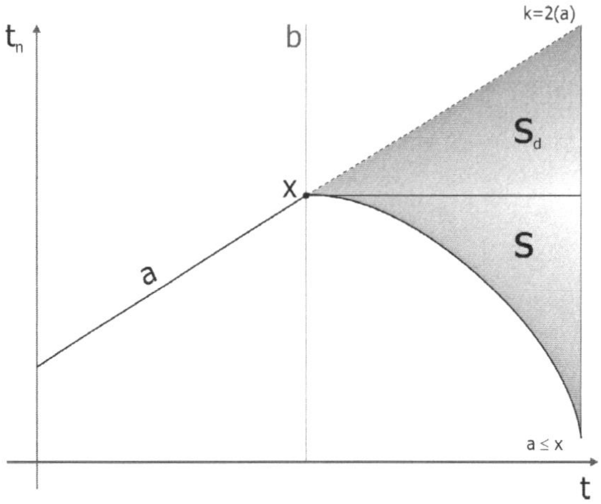

Fig. 1. Cargo diversification methodology

This figure illustrates critical point of cargo volume growth (x), when it is necessary to diversify the cargo structure in order not to reduce the amount of cargo. It is important to appreciate the right moment when the process of cargo diversification has to start, because it takes some time and the result is not immediate. Several risks that need assessment exist, when the diversification process is initiated:

- Political, economic and social conditions
- Technological solutions
- Legislation and taxation policy
- Infrastructure and logistics

- *Political support* is the decisive factor for cargo diversification, although minor, political risks exist, becoming especially topical during economic crisis conditions, which can affect the fluctuations of currency exchange rate, as well as international trade policy.
- *The economic situation* in the country determines the availability of financial resources and purchasing power, inflation, as well as currency exchange rates. All these are factors that significantly affect the proportion of international trade.
- *Social conditions* or the lifestyle of the residents. Under the conditions of diversification this factor is most directly affected by the advertising policy and public relations.
- *Technological solutions* or innovations considerably reduce the life cycle of products, which causes the necessity for investment, research and development.
- *Legislation and taxation policy*, it is important to assess, whether the created conditions are favourable for the initiation of a diversification process, whether the existing legislation creates obstacles for the launching of new products and services on the market.

Optimisation *of infrastructure and logistics* is the principal pre-requisite for the attraction of new groups of cargoes to your market segment. Serious market research or regular monitoring of the target market must be performed in order to initiate the diversification process, which is best implemented by participation in trade missions [10], industrial trade fairs or conferences, and supplementing the aforementioned data by publicly accessible information and statistical analysis. In order to initiate a diversification process, a diversification strategy [8] must be developed, which must be performed at the highest level of policy planning in the context of the state. This strategy must comply with three main criteria:

1. Technical resources;
2. Financial resources;
3. Sales volumes.

Only if these macro-economic processes are considered, the re-distribution of market segments can be planned. [3] The following external factors must be considered to successfully start the development of the strategy, which can affect the diversification result:

1. Global economy trends - cargo volumes and market prices, topical transport corridors, etc.
2. International and political environment – important international trade trends and relations with target market countries;
3. Government support – a decisive factor to initiate the process of cargo diversification.

The author offers the following formula, which has been opportunted to evaluate the quantity of cargo diversification characteristics, based on Figure 1:

$$S_d = k \times a - S_{a\geq 1}^t \qquad (1)$$

where
S_d- cargo diversification
k – cargo increase coefficient
t – time period, within which cargo increase is planned to occur
a – growing trend of cargo volume

In this formula is important to choose the optimal cargo increase coefficient (k), based on the time period (t) required for cargo diversification. This is a function that needs to be fulfilled in order to assess the potential for diversion of cargo:

$$S_d = \sum_{a\geq 1}^{t} (I + E + T) \qquad (2)$$

where
S_d- cargo diversification
I – investment for the development of infrastructure
E – entry into new markets
T – optimisation of legal status

These are the conditions that must be met in order to make it possible and successful to cargo diversification. It is necessary to make the investment for the development of infrastructure, to look for new markets for goods and services and there is a need to optimisation of legal status. Moreover, the studies completed so far have shown that even though specialisation has a positive effect on the results of seaport activity [1], excessive specialisation, understood as strong dependence on several kinds of cargoes, may be detrimental not only to the ports, but also to the local economy dependent on the port [6]. In turn, diversity of port activity unambiguously stimulates the port's growth and development [4]. On the other hand, according to some studies [6], the global shipping network is strongly dependent on the more diversified transport nodes which in turn take over the greatest traffic and show higher connectivity.

Diversification can be considered a useful business development tool for companies in any sector and location of the economy. However, you have to keep in mind that there is no recipe for successful diversification. It depends on multiple internal and external factors for each company, which should be carefully studied and taken into consideration when developing a diversification strategy. It is also very important to implement and update this strategy according to the dynamically changing conditions and strong competition under which business usually operates.

4. Conclusion

1. The problem is the lack of methods for assessing the impact of industrial transformation and the level of urban-economic development of port cities. It can be attributed to three performance indicators of companies: profitability, operational capacity and development capacity. The creation of cargo diversification methodology addresses the possibility of timely assessment of ways to identify cargo diversification deadlines.
2. Ports considerably differ in terms of their productive capacity. Currently, the cargo structure significantly differs at different ports. Differences in capacity cause a significant drop in efficiency – routes become longer as a result of significant deviations in them, the transportation requires more time, the distances covered by sea and road last longer and, consequently, more transport emissions and transport overload occur, which harms residents and the economy.
3. Diversification of cargo is the last alternative to implement, where the existing market distribution does not ensure sufficient generation of profit.
4. The decisions regarding cargo structure diversification must be made promptly, as soon as the first signs bear evidence that the critical point of cargo volume growth has been reached, without waiting for the moment when cargo volumes start to fall. Therefore, ports must promptly assess the sustainability of their cargo structure and possibilities of diversification of the cargo structure and services provided.
5. Diversification of risks is required, since the expansion of the market, as well as customer range, makes the company or a market segment independent of the economic situation in the local market. In order to initiate a diversification process, a diversification strategy must be developed, which must be performed at the highest level of policy planning in the context of the state. This strategy must conform to three main criteria: technical resources, financial resources, sales volumes.
6. Certainly, for future research, other diversification methods idenfication properties of these materials should be analysed to better understand the behaviour of the diversification methods using different models with the aim of their application for other economics applications.

5. Acknowledgments

ERDF project No. 1.1.1.2/VIAA/1/16/061 "Long-term impact of blue economy on the increase of the competitiveness of ports in the Baltic Sea region"

6. References

[1] Adler, N.; Hirte, G.; Kumar, S.; Niemeier, H.M. (2022)The impact of specialization, ownership, competition and regulation on efficiency: A case study of Indian seaports. Marit. Econ. Logist. 24, 507–536.

[2] Ansoff, H. I. (1958) A Model for Diversification., Management Science, vol. 4, no. 4, pp. 392–414. JSTOR,. Accessed 31 July 2023., pp.399-400.

[3] Barnett W., (1988) Our Steps to Forecast Total Market Demand., Harvard Bussiness Review., Available at: https://hbr.org/1988/07/four-steps-to-forecast-total-market-demand, Accessed 14 April 2023.

[4] Bun, M.J.G.; El Makhloufi, A. (2007) Dynamic externalities, local industrial structure and economic development: Panel data evidence for Morocco. Reg. Stud., 41, 823–837.

[5] Ding, Z.Y.; Jo, G.S.; Wang, Y.; Yeo, G.T. (2015) The Relative Efficiency of Container Terminals in Small and Medium-Sized Ports in China. Asian J. Shipp. Logist., 31, pp.231–251.

[6] Ducruet, C. (2020) Port specialization and connectivity in the global maritime network. Marit. Policy Manag., 49, 1–17.

[7] Europe's Seaports 2030: Challenges Ahead. European Commission, (2013), Available at: https://ec.europa.eu/commission/presscorner/detail/en/MEMO_13_448 , Accessed 2 Februar 2023.

[8] Henryanto C., (2015) Unrelated Diversification Strategies by Port Authorities: a Case Study based Comparison., pp.6-7.

[9] Kos S., Samija S., Brcic D., (2012) Multimodal transport in the function of port system containerazition developmen, Available at: https://www.researchgate.net/publication/236724372_Multimodal_transport_in_the_function_of_the_port_system_containerization_development, Accessed 2 March 2023.

[10] Marciano A., (2019) Image Analysis and Registration Methods for Cargo and vehicles X-Ray Imaging: Available at: https://theses.hal.science/tel-02044892/file/TheseFinale-MARCIANO.pdf , Accessed 12 April 2023.

[11] Monios, J. (2017) Cascading feeder vessels and the rationalisation of small container ports. J. Transp. Geogr., 59, 88–99.

[12] Notteboom, T.E.; Winkelmans, W. (2001) Structural changes in logistics: How will port authorities face the challenge? Marit. Policy Manag., 28, 71–89.

[13] Oniszczuk-Jastrzabek, A.; Czermański, E.; Dębicka, O.; Czuba, T. (2019) Globalization process in the maritime transport-causes, symptoms and effects. Ann. Univ. Apulensis Ser. Oeconomica, 1, 65–74.

[14] Pallis, A.A.; de Langen, P.W. (2010) Seaports and the structural implications of the economic crisis. Res. Transp. Econ., 27, 10–18.

[15] Peshko M., Batrakov P., Lasitsa A., Khomchenko V., (2022) Technological Wear Influence Analysis on the Decrease in the Efficiency of a Closed Loop Control of Heat Exchange Equipment.,., Proceedings of the 33rd International DAAAM Virtual Symposium "Intelligent Manufacturing & Automation" Volume 33, No.1

[16] Pluciński M., Kotowska I., Mańkowska M., Filina-Dawidowicz L., (2023) Research on Diversification Strategies of Terminal Operators—Evidence from Polish Seaport Sustainability 2023, 15(7), 5644.

[17] Rahman, N.S.F.A.; Ismail, A.; Othman, M.K.; Roslin, R.A.M.; Lun, Y.H.V. (2018) Decision making technique for analysing performance of Malaysian secondary ports. Int. J. Shipp. Transp. Logist., 10, 468–496.

[18] Strategy Train [Onlain platform] Available at: http://www.strategy-train.eu/ , Accessed 23 March 2018.

[19] Sun Y., Zhang S., Wu S., (2022) Impact of Ports' Diversification-Driven Industrial Transformation on Operating Performance: Regulatory Effect of Port Cities' Urban Economic Development Level., Water 2022, 14(8), 1243.

[20] Tongzon, J.L. (1995) Determinants of port performance and efficiency. Transp. Res. A, 29, 245–252.

[21] Wang, Z.; Subramanian, N.; Abdulrahman, M.D.; Cui, H.; Wu, L.; Liu, C. (2017) Port sustainable services innovation: Ningbo port users' expectation. Sustain. Prod. Consump., 11, 58–67.

DOI: 10.2507/34th.daaam.proceedings.064

A CONCEPTUAL MODEL FOR OPEN U-LEARNING PLATFORM

Pavlinka Radoyska* & Irina Topalova

This Publication has to be referred as: Radoyska, P[avlinka] & Topalova, I[rina] (2023). A Conceptual Model for Open U-Learning Platform, Proceedings of the 34th DAAAM International Symposium, pp.0491-0495, B. Katalinic (Ed.), Published by DAAAM International, ISBN 978-3-902734-41-9, ISSN 1726-9679, Vienna, Austria
DOI: 10.2507/34th.daaam.proceedings.064

Abstract

Modern teaching methods applied in a university environment largely determine the quality and effectiveness of the educational process. The choice of a certain method and its application is left to the respective educational institution, which must make the right choice, according to the specifics of its educational programs and goals. In this paper a comparative review of modern e-learning, m-learning and u-learning methods is presented. Their main characteristic parameters are exposed. An open conceptual model for the u-learning platform is proposed. The model is focused on using inside and external, internet-based learning resources and is based on Artificial intelligence to offer the most proper learner-centered learning.

Keywords: Teaching methods; Educational process; Educational platform; E-learning; U-learning.

1. Introduction

The learning process has two main components: learning strategies and technologies used in learning. More and more new technologies are being incorporated into the learning process due to computers and the Internet: new writing technologies, new communication technologies, new visualization and storage technologies, etc. [1]. New learning paradigms are emerging as a result of technological developments. The paradigms of e-learning, m-learning and u-learning are of greatest interest from a technology perspective. The paradigm for computer-based learning emerged as early as the middle of the 20th century. The concept of e-learning appeared at the end of the 20th century, provoked by the emergence of the Internet. Learning Management Systems (LMS) and virtual classrooms began to appear. The real boom in e-learning was at the beginning of the 21st century when many universities began to offer online courses.

The main characteristics of the e-learning are clearly described in [2]. It can be summarized that e-learning is aimed at offering anytime, anywhere learning content that enables learners to follow their own pace of learning, receive feedback and test the level of knowledge acquired, to communicate synchronously and asynchronously with teachers and other learners. LMS is used as a platform to implement e-learning. The learning materials are locked in the selected LMS and limited by the resources published in it. An LMS allows for some customization. It allows different learning paths to be followed, which are however limited by the resources of the portal used. Additional forms of interactivity can also be built into modern LMSs. In addition to the standard discussion boards, tools for voice-audio conference connections are easily embedded. It can be summarized that e-learning is aimed at offering anytime, anywhere learning content that enables learners to follow their own pace of learning, receive feedback and test the level of knowledge acquired, to communicate synchronously and asynchronously with teachers and other learners.

LMS is used as a platform to implement e-learning. The learning materials are locked in the selected LMS and limited by the resources published in it. An LMS allows for some customization. It allows different learning paths to be followed, which are however limited by the resources of the portal used [3]. The term mobile learning (m-learning) began to be used in the late 1990s with the appearance of the first mobile phones, but the real experiments began at the beginning of the 21st century with the appearance of more modern mobile devices, such as PDA devices (Personal Digital Assistants) and early smartphones. M-learning extends the possibilities of e-learning by having almost the same features as e-learning but allowing access from mobile devices. This improves access to learning resources. The use of mobile devices in m-learning leads to some problems. For example, smaller display sizes, limited capabilities of operating systems, variety of operating systems, memory limitation, and security threats [4].

Ubiquitous learning (u-learning) [5] began to be talked about at the beginning of the 21st century. U-learning can be defined as a learning environment that is supported by mobile and embedded computers and used for formal and informal learning. It provides learners with content and interaction anytime, anywhere, similar to mobile learning. But it is characterized by a high degree of personalization - the learning materials are in different formats (text, video, collaborative work, games, etc.) and are tailored to the current learning goals, interests and preferences of the learner, cognitive characteristics, learning history and current state of his competence in the field of study, the characteristics and peculiarities of the location, the technology used as a learning environment and the context of the situation in which the learning takes place [6]. We can highlight several significant features of u-learning:

- Inclusion of Social Media and User-Generated Content (YouTube, Wikipedia).
- The integration of artificial intelligence (AI) and machine learning (ML) algorithms into educational technology to build adaptive learning systems.
- Inclusion of Augmented Reality (AR) and Virtual Reality (VR). The advancements in AR and VR technologies opened up new possibilities for immersive and interactive learning experiences, allowing learners to be engaged with content in a more hands-on way.
- Internet of Things (IoT) and Smart Environments. The IoT brought about the concept of smart environments, where everyday objects and spaces are connected to the internet. This has the potential to create highly interactive and context-aware learning experiences.
- Gamification and Education Technologies Start-ups. The incorporation of gaming elements into educational experiences, as well as the proliferation of educational technology start-ups, brought new approaches to engaging learners and making education more interactive.

The purpose of this paper is to propose a conceptual model for open u-learning platform. The conceptual model represents the main components and the implementation requirements of the main tasks, faced by an open u-learning platform. The next part of the paper is organized as follows: in Section 2 we give a comparison between e-learning, m-learning and u-learning features. In Section 3 we present the most popular u-learning platforms. In Section 4 the proposed conceptual model for open u-learning platform is introduced. In Section 5 the conclusion statements and outline for further development of the work are discussed.

2. Comparison between u-learning, e-learning and m-learning

The most important parameters of current paradigms for digital learning are Accessibility, Flexibility, Collaboration, Interactivity, and Personalization. Table 1 represents the comparison of the three learning paradigms, with respect to these parameters.

Parameter	e-learning	m-learning	u-learning
Accessibility	At any time and from any place.	Anytime, anywhere and from different devices, including mobile.	Anytime, anywhere and from different devices, including mobile.
Flexibility	Pre-prepared content is available, which can be in different forms: text, video, and quizzes. The learner can determine his learning path within the proposed resources.	As with e-learning, but the content is adapted to the capabilities of mobile devices.	Resources from social networks, games, AR and VR, and IoT are added to the possibilities characteristic of m-learning. The learning path is formed using AI and ML.
Collaboration	Collaboration is often facilitated through online platforms and tools like: discussion forums, group projects, virtual	To all e-learning collaboration tools are added discussion boards, instant	Collaboration is like m-learning but can occur in a wide range of settings, including physical locations, online spaces,

	classrooms, and collaborative document editing.	messaging, and video conferencing.	and through various devices.
Interactivity	Is often facilitated through multimedia elements, quizzes, simulations, and other interactive activities. Learners engage with the content by clicking, typing, selecting options, and receiving immediate feedback.	Takes advantage of the touchscreens, sensors, and other features of mobile devices to provide interactive experiences. This can include gestures like swiping, tapping, and pinching to zoom, as well as utilizing the device's camera or microphone for activities like taking pictures, recording audio, or participating in video discussions.	It can take various forms, from virtual simulations to physical experiments, depending on the context.
Personalization	It is achieved by adaptive learning algorithms, learner profiles, and intelligent content recommendation systems. E-learning platforms may track learner progress and behaviour to provide customized content suggestions, adapt the difficulty of exercises, or offer targeted resources to address areas where a learner might be struggling.	Mobile apps can track user behaviour, preferences, and progress. It can use this data to deliver content that aligns with individual learning goals. Additionally, features like push notifications can be used to remind learners of upcoming assignments, assessments, or activities that are relevant to their specific learning journey.	Personalization in u-learning is about adapting learning experiences to fit the specific context and preferences of the learner at any given moment. For example, a u-learning system might adjust the content and format of a lesson based on whether the learner is using a laptop at home, a mobile device on the go, or participating in a hands-on activity at a physical location.

Table 1. Comparison between u-learning, e-learning and m-learning

3. U-learning platforms

U-learning is a learning paradigm and does not require a specialized learning platform. It can be implemented with more or less difficulty in any of the available LMS platforms. An implementation integrated into the Moodle platform is presented in [7]. In [8], the possibilities of popular LMS platforms such as Wikipedia, MOODLE, Web 2.0, Web 3.0 and Blackboard for the implementation of u-learning are discussed. It is understood that MOODLE, Web 2.0, Web 3.0, and Blackboard do not include social networks and do not provide the possibility of free access to educational resources, as well as the free possibility of creating them. Only Web 3.0 makes it possible to easily include social networks in the learning process.

In [9], the effects of implementing IoT in education are discussed and how u-learning that includes IoT can increase the quality of learning by enabling much more realistic experiences for students. It is argued that with the implementation of IoT, u-learning has the potential to replace traditional classroom learning. Various methods have been developed to measure the success of information systems used for training, which could be used to evaluate the effectiveness of the system after it has been built. Models for Measurement Success are presented in [10]. They have focused on the performance of the following characteristics of u-learning - 1) context-awareness, 2) interactivity, 3) personalization, and 4) flexibility.

In [11], a model for U-Learning in higher education is presented, taking into account the two aspects of learning: pedagogical and technological. A U-CLX model is presented to measure the effectiveness of U-Learning in four dimensions: time, place, environment and context. In [12] it is pointed out that the main factors for the effectiveness of digital training are 1) university level factors; 2) teacher level factors; 3) student level factors.

4. A conceptual model for open u-learning platform

We propose a conceptual scheme for building an open platform for ubiquitous learning. At the highest level, the platform includes 4 modules: Participant Profile, Learning Content Source Profile, Own Resources and Management Module (Fig. 1). In the Participant Profile module, information is stored about all participants registered in the system, who at one moment have the role of learners, at another - the role of content authors, and at others - participants in joint projects. The Participant Profile module has 3 nested modules.

- The first nested module contains static data about the participant - personal data, contact data, and account.
- The second nested module contains dynamic information that is regularly updated by the participants (topics of interest, topics on which he works, preferences for the format of the offered training materials, participation in collaborative work groups, a collection of acquired certificates, etc.)
- The third nested module contains dynamic information acquired by an analysis engine (based on processing statistical data and application of artificial intelligence). This information is updated based on the participant's activity and includes information about preferred sources of information, a comparison with the profiles of other participants, and suggestions for inclusion in various discussion groups, groups working on a joint project, etc.

The second module (Own Resources) includes four nested modules:

- e-learning platform with its own training resources;
- blogs of participants on topics of their choice;
- groups of interest;
- joint projects on which groups of participants work;
- a module with AI that makes rating evaluations of learning resources, published blogs, topics of joint projects, determining their adequacy, level of knowledge offered and their association with different fields of knowledge.

The third module (Profile of Learning Content Sources) aims to create a database of information sources and their characteristics to more quickly and adequately offer these resources to participants. The fourth module controls the operation of the system. Its task is to coordinate the work of the first three modules and make the connections between them.

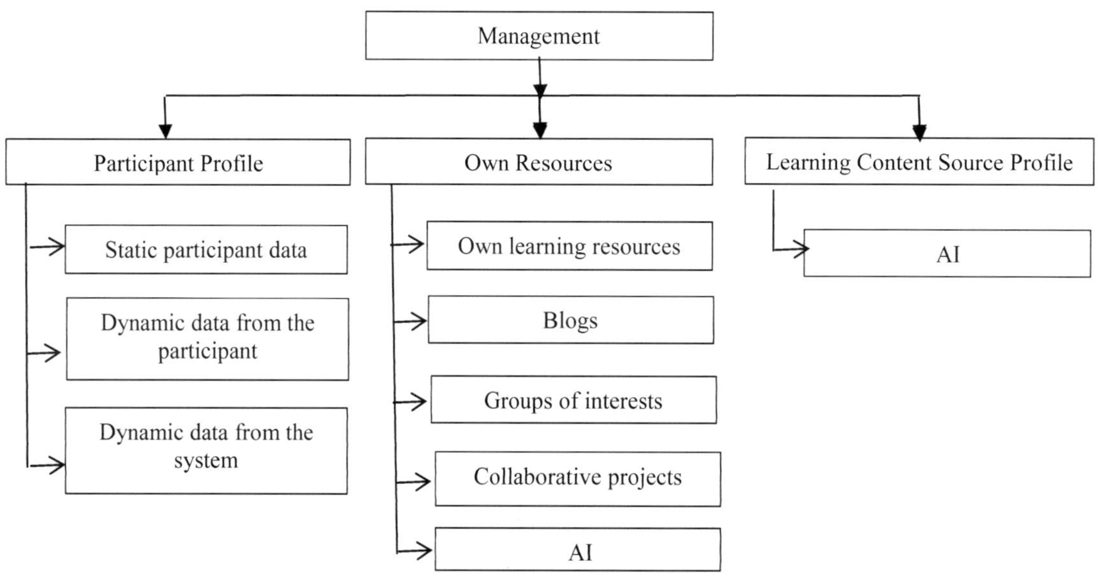

Fig. 1. The diagram of conceptual model for ubiquitous learning open platform

5. Conclusion

The advantages of the proposed conceptual model for open ubiquitous learning avoids some serious problems related to the learning process and its adaptation to the needs of the learner. Factors such as university level, teacher level, and student level can successfully be ignored.

- The level of the university is not decisive. Learning resources are searched all over the Internet, evaluated and offered according to the needs of the specific student. Thus, they are neither very difficult nor very easy to learn but are at a level corresponding to the preparation of the particular student.
- The level of the teacher is not decisive. Retrieved resources are evaluated by instructor and course rating, and resources with low ratings are not recommended.

- The level of the student is also not a factor. The suggested materials are tailored to his level. If the student starts a course and finds that his level is not adequate, he is not obliged to finish it but can choose another course on the same topic.

The disadvantages of the proposed model can be summarized in the following points:

- Great search and analysis complexity. There is no single standard for describing the courses offered (level of complexity, included components, resources, etc.) and for their rating.
- Accessibility issues to closed learning systems that may contain quality courses. This can be avoided by developing protocols for information exchange.
- The validity and recognition of certificates issued by one educational institution in another educational institution.

By implementing this model in the educational process, students and external learners will have access to the best and suitable for their learning style open resources for acquiring knowledge and improving their qualifications. They will be able to create their own resources with original content or with links to other resources. We believe that in this way the learning process will be better adapted to the needs of each learner, learning outcomes will be improved and new branches of science will be rapidly incorporated. The implementation of this model requires the efforts of a team of programmers and AI specialists. For the future, we will focus on creating the backbone of the platform, creating learner profiles, discovering and providing open resources depending on that profile. We plan to involve the students from the University of Telecommunication and Post, Sofia, to test the functionalities.

6. References

[1] Aparicio, M., Bacao, F., & Oliveira, T. (2016). An e-Learning Theoretical Framework. Educational Technology & Society, 19 (1), 292–307.

[2] Budic, H., Horvat, D., Ruzic, I. (2022). ICT in Education Development, Proceedings of the 33rd DAAAM International Symposium, pp.0017-0023, B. Katalinic (Ed.), Published by DAAAM International, ISBN 978-3-902734-36-5, ISSN 1726-9679, Vienna, Austria

[3] Perutka, K., Vymazal, M. (2021). Application of Game-based Learning in MATLAB using Object-oriented Programming, Proceedings of the 32nd DAAAM International Symposium, pp.0022-0032, B. Katalinic (Ed.), Published by DAAAM International, ISBN 978-3-902734-33-4, ISSN 1726-9679, Vienna, Austria

[4] Sarrab, M., Elgamel, L., Aldabbas, H. (2012). Mobile learning (m-learning) and educational environments. International journal of distributed and parallel systems, 3(4),

[5] Leonor Adriana Cárdenas-Robledo, Alejandro Peña-Ayala (2018). Ubiquitous learning: A systematic review, Telematics and Informatics, Volume 35, Issue 5, 2018, Pages 1097-1132, ISSN 0736-5853

[6] Kinshuk, Graf, S. (2012). Ubiquitous Learning. In: Seel, N.M. (eds) Encyclopedia of the Sciences of Learning. Springer, Boston, MA, pp 3361–3363,

[7] I Kadek Suartama, Punaji Setyosari, Sulthoni, Saida Ulfa (2020). Development of Ubiquitous Learning Environment Based on Moodle Learning Management System, International Journal of Interactive Mobile Technologies (iJIM) – eISSN: 1865-7923, VOL. 14, pp.182-204

[8] Aljawarneh, S.A. (2020). Reviewing and exploring innovative ubiquitous learning tools in higher education. Journal of Computing in Higher Education, vol. 32, pp. 57–73

[9] El-Haggar, Nahla, Lobna Amouri, Albandari Alsumayt, Fatemah H. Alghamedy, and Sumayh S. Aljameel. (2023). "The Effectiveness and Privacy Preservation of IoT on Ubiquitous Learning: Modern Learning Paradigm to Enhance Higher Education" Applied Sciences 13, no. 15: 9003.

[10] Stefanovic, D., Spasojevic, I., Havzi, S., Lolic, T., Ristic, S. (2020). Information Systems Success Models in the E-Learning Context : A Systematic Literature Review, Proceedings of the 31st DAAAM International Symposium, pp.0555-0564, B. Katalinic (Ed.), Published by DAAAM International, ISBN 978-3-902734-29-7, ISSN 1726-9679, Vienna, Austria

[11] Ramírez Villegas, Gabriel M., César A. Collazos, and Jaime Díaz. (2023). "A Model to Measure U-Learning in Virtual Higher Education: U-CLX" Applied Sciences 13, no. 2: 1091.

[12] Almazova, N., Rubtsova, A., Krylova, E., Barinova, D., Eremin, Y., Smolskaia, N., (2019). Blended Learning Model in the Innovative Electronic Basis of Technical Engineers Training, Proceedings of the 30th DAAAM International Symposium, pp.0814-0825, B. Katalinic (Ed.), Published by DAAAM International, ISBN 978-3-902734-22-8, ISSN 1726-9679, Vienna, Austria

DOI: 10.2507/34th.daaam.proceedings.065

IMPROVING THE PRINT QUALITY OF AN ADDITIVE-CAPABLE ROBOTIC CELL BY OPTIMIZING THE COOLING UNIT

Omar Ali, Ali Aburaia, Mohamed Aburaia & Kemajl Stuja

This Publication has to be referred as: Ali, O[mar]; Aburaia, A[li]; Aburaia, M[ohamed] & Stuja, K[emajl] (2023). Improving the Print Quality of an Additive-Capable Robotic Cell by Optimizing the Cooling Unit, Proceedings of the 34th DAAAM International Symposium, pp.0496-0504, B. Katalinic (Ed.), Published by DAAAM International, ISBN 978-3-902734-41-9, ISSN 1726-9679, Vienna, Austria
DOI: 10.2507/34th.daaam.proceedings.065

Abstract

Robotics and additive manufacturing are revolutionizing the manufacturing industry by enabling automated and personalized production. By combining the two technologies, it is possible to print large print objects such as a canoe for a person. This paper examines the importance of cooling in the additive manufacturing and how it can affect the shape, accuracy and surface finish of the printed part. It also discusses the use of an industrial robot to fabricate additive objects and the use of polylactide granules as the printing material. The paper addresses several factors that can affect the print quality of a 3D printer, specifically addressing melted layer aeration and cooling. The paper concludes with a discussion of a new cooling system implemented and its potential for further improvement. This new cooling system has made improvements both in terms of measured values and visually, but there is still potential for further improvement.

Keywords: Additive manufacturing; Extrusion process; Cooling; Industrial robot; Polylactide.

1. Introduction

Additive manufacturing (AM) processes have experienced a real upswing in recent years. Du to the simple applicability of additive manufacturing, it has found its way into the private sector in addition to industrial applications. FFF (Fused Filament Fabrication) is the most commonly used technology. [2]

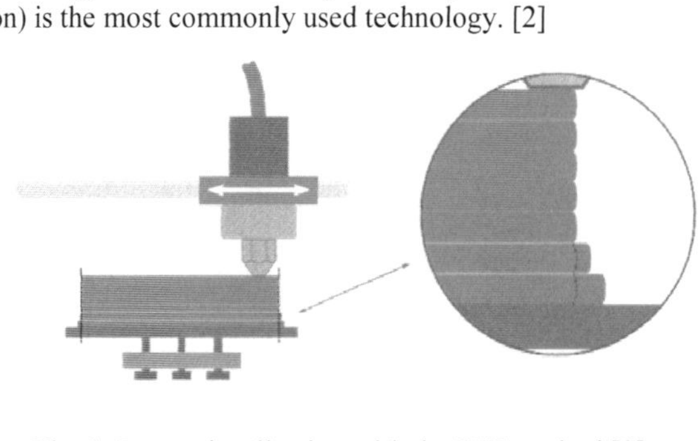

Fig. 1. Layer visualization with the FFF method [3]

FFF printing is a process in which thin layers of melted plastic filament are stacked to create a 3D object [4]. Cooling plays a crucial role in this process as the molten filament cools quickly and needs to solidify to retain the shape of the printed part. If the filament cools too quickly, it can cause warping and warping, negatively affecting the printed part's shape and accuracy. Inadequate cooling can also cause the material to contract unevenly, resulting in an uneven surface and poor lay-up. On the other hand, sufficient cooling can help the material contract evenly and form a smooth surface. [5]

Adequate cooling can also improve the strength and durability of the printed part as it helps the material cool in its defined shape. It is important to choose the right cooling strategy for each material used. For some materials, it is recommended to carry out the cooling as soon as possible to avoid distortion. For other materials, however, it may make sense to slow down the cooling to achieve a more even contraction.[6] The cooling unit is of greater importance as soon as an attempt is made to print more complex structures. That's because inaccuracies during the printing process can have a major impact on the final product. Specifically, this paper deals with an industrial robot that produces additive objects and offers the possibility of producing vertically printed objects. The printing system is divided into three main components, an industrial robot with six degrees of freedom (ABB IRB 2400/16) [7] and controller, the heated glass printing platform from Prangl [8] and finally the extruder from Noztek Xcalibur [9] with filament transport and cooling unit.

The printer to be examined uses PLA (polylactide) granules as the printing material. To be able to optimize the cooling system, the material used for printing plays a key role. PLA is a technical biopolymer based on lactic acid, which belongs to polyesters and is biodegradable [10]. Accordingly, PLA is obtained from regenerative sources and is therefore a biocompatible raw material, which can be composted under ideal conditions. However, an additive is usually added to the PLA, which specifically changes properties such as strength, elongation at break or color [11] Due to its low glass transition temperature, PLA is considered the plastic in the 3D printing sector, which is the easiest to process. The demands on the printer are very low here since a temperature-controlled construction space and a heated construction platform can be dispensed with if a good surface finish is not necessary[12]. In research and industry, the number of 3D printing robots is increasing. However, the print quality of the 3D printer is an important factor in the successful use of this technology. [13] To improve the print quality of the 3D printing robot, some steps can be taken.

Ventilation and cooling of the filament: Good ventilation and fast cooling of the filament can improve the print quality of the 3D printer. At the same time, good ventilation can help dissipate excess heat and keep the print room cooler. Faster cooling of the filament can also lead to a smoother surface and thus better print quality. [14]

Extruder temperature: Choosing the right extruder temperatures is an important factor in print quality. The optimal extruder temperature varies depending on the material and filament manufacturer. However, it is important to note that a higher temperature can result in a faster print speed and lower print quality. [15]

In summary, improving ventilation and thus cooling the filament is an important factor for better print quality. It is important to determine the optimal temperature and ventilation conditions for the filament used and the print engine to achieve the best possible print quality.

2. State of the Art

Today, robotic additive manufacturing is emerging as a potential solution to increase manufacturing flexibility. This technology makes it possible to change the orientation of the material application unit during printing, producing complex parts with the optimized material distribution. [16]

However, in technology-leading countries in general, the additive manufacturing process for the production of end products is in its infancy. A national spread is not to be expected in the next 10 years. On the one hand, these processes do not yet have the technical performance characteristics that are required for industrial series production. On the other hand, traditional industrial series production, especially in the area of medium and large series, is characterized today by a high level of economic efficiency, which cannot be solved with additive manufacturing processes in the foreseeable future, despite current technical problems. This shows the potential of additive manufacturing processes today, especially in highly specialized applications for single and small-series production.[17] Nevertheless, the market for additive manufacturing shows a growth of 30 to 40 per cent every two years [18].

Robot arms are ubiquitous in industrial production and help companies to become more economical, and robots reduce production losses through their perfection. The combination of robotic arms and additive manufacturing allows for larger and more complex printed parts that are not possible with a traditional 3D printer. [19]. There are current robots in research that produce additively manufactured parts. The EP (extrusion process) method is mostly used here. This is because a large number of printable materials and supported functionality are possible. The EP simply enables object printing and a wide variety of physical properties such as flexibility, impact strength, compression, tension, thermal resistance, and electrical conductivity can be achieved. In general, the final object that comes out of the 3D printing machine does not require any finishing, cleaning or necessary sanding, but this only applies if no support is used in the production phase. [20]

Like any other process, EP printing has advantages and disadvantages. The advantages are inexpensive, quite stable production of objects. However, numerous 3D printing processes offer a finer surface, such as the stereolithography process. The process was developed and launched by Formlabs in 2012. The stereolithography process prints with resin, which has a significantly higher resolution and finer surface than can be expected from FML processes. [10]. The disadvantages of EP printing cannot be eliminated, but it is possible to keep them as small as possible by optimally setting the printing parameters so that they are minimally noticeable.

The dimensional and mechanical properties of a printed 3D model are affected by several printing parameters. A ventilation system that moves with the printhead was designed and installed by Lee and Liu [11] for an FFF printer to control the cooling of a print model. The quality of the model printed with PLA, including the dimensions and mechanical properties, was studied for different cooling air velocities.

It was found that the cooling air velocity had different effects on the dimensional properties and mechanical strength of the printed model. More specifically, a higher cooling rate resulted in better geometric accuracy but lower mechanical strength. At the highest and lowest cooling speeds of 5 m/s and 0 m/s, respectively, the tensile strengths of the printed models differed by a factor of 4. To determine an appropriate cooling air speed setting for each specific print material, a design model was proposed. The pressure parameters determined were used in the manufacture of a Rubik's Cube as an example. The composite cube showed satisfactory performances in terms of both dimensional properties and mechanical function. Therefore, the cooling air velocity can be used as an additional control parameter in 3D printing technology. [21]

Thus, the quality of the next layers or their overhangs can be improved using a higher airflow and the associated cooling [22]. It is also worth mentioning that the layer thickness in combination with powerful cooling has a significant influence on the bending properties of the printed part. Different layer thicknesses were analysed (0.1, 0.2, 0.3, 0.4, 0.5 [mm]). It was found that parts with a layer thickness of 0.4 to 0.5 mm had the greatest flexural strength and that the tensile strength first decreased and then increased with increasing layer thickness. [23] As the new hot layer is applied, the underlying layer reflows. The more of the cooled material is melted and the faster the upper layer is cooled, the better the layers bond. A better layer connection increases the resilience of the printed object. H. Bikas, P. Stavropoulos and G. Chryssolouris [24] also came to this conclusion.

Another important parameter of the EP process is the relationship between print speed and the volume of material that is extruded through the extruder. These are crucial for a nice print result. For example, extruding too much material at too slow a speed will force excess material outward. This hurts the surface quality of the print object. An uneven relationship between print speed and material output can also cause material to deviate by as much as 20 per cent. [25]

On the other hand, the cooling process is an essential factor when it comes to overhanging or oblique print components. After an investigation in which an air-cooling system with different airflow speeds was used, it was found that with a low airflow compared to none at all, a temperature reduction of the PLA of 40 per cent could be achieved. A faster cooling process leads to better overhangs. Consequently, this results in better print quality. On the one hand, the material cannot deform without cooling. On the other hand, excessive ventilation during printing will damage the printed object. The damage occurs when the soft material is pushed away by the airflow when the new layer is printed on or hardens too quickly so that a fusion with the lower layer only takes place to a limited extent. As a result, excessive cooling leads to a deterioration in print quality and a low resilience of the printed parts. [26]

In addition, cooling too slowly will result in material distortion and this will result in a failed print. The material distortion of the printing part, during printing, causes partial or complete lifting of the printing part of the printing plate, due to the internal stresses in the material, which increase with slow cooling. It's also worth noting that warping only happens after some time. This result also corresponds to earlier studies by Sood et al. [27].[28] However, it is worth noting that the EP process can handle a variety of different materials that have different properties. Therefore, the cooling parameters must be set accordingly depending on the printing material [11].

Fig. 2. A printed object with cooling on the left, a printed object without cooling on the right [3]

Fig. 2 illustrates the effects on the printed object if it is not cooled. The picture on the right shows a layer application without cooling the filament. A material distortion can be seen here. The picture on the left shows the same system with cooling. Here you can see that no distortion has taken place in the printed object.

3. Materials and Methods

To achieve better air cooling, four parameters can play a crucial role. These are air throughput, airspeed, air temperature and humidity.

Airflow: Airflow is an important factor affecting the cooling performance. A higher airflow increases the cooling capacity, but also energy consumption. It is important to choose a balanced airflow that achieves an optimal cooling effect. [30]

Airspeed: The airspeed can also be adjusted to achieve an optimal cooling effect. Higher air velocity can result in faster cooling but also higher noise levels. [31]

Air temperature: Air temperature can also be an important factor, as cooler air provides better cooling. However, it is important to note that air that is too cold can increase energy costs. It is important to determine the appropriate air temperature for the specific application to achieve effective cooling without wasting energy unnecessarily. [32]

Humidity: Humidity can also play a role in air cooling, particularly in material properties and part quality. High humidity can lead to material deformation and contamination. It is important to monitor and regulate appropriate humidity levels to ensure optimal print quality. [32]

Fig. 3 illustrates the currently installed cooling system. Importantly, the area around the nozzle must be free of parts that could cause collisions during printing. To that end, adjustments have been made to address these issues. To define the one area in which components such as the cooling system may be located, a maximum angle for the printing process must be defined. Note that this angle only represents the difference between one simplex print and the next simplex print. The actual maximum pressure angle at which the material will not adhere to the previous plane is determined by empirical testing. Since a cooling system is required to prevent the liquid material from flowing away from the object, it must be mounted as close to the extruder as possible to cool the freshly extruded material without affecting the actual printing process. Thus, a maximum angle difference of 30° has been defined between the two pressure directions, in which the cooling system can be accommodated. On the other hand, for a practical cooling system, more complex shapes must be created to further cool the extruded material without affecting the oblique printing process. The area of the collision cone must be used here. Compressed air is used as the source for sufficient cooling of the carrier material. Therefore, the goal of the cooling section is to supply and direct the actual air. For this purpose, a component was developed that blows compressed air from three sides at a distance of 120° onto the material exit point of the nozzle. In addition, this 120° approach reduces possible problems with single-sided cooling systems, such as B. Excessive force on printed parts that can damage fine structures. Additionally, support rings for the three air outlets keep them stable when air flows through them. Due to the structural complexity of this particular air delivery system, Selective Laser Sintering (SLS) was used to fabricate the ventilation system components. The disadvantage of this system is that the materials used have melting points similar to PLA. Each airline is inside the cone as shown in Fig.3, which is not a problem when printing at an angle. [1]

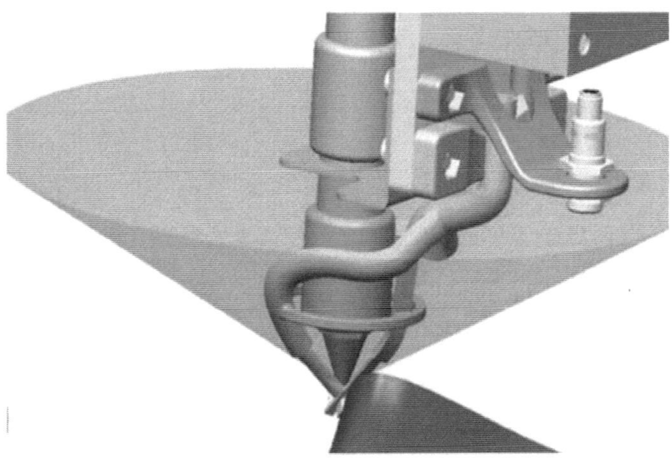

Fig. 3. Currently installed cooling within the conical collision-free sector [1]

xTo be able to properly evaluate the built-in cooling system, the above parameters must be measured and calculated. In this way, the ACTUAL values of the system can be determined to enable a comparison of the parameters between the old and new system in the event of a redesign with a new system.

In order to measure air cooling, various parameters must be recorded like air speed and air temperature. To measure the air temperature and humidity, a psychrometer from the brand RS PRO [33] is used, which can measure hot air up to 100 degrees. An anemometer of the DT-8880 from RS PRO [34] is used to measure air speed, which can measure air speeds of up to 25 m/s and also offers the possibility of measuring air temperatures of up to 50 degrees. Furthermore, it is necessary to measure and record the temperature of the freshly applied layer a few cm after extrusion in order to be able to make a comparison between the old and new cooling systems. In addition, there is an RS-833 infrared thermometer from RS PRO [35] that can measure up to 750 degrees. To do this, a device must be installed that enables the sensor to be mounted a few cm after the nozzle. It should be noted that a measurement is only possible when applying a straight layer.

To ensure consistent print conditions, a print model constructed with Fusion 360 serves as a test object. This is shown in Fig. 4 and has dimensions of 30x30x30 mm. In addition, a chamfer, and rounding of 30 mm each are integrated to include more complex geometries in the test.

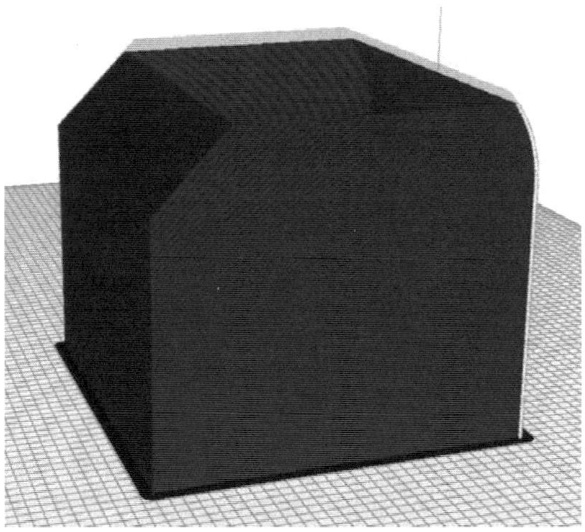

Fig. 4. Print model in CAD

Fusion 360 software is also used for the redesign of the ventilation unit to design new concepts and print them out with a 3D printer (Prusa i3). The generation of the airflow is an important factor of the system and can significantly affect various parameters such as humidity and temperature. Before the practical part is carried out, it is necessary to select suitable fans for the cooling system. For this purpose, three different axial fans are compared with each other by measuring their maximum air velocity under current. The results of this comparison are shown in Table 1.

Fan	TITAN PWM	FAN 2510	EBM Papst
Dimension [mm]	120x120x25	25x25x10	80x80x25
Operatingvoltage [V/DC]	12	12	12
Rotational speed [U/min]	2100	9600	3100
Air flow rate [m³/h]	36,05	5,1	65
Measured air velocity [m/s]	3,37	1,83	7,10

Table 1. Metrics of TITAN PWM [36], FAN 2510 [37] and EBM Papst [38]

Due to the limited installation space available for the cooling system, only a fan with maximum dimensions of 120x120 mm could be used. The EBM-Papst fan achieved the best measurement results and was smaller in its dimensions than the TITAN fan.

The FAN 2510 fan was the smallest in the comparison but achieved insufficient values for the application. Based on these results, the cooling system was implemented with three EBM Papst fans. To operate the fans, an inverter from TDK-Lambda [39] was used to convert 230V AC to 24V DC. In addition, an adjustable DC/DC Stepdown Modul from DEBO [40] was used to achieve the required 12V DC. The output voltage could be adjusted by a regulator in the module to control the operation of the fans. Fig. 5 illustrates the circuit.

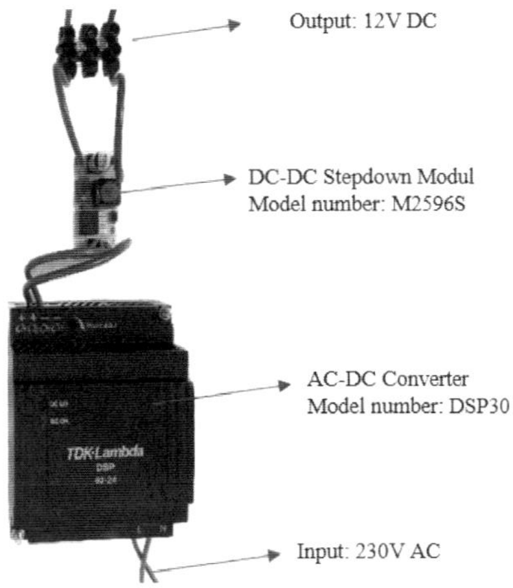

Fig. 5. Electrical construction

4. Results

Fig. 6 shows the finished construction of the cooling system, consisting of a bracket with three air ducts. Each air duct contains an EBM Papst fan. The air ducts have an elongated design to avoid sucking in warm ambient air. The walls of the air ducts are 5 mm thick and consist of two 1 mm thick outer walls with a 3 mm gap between them. This prevents heat transfer from the outer walls to the inner walls to avoid heating of the cooling air.

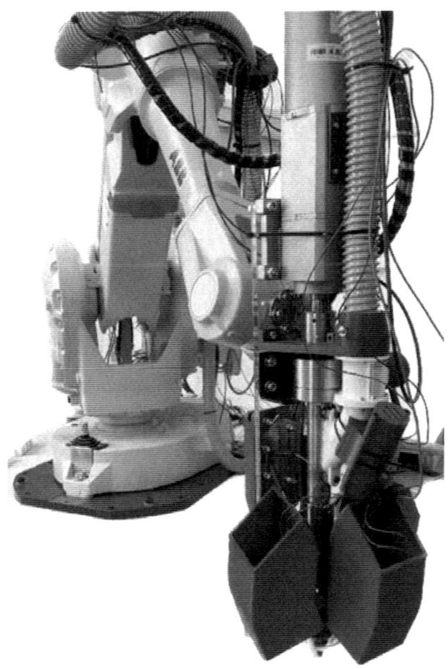

Fig. 6. Robot with the new cooling system

Fig. 7 shows the test cube printed with the new cooling system compared to the test cube printed with the old cooling system. A significant improvement in the surface finish of the test cube is visually apparent. The uneven surface of the old test cube is caused by the fact that the hot PLA does not solidify immediately after application and flows to the side. The rapid cooling of the PLA by the new cooling system prevents this trickling and thus produces a smoother surface. The measurement results also show a significant improvement.

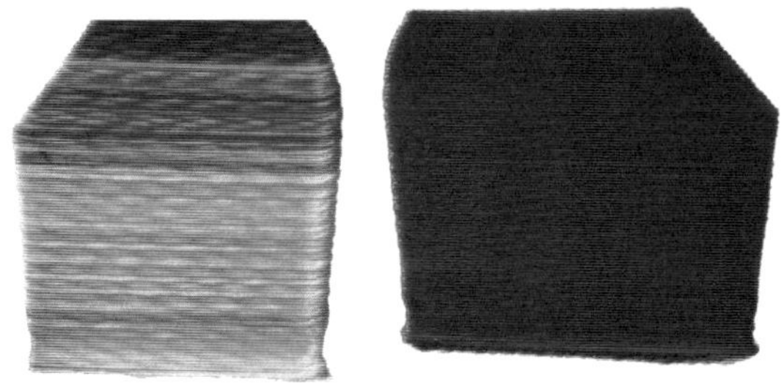

Fig. 7. Printing difference between old cooling (left) and new cooling system (right)

The new cooling system achieves an air velocity of 6.68 m/s and an air temperature of 26.1°C immediately below the nozzle tip. In comparison, the old cooling system achieves only 2.45 m/s at an air temperature of 26.2°C under the same measuring conditions. When measuring the air velocities, it is found that the old cooling system generates strong air turbulence. This can be observed on the propeller of the anemometer, which rotates irregularly. This could be due to uneven distribution of the air in the three air ducts. The new cooling system, on the other hand, achieves a constant air flow below the Nozzle. When measuring the freshly applied layer with the old cooling system, the average temperature was 192.7 degrees. This value was determined from ten measurements. The new cooling system, on the other hand, produced an average temperature of 158.3 degrees. It should be noted, however, that the PLA warmed up somewhat after the air flow moved away from the measuring point. A plausible explanation for this could be that the inner hot material heats up the surface somewhat afterwards.

5. Conclusion and outlook

With the new cooling system, improvements were achieved both in terms of the measured values and visually. Nevertheless, there is still potential for improvement. The switch to electrically driven axial fans has brought about an enormous improvement. However, the old cooling system with the compressed air system was not used. A new concept combining both variants - axial fans and compressed air - could bring about a further improvement in the cooling system and should be investigated further. In addition, there is currently no communication between the cooling system and the robot or extruder. For example, the cooling system could communicate with the robot controller to turn it on and off and integrate control of the cooling system. For example, the cooling system could ramp up the fans to maximum speed during overhangs and vertical pressure and ramp down the cooling system during horizontal pressure. However, this would require new circuitry and some consideration of the interface between the robot controller and the cooling system.

6. References

[1] C. Hofbauer, „Design object manufacturing and optimization of a six-axis robotic 3D-printing station“, Vienna, 2022.

[2] Simon ILLIG, Entwicklung einer 5-Achs-Fertigungsmaschine für kinematische Untersuchungen beim FDM.

[3] Yasin ÇAPAR, 3D Printer – Yasin ÇAPAR. [Online]. Available at: https://yasincapar.com/portfolio/3d-printer/ (Accessed on: 6. Januar 2023).

[4] H. Gohari, A. Barari und H. Kishawy, „Using Multistep Methods in Slicing 2 ½ Dimensional Parametric Surfaces for Additive Manufacturing Applications“, IFAC-PapersOnLine, Jg. 49, Nr. 31, S. 67–72, 2016, doi: 10.1016/j.ifacol.2016.12.163.

[5] M. Choudhary, S. Mukherjee und P. Kumar, „Analysis and optimization of geometry of 3D printer part cooling fan duct“, Materials Today: Proceedings, Jg. 50, S. 2482–2487, 2022, doi: 10.1016/j.matpr.2021.10.444.

[6] M. F. Afrose, S. H. Masood, P. Iovenitti, M. Nikzad und I. Sbarski, „Effects of part build orientations on fatigue behaviour of FDM-processed PLA material“, Prog Addit Manuf, Jg. 1, 1-2, S. 21–28, 2016, doi: 10.1007/s40964-015-0002-3.

[7] Robotics, IRB 2400 | ABB Robotics. [Online]. Available at: https://new.abb.com/products/robotics/robots/articulated-robots/irb-2400 (Accessed on: 17. Februar 2023).

[8] B. Prangl, „Optimierung einer 6-Achs Fused Granular Fabrication Station“, UAS Technikum Wien, Wien, 2021.

[9] Noztek Extrusion Systems, Noztek Xcalibur Hot Melt Desktop Extruder - Noztek. [Online]. Available at: https://noztek.com/product/xcalibur/ (Accessed on: 17. Februar 2023).

[10] P. Fastermann, 3D-Drucken: Wie die generative Fertigungstechnik funktioniert, 2. Aufl. Berlin, Heidelberg: Springer Berlin Heidelberg; Imprint; Springer, 2016.

[11] Mario Christian Dohr, „Prozessentwicklung zur Optimierung von additiv gefertigten Urmodellen für den Metallguss“, FH Campus Wien, Wien, 2017.

[12] P. Elsner, P. Eyerer und T. Hirth, Hg., Kunststoffe: Eigenschaften und Anwendungen, 8. Aufl. Berlin, Heidelberg: Springer Berlin Heidelberg, 2012.

[13] Fraunhofer-Institut für Produktionstechnik und Automatisierung; Deutsche Gesellschaft für Robotik; International Symposium on Robotics, Use of Industrial Robots in Additive Manufacturing - A Survey and Feasibility Study: 2 - 3 June 2014 parallel to Automatica, Munich, Germany. Berlin, Offenbach: VDE-Verl., 2014.

[14] Martin, „PLA Filament | Kompletter Guide: Einstellungen, Haftung, ...", The 3D Printer Bee, 8. Okt. 2021, 2021. [Online]. Available at: https://the3dprinterbee.com/de/pla-filament-guide/. Accessed on: 14. Februar 2023.

[15] 3Faktur, Filament 3D-Druck (FDM/FFF): Das Verfahren - 3Faktur. [Online]. Available at: https://3faktur.com/3d-druck-materialien-und-verfahren/filament-3d-druck-fdm-fff-verfahren/ (Accessed on: 14. Februar 2023).

[16] J. Lettori, R. Raffaeli, P. Bilancia, M. Peruzzini und M. Pellicciari, „A review of geometry representation and processing methods for cartesian and multiaxial robot-based additive manufacturing" (En;en), Int J Adv Manuf Technol, Jg. 123, 11-12, S. 3767–3794, 2022, doi: 10.1007/s00170-022-10432-8.

[17] C. Caviezel, R. Grünwald, S. Ehrenberg-Silies, S. Kind, T. Jetzke und M. Bovenschulte, „Additive Fertigungsverfahren (3-D-Druck)", 2017.

[18] Statista, Global 3D printing industry market size | Statista. [Online]. Available at: https://www.statista.com/statistics/315386/global-market-for-3d-printers/ (Accessed on: 29. Dezember 2022).

[19] I. C. o. I. E. a. Applications, Hg., 2018 5th International Conference on Industrial Engineering and Applications: ICIEA 2018 : April 26-28, 2018, Singapore. Piscataway, NJ: IEEE Press: IEEE Press, 2018.

[20] G. Fazzini et al., „Print On Air: FDM 3D Printing Without Supports" in 2019 II Workshop on Metrology for Industry 4.0 and IoT (MetroInd4.0&IoT), Naples, Italy, 2019, S. 350–354, doi: 10.1109/METROI4.2019.8792846.

[21] C.-Y. Lee und C.-Y. Liu, „The influence of forced-air cooling on a 3D printed PLA part manufactured by fused filament fabrication", Additive Manufacturing, Jg. 25, S. 196–203, 2019, doi: 10.1016/j.addma.2018.11.012.

[22] Arne Rotermund, „Untersuchung des Einflusses von Prozessparametern bei der additiven Fertigung auf akustische Materialeigenschaften", Deutsches Zentrum für Luft- und Raumfahrt, Braunschweig, 2017.

[23] A. Nugroho, R. Ardiansyah, L. Rusita und I. L. Larasati, „Effect of layer thickness on flexural properties of PLA (PolyLactid Acid) by 3D printing", J. Phys.: Conf. Ser., Jg. 1130, S. 1–10, 2018, doi: 10.1088/1742-6596/1130/1/012017.

[24] H. Bikas, P. Stavropoulos und G. Chryssolouris, „Additive manufacturing methods and modelling approaches: a critical review", Int J Adv Manuf Technol, Jg. 83, 1-4, S. 389–405, 2016, doi: 10.1007/s00170-015-7576-2.

[25] L. Santana, J. Lino Alves und A. Da Costa Sabino Netto, „A study of parametric calibration for low cost 3D printing: Seeking improvement in dimensional quality", Materials & Design, Jg. 135, S. 159–172, 2017, doi: 10.1016/j.matdes.2017.09.020.

[26] N. Michaelis, „Experimentelle Untersuchung elastokalorischer Kühlprozesze: konvektive Thermodynamik, latente Wärme und Materialzustandsüberwachung", Saarbrücken, 2020.

[27] A. K. Sood, R. K. Ohdar und S. S. Mahapatra, „Parametric appraisal of mechanical property of fused deposition modelling processed parts", Materials & Design, Jg. 31, Nr. 1, S. 287–295, 2010, doi: 10.1016/j.matdes.2009.06.016.

[28] A. Saluja, J. Xie und K. Fayazbakhsh, „A closed-loop in-process warping detection system for fused filament fabrication using convolutional neural networks", Journal of Manufacturing Processes, Jg. 58, S. 407–415, 2020, doi: 10.1016/j.jmapro.2020.08.036.

[29] File:3D printing calibration part-cooling fan airflow.png - Wikimedia Commons (Accessed on: 15. Februar 2023).

[30] X. Song et al., „Advances in additive manufacturing process simulation: Residual stresses and distortion predictions in complex metallic components", Materials & Design, Jg. 193, S. 108779, 2020, doi: 10.1016/j.matdes.2020.108779.

[31] A. Wibolo, A. A. I. Made und Sudirman, „Experimental study of air velocity effect of heat transfer in distillation condenser", IOP Conf. Ser.: Mater. Sci. Eng., Jg. 1175, Nr. 1, S. 12014, 2021, doi: 10.1088/1757-899X/1175/1/012014.

[32] J. Seifert, „Grundlagen Thermodynamik" in Flächenheiz- und Flächenkühlsysteme, Springer Vieweg, Wiesbaden, 2021, S. 13–31, doi: 10.1007/978-3-658-33735-3_2.

[33] RS PRO Psychrometer LCD hinterleuchtet +100 °C, 100 % RH | RS. [Online]. Available at: https://at.rs-online.com/web/p/feuchtemessgerate/1938695?cm_mmc=AT-PLA-DS3A-_-google-_-CSS_AT_DE_Mess-_und_Pr%C3%BCftechnik_Whoop-_-(AT:Whoop!)+Feuchtemessger%C3%A4te-_-1938695&matchtype=&aud-772940708119:pla-337684121950&gclid=CjwKCAiA9NGfBhBvEiwAq5vSy0IuxqPWywM0UEy3GJbxKfladYZu5kdlJGto06N6G0uZVVOufAdfZxoCVL0QAvD_BwE&gclsrc=aw.ds (Accessed on: 21. Februar 2023).

[34] RS PRO DT-8880 Anemometer bis 25m/s +50°C max.Temp., 0.01m/s, Luftstrom, Luftgeschwindigkeit, Temperatur | RS. [Online]. Available at: https://at.rs-online.com/web/p/anemometer/1938699?cm_mmc=AT-PLA-DS3A-_-google-_-CSS_AT_DE_Mess-_und_Pr%C3%BCftechnik_Whoop-_-(AT:Whoop!)+Anemometer-_-1938699&matchtype=&aud-828197004210:pla-310381506412&gclid=CjwKCAiA9NGfBhBvEiwAq5vSyyNLfPRei3gfwvuHY6UDTi1-abTQ719sS9BASosKGhx8uhddeNoEAxoCZfEQAvD_BwE&gclsrc=aw.ds (Accessed on: 21. Februar 2023).

[35] RS-833 Infrarot-Thermometer 12:1, bis +650°C, Celsius/Fahrenheit | RS. [Online]. Available at: https://at.rs-online.com/web/p/infrarot-thermometer/2068741?cm_mmc=AT-PLA-DS3A-_-google-_-CSS_AT_DE_Mess-_und_Pr%C3%BCftechnik_Whoop-_-(AT:Whoop!)+Infrarot-Thermometer-_-2068741&matchtype=&aud-

813230962291:pla-311511969494&gclid=CjwKCAiA9NGfBhBvEiwAq5vSy59t0nlQxJOM9cUNtztmaA9Ruo0oBS9TSzvgvghB5NMczSE_KsJaoRoCa1sQAvD_BwE&gclsrc=aw.ds (Accessed on: 21. Februar 2023)

[36] reichelt elektronik GmbH &amp und Team, Co. KG Internet, TITAN TFD-12025P - Titan 12V PWM Lüfter, Z-Lager, 120 mm. [Online]. Available at: https://www.reichelt.at/at/de/titan-12v-pwm-luefter-z-lager-120-mm-titan-tfd-12025p-p338254.html?r=1 (Accessed on: 4. August 2023).

[37] reichelt elektronik GmbH &amp und Team, Co. KG Internet, FAN-2510 12V - Lüfter, 12 V DC, 25 x 25 x 10 mm, U/Min: 9600. [Online]. Available at: https://www.reichelt.at/at/de/luefter-12-v-dc-25-x-25-x-10-mm-u-min-9600-fan-2510-12v-p11003.html?PROVID=2807&gclid=Cj0KCQjw756lBhDMARIsAEI0Agkpc43r_n5H_lQV8XQFjV3y4sZvroXPK94E5bQLS2sqnNY5P0wWebkaAuLEEALw_wcB (Accessed on: 4. August 2023).

[38] EBM Papst 8412NG Axiallüfter 12 V/DC 65 m³/h (L x B x H) 80 x 80 x 25.4 mm kaufen. [Online]. Available at: https://www.conrad.at/de/p/ebm-papst-8412ng-axialluefter-12-v-dc-65-m-h-l-x-b-x-h-80-x-80-x-25-4-mm-1926386.html?gclid=CjwKCAjw4ZWkBhA4EiwAVJXwqZAyIbOUDUM4nS2E_oInan5WwPnSkRJnGgsPXFDyuLIPQeoWahcbeRoCR9YQAvD_BwE&utm_source=google&utm_medium=organic&utm_campaign=shopping&ef_id=CjwKCAjw4ZWkBhA4EiwAVJXwqZAyIbOUDUM4nS2E_oInan5WwPnSkRJnGgsPXFDyuLIPQeoWahcbeRoCR9YQAvD_BwE%3AG%3As&refresh=true (Accessed on: 4. August 2023).

[39] Distrelec Austria, DSP-30-24 | TDK-Lambda Netzteil für DIN-Schiene, 83%, 24V, 1.3A, 31.2W, Einstellbar | Distrelec Österreich. [Online]. Available at: https://www.distrelec.at/de/netzteil-fuer-din-schiene-83-24v-3a-31-2w-einstellbar-tdk-lambda-dsp-30-24/p/16983488?cq_src=google_ads&cq_cmp=20016436989&cq_con=&cq_term=&cq_med=pla&cq_plac=&cq_net=x&cq_pos=&cq_plt=gp&gclid=Cj0KCQjwoK2mBhDzARIsADGbjeqh8qmwozRNicWt2fARtgyV72z5o8zx8FTUB3oi8HrE2BestOqoCzYaArZvEALw_wcB&gclsrc=aw.ds (Accessed on: 4. August 2023).

[40] reichelt elektronik GmbH &amp und Team, Co. KG Internet, DEBO DCDC 2.5W - Entwicklerboards - Spannungsregler 2,5 W, DC/DC-Wandler, M2596S. [Online]. Available at: https://www.reichelt.at/at/de/entwicklerboards-spannungsregler-2-5-w-dc-dc-wandler-m2596s-debo-dcdc-2-5w-p282576.html?PROVID=2807&gclid=Cj0KCQjwoK2mBhDzARIsADGbjepXOGhb1Nbzz1Ct4sq5eKV4QICUlbRep54Od7xShWurrQYyce5U_W4aAtxoEALw_wcB (Accessed on: 4. August 2023)

DOI: 10.2507/34th.daaam.proceedings.066

CALCULATION OF THE EFFECTIVENESS OF A DEHUMIDIFIER AND INTEGRATION OF A NEW MODEL OF A DESICCANT WHEEL INTO TRNSYS

Haris Lulic, Sadjit Metovic & Almira Softic

This Publication has to be referred as: Lulic, H[aris]; Metovic, S[adjit] & Softic, A[lmira] (2023). Calculation of the Effectiveness of a Dehumidifier and Integration of a New Model of a Desiccant Wheel into TRNSYS, Proceedings of the 34th DAAAM International Symposium, pp.0505-0514, B. Katalinic (Ed.), Published by DAAAM International, ISBN 978-3-902734-41-9, ISSN 1726-9679, Vienna, Austria
DOI: 10.2507/34th.daaam.proceedings.066

Abstract

Rotating wheel dehumidification is a relatively new air cooling technique used in air conditioning, insufficiently researched and applied in practice. The system's effectiveness depends significantly on the desiccant wheel's properties and the moist air operating parameters. Modern software and simulations with defined inputs, can predict process air conditions required for air conditioning as long as each component characteristics of the system are known. In this work, an existing component in TRNSYS was used to simulate the desiccant wheel operation, which for stationary operating conditions and predefined efficiency aligns well with experiments. However, for a time-varying process, significant deviations of the results are observed, i.e. when the entered wheel efficiency does not correspond to every condition of the air entering the wheel. Considering that TRNSYS requires the definition of the desiccant wheel's efficiency before simulation start, treating them as fixed parameters, with the existing model component located in its database, it is impossible to carry out the simulation for time-varying conditions with satisfactory accuracy. A modified desiccant wheel model was developed and integrated into TRNSYS, which allows changing the functions of potential F_1 and F_2 and efficiencies ε_{F1} and ε_{F2} during simulations, for time-varying parameters of the humid air inlet state.

Keywords: Desiccant; Dehumidification; Numerical Simulation; Adsorption; Desorption 228.

1. Introduction

Climate change and increased demands for thermal comfort have caused a greater demand for cooling energy worldwide, not only in areas with tropical climates but also in areas with temperate climates, like most European countries. This results in a significant increase in electricity consumption for the operation of compressor cooling devices, which consequently causes an unwanted increase in the use of fossil fuels and nuclear energy, and threatens the stability of power networks. In their work, Ahmović et al. [1] demonstrated the increase in average annual temperatures in continental climates attributed to climate change. As an alternative to existing conventional cooling systems, new solar cooling technologies are being developed and deployed, especially for comfort air conditioning when the time period required for cooling coincides with the maximum intensity of solar energy.

Solar air-condition units work on the principle of desiccant cooling, with the desiccant wheel based on solid desiccant appears as a key component of the system. The primary task of the desiccant wheel is to reduce the incoming air humidity in areas with high humidity, and bring it closer to the projected indoor air levels. On the other hand, in moderate climate, the desiccant wheel has the task of additionally dehumidifying the outside air in order to increase the capacity for an additional humidifier, which at the same time cools the air entering the room.

The process of dehumidification in the desiccant wheel depends both on the construction and composition of the wheel itself, and on the operating conditions of the system: desiccant wheel rotation speed, air flow, outside air temperature and humidity, and regenerative air temperature and humidity. Ma at al. [2], as well as Lulić at al. [3] have shown through their research that in solar cooling systems the greatest destruction of energy occurs at the desiccant wheel. In both papers, it is concluded that optimizing the desiccant wheel's working parameters, primarily regenerative air flow and temperature, as well as by improving the characteristics of the solid desiccant, it is possible to significantly increase the efficiency of the system component itself, and reduce the increase in entropy and increase the overall degree of efficiency of the system. The effectiveness of the dehumidifier can be expressed through two coefficients expressed as a function of the temperature and humidity of the air currents before and after the desiccant wheel. The values of the efficiency coefficients, considering the non-ideal system and processes, represent modified functions of potential based on the foundations of mathematical modeling of transport processes in the desiccant wheel. Howe [4] and Jurinak [5] developed these functions of potential in their papers, investigating the propagation of a wave front through the filling of a desiccant rotating wheel. Using the methods of nonlinear algebra [6], functions of potential $F_i(t,x)$ are defined as approximate nonlinear functions of temperature and absolute humidity, applying the same analogy as when modeling a sensible heat exchanger.

Momoi at al. [7], conducting a numerical calculation of the desiccant wheel model, found that using the values of the dehumidifier coefficients, the values of temperature and relative humidity of the air flow at the exit from the desiccant wheel can be predicted for different values of the regenerative air parameters. At the same time Bourdoukan at al. [8] using the experimental results and the results obtained by simulating the operation of the desiccant wheel model established a significant sensitivity of the change in the parameters of the process air depending on the change in the value of the efficiency coefficients of the dehumidifier. The biggest deviations occur in cases of high outdoor air humidity. Similar research was carried out by Norazam at al. [9] who, using a validated numerical method, determined the moisture removal capacity of a desiccant wheel based on solid desiccant and determined its thermal effectiveness and dehumidification effectiveness. The results obtained numerically show satisfactory deviations in terms of predicting the temperature and humidity of the process air (1.05% and 8%, respectively) compared to the experimental results. These studies are of particular importance when using software tools for simulating the operation of solar air conditioning systems (e.g. TRNSYS) that have a predefined component of the desiccant wheel with constant efficiency coefficients.

The performance of the entire solar air conditioning system largely depends on the characteristics of the dehumidifier. Nawaz at al. [10] conducted research with the aim of determining the basic characteristics of solid desiccant, specifically silica-aerogel, during system operation in non-stationary conditions (transient behavior/conditions) with different parameters of moist air at the inlet and outlet of the desiccant wheel. The paper emphasized that the adsorption capacity differs from the desorption capacity, and that the adsorption capacity is higher in samples with smaller pores in the structure. Zafar at al. [11] experimentally investigated the characteristics of a desiccant wheel with a solid desiccant (silica gel) at different parameters and ambient air flow. The results indicate that the effectiveness of dehumidification increases with increasing ambient air temperature up to a certain value, and then significantly decreases. In the paper, it is stated that the best characteristics of the desiccant wheel with silica gel filling and constant rotation speed, are shown at inlet air speeds of 3.5m/s with sensible temperature of 34°C. Narayanan at al. [12] in their research showed a change in the efficiency of the dehumidification process depending on the speed of rotation of the desiccant wheel with silica gel filling. In the case of a higher rotation speed, the contact time of moist air with the desiccant is reduced, which results in a higher humidity of the outgoing air. At the same time, higher speeds reduce the time of the regeneration process and result in a reduced performance of the desiccant wheel in the next cycle. On the other hand, low rotation speeds of the desiccant wheel significantly reduce its efficiency.

This paper describes in detail the working principle of a desiccant wheel with solid material filling (silica-gel), as well as a description of the desiccant wheel model given in TRNSYS. The theoretical bases for determining the function of potential as well as the efficiency coefficients of the desiccant wheel are given, as well as their values determined for the known configuration of the desiccant wheel. Data on the characteristics of the desiccant wheel and the tests carried out were taken from the equipment manufacturer. The obtained values of efficiency coefficients are integrated in TRNSYS by modifying the basic component of the desiccant wheel. In this way, during the simulation period, the efficiency values do not remain constant, as is the case with the configuration of the basic component, but change their value in accordance with the changing parameters of the process and regenerative air. The new component has been tested in TRNSYS and shows agreement with the results given by the basic component in operation with fixed values of wheel efficiency for constant values of inlet air parameters.

2. Air dehumidification process using desiccant wheel with solid sorption material

The process of air drying takes place in a rotating wheel, which in design resembles a rotary regenerative heat wheel, with the fact that in this case the filling of the wheel is made of a solid desiccant capable of receiving large amounts of water vapor from moist air.

Adsorption material occupies, by volume, the largest part of the interior of the wheel, while only a small part belongs to the structural part of the filling (Fig.1).

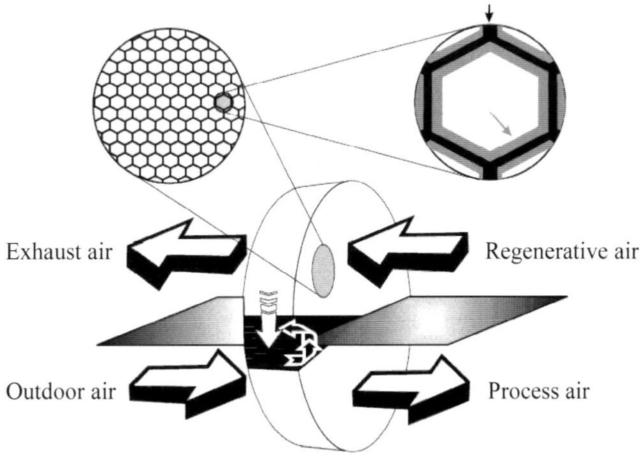

Fig. 1. Schematic of a desiccant wheel with solid sorption material [13]

In the desiccant cooling process, the purpose of the rotating desiccant wheel is to sufficiently dry the outside air so that the evaporative cooler can be used in the continuation of the process. Moisture from the moist air adsorbed in the rotating wheel in the process air stream must be transferred to the waste air on the regenerative side of the rotating wheel, so that the process can proceed without interruption. The temperature of the process air will rise at the exit from the desiccant wheel as a consequence of the adsorption process in the desiccant, i.e. the release of latent heat during phase change of the water vapor in the air. The amount of heat released by adsorption depends on the amount of water present in the desiccant and the amount of heat accumulated in the structural part of the filling of the desiccant wheel. In general, it can be said that a smaller amount of water contained in the adsorption material enables the release of a larger amount of latent heat, i.e. a higher temperature increase of the process air. On the other hand, the accumulated heat in the construction of the wheel increases its temperature, and later the temperature of the process air, which occurs as a result of convection. From this it is evident that both the water content in the dehumidifier and the temperature of the structural part of the wheel largely depend on the regeneration temperature and vice versa. The increase in the temperature of the process air during the drying process in the rotating wheel is the result of the effect of various influencing parameters such as: heat and mass transfer coefficients in the given conditions, specific sorption characteristics of the desiccant, the ratio of the volumes of the active filling and the constructive structure of the wheel, the amount and operating parameters of the air, wheel dimensions and wheel rotation speed. From this it can be seen that the correlation between the dehumidification process and the temperature increase is really complex, posing a challenge in modeling such systems.

3. Existing model of rotating desiccant wheel in TRNSYS

Dehumidifiers are devices that extract moisture from the air using a drying process, without additional cooling of moist air below the dew point temperature, which results in the extraction of water vapor from the air by the condensation process. These devices use the ability of hygroscopic materials (solid desiccants such as silica gel) to adsorb water particles on their surface. In the adsorption process, a thin molecular layer of one substance, in this case water vapor from the process air, is adsorbed on the surface of another substance, that is, a solid dehumidifier (silica gel). Over time, the structure of the dehumidifier becomes saturated, which loses its function and must be regenerated in the drying process. There are two types of dehumidifiers with filling made of solid material, which are most often used in practice. The first appears in the form of a dehumidifier cushion over which moist process air flows until the dehumidifier material becomes completely saturated. Regenerative air is periodically allowed to flow over the cushion with the aim of drying it. In another dehumidifier configuration, process and regeneration air flow simultaneously over a rotating wheel with solid sorption material, drying it on one part of the surface and wetting it on the other. Regenerative air is taken from the environment, in most cases, and heated with the aim of reducing the relative humidity. Therefore, an additional heat source is required for the air cooling process using the dehumidification technique.

For the purposes of performing simulations of the cooling process by dehumidification, a model of a rotating desiccant wheel filled with silica gel was created in TRNSYS, whose characteristics are defined by the functions of potential F_1 and F_2. The model determines the required temperature of regenerative air, which has the same value of absolute humidity as the ambient air, and which should dehumidify the process air at the exit from the rotating wheel to the set value of absolute humidity. The mass flow values of the process and regenerative air are the values set at the beginning of the calculation. Figure 2 shows a schematic view of the rotating desiccant wheel model used in TRNSYS.

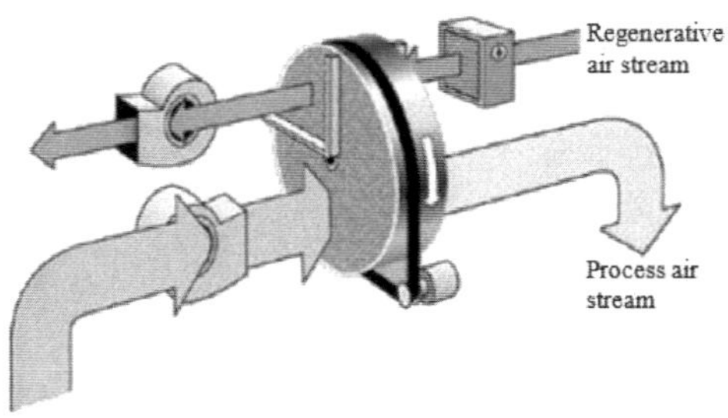

Fig. 2. Schematic of rotating desiccant wheel model in TRNSYS [14]

The ideal adsorption process that takes place from the Pin state to the Pout state is shown in the psychrometric T-x diagram in Figure 3. The process air of a certain state enters the dehumidifier. Due to the fact that water vapor is separated from the process air and adsorbed in the sorption material, the heat released during this process is transferred to the surrounding air stream. The process air leaves the dehumidifier with a lower amount of moisture and a higher temperature than it had at the entrance. In the same diagram (Figure 3), the ideal regeneration process from R_{in} to R_{out} is also shown. It is assumed that the process and regenerative air come from the same environment, that is, they have the same absolute humidity values, and that the regenerative air is heated receiving only sensible heat. Regenerative and process air pass through the rotating wheel in countercurrent. The saturated sorption material is exposed to a stream of warm and relatively dry regenerative air, which receives a certain amount of water from the dehumidifier while releasing a certain amount of heat. Regenerative air leaves the rotating wheel at a lower temperature and with a higher amount of moisture compared to the inlet conditions. As can be seen in Figure 3, an ideal dehumidifier adiabatically dries the process air and transfers a certain amount of moisture to the regenerative air. The actual sorption and desorption processes do not take place at a constant enthalpy, since the values of humidity, temperature and enthalpy in different layers of the sorption material, contained in the rotating wheel, are transmitted in the form of waves along the depth of the filling, changing their values at the same time. The dehumidifier is expected to be completely dry at the start of each process. The moist air current moves over the filling of the wheel, delivers a certain amount of water vapor and increases the moisture content in the dehumidifier. The first filling layers of the dehumidifier, which are in contact with moist air, become saturated and unable to carry out the process of drying the air, thus the parameters of the exit state of the air from the rotating wheel approach the input parameters. However, the complete filling of the wheel does not saturate at the same moment, but only the first layer, which is in contact with the stream of moist air. Adsorbed water particles from the first layer of sorption filling, in the form of waves, are transported slowly towards the inner layers, i.e. leaving behind always saturated layers of filling, they move towards unsaturated layers. The process of transporting moisture through the sorption material happens faster if there is air flow through the rotating wheel. If there is no process air flow, the air remaining in the dehumidifier reaches an equilibrium state, the process of heat and mass transfer stops, until the dehumidifier is completely dry. Through the flow of process air, "trapped" warm and humid air is released from the dehumidifier. If the dehumidification process is performed with a rotating wheel, the speed of rotation is set to allow maintaining the values of the parameters of the air outlet state within the given limits. If the speed of rotation is lower, the transport of moisture between the filling layers is higher, that is, the speed of the wave front is higher. By increasing the rotational speed of the desiccant wheel, the speed of the wave front decreases and prevents complete saturation of the filling material.

In this regard, during the investigation of the propagation of the wave front through the filling of the dehumidifier, the functions of potential F_1 and F_2 were developed. With the aim of determining the explicit values of the functions of potential, an expression is defined in which the coefficients $a_{i,j}$ determined numerically and valid only for a specific combination of the working medium that is being dehumidified and the sorption material used. The functions of potential are given by the expression:

$$F_j(t,x) = \frac{\left(\frac{T}{273,15}\right)^{a_{1,j}}}{a_{1,j}} + a_{2,j}\frac{x^{a_{3,j}}}{a_{3,j}} \tag{1}$$

Using the non-linear regression technique, the coefficients $a_{i,j}$ were determined for the combination moist air-silica gel, and their values are given in Table 1.

$a_{i,j}$	j=1	j=2
$a_{1,j}$	-1,4899998	1,4899998
$a_{2,j}$	3,7463964	-0,0898325
$a_{3,j}$	0,8624176	0,0796875

Table 1. Values of coefficients ai,j for silica gel, [4]

The graphical interpretation of the functions of potential F_1 and F_2 represents the developed curves, which show the smallest deviation from the values of the wavefront propagation characteristics for a given adsorbent. Figure 4 shows the isopotential lines for silica gel in the psychrometric T-x diagram, where the deviation of the isopotential lines in shape compared to the lines of constant enthalpy and relative humidity is clearly seen.

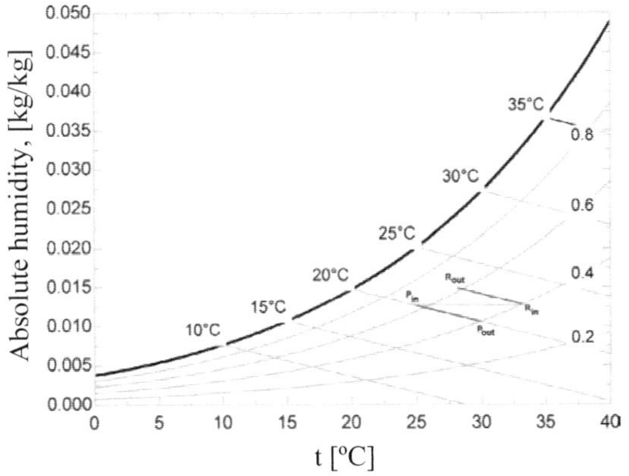

Fig. 3. Scheme of the ideal air dehumidification process using the drying technique

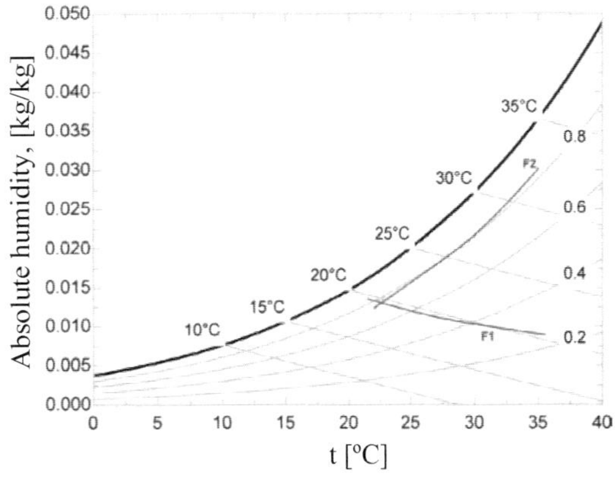

Fig. 4. Graphical representation of the values of the functions of potential F_1 and F_2

According to the expression for F_j and the coefficients given in Table 1, the values of the functions of potential for silica gel are determined according to the expressions:

$$F_{1,j} = \frac{-2865}{T_i^{1,490}} + 4{,}344 \cdot x_i^{0,8624} \tag{2}$$

$$F_{2,j} = \frac{T_i^{1,490}}{6360} - 1{,}127 \cdot x_i^{0,07969} \tag{3}$$

In these expressions, air temperature is included in [K], and absolute humidity in [kg/kg]. Using isopotential curves F_1 and F_2, Figure 5 shows a graphic representation, in a psychrometric T-x diagram, of the process of dehumidifying process air using the drying technique in a rotating wheel. The process air enters the dehumidifier with the state marked P, is heated and dehumidified along the potential line F_{1P} and exits the dehumidifier with the state D*. On the regenerative side of the wheel, air enters the dehumidifier with state R, cools and humidifies along the potential line F_{1R}, and leaves the rotating wheel with a state corresponding to the intersection of the potential lines F_{2P} and F_{1R}. The functions of potential F_1 and F_2 can be further modified, considering the non-ideality of the system and process, with two values of the efficiency coefficients ε_{F1} and ε_{F2}, which, according to [15], are defined as:

$$\varepsilon_{F1} = \frac{F_{1,D} - F_{1,P}}{F_{1,R} - F_{1,P}} \tag{4}$$

$$\varepsilon_{F2} = \frac{F_{2,D} - F_{2,P}}{F_{2,R} - F_{2,P}} \tag{5}$$

where D is the actual state of the outlet air from the rotating wheel, and D* is the state of the outlet air from the dehumidifier in the case of an ideal process.

Figure 6 gives a graphical representation of the actual dehumidification process in a rotating wheel in a psychrometric T-x diagram using isopotential lines. The rotary wheel dehumidifier component model in TRNSYS uses an iterative process to determine the actual state of the outlet air (point D). The dehumidifier is not limited by capacity, which means that the set state of air humidity at the outlet can be achieved whenever the dehumidifier is in operation. The model determines the temperature of regenerative air at the entrance to the dehumidifier, for a defined value of absolute humidity at the exit. As input data in the modeling process of this component, the following values must be provided: mass flow of air streams of process and regeneration air, state parameters of the process air (dry bulb temperature and absolute humidity) at the entrance to the dehumidifier, the desired value of the absolute humidity of the process air at the exit from the rotating wheel and the value of the absolute humidity of the regenerative air at the entrance.

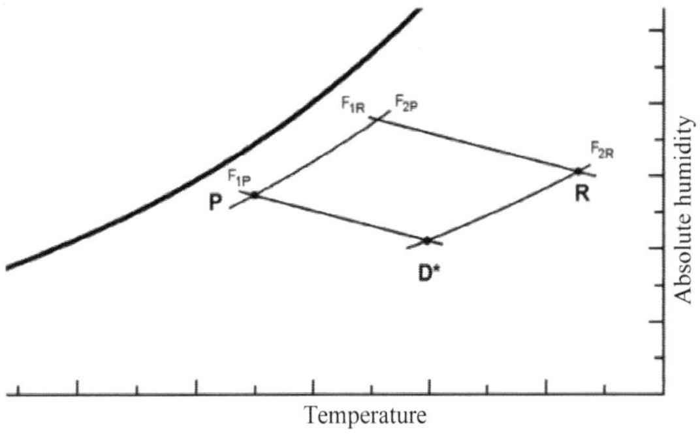

Fig. 5. Graphic representation of the dehumidification process using isopotential lines

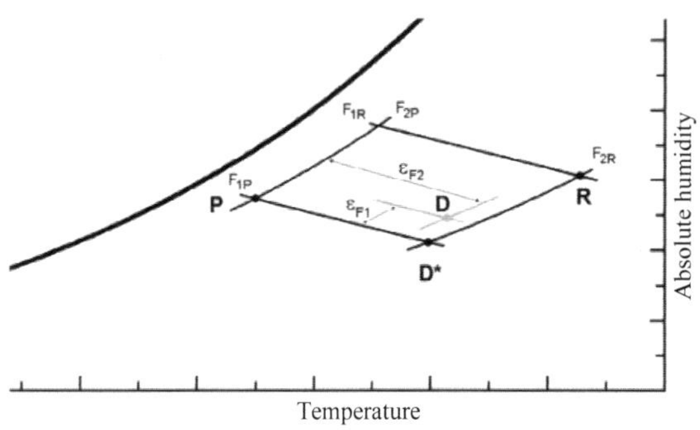

Fig. 6. Efficiency of the functions of potential F_1 and F_2 during the actual dehumidification process

The model first calculates the values of the functions of potential F_1 and F_2, for the input state of the process air. Considering that the value of the function of potential F_1 is the same for states P and D*, the output state of the process air is completely determined. When the value of the temperature for the state D* is known, the value of the function of potential F_2 at the point D* can be determined. Given that points D* and R are connected by an isopotential line, which means that the function of potential F_2 has the same value at points D* and R, the required temperature of the regenerative air can be determined. Based on the known values of the efficiency coefficients of the functions of potential (ε_{F1} and ε_{F2}), modified values of air parameters of state D* are calculated, which correspond to the function of potential at point D.

Using the obtained values for the functions of potential at point D, the model determines the appropriate value of the absolute humidity of the process air at the exit from the dehumidifier by an iterative procedure. The next procedure in the process of modeling this component of the system is the determination of the exit temperature of the process air, based on the achieved required value of absolute humidity. The condition of the regenerative air at the exit from the rotating wheel is determined using the mass balance of the air currents entering the dehumidifier and the new calculated values of the parameters of the process air at the inlet and outlet, as well as the input state parameters of the regenerative air.

4. Determination of efficiency and functions of potential F_1 and F_2 according to known characteristics of rotating, desiccant wheel

The dehumidification capacity for a specific rotating wheel depends on several factors, primarily on the production characteristics of the wheel itself, operating conditions, and mass flow of process and regenerative air. The most important characteristics of a dehumidifier in the form of a rotating wheel, which must be known before the modeling procedure, are certainly: the type and characteristics of the sorption material, the ratio of active and inactive filling structures, physical characteristics of the wheel structure and its dimensions, recommended rotation speeds at which the best cooling effect and dehumidification for given conditions and amounts of process and regenerative air is achieved. In this work, a desiccant wheel with characteristics given in Table 2 was used during the modeling of the solar air conditioning system.

For different operating conditions occurring during simulations of the created model, the values of the functions of potential and their efficiencies were determined. The data used in calculations were taken from the equipment manufacturer. The results are presented tabularly and graphically in Mollier's i-x diagram. Table 3 gives the calculated values of the functions of potential F_1 and F_2, as well as the efficiencies ε_{F1} and ε_{F2}, for the dry bulb temperatures of process and regenerative air of 19°C and 80°C, with variations of the absolute humidity of the process air at the inlet.

Sorptive material – active part of the wheel infill		Silica gel
Wheel filling structure		Honeycomb
Load-bearing part of the wheel infill – the inactive part of the infill		Fiberglass
Wheel thickness	mm	200
Surface of the front of the wheel	m^2	2,25
Percentage of the area through which the process air flows	%	58
Percentage of the area through which the regenerative air flows	%	42
Wheel housing width	mm	2010

Table 2. Characteristics of the desiccant, rotating wheel

Air stream			t	x	Functions of potential		Efficiency	
			[°C]	[g/kg]	F_1	F_2	ε_{F1}	ε_{F2}
Process air		in	19,0	6,00	-0,5545	-0,0080		
Process air	a	out	35,2	2,41	-0,5362	0,1067		
Regenerative air		out	51,2	21,40	-0,3618	0,0371	0,0856	0,6186
Process air		in	19,0	7,00	-0,5470	-0,0173		
Process air	b	out	36,4	2,86	-0,5292	0,1018		
Regenerative air		out	49,1	22,40	-0,3605	0,0257	0,0865	0,6115
Process air		in	19,0	8,00	-0,5396	-0,0254		
Process air	c	out	37,4	3,36	-0,5224	0,0965		
Regenerative air		out	47,3	23,30	-0,3592	0,0160	0,0871	0,6012
Process air		in	19,0	9,00	-0,5324	-0,0326		
Process air	d	out	38,4	3,89	-0,5154	0,0920		
Regenerative air		out	45,6	24,10	-0,3584	0,0070	0,0889	0,5935
Process air		in	19,0	10,00	-0,5260	-0,0392		
Process air	e	out	39,3	4,46	-0,5085	0,0876		
Regenerative air		out	44,1	24,80	-0,3578	-0,0008	0,0913	0,5854
Process air		in	19,0	11,00	-0,5183	-0,0451		
Process air	f	out	40,1	5,06	-0,5017	0,0833		
Regenerative air		out	42,7	25,50	-0,3570	-0,0082	0,0936	0,5772
Process air		in	19,0	12,00	-0,5114	-0,0506		
Process air	g	out	40,8	5,69	-0,4951	0,0791		
Regenerative air		out	41,4	26,20	-0,3560	-0,0151	0,0959	0,5689
Process air		in	19,0	13,00	-0,5045	-0,0557		
Process air	h	out	41,4	6,36	-0,4885	0,0748		
Regenerative air		out	40,2	26,80	-0,3554	-0,0214	0,0985	0,5599
Process air		in	19,0	13,80	-0,4991	-0,0595		
Process air	i	out	41,9	6,92	-0,4830	0,0717		
Regenerative air		out	39,3	27,20	-0,3553	-0,0259	0,1022	0,5538

Table 3. Calculated values of efficiency and functions of potential

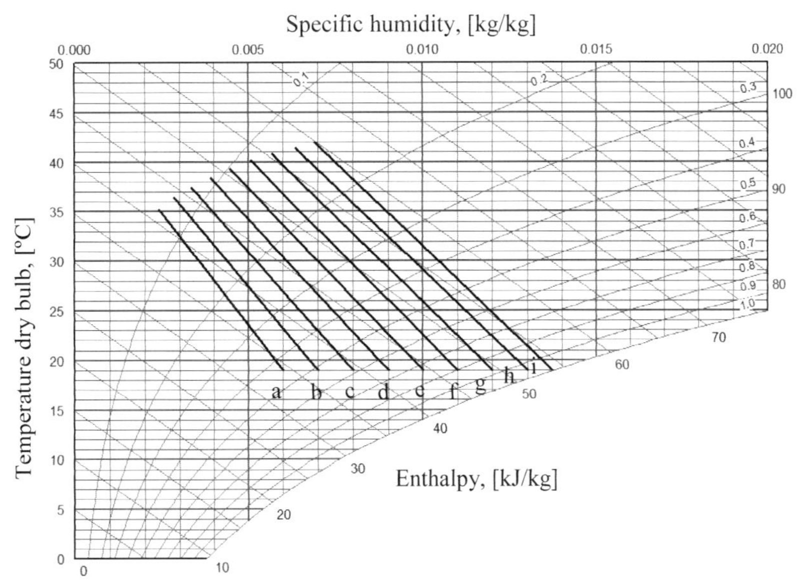

Fig. 7. Graphic representation of the dehumidification process for different input parameters of the process air

Figure 7 shows a graphic representation of the dehumidification process using the drying technique in a rotating wheel, the characteristics of which are given in Table 2 and which works with the state parameters of the process and regenerative air shown in Table 3. In the calculation, a constant value of the absolute humidity of the regenerative air of 15 g/kg was taken, that is, the maximum value of the absolute humidity that appears in the given climatic conditions during the summer period of operation of the dehumidifier. The diagram clearly shows the deviation of the actual drying process in the rotating wheel from the adiabatic, isoenthalpic process.

5. Modification of an existing TRNSYS component of a rotary desiccant wheel

When simulating the desiccant rotating wheel, more precisely the component of TRNSYS named Type683, as a fixed data that must be entered at the beginning of the calculation is the value of the efficiency ε_{F1} and the efficiency ε_{F2}, which remain unchanged during the entire simulation time, regardless of the change in the parameters of the state of the moist air in front of and behind the wheel. Considering that the efficiencies are defined by the functions of potential F_1 and F_2, which are directly dependent on the temperatures of the process and regenerative air, the simulation of the model at constant values of the efficiency of the rotating wheel gives satisfactory results in the case of constant parameters at the entrance to the dehumidifier. However, if it happens that outside air directly enters the desiccant rotating wheel, the simulation results show a certain deviation from the expected values of the output parameters of the process and regenerative air. For this reason, in this paper, a modification of the existing TRNSYS component of the desiccant rotating wheel was made, which during the simulation can change the efficiency values depending on the input state parameters of the moist air and maintain the values of the output properties within the limits of satisfactory accuracy. The calculated efficiency values are entered into the new model, which during the simulation interpolates the entered data and assigns to the rotating wheel the values of efficiency that belong to the parameters of the process air at the inlet.

5.1 Integration of the modified rotating wheel component in TRNSYS

Preparing and creating a new component in TRNSYS consists of two phases. The first phase involves modeling the actual processes that occur in the operation of the component and defining the mathematical model, while the second phase mainly refers to the collection and use of fixed data obtained from the equipment manufacturer, with the aim of better describing the characteristics of the given component. In this case, the first phase was already done when creating the basic component, and the modification of the component was connected to the second phase of component creation in TRNSYS and followed the existing constitutive relations and supplemented the model. The goal was to establish a component that will work equally with other TRNSYS components. The program that defines the mode of operation of the new component was written in FORTRAN and compiled for TRNSYS. The component has been tested in TRNSYS and shows agreement with the results given by the base component in operation with fixed wheel efficiency values.

The new component enables the change of the efficiency value of the desiccant wheel during the simulation, and for the given parameters of the process air during the simulation it assigns the rotating wheel the corresponding values, while the basic model of the component works with the same efficiency values for the entire simulation period, and at the same time the parameters of the state of the process air are variable. Figure 8 shows the change in the efficiency of the desiccant rotating wheel that works according to the new model during the process of simulating the operation of solar air conditioning. It can be seen from the diagram that the efficiency values are not constant, but vary depending on the changes in the parameters of the state of the process and regenerative air.

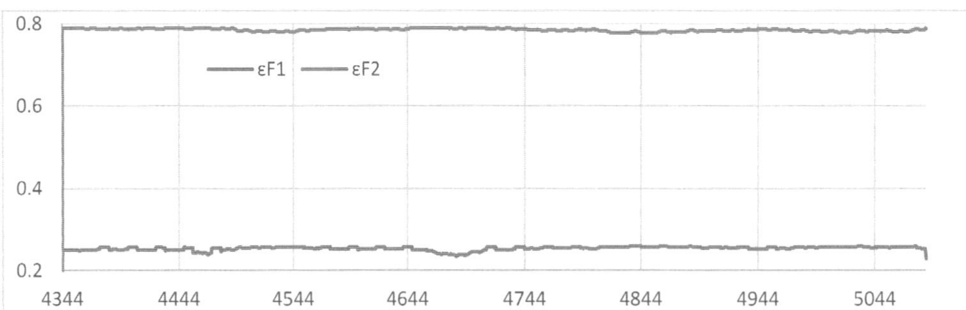

Fig. 8. Values of the efficiency functions of the desiccant rotating wheel during the simulation period

5.2 Validation of operation of the modified component of the desiccant rotating wheel

For the evaluation of performance and verification of calculated values using the new component in TRNSYS, a simple and limited simulation was employed. To test the new component, individual simulations of the existing component were conducted for 18 different combinations of process and regenerative air input parameters, with constant values for the wheel efficiency.

In doing so, the idea was to show the agreement of the output parameters of the state of the air from the desiccant rotating wheel obtained during the simulation with the experimental values during the operation of the wheel in stationary conditions, i.e. for one set of input parameters of the state of the process and regenerative air and the corresponding values of the wheel efficiency (Figure 9).

The obtained simulation results were compared to the results of dynamic simulation of the new component, and they proved to be identical. It's worth noting that in the simulation of the new component, the input data set was formatted in the form of temperature and absolute humidity of the process air at the wheel's inlet, which change over time, with each time interval representing one of the mentioned 18 combinations of moist air state parameters. The printouts after a certain time interval showed that, for the given values of the parameters of the state of the process air in that interval, the assigned values of the efficiency of the wheel coincide with the values of the fixed parameters entered before the start of the individual simulations on the existing model for the same values of temperature and absolute humidity of the process and regenerative air. It should be emphasized that the new model performs interpolation of the data values at its disposal with the aim of determining the efficiency of the wheel as accurately as possible in each time period of the simulation.

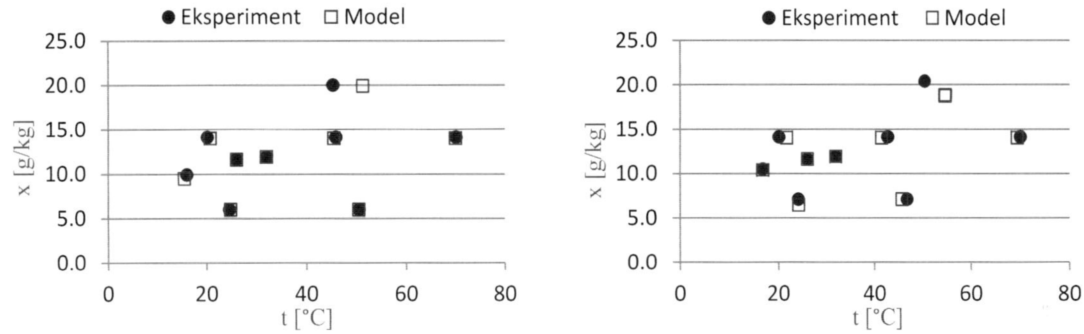

Fig. 9. Deviations between the simulation results obtained in TRNSYS and the values obtained through experiments, according to [16] (left) and [17] (right).

6. Conclusion

In the process of modeling a solar air conditioning system using the dehumidification process, the most challenging step is modeling the rotating wheel and determining the efficiencies ε_{F1} and ε_{F2}, whose values describe the cumulative effect of various influential parameters in the complex and simultaneous heat and mass transfer process. Considering that the values of ε_{F1} and ε_{F2} represent the efficiencies of the functions of potential F_1 and F_2, which are directly dependent on the input parameters of the process and the regenerative air, it can be concluded that special attention should be directed to the accurate input of these values when developing a model for a solar air conditioning system.

The results obtained by simulating basic models of solar cooling by dehumidification under steady-state conditions represent various values of the state parameters of moist air during the process and show good agreement with experimentally determined values of air parameters for the same conditions. During the simulation of the model under time-varying input parameters, there was a deviation in the output parameters of the process and regenerative air from the rotating wheel, as the simulation program TRNSYS uses a component that works with fixed parameters of the functions of potential F_1 and F_2, as well as the efficiencies ε_{F1} and ε_{F2}. The new desiccant wheel model developed in this work and integrated into the TRNSYS database allows for changes in the wheel efficiency values depending on the variation of the moist air state parameters at the inlet over time, thereby increasing the accuracy of the simulation.

In future research, activities should be focused on improving the characteristics of the desiccant wheel and reducing the regeneration temperature. This primarily refers to the invention of new sorption materials with a higher degree of adsorption and a lower production price. In this way, the efficiency of the desiccant wheel and the entire system will increase, and the price reduction will enable the application of multi-stage dehumidification, which reduces the regeneration temperature.

7. References

[1] Ahmovic, I.; Dzaferovic, E. & Hadziahmetovic, H. (2018). Analysis of Temperature Data in the Area of the City of Sarajevo (Bjelave) in Period From 2000-2010, Proceedings of the 29th DAAAM International Symposium, pp.0260-0266, B. Katalinic (Ed.), Published by DAAAM International, ISBN 978-3-902734-20-4, ISSN 1726-9679, Vienna, Austria, DOI: 10.2507/29th.daaam.proceedings.037

[2] Ma, Z.; Ren, H. & Sun, Z. (2019). Energy and exergy analysis of a desiccant cooling system integrated with thermal energy storage and photovoltaic/thermal-solar air collectors, Science and Technology for the Built Environment, Online First, 1-16

[3] Lulić, H.; Metović, S. & Softić, A. (2023). Exergy analysis of a solar assisted desiccant cooling system, 14th Days of BHAAAS in Bosnia and Herzegovina, International Conference on Innovative and Interdisciplinary Applications of Advanced Technologies – IAT 2023, Tuzla, Bosna and Hercegovina, June, in press

[4] Howe, R.R. (1983). Model and Performance Characteristics of a Commercially-Sized Hybrid Air Conditioning System Which Utilizes a Rotary Desiccant Dehumidifier, MS Thesis, University of Wisconsin – Madison

[5] Jurinak, J.J. (1982). Open Cycle Desiccant Cooling – Component Model and System Simulations, PhD Thesis, University of Wisconsin – Madison

[6] Banks, P.J.; Close, D.J. & Maclaine-Cross, I.L. (1970). Coupled Heat and Mass Transfer in Fluid through Porous Media – an Analogy with Heat Transfer, Proceedings of the fourth international heat transfer conference, 1970., Versailles, vol. VII, pp. 1-10. Elsevier publishing company, Amsterdam

[7] Momoi, Y.; Yoshie, R.; Satake, A. & Yoshino, H. (2010). Calculation of Dehumidification Coefficients for Numerical Simulation of Desiccant Wheel, Proceedings of the 31st AIVC Conference, Seoul, Korea, https://www.aivc.org/sites/default/files/3B-2.pdf

[8] Bourdoukan, P.; Wurtz, E.; Joubert, P. (2010). Comparison between the conventional and recirculation modes in desiccant cooling cycles and deriving critical efficiencies of components, Energy, Volume 35, Issue 2, Pages 1057-1067

[9] Norazam, A.S.; Kamar, H.M.; Kamsah, N. & Alhamid, M.I. (2018). Simulation of adsorption process in a rotary solid desiccant wheel, The 10th International Meeting of Advances in Thermofluids, 16–17 November 2018, Bali, Indonesia (IMAT 2018), AIP Conference Proceedings, Volume 2062, Issue 1, Print ISSN 0094-243X, AIP Publishing, 2019, https://doi.org/10.1063/1.5086547

[10] Nawaz, K.; Abdelaziz, O. & Jacobi, O. (2021). Characterization of Solid Desiccant for Dehumidification Applications, 18th International Refrigeration and Air Conditioning Conference at Purdue, May 24-28, 2021., Proceedings from 2021., Paper 2217, pp 2694, Page 1-9, https://docs.lib.purdue.edu/iracc/2217

[11] Zafar, A. & Taliv, H. (2022). Experimental Analysis and Performance Evaluation of Solid Desiccant Dehumidifier Using Silica Gel, International Journal of Analytical, Experimental and Finite Element Analysis, Volume 9, Issue 1, March 2022, pp 1 – 8, https://doi.org/10.26706/ijaefea.1.9.20211202

[12] Narayanan, R.; Saman, W.; White, S. & Goldsworthy, M. (2010). Modeling and parametric analysis of silica-gel desiccant wheels. Refrigeration Science and Technology Proceedings 2010-2, ISBN-139781617387845, 9th IIR-Gustav Lorentzen Conference on Natural Working Fluids 2010, Sydney, Australia, 12-14 April 2010, https://hdl.handle.net/10018/1023729

[13] Lindholm T. (2000). Evaporative and Desiccant Cooling Techniques, Ph.D. Dissertation, Department of Building Services Engineering, Chalmers University of Technology, Goeteborg, Sweden, ISBN 91-7197-904-2

[14] TRNSYS user´s manual (2004)., Solar Energy Labs, University of Wisconsin, Madison

[15] Schultz K.J. (1983). The Performance of Desiccant Dehumidifier Air-Conditioning Systems Using Cooled Dehumidifiers, MS Thesis, University of Wisconsin – Madison

[16] Höfker G. (2001)., Desiccant Cooling with Solar Energy, Ph.D. Thesis, Institute of Energy and Sustainable Development, De Montfort University Leicester, Department of Building Physics, Fachhochschule Stuttgart, Hochschule fur Technik

[17] Maalouf C.; Wurtz E. & Allard F. (2006). Evaluation of the cooling potential of a desiccant evaporative cooling system using the SimSPARK environment, UTAP, University of Reims, Campus du Moulin de la Housse, Reims, France, Published 2008

DOI: 10.2507/34th.daaam.proceedings.067

ERRORS EVALUATION OF WEIGHING PLATFORMS WITH ARRANGEMENTS OF 2, 3 AND 4 FORCE SENSORS

Dmitrii Smirnov, Andrei Vorotnikov & Yuri Poduraev

This Publication has to be referred as: Smirnov, D[mitrii]; Vorotnikov, A[ndrei] & Poduraev, Y[uri] (2023). Errors Evaluation of Weighing Platforms with Arrangements of 2, 3 and 4 Force Sensors, Proceedings of the 34th DAAAM International Symposium, pp.0515-0520, B. Katalinic (Ed.), Published by DAAAM International, ISBN 978-3-902734-41-9, ISSN 1726-9679, Vienna, Austria
DOI: 10.2507/34th.daaam.proceedings.067

Abstract

Force sensors are used in many fields of activity as components for measuring forces applied to them or measuring weights installed on them. To measure the weight of large objects, weighing platforms are used, which require the mutual placement of several force sensors. When designing weighing platforms, it is necessary to implement solutions to various scientific problems. In particular, this is an increase in the measurement range of the platform while maintaining sensitivity. Or is it achieving greater stability of the weighing platform by increasing the number of support points. In addition, when installing force sensors in weighing platforms, additional errors arise due to the following factors: displacement of the position of the applied force on the platform relative to the measuring centre, inaccuracy of sensor installation (deviations in the shape and size of the surfaces of force sensors and the surfaces on which they are installed), as well as differences in the characteristics of the sensors. This paper discusses 3 sensor arrangements for the measuring platform: with 2, 3 and 4 force sensors respectively. An evaluation was made of the errors that arise when using these arrangements.

Keywords: force sensor; weighing platform; errors evaluation; sensor calibration

1. Introduction

Currently, force sensors are used for both domestic, industrial and medical purposes. Also in the development of robotics, force-torque sensing takes a special place [13]. Basically, it is needed either in joints of robots or on end-effector, as well as weighing platforms for studying robotic technological processes. Often, designs use several force sensors simultaneously to increase the range of measured values while maintaining sensitivity or to weigh large objects. For example, a common arrangement of sensors is a "rectangle", where the sensors are located along the edges of the measuring platform, which makes it more stable (Fig. 1.a). This arrangement is quite widely used; examples include: floor, platform, automobile or food scales. For example, work [9] describes the design of portable vehicle scales powered by solar energy for use in quarry conditions. In [10], this arrangement is used to weigh the hive, which helps beekeepers make observations. In [11], to assess the distribution of strawberry yield on a field, using a berry picking trolley equipped with a system of 4 force sensors.

In addition to this application, scientists are making attempts to use this arrangement for medical purposes. For example, as in work [2], where 4 force sensors are located at the edges of the bed under its legs and are used to determine the characteristics of human movements in sleep and their recognition based on sensor data. [3] integrated sensors into a hospital bed to monitor respiratory rate and tidal volume, and [4] described estimating a patient's heart rate using force sensors signals. Also, there are weighing platforms for monitoring the patient's gait, which use 6-component sensors to determine the magnitude and direction of forces and moments [5]. This article will discuss single-component single point type sensors.

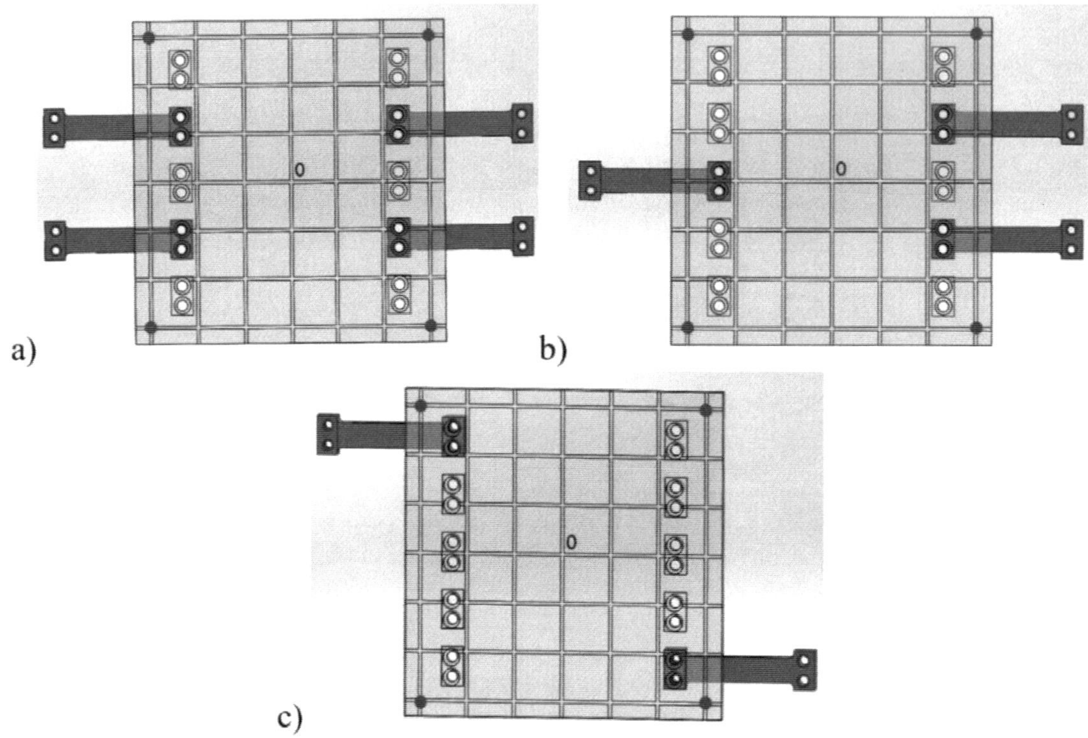

Fig. 1. Illustration of weighing platforms with arrangements of 2, 3 and 4 force sensors

The triangle design is often used to weigh bins of bulk materials, such as in [6] for a rice vending machine. However, an equilateral triangle is chosen, and the mass is always applied at the centre of this triangle. There is also an application for estimating engine thrust, where the air flow pressure is estimated using this arrangement [7]. In this work we will consider the diagram of an isosceles triangle (Fig. 1.b).

The final one is a diagram with two force sensors, the fastenings of which are located at the edges of the weighing platform (Fig. 1.c). This arrangement was used in [1] to determine the force acting on the motor rod. And also in the design of medical scales for weighing infants [8], where the error value of the eccentric application of force remained within the permissible error and amounted to 0.04% of the impact force at the nominal mass (+ - 2 grams with a nominal load capacity of the sensor of 10 kg). The purpose of this work is to evaluate the errors of weighing platforms using several force sensors and the applicability of the considered arrangement s.

2. Materials and methods

To read voltages from force sensors, an ADS1256 analogue-to-digital converter is used, which has a 24-bit capacity and 8 channels, which allows you to connect 4 independent force sensors simultaneously and take their data with an accuracy of up to 298 nanovolts at a supply voltage of 5V. ADS1256 works in conjunction with Arduino Uno, which communicates with the ADC via the SPI interface for subsequent post-processing and transmission of measurement results to the serial port interface window. The schematic electrical diagram is shown in Figure 2. Post-processing consisted of converting the data from the ADC into voltage (mV) and, after calibration, into mass (grams). The reading frequency was 1 kHz. In order to minimize errors associated with spikes and conversion of analogue values using an ADC, average data over 400 values were evaluated.

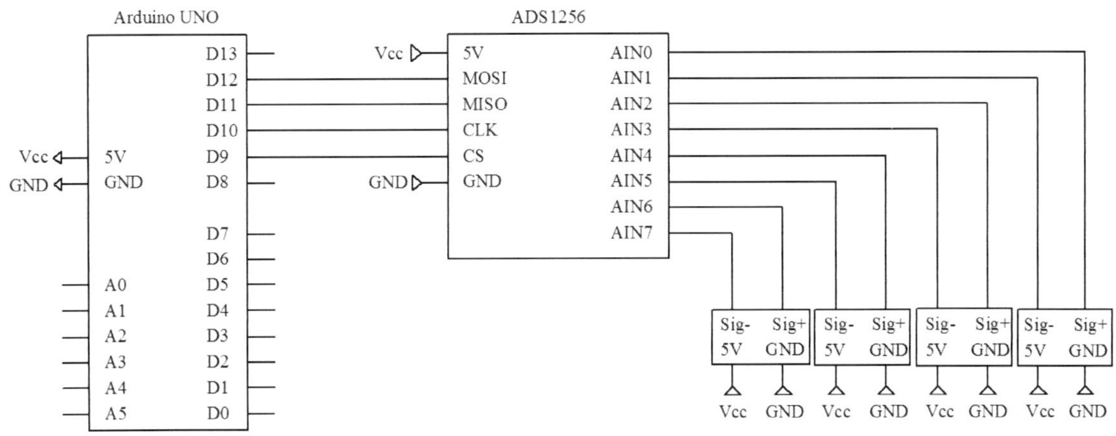

Fig. 2. Schematic diagram of connecting the components of the weighing platform

To conduct an evaluation of the errors of weighing platforms, it is necessary to have sensors that have been calibrated in order to correctly assess the resulting errors and compare them with the existing characteristics of the sensors. Therefore, the initial stage of scientific research was the calibration of each sensor separately. To do this, each sensor was gradually loaded up to 110% of the nominal load capacity, which is 100 grams, and unloaded using standard weights of accuracy class M1 (Fig. 3.). To obtain the coefficients for converting voltage, expressed in mV, into units of mass - grams, an approximation was carried out using the least squares method by the following minimization function:

$$\sum_{i=1}^{n}(U_i-(am_i+b))^2 \to \min$$

where U_i is the voltage from a separate sensor when applying mass m_i, where i is the index number of the measurement, n is the number of measurements, a and b are approximation coefficients. As a result of calibration, the following sensor characteristics were obtained, shown in Table 1 below, for 2 cycles of 7 measurements during loading and unloading.

Number of sensors	**1**	**2**	**3**	**4**
Nominal load capacity [g]	100			
Zero balance %	0,027	0,033	0,008	0,005
Nonliniarity %	0,108	0,153	0,048	0,169
Hysteresis U %	0,045	0,014	0,010	0,072
Maximum deviation [g]	0,108	0,153	0,048	0,169
RCP mv/v	1,274	1,252	1,215	1,244
Standard deviation of sensor data after calibration from the applied mass [g]	0,029	0,042	0,013	0,054
Systematic error U%	0,031	0,076	0,009	0,133

Table 1. Sensor characteristics obtained as a result of calibration.

Figure 3 illustrates the calibration process force sensors using reference weights. Also, the right figure 3 shows the results for all sensors, where dots indicate experimental data, and solid colored lines show straight lines based on the found coefficients after the approximation.

As you can see, although the sensors are the same, they have different characteristics, which can affect the values measured by the weighing platform. When using certified sensors, these characteristics are indicated in the product passport or in the calibration certificate.

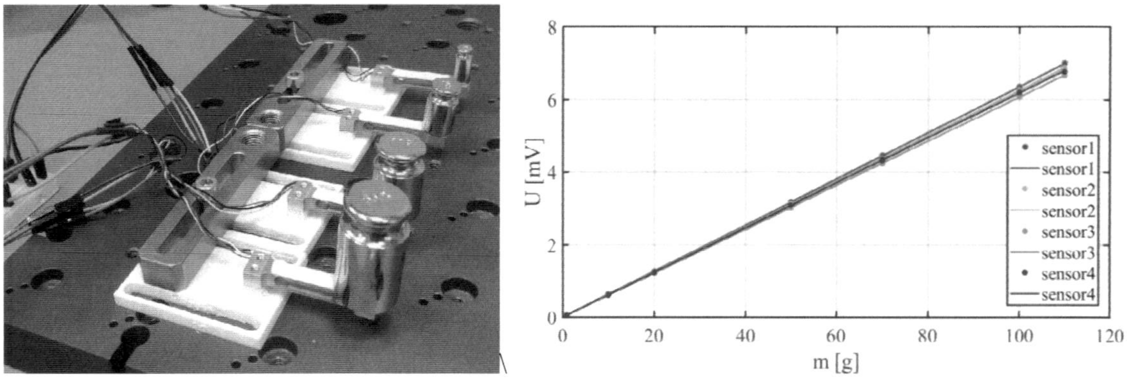

Fig. 3. Calibration process using standard weights (left) and calibration curves obtained after fitting (right)

Initial calibration was also used during installation to ensure that the maximum permissible force on the sensor was not exceeded, which could lead to its failure. And also for leveling the platform, focusing on the data. After all, if the mass of the platform is evenly distributed across the sensors, then we can say that it is leveled. When multiple sensors are used under one platform, the mass applied to it is distributed among the sensors. Accordingly, to evaluate the applied forces, it is necessary to add up the sensor data. Installation of sensors on the same platform makes them structurally interconnected, which leads to the appearance of a complex error, similar to that described in [12], i.e. consisting of several components:

$$F_{\Sigma} = \sum_{i=1}^{n} F_i + \Delta_{sen} + \Delta_{montage} + e$$

where F_{Σ} is the resulting force corresponding to the mass applied to the platform, taking into account errors (Δ_{sen} is the error caused by the presence of differences in the characteristics of the sensors, e is the eccentricity error (Fig. 4. on the right), caused by the displacement of the position of the weight on the measuring platform relative to the measuring centre, $\Delta_{montage}$ is the error arising due to deviations in the shape and size of the surfaces of the force sensors and the surfaces on which they are installed), F_i – data of the i-th sensor after calibration.

To evaluate the effect of the above errors, two experiments were carried out for all arrangement s. In the first it was assessed. To do this, the platform was sequentially loaded and unloaded in the same way as was done when calibrating each sensor separately earlier, it is important to note that the loads were located only in the centre of the platform. And in the second experiment, it was evaluated, where only a weight of 100 grams was used (which corresponds to the value of the nominal load capacity, in order to achieve the largest error value, since it was found that the error increases with increasing applied mass). For this purpose, a grid was outlined in advance (Fig. 1), at the intersection points of which (49 positions), weights were installed.

Based on the data obtained during the first experiment, the values of the maximum error and the standard deviation of the data of the resulting force of the sensors relative to the applied mass were calculated (Table 2). A similar assessment was carried out in [14]

Number of sensors	**2**	**3**	**4**
Maximum error [g]	3,6573	1,2615	0,5711
Standart diviation [g]	1,276	0,432	0,180

Table 2. Numerical values of the maximum error and standard deviation of the data of the resulting force of the sensors from the value of the applied mass for various sensor layouts

To reduce this error, the already assembled sensor system was recalibrated using the following algorithm:
1. Removing data in 4 extreme positions on the platform, indicated in Figure 1 and Figure 4 by red dots. – the value of the resulting force in the upper left position with coordinates (-3;3) when mi mass is applied; respectively – in the lower left (-3;-3), – in the lower right (3;-3), – in the upper right (3;3).
2. Approximation of average values from 4 positions relative to the applied mass mi using the least squares method:

$$\sum_{i=1}^{n}(\overline{F_{\Sigma_i}} - (a_{\Sigma}m_i + b_{\Sigma}))^2 \rightarrow \min$$

3. Entering the obtained coefficients a_{Σ} and b_{Σ} to obtain the resulting force corresponding to the mass applied to the platform:

$$F_{\Sigma cor} = \frac{(F_{\Sigma} - b_{\Sigma})}{a_{\Sigma}}$$

This algorithm made it possible to reduce the influence of error. As a result, when installing weights in the centre of the platform, the following deviations were obtained, shown in Table 3. Which are comparable with the data given in Table 1.

Number of sensors	**2**	**3**	**4**
Maximum error [g]	0,2442	0,3624	0,2125
Standart deviation [g]	0,0532	0,0717	0,0666

Table 3. Numerical values of the maximum error and standard deviation of the data of the resulting force of the sensors from the value of the applied mass for various sensor layouts after recalibration

To evaluate the effect of error e, a second experiment was conducted. For this purpose, a weight weighing 100 g was used and was located at the intersection points of the intended grid on the platform (Fig. 1), which made it possible to visualize the distribution of error e relative to the average value from all positions on the platform (in order not to take into account the effect of other errors) depending on the location of the load on platform (Fig. 4).

$$e_j = F_{\Sigma j} - \overline{F_{\Sigma j}}\ ,$$

where j is the serial number of the resulting force at different positions on the platform (marked purple in the fig- 4).

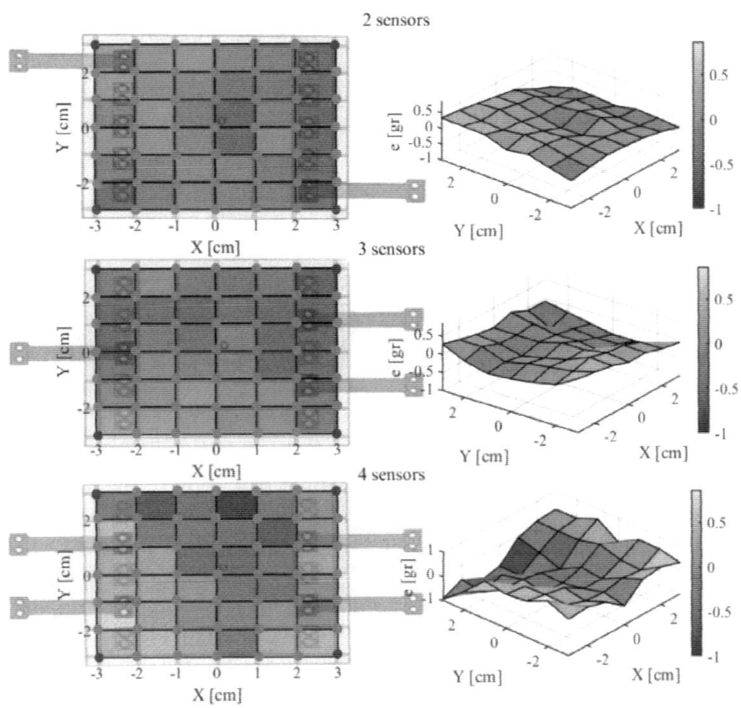

Fig. 4. Illustration of the distribution of error *e* relative to the average value from all positions on the platform depending on the location of the load for all arrangements.

The results show that a shift in the position of the weight on the measuring platform relative to the measuring centre leads to significant errors in the data up to 1 gram, which is 1% of the nominal load capacity of the sensors. Moreover, increasing the number of supports does not solve this problem, but on the contrary leads to an increase in error. It is important that the resulting error is unevenly distributed over the measuring surface. The relationship between the distribution of this error and the geometric dimensions between the supports will be investigated in future work. After all, if we identify the systematic component of such a phenomenon, then it will be possible to design weighing systems of any size with a sense of the error with which they will carry out measurements.

3. Conclusion

In this work, various arrangements of 2, 3 and 4 force sensors for weighing platforms were considered. In the course of the work, it was found that the error that occurs when several sensors are used together, the connection between each other is constructive, complex and conditional, takes into account the following factors: placing weight measurements on the measuring platform relative to the measuring centre, inaccuracy of sensor measurements (deviations in the shape and size of the characteristics force sensors and the characteristics they provide), as well as differences in sensor characteristics. The first will be considered in subsequent works: the study of the error arising due to the displacement of dimensions relative to the centre depending on the geometric dimensions between the supports, which will make it possible to design weighing platforms with the size of this error. The second can lead to the creation of additional bending moments, leading to increased errors and a more labour-intensive assembly process. Supposedly, the performance issue can be corrected by using additional soft spacers, which will be tested in subsequent work. The latter error introduced by analogue conversions in sensor characteristics can be reduced by retransmission after the sensor system has been assembled. It can also be said that the presented diagrams represent consistency for use.

4. References

[1] Belitsky E., Solovyev M., Kovalsky V., Vorotnikov A., Kordonsky A., Grin A., Poduraev Yu. (2023) "Mechatronic linear movement device for constructing surfaces of complex shapes with the possibility of force measurement", Robotics and Technical Cybernetics. No. 11 (2)., pp. 85-93. DOI: 10.31776/rtcj.11201 – Text: electronic.

[2] Adami, A. M., Hayes, T. L., Pavel, M., & Singer, C. M. (2006, January). Detection and classification of movements in bed using load cells. In 2005 IEEE Engineering in Medicine and Biology 27th Annual Conference (pp. 589-592). IEEE..

[3] Jung, H., Kimball, J. P., Receveur, T., Gazi, A. H., Agdeppa, E. D., & Inan, O. T. (2022). Estimation of tidal volume using load cells on a hospital bed. IEEE Journal of Biomedical and Health Informatics, 26(7), 3330-3341.

[4] Jung, H., Kimball, J. P., Receveur, T., Agdeppa, E. D., & Inan, O. T. (2021). Accurate ballistocardiogram based heart rate estimation using an array of load cells in a hospital bed. IEEE Journal of Biomedical and Health Informatics, 25(9), 3373-3383.

[5] Payton, C. J., & Burden, A. (Eds.). (2017). Biomechanical evaluation of movement in sport and exercise: the British Association of Sport and Exercise Sciences guide. Routledge.

[6] Yuliandoko, H., Panduardi, F., Lusi, N., & Wibowo, S. (2023, February). Precision Rice Vending Machine by Using Multiple Load Cell and IoT Based. In 2023 International Conference on Computer Science, Information Technology and Engineering (ICCoSITE) (pp. 296-300). IEEE.

[7] Baer, P. H. (2008). 3 Load Cell Tumble Meter Development (No. 2008-01-3004). SAE Technical Paper.

[8] Новиков, Ю. С., & Ситдиков, Ф. Г. (2001). Весы медицинские электронные. // Novikov Yu. S.& Sitdikov F. G. (2001) Medical electronic scales.

[9] Zamri, N. S. M., Mohamad, K. A., Alias, A., Razali, E. F., & Nordin, M. S. (2022). Development of A Solar-Powered Weighbridge System Using Load Cells in Quarry Site. Evolution in Electrical and Electronic Engineering, 3(1), 402-408..

[10] Zacepins, A., Pecka, A., Osadcuks, V., Kviesis, A., & Engel, S. (2017). Solution for automated bee colony weight monitoring. Agronomy Research, 15(2).

[11] Anjom, F. K., Vougioukas, S. G., & Slaughter, D. C. (2018). Development and application of a strawberry yield-monitoring picking cart. Computers and electronics in agriculture, 155, 400-411.

[12] Vorotnikov, A. A., Poduraev, Y. V., & Romash, E. V. (2015). Estimation of error in determining the centres of rotation of links in a kinematic chain for industrial robot calibration techniques. Measurement Techniques, 58, 864-871.

[13] Solovyev, M., Andrei, V., Klimov, D., Vladislav, K., & Poduraev, Y. (2017). Control system of the Articulated Arm Braking Mechatronic Machine (AABMM). Annals of DAAAM & Proceedings, 28.

[14] Vorotnikov, A., Romash, E., Isaev, A., Bashevskaya, O., Bianchi, G., & Poduraev, Y. (2016, January). Uncertainty estimation of axes direction determination of industrial robot using an ellipsoid concentration model. In Proceedings of the 27th DAAAM International Symposium (pp. 0480-0486).

DOI: 10.2507/34th.daaam.proceedings.068

CREATION OF A FRAMEWORK FOR AN EMPATHY-BASED IMPROVEMENT OF A HYBRID WORKING SYSTEM

Tobias Möltner, Kemajl Stuja, Mohamed Aburaia & Ali Aburaia

This Publication has to be referred as: Tobias, M[oeltner]; Kemajl, S[tuja]; Aburaia, M[ohamed] & Aburaia, A[li] (2023). Creation of a Framework for an Empathy-Based Improvement of a Hybrid Working System, Proceedings of the 34th DAAAM International Symposium, pp.0521-0525, B. Katalinic (Ed.), Published by DAAAM International, ISBN 978-3-902734-41-9, ISSN 1726-9679, Vienna, Austria
DOI: 10.2507/34th.daaam.proceedings.068

Abstract

Collaborative robotics is changing the way we work, making it safer, more efficient, and more productive, by merging human and robot strengths. The increasing utilization of hybrid working systems has led to a need for improved methods of operator stress detection and mitigation. This thesis proposes a framework for an empathy-based improvement of hybrid working systems. The framework uses galvanic skin response (GSR) measurements, captured with the physiological data sensor "Empatica E4", to detect moments of stress (MOS) in the operator. Subsequently, a stress-level, based on the number of MOS within a predefined duration, is estimated. Therefore, the framework's reactive mechanism adapts the collaborative robot's (cobot) manipulation speed and task allocation by sending the corresponding signals to the robot control. The aim of this adaptation is to reduce the operator's stress-level while maintaining or increasing productivity. The framework was evaluated with a platform for assembling a miniature robot. Results showed that the framework was able to effectively detect moments of stress and adapt the cobot's manipulation speed accordingly. The framework could be used to reduce operator stress and improve productivity in a variety of industrial applications.

Keywords: Collaborative robotics; Empathy; Stress detection and reduction; Reactive mechanism.

1. Introduction

Cobots enhance the classical approach of industrial robots by making it possible for humans and robots to work on a task simultaneously. The growing importance of concurrent utilization of humans and robots is leading to the development of dynamic interfaces to accommodate individual workers' capabilities [1]. Understanding the human's state, particularly stress, is crucial, and this work aims to design an interface that collects physiological data to assess stress-levels and establish a reactive interaction, referred to as empathy, between the robot and the human. Human-robot collaboration poses challenges such as safety risks, data privacy and security, and limitations in payload and reach. However, one of the biggest challenges is to keep the productivity level consistent, as human behavior is unpredictable and can be affected by factors such as stress and mood. Thereby, the interpretation of the mental state of the operator and a corresponding reaction becomes a key factor. Recent developments are emphasizing the demand for a corresponding non-invasive reactive mechanism that is affected by human manipulation to a minimal extent [2].

That is why authors have dealt with the problem of finding a suitable solution for interpreting the operator's mental state. The paper [3] classifies five existing State of the Art approaches for the operator's emotional state assessment. It concludes that each existing approach is based on "facial recognition", "speech recognition", "body language", "electrocardiography (ECG)" or electroencephalography (EEG). In the paper [4] the authors investigate the effectiveness of a human sensor-based emergency stop interface designed to respond promptly when a human operator detects or anticipates a potential emergency in the context of Human-Robot Collaboration. The proposed approach utilizes a mobile EEG to detect potential emergencies.

With the BAZAR Cherubini et al. merge the vision and sensing-based approach with a reactive system for power and force limiting with torque sensors within the robot's structure [6]. It is deployed in smart logistics use cases like carrying specific car parts to autonomous ground vehicles for assembling. Since the examined literature shows that the challenge of creating a secure and tamper-free interpretation of the mental state of the human remains, this work suggests a novel approach using GSR/electrodermal activity (EDA). To provide a comprehensive overview of the work's proceedings, the paper is organized as follows. Before depicting the context of the work, the problem and the starting point is outlined. To examine the effectiveness of the framework, a subset is applied in a laboratory setup. Section 3 displays used materials and methods. In section 4, the concept of the framework is shown. The analysis of the results is presented in section 5. In section 6, a concluding reflection of the implementation is presented and recommendations for further research are given.

2. Problem statement

Since 2015, the University of Applied Sciences Technikum in Vienna has been developing low-cost robots made of polyamide for educational purposes [7]. This robot series bears the name of the developer, Mo-Robot or Morobot. In the meantime, however, they have been further modified and optimised for automated production (Design for Automation). One task in this production line is assembly. As can be seen in Figure 1. a), this task consists of several steps. Although all parts have been optimised for automated assembly, hybrid applications (robot and human - Figure 1, b and c) are also being tested as part of the research. These hybrid working systems are designed to combine the high performance of robotic systems with the energy efficiency of humans. In addition, humans are able to check the functionality of the products with little effort (Figure 1, d and e).

The evaluation of concept variants (robot-robot, human-human and robot-human) according to the methodical system in [8] and based on customer requirements has shown that this variant is the optimal one for a given batch size. In addition, threatening circumstances such as climate change and the scarcity of the earth's fossil resources are leading to greater energy awareness in society. Energy efficiency has therefore become an integral part of this work. However, the interaction between the technical part of the system (the robot) and the biological part (the human) is a major challenge in maintaining a constant level of productivity. In addition, the unpredictability of human behaviour (stress, mood swings, etc.) has an immense impact on the overall efficiency of the hybrid work system. It is therefore necessary to design a reactive system between the robot and the human.

This reaction is called "empathy" and must be decomposed into well-defined correlations between the participants of the system. Each correlation should be defined as an input variable (e.g. stress moment level) for the reactive part of the system (robot system). This means that the rate of work is set by the human, e.g. in case of fatigue the manipulator's speed should be reduced resp. the manipulator should react with emphatic behaviour.

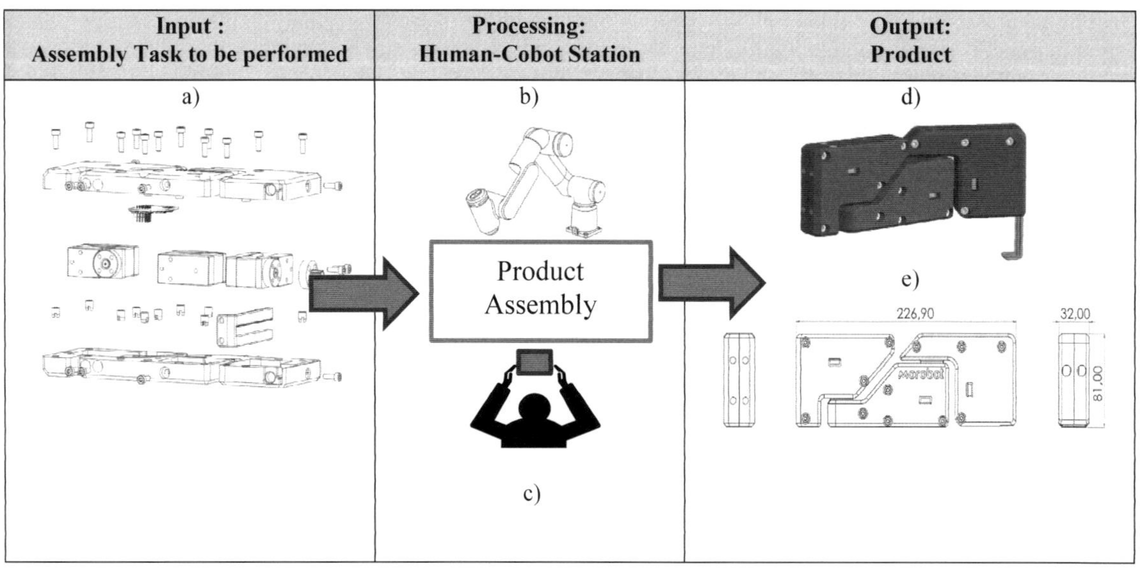

Fig. 1. Human-Cobot Station- Assembly Task.

3. Material and Methods: Processing Station

The proposed structure of the subset, which is shown in figure 2, encapsulates a specific use case scenario where a miniature robot (morobot) is assembled by the human operator and the cobot. The reactive mechanism is the controlling parameter of the PLC program so that the speed of the cobot can be adapted based on the stress level of the operator. An experiment with ten participants was conducted in order to validate the framework.

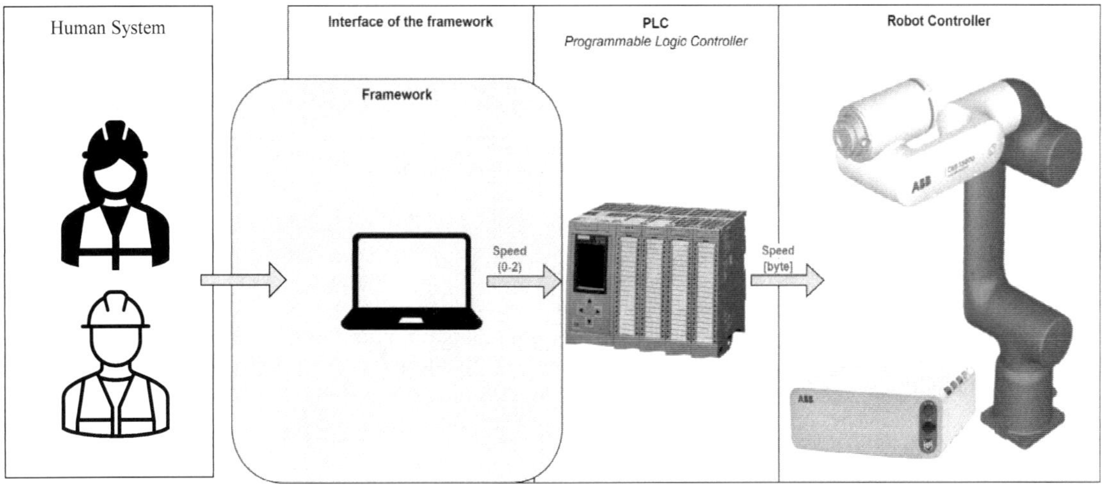

Fig. 2. Structure of the application of the framework's subset

To allow validation of the framework, each operator's stress-level and the corresponding timestamp were captured during the morobot assembly. Due to the window size and the fact that a maximum number of one MOS can be captured per window, stress-level values from 0-2 are provided every 10 seconds.

4. Conception of Collaboration-Framework

As there is currently no available approach for a reaction toward the emotional state of the operator without using vision-based or hardware-expensive equipment, research on possible empathetic approaches was conducted. As a consequence of that, a rule-based algorithm for the detection of MOS suggested in [5] was called into play. With the help of that algorithm a framework, which is composed of three sequential phases that are repeated cyclically (see figure 1), was constructed.

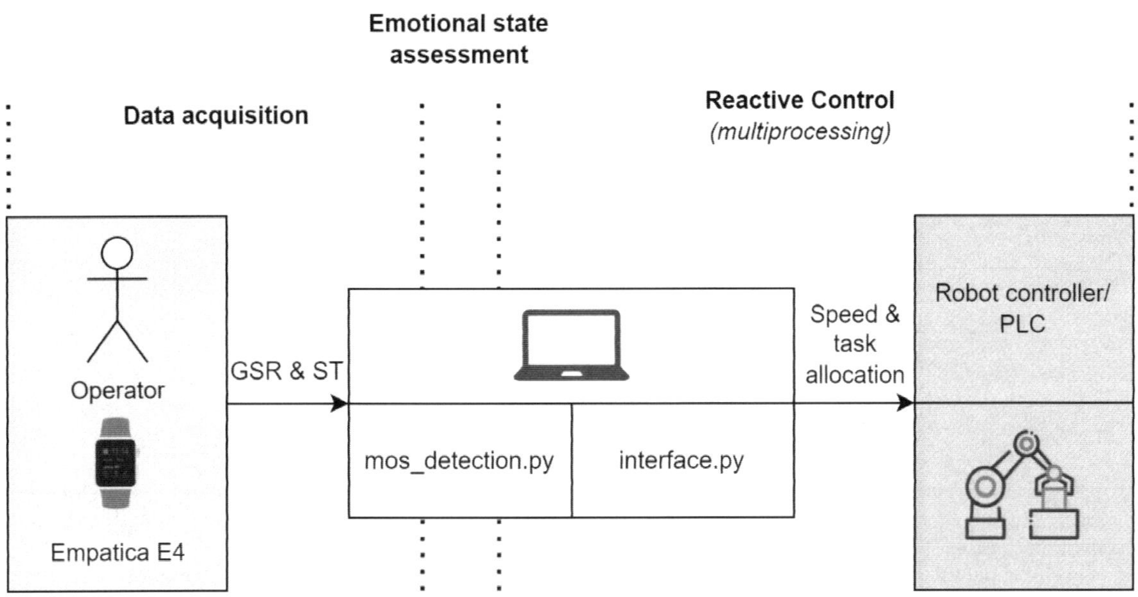

Fig. 3. Structure of the framework and its phases

In the first phase, the physiological data of the operator is captured utilizing the "Empatica E4" sensory module and setting up a Bluetooth-based stream with the USB dongle "BLED112". Subsequently, a PC, serving as an interface, pre-processing the data and estimates a stress-level, based on the changes in the GSR data. This is done with the help of the rule-based MOS detection algorithm suggested in research of Kyriakou, Moser et al. [6]. The proceedings of both works make use of the GSR-value propagation over time and investigate the magnitude and the corresponding derivative. Since the interface is continuously collecting data, a moving window approach was chosen in order to be able to pre-process and make use of the data during runtime. Establishing a window size of 10s ensures that the defined rules of the MOS detection can be applied, and the reactive control's latency is kept to an acceptable extent. By counting the MOS within the last and its previous window, a stress-level from 0 to 2 is estimated.

- ***Case 1*** - The number of MOS in the present and past window is 0; **The operator is not stressed**.
- ***Case 2*** - The number of MOS in the present and past window is 1; **The operator is extremely stressed**.
- ***Case 3*** - The number of MOS in the present window is 1, and the number of MOS in the past window is 0; **The operator is moderately stressed.**
- ***Case 4*** - The number of MOS in the present window is 0, and the number of MOS in the past window is 1; **The operator is moderately stressed.**

According to the estimated level of stress, being an interpretation of the mental state of the operator, the interface sends signals for adapting the speed and task allocation of the collaborative robot. For the purpose of establishing parallel communication between sensor-PC and PC-cobot, multiprocessing for the PC's Python modules was implemented. Thereby, the communication for data acquisition and reactive control can be maintained simultaneously. The following figure gives a comprehensive overview of the implemented processes and the required organization of units and connections for establishing the functionality of the framework.

Furthermore, this reactive mechanism was validated with a subset located in the facilities of "UAS Technikum Wien". In the experiment each of the ten participants had to assemble a miniature robot (morobot) with a predefined task allocation between the human and the cobot "CRB 1500". To apprehend the effect of the presence or absence of the reactive control, each participant assembled the morobot with and without it being active in an arbitrary order.

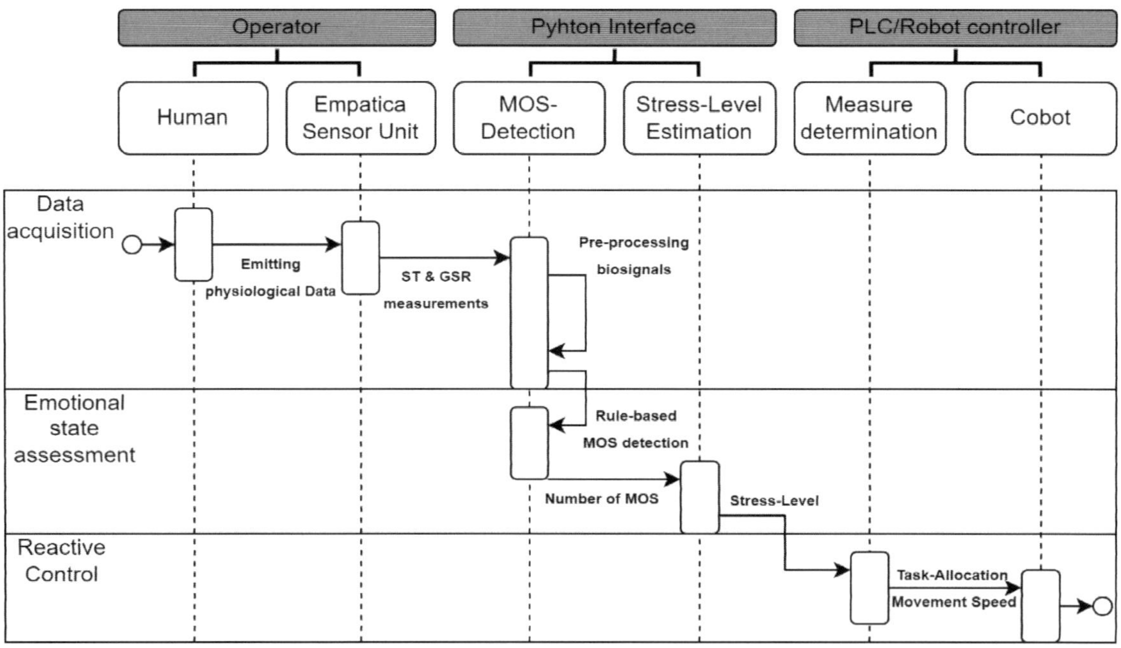

Fig. 4. Units and connections of the framework

5. Results

The results showed that the framework was able to adapt to the speed of the cobot and thereby change the Duty Cycle of the assembly process. Furthermore, a change in the workload of both the cobot (WLC) and the human (WLH) with the reactive mechanism enabled or disabled can be observed. However, the effect of the presence of the reactive mechanism on the well-being of the operator was difficult to examine. As table 1 also shows, due to the sequential process order of the subset, a validation of the positive impact of the framework's presence on the operator's mean stress level (OPML) was strenuous to accomplish.

Parameter	Enabled	Disabled	Difference[%]
OMSL	0.5	0.38	24.23
WLC	7.27min	7.52min	10.32
WLH	7.42min	7.7min	4.64
Duty Cycles	18.67min	16.83min	9.86

Table 1. Parameter of the validation process and relative comparison

6. Conclusion and Future Work

The proposed framework introduces a novel approach to reacting to the operator's mental state by capturing MOS, estimating stress levels, and implementing empathy-based improvements in a hybrid working system. Unlike existing solutions, this framework offers a non-invasive, difficult-to-manipulate algorithm for assessing operator well-being. It suggests actions to enhance the emotional state of human operators during collaborative tasks, potentially increasing productivity and reducing Duty Cycle. By fostering empathy between cobots and human workers in the manufacturing environment, this framework aims to improve job satisfaction and overall well-being, marking a crucial step toward a more harmonious and productive future in manufacturing. However, the performed research's validation is limited to the utilized subset consisting of equipment present in the facilities of the "UAS Technikum Wien". Therefore, the validation proceedings leave gaps for improvements.

In terms of future work, several opportunities for improvement and research gaps have emerged. Validation of the framework with a customized or substantially modified subset within the "Digitale Fabrik" of the UAS "FH Technikum Wien" could enhance result significance. Exploring suitable cutoff frequencies for MOS detection and adjusting statistical power calculation parameters, such as effect size and sample size, could lead to more accurate evaluations. Additionally, leveraging geo-data from "Empatica's" E4 sensory module could establish a novel safety mechanism for collaborative setups. Addressing challenges posed by metal implants in data capture and investigating such scenarios further would provide valuable insights. In an evolving work landscape, this framework not only envisions empathy-based hybrid workstations but also sets the stage for a transformative paradigm where technology and human connection seamlessly shape the future of collaboration and productivity.

7. Acknowledgments

Our special thanks go to the University of Applied Sciences Vienna -FH Technikum Wien- for providing the equipment and the laboratory as well as for sponsoring this work. Furthermore, we would like to thank the research team of the Paris Lodron University Salzburg and Fraunhofer Austria Research GmbH for providing knowledge and support.

8. References

[1] Arash Ajoudani, Andrea Maria Zanchettin, Serena Ivaldi, Alin Albu-Schäffer, Kazuhiro Kosuge, and Oussama Khatib (2018). Progress and prospects of the human–robot collaboration. Autonomous Robots, 42:957–975.

[2] Inkulu, A. K., Bahubalendruni, M. R. & Dara, A. (2022). Challenges and opportunities in human robot collaboration context of industry 4.0-a state of the art review. Industrial Robot: the international journal of robotics research and application, 49(2), 226–239.

[3] Martin Karl Moser, Bernd Resch, and Maximilian Ehrhart (2023). An individual-oriented algorithm for stress detection in wearable sensor measurements. IEEE Sensors Journal.

[4] Achim Buerkle, Thomas Bamber, Niels Lohse, and Pedro Ferreira (2021). Feasibility of detecting potential emergencies in symbiotic human-robot collaboration with a mobile eeg. Robotics and Computer-Integrated Manufacturing, 72:102179.

[5] Kalliopi Kyriakou, Bernd Resch, Günther Sagl, Andreas Petutschnig, Christian Werner, David Niederseer, Michael Liedlgruber, Frank Wilhelm, Tess Osborne, and Jessica Pykett (2019). Detecting moments of stress from measurements of wearable physiological sensors. Sensors, 19(17):3805.

[6] Andrea Cherubini, Robin Passama, Benjamin Navarro, Mohamed Sorour, Abdellah Khelloufi, Osama Mazhar, Sonny Tarbouriech, Jihong Zhu, Olivier Tempier, André Crosnier, et al (2019). A collaborative robot for the factory of the future: Bazar. The International Journal of Advanced Manufacturing Technology, 105:3643–3659.

[7] Stuja K., Bruqi, M., Markl E. & Aburaia, M. (2016). Lightweight 4 -Axis Scara Robot for Education and Research, Proceedings of the 27th DAAAM International Symposium, pp. 0102-0108, B. Katalinic (Ed.), Published by DAAAM International, ISBN 978-3-902734-08-2, ISSN 1726-9679, Vienna, Austria.

[8] Stuja, K.; Katalinic, B.; Aburaia, M. & Pajaziti, A. (2021). Design of a Robot Application with Regard to Energy Efficiency, Proceedings of the 32nd DAAAM International Symposium, pp.0033-0039, B. Katalinic (Ed.), Published by DAAAM International, ISBN 978-3-902734-33-4, ISSN 1726-9679, Vienna, Austria

DOI: 10.2507/34th.daaam.proceedings.069

BENEFITS OF KNOWLEDGE MANAGEMENT IN CYBERSECURITY – INITIAL STUDY

Jan Albrecht, Petra Dostálová, Lukáš Králík & Gabriela Králíčková

This Publication has to be referred as: Albrecht, J[an]; Dostalova, P[etra]; Kralik, L[ukas] & Kralickova, G[abriela] (2023). Benefits of Knowledge Management in Cybersecurity – Initial Study, Proceedings of the 34th DAAAM International Symposium, pp.0526-0530, B. Katalinic (Ed.), Published by DAAAM International, ISBN 978-3-902734-41-9, ISSN 1726-9679, Vienna, Austria
DOI: 10.2507/34th.daaam.proceedings.069

Abstract

The work focuses on the aspects of knowledge management that could be beneficial in dealing with cybersecurity issues, specifically targeting common users with a basic knowledge in that area. To understand the connection, the paper briefly introduces knowledge management itself, capturing the essential information. The work explains how technological innovation changes the value of intangible assets that are closely related to cybersecurity, followed by the description of threats in the cybersecurity field, characterizing its categories, and giving current examples. The importance and potential benefits of knowledge management are often overlooked, making it much harder for many organizations to deal with various types of threats. The work presents how valuable knowledge management can be and how to use it to deal with the problem of avoiding and mitigating cybersecurity threats with their consequences, depicting the cycle of knowledge management, and describing its individual parts.

Keywords: Knowledge management; Cybersecurity; Technological innovation; ISMS; Continuous improvement.

1. Introduction

During the last decade, the world has been witnessing a major increase in technological innovation, which is exponentially rising. The improvement and exploration of new possibilities is not a problem on its own, on the contrary, there are plenty of useful outcomes that can help in many ways and be an enormous help in a lot of areas (healthcare, cybersecurity, predictions, etc.). New possibilities, often intentionally meant to be used mainly for beneficial outcomes, have a great potential of being misused by attackers. The rise of technological innovation has been there for some ages now and it is a natural way of evolution [1], [2], [3]. By acknowledging that fact, organizations and people should not only think about the possible benefits with positive outcomes, but also about the possibilities of misuses and what threat can they pose [3]. As Talet (2017) mentions, knowledge management (KM) alongside the potential risks requires to be taken very seriously in the matter of obtaining the actual and relevant information of possible threats [4]. By doing so, organizations can be prepared to respond to new threats, reducing the risk of them causing some crucial harm. One of the vulnerable parts of an organization and its cybersecurity measures is usually the human factor. When people within are not educated enough about the possible threats and consequences, they are not able to properly deal with the issues, even if the problem is only at the start and could be manageable with the right knowledge [5], [6].

In some areas it could be hard to use KM, however, cybersecurity on the level of the common user is one of the examples where the power of knowledge and its sharing can be a crucial factor for a fast and appropriate response to threats [4], [6]. The paper aims to introduce the benefits of using KM in cybersecurity on the level of the common user and suggests how should be the KM implemented into the organization's processes to be most effective.

2. The role of Knowledge management

In the previous ages, visible assets were the most important ones to be taken care of, today we live in an era, where non-material assets are becoming more and more valuable, hence their need for protection is also on the rise. Personal experience, relationships, talents, skills, subconscious, and many more were there even before, but their importance was overshadowed by the necessity to protect tangible assets such as material, human resources, premises, etc [4], [5]. Nowadays, when technology rises and with it, the amount of information, data, and knowledge, makes their value for dealing with threats (among other fields) greater than before. However, their value does not come only from their acknowledgment, but mainly from the way of their efficient gathering, storage, and sharing [5], [7], [8].

The Knowledge Management standard is ISO 30401:2018 and includes requirements, sets of best practices, processes, and many other topics related to KM [9]. As it functions as a great guidance tool for the implementation of KM principles, some organizations prefer to use frameworks such as the Information Technology Infrastructure Library (ITIL), Control Objectives for Information and Related Technologies (COBIT), AGILE, and many others. These tools mentioned are not primarily KM tools, but they do include the knowledge principles in their structure, making them suitable for that application, whilst the KM implementation on its own can be perceived as redundant for them [9], [10].

Knowledge management is working with more types of knowledge and the correct application should include the combination of them all. The main difference between them is in their expression, the possibility of their documentation, and the way people use and remember them. Their more thorough description follows [4], [5], [11], [12]:

- **Tacit knowledge** – This type can include the general knowledge of an individual, that one is usually not able to fully describe and document it. It can be sometimes characterized as personal knowledge, that people use naturally based on their subconscious, judgment, intuition, experiences, and understanding. The goal of KM is to try to add important information to the people's common knowledge and reactions.
- **Implicit knowledge** – Sometimes known as a subpart of tacit knowledge is the implicit one. They are better for description but are still considered as more personal data, gained by time and experience. Sharing and explaining are usually done by seeing the action in the process so others can see it, remember it, and apply it by themselves.
- **Explicit knowledge** – This type of knowledge is pretty normal for their easy description because they consist of known facts, standards, manuals, and clear explanations which should be understood the same by everyone. Their ease of expression makes them used for tangible documentation, manuals, and instructions.

3. Cybersecurity threats

The era of technological innovation and with it the related rise of cyber technologies used by a vast area of people brings certain advantages as well as threats and risks. With the contrast of many other kinds of securities and their threats, this type usually targets intangible assets, which can make it way more difficult to not only make the needed preparation for it but also to deal with them once they occur [13], [14], [15]. This area can divide the threats into two categories, intentional and unintentional, where both pose different risks and they are more thoroughly discussed in the sections below [13]. The major goal of that part is not to list all the potential threats, but to present the various types of threats associated with cybersecurity and to point out the necessity of its acknowledgment.

3.1 Intentional threats

These are usually the outcomes of an attack executed by an attacker with the intention of causing harm of any kind. The harm in this area does not usually mean endangering the health of someone (but even these types can happen, e.g. targeting some functions of an automated vehicle, making it a potential tool to cause damage not only to the people inside, but also to others), but more often can be seen in the form of attacks that are intended to steal, modify, or delete some information, whether for the advantageous use (so the attackers can have some kind of leverage against them), or for destructive purpose. A few selected examples of these attacks are listed below [2], [3], [13], [14], [15]:

- Malicious actors;
- Social Engineering (Phishing, Vishing, Pharming, etc.);
- Ransomware;
- (Distributed) Denial of Service attacks;
- Cyber Espionage;
- Insider threats [16], [17];
- Zero-day attacks.

3.2 Unintentional threats

Technology innovation is much faster than before and that also raises the number of possibilities not only for attackers to execute their intended attack, but also for some issues from outside or within the system. These threats are hard to prepare for and they can be usually caused by [2], [6], [13], [15]:

- Human error;
- Risky behaviours due to a lack of knowledge about security;
- Software vulnerabilities, flaws (making opportunity for attacks, e. g. zero-day attacks);
- Wrong configurations of systems;
- Misuse of tools, software, hardware, etc.
- Failure to keep the software or hardware up-to-date;
- Threats caused by weaknesses and failures of third-party subjects.

Another category could describe threats caused by Natural threats such as floods, thunderstorms, earthquakes, etc. [14], but since they are the same for many security fields, their description in this paper would be only a piece of redundant information.

4. The benefits of the use of Knowledge Management in Cybersecurity

Information and knowledge are the most important tools in dealing with the security of intangible assets, which is one of the most targeted areas in cybersecurity. However, there can be a common misunderstanding between just having the information and being able to use it well [4], [5]. An organization can have plenty of data about cybersecurity and many other related areas, but with the lack of knowledge, right distribution, presentation, and application, the whole process of their collection could be wasted. No matter the technological innovation phase, people are the most significant units in an organization to be well educated about all possibilities of the potential threats and risks included [4]. Cybersecurity deals with a lot of information and uses different types of SW and HW, but there is always someone who operates with, works with them, or just has access to them, making them a vulnerability, a weak point of the system itself. The common saying states that the system is only as strong as its weakest unit, then organizations have to always concentrate on the education and experience of their employees.

4.1 The cycle of knowledge management

As are the types of knowledge mentioned above, the process of their implementation is also an important aspect. The whole cycle described in this part is similar to the widely used Plan-Do-Check-Act matrix (known as PDCA or Deming matrix), which is used for continuous improvement (mainly in business process management) [13]. By understanding the continuous cycle and its benefits, the individual parts of KM were then incorporated into a continuous process which consists of 4 parts. The cycle captures what steps should be performed to set (and maintain) adequate knowledge of people about potential threats and how to respond to them. The individual parts are briefly described below [4], [5]:

- **Threat awareness** – The initial step should be to acknowledge all potential threats that can pose a risk to the subject, collect all the possible and relevant information, and create a well-constructed, and understandable document, that introduces all that is necessary to the involved group of individuals. Thorough threat awareness is a crucial part of all other steps.
- **Incident response** – Once good knowledge about the potential threats is established, everyone related to the area of interest should be able to know how to respond based on the threat they deal with. This mostly includes the documentation of the policies, best practices, and procedures, that should be done to achieve the best possible outcome of the occurred situations.
- **Knowledge sharing** – The communication between individuals in an organization is crucial for creating tacit/implicit knowledge. Even when the information about threats and responses should be already presented in the general threat awareness, sharing information between others can help to maintain that knowledge and understand it not only as something necessary to know but also as a custom that should be kept to establish better conditions at work.
- **Training/Education** – From the first step, there should be created documentation about the potential threats, and the suitable responses to them are well described for all interested individuals. However, periodical training and education should be involved to ensure that they know all the important information for sure, and mainly, that they understand them correctly.

To ensure that all the knowledge about the threats and the responses to them are all the time kept up to date, and by doing so is maintained the level of cybersecurity to the intended level, the whole process should be continuously updated and improved [4], [13]. That step should always check if there are any possible changes to improve the knowledge about the potential threats, and if so, it should be incorporated into the whole process. The whole cycle of knowledge management is pictured in Figure 1.

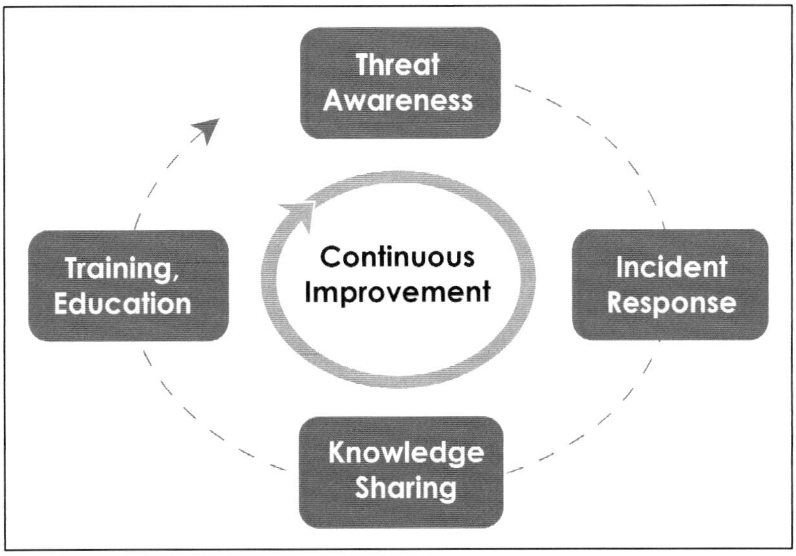

Fig. 1. Knowledge management cycle in the area of cybersecurity.

With the suggestions and steps described, organizations can improve their management of knowledge, and with it their overall cybersecurity measures and preparations. Since each organization has individual needs and capabilities, the way of implementation can vary. The first step should always be a thorough analysis of their current knowledge management, highlighting the main points that need to be changed. That part is highly recommended to consult with an expert from the field (internal/external) and with all the related and responsible people of the organization. If everything proceeds successfully, the expected outcome should make the organization better prepared to deal with many potential threats and change the behaviour of individuals within the organization, making them more responsible, since they are now well-informed about the potential consequences and how to prevent them. However, the preparation, well-executed analysis, and plan of implementation should not be underestimated. Poor or hasty implementation can result in no improvement in security, and may even lead to a negative response from employees.

5. Conclusion

The topic of knowledge management and its benefits, specifically in cybersecurity is sometimes a very overlooked point of view. Knowledge management serves as a universal tool, which can be slightly modified to be used in many different areas based on its needs. Cybersecurity is highly related to intangible assets, making them the prime target of cyber-attacks. To appropriately react to that fact, the necessity of their protection needs to have a high priority as well, in order to sufficiently protect the most valuable of them. For that, the knowledge of the potential threats, the measures, and responses to them are greatly important to mitigate not only their risks of occurrence but also to help with the potential consequences.

The article briefly describes knowledge management on its own, followed by the diversification of the knowledge types, where understanding and the right incorporation of Tacit, Implicit, and Explicit knowledge are the core parts of the successful application of knowledge management. The work suggests what four steps should be included in the cycle of knowledge management for cybersecurity, stressing the importance of continuous improvement, which is nowadays more important due to the rapid changes in the technological landscape. What one wants to achieve by the right implementation of KM within the organization is to elevate individuals' knowledge to a level where they are aware of potential threats and their consequences, enabling them to respond adequately to such threats. Additionally to that, the other focused area is, to change their behaviour to not only gather information and knowledge just because the regulations and policies say so but also because they know what damage it could eventually cause if they did not know how to respond to some threats.

The possibilities for further research can vary in multiple directions. The process of implementation of the KM cycle in organizations offers many opportunities to study, from the process on its own, methods of implementation, up to the outcomes it could bring. Another point of view about that study could explore the options of education about the knowledge among individuals and their different effectiveness.

6. Acknowledgments

This work was supported by the Internal Grant Agency of Tomas Bata University - Faculty of Applied Informatics under project No. IGA/FAI/2023/003.

7. References

[1] O. V. Syuntyurenko (2022), Predicting Potential Threats and Megarisks in Information Technology Development, Scientific and Technical Information Processing, Vol. 49, no. 1, 48–59, doi: https://doi.org/10.3103/S0147688222010130

[2] Prague Proposals on Cyber Security of EDTs (2021), https://www.nukib.cz/download/Prague_Proposals_on_Cyber_Security_of_EDTs.pdf

[3] D. F. Reding and J. Eaton, Science & Technology Trends 2020–2040 (2020), https://www.nato.int/nato_static_fl2014/assets/pdf/2020/4/pdf/190422-ST_Tech_Trends_Report_2020-2040.pdf

[4] Talet, A. N. (2018), The Role of Knowledge Management with Risk Management for Information Technology Projects Risk Assessment. International Journal of Environment and Sustainability [online]. vol. 6, no. 2. ISSN 1927-9566.

[5] Milton, N. J. and Lambe P. (2020), The knowledge manager's handbook: a step-by-step guide to embedding effective knowledge management in your organization. Second edition. London, United Kingdom: Kogan Page Limited. ISBN 9780749484613. Accessed: http://search.ebscohost.com/login.aspx?direct=true&scope=site&db=nlebk&AN=2258716&authtype=ip,shib&custid=s3936755

[6] Rahim, A. S.; Widodo, P.; Reksoprodjo, A. H. & Alsodiq, A. (2023). Identify Cyber Intelligence Threats in Indonesia. International Journal Of Humanities Education and Social Sciences, 3(1), doi: https://doi.org/10.55227/ijhess.v3i1.426

[7] Paulova, I.; Vanova, J.; Rusko, M.; Hekelova, E. & Kralikova, R. (2017). Knowledge Managements for Improvement the Competitiveness of Organization, Proceedings of the 28th DAAAM International Symposium, pp.1221-1226, B. Katalinic (Ed.), Published by DAAAM International, ISBN 978-3-902734-11-2, ISSN 1726-9679, Vienna, Austria, DOI: 10.2507/28th.daaam.proceedings.170

[8] Mohapatra, S.; Agrawal, A. and Satpathy, A. (2016). Designing Knowledge Management Strategy. In: Designing Knowledge Management-Enabled Business Strategies. Management for Professionals. Springer, Cham. https://doi.org/10.1007/978-3-319-33894-1_5

[9] ISO 30401:2018 (2018) Knowledge Management Systems – Requirements, International Organization for Standardization, Geneva, preview: https://www.iso.org/obp/ui/en/#iso:std:68683:en

[10] Knoco Ltd (2021), Global Survey of Knowledge Management 2020-2021. Online. ©KNOCO LTD 2021. Knowledge management consultants. 2021. Accessed: https://www.knoco.com/knowledge-management-survey.htm.

[11] Rowley, J. (1999), "What is knowledge management?", Library Management, Vol. 20 No. 8, pp. 416-420. https://doi.org/10.1108/01435129910291175

[12] Davies, M. (2015), Knowledge–Explicit, implicit and tacit: Philosophical aspects. International encyclopedia of the social & behavioral sciences, v. 13: 74-90.

[13] IEC/ISO 27001:2022 (2022), Information security management systems - Requirements, International Electrotechnical Commission, Geneva, preview: https://www.iso.org/obp/ui/#iso:std:iso-iec:27001:ed-3:v1:en

[14] Jerab, D. A. (2023), How to Deal With Risks and Threats, SSRN Electronic Journal, doi: http://dx.doi.org/10.2139/ssrn.4544513

[15] Sheth, A.; Bhosale, S.; Kurupka, F. (2021), Research Paper on Cyber Security. Contemporary Research in India (ISSN 2231-2137. Special Issue: April 2021.

[16] Probst, Christian & Hunker, Jeffrey & Gollmann, Dieter & Bishop, Matt. (2010), Insider Threats in Cyber Security, doi: dx.doi.org/10.1007/978-1-4419-7133-3

[17] Efijemue, O.; Ejimofor, I. and Owolabi, O. S. (2023), Insider Threat Prevention in the US Banking System. International Journal on Soft Computing [online]. 2023, vol. 14, no. 3, s. 17-28. ISSN 2229-7103, doi: 10.5121/ijsc.2023.14302

DOI: 10.2507/34th.daaam.proceedings.070

JIG FOR CLAMPING THE CHUCK DURING MILLING

David Sobora & Michal Povolny

This Publication has to be referred as: Sobora, D[avid] & Povolny, M[ichal] (2023). Jig for Clamping the Chuck During Milling, Proceedings of the 34th DAAAM International Symposium, pp.0531-0536, B. Katalinic (Ed.), Published by DAAAM International, ISBN 978-3-902734-41-9, ISSN 1726-9679, Vienna, Austria
DOI: 10.2507/34th.daaam.proceedings.070

Abstract

Nowadays, production processes are very demanding to constantly improve productivity, reduce the energy and ecological load and strive for the highest possible efficiency of the entire process. One of these aspects is precisely the reduction of secondary times when clamping the workpiece or fixtures. For example, clamping a universal chuck when milling on the table of classic machine tools or multi-axis milling centres. This issue is also related to many other companies that deal with piece or small-batch production of milled flat surfaces on rotary components. The aim of this article is to present a clamping jig that improves the clamping of chucks on the machine table, but at the same time it can be used for clamping other components or fixtures. The article deals with the principle of the product's function and the possibilities of its use in practice.

Keywords: Universal chuck; Clamping; Milling; Clamping jig

1. Introduction

At the Department of Machining Technology of the Faculty of Mechanical Engineering of the University of West Bohemia in Pilsen, a requirement arose for a solution to the problem of clamping a universal chuck during milling on the table of conventional machine tools. This problem is also faced by many other companies involved in piece or small batch production of milled planar surfaces on rotating components. The universal chuck is used for clamping rotary components mainly on lathes. However, in many cases the use of a chuck is necessary for milling flat surfaces on rotating components, where the universal chuck is also widely used. It simplifies workpiece alignment and at the same time ensures a sufficiently rigid clamping [1], [2], [3]. The universal chuck is usually clamped to the milling machine table by two or three clamps secured by screws in the T-slots. This method of clamping can also be used in combination with a rotary table, mostly used on manual machine tools. Since the shape and design of the universal chuck is primarily adapted for clamping to the lathe spindle, clamping the chuck to the milling machine table presents several problems [1], [2], [3], [4], [12]. The aim of this bachelor thesis is to propose a design solution-fixture for simple, rigid, and fast clamping of universal chucks on milling machine tables. This task was directly assigned by the Department of Machining Technology of the University of West Bohemia in Pilsen.

2. Determination of the requirements for the new product design

The design of the universal chuck is mainly adapted for turning, where the chuck rotates together with the workpiece, or the main cutting motion is performed by the workpiece.

It is clamped by means of screws for which threaded holes are formed on the rear wall of the chuck. This assembly is clamped to the flange, which is part of the lathe spindle, by means of these screws. However, in a number of cases, the use of a universal chuck is necessary in milling, where the chuck must also be clamped, but in this case on the machine table [1], [5], [6], [7], [12], [13]. In the current state, the chuck is clamped in small batch and piece production mainly by two or three flat or adjustable clamps supported by washers that approximate the height of a universal chuck.

In some cases, the machine operator has made special clamping stones for a specific chuck. This method of clamping is much faster compared to using flat chucks. The big disadvantage of this method is that it is limited to one specific height or can only be used to clamp one type of chuck [5], [6], [7], [12], [13].

Description of the process of simultaneous clamping of the chuck with flat clamps on the machine tool table

- Placing the chuck on the machine table
- Insertion of 2 T-nuts into the table grooves
- Assembling 2 washers close to the height of the chuck
- Placing the flat clamp on the chuck and washer
- Pushing the bolt through the clamp and screwing it into the T-nut
- Comparison of chucks and clamps on the machine table
- Tightening the screws

Fig. 1. Clamping the universal chuck with flat clamps

According to the experience of the machine tool operator, the conventional method of clamping by means of clamps is laborious and lengthy, due to which the lead time for clamping the chuck is significantly extended, which is approximately 5 minutes. A further complication is the protruding part of the screw above the chuck and the long end of the chuck which is remote from the chuck and its length is not precisely defined. These facts complicate work especially on 5-axis and multi-axis milling centres. When tilting and rotating the table, care must be taken to ensure that the tool does not hit the protruding screw above the chuck and the protruding end of the chuck next to the chuck when machining or travelling in a rapid feed.

Fig. 2. Another way of clamping the universal chuck with clamping stones

Determination of points for a new clamping solution

a) **Time-consuming clamp assembly**: The new clamping design should be easy and quick to assemble, thus significantly reducing lead times. Consequently, the financial costs of clamping the chucks will also be reduced.
b) **Incorrect chuck alignment**: In the case of chuck clamping, it is very difficult and laborious to set the chuck in the same position when re-clamping. The new design should make it more accurate and easier to position the chuck when re-clamping.
c) **Laborious assembly of the pad under the clamp**: For the correct functioning of the flat clamp, it is necessary to support the opposite side, in relation to the clamped part, so that both ends lie at the same height from the machine tool table surface. The new clamping method could ideally eliminate this action or at least reduce the laboriousness of assembling the washer.
d) **Protruding screw above the clamp**: In the case of a flat clamp, the clamping force is exerted by tightening a bolt or nut (when using a clamp bolt), the head of which, however, protrudes above the clamp. This is a complication especially on multi-axis machine tools when tool passes and workpiece tilting. As machining centres are now a common part of production, the new design should ideally not protrude too much or only minimally.
e) **Preservation of partial universality**: In this case, this feature is perfectly fulfilled by clamping with flat clamps and it would be advisable to maintain it when designing a new fixture. After consultation with the machine operator, it was found that the most used types of universal chucks clamped on the milling machine table are: 80, 100, 160, 200, 250. As mentioned above they differ not only in outer diameter but also in width. That is why the new design of the jig should be applicable in this range of universal chucks.

3. Design of the adjustable jig

One suitable solution for clamping the universal chuck is by means of an adjustable jig, which is guided by a flat guide. The name Cobra was chosen from the upper part of the jig, which resembles the head of a snake. The clamping force is exerted by tightening a screw passing through the jig [6], [8], [9], [10], [11]. The fixture consists of sliding member 1, fixed member 2, and screw 3, member 4 is a universal chuck TOS IUS200 whose height is 76mm. The individual parts are better seen in Figure 3, where the individual parts are shown. [6], [8], [9], [10], [11].

The clamping surface on member 1 is intentionally not milled perpendicular to the guide, it is milled at a 4° angle due to the clearance in the flat guide and the clearance between the T-slot and the T-nut. This design feature ensures that when the bolt is tightened and member 1 is tilted, the clamping force at the end of this face is still perpendicular to the milling machine table and does not "push" the chuck out of the jig. Another design element on member 1 is the U-shaped milled area, which is designed to centre the chuck in the axis of the jig, thus simplifying the adjustment during re-clamping [6], [8], [9], [10], [11], [13].

The product provides chuck clamping from a height of 45mm to 82mm, or the most used chucks clamped on milling machine tables from 80mm to 250mm diameter (6 chuck bits). It can also be combined for different sizes of T-slots in machine tables. Depending on the size of the T-slot, the size of the clamping screw can also be varied, as well as the size of the entire jig [6], [8], [9], [10], [11], [13]. This jig can be used to clamp any workpiece that has a plane surface height from the machine table within the jig's adjustability range. To increase the clamping range, it would be a good solution to produce this fixture in sets.

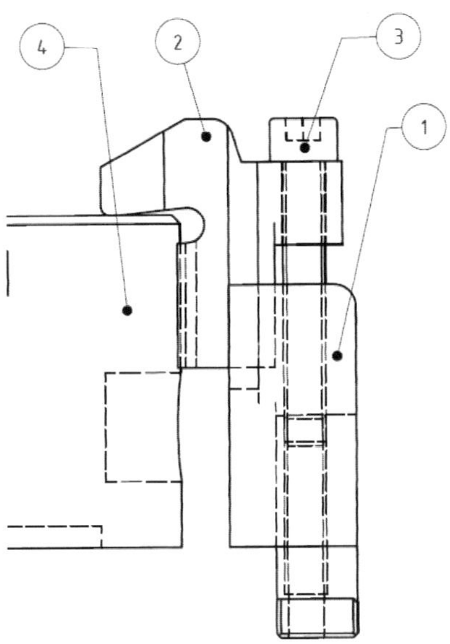

Fig. 3. Adjustable jig

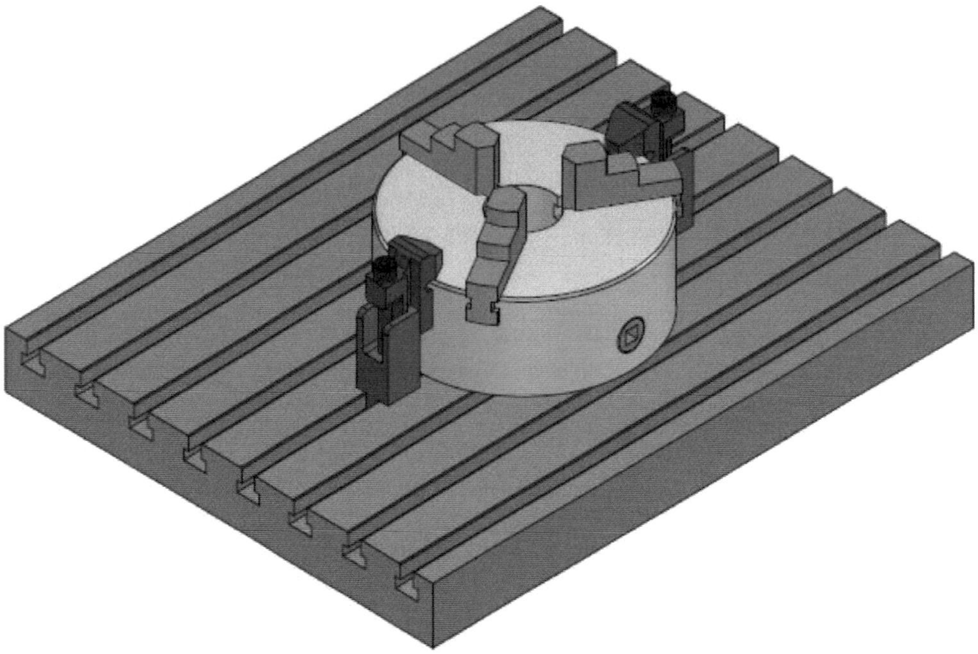

Fig. 4. Model of the adjustable jig with chuck

4. Design of the adjustable jig for tighter clamping

To ensure a tighter clamping or to be used as a stop in combination with the above-mentioned fixture, a variant of the adjustable clamping fixture with an extended seating area to the machine table and a hole for an M10 screw, which is tightened by means of a conventional T-nut to the machine table, was also designed. The sliding member and the hexagon socket head screw are identical to the previous design. The assembly of this type of fixture is shown in Fig. 5. The main disadvantage of this type of fixture is the increased clamping time for tightening the two additional M10 screws, however, this fixture can be used without the M10 screws in the same manner as the previous adjustable fixture design. However, due to the extended seating area, the tilting of the entire jig is reduced when tightening the M10 hexagon socket head screw. As a result, this type of jig has a more versatile application compared to the previous type.

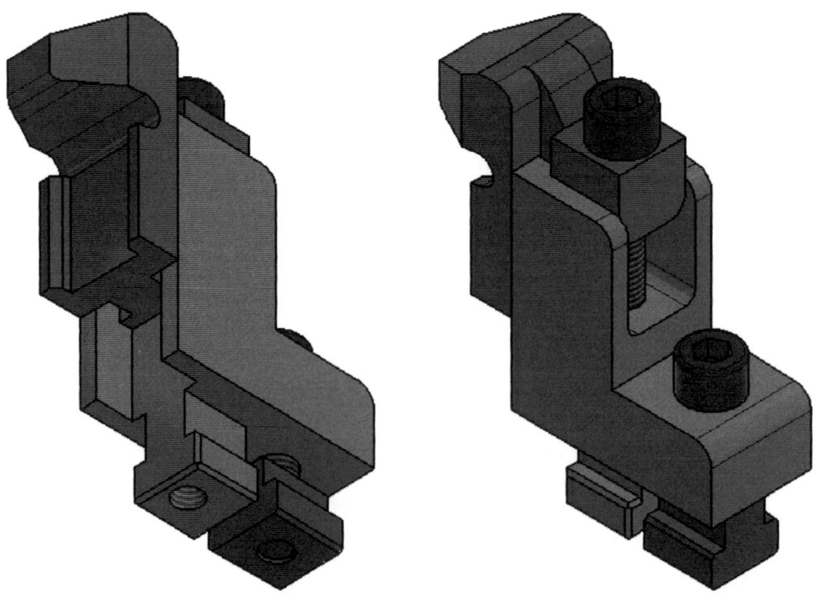

Fig. 5. Adjustable jig for tighter clamping

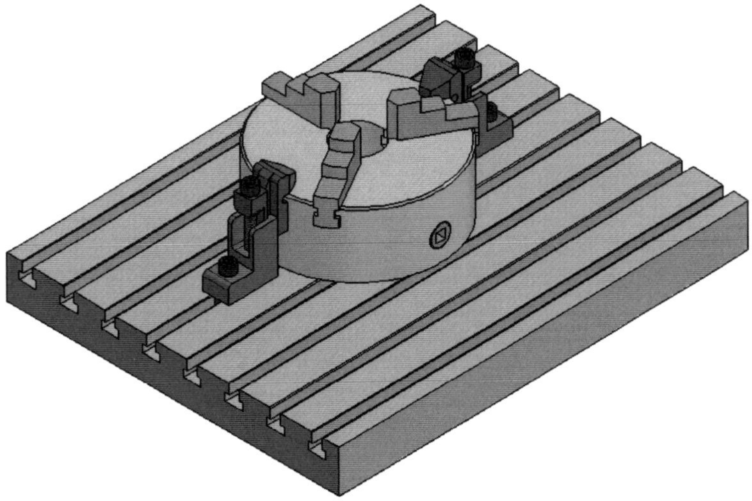

Fig. 6. Model of the adjustable jig with chuck for tighter clamping

5. Prototype production using 3D printing

To produce the prototype of the selected product, a currently widely used technology was chosen, namely 3D printing of plastics. Figure 7 and Figure 8 below show the fixture assembly produced by 3D printing that clamps the TOS IUS 160 universal chuck. Unfortunately, using this manufacturing method, the correct manufacturing tolerances of the fixture cannot be determined because this type of 3D printing of plastic is not accurate enough, the plastic curls due to temperature, and the roughness of the seating surfaces is also much worse than indicated in the drawing. However, to introduce the idea and function of the design, this prototype is quite sufficient [6], [8], [9], [10], [11].

Fig. 7. Universal chuck TOS 160 with prototypes of adjustable jig

Fig. 8. Detail of the model of the adjustable jig

6. Conclusion

The article discusses a practical solution to address the challenges associated with clamping universal chucks for milling on conventional machine tools. This issue is relevant not only to academic institutions, such as the University of West Bohemia in Pilsen but also to numerous companies involved in piece or small-batch production of milled planar surfaces on rotating components. The existing method of clamping universal chucks with flat clamps is time-consuming and has several limitations, including difficulty in chuck alignment, laborious assembly of support pads, and protruding screws that can interfere with the machining process, especially on multi-axis milling centres. To overcome these challenges, a new clamping solution called the "Cobra" jig was proposed, which aims to simplify, expedite, and improve the clamping process. The "Cobra" jig is designed to be adjustable, accommodating a range of universal chuck sizes and heights, making it a versatile solution for a variety of milling applications. Additionally, a variant of the adjustable clamping fixture was introduced for tighter clamping and as a stop, enhancing its flexibility and usability. To validate the design concept, a prototype was produced using 3D printing, although it had some limitations in terms of manufacturing tolerances and surface quality. Nonetheless, it effectively showcased the functionality of the design. This article highlights the importance of innovation in manufacturing processes to improve productivity, reduce setup times, and enhance efficiency. The "Cobra" jig represents a step forward in addressing clamping challenges in milling operations, offering a promising solution for both academic and industrial settings. Further development and refinement of the design may lead to significant improvements in the clamping processes for universal chucks, benefiting the broader manufacturing community.

7. Acknowledgements

The article contribution has been prepared under project SGS-2022-007 - Research and Development for Innovation in Engineering Technology – Machining Technology IV.

8. References

[1] Chvala, B. & Votava, J. (1980). Jigs for machining and feeding equipment. ČVUT. Praha

[2] Novotny, K. & Zemcik O. (2015). Preparations and tools, skriptum, Brno.

[4] Rasa, J & Gabriel, V. (2005) Engineering Technology 3. Vol. 1, 2. Praha: Scientia.

[5] "Clamping and other fixtures", www.presston.cz, 2023. [Online]. Available: https://www.presston.cz/konstrukce/pripravky-upinaci-a-jine/. [Accessed: 2023-03-09].

[6] Sykora, J. (2020). Additive Manufacturing of Unmanned Aerial Vehicle, Annals of DAAAM for 2020 & Proceedings of the 31st International DAAAM Symposium, Published by DAAAM International, Vienna, 2020, ISBN 978-3-902734-29-7,

[7] Povolny, M; Fulemova, J; Rehor, J & Zatloukal, T. (2019). Study of Clamping System Plasticity for Reaming Tools, Annals of DAAAM for 2019 & Proceedings of the 30nd International DAAAM Symposium, Published by DAAAM International, Vienna, 2019, ISBN 978-3-902734-22-8,

[8] www.ust.fme.vutbr.cz, (2004). Humar, A., Machining technology - 2nd part. [Accessed: 2023-03-09]

[9] "TOS Svitavy, a.s.", in www.tos.cz, [Online]. [Accessed: 2023-03-09]

[10] Principles and technologies of 3D printing, www.mmspektrum.com, (2022). [Online]. [Accessed: 2023-04-18].

[11] Introduction to FDM 3D printing, 3D Hubs. [Online]. Available: https://www.3dhubs.com/knowledgebase/introduction-fdm-3d-printing/. [Accessed: 2023-03-09]

[12] Povolny, M., Zatloukal, T., & Rehor, J. (2021). Construction of Tool for Machining Hard Materials. In B. Katalinic (Ed.), DAAAM Proceedings (1st ed., Vol. 1, pp. 0299–0305). DAAAM International Vienna. https://doi.org/10.2507/32nd.daaam.proceedings.043

[13] LI, E; ZHOU, J; YANG, Ch; WANG, M & ZHANG, S. Clamping force prediction based on deep spatio-temporal network for machining process of deformable parts. Online. Scientific Reports. 2023, roč. 13, č. 1. ISSN 2045-2322. Available: https://doi.org/10.1038/s41598-023-33666-2. [cit. 2023-10-15].

DOI: 10.2507/34th.daaam.proceedings.071

DEVELOPMENT OF A HIGH-PERFORMANCE TOOL FOR SUPERALLOY MACHINING (HRSA)

Jan Mezulianik, Michal Povolny & Lubos Kroft

This Publication has to be referred as: Mezulianik, J[an]; Povolny, M[ichal] & Kroft, L[ubos] (2023). Development of a High-Performance Tool for Superalloy Machining (HRSA), Proceedings of the 34th DAAAM International Symposium, pp.0537-0543, B. Katalinic (Ed.), Published by DAAAM International, ISBN 978-3-902734-41-9, ISSN 1726-9679, Vienna, Austria
DOI: 10.2507/34th.daaam.proceedings.071

Abstract

The main objective of this article is to develop a new specialized cutting tool capable of efficiently machining difficult-to-machine materials, such as martensitic stainless steels (e.g., AISI 440C, AISI 431, AISI 630), Inçonel® 718 in AMS 5663 condition, Mionic 80, and Titanium Alloy Grade 5. These materials are commonly used in industries such as aerospace, space, energy, and medical. The article is focused on creating a complete tool assembly, which includes a new cutting insert with modified geometry and a special holder allowing for effective cooling of the cutting insert. The first part concentrates on the development and testing of the cutting insert, while the second part is dedicated to the development and testing of the entire tool assembly.

Keywords: Durability enhancement; turning tools; cutting edge adjustment; additive manufacturing

1. Introduction

This research is focused on the development of a new tool for efficient machining of hard-to-machine materials in industrial sectors such as aviation, aerospace, energy, and the medical industry. The research aims to develop a complete toolset, including a new cutting insert with modified geometry and a special holder that allows for effective cooling of the cutting edge. The research is a joint initiative of BONAR a.s. and the University of West Bohemia in Pilsen, Faculty of Mechanical Engineering. BONAR a.s. and the University of West Bohemia in Pilsen will collaborate through an "effective cooperation" arrangement, with the University of West Bohemia being the research partner and responsible for part of the research funding. The goal of this research is to provide an innovative solution for efficient machining of hard-to-machine materials [1], which have wide applications in various industrial sectors, including aviation [2,3]. It is expected that the new tool will increase cutting performance by at least 20 % compared to existing tools and extend the lifespan of cutting inserts [4]. This should lead to increased productivity and reduced machining costs for these challenging materials. The research consists of two phases. The first phase is focused on the development and optimization of the flank geometry of the cutting insert to reduce the contact area and increase the lifespan of the inserts [5,6]. The second phase of the research is focused on the development of a toolset with a cooling system that allows for effective cooling of the cutting insert during machining.

It is expected that the new type of cutting insert and toolset will bring significant benefits to customers in progressive industrial sectors [7,8]. Increased cutting performance and extended insert life are anticipated, leading to increased revenues in this market segment. The research also has significant export potential since most customers are located outside the Czech Republic. The research's manufacturing processes will include innovative approaches such as the use of metal 3D printing for producing the tool holder and cooling jets. This will enable the application of design elements that are difficult to achieve with conventional methods. The aim of the research described in this article is, therefore, to provide an innovative solution for more efficient and high-performance machining of hard-to-machine materials in the aviation and aerospace industries.

2. Stage 1: Development of the cutting insert

Description of the development of the cutting insert in the first phase of the research: The first phase of the research was focused on the development of a cutting insert for efficient machining of hard-to-machine materials. The primary objective of this phase was to make design modifications to the flank of the cutting insert to reduce the contact area between the flank of the insert and the workpiece surface [9].

These geometry modifications of the insert were achieved through grinding. Reducing the contact area resulted in a reduction in the heat generated due to friction between the insert and the workpiece [10]. It was also intended to increase the lifespan of the cutting insert because less heat was transferred into the tool's cutting edge and was better dissipated from the cutting zone. Another benefit of reducing the contact area was a decrease in friction and its impact on the wear of the insert's coating or base material, depending on the specific design of the insert.

As part of the development of the cutting insert, design modifications were carried out to achieve the desired parameters. Modifications to the geometry of "standard" replaceable inserts were designed to improve performance parameters. For each type of insert intended for specific machining operations (roughing to finishing), suitable cutting conditions were selected, including cutting speed, feed rate, and cutting depth. These conditions were combined with other factors such as cooling, coating, etc., to achieve optimal insert performance.

Fig. 1. Models of ground W-Type RCIs

During the testing phase, experiments were conducted with the designed variants of cutting inserts and compared to standardized inserts. The main parameters monitored during these experiments included insert wear and the achieved surface quality, including dimensional accuracy and surface roughness. For some of the tests, a single type of material was used, while for other insert variants, statistical analysis was performed based on the results of physical experiments.

Fig. 2. Replaceable cutting insert with ground flank

Based on the evaluation of the conducted experiments, optimization of the characteristics of the designed inserts was carried out to achieve the desired performance parameters. The goal was to increase cutting performance during machining by approximately 20 % and extend the lifespan of the insert.

A special type of negative geometry cutting edge was designed to enhance cutting performance, enable the double-sided use of the insert, and create a groove for focused cooling with a cooling fluid. The development of the tool holder that allows for this cooling was part of the second phase of the research.

3. Stage 2: Development of the Tool Assembly

In this phase of the research, the goal was to develop a special tool assembly with a cooling system for the flank of the replaceable cutting inserts (RCI). Two variants of holders (tools) were developed in this phase. One technical solution was developed by BONAR a.s., and the other, different solution was developed at the Regional Technological Institute within the Faculty of Mechanical Engineering at the University of West Bohemia in Pilsen. Both teams (BONAR and the University of West Bohemia) applied different approaches to the solution and subsequently evaluated which variant was better and achieved better results. In the initial phase of holder development, the construction of various competing tool holders was examined with the aim of delivering the process fluid as close as possible to the flank part of the RCI.

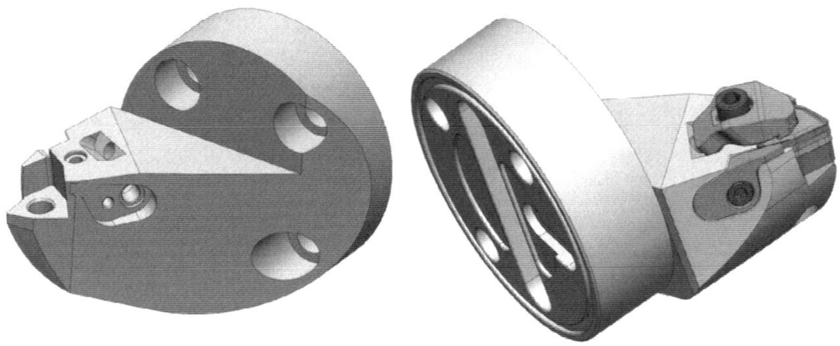

Fig. 3. Model of W-Type RCI holder

Based on the design concept, prototypes of each RCI holder were manufactured and subsequently tested, both statically (i.e., for fluid routing and flow) and later during actual machining operations. Based on these tests, the design solution was optimized for the subsequent full-scale experiment.

Fig. 4. Prototype holder for W-Type RCI

The next phase of the tool assembly development stage involved testing on selected workpiece materials. Since this innovative cooling tool assembly was being compared to tools tested in the previous stage, this phase was quite dynamic, with a primary focus on fine-tuning the delivery of the process fluid. The developed holder aims to provide the supply and distribution of cooling fluid pressurized to over 40 bar into the groove on the flank of the RCI, thus ensuring more intensive and even cooling and further enhancing machining performance. In the final phase of this stage, samples of the tool assembly were deployed in selected industrial enterprises primarily involved in manufacturing of components for the aviation industry. This allowed for the verification of the results obtained during the research phase.

4. Experimental Testing

Experimental testing was conducted simultaneously for both stages, i.e., for the development of the cutting insert and the development of the tool assembly. In the first phase of the experiment, the influence of the ground flank on the lifespan of the RCI was observed. With a focus on ensuring the highest test reliability (for example, to eliminate the influence of changing bar diameter [11]), the experiment was carried out in face turning. Cutting conditions were kept constant, with the only variable being the angle of the ground groove on the flank of each RCI. The experiment was conducted under the parameters mentioned below.

The objective of the durability testing experiment was to compare the innovated holder and RCIs with a ground flank to conventionally supplied holders and conventional RCIs. The same workpiece material as in the previous case of ground flank testing was chosen for this experiment. Machining was performed during face turning to ensure a constant tool path and unaffected results due to diameter changes. For this experiment, three types of tools were selected:

Tool I: Classic tool holder with an RCI without a ground flank. This tool served as a control group representing the standard configuration commonly used.

Tool II: Classic tool holder with an RCI with a ground flank. This tool had a ground replaceable cutting insert, which had the potential to improve its performance.

Tool III: Innovated W-type tool holder with an RCI with a ground flank. This tool represented the innovated tool assembly with optimized properties and a modified W-type RCI.

Cutting conditions remained constant and were set as follows: cutting speed (Vc) was 35 m/min, feed rate (f) was 0.2 mm/rev, depth of cut (ap) was 1.5 mm, and high-pressure cooling was used. These parameters ensured consistent cutting conditions for all tested tools. The experiment monitored the overall impact of the ground flank and innovated cooling on the lifespan of the RCIs. The volume of material removed was also evaluated for each tool type. This approach allowed assessing the effectiveness and advantages of using ground flank and tool with innovated cooling.

Fig. 5. Innovated tool during experiments

Experiment Parameters:

Workpiece:

- Inconel 718

Tools:

- Tool I: Classic tool holder + RCI without a ground flank
- Tool II: Classic tool holder + RCI with a 7° ground flank
- Tool III: Innovated W-type tool holder + RCI with a 7° ground flank

Cutting Conditions:

- Vc = 35 m/min (cutting speed)
- f = 0.2 mm/rev (feed rate)
- ap = 1.5 mm (depth of cut)

High-pressure cooling

Tool I										
cut	D [mm]	L [mm]	ΔL [mm]	t [min]	Δt [min]	vf [mm/min]	n [ot/min]	ap [mm]	fo [mm/ot]	vc [m/min]
1	165	85	85	6,294	6	13,504	67,520	1,5	0,2	35
2	165	85	170	6,294	13	13,504	67,520	1,5	0,2	35
3	165	85	255	6,294	19	13,504	67,520	1,5	0,2	35
4	165	85	340	6,294	25	13,504	67,520	1,5	0,2	35
5	165	85	425	6,294	31	13,504	67,520	1,5	0,2	35

Tool II										
cut	D [mm]	L [mm]	ΔL [mm]	t [min]	Δt [min]	vf [mm/min]	n [ot/min]	ap [mm]	fo [mm/ot]	vc [m/min]
1	165	85	85	6,294	6	13,504	67,520	1,5	0,2	35
2	165	85	170	6,294	13	13,504	67,520	1,5	0,2	35
3	165	85	255	6,294	19	13,504	67,520	1,5	0,2	35
4	165	85	340	6,294	25	13,504	67,520	1,5	0,2	35
5	165	85	425	6,294	31	13,504	67,520	1,5	0,2	35
6	165	85	510	6,294	38	13,504	67,520	1,5	0,2	35

Tool III										
cut	D [mm]	L [mm]	ΔL [mm]	t [min]	Δt [min]	vf [mm/min]	n [ot/min]	ap [mm]	fo [mm/ot]	vc [m/min]
1	165	85	85	6,294	6	13,504	67,520	1,5	0,2	35
2	165	85	170	6,294	13	13,504	67,520	1,5	0,2	35
3	165	85	255	6,294	19	13,504	67,520	1,5	0,2	35
4	165	85	340	6,294	25	13,504	67,520	1,5	0,2	35
5	165	85	425	6,294	31	13,504	67,520	1,5	0,2	35
6	165	85	510	6,294	38	13,504	67,520	1,5	0,2	35
7	165	85	595	6,294	44	13,504	67,520	1,5	0,2	35
8	165	85	680	6,294	50	13,504	67,520	1,5	0,2	35
9	165	85	765	6,294	57	13,504	67,520	1,5	0,2	35

Table 1. Experiment Progress Table

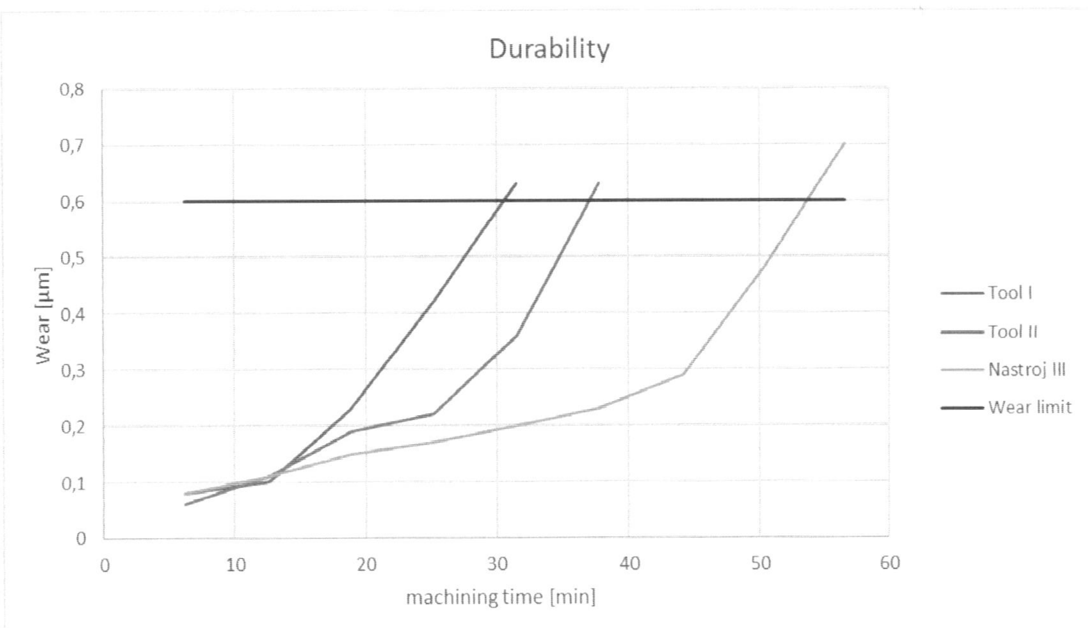

Fig. 6. Durability graph for tested tools

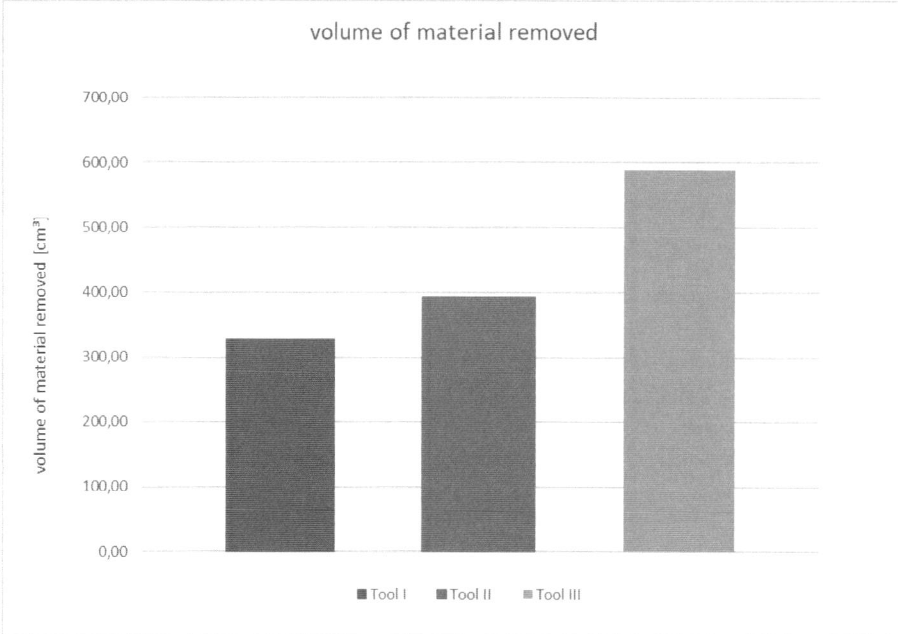

Fig. 7. Volume of removed material graph

The results clearly indicate that the combination of a ground RCI with a conventional tool has a minimal impact on the lifespan but shows a more uniform increase in wear. When using a ground RCI in combination with the innovated tool holder, almost double the lifespan was achieved, and 30 % more material was removed compared to using a conventional tool with a standard RCI. The experiments have thus clearly demonstrated the high potential of using ground RCIs and optimized cooling in machining hard-to-machine materials such as titanium, Inconel, and others.

5. Conclusion

The outcome of this research is an innovative solution for more efficient and high-performance machining of hard-to-machine materials in the aviation and aerospace industries, as well as in other industrial sectors. The research, jointly conducted by BONAR a.s. and the Faculty of Mechanical Engineering at the University of West Bohemia in Pilsen, focused on the development of a new tool assembly comprising a cutting insert with optimized geometry and a special holder enabling effective cooling of the cutting edge. During the research, the first stage was dedicated to the development and optimization of the geometry of the cutting insert's back to reduce the contact area and extend the lifespan of the inserts. The second stage focused on the development of the tool assembly with a cooling system that allowed effective cooling of the cutting insert during machining. The third step involved the execution of experiments that demonstrated the achievement of the initially set goals and expectations.

It is expected that the new type of cutting insert and tool assembly will bring significant benefits to customers in the aviation and aerospace industries, such as increased cutting performance and extended insert lifespan. This would result in higher productivity and reduced costs associated with machining these challenging materials. The research outcome also represents significant export potential since most potential customers are located outside the Czech Republic. Innovative manufacturing processes, such as the use of metal 3D printing for tool holder production, allowed for the application of design elements that were challenging to achieve with conventional methods.

In summary, this research has successfully delivered an innovative solution for efficient and high-performance machining of hard-to-machine materials in the aviation, aerospace, and other industrial sectors. The new tool assembly, developed in collaboration between BONAR a.s. and the University of West Bohemia in Pilsen, brings substantial benefits to customers, enhances their productivity, and increases their competitiveness in the market. In the future this topic of specialized RCI micro-geometry and focused cooling might be extended with other shapes of RCI and geometry specific for other machined materials. This will broughden the applicability of these special turning tools on the industrial market.

6. Acknowledgments

The article contribution has been prepared under project SGS-2022-007 - Research and Development for Innovation in Engineering Technology – Machining Technology IV.

7. References

[1] Bakša, T., Kroupa, T., Hanzl, P., & Zetek, M. (2015). Durability of Cutting Tools during Machining of Very Hard and Solid Materials. Procedia Engineering, 100, 1414–1423. https://doi.org/10.1016/j.proeng.2015.01.511

[2] Baksa, T., Schornik, V., Adamek, P., & Zetek, M. (2016). Machining of Inconel 718 Using Uncoated Cutting Tools With Different Cutting Edge Quality. In B. Katalinic (Ed.), DAAAM Proceedings (1st ed., Vol. 1, pp. 0441–0446). DAAAM International Vienna. https://doi.org/10.2507/27th.daaam.proceedings.065

[3] Clares Rodriguez, J. M., Vazquez Martinez, J. M., Gomez-Parra, A., Puerta Morales, F. J., & Marcos, M. (2016). Experimental Methodology for Evaluating Workpieces Surface Integrity in Dry Turning of Aerospace Alloys. In B. Katalinic (Ed.), DAAAM Proceedings (1st ed., Vol. 1, pp. 0849–0855). DAAAM International Vienna. https://doi.org/10.2507/26th.daaam.proceedings.118

[4] Daniyan, I. A., Mpofu, K., Adeodu, A. O., & Oyesola, M. (2023). Hard Turning Operation of Alloy Tool Steel (AISI D3) Using Cubic Boron Cutting Tool. 2023 14th International Conference on Mechanical and Intelligent Manufacturing Technologies (ICMIMT), 153–157. https://doi.org/10.1109/ICMIMT59138.2023.10199225

[5] Jurina, F., Vopat, T., Kuruc, M., & Simna, V. (2019). The Tool Wear Observation of Milling Tools in High Feed Machining of Hardened Steels. In B. Katalinic (Ed.), DAAAM Proceedings (1st ed., Vol. 1, pp. 0753–0757). DAAAM International Vienna. https://doi.org/10.2507/30th.daaam.proceedings.103

[6] Sykora, J., & Syrovatka, S. (2022). Influence of Tool Geometry on the Process of Turning Ti-6Al-4V. In B. Katalinic (Ed.), DAAAM Proceedings (1st ed., Vol. 1, pp. 0336–0343). DAAAM International Vienna. https://doi.org/10.2507/33rd.daaam.proceedings.047

[7] Povolny, M., Zatloukal, T., & Rehor, J. (2021). Construction of Tool for Machining Hard Materials. In B. Katalinic (Ed.), DAAAM Proceedings (1st ed., Vol. 1, pp. 0299–0305). DAAAM International Vienna. https://doi.org/10.2507/32nd.daaam.proceedings.043

[8] Marsalek, O., Horak, L., Syrovatka, S., & Zatloukal, T. (2022). Innovation in the Problematics of Turning with a Linear Cutting Edge. In B. Katalinic (Ed.), DAAAM Proceedings (1st ed., Vol. 1, pp. 0365–0370). DAAAM International Vienna. https://doi.org/10.2507/33rd.daaam.proceedings.051

[9] Majerik, J., Barenyi, I., & Sedlak, J. (2019). Investigation of Flank Wear when Finish Turning of High Strength Steel with CBN Cutting Insert. In B. Katalinic (Ed.), DAAAM Proceedings (1st ed., Vol. 1, pp. 0215–0222). DAAAM International Vienna. https://doi.org/10.2507/30th.daaam.proceedings.028

[10] Vopat, T., Straka, R., Kuruc, M., & Peterka, J. (2022). The Effect of Cutting Edge Radius Sizes on Tool Life in Machining Nickel Alloy Inconel 718. In B. Katalinic (Ed.), DAAAM Proceedings (1st ed., Vol. 1, pp. 0180–0187). DAAAM International Vienna. https://doi.org/10.2507/33rd.daaam.proceedings.025

[11] Zetek, M., Vozar, V., Baksa, T., & Zetkova, I. (2017). Influence of the Milling Cutter Diameter on the Cutting Tool Life when Machining Inconel 718. In B. Katalinic (Ed.), DAAAM Proceedings (1st ed., Vol. 1, pp. 0417–0422). DAAAM International Vienna. https://doi.org/10.2507/28th.daaam.proceedings.058

DOI: 10.2507/34th.daaam.proceedings.072

NUMERICAL MODEL FOR IDENTIFICATION OF FIBER-REINFORCED CONCRETE PARAMETERS FROM THE CRACK-MOUTH OPENING DISPLACEMENT

Ivica Kožar & Tea Sulovsky

This Publication has to be referred as: Kozar, I[vica] & Sulovsky, T[ea] (2023). Numerical Model for Identification of Fiber-Reinforced Concrete Parameters from the Crack-Mouth Opening Displacement, Proceedings of the 34th DAAAM International Symposium, pp.0544-0550, B. Katalinic (Ed.), Published by DAAAM International, ISBN 978-3-902734-41-9, ISSN 1726-9679, Vienna, Austria
DOI: 10.2507/34th.daaam.proceedings.072

Abstract

Determining material parameters is crucial for understanding their physical properties, improving the design process, and optimizing costs. Due to the challenges of measuring parameters in fibre-reinforced concrete, numerical models and inverse analysis are employed for parameter identification. Laboratory tests, including bending and compressive strength tests, were conducted on concrete samples, recording three parameters: vertical displacement, crack mouth opening displacement, and applied force. The laboratory test results served as a reference for developing a numerical model of micro-reinforced concrete, which generated simulated data used as input for an inverse model based on the Levenberg-Marquardt method. Model verification involved comparing the results with the data obtained from the laboratory tests.

Keywords: fiber-reinforced concrete; mathematical model; inverse analysis; parameter estimation

1. Introduction

The identification and quantification of parameters for engineering materials play a significant role in improving the understanding of their physical properties, enhancing design methods, and optimizing costs. The process of identifying parameters often requires numerous laboratory tests that can be time-consuming, financially draining, and often inadequate. In the case of fibre-reinforced concrete, characterized by the replacement of traditional steel reinforcement with short fibres, the parameters describing the concrete mixture, steel fibres, and their mutual interaction are often physically immeasurable values.

One of the methods for indirectly determining unknown parameters is the use of computer models that can replace costly experimental tests. However, the heterogeneity of concrete presents a significant limitation when describing this material using numerical methods [1], which is further complicated by the addition of fibres to the mixture. In previous research, two approaches have been used to address this problem: the application of an appropriate finite element model where fibres are discretized and placed along the edges of finite elements [2], and the use of the Fiber Bundle Model (FBM) for composite materials [3].

This paper presents a new deterministic numerical model that, through inverse analysis based on the Levenberg-Marquardt algorithm, identifies the material parameters of fibre-reinforced concrete that affect the overall response of the finished concrete element.

2. Laboratory testing

To gain insight into the behaviour of beams made of fibre-reinforced concrete, laboratory tests were conducted. Two tests were performed: three-point bending of beams and testing of the compressive strength of cubes. All test specimens were made using self-compacting concrete. The compressive strength test was carried out on standard concrete samples with dimensions of 150x150x150 mm by applying a load until the cube failure, following the standard for determining the compressive strength of cured concrete test specimens [4]. The measurement results are presented in Fig. 1. and their mean value was taken as the reference value for the compressive strength of concrete.

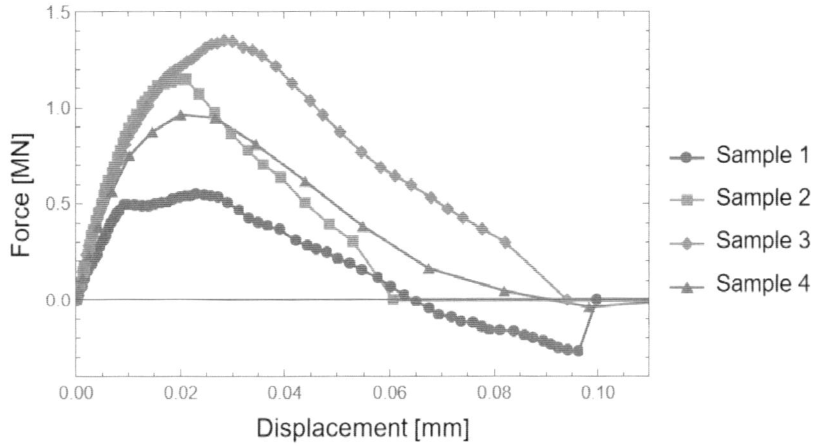

Fig. 1. Concrete's force-displacement diagram

Samples for three-point bending tests were categorized into three groups: beams without fibres, beams with regular fibres, and beams with rough-surfaced fibres. When preparing fibre-reinforced concrete, fibres are usually added during the mixing of wet ingredients, resulting in their homogeneous but unpredictable distribution within the finished mixture. To eliminate the uncertainty of fibre positions within test specimens, fibres were placed at predefined positions using auxiliary elements before pouring the concrete mixture. After 28 days, the auxiliary elements were removed, leaving a "notch" on the beam that predefined the crack position and facilitated the observation of fibre behaviour during bending. All three-point bending tests were conducted with displacement control until the beam failure, following the standard for determining the flexural strength of cured concrete test specimens [5]. Measured quantities included vertical displacement, crack mouth opening displacement (horizontal displacement), and applied force. The test results, which were later used for the validation of the numerical model, are shown in Fig. 2. to 4.

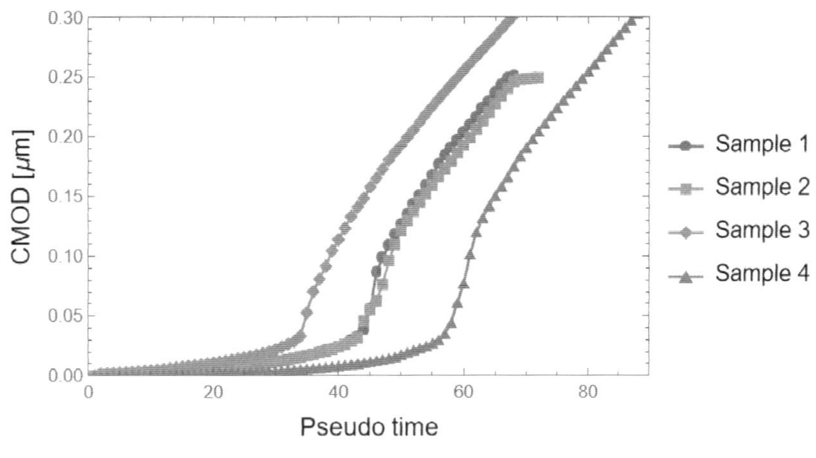

Fig. 2. Crack-mouth opening displacement diagram for beams without fibres

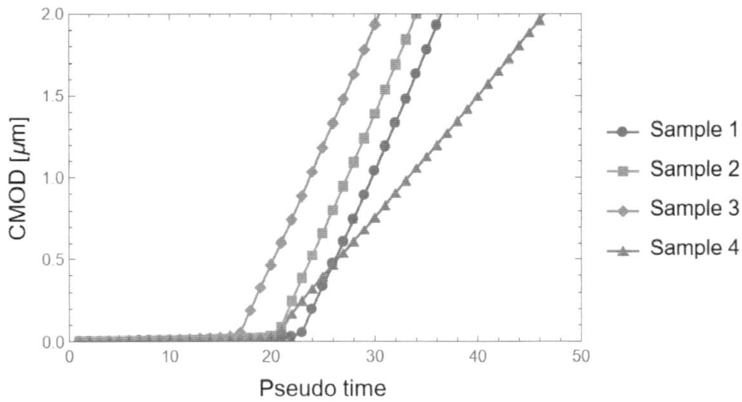

Fig. 3. Crack-mouth opening displacement diagram for beams with regular fibres

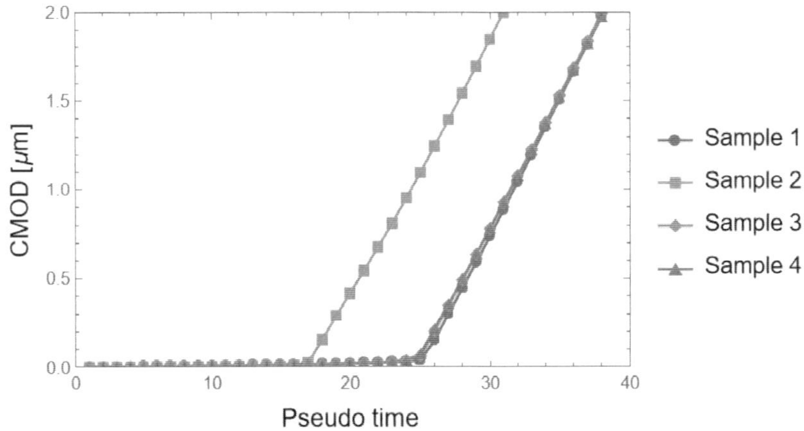

Fig. 4. Crack-mouth opening displacement diagram for beams with rough-surfaced fibres

3. Crack-mouth opening displacement

Crack Mouth Opening Displacement (CMOD) is the horizontal deformation of the beam that occurs during bending, measured at the bottom of the crack, as shown in Fig. 5.

Fig. 5. Three-point bending test configuration with a crack mouth opening displacement gauge installed

Among all the results obtained from laboratory testing, CMOD is the only parameter that directly relates to the cross-sectional and longitudinal aspects of the beam. By analyzing this parameter, it is possible to gain insight into the influence of material parameters, such as those describing fibres in the concrete, on the global response of the beam under loading.

4. Mathematical model of fibre-reinforced concrete

The mathematical model of fibre-reinforced in this research is a forward model based on principles laid out by Menke (2018) [6] and it aims to reproduce certain beam behaviour parameters that relate to material parameters and experimental results and is suitable for the formulation of an inverse model [7]. As the heterogeneity of the fibre-reinforced concrete represents a great challenge for numerical modelling, the material is simplified by describing the concrete and the fibres individually, assuming homogeneity of each. Concrete is divided into layers, following the approach introduced by Kožar et al. in their work from 2021 [8], and is mathematically described using the force-displacement diagram:

$$f_b(x, a_t, b_t, a_c, b_c) = \begin{cases} a_c x E_b exp(-b_c x) & if x < 0 \\ a_t x E_b exp(-b_t x) & if x \geq 0 \end{cases} \tag{1}$$

Where a_t and b_t are the parameters describing the behaviour of concrete in tension, a_c, and b_c are the parameters describing the behaviour of concrete in compression, and E_b is the modulus of elasticity of concrete. The equation relating displacement and force for steel fibres is as follows:

$$f_a(x, F_t, E_u, E_d) = \begin{cases} x E_b exp(-b_c x) & if x < 0 \\ x E_u & if x > 0 \wedge x < x_{elast} \\ (x - x_{elast}) E_d & if x \geq x_{elast} \wedge x < x_{limit} \\ 0 & if x > x_{limit} \end{cases} \tag{2}$$

The parameters describing the behaviour of fibres are the ultimate force F_t and the modulus of elasticity of steel fibres under loading E_u and unloading E_d. Equations (1) and (2) are independent of each other but are related within the equilibrium equations:

$$\begin{gathered} F(\epsilon, \kappa) = \Delta h \textstyle\sum_{i=1}^{sloj} f_b[(h_i - \epsilon h) tg(\kappa)] + \Delta a \, f_a(h_a - \varepsilon \, h) = 0 \\ M(\epsilon, \kappa) = \Delta h \textstyle\sum_{i=1}^{sloj} (h_i - \epsilon h) f_b[(h_i - \epsilon h) tg(\kappa)] + \Delta a \, (h_a - \varepsilon \, h) \, f_a(h_a - \varepsilon \, h) = 0 \end{gathered} \tag{3}$$

Where the upper equation represents the force balance, and the lower equation represents the moment balance. From the equilibrium equations, the position of the neutral axis of the beam ε and the crack mouth opening displacement angle κ can be determined. These two values are then used in the expression for the crack mouth opening displacement:

$$d_{cmod} = (1 - \varepsilon) \cdot \mathrm{h} \cdot tg(\kappa) \tag{4}$$

Where h represents the height of an individual layer of the beam, the results obtained through numerical modelling of expression (4) are presented in Fig. 6.

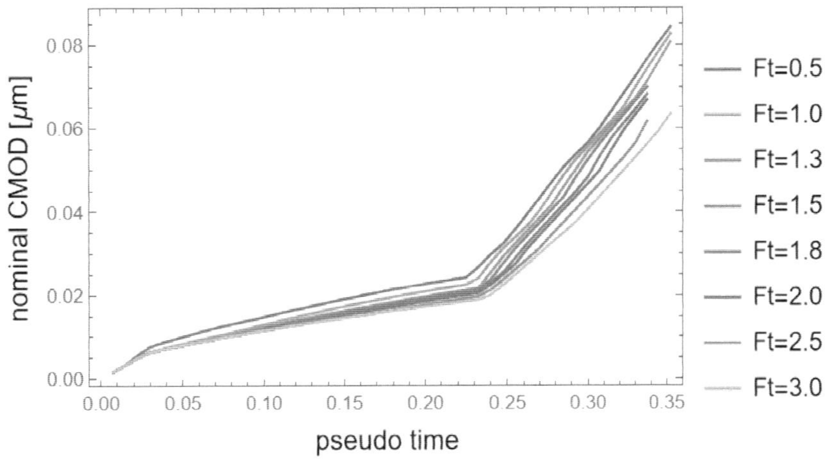

Fig. 6. CMOD Results for E_b = 40 GPa and F_t values ranging from 0.5 kN to 3.0 kN

Obtained results from the forward model were successfully validated using the laboratory testing data, as well as the crack-mouth opening displacement results generated using the data-driven stochastic model formulated by Kožar et al. in their work from 2021 [9].

5. Inverse model

Levenberg-Marquardt Method is an iterative technique for solving nonlinear least squares problems to estimate parameters. This method iteratively adjusts a function's parameter until it finds a value that minimizes the error between predicted and actual data. The general notation for the sum of squares in the equation describing the model function fitting, $\hat{y}(x, p)$, to the independent variable t and a vector of n parameters p, given a set of m data points (t_i, y_i), is:

$$SS = \sum_{i=1}^{n}(y_i - \hat{y}_i)^2 \tag{5}$$

Where y_i is the known (measured) data, and $\hat{y}_i$ is the target function. If $\hat{y}(x, p)$ is nonlinear in the model parameters p, the minimization process becomes iterative.

$$\frac{\partial ss}{\partial p} = 0 \tag{6}$$

Substituting expression (5) into (6) yields:

$$\frac{\partial ss}{\partial p} = -2\sum_{i=1}^{n}(y_i - \hat{y}_i)\frac{\partial \hat{y}_i}{\partial p} \tag{7}$$

Where $\frac{\partial \hat{y}_i}{\partial p}$ is the Jacobian matrix of dimensions mxn, representing the local sensitivity of the function to variations in parameter p. This is also known as the sensitivity coefficient at each measurement point χ_p and can be expressed as:

$$\sum_{i=1}^{n}(y_i - \hat{y}_i)\chi_p = 0 \tag{8}$$

The perturbed value of the parameter with each iteration is denoted as Δp. The target function with $p + \Delta p$ is then written as:

$$\hat{y}_i(p + \Delta p) = \hat{y}_i(p) + \frac{\partial \hat{y}_i}{\partial p}\Delta p \tag{9}$$

When (9) is substituted back into (8), the perturbation value of parameter p is obtained as:

$$\sum_{i=1}^{n}\left(y_i - \hat{y}_i(p) - \frac{\partial \hat{y}_i}{\partial p}\Delta p\right)\chi_p = 0 \tag{10}$$

From expression (10), the value of the perturbation of the parameter p is obtained as:

$$\Delta p = \frac{\sum_{i=1}^{n}(y_i - \hat{y}_i(p))\chi_p}{\sum_{i=1}^{n}{\chi_p}^2} \tag{11}$$

6. Determination of the material parameters Eb and Ft

The previously described method is applied to determine the values of the modulus of elasticity of concrete E_b and the threshold force of the steel fibre Ft. Expression (11) is rewritten as:

$$\Delta E_b = \frac{\sum_{i=1}^{n}(y_i - d_{cmod}(E_b, F_t))\chi_{E_b}}{\sum_{i=1}^{n}(\chi_{E_U})^2} \tag{12}$$

for the minimization with respect to E_b, and

$$\Delta F_t = \frac{\sum_{i=1}^{n}(y_i - d_{cmod}(E_b, F_t))\chi_{F_t}}{\sum_{i=1}^{n}(\chi_{F_t})^2} \tag{13}$$

for the minimization with respect to F_t. The calculation begins with initial values $E_{b,0}$ and $F_{t,0}$. Expression (4) is used as the target function, and sensitivity coefficients obtained from expression (8) are:

$$\chi_{E_{b,1}} = \frac{y_1 - d_{cmod}(E_{b,0}, F_{t,0})}{\Delta E_b} \tag{14}$$

for the parameter E_b and

$$\chi_{F_{t,1}} = \frac{y_1 - d_{cmod}(E_{b,0}, F_{t,0})}{\Delta F_t} \quad (15)$$

for the parameter F_t. Consequently, the expressions for the perturbation of both parameters are as follows:

$$\Delta E_b = \frac{\sum_{i=1}^{n}\left(y_i - d_{cmod}(E_{b,0}, F_{t,0})\right)\chi_{E_{b,1}}}{\sum_{i=1}^{n} {\chi_{E_{b,1}}}^2} \quad (16)$$

for parameter E_b, and

$$\Delta F_t = \frac{\sum_{i=1}^{n}\left(y_i - d_{cmod}(E_{b,0}, F_{t,0})\right)\chi_{F_{t,1}}}{\sum_{i=1}^{n} {\chi_{F_{t,1}}}^2} \quad (17)$$

for parameter F_t. Adding the initial assumption to expressions (16) and (17) for the respective parameters yields the final iteration values:

$$F_{t,1} = F_{t,0} + \Delta F_t \quad (18)$$

$$E_{b,1} = E_{b,0} + \Delta E_b \quad (19)$$

The process is then repeated with a new iteration and continues until local minimum values are found where both parameters converge. This procedure was applied to the data from Fig. 1., with E_b = 45.0 GPa and F_t = 1.8 kN. Through the inverse analysis, satisfactory results were obtained, with the resulting values of these parameters being E_b = 43.12 GPa and F_t = 1.71 kN, converging within 20 iterations. The resulting values are deemed satisfactory and they validate the inverse procedure.

The Levenberg-Marquardt algorithm's success in finding optimal parameter estimates is sensitive to the choice of initial parameter values which influences the model's convergence. Future work will focus on enhancing the robustness of the initial parameter estimation process to mitigate this sensitivity and improve accuracy, as well as conducting a sensitivity analysis of the obtained parameters. Such analysis aims to determine the individual components and coefficients' influence on the optimization problem's solution and identify those with the most significant impact on the behaviour of the concrete element [10].

7. Conclusion

In this study, challenges of identifying material parameters in fibre-reinforced concrete were addressed, a task hampered by the limitations of traditional laboratory testing. A deterministic numerical model which employs the Levenberg-Marquardt algorithm for inverse analysis was introduced. This novel approach efficiently and accurately estimated material parameters, which was specifically presented on the modulus of elasticity of the concrete (E_b) and the threshold force of steel fibres (F_t). The results mathematically matched the values obtained through laboratory experiments, confirming the viability of the method.

Through this research, the problem of efficient parameter determination using a deterministic model was solved. This effectively eliminates the need for extensive, resource-intensive laboratory tests, reducing both time and cost and paves the way for further optimization and practical applications of fibre-reinforced concrete. Future work will focus on refining the inverse model and conducting a comprehensive sensitivity analysis of material parameters, with the goal of enhancing the performance and reliability of these materials in engineering applications.

8. Acknowledgments

This work was supported by project HRZZ 7926 "Separation of parameter influence in engineering modelling and parameter identification".

9. References

[1] Wang, S. R., Zhao, J. Q., Wu, X. G., Yang, J. H., & Liu, A. (2021). Meso-Scale Simulations of Lightweight Aggregate Concrete Under Impact Loading. International Journal of Simulation Modelling (IJSIMM), 20(2), pp. 291-302. doi:10.2507/ijsimm.20(2).487

[2] Smolčić, Ž., & Ožbolt, J. (2014). Meso scale model for fiber-reinforced-concrete: Effective bond-slip relationship of fibers. Proceedings of works (Faculty of Civil Engineering, University of Rijeka), 17, 197-212. https://doi.org/10.32762/zr

[3] Sampson, W. W. (2009). Modelling Stochastic Fibrous Materials with Mathematica. (Engineering Materials and Processes). Springer Nature., London, United Kingdom

[4] Testing hardened concrete – part 3: Compressive strength of test specimens (en 12390-3:2019), 2019, doi: 10.3403/30360097U

[5] Testing hardened concrete – part 5: Flexural strength of test specimens (en 12390-5:2019), 2019, doi: 10.3403/30360073U

[6] Menke, W. (2018). Geophysical Data Analysis (4th Edition). Academic Press Inc. https://doi.org/10.1016/B978-0-12-813555-6.00001-0

[7] Kožar, I., Malić Torić, N., & Rukavina, T. (2018). Inverse model for pullout determination of steel fibers. Coupled Systems Mechanics, 7(2), 197-209. https://doi.org/10.12989/csm.2018.7.2.197

[8] Kožar, I., Bede, N., Mrakovčić, S., & Božić, Ž. (2021). Layered model of crack growth in concrete beams in bending. Procedia Structural Integrity, 31, 134-139. https://doi.org/10.1016/j.prostr.2021.03.022

[9] Kožar, I., Bede, N., Bogdanić, A., & Mrakovčić, S. (2021). Data-Driven Inverse Stochastic Models for Fiber Reinforced Concrete. Coupled Systems Mechanics, 10(6), 509-520. https://doi.org/10.12989/csm.2021.10.6.509

[10] Ganin, P., Kobrin, A., Shilin, D., Moskvin, V., & Shestov, D. (2019). Synthesis of Real-Time Control Systems for Multilink Industrial Robots Based on Hybrid Neural Network Approach of Solution Inverse Kinematics Problem, Proceedings of the 30th DAAAM International Symposium, pp.0513-0517, B. Katalinic (Ed.), Published by DAAAM International, ISBN 978-3-902734-22-8, ISSN 1726-9679, Vienna, Austria

DOI: 10.2507/34th.daaam.proceedings.073

ASSESSING UNITY'S APPLICABILITY FOR CREATING DIGITAL TWINS BY CREATING A MULTI-ROBOT DIGITAL TWIN

Elisa Bauer, Ali Aburaia, Horst Orsolits & Mohamed Aburaia

This Publication has to be referred as: Bauer, E[lisa]; Aburaia, A[li]; Orsolits, H[orst] & Aburaia, M[ohamed] (2023). Assessing Unity's Applicability for Creating Digital Twins by Creating a Multi-Robot Digital Twin, Proceedings of the 34th DAAAM International Symposium, pp.0551-0556, B. Katalinic (Ed.), Published by DAAAM International, ISBN 978-3-902734-41-9, ISSN 1726-9679, Vienna, Austria
DOI: 10.2507/34th.daaam.proceedings.073

Abstract

A Digital Twin (DT) replicates physical entities virtually and accurately mirrors their behaviour and condition. The number of Scopus database publications mentioning "digital twin" in their abstract increased from 51 to 2381 between 2016 and 2020. With the growing demand for DTs, there is a need for software to create and run them. While Unity has often been used for DT creation in research, its suitability for this purpose remains unexplored. This thesis aims to investigate Unity's suitability by creating a DT of a factory. The DT includes Kuka and UR5 robots, as well as linear and rotary actuators. Communication between the DT and physical system is established using ROS. Tests were conducted to measure the ROS connection latency and the accuracy of robot movements. Latency was tested under different scenarios: regular operation, reduced GPU power, and additional CPU load. Although latency was generally low, additional CPU load had a significant influence on it. The accuracy and repeatability of robot movements were assessed using the ISO 9283:1998 standard as a guide. Compared to their physical counterparts, the robot's movements exhibited significantly worse accuracy and repeatability by orders of magnitude. The robots controlled using the ArticulationBody script showed large positioning errors.

Keywords: Digital Twin; ROS; Unity

1. Introduction

Industry 4.0 has become a transformative global influence, revolutionizing industries. Deloitte's 2019 survey revealed that 74% of executives anticipated its substantial impact, 88% of leaders in other industries and 92% of IT executives agreed on its influential role [1]. Using Cyber-Physical Systems and Artificial Intelligence, Industry 4.0's implementation enhances flexibility, resource efficiency, and product adaptation to individual consumer needs [2]. A key factor to achieving these benefits is the concept of Digital Twins (DT). A DT virtually replicates a physical product, machine or system to mimic their condition and behaviour as closely as possible through an array of simulations [3]. The goal of a DT is to mirror the behaviour and condition of a physical counterpart. To achieve this one DT can consist of multiple simulations [3]. The Unity engine includes multiple relevant simulation features, like the Nvidia PhysX physics engine and allowing customization via C# scripts. Unity Technologies further supports developers in the robotics field with additional packages geared towards robotics, including robot control and Robot Operating System (ROS) integration [4]. These features make Unity an attractive option to researchers.

The result is that the Unity game engine is already commonly employed in academic DT simulations. Many of these focus on visualization and did not evaluate the simulation accuracy of Unity in their studies [5–7]. The aim of this work is to analyse the accuracy of Unity's simulations regarding robot movement and the ROS connection.

2. Materials and methods

The physical and digital systems consist of three Kuka KR 6 R900 sixx robots, two UR5 robots mounted on linear rails, a Festo tripod EXPT-45, and an eXtended Transport System by Beckhoff in an oval configuration, featuring six movers. There are also various small linear and rotational actuators, see Fig. 1. The DT was developed using Unity version 2021.3.8f1. All throughout development and testing a PC equipped with an 8-Core Ryzen 7 5800X CPU running at 3.8 GHz, an Nvidia Geforce RTX 3070 GPU, and 32GB of RAM was used.

Fig. 1. View of the DT in Unity including the Kuka, UR5 and Delta robots, the eXtended Transport System and, shown in red, the various smaller moving components.

In order to measure the movement accuracy, it is divided into two components, pose accuracy and timing accuracy. The timing is given by the data of the physical twin. The connection between the physical and DT is realised using ROS. For the timing the latency of the ROS connection is examined. The pose accuracy and repeatability of the robots is measured using the standard ISO 9283:1998 as a guide [8].

3. Practical realisation

To measure the variables mentioned above a DT was created. The models of the factory were available as SolidWorks files. Two methods were used to convert the models from CAD files to formats which can be imported into Unity. The first method, visualised in Fig.2., was used for the robots and actuators, except for the Delta robot. They were exported from SolidWorks as Unified Robotics Description Format (URDF) files using the URDF Exporter Addon. The URDF files include the links, joints and geometry, which were set in the Addons interface in SolidWorks. The URDF files were then imported into Unity using the URDF Importer package. The importer creates the robots as nested GameObjects with the ArticulationBody script attached to each link. The joint settings are automatically taken from the URDF file.

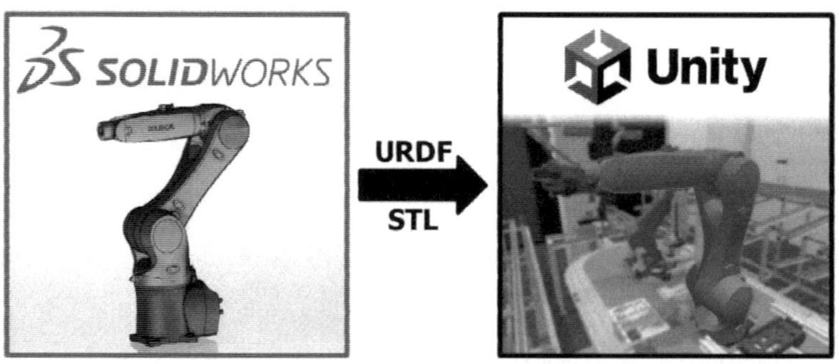

Fig. 2. Visualisation of the software and file types to import the robots into Unity.

In the second method, shown in Fig. 3., the models were exported from SolidWorks as Virtual Reality Modelling Language (VRML) files. The files were then imported into Blender and optimised, by:

- Combining the meshes into meaningful objects.
- Setting the origins.
- Cleaning up the geometry.
- Deleting unnecessary geometry.
- Reducing the vertex count.
- Removing excess materials.
- Assigning new materials.

Once the geometry was optimised the models were exported as FBX files, in this format the models were imported into Unity.

Fig. 3. Visualisation of the software and file types to import other geometry into Unity.

The connection to the physical twin was established using the ROS TCP Endpoint node in combination with the ROS TCP Connector package in Unity. The physical twin publishes the data which is used to control the movement in the DT. The robots in the DT move according to the data immediately after receiving it. For the robots and actuators, the joint angles are published, which is why they are controlled using the ArticulationBody package. The package lets one set the angles of rotatory joints and distances for linear joints. The movers' data is their distance from their home position along the rail. Their position in the world coordinate system is calculated using a bézier curve along the path. The Delta robot's data is the position of its end effector in the robot's coordinate system. The position is converted into world coordinates using build in Unity functions. The Delta robot's end effector and the movers are GameObjects and can be moved to the correct position using the goal position in the world coordinate system. The DT can be set to automatically connect to a set ROS topic on start or not. If the ROS connection is not automatically established, it can be started using the UI. The UI also has options to move the robots manually, display the robots' status and change the camera angle.

4. Experiments

Two experiments were conducted, one to find the latency of the ROS connection and one to find the accuracy and repeatability of the robots. The way the two experiments were designed is presented in the following chapters.

4.1. Ros Connection

To measure the latency the published data was recorded in a rosbag file. The file contained 157509 messages and lasted 1224 seconds. The file was played back and the times, at which the messages were received in Unity, were recorded. The recorded data was then compared to the time at which the data was published. This test was conducted for six scenarios:

- Regular operation
- Without simulating the robots' movements
- With the GPU power limited to 80%
- With the GPU power limited to 20%
- Adding an extra load of 30% on the CPU
- Adding an extra load of 60% on the CPU

The first two scenarios are designed to give a benchmark and to test the effect of computational load in Unity on the connection. The other two sets are designed to gauge the influence of the GPU and CPU respectively. Each scenario was recorded two times with no other programmes running on the PC. The scenarios were designed to test whether running the simulation in Unity or the hardware has an effect on the latency.

4.2. Accuracy and Repeatability

The second experiment was testing the positioning and orientational accuracy and repeatability of the robots and movers. The standard ISO 9283:1998 was used as the base to design the methods of measuring and calculating the robots accuracy and repeatability [8]. The methods of the standard could not be followed completely as they are designed for physical systems. The standard defines five measurement points using an ISO Cube. The points are located on the diagonals of the measurement plane. For 6-axis robots the plane lies diagonally in the cube connecting one lower side with the opposite upper side. The cubes are placed in front of the robots within their workspace, see Fig. 4.

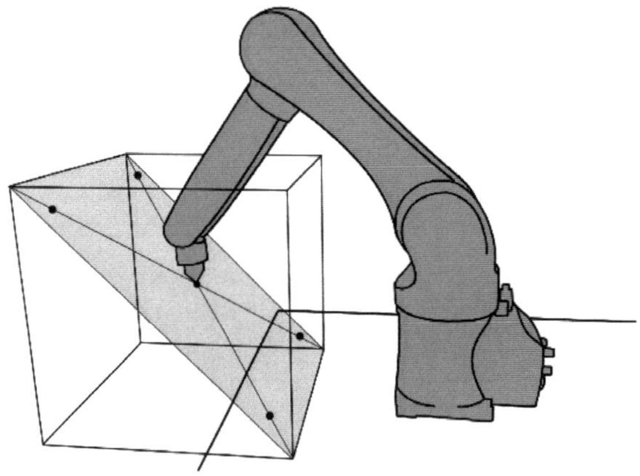

Fig. 4. The ISO Cube in front of a Kuka robot

In one measurement cycle the robot moves to each point and stops while the measurement of the end effectors position and orientation is recorded. The BioIK package was used to move the robots by end effector position rather than joint angles. The resulting joint angles for each point were saved and used to move the robot during the measurement cycles. As mandated by the standard 30 measurement cycles as well as a 0 cycle, which is not recorded, were performed. The positional and orientational accuracy and repeatability for each point were calculated as defined in the standard.

5. Results

For the ROS connection the first tests, running with and without the robots, was designed to evaluate if higher workload on the Unity application has a significant impact on the connection speed. The results are presented below in Table 1.

	With robots	With robots	Without robots	Without robots
Mean Error [ms]	16.07	16.65	27.82	19.94
Max. Error [ms]	89.98	55.36	50.49	54.05
Standard Deviation [ms]	11.08	9.49	10.44	10.02

Table 1. The mean and maximum error and the Standard Deviation in milliseconds for the testcases with no hardware limitations.

The average latency remained below 30ms for the tests without hardware constraints, for both scenarios: with and without robots. Although the maximum error was slightly higher in tests involving robots, the standard deviation remained below 11ms in all cases. Considering the application's frame rate is approximately 60 frames per second, equivalent to 16.67ms per frame, the observed 3-7ms difference was considered insignificant. Furthermore, since the tests with robots showed slightly lower errors, it can be concluded that the simulation does not adversely affect latency. These two test sets were established as benchmarks for the scenarios involving hardware limitations. Subsequent tests were performed with the robot simulation. Specifically, the GPU was restricted to 80% and 60% power consumption, as indicated in Table 2.

	80% GPU	80% GPU	60% GPU	60% GPU
Mean Error [ms]	16.89	30.19	9.53	33.73
Max. Error [ms]	53.87	53.99	81.41	57.24
Standard Deviation [ms]	9.53	10.58	8.19	10.07

Table 2. The mean and maximum error and the Standard Deviation in milliseconds for the testcases with GPU limitations.

The outcomes from both sets of tests involving the GPU remained relatively consistent with those conducted without any constraints. The maximum error and standard deviation in both sets were comparable to the benchmark tests. Given that the increase in the mean was only around 5ms compared to the tests without constraints, and the maximum error was either lower or similar to the benchmark, it can be inferred that GPU performance has very little impact on latency. Subsequently, the CPU was subjected to testing under additional loads of 30% and 60%, and the recorded results are presented in Table 3.

	30% CPU	**30% CPU**	**60% CPU**	**60% CPU**
Mean Error [ms]	27.19	12.51	22245.49	20526.7
Max. Error [ms]	53.30	186.38	39624.55	36995.82
Standard Deviation [ms]	10.49	10.40	11054.03	10453.98

Table 3. The mean and maximum error and the Standard Deviation in milliseconds for the testcases CPU limitations.

In the first set of tests with additional 30% CPU load, the results remained consistent with previous tests for the mean and maximum error. Although plotting the difference for each messages shows a shift towards higher differences. However, when the CPU load was increased to 60%, a significant difference was observed. Latency reached a maximum of 396.24 seconds, and the standard deviation for both runs exceeded 10,000ms, in contrast to the consistent 10ms observed in all other runs. Fig. 5. illustrates how the difference increased with each message, with errors accumulating over the test duration. This test clearly demonstrates the substantial impact of CPU load on latency. The slight difference observed in the plotted error with smaller additional loads implies that even minor increases in CPU load or limited CPU power can have a noticeable effect.

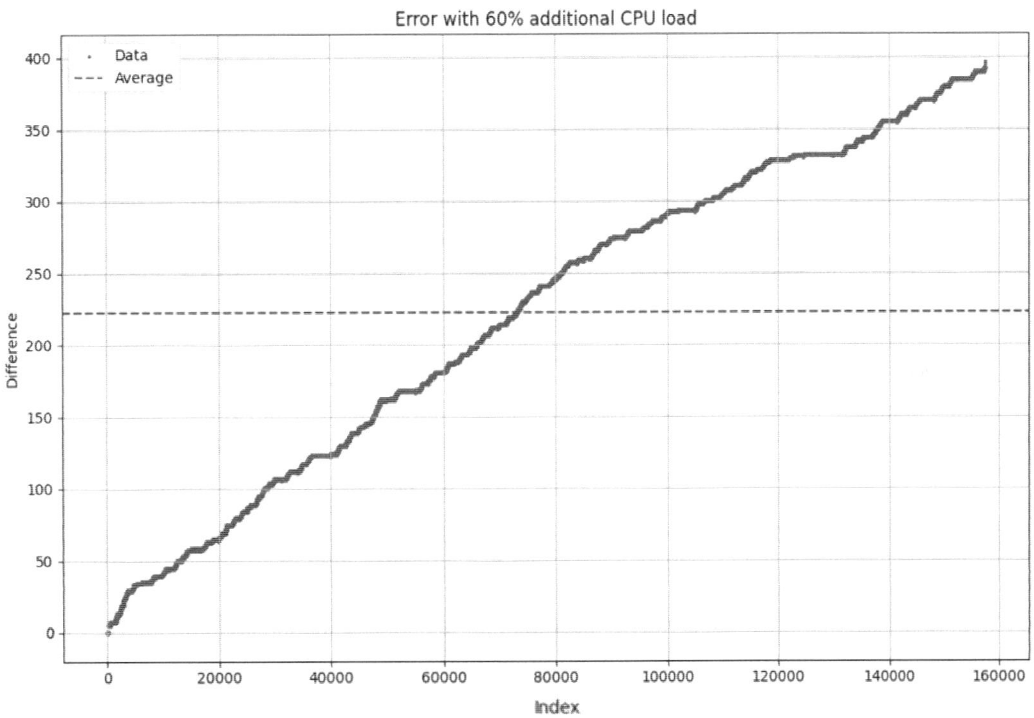

Fig. 5. Test case with an additional load of 60% on the CPU. The difference, in seconds, is plotted over the duration of the recording, measured in number of messages. The average is shown by a red dotted line.

The results of the accuracy and repeatability measurements shows that the Kuka and UR5 robots performed worse than their physical counterparts, in all categories except for orientational repeatability. The average positional accuracy (APp) of the Kuka robots lied between 14.4mm and 114.2mm. The average orientational accuracy (APa, APb and APc) were -0.28328°, -0.28885° and -0.01599° respectively. In comparison to the pose accuracy of between 0.5mm and 0.7mm, which was measured on a physical KUKA model AGILUS KR 6 R900 sixx without prior calibration [9]. The positional repeatability (RP_l) of the robots ranged from 0.5mm to 5.6mm, which is up to 3 orders of magnitude greater than ± 0.03 mm, which is the pose repeatability given by the manufacturer [10]. The findings suggest that the accuracy is influenced by the end effector. This is evident from the minor 0.4mm difference in accuracy between the two robots with identical end effectors, with greater differences between robots with different end effectors.

This indicates that the accuracy is not solely determined by the number of links or types of joints, but also by the geometry of the links. The UR5 robots had a similar result with an average APp across all points for both robots of 82.5mm. The repeatability was however a lot better than of the Kuka robots with an average of 0.021mm. The manufacturer of the UR5 robots specifies a repeatability of ± 0.1 mm [11]. However, the orientational repeatability for all robots was high, the results are in the range of 3.8e-14° to 1.9e-12°.

The high APp of the robots can be explained by the method of evaluation. The commanded TCP position was defined by the ISO Cube within the world coordinate system. The robot's movement given by the TCP position was executed using the BioIK package to record the joint positions. During this process, when converting angles from the BioIK script to ArticulationBodies, rounding errors might impact the resulting TCP position. These errors only affect the measurement process but do not impact the DT during regular operation. Because during regular operation the angles are provided by the physical twin's data. However, such errors cannot account for the low positional repeatability observed in the Kuka robots. The ArticulationBody script moves the robots by applying forces to the joints, which should let the joint reach the set position. Various factors, such as collisions, inertia, and controller settings, can influence the accuracy of these movements. To achieve the most precise results, values need to be adjusted to each situation, as they cannot all be set to manufacturer specifications due to the absence of a standardized unit. The Movers and the Delta robot are not controlled using this method. The Delta robot's end effector is positioned, using the Transform script, according to the data provided by ROS, while the robot's links move using an inverse kinematic script. Since the end effector is accurately set to its desired location, there is no disparity between the commanded and executed poses. The measurement accurately reflects this precision. APp and RPl for these robots are 0.0mm. Because the end effector doesn't rotate, there is no orientational accuracy or repeatability to consider. This is in contrast to the physical robot, which boasts an accuracy of 0.5mm and a repeatability of ± 0.1mm [12].

6. Conclusion

The connection via ROS with the mentioned packages provides a fast connection for small message sizes. The Connection is reliable as all messages were received by the Unity subscriber. The computational load of the Unity application and the GPU have no significant effect on the connection speed. The latency is however highly affected by the CPUs performance. Even with smaller additional loads of 30% plotting the error for each point showed a shift towards higher errors. At 60% additional load the error becomes greater for each message. The positional accuracy and repeatability for the robots controlled using the ArticulationBody script is lower than, what the manufacturer of the robots claim, their physical counterparts can achieve, by orders of magnitude. Although there are many settings and variables which can be changed in the ArticulationBody script. A more in-depth study of the effects of these variables on the accuracy is necessary.

7. References

[1] N. Farkas-Mills, The Fourth Industrial Revolution: At the intersection of readiness and responsibility. [Online]. Available: https://www2.deloitte.com/content/dam/Deloitte/uk/Documents/energy-resources/deloitte-uk-fourth-industrial-revolution.pdf (accessed: Jul. 27 2023).

[2] H. Kagermann, W. Wahlster, and J. Helbig, Recommendations for implementing the strategic initiative INDUSTRIE 4.0. Final report of the Industrie 4.0 Working Group: Forschungsunion, acatech, 2013. Accessed: Jul. 27 2023.

[3] S. Haag and R. Anderl, "Digital twin – Proof of concept," Manufacturing Letters, vol. 15, pp. 64–66, 2018, doi: 10.1016/j.mfglet.2018.02.006#.

[4] Unity Technologies, Unity-Robotics-Hub. [Online]. Available: https://github.com/Unity-Technologies/Unity-Robotics-Hub (accessed: Jul. 10 2023).

[5] F. Tao, B. Xiao, Q. Qi, J. Cheng, and P. Ji, "Digital twin modeling," Journal of Manufacturing Systems, vol. 64, pp. 372–389, 2022, doi: 10.1016/j.jmsy.2022.06.015.

[6] S. Xu, S. Moore, and A. Cosgun, "Shared-Control Robotic Manipulation in Virtual Reality," May. 2022. [Online]. Available: http://arxiv.org/pdf/2205.10564v1

[7] G. Garg, V. Kuts, and G. Anbarjafari, "Digital Twin for FANUC Robots: Industrial Robot Programming and Simulation Using Virtual Reality," Sustainability, vol. 13, no. 18, p. 10336, 2021, doi: 10.3390/su131810336.

[8] Manipulating industrial robots - Manipulating industrial robots — Performance criteria and related test methods, 9283:1998, International Organization for Standardization, 1998.

[9] D. Willi and S. Guillaume, "Calibration of a Six-Axis Robot for GNSS Antenna Phase Center Estimation," J. Surv. Eng., vol. 145, no. 4, 2019, doi: 10.1061/(ASCE)SU.1943-5428.0000291.

[10] KUKA Deutschland GmbH, KR 6 R900 EX. [Online]. Available: https://www.kuka.com/-/media/kuka-downloads/imported/8350ff3ca11642998dbdc81dcc2ed44c/0000293617_en.pdf?rev=aaaf08e92a6f44dbb1ad5c48467d59ae&hash=57ED900A4C91C2BE6ADD02875DDBC9DF (accessed: Aug. 1 2023).

[11] Universal Robots A/S, UR5 Technical specifications: 6-axis robot arm with a working radius of 850 mm / 33.5 in. [Online]. Available: https://www.universal-robots.com/media/50588/ur5_en.pdf (accessed: Aug. 1 2023).

[12] Festo Corporation, Electrical Tripod EXPT. [Online]. Available: https://www.festo.com/net/SupportPortal/Files/13547/EXPT-PSI-US.pdf (accessed: Aug. 1 2023).